The Wiley Blackwell Companion to Political Geography

Wiley Blackwell Companions to Geography

Wiley Blackwell Companions to Geography is a blue-chip, comprehensive series covering each major sub discipline of human geography in detail. Edited and contributed by the disciplines' leading authorities each book provides the most up to date and authoritative syntheses available in its field. The overviews provided in each Companion will be an indispensable introduction to the field for students of all levels, while the cutting-edge, critical direction will engage students, teachers, and practitioners alike.

Published

The Wiley Blackwell Companion to Tourism
Edited by C. Michael Hall

A Companion to Feminist Geography
Edited by Lise Nelson and Joni Seager

A Companion to Environmental Geography
Edited by Noel Castree, David Demeritt, Diana Liverman, and Bruce Rhoads

A Companion to Health and Medical Geography
Edited by Tim Brown, Sara McLafferty, and Graham Moon

A Companion to Social Geography
Edited by Vincent J. Del Casino Jr., Mary Thomas, Ruth Panelli, and Paul Cloke

The Wiley-Blackwell Companion to Human Geography
Edited by John A. Agnew and James S. Duncan

The Wiley-Blackwell Companion to Economic Geography
Edited by Eric Sheppard, Trevor J. Barnes, and Jamie Peck

The Wiley-Blackwell Companion to Cultural Geography
Edited by Nuala C. Johnson, Richard H. Schein, and Jamie Winders

The Wiley Blackwell Companion to Tourism
Edited by Alan A. Lew, C. Michael Hall, and Allan M. Williams

The Wiley Blackwell Companion to Political Geography
Edited by John Agnew, Virginie Mamadouh, Anna J. Secor, and Joanne Sharp

The Wiley Blackwell Companion to Political Geography
Edited by Virginie Mamadouh, Natalie Koch, Chih Yuan Woon, and John Agnew

Also available:

The New Blackwell Companion to the City
Edited by Gary Bridge and Sophie Watson

The Blackwell Companion to Globalization
Edited by George Ritzer

The Handbook of Geographic Information Science
Edited by John Wilson and Stewart Fotheringha

The Wiley Blackwell Companion to Political Geography

Second Edition

Edited by

Virginie Mamadouh, Natalie Koch,
Chih Yuan Woon, and John Agnew

WILEY Blackwell

This edition first published 2025

Edition History
John Wiley & Sons Ltd. (1e, 2015)

Registered Offices
John Wiley & Sons, Inc., 111 River Street, Hoboken, NJ 07030, USA
John Wiley & Sons Ltd, New Era House, 8 Oldlands Way, Bognor Regis, West Sussex, PO22 9NQ, UK

For details of our global editorial offices, customer services, and more information about Wiley products visit us at www.wiley.com.

The manufacturer's authorized representative according to the EU General Product Safety Regulation is Wiley-VCH GmbH, Boschstr. 12, 69469 Weinheim, Germany, e-mail: Product_Safety@wiley.com.

Library of Congress Cataloging-in-Publication Data
Names: Mamadouh, Virginie, 1963– editor | Koch, Natalie editor | Woon, Chih Yuan editor | Agnew, John A. editor
Title: The Wiley Blackwell companion to political geography / Edited by Virginie Mamadouh, Natalie Koch, Chih Yuan Woon, John Agnew.
Other titles: Wiley-Blackwell companions to geography
Description: Second edition. | Hoboken, NJ : Wiley-Blackwell, 2025. | Series: Wiley Blackwell companions to geography | Originally published as: A companion to political geography. | Includes bibliographical references and index.
Identifiers: LCCN 2024062223 (print) | LCCN 2024062224 (ebook) | ISBN 9781119753971 cloth | ISBN 9781119753896 adobe pdf | ISBN 9781119753919 epub
Subjects: LCSH: Political geography | Geopolitics
Classification: LCC JC319 .C646 2025 (print) | LCC JC319 (ebook) | DDC 320.1/2–dc23/eng/20250312
LC record available at https://lccn.loc.gov/2024062223
LC ebook record available at https://lccn.loc.gov/2024062224

Cover Design: Wiley
Cover Image: Photo by Natalie Koch, Mural by Máisel López Valdés

Set in 9/12.5pt Sabon by Straive, Pondicherry, India

Printed and bound by CPI Group (UK) Ltd, Croydon CR0 4YY
C9781119753971_240425

Contents

List of Figures and Tables

Figures

Table

Cover Concept

Natalie Koch

The picture on the cover was taken when I was visiting the Katara Cultural Village in Doha in February 2023, shortly after the 2022 FIFA Men's World Cup was hosted in Qatar. Opened in 2010, Katara was a government project to position the State of Qatar as a "cultural beacon" and to create spaces for new forms of cultural exchange. Katara regularly hosts special events, including for the country's annual National Day celebrations, which for the past 10 years have emphasized the unity of Qatari citizens and non-Qatari residents. The football mural painted on this Katara wall ahead of the FIFA World Cup is the work of the Cuban artist Máisel López Valdés and is entitled *Los amigos cubano y catarí* (The Cuban and the Qatari friends). It illustrates how the venue was folded into the state's celebratory story of global exchange at the 2022 event.

It also shows how the nationalist story of citizen/noncitizen friendship was an essential element of that positive story – the two boys serving as icons of both groups. But just as FIFA so often does in promoting the men's championship as its premier event, women and girls are quietly excluded from this nationalist image. They are perhaps subtly appealed to as mothers, but agency is ceded to the young boys who will grow up to embody the ideals indexed here – of friendship, cooperation, joy, and delight in common experiences like football. The Katara mural thus highlights this book's concern with what is visible *and* invisible in how political geographers approach politics, space, and identity across many scales and contexts. And it reflects our goal of provoking deeper questions about what immediately meets the eye.

List of Contributors

Paul C. Adams
Department of Geography and the Environment
University of Texas at Austin
Austin, TX, USA

John Agnew
Department of Geography
University of California
Los Angeles, CA, USA

Anne-Laure Amilhat Szary
Université Grenoble Alpes
CNRS
Sciences Po Grenoble – UGA
Pacte
Grenoble, France

Joshua Barkan
Department of Geography
University of Georgia
Athens, GA, USA

Silvia Binenti
Department of Conflict and Development Studies
Ghent University
Ghent, Belgium

Fabien Cante
Department of Geography
University College London
London, UK

Amaël Cattaruzza
French Institute of Geopolitics
Université Paris 8
Saint-Denis, France

Karen Culcasi
Department of Geology and Geography
West Virginia University
Morgantown, WV, USA

Cristina Del Biaggio
Pacte
Université Grenoble Alpes, CNRS, Sciences Po Grenoble
Grenoble, France

Jason Dittmer
Department of Geography
University College London
London, UK

Patricia Ehrkamp
Department of Geography
University of Kentucky
Lexington, KY, USA

Hilary Faxon
W.A. Franke College of Forestry and Conservation
University of Montana
Missoula, MT, USA

Md Azmeary Ferdoush
Department of Geography and Environment
Loughborough University
Loughborough, UK

Jennifer L. Fluri
Department of Geography
University of Colorado
Boulder, CO, USA

Benjamin Forest
Departments of Geography and Political Science
McGill University
Montreal, Canada

Coleen Fox
Department of Geography
Dartmouth College
Hanover, NH, USA

Po-Yi Hung
Department of Geography
National Taiwan University
Taipei, Taiwan

Alex Jeffrey
Department of Geography
University of Cambridge
Cambridge, UK

Shaun Johnson
Department of Geography
University of Kansas
Lawrence, KS, USA

Andrew E.G. Jonas
School of Environmental Sciences
University of Hull
Hull, UK

Halie Kampman
Department of Geography
Pennsylvania State University
University Park, PA, USA

Brian King
Department of Geography
Pennsylvania State University
University Park, PA, USA

Kendra Kintzi
Cornell Atkinson Center for Sustainability
Cornell University
Ithaca, NY, USA

Natalie Koch
Geography and the Environment Department
Maxwell School of Citizenship and Public Affairs
Syracuse University
Syracuse, NY, USA

Kirsten Koop
Pacte
Université Grenoble Alpes, CNRS, Sciences Po Grenoble
Grenoble, France

Sara Koopman
School of Peace and Conflict Studies
Kent State University
Kent, OH, USA

Yu-Chia Lin
Department of Geography
National Taiwan University
Taipei, Taiwan

Andrew M. Linke
School of Environment, Society and Sustainability
University of Utah
Salt Lake City, UT, USA

Virginie Mamadouh
Department of Geography, Planning and International Development Studies
University of Amsterdam
Amsterdam, The Netherlands

Andrea Marston
Department of Geography
Rutgers University
New Brunswick, NJ, USA

Lauren L. Martin
Department of Geography
University of Durham
Durham, UK

Sami Moisio
Department of Geosciences and Geography
University of Helsinki
Helsinki, Finland

Martin Müller
Department of Geography and Sustainability
University of Lausanne
Lausanne, Switzerland

Luca Muscarà
Department of Economics
University of Molise
Campobasso, Italy
Pacte
Université Grenoble Alpes, CNRS, Sciences Po Grenoble
Grenoble, France

Camille Noûs
Laboratoire Cogitamus
www.cogitamus.fr

John O'Loughlin
Institute of Behavioral Science
University of Colorado
Boulder, CO, USA

Anssi Paasi
Geography Research Unit
University of Oulu
Oulu, Finland

Carine Pachoud
AGIR – INRAE
University of Toulouse
Toulouse, France

Joe Painter
Department of Geography
Durham University
Durham, UK

Shaina Potts
Department of Geography
University of California Los Angeles
Los Angeles, CA, USA

Joseph S. Robinson
Independent scholar

Michael Samers
Department of Geography
University of Kentucky
Lexington, KY, USA

James D. Sidaway
Department of Geography
National University of Singapore
Singapore

Chris Sneddon
Departments of Geography and Environmental Studies
Dartmouth College
Hanover, NH, USA

Karen E. Till
Department of Geography
Maynooth University
Maynooth, Ireland

Violante Torre
Department of Geography and Sustainability
University of Lausanne
Lausanne, Switzerland

Clotilde Trivin
Department of Geography and Sustainability
University of Lausanne
Lausanne, Switzerland

Barney Warf
Department of Geography
University of Kansas
Lawrence, KS, USA

Philippa Williams
School of Geography
Queen Mary University of London
London, UK

Chih Yuan Woon
Department of Geography
National University of Singapore
Singapore

Herman van der Wusten
Human Geography
Planning and International Development
University of Amsterdam
Amsterdam, The Netherlands

Notes on Contributors

Paul C. Adams is professor of geography at University of Texas at Austin. His visiting appointments have included associate professor II at the University of Bergen (2015–2020), and Ander Visiting Professor in Global Media Studies, Karlstad University (2016–2017). He has also held Fulbright Fellowships to University of Bergen (2010), and McGill University and University of Montreal (2001). His books include *Geographies of Media and Communication* (Wiley-Blackwell, 2009) and *Disentangling: The Geographies of Digital Disconnection* (Oxford University Press, 2021, with André Jansson). He is founder of the Media and Communication Specialty Group of the American Association of Geographers and currently serves as the Human Geography Editor for *Annals of the American Association of Geographers*. His research interests lie in communication processes as revealed through various geographical frameworks.

John Agnew is Distinguished Professor of Geography and Italian at UCLA. He was a co-editor of two previous editions of this *Companion*. He is a Fellow of the British Academy and a former president of the American Association of Geographers and the Regional Studies Association. In 2019 he received the Vautrin Lud Prize in Geography. His most recent book is *Hidden Geopolitics: Governance in a Globalized World* (Rowman & Littlefield, 2022).

Anne-Laure Amilhat Szary is a full professor at Grenoble-Alpes University and honorary member of the Institut Universitaire de France. A political geographer dedicated to border studies, her research questions the interrelations between art and culture in and about contested places. She is a founding member of the anti-Atlas of borders collective (http://www.antiatlas.net/en/), an art-science project.

Joshua Barkan is an associate professor of geography at the University of Georgia, where he studies the role of law in shaping geographical political economy.

His first book, *Corporate Sovereignty: Law and Government under Capitalism* (University of Minnesota Press, 2013), examined corporations as political institutions. His current research explores the long history of concession agreements in the transformation of political space.

Silvia Binenti is a PhD student in the Department of Geography at University College London and a teaching assistant at Ghent University. Her research combines geographical and anthropological insights to look at the geopolitics of everyday objects. Her current research focuses on political T-shirts in Italy.

Fabien Cante is lecturer in urban and development geography at University College London. His ethnographic research in Abidjan, Côte d'Ivoire, examines the politics of peace-building from the perspective of everyday life, urban atmospheres, and the media infrastructures that shape them. He writes and teaches across geography, anthropology, and cultural studies, centering the Black Atlantic as a comparative frame and space of critical theory.

Amaël Cattaruzza is a professor at the Institut Français de Géopolitique, Université Vincennes-Saint-Denis Paris 8. He studies contemporary conflictuality from various perspectives (including borders and conflicts, postconflict, military geography) and is currently particularly interested in the geopolitics of digital space as part of the GEODE (Geopolitics of the Datasphere) research group and the IFGLab laboratory. From 2020 to 2024, he was President of the French Geography Committee.

Karen Culcasi is associate professor of geography at West Virginia University. Her recent book, *Displacing Territory: Syrian and Palestinian Refugees in Jordan* (University of Chicago Press, 2023), won the 2023 Meridian Book Prize for Outstanding Scholarship. She is currently working on a project on anti-Muslim discrimination in rural West Virginia. She teaches courses on Political Geography, Geographies of the "Middle East" (though she prefers the term "Southwest Asia and North Africa"), Digital Cartography, Geopolitical Theories, and Geographic Thought.

Cristina Del Biaggio, affiliated to the Pacte research center, is assistant professor in geography at the Université Grenoble Alpes. She analyzes the geographical, political, and social dimensions of migration. Her research currently focuses on the conditions under which migrants cross the Alpine borders. Adopting a forensic architecture approach, she co-investigated on the death of a migrant woman at the French–Italian border. She is co-leading a project aiming at building a unified database of deaths at Europe's external and internal borders.

Jason Dittmer is professor of political geography at University College London. His research focuses on the everyday dimensions of geopolitics, from popular culture to diplomacy. His current research focuses on heritage in Gibraltar.

Patricia Ehrkamp is Arts & Sciences Distinguished Professor in the Department of Geography at the University of Kentucky. Her research examines contemporary processes of refugee geopolitics, trauma, immigration, citizenship, and democracy in the United States and Europe.

Hilary Faxon is an assistant professor of environmental social science at the W.A. Franke College of Forestry and Conservation at the University of Montana. Her research interests include struggles for land and democracy in Myanmar and environment, development, and technology in Southeast Asia.

Md Azmeary Ferdoush is lecturer in geography and environment at Loughborough University. He is the author of *Sovereign Atonement: Citizenship, Territory and the State at the Bangladesh-India Border* (Cambridge University Press, 2024) and co-editor of *Borders and Mobility in South Asia and Beyond* (Amsterdam University Press, 2018). He was named Stanley D. Brunn Early Career Scholar by the Political Geography Specialty Group of the American Association of Geographers in 2023.

Jennifer L. Fluri is a professor of geography at the University of Colorado Boulder. She is a feminist political geographer whose research has focused on gender, geopolitics, and economic development in Afghanistan during the nearly 20-year US occupation. With the US withdrawal from Afghanistan in 2021, her research transitioned to an examination of the gendered experiences of Afghan evacuees, refugees, and asylum seekers in the United States and Canada, and the continuation of women-led enterprises in Afghanistan. She has authored over 40 peer-reviewed articles and book chapters, is co-author of three books, and has co-edited two books. She is the co-founder and co-director of the Housing Research and Education Center, which works to inform political policy, amplify community voices, support agencies working on housing access and affordability, and build relationships with community stakeholders and residents to strengthen housing access for all.

Benjamin Forest is an associate professor of geography at McGill University and is affiliated with the Department of Political Science, Centre for the Study of Democratic Citizenship, and McGill Institute for the Study of Canada. He received his PhD from UCLA in 1997 and taught at Dartmouth College from 1998 to 2006. His current research examines redistricting and political community, the political representation of ethnic minority groups and women, the politics of memory and identity, and various issues of governance.

Coleen Fox is a senior lecturer in the Geography Department at Dartmouth College. Her research examines transnational environmental governance, the political ecology of river restoration, and conservation politics.

Po-Yi Hung is a professor of geography at National Taiwan University. He uses food, agriculture, and fisheries as lenses to look into border and territoriality, mobility and infrastructure, nature and society. He has conducted research in Taiwan, China, the Highlands of Southeast Asia, and Mauritius in the Indian Ocean. He is the author of *Tea Production, Land Use Politics, and Ethnic Minorities: Struggling over Dilemmas in China's Southwest Frontier* (Palgrave Macmillan, 2015).

Alex Jeffrey is professor of political and legal geography at the University of Cambridge. His latest is book is *The Edge of Law: Legal Geographies of a War Crimes Court* (Cambridge University Press, 2020).

Shaun Johnson is a PhD student in geography at the University of Kansas. He earned his BS in Geography and Political Science from Illinois State University and his MA in Geography from the University of Kansas. His research interests include electoral and political geographies and the study of digital networks of care.

Andrew E.G. Jonas is professor of human geography at the University of Hull. His PhD is from The Ohio State University under the supervision of Kevin R. Cox. His co-authored and edited books include *The Urban Growth Machine: Critical Perspectives Two Decades Later* (SUNY Press, 1999), *Interrogating Alterity* (Ashgate, 2010), *Territory, State and Urban Politics* (Ashgate, 2012), *Urban Geography: A Critical Introduction* (Wiley-Blackwell, 2015), *Handbook on Spaces of Urban Politics* (Routledge, 2018), and *Handbook on Changing Geographies of the State* (Edward Elgar, 2020).

Halie Kampman is a postdoctoral scholar in the Department of Geography at the Pennsylvania State University. Her current work focuses on the opioid epidemic in Appalachia, where she collaborates with the interdepartmental Consortium on Substance Use and Addiction. She earned her PhD in environmental studies from the University of California, Santa Cruz, where she led collaborative social science–oriented projects on food and health geographies in Senegal and Ghana.

Brian King is professor and head of the Department of Geography at the Pennsylvania State University. He is a broadly trained geographer whose work focuses on livelihoods, conservation and development, environmental change, and human health. Over the past decade he has worked in projects in southern Africa exploring how social and ecological systems are being transformed by HIV/AIDS. More recently, his laboratory group has examined how Covid-19 is transforming the US opioid epidemic.

Kendra Kintzi is a postdoctoral fellow with the Atkinson Center for Sustainability at Cornell University. Her research focuses on the political ecology of decarbonization, smart development, and the possibilities for just environmental governance in Southwest Asia.

Natalie Koch is professor of geography at Syracuse University's Maxwell School of Citizenship and Public Affairs. She is a political geographer who works on geopolitics, authoritarianism, identity politics, and state power in hydrocarbon-rich countries, primarily in the Arabian Peninsula. In addition to editing several collections, her monographs include *The Geopolitics of Spectacle: Space, Synecdoche, and the New Capitals of Asia* (Cornell University Press, 2018) and *Arid Empire: The Entangled Fates of Arizona and Arabia* (Verso, 2022).

Kirsten Koop is a geographer and assistant professor in geography at Université Grenoble Alpes. She works at the intersection of critical development geography, territorial sciences, and sustainability transitions studies and she is interested in the levers for societal transformations in times of global change. She is currently doing research on changing imaginaries, narratives, and practices induced by citizen initiatives in the Alps.

Sara Koopman is an assistant professor in the School of Peace and Conflict Studies at Kent State, which is a living memorial to the student peace protesters killed there by the National Guard in 1970. She is a feminist political geographer interested in the sociospatial aspects of peace and peace-building and how interlocking systems of unequal power and privilege shape those. She looks at solidarity efforts that work across both distance and difference to build peace and justice with a particular focus on global North–South international solidarity, how it can fall into colonial patterns, and how to avoid these. She is particularly inspired by organizing for peace in Colombia, in which she has long been active.

Yu-Chia Lin is a research assistant at National Taiwan University, where he earned his master's degree in geography. His research interests generally are at the intersection of political and economic geographies.

Andrew M. Linke is associate professor of geography, University of Utah. He obtained his PhD in 2013 from the University of Colorado-Boulder Department of Geography. His research interests lie in political violence, political geography, human–environment interactions, spatial statistics, and geographic information systems. He has published research articles in *Political Geography, American Sociological Review*, *Global Environmental Change*, *Scientific Reports*, *Population and Environment*, *Journal of Peace Research*, and other journals.

Virginie Mamadouh is associate professor of political and cultural geography at the University of Amsterdam. She was a coeditor of the previous edition of this *Companion*. She has published on geopolitics, globalization, diplomacy, and

European integration and she served as the chair of the Commission on Political Geography of the International Geographical Union.

Andrea Marston is assistant professor of geography at Rutgers University. Her research explores the political economics and cultural politics of resource governance in Latin America. She is author of *Subterranean Matters: Cooperative Mining and Resource Nationalism in Plurinational Bolivia* (Duke University Press, 2024).

Lauren L. Martin is associate professor of political geography in the Department of Geography at Durham University. She has published research on immigration detention, carceral economies of migration control, and humanitarian technology. Her current research explores the politics of technological change in refugee humanitarianism and refugee reception in postindustrial cities.

Sami Moisio is professor of spatial planning and policy in the Department of Geosciences and Geography at the University of Helsinki. His recent books include *Geopolitics of the Knowledge-Based Economy* (Routledge, 2018) and *The Urban Field: Capital and Governmentality in the Age of Techno-monopoly* (co-authored with Ugo Rossi, Agenda Publishing, 2024).

Martin Müller is an urban and political geographer and professor in the Department of Geography and Sustainability at the University of Lausanne. In addition to maintaining a theoretical interest in the overlaps of and differences between actor-network theory and assemblage thinking, his current research looks at the skewed geopolitics of knowledge production and at cities in the global world. Using the concept of the "Global Easts," he seeks to take global theory beyond the binaries of North and South and expand the horizons of decolonial theory.

Luca Muscarà is professor of political geography, geopolitics, globalization, and sustainable development at Università del Molise, Department of Economics. An Honorary Fellow of the Société de Géographie, Paris, he is a research associate at Pacte, Laboratoire de Sciences Sociales, Grenoble, from where he co-developed an exhibition on Jean Gottmann's photo archives. His recent historical research includes works on David Lowenthal and Michael Walzer.

Camille Noûs was born in France in 2020, embodying the community's contribution to research work in the form of a collective signature. This co-signature asserts the collaborative and open nature of the production and dissemination of knowledge, under the control of the academic community. Camille Noûs is a member of the multidisciplinary, interdisciplinary, and transdisciplinary Cogitamus research group. By co-authoring this text with Camille Noûs, the authors of Chapter 9 are endorsing the values promoted by this collective signature. For more information, see: https://www.cogitamus.fr/camilleen.html.

John O'Loughlin is College Professor of Distinction in Geography and a Fellow of the Institute of Behavioral Science at the University of Colorado-Boulder. He obtained his PhD in 1973 from the Pennsylvania State University. He was editor of *Political Geography* 1981–2015 and *Eurasian Geography and Economics* 2001–2017. His research interests are in the geographies of conflict, including the relationship between climate change and conflict in Sub-Saharan Africa, especially Kenya, and territorial conflicts in the former Soviet Union.

Anssi Paasi is emeritus professor of geography at the University of Oulu. He is the author of *Territories, Boundaries and Consciousness* (Wiley, 1996). His coedited books include *The Sage Handbook of Human Geography* (Sage, 2014), *The New European Frontiers* (Cambridge Scholars Publishing, 2014), *Regional Worlds: Advancing the Geography of Regions* (Routledge, 2015), *Handbook on the Geographies of Territories and Regions* (Edward Elgar, 2018), and *Borderless Worlds for Whom?* (Routledge, 2019).

Carine Pachoud has a Master's degree in agronomy and a PhD in geography. She holds a junior professor chair at France's National Research Institute for Agriculture, Food and Environment (INRAE) in Toulouse. She analyses societal transformations from a territorial perspective and develops participatory approaches to accompany transformations, focusing on mountain areas.

Joe Painter is professor of geography at Durham University, where he teaches urban and political geography. He works on geographies of state formation, urban and regional governance, European integration, territory, and migration. His books include *New Borders: Hotspots and the European Migration Regime* (Pluto Press, 2019) and *Rethinking Territory* (Edward Elgar, forthcoming).

Shaina Potts is an associate professor of geography and global studies at UCLA. She is an economic, political, and legal geographer with a focus on the articulation of transnational political economy, geopolitics, and law. She is especially interested in the mutual constitution of technical, legal, and economic practices, on the one hand, and sovereignty, empire, and territoriality, on the other.

Joseph S. Robinson is a political and cultural geographer whose work sits at the intersection of memory studies, transitional justice, trauma studies, peace and conflict studies, and the study of time and society. He received his PhD in Geography from Maynooth University in 2020 and is the author of *Transitional Justice and the Politics of Inscription: Memory, Space, and Narrative in Northern Ireland* (Routledge, 2018).

Michael Samers, who obtained his DPhil from Oxford University, is professor of geography at the University of Kentucky, having previously held positions at the

Universities of Liverpool and Nottingham. A Fulbright Scholarship awardee in France, and former co-editor of *Geoforum*, he is the author of *Migration* (with Michael Collyer, Routledge, 2017, 2nd ed.), *Cities, Migration and Governance: Beyond Scales and Levels* (with Felicitas Hillmann, Routledge, 2024) and *Migration and Nationalism* (with Jens Rydgren, Edward Elgar, 2024).

James D. Sidaway has served as professor of political geography at the National University of Singapore since 2012. He previously taught at the University of Amsterdam and several universities in the UK. Drawing from his broad interests in political geography, James's longstanding fascination with the history and philosophy of geographical thought and the relationship between geography and a range of traditions in area studies have recently led him to studies in decolonial Muslim geographies.

Chris Sneddon is a professor in the Departments of Geography and Environmental Studies at Dartmouth College. His research focuses on environmental politics, the geopolitics of development, water conflicts, the politics of environmental knowledge production, and the hydropolitics of the Mekong River basin.

Karen E. Till is professor of cultural geography at Maynooth University and a Member of the Royal Irish Academy. She collaborates with diverse publics on initiatives supporting place-based memory-work and care in cities wounded by state-perpetrated violence. Her publications include *The New Berlin: Memory, Politics, Place* (University of Minnesota, 2004) and the *Traveller Community Mapping Coolock Story Map: Storied Places of Belonging and Unbelonging*, co-authored with Pavee Point Traveller and Roma Centre, TravAct Coolock, and Maynooth Geography.

Violante Torre is a PhD candidate and teaching assistant in urban geography in the Department of Geography and Sustainability at the University of Lausanne. Her doctoral research project focuses on decentering iconic buildings by understanding them as made, remade, and renegotiated on the ground through human–material relations, everyday practices, and affective experiences. Her work focuses on Istanbul's Hagia Sophia and its latest conversion from museum to mosque, centering city dwellers' experiences of symbolic and conflictive urban space.

Clotilde Trivin is a PhD candidate in urban geography in the Department of Geography and Sustainability at the University of Lausanne. Her Swiss National Science Foundation–supported doctoral project investigates large-scale cultural buildings as "affective buildings," exploring both how their sociomaterial entanglements are experienced and assembled as well as enmeshed in broader urban assemblages. Drawing on more-than-human geographies and architectural geographies, her project examines the ways in which the European Solidarity Centre

(Gdańsk, Poland) is materially assembled, affectively experienced, and politically mobilized to make and unmake urban imaginaries.

Barney Warf is a professor of geography at the University of Kansas. His research and teaching interests lie within the broad domain of human geography. He has studied a wide array of topics, particularly information technologies as viewed through the lens of political economy and social theory. He maintains an active interest in political geography, including elections, voting technologies, the US Electoral College, and the geographies of Donald Trump. He also edits *Geojournal.*

Philippa Williams is professor of geography in the School of Geography, Queen Mary University of London. She researches, teaches, and writes on experiences of the state, citizenship, violence/nonviolence, everyday peace, and the everyday geopolitics of digital India. Philippa is co-author of *Privacy Techtonics: Digital Geopolitics, WhatsApp and Democracy in India* (Bristol University Press, 2025).

Chih Yuan Woon is an associate professor at the Department of Geography, National University of Singapore. He is a political geographer who has research interests in critical geopolitics, postcolonial geographies, and issues to do with peace, nonviolence, and security. Currently he is examining the geopolitical and geoeconomic implications of China's "rise," contextualizing such debates through the Belt and Road Initiative as well as China's ongoing interest and interventions in the polar regions. He has served as the co-editor of *Territory, Politics, Governance* (2019–2023) and the *RGS-IBG Book Series* (2020–2023) and sits on the editorial boards of *Critical Military Studies*, *Environment and Planning C: Politics and Space*, *Geography Compass* (Political Geography Section), *Territory, Politics, Governance* and *The Singapore Journal of Tropical Geography.*

Herman van der Wusten is professor emeritus of the University of Amsterdam, where he taught political geography. He has studied Dutch electoral politics and has also been active in the field of international relations. His current main interests are political center formation in Europe, EU governance, and diplomacy.

Chapter 1

Introduction

Virginie Mamadouh[1], Natalie Koch[2], Chih Yuan Woon[3], and John Agnew[4]

[1]Department of Geography, Planning and International Development Studies, University of Amsterdam, Amsterdam, The Netherlands
[2]Geography and the Environment Department, Maxwell School of Citizenship & Public Affairs, Syracuse University, Syracuse, NY, USA
[3]Department of Geography, National University of Singapore, Singapore
[4]Department of Geography, University of California, Los Angeles, CA, USA

On July 26, 2024, the opening ceremony of the Olympic Games in Paris showed the usual tension between the celebration of the movement's values of "excellence, respect, and friendship" on the one hand and nationalism-infused competition between states on the other. Yet the opening was also innovative: for the first time, the parade of national sports delegations was staged outside a stadium, replaced by a fleet of vessels of various shapes and sizes crossing the city on the Seine River. In this way, Paris's urban spaces were used to invite the Games' global audience to feel included in the spectacle to a new and remarkable degree. But political divides were not cast off in the Seine display. As with all events since the modern Olympics began in 1896, the alleged political neutrality of the Games was ambiguous: national sports federations are organized territorially but some are states and others are not. The US territory Puerto Rico, for example, has had its own Olympic team since 1948, and in 2016 the International Olympic Committee introduced a Refugee Olympic Team. And as in years before, 2024 saw certain delegations deprived of their national flag. These anomalies originated in international sanctions (e.g., against Russia and Belarus) or national vetoes (e.g., the People's Republic of China accepting Taiwan's participation as "Chinese Taipei," but with a strict ban on all Taiwanese flags and symbols at Olympic events and surrounding public spaces).

The intersection of sports and politics is well established (Koch 2017). Mega sporting events like the Olympics vividly demonstrate how national sentiments are

The Wiley Blackwell Companion to Political Geography, Second Edition.
Edited by Virginie Mamadouh, Natalie Koch, Chih Yuan Woon, and John Agnew.

galvanized among sports participants and supporters alike. Sometimes they can be the source of national unity; other times, they can kindle or reignite national divides. A month before the Olympics the 2024 UEFA European Football Championship tournament saw the spark of various nationalist flashpoints. Fans from Serbia screamed about Kosovo during their games, even though Kosovo was not a participant. They were remembering the still-rankling secession of largely Albanian Kosovo as they attended what was supposed to be a football game, not a political rally. Scottish fans played nice everywhere, but when their team was eliminated in a loss to Germany, England fans celebrated in excess. The football festivities in the summer of 2024 remind us that Europe's football fans, self-selected as they may be, firmly grasp the nationalist sentiments that still haunt Europe nearly 80 years after the end of World War II. The irony is that many of the players of Europe's national teams are first- and second-generation immigrants, whose very presence in Europe is an issue that excites contemporary nationalism more than the traditional animosities, such as England versus Germany.

Pundits and academics alike have asserted that the recent revival of nationalist sentiments is a response to the excesses of globalization, including the decline of the welfare state, the loss of supposedly preferable manufacturing jobs, and the seeming collapse of national-ethnic homogeneity in the face of dramatic increases in international migration – at least in the United States and Western Europe. Nostalgia for lost golden ages, such as the 1950s postwar boomtimes in the United States and Germany, and waning hope for a brighter future in the local places people call home, have also been identified as significant factors. Fictive as it may be, reinstating national territorial sovereignty is cast as a solution in this reactionary nationalist storyline. In the United States this has manifested in the "Make America Great Again" campaign of the far-right former president and president-elect Donald Trump, and in the United Kingdom it fueled the ill-fated Brexit decision, which saw Britain leave the supranational community of the European Union (EU).

This trend to what can be called "national populism" mobilizes particular geopolitical imaginaries in the United States and Western Europe, but it is far more widespread. From India to Brazil to Russia, Argentina, Israel, South Africa, and China, similar sentiments are on display. The commonalities include increasing hostility to international cooperation and the so-called liberal international order, suspicion about foreign investors and foreign aid, anger toward US hypocrisy about the integrity of only *some* national borders, fear of foreign interference in national affairs, and defensiveness against liberal cultural trends emanating from the so-called West that challenge "traditional" norms about gender, religion, and social hierarchies.

The reactionary spirit of today's national populism is hardly new. But it has taken on new life following a series of major geopolitical shifts, including the global financial crisis of 2008–2009, the Arab Spring movements beginning in late 2010, the 2015 refugee crisis in Europe, and the Covid-19 pandemic. In each case, populist politicians reacted to events by playing to the anger and resentments of populations experiencing poor employment prospects, facing corrupt elites cashing in on political ties, and threatening cultural shifts. And in many cases, immigration and territorial borders became an easy focus for populist attacks, even though populists seldom acknowledge that contemporary migration stems from the legacies of colonialism and neocolonial interventionism, and the authoritarianism of governments and corporations alike.

For example, the 2015 influx of refugees in Europe was largely the result of the Arab Spring uprisings across the Middle East and North Africa – some of which toppled US-backed authoritarian regimes (as in Egypt and Tunisia) and others of which led to prolonged civil war (as in Syria and Yemen), but none of which has since brought democracy and civil liberties, let alone economic prosperity, to their countries. This failure is not a failure of vision on the part of the Arab democracy activists: their fight may have started by attacking an oppressive regime, but they found themselves fighting a structure of oppression facilitated by Western empire, past and present, in the region. This includes a long history of militarism, most recently reflected in the disastrous results of the United States' interventions in Iraq and Afghanistan in the so-called War on Terror (from 2002 onward).

But as with the British and French colonial undertakings in the region in the previous century, American war-making in some places was accompanied by supporting their autocratic neighbors, as in Egypt. Yet European narratives about the 2015 immigrant "invasion" did not reflect on these imperial histories and presents, instead preferring sensational stories that blamed the victims. Similar victim-blaming attacks on migrants and asylum seekers have been seen at highly politicized border crossing points between the United States and Mexico, Myanmar and Bangladesh, Syria and Turkey, Haiti and the Dominican Republic, and Sudan and Ethiopia. Politicians fueling populist agitation consistently overlook the fact that the individuals crossing these borders are fleeing civil war, genocide, unemployment, and failed states – and sometimes all of these – instead framing their national communities as righteous defenders of their homelands. Nationalist protectionism, enacted through xenophobic rhetoric and harsh border security measures, was already on the rise before the Covid-19 pandemic began in early 2020, but that gave yet more fuel to this fire.

Covid-19 showed that planetary problems invariably cross territorial borders, but many political leaders around the world intensified their commitment to the dream that closed borders were the solution. The result was ill-coordinated global responses that led to a set of disparate and ultimately deadly outcomes. In many countries, border closures revived the sense of being alone in the world and increased calls for national autonomy and even autarchy, rather than encouraging a deeper sense of global community. This spur to reterritorialization was enhanced by the perception of vulnerability to the vagaries of global supply chains shipping goods halfway around the world, including those necessary for dealing with Covid-19 like personal protective equipment. On top of disturbed supply chains due to work and transport stoppages, one of the most iconic pictures from the pandemic is of the gigantic container ship the *Ever Given* stuck in the Suez Canal in March 2021. Behind it, hundreds of other ships sat waiting to make the trip through the canal, while hundreds of others rerouted to older, longer shipping lanes around Africa. The call to domesticate production was not an unsurprising result of both the supply-chain backups and the price increases that followed. This period was a turning point when retreating behind familiar borders looked like a solution for many.

The wars erupting in Ukraine in 2022 and Gaza in 2023 would also seem to suggest a revival of nationalist imperatives enacted at territorial borders. Both can be viewed as border disputes, but they are also much more than that. Each involves competing claims to parts or the whole of some other national territory. There are many places where such conflicts could now erupt, for example between China and Taiwan, the two Koreas, Sudan and South Sudan, or Venezuela and Guyana. There also many devastating so-called civil wars, often with external sponsors, such as those in Myanmar, Yemen, and Sudan, as well as countries in which the writ of central government does not hold in regions and localities dominated by narco-gangs and insurgents, such as Mexico, Haiti, Ecuador, and Colombia. All of this is occurring in a geopolitical context in which the United States is politically polarized over its precise role in intervening in international territorial disputes, but is nonetheless wary of ceding its global leadership to states like China and Russia.

Territorial questions continue to be at the center of geopolitical metanarratives today, but economic questions are never far away. For example, economic hegemony is a fixture of the perennial prophecies of a "new Cold War" between the United States and China (Schindler et al. 2024). Such speculations attend to recent shifts away from the hyperglobalization of the period 1990–2010, but the fact is that the world remains economically and socially integrated to an extent that would have been unthinkable 100 years ago. And the United States and China remain economically entangled in ways that the Soviet Union and the United States never were during the Cold War from 1947 to 1991. Capital and information flow relatively freely worldwide. If political elites from the United States or any other country tried to impose major limits on the open world economy, it would be costly for all concerned – not just for consumers facing steep price hikes resulting from tariffs and shortages due to poorly performing domestic producers, but also for the bottom lines of companies that have built their strategic-economic advantage and profitability on globalizing their operations. Even as sanctions have become a favorite tool of American and Western indirect interventionism (for example, sanctioning Russian politicians and companies instead of getting militarily involved in the war in Ukraine), they are fundamentally designed to maintain the global economy's status quo of deep integration.

Massive planetary problems must be addressed through global cooperation, ranging from climate change to global health, food and water security, satellite infrastructure, telecommunications and artificial intelligence (AI), labor rights, and more. But we are in a fraught era in which a contest between "nationalists" and "globalists" is underway. So far, the outcome is unclear. Recent elections from Brazil and India to France and Britain suggest a standoff in which the hard-core nationalists have had some setbacks. Even in the face of rising nationalisms, the EU remains popular across most of its member states, partly perhaps because of the failed example of Brexit. According to the pollster YouGov in June 2024, a referendum on membership in the EU would result in a crushing victory for "Remain" in each of the largest member states. And as the far-right Alternative für Deutschland (AfD) party expanded its electoral victories across Germany in 2024, millions of anti-AfD protesters responded by filling the streets of nearly every German city, chanting *Nie wieder ist jetzt!* ("Never again is now") and holding signs reading *Faschisten sind keine Alternative für Deutschland* ("Fascists are *no* alternative for Germany"). It is in this context that future political geography might well be made and what the chapters in this book hopefully offer in understanding how we got to this point.

Traditions and Transitions in Political Geography

This book is a new edition of the *Wiley Blackwell Companion to Political Geography* published a decade ago (Agnew et al. 2015). Two former editors (Agnew and Mamadouh) and two new editors (Koch and Woon) make up the editorial team of this volume and it is hoped that this change will bring different synergies, experiences, and knowledges to approaching and thinking about (contemporary/recent developments in) political geography and engage a more diverse set of authors. We also bridge generations, as our training in political geography fell both before and after two key junctures for the field: the Cold War and the critical theory turn. The term "political geography" was coined in 1751 by the French economist and statesman Jacques Turgot, and inscribed in academia through Friedrich Raztel's seminal text *Politische Geographie* in 1897, but the discipline was truly revolutionized after the 1980s with the end of the Cold War and the reimagination of political geography through critical theory.

Changes in the field had deeper roots, of course, as political geography, born at the time of interimperial rivalries and developed in the interwar period, was strongly curtailed after its implication in the conduct of World War II (see Muscarà, this volume; Sidaway, this volume). Seeds of change were first seen in the 1960s, when a renewed political geography was premised on developing a spatial science of electoral geography (see Johnson and Warf, this volume). By the 1980s, however, critical approaches, including Marxism, postmodernism, and Immanuel Wallerstein's world systems theory, began to enrich the subdiscipline by bringing into view questions of power, inequality, and social justice (Taylor 1982; Smith 1984; Flint & Taylor 2024). As political geography research gained momentum, debates and theorizations around borders, territory, political identity, power, and resistance started to flourish. At the same time, geopolitics was revived in policy-making and academic circles in the US through the politicking and visions of Henry Kissinger and Zbigniew Brzezinski in the 1970s and 1980s.

The "revival" of geopolitics, as Leslie Hepple (1986) called it, was instrumental in shaping geographical engagements with the topic in the 1990s, albeit from a critical perspective that focused on "disrupting the hegemonic practices of statecraft, challenging the taken-for-granted specifications of the world in various mappings by elites and academics" (Dalby 2010: 281; see also Ó Tuathail 1996; Ó Tuathail & Agnew 1992). This strand of critique proliferated, producing new fields of study including popular geopolitics (Sharp 1993; Dittmer 2010), feminist geopolitics (Dowler & Sharp 2001; Dixon 2015), alter-geopolitics (Koopman 2011), and subaltern and liminal geopolitics (Sharp 2013; McConnell 2017). These approaches all reflect the infusion into political geography of feminist and postcolonial/decolonial theories, which have foregrounded the voices and subjectivities of marginalized groups. The feminist perspective that today characterizes much of political geography – especially visible in two of the newest textbook surveys of the subfield (Smith 2020; Squire & Jackman 2023) – explicitly challenges the masculinist biases in the subdiscipline and reorients the field from state-centric analyses to consider a wider, multiscalar set of issues, including everyday spatial practices.

In fact, for over three decades now political geographers have worked to show that the state is not a mere container of political processes and events (Agnew 1994; Paasi 1996). *How* they do so takes many forms. For instance, research on state capitalism interrogates the geoeconomic agency of the state (and corporate allies), which has produced the geopolitical, economic, and financial processes pertaining to the secular development of global capitalism (Alami & Dixon 2020; Mallin & Sidaway 2024; Potts, this volume), as well as the diverse ways in which state and nonstate alliances are shifting in the face of the climate crisis, such as greening state capitalism (Koch 2022). Other political geographers have eschewed interrogating solely the nature of the state, instead zooming in on how state effects have permeated space, place, and society through quotidian and prosaic means (Painter 2006; Benwell & Dodds 2011; Jeffrey 2013). The state, in this perspective, is "peopled," (re)produced by different actors and their networks, rather than being an abstract, amorphous political entity (Medby 2018).

Territory likewise continues to animate a great deal of research in political geography, moving well beyond traditional approaches that considered it as part of a great state–nation–territory triad (Gottmann 1973; Knight 1982; Paasi 1996) to open up deeper reflections on its diverse logics, expressions, and constitution. Beginning from Thongchai Winichakul's (1994) seminal book *Siam Mapped*, political geographers not only acknowledge territory to be a Western construct, but also aim to spotlight alternative, contextual, and Indigenous approaches to the concept (Daigle 2016; Palmer 2020). The multidimensional nature of borders also continues to be a touchstone of contemporary political geography. Despite assertions of globalization and the "borderless world," political geographers argue that borders are important to the process of territorialization. They symbolically perpetuate meaning and physically shape the mobilities of people. But rather than seeing borders as a "thing," scholars now emphasize the *processes* of bordering and related questions of agency, motivations, and power relations (see Amilhat Szary, this volume).

The compound nature of territory has likewise been given closer study (Billé 2019), especially on how sovereign power and control can extend volumetrically and vertically into spaces such as oceanic realms (Steinberg & Peters 2015), outer space (Klinger 2020), and the varying depths of the subterranean (Squire & Dodds 2020). Cities have also become an increasingly important lens through which to examine traditional questions related to borders, territory, political movements, and geopolitics. Political geographers writing on urban geopolitics, for example, have attended to the unique spatialization of conflict in cities (Fregonese 2017; Rokem & Boano 2018) as well as the flows of state power through global city networks, forging new North–South and South–South connections (Koch 2013). Cities are also important places where infrastructure projects are realized and contested, including surrounding sports mega-events like the Olympics and the World Cup (Müller 2017). Well beyond the urban scale, political geographers have joined other social scientists in advancing critical research on infrastructure more broadly, addressing the local, regional, and transnational power dynamics of new projects ranging from roads to rail, ports, dams, and digital infrastructure.

Moreover, the rapid advance of AI and digital infrastructure in diverse domains of everyday life has given renewed attention to the role of the digital and nonhuman. This has in turn raised pertinent questions about the role of human agency in the age of automation/AI, as well as the challenges in attributing moral responsibilities to different (geo)political processes and outcomes (Amoore 2020; Erskine 2024). But rather than fetishizing technological systems and innovations, interesting political geography scholarship has also demonstrated how glitches, in the form of small-scale disorientations in digital urban mediations, can also provide the possibilities for resistant politics and movements to advance more fair and just technological futures (Leszczynski & Elwood 2022). Much of the new research on cyberspace and technology builds from and extends political geography's much longer engagement with geospatial tools – both using geographic information systems (GIS) and remote sensing to map and analyze various political patterns, and critically reflecting on the ethics of using these technologies in our research (O'Loughlin 2005; Bennett et al. 2022; see Culcasi, this volume).

The jointly political and ethical challenges of today's digital environment parallel the persistent sensitivity that the field of political geography has to changes happening outside academia. Political geographers are grappling with many of the contemporary issues flagged in the opening section of this chapter, which are also engaged by this volume's authors. The thematic and topical foci of the first edition of the text remain relevant today, but they have been updated to reflect new trends. Debates about war and peace have not escaped the intellectual radar of political geography and have become even more pertinent in light of today's unfolding landscape of war (Murphy 2022; Agnew 2024). The ever-growing impacts of the climate crisis on security, migration, and resources demand urgent attention, as do the implications of the Covid-19 pandemic on the politics of health and wellbeing.

Keeping up with the dynamic global context of our research has led political geographers to embrace novel methodological techniques and fieldwork. Conventional research methods such as interviews, ethnography, and surveys remain important, but the increasing digitalization of society has encouraged political geographers to explore more web-based methods, such as online interviews and virtual ethnography (Duggan 2017; Murrey 2023). Virtual tools proved extremely valuable for scholars continuing their research when widespread Covid-19 lockdown conditions made in-person fieldwork challenging or impossible – and their usefulness is still being discovered even after the end of the pandemic. Thanks to its record of embracing theoretical and empirical diversity, political geography is well positioned to continue capitalizing on such methodological innovations, virtually, in person, and in all the spaces in between.

Outline of the Book

The aim of each section of this book is not to be comprehensive, but to open up sufficient intellectual space for readers to reflect on the current and future agendas around political geography. Some chapters were updated from the 2015 edition while others are completely new. We have been careful also not to expand the volume beyond limits. This means that some themes were removed to make room for new ones, while others have been combined or reorganized. We have balanced old and new contributors, engaging early career researchers and diversifying orientations and localizations. The overall structure of the volume conceived with Anna Secor and Joanne Sharp has been maintained, although we have renamed the last section and added a new opening section.

The first section is "Framing Political Geography," consisting of four chapters that outline the scope and remit of political geography, as well as the field's imbrication with multiply scaled politics and power relations. Chapter 2 on **Political Geography in Historical Context** by Luca Muscarà and Chapter 3 on **The Politics of Political Geography** by James Sidaway trace the diverse histories and political trajectories of the subfield. In reflecting on the interface of historical and political geographies, as well as the significance of translation in an increasingly globalized field, both chapters remind us that the process of establishing disciplinary identity has always been contested. There are different ways of "seeing" and "knowing" the political and there should be critical reflections on what gets included or excluded in dominant narratives of political geography. These themes are continued in Chapter 4, **Geopolitics of Knowledge** by Anssi Paasi, which extends this point by underscoring the Euro-American dominance in political geography scholarship. He attributes this to wider characteristics associated with the neoliberal university, academic capitalism, and the geopolitics of knowledge – which all frame what counts as legitimate knowledge in our discipline today. Karen Culcasi, in Chapter 5 on **The Powers of Maps**, likewise engages questions of geopolitics and the history of geographical knowledge practices by focusing on maps and map-making as both method and object of study. Although map-making has long been neglected in political geography, Culcasi contends that maps are more than just texts for political geographers to analyze: the many powers of maps can be valuable for political geographers to engage with more directly, such as through novel counter-mapping initiatives.

The second section, "(Contested) Key Concepts," addresses concepts that we consider to be foundational to the study of political geography: power, state, sovereignty, territory, scale, boundaries/borders, citizenship, security, conflict, peace, and memory. Rather than outlining discrete definitions of these concepts, contributors are united in their efforts to demonstrate how each idea is complex and multifaceted in nature, and contested from different intellectual and geographical lineages. They therefore draw out the traditions of each concept, tracing how political geographers have engaged with these notions more recently as well as foregrounding their analytical possibilities for future research agendas. In Chapter 6 on **Power**, Joe Painter and Natalie Koch begin by noting that power was undertheorized in political geography for many years, but is now a central keyword in the field. In joining theoretical discussions inspired by the work of notable theorists, political geographers today have avoided reifying power as a "thing" to be possessed, and instead offer critical insights about the diverse practices, relations, and spatialities of power. In Chapter 7 on **The State**, Alex Jeffrey focuses on the relationship between territory and power. He goes on to contrast critical and radical approaches to conceptualizing the state with traditional understanding. In Chapter 8 on **Sovereignty**, Joshua Barkan discusses the ambiguities of state sovereignty and more recent forms of sovereign power. Cristina Del Biaggio, Kirsten Koop, Carine Pachoud, and Camille Noûs in Chapter 9 on **Territory** contrast conceptions of territories as political and social constructs in francophone and anglophone geographies, before discussing recent attempts in South and North America to decolonize the concept. In Chapter 10 on **Scale**, Andrew Jonas examines how political geographers use scale to reflect on territorial hierarchies of political organization and power relations, encompassing a truly diverse set of approaches. And in Chapter 11 on **Borders**, Anne-Laure Amilhat Szary likewise examines a plural tradition in political geographical research. Tracking the past and present of border studies, she shows how scholars are forced to reckon with new spatial practices, which are constantly challenging basic assumptions about the material and symbolic meaning of borders and border practices.

The following five chapters in this section introduce concepts that are less explicitly geographical although definitively spatial. In Chapter 12, **Citizenship**, Patricia Ehrkamp shows how citizenship regimes reflect a mode of stretching rights and political community across space, and how political geographers have examined the many scales and practices of the making of citizens and their political subjectivities, as well as citizenship. In Chapter 13 on **Security**, Lauren Martin considers its many intersections with state power, livelihoods, environment, economy, migration, and climate change. Her critical review also addresses some of the key approaches used in political geographies of security, ranging from constructivist and performative theories and securitization to governmentalities, security assemblages, and ontological security. In Chapter 14 on **Conflict**, Amaël Cattaruzza discusses the evolution and expansion of the notion of conflict in political geography, questioning more specifically the "return of war" and exploring with cyberwar the influence of the sociotechnical context on the spatial materialization of conflicts. But in Chapter 15 on **Peace**, Fabien Cante and Philippa Williams show how political geographers have been increasingly concerned with the concept and practice of peace in the past 15 years. As the global political landscape continues to be fraught with violence, they raise urgent questions about what conflict and peace mean to different populations at a range of geographical scales, and how political geographers might foster more convivial and harmonious relationships and coexistences on the ground. Finally, in Chapter 16 on **Place and Memory**, Karen Till and Joseph Robinson focus on the contested politics to memory, including how certain "pasts" are publicly remembered whereas others are downplayed or ignored. What these remembrance practices achieve politically and how they can be harnessed toward more equitable futures are just some inquiries of interest to political geographers.

The third section, "Approaching Political Geography," shifts focus from the individual concepts themselves to how geographers have understood key processes. More specifically, the chapters featured here explore different epistemological traditions and approaches that have undergirded how concepts have been dealt with and mobilized in political geography. In Chapter 17 on **Spatial Analysis**, Andrew Linke and John O'Loughlin introduce the significance of context (place) and spatial dependencies (spillover effects) in the field. Increased data accessibility and new analytical tools have given spatial analysis a renewed focus, providing novel opportunities to examine geographical distributions of political phenomena. By contrast, the relationship between state and capital (between politics and economics) is a longstanding concern for political geography insofar as there is a close link between the changing regimes of capital accumulation and states' spatial patterning. This relationship, which Shaina Potts considers under the broad rubric of **Political Economy** in Chapter 18, points to a variety of approaches to conceptualize ongoing transformations in state–capital relations and their consequences for sovereignty, territoriality, and geopolitics in an interconnected world.

One of the most significant retheorizations within political geography over the past few decades revolves around infusing more critical perspectives into geopolitics, as discussed by Sami Moisio in Chapter 19, **Geopolitics and Critical Geopolitics.** The field of critical geopolitics has extended the intellectual foci of geopolitics toward the realm of representations and discourses – problematizing powerful geographical framings of world politics and statecraft. Feminist and postcolonial/decolonial interventions have proliferated after the publication of the previous edition of this collection. They remain committed to broadening our understandings of "where" the political is located, drawing attention to the everyday, underrepresented, and marginalized spaces through which political geographies are constantly made and remade. In Chapter 20 on **Feminist Political Geography**, Jennifer Fluri provides an overview of the themes beyond the initial interest in foregrounding women's roles in politics. These range from clusters around gender, nation, and state to corporeal geopolitics and body politics, domestic spaces and informal politics, and feminist approaches to conflict and security. Chih Yuan Woon's Chapter 21 on **Postcolonial and Decolonial Political Geographies** maps out the development of these debates and research agendas in political geography. He also reflects on the epistemological, ontological, and methodological possibilities offered by the idea of a "pluriverse" to confront imperial and neocolonial geographies.

After setting out the key concepts and main approaches, the fourth section of the Companion, "Doing Political Geographies," explores the diverse sites and scales where political processes play out. In Chapter 22 on **Electoral Geographies**, Shaun Johnson and Barney Warf review the evolution of this subdiscipline, which was once dominated by empiricist and positivist approaches and was a major site of spatial analysis. More recently, political geographers have examined voting as one of the most formal and institutionalized spheres and processes that enable state elites to gain and obtain power. But, as Johnson and Warf show, it is also a powerful channel for reproducing political identities and senses of belonging. Benjamin Forest too addresses this theme of identity politics in Chapter 23 on **Nationalism and Nation-Building.** After reviewing the types of nationalism, he outlines the relationship of language, religion, and citizenship with national identity. In Chapter 24 on **Multilevel Governance,** Herman van der Wusten considers the diverse political geographies of state organization, including federalism as well as local and regional government in unitary states. Although the multilevel governance analytical frame was initially developed in political science and policy analysis to conceptualize policy-making in the EU, he shows its wider application to other contexts around the world.

In Chapter 25 on **Social Movements,** Sara Koopman examines how geographers brought space to social movement thinking and praxis. Drawing on the feminist concern with small "p" politics, social movement research is concerned with questions of embodied practices of building solidarity to form communities of resistance, as well as questions of *who* is included and represented in different forums and polities. These issues are also taken up in Chapter 26, **Identity Politics** by Azmeary Ferdoush. Although identity politics is often approached as a politics of exclusionary categories (both social and spatial), he examines practices of inclusion as well, drawing examples from religion and sexuality. But of course, social movements and identity politics are not necessarily only played out in "real" and "physical spaces." Crucially, as Paul Adams shows in Chapter 27 on **Digital Media**, new identities and communities can be formed in "alternative" (virtual) spaces, potentially facilitating activism networks to emerge across vast geographical distances and allowing for the dissemination of images of oppression and resistance to an accessible audience. New technological interfaces, he shows, also differentially impact multiple scales of sovereignty. Closing this section, Michael Samers highlights similar tensions around sovereignty in Chapter 28 on **Migration.** He specifically shows how debates on *international* migration are imbricated debates about governance at many scales, ranging from the local to the supranational – while also warning that such spatial metaphors of scale can hinder critical thinking about the dynamic nature of migration today.

The fifth and final section, "Material and Environmental Political Geographies," covers the material turn in political geography, as well as burgeoning scholarship on the environment in the field. Chapter 29 on **Material Political Geographies**, by Violante Torre, Clotilde Trivin, and Martin Müller, opens the section by outlining how political geographers have investigated the many entanglements between humans and the material world. Although the "cultural turn" has reoriented the subdiscipline toward the power of representations in shaping the political world, it is crucial not to lose sight of matter and materialities that are caught up in actor-networks/assemblages. Silvia Binenti and Jason Dittmer likewise engage an assemblage approach in Chapter 30 on **Popular Culture** in political geography. Using examples of consumerism and humor, they point to the everydayness of popular culture as emerging out of the assemblage of relations that link bodies, media artifacts, and other objects and materialities. The next chapters spotlight the role of nature and natural environments in driving different kinds of political trajectories, especially in the era of the Anthropocene. In Chapter 31 on **Natural Resources**, Andrea Marston shows how these become recognized as such and with what political effects. She discusses the critical interface between resource geography and political geography through recent literature on extractivism and the plantation. Hilary Faxon and Kendra Kintzi, in Chapter 32 on **Geopolitical Ecologies**, consider the intersections of political geography and political ecology, a heterogenous field committed to critical theory, grounded methodologies, and spatial justice. They show how research from a geopolitical ecology perspective can shed light on how new forms of value and authority emerge beyond the state.

Material political geographies can be extended to the realms of water, health, and the ocean, whereby the territorial management of maritime spaces, water resources, and deadly viruses has tremendous consequences for the survival, precarity, and mobility of different groups of people in everyday life. Chris Sneddon and Coleen Fox discuss political geographical contributions to the study of **Water Politics** in Chapter 33, showing how these power relations are about both the power to govern, control, and manipulate water access and the power to confer meaning to water. In Chapter 34 on **Health Geographies,** Brian King and Halie Kampman investigate the relationship between politics and human health. In tracking the evolution from medical to health geography, they demonstrate how this transition has contributed to political geographical understandings of health governance, disease management, and food politics. Finally, in Chapter 35 on **Ocean Geographies,** Po-Yi Hung and Yu-Chia Lin introduce diverse approaches to conceptualizing the spatial politics of the ocean and highlight four themes of relevance: disputed islands and oceanic gray zones, asylum seekers and migration, fisheries and infrastructure, and high seas governance and science studies.

The overarching goal of this new *Companion* represents a fresh look at the subdiscipline of political geography and how it can be mobilized to investigate, interpret, and understand our changing world. There are many ways to classify, thematize, and divide a subdiscipline and hence, by definition, there are limits to every chapter outline. In organizing this collection into distinct sections, we showcase both theory and concepts with topical and empirical investigations to signal the dynamism and vibrancy of the field.

References

Agnew, J. (1994). The territorial trap: The geographical assumptions of international relations theory. *Review of International Political Economy* *1*(1): 53–80.

Agnew, J. (2024). The language of intractability and the Gaza war: Conflating anti-Semitism and anti-Zionism is historically problematic and misses how much contemporary Israel has become a role model for ethno-nationalists worldwide. *Human Geography* *17*(2): 203–209.

Agnew, J., Mamadouh, V., Secor, A., & Sharp, J. (eds) (2015). *The Wiley Blackwell Companion to Political Geography*. London: Wiley Blackwell.

Alami, I., & Dixon, D. (2020). The strange geographies of "new" state capitalism. *Political Geography* *82*: 102237.

Amoore, L. (2020). *Cloud Ethics: Algorithms and the Attributes of Ourselves and Others*. Durham, NC: Duke University Press.

Bennett, M., Chen, J., Alvarez León, L., & Gleason, C. (2022). The politics of pixels: A review and agenda for critical remote sensing. *Progress in Human Geography* *46*(3): 729–752.

Benwell, M., & Dodds, K. (2011). Argentine territorial nationalism revisited: The Malvinas/Falklands dispute and geographies of everyday nationalism. *Political Geography* *30*(8): 441–449.

Billé, F. (ed.) (2019). *Voluminous States: Sovereignty, Materiality, and the Territorial Imagination*. Durham, NC: Duke University Press.

Daigle, M. (2016). Awawanenitakik: The spatial politics of recognition and relational geographies of indigenous self-determination. *The Canadian Geographer/Le Géographe canadien* *60*(2): 259–269.

Dalby, S. (2010). Recontextualising violence, power and nature: The next twenty years of critical geopolitics. *Political Geography 29*: 280–288.

Dittmer, J. (2010). *Popular Culture, Geopolitics, and Identity*. Lanham, MD: Rowman & Littlefield.

Dixon, D. (2015). *Feminist Geopolitics: Material States*. Burlington, NJ: Ashgate.

Dowler, L., & Sharp, J. (2001). A feminist geopolitics? *Space and Polity 5*: 165–176.

Duggan, M. (2017). Questioning "digital ethnography" in an era of ubiquitous computing. *Geography Compass 11*(5): e12313.

Erskine, T. (2024). AI and the future of IR: Disentangling flesh-and-blood, institutional and synthetic moral agency in world politics. *Review of International Studies 50*(3): 5340559.

Flint, C., & Taylor, P.J. (2024). *Political Geography, World-Economy, Nation-State and Locality*, 8th edn. London: Routledge.

Fregonese, S. (2017). Affective atmospheres, urban geopolitics and conflict (de)escalation in Beirut. *Political Geography 61*: 1–10.

Gottmann, J. (1973). *The Significance of Territory*. Charlottesville, VA: University Press of Virginia.

Hepple, L. (1986). The revival of geopolitics. *Political Geography Quarterly 5*: S21–S36.

Jeffrey, A. (2013). *The Improvised State: Sovereignty, Performance and Agency in Dayton Bosnia*. London: Wiley-Blackwell.

Klinger, J. (2020). Critical geopolitics of outer space. *Geopolitics 26*(3): 661–665.

Knight, D. (1982). Identity and territory: Geographical perspectives on nationalism and regionalism. *Annals of the Association of American Geographers 72*(4): 514–531.

Koch, N. (2013). Why not a world city? Astana, Ankara, and geopolitical scripts in urban networks. *Urban Geography 34*(1): 109–130.

Koch, N. (ed.) (2017). *Critical Geographies of Sport: Space, Power and Sport in Global Perspective*. New York: Routledge.

Koch, N. (2022). Greening oil money: The geopolitics of energy finance going green. *Energy Research & Social Science 93*: 102833.

Koopman, S. (2011). Alter-geopolitics: Other securities are happening. *Geoforum 42*(3): 274–284.

Leszczynski, A., & Elwood, S. (2022). Glitch epistemologies for computational cities. *Dialogues in Human Geography 12*(3): 361–378.

Mallin, F., & Sidaway, J. (2024). Critical geoeconomics: A genealogy of writing politics, economy and space. *Transactions of the Institute of British Geographers 49*(1): e12600.

McConnell, F. (2017). Liminal geopolitics: The subjectivity and spatiality of diplomacy at the margins. *Transactions of the Institute of British Geographers 42*(1): 139–152.

Medby, I. (2018). Articulating state identity: "Peopling" the Arctic state. *Political Geography 62*: 116–125.

Müller, M. (2017). How mega-events capture their hosts: Event seizure and the World Cup 2018 in Russia. *Urban Geography 38*(8): 1113–1132.

Murphy, A. (2022). The modernist territorial order and the Ukraine crisis. *Political Geography 97*: 102667.

Murrey, A. (2023). A decolonial political geography of resistance and digital infrastructural harm in Cameroon and Ethiopia. *Globalizations 20*(6): 967–993.

O'Loughlin, J. (2005). The war on terrorism, academic publication norms, and replication. *Professional Geographer 57*(4): 588–591.

Ó Tuathail, G. (1996). *Critical Geopolitics: The Politics of Writing Global Space*. Minneapolis, MN: University of Minnesota Press.

Ó Tuathail, G., & Agnew, J. (1992). Geopolitics and discourse. *Political Geography Quarterly 11*: 190–204.

Paasi, A. (1996). *Territories, Boundaries, and Consciousness: The Changing Geographies of the Finnish-Russian Boundary*. New York: John Wiley & Sons.

Painter, J. (2006). Prosaic geographies of stateness. *Political Geography 25*(7): 752–774.

Palmer, M. (2020). Rendering settler sovereign landscapes: Race and property in the empire state. *Environment and Planning D: Society and Space 38*(5): 793–810.

Rokem, J., & Boano, C. (2018). *Urban Geopolitics: Rethinking Planning in Contested Cities*. New York: Routledge.

Schindler, S., Alami, I., DiCarlo, J., et al. (2024). The second cold war: US-China competition for centrality in infrastructure, digital, production, and finance networks. *Geopolitics 29*(4): 1083–1120.

Sharp, J. (1993). Publishing American identity: Popular geopolitics, myth and the reader's digest. *Political Geography 12*(6): 491–503.

Sharp, J. (2013). Geopolitics at the margins? Reconsidering genealogies of critical geopolitics. *Political Geography 37*: 20–29.

Smith, N. (1984). *Uneven Development*. London: Blackwell.

Smith, S. (2020). *Political Geography: A Critical Introduction*. Hoboken, NJ: Wiley-Blackwell.

Squire, R., & Dodds, K. (2020). Introduction to the special issue: Subterranean geopolitics. *Geopolitics 25*(1): 4–16.

Squire, R., & Jackman, A. (2023). *Political Geography: Approaches, Concepts, Futures*. Berkeley, CA: Sage.

Steinberg, P., & Peters, K. (2015). Wet ontologies, fluid spaces: Giving depth to volume through oceanic thinking. *Environment and Planning D: Society and Space 33*(2): 247–264.

Taylor, P.J. (1982). A materialist framework for political geography. *Transactions of the Institute of British Geographers NS7*: 15–34.

Winichakul, T. (1994). *Siam Mapped: A History of the Geo-Body of a Nation*. Honolulu: University of Hawaii Press.

Framing Political Geography

Chapter 2

Political Geography in Historical Context

Luca Muscarà[1,2]

[1]*Department of Economics, University of Molise, Campobasso, Italy*
[2]*Pacte, Université Grenoble Alpes, CNRS, Sciences Po Grenoble, Grenoble, France*

> History is that certainty produced at the point
> where the imperfections of memory
> meet the inadequacies of documentation.
>
> —Julian Barnes, 2011

To understand the making of political geography, and avoid the *view from nowhere* (Nagel 1986; Shapin 1998), its authors and theories need to be placed in historical context. Yet the past *is* a foreign country, and it is impossible to fully project ourselves into the mindsets and intellectual atmospheres of bygone times (Lowenthal 2015). This is even more the case as each generation rewrites at least part of political geography's history (Keighren et al. 2012) for reasons of disciplinary identity, often manipulating the past for some present gain, either due to ideological bias and/or to reach some political or academic advantage (Johnston & Sidaway 2016).

This narrative account of the history of political geography was inspired by confronting Koch's (2014) warning on the relative historical amnesia of the younger generations (as if nothing predating the Cold War's paradigm shift mattered today) with Lowenthal's (2015) call that intergenerational transmission is what keeps a society together – including academic communities like that of political geography. The modern origins of political geography are in the period of interimperial rivalries (1870–1945), between the Franco-Prussian War and World War II. Indeed, wars offer a convenient periodization, as they play a defining role in the making of political geography, but focusing on their discontinuities requires

The Wiley Blackwell Companion to Political Geography, Second Edition.
Edited by Virginie Mamadouh, Natalie Koch, Chih Yuan Woon, and John Agnew.

simultaneously acknowledging the existence of threads of continuity (Mercier 2001). If revisiting the discipline's plural histories is inevitably a selective enterprise, tracking the genealogy of some earlier threads promotes a deeper historical understanding of our troubled times.

As modern political geography emerged as an academic subfield within a unitary geography strongly connected to history, it is useful to take a longer perspective. That is, we should also consider the Industrial Revolution with its many implications, the rise of nationalism (and internationalism), and the evolution of colonialism into imperialism, including reference to scientific developments such as those occurring in cartography and the spread of Social Darwinism. These threads provide the context of modern political geography and are intertwined in a complex web that Agnew (2002) aptly characterized as "naturalized knowledge." Since World War II the field has undergone deep changes and political geography continues to be in the making (Agnew & Muscarà 2012). Following a period of eclipse at the beginning of the Cold War, those early pioneers of political geography were increasingly challenged by critical theory for their Western and state centrism, nationalism, imperialism, racism, and, more recently, masculinism. Relating some old ideas to current morality is difficult, but such judgments do not exhaust the interest of the historian, for historical understanding requires an effort of imagination to comprehend the past in its otherness and alterity. This is true for political geography in particular, as some of its foundational ideas have crossed generations as well as divisions of (intellectual) labor, and at times seem to return to haunt our present.

The Russian aggression in Ukraine (started in February 2022 yet ongoing since 2014) and Israel's war with Hamas/Hezbollah (after the massacre of October 7, 2023) involve two nuclear powers and are fought in areas that have tragically seen armed violence for a century or longer. For all their recent causes and the differences between present and past, some of their roots are to be found in history. Once peace returns, these wars may rewrite parts of the world's geopolitical and geoeconomic map in a change comparable to that introduced by the end of the Cold War, affecting the evolution of political geography.

Early Unity of History and Geography

Political geography existed before its name was invented, within a unitary conception of geography that since ancient Greece has included both history and science. If the former was a literary and artistic description of more or less exotic creatures, places, and events, a scientific representation of the Earth as three-dimensional geometric space was constructed on astronomical observation and mathematics (Muscarà 2019). This dual nature of geographical knowledge was already seen as useful to power. Strabo, the Greek geographer, philosopher, and historian who lived during Rome's transition from Republic into Empire, considered geography as oriented in its entirety to the practice of government.

Geography and science flourished in the southern Mediterranean during the Umayyad and Abbasid caliphates, when medieval Europe was dominated by a place-based view of sovereignty in the fragmentation and isolation of political authority based on feudal relations of individual allegiance (Gottmann 1973), within the unitary frame of Christendom. There were no linear political boundaries then, but nonlinear linguistic ones (Hale 1993: 33). Only after the fifteenth-century rediscovery of Ptolemy's *Geographia* prompted the understanding of a continuous global ocean, together with other developments, was the way open to oceanic exploration. Sponsored by the rulers of the time, it would originate Europe's "Age of Reconnaissance" (Parry 1963), deepening the relationship between geographical knowledge and power.

A space-based conception of political authority appeared as early as 1494, as the Pope extended the Holy Roman Empire across the Atlantic and the Treaties of Tordesillas partitioned the "islands" of the New World between Spain and Portugal along a pole-to-pole meridian. Still, within Europe, territory would long continue to be seen in the medieval fashion as a series of differentiated places. Even the 1648 Treaties of Westphalia, traditionally considered as the birth of the modern territorial state system, had no maps: sovereignties appeared as lists of place names (Branch 2014).

With the Scientific Revolution, geography and cartography would contribute to shaping the modern European state, augmenting the power of the military and civilian administration, beginning in France with Louis XIV's centralized rule. There, Vauban introduced the notion of an exclusive territory and the Cassinis created the first scientific map of a state based on geodetic triangulation. Thus it is not surprising that the expression *géographie politique* first appeared in Paris, as the title of a book planned by Anne Robert Jacques Turgot (1727–1781). As its extensive table of contents shows (Turgot 1751), his project was ambitious: a historical study of physical geography's influence on the distribution of peoples on the globe and its division into states, including demographic, economic, social, and juridical conditions to inform the art of government. But it never saw the light of day, as Turgot took on governmental positions and eventually became an influential political economist.

The best example of the unity of geography at that time is perhaps Alexander von Humboldt (1769–1859), whom Goethe considered the prototypical modern scientist. As the Enlightenment reached Prussia, Humboldt developed his holistic philosophy of nature encompassing science, art, and literature. In it, everything living on Earth shared the same vital principle (*Lebenskraft*) and was to be studied in its geographical associations in specific places, anticipating the notion of ecosystems (Vallino 2021). An explorer and polymath, Humboldt contributed scientific discoveries to dozens of fields in the natural sciences. Moreover, believing in French revolutionary ideals, he also contested the practice of slavery, supporting its abolition. His lesson would be foundational in the nineteenth-century process of the academic institutionalization of geography in Prussia and beyond.

Humboldt lived at a time of profound social and political change. As the American and French Revolutions shattered previous political orders, establishing the "People" as a self-conscious political unit, French nationalism and the nation-in-arms emerged in response to foreign attacks on the Revolution. With Napoleon, the missionary character of this idealist nationalism waged war on other European countries, triggering in turn a nationalist resistance that established the principle of national self-determination. The rights of nations were seen as the same as individual rights (such as freedom from interference and oppression) in the 1848 upheavals, but once Germany and Italy were unified, their utopian nationalism would turn into integral nationalism and the new states would aggressively compete with the established powers for territory and resources (Hertz 1950).

This evolution of nationalism was coupled with cartographic developments. In the eighteenth century, "natural" borders were often seen as the ideal national boundaries (as if nature itself provided each community with its "natural" place on Earth). In 1815, the Treaty of Paris was the first major treaty to include a *map* of sovereignties. If geometric cartography had already been applied in the colonies, in Europe it would produce a linearization of political borders that gave birth to the cartographic "nation-state," whose sovereignty was now scientifically grounded on territory: "The linear boundary circumscribed the territorial claim completely, while the spatial homogeneity within those lines obscured other forms of authority" (Branch 2014: 137–138). Remaining forms of nonterritorial authority such as feudal rights, not representable cartographically, lost legitimacy and were removed from the practice of European international politics, while national identities, conceived as homogenous and bounded by the state, were projected on territory with the accuracy of scientific cartography.

Meanwhile, the Industrial Revolution had rapidly spread to Europe and beyond. Technological innovation in manufacturing changed national economies and societies in each industrializing country. As the demographic transition set in, and national populations rapidly grew, rural–urban migration affected the inner balance of states. Pressure on territory increased, while surface and underground railways sped up movement within states and cities, and telegraph and electricity, oceanic steamships, and cables connected the colonial empires with the metropoles. The resulting space–time compression accelerated history, revolutionizing ways of life. For Europe's newly industrialized states to thrive, every national economy needed raw materials and new markets, largely found in the colonies, but London (and Paris) already commanded most of those. As the new players transitioned to industrial powers searching for more resources and territory, geographical learned societies, already established as alliances of explorers, traders, and the military, would no longer suffice. A "new geography" was required to provide each state and its citizens with an understanding of its identity and place at the global scale.

Modern Geographers in the Age of Empires

While geography often emphasized the planetary viewpoint, its "evolution has partaken heavily of the particular national, cultural and environmental traditions and outlooks in which its leading scholars were nurtured" (Hooson 1983: 157). Political geography coemerged with modern geography at the end of the nineteenth century as they were institutionalized in national education systems, and most geographers, however committed to an "objective" geography, responded to the atmosphere and concerns of their countries and empires with ideas rooted in their understanding of their respective nations and their environment. In the Humboldtian tradition, physical and human geography continued to be conceived as unitary, and the latter would remain well connected to history until World War II.

Nature occupied a dominant position to explain social and political change: after 1859, Darwin revolutionized biology and European science, and Social Darwinists developed ideas of struggle for survival that were applied to states in their imperial scramble. Confronted with the remote tropical and subtropical worlds, geographers and ethnologists explained to English-speaking imperialists the relation between climate and the world's human variety, developing the "climate's moral economy" (Livingstone 1992: 221). Such environmental determinism was perhaps reassuring in reaction to all the technological changes: evidence of different degrees of technological development justified racist and/or paternalistic policies (such as the French *mission civilisatrice*) toward nonwhite colonized peoples (Meyer 2020).

With a background in the natural sciences, German geographer Friedrich Ratzel (1844–1904) believed in the general unity of organic life on Earth and was influenced by Social Darwinism. First in his *Anthropogeographie* (1882–1891), then in *History of Mankind* (1885), he extended notions of natural selection to human history as he thought that "human society and politics were subject to the same natural laws" (Klinke & Bassin 2018: 55). His seminal book *Politische Geographie* (1897) started the discipline of political geography. His organic view of the state considered it a higher form of biological evolution (Bassin 1987). Yet, realizing that the state was an "imperfect organism," Ratzel made up for its lack of inner connectivity with the spiritual force of its *Volk*, nurtured by its attachment to the soil. He underlined the importance of the state's geographical position and large territory. In an article on biogeography, Ratzel claimed that just as any organism grows and extends its spatial range, so too does the state, by territorially expanding its *Lebensraum*: "Just as the struggle for existence in the plant and animal world always centres about a matter of space, so the conflicts of nations are in great part only struggles for territory" (Ratzel 1897–1898: 458).

The Berlin Conference on European colonization and trade in Africa (1884–1885) triggered Ratzel's interest in political geography and he supported "the German acquisition of an overseas colonial empire" (Bassin 1987: 126–127). Still, the extractive system of formal empire soon met its limits as no more "empty" spaces were available for colonization by the newly industrializing states. Territorial expansion thus became a "zero-sum game" and interimperial competition exploded. Ratzel's political geography would become a model for the young German state after his death, as it fit well into the German intellectual context that was then actively promoting a military and competitive view of neighboring imperial powers: France, Russia, and Britain (Fischer 1967). His naturalized knowledge would influence the development of political geography in Europe and the United States through the work of his American student Ellen Churchill Semple (1863–1932). Since most of Ratzel's work was never translated into English, its interpretation in the Anglophone world was long dominated by Semple's geographically determinist version of Ratzel.

In France, the defeat in the Franco-Prussian War (1870–1871) was explained through the Prussian military's superior geographical knowledge of French territory, prompting a reform of national education in the Third Republic that gave geography a prominent place. A coherent program from primary school to university was developed by Paul Vidal de la Blache (1845–1918), his student Lucien Gallois (1857–1941), and others (Berdoulay 1981). A national school of geography spanning four generations developed around the regional paradigm emerging from their study of the interactions between societies and their natural environment. Faithful to Humboldt's unitary view of geography, Vidal shared Ratzel's naturalistic foundations of geography, but he saw the state as historical, not biological. France emerged from the harmony of its regions, whose "personality" Vidal explained through the cultural/economic relationship between local communities and their habitats: physical geography offered a range of possibilities that each community would value differently, developing locally specific *genres de vie*, or ways of life.

Vidalian geography also recognized the importance of human movement, comprising cities and their networks, as well as broader, supranational *cultural* regions across the world (Mercier 2001). For Vidal, the challenge of understanding the geohistorical and social transformations brought about by the nineteenth century's technological revolution required everyone to study the geography not just of one's homeland but of the entire planet. This shifted the inside/outside frontier from the state to the planet by articulating scales beyond the national one (Robic 2020). With World War I, Vidal and the Vidalians distanced themselves from Ratzel, developing their possibilist version of political geography.

In Britain, Halford Mackinder (1861–1947) was the leading advocate of British geographical education. His political geography naturalized world history by anchoring it in the physical map of the Earth: the oceans and the landmasses were its stage, and size, morphology, and geographical position determined history. As a lifelong supporter of British imperialism, Mackinder developed a geostrategic model to guarantee the British Empire's continued global supremacy, in view of the challenges it faced from Russia and Germany. In discerning between sea and land powers, he was influenced by both Ratzel and the American naval historian Alfred Mahan (1840–1914). But while Ratzel supported the development of a German oceanic navy, Mackinder was more concerned with challenges coming from the impact of railways on the Eurasian continent. Thus he redrew the global political map in three areas: the "Heartland" (an inner Eurasian core inaccessible from the sea), the "Outer Crescent" (an oceanic belt of maritime states held together by maritime navigation), and the "Inner Crescent" (an intermediate coastal continental area, partially accessible from the sea and the Heartland).

Mackinder's political geography was thus enclosed in a simple formula: controlling the Heartland, the strategic "pivot" of the World-Island, was the key to global supremacy. This dictum was a direct response to fears that the British Empire could be threatened by Russia and Germany. First articulated in 1904, this Heartland theory was adjusted in 1919 and 1943 (see later discussion) according to changing geopolitical realities. Like other forms of geographical determinism in early political geography, Mackinder's ideas have long been rejected by critical political geographers. Yet his impact on American strategic studies continued and remains a persistent dogma in military and statecraft circles.

A relevant exception to Mackinder's and Ratzel's state-centered environmental determinism (and politically distant from Vidal) were Frenchman Elisée Reclus (1830–1905) and his Russian friend Pyotr Kropotkin (1842–1891), both exiled for their anarchist views. Sharing a deep human sympathy for the people of all countries, their geographies were devoid of any Social Darwinism. Kropotkin even countered Darwin, stressing the lesson of cooperation rather than competition in the natural world (Hooson 1983).

Between Two World Wars

World War I (1914–1918) resulted from a fatefully shortsighted combination of aggressive nationalism, militarism, and industrialism. It introduced a major discontinuity in world history and in the history of political geography. In every warring country, geographers provided cartographic services to their armies and helped "policy-makers understand wartime commodity requirements" (Wilbanks 1985: 5–6). A few even contributed to military strategy and/or foreign policy. With German aggression again seen as a threat in France, French geographers distanced themselves from Ratzel. For example, Camille Vallaux denounced the imperialist tone of Ratzel's writings, and Vidal in 1917 attacked Ratzel's notion of *Raum* as an expression of the "hegemonic will of an aggressive power" (Robic 2020: 81), which ridiculed the principle of self-determination of peoples on which US president Woodrow Wilson would base his "scientific peace" plan for the postwar period.

In the United States, geography had been influenced by the German tradition (Wright & Koch 2009). As the country entered the war, political geography gained prominence with Isaiah Bowman (1878–1950), who would develop a different thread. Sharing Wilson's liberal idealism, he actively supported his newly engaged foreign policy as chief territorial specialist at the Versailles Treaty (1919). Bowman contributed to the idea of a scientific peace: each problem required the most rational solution, reached by analyzing extensive geographical information, either coming from the American Geographical Society's library or from extensive fieldwork, (such as that carried out for the occasion by Columbia University's geographer Douglas Johnson). In Versailles they collaborated with other geographers, especially the French and British. As 3000 miles of new borders had to be drawn, geographers from every delegation were mobilized and American geographers produced thousands of maps for the carving out of new states in the former realm of the three defeated empires (the German, Austrian-Hungarian, and Ottoman Empires). Ideally, the "new world order" would have resulted from applying the principle of national self-determination to the successor states. In practice, the "Big Four" (the United States, the United Kingdom, France, and Italy) created a *cordon sanitaire* in eastern Europe: a north–south string of "buffer" states to separate Germany and Russia and prevent their future territorial expansion east and west, respectively.

Other geographers sought to shape the postwar political order at Versailles. For his part, Mackinder was not in the British delegation, but having insider knowledge of the British government's plans he tried to influence the Peace Treaty by extending the Heartland to eastern Europe (Mackinder 1919). He led a 1920 governmental mission to southern Russia in the Civil War (1917–1922), after which he advocated for a double *cordon sanitaire* with an independent Belarus, Ukraine, and the Caucasus, albeit ineffectively. Despite his realist credentials, he also supported the League of Nations "as a means of transcending the awful realities of a world based on *Realpolitik*" (Ashworth 2011: 281). Overall, the Peace Treaty was an eye-opener to geographers: politics ruled beyond science. Geographers from different nationalities called in vain for preserving economic and urban transboundary regions, while the military, politicians, and diplomats saw borders as untouchable. Ultimately, the result was that national schools of geography were reinforced while international collaborations were also reshaped.

Lucien Febvre's *La Terre et l'evolution humaine* (1922) historicized Vidalian geography by characterizing its view of the human–environment nexus as "possibilist," in opposition to the deterministic geographies of Ratzelian inspiration. The discontinuity introduced by the war would be reflected in books such as Albert Demangeon's *Le déclin de l'Europe* (1920), which emphasized the shift in world power from Europe to America, as well as Bowman's *The New World: Problems in Political Geography* (1921), which presented an inventory of 35 countries and world regions with a problem-solving approach for the US national interest. Yet the prewar traditions and ways of imagining political geography would persist differently in Europe. For example, Vidal's legacy was developed by his former students, who continued to trace the "delicate, lace-like character of the interrelations between geographical environment, the psychology of peoples, and international politics" (Gottmann 1942: 205), best exemplified in the work of André Siegfried (1875–1959).

Meanwhile German nationalists, frustrated by defeat and the heavy territorial and economic reparations imposed at Versailles, returned to Ratzel's work through the mediation of the Swedish political scientist Rudolf Kjellén (1864–1922), who coined the term *Geopolitik*. While German geographers were not invited to the newly established International Geographical Union, they continued Ratzel's work after the war, extending his organic view of the state and notion of *Lebensraum*, and developing a pan-regional global model (O'Loughlin & Van der Wusten 1990). It is through their work that Ratzel's ideas reached Adolf Hitler: in 1925, the Munich-based geographer Karl Haushofer visited Hitler in prison and gifted him a copy of Ratzel's *Politische Geographie*. Haushofer worked to popularize grand geopolitical thinking through his journal *Zeitschrift für Geopolitik* (1924–1944). Here and elsewhere, he combined Ratzel's ideas, in their articulation by Kjellén, and Mackinder's Heartland model, providing an apparently scientific justification for Hitler's territorial expansionism in eastern Europe (while their Japanese translation would do the same for imperial Japan's in Asia). Yet in France, geographers consistently questioned *Geopolitik*'s scientific credibility: Albert Demangeon (1932, 1939) and Jacques Ancel (1936) attacked its method as "simple graphical planning on the map." And in Germany, Karl Wittfogel (1957 [1929]) criticized its lack of attention to economic issues and aimed at replacing nationalist geopolitics with Marxist geopolitics.

Meanwhile, the Soviets' 1917 October revolution and the Russian Civil War (1918–1920) produced a wave of Russian political refugees. In Paris, this included émigré geographer Pëtr Savitsky, who promoted an alternate geopolitical vision within the "Eurasianist" movement. It was a determinist geography centered on the Eurasian character of Russia, in reaction to its Westernization under the Romanov dynasty and to communism (Laruelle 2008). After the collapse of the Union of Soviet Socialist Republics (USSR), both Eurasianism and *Lebensraum* geopolitics resurfaced in Aleksandr Dugin's support for a Eurasian Economic Union, the Russian annexation of Crimea (Klinke & Bassin 2018), and the subsequent full-scale invasion of Ukraine in 2022.

The interwar period also saw developments in the scientific internationalist thread following the 1920 establishment of the League of Nations. Its International Institute for Intellectual Cooperation (IIIC), for example, promoted the 1932 "Why war?" exchange between Albert Einstein and Sigmund Freud. It also held international conferences in Paris in 1937–1938, in a late failed attempt to find a peaceful solution to the problem of the markets and distribution of raw materials raised by the so-called *Have-not* countries: Germany, Italy, and Japan (Dennery et al. 1939). Ultimately, as the League of Nations was unable to maintain the prevailing order after the Japanese invasion of Manchuria (1931), Fascist Italy's war on Ethiopia (1935–1936), and Nazi Germany's occupation of Austria and Czechoslovakia (1938), this internationalist spirit did not prevail.

If the 1939 pact between Nazi Germany and the USSR was *Geopolitik*'s top achievement, it became obsolete soon after when Hitler invaded Poland and World War II began. During this war, "the demand for geographers as specialists—in foreign area studies and transportation—simply exploded" (Wilbanks 1985: 5–6). After the United States entered the war in 1941, there were more geographers in Washington than in all American universities, providing cartography, photo-interpretation, and area reports to the Army's geographical services and military intelligence (Harris 1997). Scholars also tried to contribute intellectually to their countries' war efforts. Since Goebbels' propaganda portrayed *Geopolitik* as the key to early Nazi success, geographers (including Bowman, Hans Weigert, Derwent Whittlesey, and Jean Gottmann) analyzed it for the American public and scholars. In their reading, *Geopolitik* went well beyond military theory to become a science of war, under a sort of deification of the state (Gottmann 1947).

As part of this intellectual counteroffensive, Bowman invited Mackinder to revisit his Heartland theory once more in 1943. The British geographer, who predicted "the war between Russia and Germany for the control of Eastern Europe" (Blouet 2001: 14), claimed his Heartland theory to be even more useful, since a USSR victory over Germany would make it the "greatest land Power on the globe" (Mackinder 1943). Only an alliance of the United States, Britain, and France would keep Germany in check, he argued, hinting at a defense strategy centered on the North Atlantic powers that echoed the vision of contemporaries like Walter Lippmann (1943) and Nicholas Spykman (1944), perhaps influencing the 1949 creation of the North Atlantic Treaty Organization (NATO). While the war proved the enduring appeal of Mackinder's Heartland model, Eurasia's geo-strategic centrality was challenged by Spykman, a Dutch-American political scientist. He recentered the world map on the United States, arguing that control over its "rimland" was more crucial than the Eurasian landmass (Spykman 1944).

The Cold War's Paradigm Shift

At the Yalta conference (1945), Joseph Stalin, Franklin D. Roosevelt, and Winston Churchill agreed that the new borders would follow the respective position of Allied armies when the war ended; thus political geographers played no comparable role to that in the 1919 Versailles Treaty. Eastern Europe, including the "buffer" states of the former *cordon sanitaire*, would become part of the USSR's sphere of influence, as if Stalin had also learned Mackinder's lesson. Bowman supported the American plan for a new postwar order based on a new international organization to overcome the failure of the League of Nations. He also advised on refugee resettlement and supported the process of decolonization, so that once the former colonies became independent they could be open markets to US firms and sustain economic growth (Smith 2003).

Nevertheless, the USSR's different understanding of the international organization, and its development of nuclear weapons after 1949, would turn the Soviet and American superpowers' reciprocal diffidence into open hostility, intensified by the concomitant communist revolution in China. As the Cold War set in with its bipolar order, the two opposing blocks would compete to secure loyalty from the new independent states, while confronting each other through proxies in regional wars in Korea, Vietnam, and beyond. These conflicts expressed the US policy of *containment*, proposed by American diplomat George Kennan, which became a central tenet of American policy toward the USSR during the Cold War. Although Kennan's doctrine was not based on any reading of Mackinder's Heartland theory, it nonetheless revived the Heartland concept in subsequent American foreign policy toward the Soviet Union, becoming a fixture of Washington's military and political culture (Hooson 2001: 234).

The Cold War was marked by an intensified culture of militarism especially in the United States and USSR, including the arms race, the proliferation of nuclear arsenals, and the space race – all of which required huge financial efforts and a parallel scientific and industrial gigantism (Galison & Hevly 1992). Large-scale research projects saw many scientists and scholars adhering uncritically to government projects, including many geographers who embraced the "quantitative turn." Only in the 1970s did they begin to take a critical stance, seeding a paradigm shift in Anglo-American geography (Johnston & Sidaway 2016).

Increasingly abstracted from its biophysical base, human and physical geography split into two separate subfields. The resulting disunity weakened geography's holistic ambitions (Castree 2011: 289), while competition with academic disciplines with "harder" scientific credentials produced a turn toward specialization and fragmentation. To become more "scientific," human geography also severed its ties to history and distanced itself from regional geography, trying to become "a law-seeking discipline concerned with universal laws, not regional specifics" (Livingstone 1992: 355).

As nuclear weapons implied the possibility of absolute destruction, naturalized knowledge became obsolete (Muscarà 2015). Still, in the public perception, political geography was assimilated to *Geopolitik* and experienced a temporary eclipse. While in Europe most geographers were busy with reconstruction, in the United States their role as an "aid to statecraft" was taken by specialists in international relations and area studies. Meanwhile, fear of communism and of a nuclear war with the USSR triggered a Red Scare that also affected the atmosphere in American universities. After Senator Joseph McCarthy attacked sinologist Owen Lattimore in 1950, the adjective "political" became too risky for many (Harvey 1984).

The best exception to this 1950s eclipse of political geography was Jean Gottmann (1915–1994), the first cosmopolitan geographer of the twentieth century (Hooson 1983). Differently from most of his contemporaries he moved beyond a singular national rootedness. A Jewish-Russian refugee and naturalized French citizen, he was trained in the Vidalian school and became Demangeon's youngest assistant (1936–1940). But with the Nazi invasion of France he fled to the United States and, after two decades of untenured "Atlantic transhumance," in

1968 he settled in Oxford. During the Cold War he anticipated the "cultural turn" in political geography, introducing the psychological dimension (Gottmann 1952) after the thread developed by Siegfried and Demangeon. He also pioneered the urban scale in political geography, introducing the notion of multiscalar urban networks in his study of the US megalopolis, which evolved the Vidalian regional method. Considered the first intellectual of geography (Agnew 2002: 18), in the 1970s he created the Research Committee on Political Geography in the International Political Science Association to overcome the prewar boundaries between the two disciplines.

Meanwhile, political geography was slowly reinvented after the "terrible blow" of the 1948 closure of Harvard's department (Smith 2003; Wright & Koch 2009). In the 1960s, a renewed orientation toward spatial analysis was seen as a way to restore the field's "scientific status." Part of geography's "quantitative turn," this was especially strong in the United States, where statistical and mathematical tools combined with the "use of a formal vocabulary to reduce complex geographical patterns to simpler relations" (Barnes 2011: 381). Spatial analysis also appealed to the military-industrial complex, which was a major source of research funding at this time. But its positivism was soon contested by some of its very proponents, due to the distance between its "sophisticated theoretical and methodological framework" and its "ability to say anything really meaningful about events as they unfold around us" (Harvey 1972: 6). These tensions notwithstanding, the spatial-analytic evolved and remains one of political geography's enduring perspectives, extending its initial focus on electoral geography to a reengagement with the spatiality of states, geopolitics, identity, and nationalism (Agnew & Muscarà 2012).

Most importantly, political geography's relation to power changed significantly in the postwar period and was increasingly shaped by politics in its new intellectual center in the United States. The Vietnam War, the violation of minorities' civil rights, and the lack of equal opportunities for women became leading issues in the United States and ordinary people increasingly questioned the nation's democratic credentials while its political elite continued to assert them in their ideological opposition to the Soviet regime. With the Baby Boom generation's coming of age, students and scholars with more diverse social backgrounds contributed to the re-politicization of universities, which appeared outside the classroom in mass protest movements. It also manifested in academic writing and inside the classroom: as everything became political, geographers took an increasingly critical stance toward social injustice, questioning racism, sexism, ethnocentrism, and political prejudice (Harvey 1974).

As the gap between governments and people widened, and opposition to US foreign policy increased, Yves Lacoste (1976) denounced geography as serving the waging of war, giving geographers more reasons for distancing themselves from direct engagement with governments' policies. Throughout the 1970s, political geographers sought to recalibrate away from the quantitative turn's positivism. In searching for new paradigms, they became more open to extradisciplinary influences and new political-economic perspectives. In the context of the Cold War's bipolar order, this was often built in opposition to the neoliberal state's agenda, as with (neo-)Marxist theories. For example, Harvey developed a spatial approach to Marx's model by theorizing capitalism's "spatial fix," Doreen Massey theorized the spatial division of labor, while Peter Taylor elaborated the spatial dimension of Immanuel Wallerstein's World Systems Analysis to show how hegemonic states systematically exploit those on the global periphery.

By the early 1980s, John Agnew was already questioning World Systems Analysis's spatiality, and later, with his "Territorial Trap," he challenged the problematic geographical assumptions of international relations theory (Agnew 1994). Others interrogated the binary geopolitical frames of the Cold War, while geographers started to gather under a variety of flags. With poststructuralism, Michel Foucault's intellectual genealogies highlighted the central relation between knowledge and power, which geographers such as Claude Raffestin (1980) and innumerable others developed geographically. With postcolonialism, inspired by cultural and literary critic Edward Said, the language of power was developed spatially by generations of postcolonial geographers (see Woon, this volume).

Concurrently, the concept of geopolitics was reintroduced by Henry Kissinger in the 1970s. And by the 1980s, geographers like Saul Cohen, Yves Lacoste, David Harvey, and Peter Taylor were also starting to use the term more frequently (Hepple 1986). But it was not fully revived as a research area in political geography until John Agnew, Gearóid Ó Tuathail, Simon Dalby, Anssi Paasi, and a handful of others introduced a new wave of studies to define a new, *discursive* approach to "critical geopolitics" (Ó Tuathail 1986, 1996; Dalby 1988, 1990; Ó Tuathail & Agnew 1992; see also Agnew 2013; Moisio, this volume).

The collapse of the USSR in 1991 triggered a renewed interest not only in geopolitics but also in the environment and nature (Castree 2011: 293). Influenced by French scholars such as Isabelle Stengers and Bruno Latour, this research rejected the division between "natural" and "artificial" as a misrepresentation of reality. If during the Cold War climate research had been appropriated by the Earth sciences (Hulme 2009), the end of the Cold War led Simon Dalby to read the renewed interest in the environment in the 1990s as resulting from the search for a new security threat, after the global security community had suddenly lost a master narrative (Dalby 1999, 2003; Deudney 1999). Only in the last 10 years have more political geographers started to pick up these themes (see Faxon & Kintzi, this volume; Marston, this volume).

This focus on the environment was quickly eviscerated after the terrorist attacks on the United States of September 11, 2001, when the global security community found its new raison d'être in terrorism, resolving this problem for the military-industrial complex. Many academic political geographers were keen to jump on the new narrative of the Global War on Terror (GWOT), almost uniformly remaining critical of the United States and approaching the GWOT as yet another opportunity to criticize US hegemony (Brunn 2004; Hannah 2006; Dalby 2007; Ingram & Dodds 2009). Since the mid-2010s, more political geographers have returned to more diverse research agendas and the environment has reentered political geography – although nature/human agencies were reversed compared to the days of environmental determinism. As a wide scientific consensus has long proved that the trend toward warming in the atmosphere and ocean temperatures that began in the Industrial Revolution has planetary consequences, Dalby (2024) urged security studies to extend the focus on firepower from its military and nuclear meanings to climate and ecological changes, within the wider politics of nature–society relations.

In the post–Cold War era, critical geopolitics has engaged a wide range of scholars and research, from critical Anthropocene theory to border studies, the history of geography, and feminist and gender geography. Although shaped as an archipelago of epistemic communities, critical geopolitics appears to be the most thriving area of contemporary political geography in the twenty-first century. Its lasting success has perhaps been granted by its lack of an overarching theory or unitary program of research on which everybody would agree. The simultaneous coexistence of and dialogue with different theoretical perspectives not only characterize contemporary political geography with dynamism and a healthy pluralism, but also seem to mirror the lack of an enduring dominant narrative since the end of the Cold War.

Conclusion

Historical research on political geography has advanced substantially in the last decades (Agnew & Muscarà 2012; Martin 2015; Johnston & Sidaway 2016; Ferretti 2021). Although history does not repeat itself, historical understanding of the relation between political geography and its contexts can enhance our comprehension of present challenges. Political geographers' focus on war and peace was renewed by the GWOT and has been reignited by the wars of the 2020s (Mamadouh 2024). Over this time, the possibility of returning to a new Cold War has often been evoked in the West – first against China, then against Russia. This seems unlikely in a world more connected than ever, and even if some commentators have noted that Western sanctions on Russia, China, Iran, and North Korea made it advantageous for those countries to do business together (Kendall-Taylor & Fontaine 2024), these countries remain very different bedfellows.

Russian president Vladimir Putin's aggression in Ukraine in 2014 brought back echoes of *Lebensraum* (Klinke & Bassin 2018) and the Russian invasion of 2022 was framed within Mackinder's Heartland theory (Hochberg & Hochberg 2023). After the 1919 US plan for a "scientific peace" failed in Versailles, Bowman (1921) concluded that there "peacemakers learned that definable problems could be measured, but that sentiment was rarely so." In this vein, it is useful to recall that Gottmann (1973) explained territory as a "psychosomatic device," the material projection of the psychological search for security and for opportunities. This applies to territorial identity conflicts such as the 2020s wars between Russia and Ukraine or between Israel and Hamas/Hezbollah. While the world struggles toward multipolarism, a new bipolar order would be psychologically regressive, even though the fears to promote it are fabricated by nationalist populist leaders (Muscarà 2020) or the industrial-military complex in need of an enemy to stay in business.

Yet security and opportunities are not necessarily in conflict, and geopolitics and globalization are not the opposite terms that they are often seen to be: a hidden geopolitics is at work within globalization itself (Agnew 2023). Today's diverse crises at the global scale underscore the continued need for effective planetary governance. To reempower the world's citizens beyond states and non-state agencies, greater imagination is needed alongside dual national and planetary citizenship (Archibugi 1992). In the present watershed for the narrative of the post-post–Cold War geopolitical "order", the history of science reminds us that order and skepticism are intertwined, and the history of political geography reminds us that geopolitical imaginaries move across generations.

References

Agnew, J. (1994). The territorial trap: The geographical assumptions of international relations theory. *Review of International Political Economy* 1(1): 53–80.

Agnew, J. (2002). *Making Political Geography*. London: Arnold.

Agnew, J. (2013). The origins of critical geopolitics. In Dodds, K., Kuus, M., & Sharp, J. (eds), *The Ashgate Research Companion to Critical Geopolitics*. Burlington, NJ: Ashgate, 19–32.

Agnew, J. (2023). *Hidden Geopolitics: Governance in a Globalized World*. Lanham, MD: Rowman & Littlefield.

Agnew, J., & Muscarà, L. (2012). *Making Political Geography*, 2nd edn. Lanham, MD: Rowman & Littlefield.

Ancel, J. (1936). *Géopolitique*. Paris: Delagrave.

Archibugi, D. (1992). Models of international organization in perpetual peace projects. *Review of International Studies* 18: 295–317.

Ashworth, L. (2011). Realism and the spirit of 1919: Halford Mackinder, geopolitics and the reality of the League of Nations. *European Journal of International Relations* 17(2): 279–301.

Barnes, J. (2011). The Sense of an Ending. London: Jonathan Cape. Spatial analysis. In Agnew, J., & Livingstone, D. (eds), *The SAGE Handbook of Geographical Knowledge*. Los Angeles, CA: Sage, 381–389.

Bassin, M. (1987). Imperialism and the nation state in Friedrich Ratzel's political geography. *Progress in Human Geography* 11(4): 473–495.

Berdoulay, V. (1981). *La formation de l'école française de géographie (1870–1914)*. Paris: CTHS.

Blouet, B.W. (2001). *Geopolitics and Globalization in the Twentieth Century*. London: Reaktion Books.

Bowman, I. (1921). *The New World: Problems in Political Geography*. Yonkers, NY: World Books.

Branch, J. (2014). *The Cartographic State: Maps, Territory and the Origins of Sovereignty*. Cambridge: Cambridge University Press.

Brunn, S.D. (2004). *11 September and Its Aftermath: The Geopolitics of Terror*. London: Frank Cass.

Castree, N. (2011). Nature and society. In Agnew, J., & Livingstone, D. (eds), *The SAGE Handbook of Geographical Knowledge*. Los Angeles, CA: Sage, 287–299.

Dalby, S. (1988). Geopolitical discourse: The Soviet Union as other. *Alternatives: Social Transformation and Human Governance 13*(4): 415–442.

Dalby, S. (1990). *Creating the Second Cold War: The Discourse of Politics*. London: Pinter.

Dalby, S. (1999). Threats from the south? Geopolitics, equity, and environmental security. In Deudney, D., & Matthew, R.A. (eds), *Contested Grounds: Security and Conflict in the New Environmental Politics*. Albany, NY: State University of New York Press, 155–185.

Dalby, S. (2003). Environmental insecurities: Geopolitics, resources and conflict. *Economic and Political Weekly 38*(48): 5073–5079.

Dalby, S. (2007). Regions, strategies and empire in the global war on terror. *Geopolitics 12*(4): 586–606.

Dalby, S. (2024). *Pyromania: Fire and geopolitics in a climate-disrupted world*. Newcastle-upon-Tyne: Agenda Publishing.

Demangeon, A. (1920). *Le Déclin de l'Europe*. Paris: Payot.

Demangeon, A. (1932). Géographie politique. *Annales de géographie 41*(229): 22–31.

Demangeon, A. (1939). Géographie politique, à propos de l'Allemagne. *Annales de Géographie 48*(272): 113–119.

Dennery, E. et al. (1939). *Le Problème des matières premières*. Conférence Permanente des Hautes Études Internationales, Paris, IIIC.

Deudney, D.H. (1999). Bringing nature back in: Geopolitical theory from the Greeks to the global era. In Deudney, D.H., & Matthew, R.A. (eds), *Contested Grounds: Security and Conflict in the New Environmental Politics*. Albany, NY: State University of New York, 155–185.

Einstein, A., & Freud, S. (1932). *Why War?* Paris: IIIC. https://archive.org/details/freud-einstein-1933-war/mode/2up (accessed November 19, 2024)

Febvre, L. (1922). *La Terre et l'évolution humaine*. Paris: La Renaissance du Livre.

Ferretti, F. (2021). History and philosophy of geography II: Rediscovering individuals, fostering interdisciplinarity and renegotiating the "margins". *Progress in Human Geography 45*(4): 890–901.

Fischer, F. (1967). *Germany's Aims in the First World War*. New York: Norton.

Galison, P., & Hevly, B. (eds) (1992). *Big Science: The Growth of Large Scale Research*. Stanford University Press: Stanford, CA.

Gottmann, J. (1942). The background of geopolitics. *Military Affairs: Journal of the American Military Institute 4*: 197–206.

Gottmann, J. (1947). Doctrines géographiques en politique. In Gottmann, J., & Mirkine-Guetzévitch, B. (eds), *Les Doctrines politiques modernes*. New York: Brentano's, 17–40.

Gottmann, J. (1952). *La Politiques des états et leur géographie*. Paris: Colin.

Gottmann, J. (1973). *The Significance of Territory*. Charlottesville, VA: Virginia University Press.

Hale, J. (1993). *The Civilization of Europe in the Renaissance*. London: Harper Collins.

Hannah, M.G. (2006). Torture and the ticking bomb: The war on terrorism as a geographical imagination of power/knowledge. *Annals of the Association of American Geographers* 96(3): 622–640.

Harris, C. (1997). Geographers in the U.S. Government in Washington, DC, during World War II. *Professional Geographer* 49(2): 245–256.

Harvey, D. (1972). Revolutionary and counter-revolutionary theory in geography and the problem of ghetto formation. *Antipode* 4: 1–13.

Harvey, D. (1974). What kind of geography for what kind of public policy? *Transactions: Institute of British Geographers* 63: 18–24.

Harvey, D. (1984). On the history and present condition of geography. *Professional Geographer* 36(1): 1–10.

Hepple, L. (1986). The revival of geopolitics. *Political Geography Quarterly* 5(S): s21–s36.

Hertz, J. (1950). Idealist internationalism and the security dilemma. *World Politics* 2(2): 157–180.

Hochberg, M., & Hochberg, L. (2023). "Confining the enemy". Mackinder's theory of containment and the conflict in Ukraine. *Naval College Review* 76(2): 7. https://digital-commons.usnwc.edu/nwc-review/vol76/iss2/7 (accessed November 19, 2024)

Hooson, D. (1983). National cultures and academic geography in an urbanizing age. In Patten, J. (ed.), *The Expanding City*. London: Academic Press, 157–178.

Hooson, D. (2001). Geography in Russia: Glories and disappointments. In Dunbar, G. (ed.), *Geography: Discipline, Profession and Subject since 1870*. Dordrecht: Kluwer, 225–243.

Hulme, P. (2009). *Why We Disagree about Climate Change*. Cambridge: Cambridge University Press.

Ingram, A., & Dodds, K. (2009). *Spaces of Security and Insecurity: Geographies of the War on Terror*. Burlington, NJ: Ashgate.

Johnston, R., & Sidaway, J. (2016). *Geography and Geographers: Anglo-American Human Geography since 1945*, 7th edn. New York: Routledge.

Keighren, I.M., Abrahamsson, C., & Della Dora, V. (2012). On canonical geographies. *Dialogues in Human Geography* 2(3): 296–312.

Kendall-Taylor, A., & Fontaine, R. (2024). The axis of upheaval. *Foreign Affairs*, April 23. https://www.foreignaffairs.com/china/axis-upheaval-russia-iran-north-korea-taylor-fontaine (accessed November 19, 2024)

Klinke, I., & Bassin, M. (2018). Introduction: *Lebensraum* and its discontents. *Journal of Historical Geography* 61: 53–58.

Koch, N. (2014). Making political geography: Placing the new generations of political geographers. *Political Geography* 38: 50–51.

Lacoste, Y. (1976). *La Géographie, ça sert, d'abord, à faire la guerre*. Paris: Maspero.

Laruelle, M. (2008). *Russian Eurasianism: An Ideology of Empire*. Washington, DC: W. Wilson Center Press.

Lippmann, W. (1943). *U.S. Foreign Policy: Shield of the Republic*. Boston, MA: Little Brown.

Livingstone, D. (1992). *The Geographical Tradition*. Oxford: Blackwell.

Lowenthal, D. (2015). *The Past Is a Foreign Country – Revisited.* Cambridge: Cambridge University Press.

Mackinder, H. (1904). *The Geographical Pivot of History.* London: Royal Geographic Society.

Mackinder, H. (1919). *Democratic Ideals and Reality: A Study in the Politics of Reconstruction.* London: Constable.

Mackinder, H. (1943). The round world and the winning of the peace. *Foreign Affairs 21*(4): 595–605.

Mamadouh, V. (2024). Geography at war, geographers at peace. In Flint, C., & Dempsey, K. (eds), *Making Geographies of Peace and Conflict.* London: Routledge, 12–36.

Martin, G.J. (2015). *American Geography and Geographers. Toward Geographical Science.* Oxford: Oxford University Press.

Mercier, G. (2001). The geography of Friedrich Ratzel and Paul Vidal de la Blache: A comparative analysis. *Geography Online* 2(1). http://www.siue.edu/GEOGRAPHY/ONLINE/mercier.htm#N_1_ (accessed November 19, 2024)

Meyer, W.B. (2020). Environmental determinism. In Kobayashi, A. (ed.), *International Encyclopedia of Human Geography*, 2nd edn. Oxford: Elsevier, 175–181.

Muscarà, L. (2015). The banality of Empire. In Agnew, J., Mamadouh, V., Secor, A.J., & Sharp, J. (eds), *The Wiley Blackwell Companion to Political Geography.* Oxford: Wiley Blackwell, 324–338.

Muscarà, L. (2019). Maps, complexity and the uncertainty of power. In Coleman, M., & Agnew, J. (eds), *Handbook on the Geographies of Power.* Cheltenham: Elgar, 362–379.

Muscarà, L. (2020). The great swindle of Nationalist Sovereigntism: On territory, psychology, and communication technologies. In Moisio, S., Koch, N., Jonas, A.E.G., et al. (eds), *Handbook on the Changing Geographies of the State.* Cheltenham: Elgar, 96–106.

Nagel, T. (1986). *The View from Nowhere.* Oxford: Oxford University Press.

O'Loughlin, J., & Van der Wusten, H. (1990). Political geography of pan-regions. *Geographical Review 80*(1): 1–20.

Ó Tuathail, G. (1986). The language and nature of the "new geopolitics" – the case of US-El Salvador relations. *Political Geography Quarterly 5*(1): 73–85.

Ó Tuathail, G. (1996). *Critical Geopolitics: The Politics of Writing Global Space.* Minneapolis, MN: University of Minnesota Press.

Ó Tuathail, G., & Agnew, J. (1992). Geopolitics and discourse: Practical geopolitical reasoning in American foreign policy. *Political Geography 11*(2): 190–204.

Parry, J.H. (1963). *The Age of Reconnaissance.* Cleveland, OH: World Publishing Company.

Raffestin, C. (1980). *Per una geografia del potere.* Milan: Unicopli.

Ratzel, F. (1897–1898). Studies in political areas. *American Journal of Sociology 3*: 279–313, 449–463

Robic, M.-C. (2020). D'une guerre l'autre. Paul Vidal de la Blache. *Inflexions 43*(1): 77–82.

Shapin, S. (1998). Placing the view from nowhere: Historical and sociological problems in the location of science. *Transactions of the Institute of British Geographers 23*(1): 5–12.

Smith, N. (2003). *American Empire. Roosevelt's Geographer and the Prelude to Globalization.* Berkeley, CA: California University Press.

Spykman, N. (1944). *The Geography of the Peace.* New York: Harcourt Brace.

Turgot, A.-R.-J. (1751). *Plan d'un ouvrage sur la géographie politique.* https://www.institutcoppet.org/oeuvres-de-turgot-025-plan-geographie-politique (accessed November 19, 2024).

Vallino, C. (2021). *I pionieri del pensiero del nostro tempo*. ETS: Pisa.

Vidal de la Blache, P. (1917). *La France de l'est*. Paris: Colin.

Wilbanks, T. (1985). Geography and public policy at the national scale. *Annals of the Association of American Geographers* 75: 4–10.

Wittfogel, K. (1957 [1929]). *Oriental Despotism: A Comparative Study of Total Power*. New Haven, CT: Yale University Press.

Wright, R., & Koch, N. (2009). Ivy League and Geography in the US. In Kitchin, R., & Thrift, N. (eds), *International Encyclopedia of Human Geography*. Oxford: Elsevier, 616–622.

Chapter 3

The Politics of Political Geography

James D. Sidaway
Department of Geography, National University of Singapore, Singapore

> New ideas are born in a variety of ways. Sometimes they come from careful study and repeated experimentation, sometimes from dreams and flashes of inspiration, and sometimes from conversation and debate. But occasionally ideas are born through the attempt, whether successful or flawed, to communicate old ideas to new audiences.
>
> —Alicia Steinmetz (2021: 83)

Neither the prior 2015 edition nor the 2003 forerunner of this *Companion* contained a chapter on the politics of political geography. However, both set out a key aspect of this in sketching political geography's foundational history and the 2015 edition opened with reference to how political geography was "a term coined by the French statesman Turgot in the eighteenth century and established by German geographer Friedrich Ratzel with his landmark volume *Politische Geographie* (1897)" (Agnew et al. 2015: 3).

Jacques Turgot (1727–1781) penned his summary proposal for a book on "*la géographie politique*" in 1751 when he was a student, before he became an administrator and then minister in pre-revolutionary France. His political geography reflected his wider writings and policy to break from feudalism and envision and advance a stagist view of progress and history (Heffernan 1994). Such notions acquired wider circulation as part of the revolutionary milieu only half a century later. As Debarbieux and Rudaz (2015: 50) comment:

> Turgot's [proposed] book thus aspired to be a general reflection on the modalities for coordinating the geography of the modern state with physical geography, guided by

The Wiley Blackwell Companion to Political Geography, Second Edition.
Edited by Virginie Mamadouh, Natalie Koch, Chih Yuan Woon, and John Agnew.

> an interest in optimizing national territoriality . . . complemented by a normative proposal, namely, that the match between natural facts and political decisions constitutes an index of modernity and an aspect of moral character.

But whereas Turgot's putative earlier use of the term is barely or only cursorily cited in most disciplinary histories, the German origin story, oft repeated in textbooks and classrooms, offers a compelling narrative that locates the subdiscipline in a zeitgeist of nationalist state-building and competing imperial projects. The establishment of the new state of Germany (out of a fragmented patchwork of principalities) in the late nineteenth century provided intellectual space, patronage, and a rationale for political geography as an expression and tool of European state-building. In cueing the coalescing rise of such status quo pressures, political geography therefore predates the modern state-centric discipline of international relations (IR), which was institutionalized after World War I.

As Mark Bassin (1987) has shown, a complex mix of Prussian-German nationalism, Hegelian philosophy, and Darwinian biology influenced Ratzel, who continues to pull interest, most recently heralded by a landmark first book-length biography in English (Klinke 2023). Crucially, German nationalists from the early twentieth century drew inspiration from Ratzel's references to national/state *Lebensraum* (living space) – an idea developed by several others and also reworked by the conservative Swedish intellectual Rudolf Kjellén, who coined the term *Geopolitik* (geopolitics) in 1899. Situating this, the introduction to the 2003 forerunner of this *Companion* described how "early twentieth century political geography was largely in thrall to the great nation-states of the time" and characterized by a "tendency to read geography in largely physical terms [that] was combined with a reductionist understanding of politics as the activities of states and their elites" (Agnew et al. 2003: 3).

Political geographers' focus on elites was, however, combined with attention to the spatial distribution and composition of populations. In turn, the nexus of nation, state, and space expressed a wider nineteenth-century biopolitics, at a moment when nationalism, empire, and power were increasingly manifest via the advent of mass (print) media, the expansion of education and roll-outs of instrumental technologies such as censuses, and cartographic and population surveys mapping the locations, boundaries, and morphology of states (the work of Ghanaian political geographer Ernest Amano Boateng (1978: 5, 11) on Africa is illustrative). Within the momentum of these evolving and subsequent decolonizing contexts, political geography served pedagogic functions – in schools and universities – impressing the nation-state as a spatial project onto a "national" population. This calling was pioneered in Germany and soon emulated in a number of other European countries, as well as in the postcolonial Americas, especially as the field has become entangled both with geopolitics in Europe, the Americas, and Asia (see Moisio, this volume; Sidaway, Mamadouh, & Power 2013) and with geoeconomics, carrying similarly multiplex transatlantic and East Asian histories, since the early twentieth century (Mallin & Sidaway 2024).

Classical political geography was thus engendered by a political frame powerfully expressed by nation/state-building in late nineteenth-century Europe, and wider imperial and interstate competition. While many of these political geographers may have been on the political right, the consequences of their work in terms of territorial claims and aspirations to planetary space were often disturbing to the promise of international order through the first half of the twentieth century. A handful of attempts to forge an alternative, socialist political geography, by Robert Dickson and Frank Horrabin in the United Kingdom (see Hepple 1999; Johnston 2000) and Karl Wittfogel in Germany (see Smith 1987; Bassin 1996), were hardly influential in policy circles. Or in other cases, like Owen Lattimore in the United States, they ran afoul of Cold War red-baiting (Harvey 1983). Prior radical alternatives had been articulated by the Russian anarchist and geographer Pyotr Alexeyevich Kropotkin (1842–1921) (see Laughlin 2016) and French anarchist Élisée Reclus (1830–1905) – who were well connected with each other and with wider scientific and political networks (Stoddart 1975; Dunbar 1978a, 1978b; Ferretti 2011, 2018; Pelletier 2013). Such activations influenced other geographers who celebrated communities and political projects outside/beyond imperial sovereignty, attesting "the complexity and heterogeneity of the European intellectual field during the imperial age" addressed by Federico Ferretti (2017: 332).

However, for some years after World War II, notably the 1950s and 1960s, the prior rationales for traditional political geography served as impetus for other disciplines, at least in much of Western Europe and North America. Newer disciplines, notably IR, thereby displaced political geography's pre-1945 role as a science of imperial governmentality. In Quinn Slobodian's (2023: 189) suggestive claims, geopolitics' "use by the National Socialists [Nazis] rightly drove it into disrespect and discredit, but . . . much geopolitical thinking travelled with émigrés across the Atlantic to serve as the founding framework of international relations and US foreign policy from Henry Morgenthau to Henry Kissinger." Slobodian makes a vital point regarding the diffusion of political geography/geopolitics into postwar, American-led IR. But he notably omits mention that, instead of disrepute, the formal tradition of geopolitics also went from strength to strength through the 1950s into the 1970s in enduringly fascist Portugal and Spain (Sidaway 2000; Sidaway & Power 2005).

Classical geopolitics also found fertile ground in Turkey, according to Pinar Bilgin (2007: 740), as "a 'scientific' perspective on statecraft, and [among] the military as an actor licensed to craft state policies (by virtue of its mastery over geopolitical knowledge)." Similar geopolitical perspectives took renewed and murderous forms in Pinochet's Chile, while mobilizing other military writers and rulers in the Argentine and Brazilian dictatorships through the 1970s (Hepple 1992; Hewitt 2001). They were reanimated in Brazil during the presidency of Jair Bolsonaro (2019–2022) when they were folded into wider far-right civilizational narratives (de Souza Chaloub 2023).

In the 1960s, however, academia's newly US-centric landscape was shaped by the concurrent vogue for positivism and quantitative analysis, which dominated much academic human geography. This saw political geography relatively sidelined, even while the substrand of electoral geography was reinforced by the development of quantitative methods (Johnson and Warf, this volume; Muscarà, this volume). Electoral geography had a somewhat autonomous track – with a 1913 *Tableau politique de la France de l'Ouest sous la Troisième République* by the French scholar André Siegfried (1875–1959) – regarded as foundational (Sanguin 1985). But in this subfield too, the quantity of work long remained modest. An introduction to a textbook on the *Geography of Elections* declared:

> Although the roots of electoral geography can be traced to the early years of this century . . . research by geographers into various aspects of voting has never been more than slight in volume. With the possible exception of French work, the heritage for current workers in electoral geography consists only of a disconnected set of isolated studies, few in number and limited in purpose. (Taylor & Johnston 1979: 21)

Two edited collections affirmed that political geography continued being researched and taught in the 1960s (Fisher 1969; Kasperson & Minghi 1969), but there was no dedicated journal. Discussions about establishing one in the United States in 1945 garnered little institutional support: it was thought there would not be enough quality submissions (Minghi 2002). Moreover, for one leading geographer associated with quantitative methods – Brian Berry, then based at the University of Chicago – political geography had become a "moribund backwater" by the late 1960s (see Antonsich et al. 2009). This fate would change over the next two decades. The politics of these evolutions across time (yielding a sense of revitalization) and across space (yielding further translations) forms the foci of the next two parts.

The Politics of Political Geography's Revitalizations

In the last two decades of the twentieth century, the Anglophone subdiscipline underwent a striking renaissance in the United Kingdom and North America. A mid-1970s textbook (Muir 1975: 1) still regarded political geography as "simultaneously one of the most retarded and most undervalued branches of geography, and one that offers the greatest potential for both theoretical and practical advance" and argued for quantitative methods as a way forward. However, the appearance of a specialist journal in 1982 – *Political Geography Quarterly*, subsequently today's *Political Geography* – marked an opportunity for a methodologically pluralistic, revitalized, and critical political geography. The journal rapidly established a reputation within and beyond academic geography, flagging the field for more political geography–focused journals and textbooks.

Within this revitalization and expansion, feminist and allied scholarship broadened the concerns of political geography, particularly in surfacing connections between "macro" issues of states and geopolitics and the "micro" – or "intimate" (Pain 2015) – politics of everyday lives as concerns central to human security; modalities of what Rachel Pain (2021) elsewhere calls "geotrauma." What Joan Nogué (1998: 35, my translation) summarized as "a much more open political geography" also yielded a sense of the field as rather unwieldy and diffuse (Brown & Staeheli 2003), or even postdisciplinary (Painter 2003). The blurring of intellectual boundaries with the widening range of theoretical inspirations from critical social thought, interdisciplinary crossovers, and diverse political orientations untethered the subdiscipline from perceived moorings in service of state-centered interests toward a primary focus on education (e.g., Raento et al. 2010; Williams et al. 2013) to better understand how geographical knowledges are constituted and put to use in political action.

Yet the legacies of the nationalist and right-wing menace of the term "political geography" endured. When a "Commission on Political Geography" was first proposed at the Twenty-Fifth Congress of the International Geographical Union (IGU, formally established in 1922) in Paris in August 1984, the Soviet delegation objected to its "fascist and Nazi connotations" (International Geographic Union Commission on Political Geography 2024). As a result, the subdisciplinary group formed in 1988 was first known as the "Commission on the World Political Map" before it was finally renamed in August 2000 (Sanguin 2013). Indeed, right-wing geopolitical discourses have persisted and are on active service beyond the academy, as Murphy et al. (2004: 619) pointed out:

> Political geographers typically invoke the term [geopolitics] with reference to the geographical assumptions and understandings that influence world politics. Outside of the academy, geopolitics often connotes a conservative or right-wing political-territorial calculus associated with the strategic designs of Henry Kissinger, Aleksander Dugin, and followers of the new *Geopolitik* in Germany.

Murphy et al. cite examples of what they term a "neoclassical geopolitics" from a range of national contexts. Post-Soviet Russia soon became a key domain for such neoclassical geopolitics, as Alan Ingram (2001) and Graham Smith (1999) detailed. Likewise, John O'Loughlin (2000: 127) described this as "high geopolitics," motivated by traditional state-centric security concerns. Reviewing the circulation of far-right French and German Cold War materials, Bassin (2023: 1861) noted "the remarkable flexibility that gives the legacy of classical geopolitics its perennial appeal and utility." At the same time that various strands of right-wing and reiterations of "neoclassical geopolitics" (Megoran 2010) were being popularized beyond the academy, the revival of the field in Anglophone academic geography in the 1980s and 1990s saw political geography influenced by critical currents, including Marxism, poststructuralism, and feminism.

However, to some observers the revitalized critical political geography remained overly focused on Anglo-American cases and related global concerns. Writing a few years after the establishment of the journal *Political Geography*, Peter Perry (1987: 6) voiced the discontent that "Anglo-American political geography poses and pursues a limited and impoverished version of the discipline, largely ignoring the political concerns of four fifths of humankind." Eleonore Kofman (1994: 437) reiterated this in the mid-1990s, decrying "the heavily Anglocentric, let alone Eurocentric, bias of political geography writing." In this aspect, political geography was not spared the critique – then and still (Chapman 2023) – being leveled at "Anglo-American" human geography and social sciences more widely.

Yet other critical schools developed in political geography in postwar Europe. German-language geographers have scrutinized the historiographies of *Geopolitik* and the complex postwar maneuvers by which German geographers established a sense of separation from the wartime and 1930s subdisciplinary past (Natter 2003; Hannah 2016; Michel 2016). Indeed, the need to challenge far-right discourse continues to animate the German subdiscipline (Mullis & Miggelbrink 2022; Kuebler & Schopper 2023). Elsewhere in postwar Europe, awareness of the complicity of political geography in wartime occupation and conciliation policies even resulted in legal action, for example with the trial of a leading academic in Denmark after his arrest in 1945 (Larsen 2015). In postwar France, Francophone political geography and geopolitics have also (re)emerged. This early critical approach was exemplified in the work of Yves Lacoste (1976) as well as scholarship associated with *Hérodote* and writings by Paul Claval (1994) and Claude Raffestin (1980; Raffestin et al. 1995; Raffestin & Butler 2012). Many of these and allied works have appeared in Italian, Portuguese, Spanish, and sometimes German translations. But dialogue between all these and Anglophone scholarship has remained somewhat limited (Fall & Rosière 2008; Fall, 2012).

In the late 1960s and through the 1970s, Jean Gottmann (1915–1994), professor at the Oxford School of Geography from 1968 to 1983, was a notable bridge between Francophone and Anglophone political geography, and an original theoretical voice at a time when the field was relatively marginal (Muscarà 2005). Writing more than 75 years ago in *Annales de Géographie*, Gottmann (1947) had called for the geographical scrutiny of *carrefours* (crossroads) in a world of flows and mobility. Although Gottmann was better known for his account of an emerging *Megalopolis* (Gottmann 1961), he made significant contributions to political geography over 40 years, including a book on *The Significance of Territory* (Gottmann 1973). Subsequently, however, the depth and quantity of exchanges between Francophone and Anglophone geography lessened. Leslie Hepple (2000: 269) thus noted:

> The Anglophone geographical and geopolitical communities tend to know of the existence of the *Hérodote* school, but there has been remarkably little engagement

> with its ideas or referencing of *Hérodote* sources. . . . This neglect appears remarkable when one starts to list some of the features of the Lacoste-Hérodote school: the early work by Lacoste in the Marxist tradition; the construction of *Hérodote*'s analysis within the radical, post-Marxist culture of Vincennes (where Foucault and Deleuze, with other influential thinkers, worked for periods); and a direct engagement with Foucault in the first issues of *Hérodote*.

Though first envisaged in 1972, *Hérodote*: *Revue de géographie et de géopolitique* was established by Yves Lacoste and Béatrice Giblin in 1976 at the then eight-year-old Université de Vincennes, which itself was set up "après les événements de Mai 68" (Lacoste 2018: 157). Yet, although a selection of papers appeared in translation (Girot & Kofman 1987), *Hérodote* was not widely cited in the Anglophone subdiscipline. Hepple attributed this inattention to the linguistic limitations of many Anglophone geographers. In tandem with this, he identified a certain linguistic imperialism on the part of the wider Anglophone academy.

Critical Anglophone political geography, like the wider Anglophone discipline, operates predominantly in a transatlantic circuit of communication, conferences, and journals in which US–UK linkages are predominant (see Paasi, this volume). Anglophone political geographers are mobilized in the Institute of Australian Geographers Political Geography Study Group, the Specialty Group on Political Geography of the American Association of Geographers, the Political Geography Research Group of the Royal Geographical Society (with the Institute of British Geographers) (RGS-IBG), plus the (formally bilingual: English and French) Commission on Political Geography of the IGU. However, their links with groups elsewhere – such as, to name but two, the Arbeitskreise Politische Geographie in Germany or the Commission de géographie politique of the Comité National Français de Géographie – are uneven.

The politics of the Israel–Palestine conflict have long reverberated through the politics of political geography (Falah 1989; Kliot & Waterman 1990). Ghazi-Walid Falah and Nadia Abu-Zahra (2022: 9), for example, outline a telling anecdote of the IGU's rejection of a 1992 membership application from a recently constituted Palestinian Geographical Society on the grounds that the main applicant on behalf of the Palestinian society was at the time residing in the United States. The IGU's response stated that "the application had to come from Palestinian geographers or institutions, residing in . . . the West Bank and Gaza Strip, where such territory 'does not overlap with Israel's sovereign territory' – that is, the 'domain' of the Israeli Geographical Society."

Two decades ago the Anglophone subdiscipline's flagship journal *Political Geography* saw one of the editors who was committed to a boycott of the Israeli academy refusing to handle a paper by two authors based in Israeli universities (Slater 2004). Such academic boycotts did not correspond with the journal's official editorial or publisher positions (O'Loughlin 2004; Waterman 2005) and the paper – a comparative analysis of Israel as an "ethnocratic" regime – was subsequently accepted for publication (Yiftachel & Ghanem 2004). Strongly held contentions on this issue have never been buried and, two decades on, in the wake of the October 7, 2023, Hamas attack and the Israeli war unleashed on Gaza, questions about who may act and speak for Palestine, how, where, and on what terms resurged at the RGS-IBG (Griffiths et al. 2024).

The Politics of Translating Political Geographies

In the light of the contexts and debates just sketched out, one productive way forward is to ask: What is lost (and gained) in translations? Translation is fraught with difficulties and possibilities. Reviewing these and how they are shaped by commercial, cultural, and geopolitical power, Susan Bassnett (2005: 397) elucidates:

> Translation is a kind of journey. It is an activity that always involves motion, it is a passage from one language to another, and hence from one culture into another. There is also always a temporal dimension, for what is written in one place, in one time, is then rendered for readers in another place and another time.

Attentive to translations as continuations and alterations to texts through space and time, Sidaway et al. (2004) explore categories and concepts that have been central to much political geography in Portuguese, Spanish, Malay, Korean, Japanese, and Urdu. They point out that the question of what happens to even quite fundamental political geography concepts, such as the state, sovereignty, borders, and territory, in different national-linguistic contexts begins to problematize universalized, taken-for-granted assumptions. Thus:

> Languages and meanings of the political are everywhere caught up in wider cosmographies and hermeneutics. These are usefully approached with an openness to political thought and meanings: recognizing that our categories, terms, and analyses deserve enrichment from relatively unaccustomed sources. (Sidaway et al. 2004: 1046)

Likewise, Claudio Ingerflom (1993) and Oleg Kharkhordin (2001) consider the significant differences in the concept of the state between Russian and West European languages. In all cases, the state emerges as a complex concept rooted in residual mystical foundations of power and tied in with national imaginaries. Critical analysis of these foundations therefore needs to move beyond the spoken and written word to consider other representations. One of the most suggestive conceptualizations has come from Thongchai Winichakul, a historian of Thai ideas of identity/sovereignty. Based on his study of Thai concepts of territory and sovereignty, Winichakul (1996: 69–70) develops the concept of the sovereign "geo-body" that "occupies a certain portion of the earth's surface which can be easily identified. It seems to be concrete to the eyes and having a long history as if it were natural, and independent from technology or any cultural and social construction." Winichakul explicates the geo-body as an effect of modern practices and technologies of mapping. In the Thai case, the earlier traditional sense of undefined frontiers and interlaced sovereign powers was transformed through contact with colonial powers who were carving out bounded territories in Southeast Asia. In response, Thai territory was mapped, demarcated, and represented as the geo-body of Siam. In turn, the "Thai-ness" of these lands was further asserted when Siam was renamed Thailand in 1939.

However, representations of geo-bodies are not confined to conventional maps. One example is the circulation in South Asia of striking representations of *Bharat Mata* (Mother India) as the geo-body of the subcontinent. Sumathi Ramaswamy (2002: 178) notes the significance of these in late colonial and postcolonial India, where

> by the middle decades of the [twentieth] century, the map and the mother's body had become interchangeable, one substituting for the other. Indeed, in 1936, when the very first "temple" to Bharat Mata was opened in Banares [Vanaasi], it did not house an image of the goddess but a marble map of India made to scale with all its topographic features shown in great detail.

It is partly these cosmo-socio-geopolitical interfaces that led Sankaran Krishna (1994) to the concept of cartographic anxiety, describing postcolonial India's fixation with territoriality, mappings, and the secure inscription of borders. Similarly, Freitag (2001: 39) points to the acts of imagining nation and community, "where spectatorship meets creation in complex interplays between bodies, visuality, discourse, institutions and figurality"; and Stephen Turner (2002: 93) contends that "the subaltern, however disempowered in the contexts of official institutions, public media, and government, certainly can and does speak, *but not necessarily to you*."

In other words, there is a potential diversity of other political geographies – other ways of seeing and knowing the political. There are other ethics of power and space to acknowledge. This applies both to sites/foci, classifications, and theoretical inspirations. In terms of the latter, James Tyner calls for a "broadening of horizons." Writing from the United States, Tyner's (2004: 341) focus on the political thought of the African-American radical Malcolm X (El-Hajj Malik El-Shabazz) makes a wider point:

> Our geographies, and especially our political geographies, remain largely distant from non-European theorists and theories. Our texts on nationalism and identities, in particular, are woefully ignorant of Pan-African nationalism and other African diasporic movements. Also largely missing are the geographies of Pan-Asian, American and Chicano movements.

Sustained attention to location, translation, language, and context offers fruitful and challenging departures. Yet, as in Tyner's engagement of Malcolm X's geographical perspective, there is something more fundamental at stake than other languages, artifacts, and narratives supplementing (and sometimes disrupting) the conventional foci of much Anglophone political geography.

First, "translations" may be incommensurable. More specifically, translation frequently speaks to, and sometimes becomes part of, concrete political struggles. Michelle Daigle (2016) traces this via how Omushkegowuk Cree ontologies of self-determination are experienced outside and/or alongside formal state and intergovernmental structures in Canada. Another example is the 1840 Treaty of Waitangi, the founding document of the present-day New Zealand state, of which two – English and Maori – versions were signed by the representative of the British crown and Maori tribal representatives, respectively. The text of this Treaty/Triti, purportedly laying the basis for the colonial settlement of Aotearoa New Zealand, still configures the terms of struggles over the rights and obligations it is seen to confer (Clavé-Mercier 2024). In essence, as Marilyn Lashley (2000: 4–5) notes:

> the English and Maori texts of the treaty differ profoundly. According to the English text, indigenous Maori ceded "all the rights and powers of sovereignty over their respective territories" . . . Maori people retained "full and undisturbed possession of their lands and estates, forests, fisheries and other properties" . . . in return for "royal protection" . . . According to the Maori text, Maori ceded *kawanatanga* (governance) *not* sovereignty and retained *rangatiratanga* (chieftainship). For Maori, this distinction is paramount.

Most of the original Maori signatories to the Treaty signed only the Maori version and disputes over its interpretation, compounded by the existence of the two versions, are vigorously contested in contemporary Aotearoa New Zealand (Clavé-Mercier 2024).

Secondly, translations may oversimplify and de/recontextualize concepts. Categories and terms that have come down to modern Anglophone political geography (and allied work in political science and IR) to interpret China, for example, represent readings of Chinese terms that were rendered through colonial narratives. An example is how the idea of a characteristic Chinese "tributary system," which has been reanimated with the rise of China in recent decades (see Wang 2021), reworks some concepts that have Western colonial frames. Hsiao-ting Lin (2009: 491) notes how a specific understanding of China relates to a series of negotiations and translations:

> more from the late nineteenth-century Qing experience of dealing with the British and the Russians than from Chinese imperial tradition. The British, in particular, introduced the terms "sovereignty" and "suzerainty" into regional diplomacy, thereby parsing the notion of imperial sovereignty and territoriality in new ways.

Finally, translations may conceal misplaced assumptions about universality. Revisiting the case of Japanese geopolitics, Atsuko Watanabe (2018: 29) argues that any "universal applicability" of geopolitics as it is translated and takes root in different milieu "actually contains, and even depends on, diversity in usage."

Critical geopolitics has been insistent that all descriptions are views from somewhere; in other words, it contextualizes geopolitical discourses and is wary of universalistic pretensions found in some classical geopolitics. But more sensitivity to the structural positionality of academic norms and suppositions is vital, including those made in a critical guise. This comes more clearly into view when it is recognized that assumptions about universality in political geography themselves tend to be products of particular circumstances.

Conclusion

An enduring preoccupation of the social sciences – and especially geography – has been to envision how connections between diverse phenomena and places are configured through universal processes, narratives, and logics. However, ideas of universal processes, being simultaneously bound up with Western dominance and always geographically specific and situated, cleave to "a particular 'lay of the land' within the specific national/cultural contours of the . . . academy" (Megoran 2024). Whereas there are expanding calls to bring decolonial perspectives and politics to political geography (Naylor et al. 2018; Radcliffe and Radhuber 2020; Woon, this volume), there is also a longer-established and related sense of critical possibility and challenge in terms of alternative meso- and macrogeographies (Powell et al. 2017). For example, building on work on political geographies of the atmosphere, subterranean geopolitics, and oceans, or what Mia Bennett and Klaus Dodds (2024) have termed "earthly volumes," Yolanda Collins (2024: 3) wants to transcend "fixed, land-centered perspectives [that] limit understandings of power and nature."

Similarly, commitments to alternative metageographies continue to inspire other "decolonial" political geography (Sidaway 2023), including engagements with "trans-areas, such as the Indian Ocean Region, Zomia, the Black Atlantic, and other diaspora geographies" (Akhter 2023: 370). Both attend to translations, such as the ways in which Arabic discourses describe what in English is termed Arabia as *Al-Jazira* – or the peninsula (see too Halliday 2002). Elsewhere, Fred Halliday (2002: 214) reminds us of the significance of vantage points in naming practices, with an example from Cold War narratives: "What the Western world calls 'the Cuba missile crisis' of October 1962 is to the Russians the 'Caribbean crisis' and to the Cubans, who point out none of the missiles involved was 'Cuban', it is *la crise de octubre*."

Yet each alternative (including the conventional Western designations) carries distinctive sets of exclusions and focus points. Reviewing the move to relational and situated political geographies, via a case study of the Comoros, Zhu and Grydehøj (2023: 7) note:

> All these choices and approaches are entangled in geographies of power. Our choice to pin the Comoros to this or that map and see it from this or that perspective inevitably brings certain sets of relations into focus while obscuring others . . . The

trouble with relational thinking is not that there is too much of it but instead that it too often situates itself outside relation while simultaneously failing to acknowledge the limits to human understanding of relational complexity.

Issues of nomenclature and focus thereby reveal more profound differences of vantage point and scales and geometries of reference. Openness to these alternatives should be productively disruptive. In such a spirit, this chapter has no neat concluding takeaways. Developing conceptual alternatives is an ongoing challenge, and political geography is richest when reworked, resituated, redeployed, and reimagined – even while we necessarily expect and acknowledge a range of differences.

In this regard, however, I will close with cautionary comments on the standard evolutionary arc in narrating the political history of political geography. In an essay bearing the same title as this one published in 2008, Guntram Herb (2008: 21) questions the tendency "to present a story of progress from a benighted past to an enlightened present . . . of a heyday characterized by racism, imperialism and war . . . followed by a period of stagnation and decline . . .. and finally a Phoenix-like [critical] revival." Herb had rightly countered any reductionist singular story. Contemporary political geography's pluralism (in topics and methods) and politically engaged themes – around, for example, anarchism and poststatism (see Ince & Barrera de la Torre 2016), Anthropocene geopolitics (see Dalby 2020), "more-than-human" political geographies (see Fleischmann 2023), gender, racializations, racisms (and anti-Blackness), anti-racist mobilizations, and sexuality (see Faria, Massaro, & Williams 2020; Chacko 2023; Hawthorne & Lewis 2023) – reflect today's political conjuncture(s). This applies in particular in Anglophone and Western contexts, where a laudable subdisciplinary pluralism is more often methodical or topical than pluralism *across* the ideological and political spectrum.

As Bassin (2023) and others cited here have charted, there is a more complex story to be excavated. Succinctly framed in his review essay on some recent political geography textbooks, Nick Megoran (2024) considers that, while the texts fail to transcend the contexts they are written from: "Their benefit as classroom texts is to demonstrate to students that there is a politics and geography to political geography: the times and places in which political geographers work indelibly mark the type of political geographical work that we produce." In different ways, these diverse authors are engaging the limits, range, and situatedness of how today's political geographies will evolve and be viewed in years to come.

Acknowledgments

I am very grateful to the editors, especially Natalie Koch and Virginie Mamadouh, as well as Nick Megoran and Vani S, for their helpful comments on earlier drafts. I am also grateful to the publisher and editors (Kevin R. Cox, Murray Low, and

Jennifer Robinson) of *The SAGE Handbook of Political Geography* (2008) for permission to rework some material from my chapter on "The Geography of Political Geography" from the *Handbook* for inclusion here. As per that 2008 chapter, I dedicate this one to the celebration of the many border crossings made with my daughter Jasmin Leila Sidaway (1997–2007).

References

Agnew, J., Mamadouh, V., Secor, A.J., et al. (2015). Introduction. In Agnew, J., Mamadouh, V., Secor, A.J., et al. (eds), *The Wiley-Blackwell Companion to Political Geography*. Malden, MA: Wiley, 1–9.

Agnew, J., Mitchell, K., & Toal, G. (2003). Introduction. In Agnew, J., Mitchell, K., & Toal, G. (eds), *A Companion to Political Geography*. Oxford: Blackwell, 1–9.

Akhter, M. (2023). Muslim peripheries: A world regional perspective. *Dialogues in Human Geography 13*(3): 367–371. doi:10.1177/204382062311917

Antonsich, M., Minghi, J., Johnston, R., et al. (2009). Interventions on the "moribund backwater" forty years on. *Political Geography 28*(7): 388–394. doi:10.1016/j.polgeo.2009.10.007

Bassin, M. (1987). Imperialism and the nation-state in Friedrich Ratzel's political geography. *Progress in Human Geography 11*(4): 473–495. doi:10.1177/030913258701100

Bassin, M. (1996). Nature, geopolitics and Marxism: Ecological contestations in Weimar Germany. *Transactions of the Institute of British Geographers NS21*(2): 315–341. doi:10.2307/622484

Bassin, M. (2023). "Everything is revealed in maps": The European far right and the legacy of classical geopolitics during the Cold War. *Geopolitics 28*(5): 1843–1867. doi:10.1080/14650045.2022.2078709

Bassnett, S. (2005). Translating terror. *Third World Quarterly 26*(3): 393–403. doi:10.1080/01436590500033628

Bennett, M.M., & Dodds, K. (2024). Earthly volumes, voluminous materialities: Working with apprehension. *Territory, Politics, Governance 12*(1): 1–11. doi:10.1080/21622671.2023.2242394

Bilgin, P. (2007). "Only strong states can survive in Turkey's geography": The uses of "geopolitical truths" in Turkey. *Political Geography 26*(7): 740–756. doi:10.1016/j.polgeo.2007.04.003

Boateng, E.A. (1978). *A Political Geography of Africa*. Cambridge: Cambridge University Press.

Brown, M., & Staeheli, L.A. (2003). "Are we there yet?" Feminist political geographies. *Gender, Place & Culture: A Journal of Feminist Geography 10*(3): 247–255. doi:10.1080/0966369032000114019

Chacko, P. (2023). Racial capitalism and spheres of influence: Australian assertions of white possession in the Pacific. *Political Geography 105*: 102923. doi:10.1016/j.polgeo.2023.102923

Chapman, E. (2023). Revocalising human geography: Decolonial language geographies beyond the nation-state. *Progress in Human Geography 47*(1): 24–42. doi:10.1177/03091325221131852

Claval, P. (1994). *Géopolitique et géostratégie: la pensée politique, l'espace et le territoire au XXe siècle*. Paris: Nathan.

Clavé-Mercier, V. (2024). "Sovereignty is still the name of the game": Indigenous theorising and strategic entanglement in Maōri political discourses. *Review of International Studies*: 1–20. doi:10.1017/S026021052400010X

Collins, Y.A. (2024). A political ecology of atmospheres: A voluminous case study of the Guiana Shield. *Political Geography 109*: 103048. doi:10.1016/j.polgeo.2023.103048

Daigle, M. (2016). Awawanenitakik: The spatial politics of recognition and relational geographies of Indigenous self-determination. *Canadian Geographies/Géographies canadiennes 60*(2): 259–269. doi:10.1111/cag.12260

Dalby, S. (2020). *Anthropocene Geopolitics: Globalization, Security, Sustainability*. Ottawa: University of Ottawa Press.

Debarbieux, B., & Rudaz, G. (2015). *The Mountain: A Political History from the Enlightenment to the Present* (trans. J.M. Todd). Chicago, IL: University of Chicago Press.

de Souza Chaloub, J.G. (2023). The geopolitical imaginary of the Brazilian ultra-right. *Bulletin of Latin American Research 42*: 539–550. doi:10.1111/blar.13470

Dunbar, G. (1978a). Élisée reclus, geographer and anarchist. *Antipode 10*: 16–21.

Dunbar, G.S. (1978b). *Élisée Reclus: Historian of Nature*. Hamden, CT: Archon.

Falah, G. (1989). Israelization of Palestine human geography. *Progress in Human Geography 13*(4): 535–550. doi:10.1177/030913258901300404

Falah, G., & Abu-Zahra, N. (2022). The evolution of Palestinian critical geography in Palestine and beyond. In Berg, L.D., Best, U., Gilmartin, M., et al. (eds), *Placing Critical Geography: Historical Geographies of Critical Geography*. London: Routledge, 9–27.

Fall, J.J. (2012). Reading Claude Raffestin: Pathways for a critical biography. *Environment and Planning D: Society and Space 30*(1): 173–189. doi:10.1068/d5411

Fall, J., & Rosière, S. (2008). Guest editorial. On the limits of dialogue between Francophone and Anglophone political geography. *Political Geography 27*(7): 713–716. doi:10.1016/j.polgeo.2008.08.003

Faria, C., Massaro, V.A., & Williams, J.M. (2020). Feminist political geographies: Critical reflections, new directions. *Environment and Planning C: Politics and Spaces 38*(7–8): 1149–1159. doi:10.1177/2399654420952419

Ferretti, F. (2011). The correspondence between Élisée Reclus and Pëtr Kropotkin as a source for the history of geography. *Journal of Historical Geography 37*(2): 216–222. doi:10.1016/j.jhg.2010.10.001

Ferretti, F. (2017). Tropicality, the unruly Atlantic and social utopias: The French explorer Henri Coudreau (1859–1899). *Singapore Journal of Tropical Geography 38*(3): 332–349. doi:10.1111/sjtg.12209

Ferretti, F. (2018). *Anarchy and Geography: Reclus and Kropotkin in the UK*. London: Routledge.

Fisher, C.A. (ed.) (1969). *Essays in Political Geography*. London: Methuen.

Fleischmann, L. (2023). More-than-human political geographies: Abjection and sovereign power. *Political Geography 107*: 102949. doi:10.1016/j.polgeo.2023.102949

Freitag, S.B. (2001). Visions of the nation: Theorizing the nexus between creation, consumption, and participation in the public sphere. In Dwyer, R., & Pinney, C. (eds), *Pleasure and the Nation: The History, Politics, and Consumption of Public Culture in India*. Oxford: Oxford University Press, 35–75.

Girot, P., & Kofman, E. (eds) (1987). *International Geopolitical Analysis: A Selection from Hérodote*. London: Croom Helm/La Découverte.

Gottmann, J. (1947). De la méthode d'analyse en géographie humaine. *Annales de Géographie 301*: 1–12. doi:10.3406/geo.1947.12424

Gottmann, J. (1961). *Megalopolis: The Urbanized Northeastern Seaboard of the United States*. New York: The Twentieth Century Fund.

Gottmann, J. (1973). *The Significance of Territory*. Charlottesville, VA: University of Virginia Press.

Griffiths, M., Hughes, S., Mason, O., et al. (2024). An open letter to the *SJTG* and the Royal Geographical Society (with IBG): The war on Gaza, the Royal Geographical Society (with IBG), and a Palestinian literary event. *Singapore Journal of Tropical Geography 45*(1): 6–17. doi:10.1111/sjtg.12527

Halliday, F. (2002). *Two Hours That Shook the World. September 11, 2001: Causes and Consequences*. London: Saqi Books.

Hannah, M.G. (2016). Innovations in the afterlife of the Cold War: German-language human geography. *Social & Cultural Geography 17*(1): 71–80. doi:10.1080/14649365.2015.1066841

Harvey, D. (1983). Owen Lattimore: A memoire. *Antipode 15*(3): 3–11. doi:10.1111/j.1467-8330.1983.tb00052.x

Hawthorne, C., & Lewis, J.S. (eds) (2023). *The Black Geographic: Praxis, Resistance, Futurity*. Durham, NC: Duke University Press.

Heffernan, M. (1994). On geography and progress: Turgot's plan d'un ouvrage sur la géographie politique (1751) and the origins of modern progressive thought. *Political Geography 13*(4): 328–343. doi:10.1016/0962-6298(94)90002-7

Hepple, L.W. (1992). Metaphor, geopolitical discourse and the military in South America. In Barnes, T., & Duncan, J. (eds), *Writing Worlds: Text, Language and Discourse in Geography*. Routledge: London, 136–154.

Hepple, L.W. (1999). Socialist geography in England: J. F. Horrabin and a workers' economic and political geography. *Antipode 31*(1): 80–109. doi:10.1111/1467-8330.00093

Hepple, L.W. (2000). Géopolitiques de Gauche: Yves Lacoste, *Hérodote* and French radical geopolitics. In Dodds, K.J., & Atkinson, D. (eds), *Geopolitical Traditions: A Century of Geopolitical Thought*. London: Routledge, 268–301.

Herb, G.H. (2008). The politics of political geography. In Cox, K., Low, M., & Robinson, J. (eds), *The Sage Handbook of Political Geography*. London: Sage, 21–40.

Hewitt, K. (2001). Between Pinochet and Kropotkin: State terror, human rights and the geographers. *Canadian Geographer 45*: 338–355. doi:10.1111/j.1541-0064.2001.tb01186.x

Ince, A., & Barrera de la Torre, G. (2016). For post-statist geographies. *Political Geography 55*: 10–19. doi:10.1016/j.polgeo.2016.04.001

Ingerflom, C.S. (1993). Oublier l'état pour comprendre la Russie? (XVIe–XIXe siècle): Excursion historiographique. *Revue des études slaves 66*(1): 125–134. doi:10.3406/slave.1994.6168

Ingram, A. (2001). Alexander Dugin: Geopolitics and neofascism in Europe. *Political Geography 20*(8): 1029–1051. doi:10.1016/S0962-6298(01)00043-9

International Geographic Union Commission on Political Geography (2024). History of IGU-CPG. https://igu-cpg.unimib.it/cpg-history (accessed February 22, 2024).

Johnston, R. (2000). City-regions and a Federal Europe: Robert Dickinson and post-World War II reconstruction. *Geopolitics 5*(3): 153–176. doi:10.1080/14650040008407699

Kasperson, R.E., & Minghi, J.V. (eds) (1969). *The Structure of Political Geography*. Chicago, IL: Aldine.

Kharkhordin, O. (2001). What is the state? The Russian concept of *gosudarstvo* in the European context. *History and Theory 40*(2): 206–240. doi:10.1111/0018-2656.00163

Klinke, I. (2023). *Life, Earth, Colony: Friedrich Ratzel's Necropolitical Geography*. Ann Arbor, MI: University of Michigan Press.

Kliot, N., & Waterman, S. (1990). The political impact on writing the geography of Palestine/Israel. *Progress in Human Geography 14*(2): 237–260. doi:10.1177/0309132 59001400203

Kofman, E. (1994). Unfinished agendas: Acting upon minority voices of the past decade. *Geoforum 25*(4): 429–443. doi:10.1016/0016-7185(94)90022-1

Krishna, S. (1994). Cartographic anxiety: Mapping the body politic in India. *Alternatives: Global, Local, Political 19*(4): 507–521. doi:10.1177/030437549401900040

Kuebler, F., & Schopper, T. (2023). Teaching geographies of the far right in Germany – conceptualising hate and fear as didactic challenges. *Journal of Geography in Higher Education 47*(5): 773–791. doi:10.1080/03098265.2021.1990226

Lacoste, Y. (1976). *La géographie, ça sert, d'abord, à faire la guerre*. Paris: Maspero.

Lacoste, Y. (2018). Vincennes et le département de géographie. *Hérodote 168*(1): 157–163. doi:10.3917/her.168.0157

Larsen, H.G. (2015). Geopolitics on trial: Politics and science in the wartime geopolitics of Gudmund Hatt. *Journal of Historical Geography 47*: 29–39. doi:10.1016/j.jhg.2014.05.031

Lashley, M.E. (2000). Implementing treaty settlements via Indigenous institutions: Social justice and detribalization in New Zealand. *The Contemporary Pacific 12*(1): 1–55. doi:10.1353/cp.2000.0018

Laughlin, J.M. (2016). *Kropotkin and the Anarchist Intellectual Tradition*. London: Pluto Press.

Lin, H.-T. (2009). The tributary system in China's historical imagination: China and Hunza, ca. 1760–1960. *Journal of the Royal Asiatic Society 19*(4): 489–507. doi:10.1017/S1356186309990071

Mallin, F., & Sidaway, J.D. (2024). Critical geoeconomics: A genealogy of writing politics, economy and space. *Transactions of the Institute of British Geographers 49*(1): 12600. doi:10.1111/tran.12600

Megoran, N. (2010). Guest editorial: Neoclassical geopolitics. *Political Geography 29*(4): 187–189. doi:10.1016/j.polgeo.2009.09.001

Megoran, N. (2024). Review essay: What's the purpose of teaching political geography? *Political Geography 113*: 103072. doi:10.1016/j.polgeo.2024.103072

Michel, B. (2016). "With almost clean or at most slightly dirty hands". On the self-denazification of German geography after 1945 and its rebranding as a science of peace. *Political Geography 55*: 135–143. doi:10.1016/j.polgeo.2016.09.008

Minghi, J. (2002). "Do not start a journal on political geography": Bowman to Whittlesey—1945. *Political Geography 21*(6): 739–744. doi:10.1016/S0962-6298(02)00025-2

Muir, R. (1975). *Modern Political Geography*. London: Macmillan.

Mullis, D., & Miggelbrink, J. (eds) (2022). *Lokal extrem Rechts: Analysen alltäglicher Vergesellschaftungen*. Bielefeld: transcript.

Murphy, A., Bassin, M., Newman, D., et al. (2004). Forum: Is there a politics to geopolitics? *Progress in Human Geography 28*(5): 619–640. doi:10.1191/0309132504ph508oa

Muscarà, L. (2005). Territory as a pyschomatic device: Gottmann's kinetic political geography. *Geopolitics 10*(1): 26–49. doi:10.1080/14650040590907721

Natter, W. (2003). Geopolitics in Germany, 1919–45: Karl Haushofer, and the *Zeitschrift für Geopolitik*. In Agnew, J., Mitchell, K., & Toal, G. (eds), *A Companion to Political Geography*. Malden, MA: Blackwell, 187–203.

Naylor, L., Daigle, M., Zaragocin, S., et al. (2018). Interventions: Bringing the decolonial to political geography. *Political Geography 66*: 199–209. doi:10.1016/j.polgeo.2017.11.002

Nogué, J. (1998). *Nacionalismo y territorio*. Lleida: Milenio.

O'Loughlin, J. (2000). Geography as space and geography as place: The divide between political science and political geography continues. *Geopolitics 5*(3): 126–137. doi:10.1080/14650040008407695

O'Loughlin, J. (2004). Academic openness, boycotts and journal policy. *Political Geography 23*(6): 641–643. doi:10.1016/j.polgeo.2004.08.002

Pain, R. (2015). Intimate war. *Political Geography 44*: 64–73. doi:10.1016/j.polgeo.2014.09.011

Pain, R. (2021). Geotrauma: Violence, place and repossession. *Progress in Human Geography 45*(5): 972–989. doi:10.1177/0309132520943676

Painter, J. (2003). Towards a post-disciplinary political geography. *Political Geography 22*(6): 637–639. doi:10.1016/S0962-6298(03)00070-2

Pelletier, P. (2013). *Géographie et anarchie. Reclus, Kropotkine, Metchnikoff*. Paris: Éditions du Monde libertaire.

Perry, P. (1987). Editorial comment. *Political geography quarterly*: A content (but discontented) review. *Political Geography Quarterly 6*(1): 5–6. doi:10.1016/0260-9827(87)90026-7

Powell, R.C., Klinke, I., Jazeel, T., et al. (2017). Interventions in the political geographies of "area". *Political Geography 57*: 94–104. doi:10.1016/j.polgeo.2016.03.004

Radcliffe, S.A., & Radhuber, I.M. (2020). The political geographies of D/decolonization: Variegation and decolonial challenges of/in geography. *Political Geography 78*: 102128. doi:10.1016/j.polgeo.2019.102128

Raento, P., Minghi, J., Cox, K.R., et al. (2010). Interventions in teaching political geography in the USA. *Political Geography 29*(4): 190–199. doi:10.1016/j.polgeo.2010.04.005

Raffestin, C. (1980). *Pour une géographie du pouvoir*. Paris: Litec.

Raffestin, C., & Butler, S.A. (2012). Space, territory, and territoriality. *Environment and Planning D: Society and Space 30*(1): 121–141. doi:10.1068/d21311

Raffestin, C. with Lopreno, D. and Pasteur, Y (1995). *Géopolitique et histoire*. Lausanne: Payot.

Ramaswamy, S. (2002). Visualising India's geo-body: Globes, maps, bodyscapes. *Contributions to Indian Sociology 36*(1–2): 151–189. doi:10.1177/006996670203600106

Sanguin, A.-L. (1985). Political geographers of the past II: André Siegfried, an unconventional French political geographer. *Political Geography Quarterly 4*(1): 79–83.

Sanguin, A.-L. (2013). Institutional resurgence and future of political geography in France since the 1970s. *L'Espace Politique 21*: 2773. doi:10.4000/espacepolitique.2773

Sidaway, J.D. (2000). Iberian geopolitics. In Atkinson, D., & Dodds, K. (eds), *Geopolitical Traditions: Critical Histories of a Century of Geopolitical Thought*. Abingdon: Routledge, 118–149.

Sidaway, J.D. (2023). Beyond the decolonial: Critical Muslim geographies. *Dialogues in Human Geography 13*(3): 337–358. doi:10.1177/20438206221121651

Sidaway, J.D., Bunnell, T., Grundy-Warr, C., et al. (2004). Translating political geographies. *Political Geography 23*(8): 1037–1049. doi:10.1016/j.polgeo.2004.05.001

Sidaway, J.D., Mamadouh, V., & Power, M. (2013). Reappraising geopolitical traditions. In Dodds, K., Kuus, M., & Sharp, J. (eds), *The Ashgate Research Companion to Critical Geopolitics*. Farnham: Ashgate, 165–187.

Sidaway, J.D., & Power, M. (2005). "The tears of Portugal": Empire, identity, "race", and destiny in Portuguese geopolitical narratives. *Environment and Planning D: Society and Space 23*(4): 527–554. doi:10.1068/d345t

Slater, D. (2004). Editorial comment: Academic politics and Israel/Palestine. *Political Geography 23*(6): 645–646. doi:10.1016/j.polgeo.2004.08.002

Slobodian, Q. (2023). The call to political geography. *New German Critique 50*(3(150)): 185–192. doi:10.1215/0094033X-10708433

Smith, G.E. (1999). The masks of Proteus: Russia, geopolitical shift and the new Eurasianism. *Transactions of the Institute of British Geographers 24*(4): 481–494. doi:10.1111/j.0020-2754.1999.t01-2-00481.x

Smith, N. (1987). Essay review. Rehabilitating a renegade? The geography and politics of Karl August Wittfogel. *Dialectical Anthropology 12*(1): 127–136. doi:10.1007/BF00734792

Steinmetz, A. (2021). Hobbes and the politics of translation. *Political Theory 49*(1): 83–108. doi:10.1177/0090591720903393

Stoddart, D.R. (1975). Kropotkin, Reclus, and "relevant" geography. *Area 7*(3): 188–190. https://www.jstor.org/stable/20001005 (accessed November 19, 2024)

Taylor, P.J., & Johnston, R. (1979). *Geography of Elections*. London: Routledge.

Turner, S. (2002). Sovereignty, or the art of being native. *Cultural Critique 51*: 74–100. doi:10.1353/cul.2002.0023

Tyner, J. (2004). Territoriality, social justice and gendered revolutions in the speeches of Malcolm X. *Transactions of the Institute of British Geographers NS29*(3): 330–343. doi:10.1111/j.0020-2754.2004.00308.x

Wang, Y.-K. (2021). The strange journey of the tributary system. *Millennium: Journal of International Studies 50*(1): 267–277. doi:10.1177/03058298211050674

Watanabe, A. (2018). Greater East Asia geopolitics and its geopolitical imagination of a borderless world: A neglected tradition? *Political Geography 67*: 23–31. doi:10.1016/j.polgeo.2018.09.005

Waterman, S. (2005). True exchanges or thought crime? *Political Geography 24*(8): 998–1001. doi:10.1016/j.polgeo.2005.06.006

Williams, A.J., Jeffrey, A., McConnell, F., et al. (2013). Interventions in teaching political geography: Reflections on practice. *Political Geography 34*: 24–34. doi:10.1016/j.polgeo.2013.03.002

Winichakul, T. (1996). Maps and the formation of the geobody of Siam. In Tønnesson, S., & Antølov, H. (eds), *Asian Forms of the Nation*. Richmond, VA: Curzon, 67–91.

Yiftachel, O., & Ghanem, A. (2004). Understanding ethnocratic regimes: The politics of seizing contested territory. *Political Geography 23*(6): 647–676. doi:10.1016/j.polgeo.2004.04.003

Zhu, X., & Grydehøj, A. (2023). Troubling the politics of island relation: The Comoros in between, on the edge, and as microcosm. *Political Geography 101*: 102838. doi:10.1016/j.polgeo.2023.102838

Chapter 4

Geopolitics of Knowledge

Anssi Paasi

Geography Research Unit, University of Oulu, Oulu, Finland

The Geopolitics of Knowledge

One of the major debates in international social science since the turn of the millennium has revolved around the hegemony of the English language and Anglophone journals. In geography, this debate has focused on several themes. The first is the expanding role of English as a lingua franca in academic publishing. For some authors, this represents linguistic imperialism and cultural colonization (see O'Neil 2017), which marginalizes other linguistic cultures and voices that have been historically significant in geography as an academic field. The second is Anglophone journals as synonymous with academic quality, resulting from the increased importance of academic ranking tables and bibliometric tools such as Web of Science and Scopus. These tools are often seen as "global" databases, but both are structurally biased against research/knowledge produced in non-Western countries and in non-English languages. Both databases are overwhelmingly dominated by Anglophone journals (Tennant 2020; see also Paasi 2005). Ranking lists strongly shape social reality and the practices of higher education institutions. Such lists have become policy instruments for the global management of academia (Kauppi & Erkkilä 2011). The third theme is the neoliberal new public management (NPM) ideology that underlines (public-sector) accountability, rankings, and achieved "results" to improve competition, efficiency, and performance.

"Competition" is definitely not a new feature in academia. Bourdieu (1975), for example, saw science as a social field characterized by power relations,

The Wiley Blackwell Companion to Political Geography, Second Edition.
Edited by Virginie Mamadouh, Natalie Koch, Chih Yuan Woon, and John Agnew.

struggles, and strategies. He argued bluntly that every scientific choice – the choice of the area of research, methods, or the place of publication – is a political investment strategy, directed toward maximization of scientific profit; that is, of "potential recognition by the agent's competitor-peers" (Bourdieu 1975: 19, 22–23). Geographer Ron Johnston (2005: 2, 4) echoed similar ideas:

> Refereed journals are at the core of the contemporary academic enterprise . . . most new knowledge, however defined, is carried in their pages. At the same time, there is intense pressure for individual academics to publish in "high-quality" journals, to meet both individual (career progression) and collective (departmental evaluation) goals . . . Along with the charisma/status of publishing in a prestigious outlet, however, there is an issue of audience. Academics want their papers to be read and then cited, to be used as exemplars in later works.

Thus, competitive rationalities that "demand" recognition and pressures from universities and government departments encourage researchers around the world toward publishing models where indexed journals and impact factors are seen as important standards of "quality" (Agnew 2009; Sheppard 2013). Instead of monographs, institutional recognition is increasingly attributed to journal articles. Competition, pressure to publish, rankings, and the instrumentalization of research undoubtedly underlie many researchers' wellbeing problems. Self-abuse, anxiety, and other mental problems have become a serious threat in contemporary academic life (Berg, Huijbens, & Larsen 2016; Hall 2023).

Brankovic, Ringel, and Werron (2018) argue that macro-level trends (neoliberal ideology, the spread of audit culture, globalization) are not enough to understand rankings, but that attention must be paid to the "carriers" who "spread the gospel," the actors who work to reproduce, market, and disseminate ideas and thereby strengthen their legitimacy. This reflects the increasing role of international policy transfer (Dolowitz & Marsh 1996), which appears in the circulation of "best practices" across global space. Global policy networks bring together advisors, consultants, policy-makers, governments, and transnational actors (Prince 2012: 189). National governments rely increasingly on foreign and transnational agencies (such as the Organisation for Economic Co-operation and Development, OECD) in their policy choices. This also affects the diffusion of "international" standards for academic work.

For the ideal of "international science," cross-border communication is a virtue and science is largely a network-based activity that occurs at and across different scales: "There are no boundaries, no walls, between the doing of science and the communication of it; communicating is the doing of science" (Montgomery 2003: 1, cited in Cronin 2005). Knowledge is never completely made in one place and then simply consumed elsewhere. It is constantly negotiated and tends to transform as it circulates (Agnew 2007; Agnew & Livingstone 2011). Despite the moral ideal of a borderless academy and free-flowing knowledge, academic communication and exchange of ideas always echo power relations and contextuality. Similarly, languages also affect the production and dissemination of scientific information. Scholars from "peripheral" writing spaces and language areas often face significant obstacles to publishing in major Anglophone, English-language journals (see Canagarajah 2002; Tietze & Dick 2009), an issue also recognized by geographers (Müller 2021). This is part of a wider problem of the "geopolitics of knowledge," a notion proposed by Walter Mignolo (2002) in his works on postcolonialism, history of capitalism, and the dominance of Western epistemologies.

The Aims and Structure of This Chapter

The following sections will examine the debate about the Anglophone hegemony in geography and the dominance of the English language. I use this debate as my research material to explore both the roles of wider material and ideological processes and the contextual factors that characterize the globalizing academic world. This debate is constantly ongoing and (political) geographers are bringing new geopolitical realms into the debate, for example the postsocialist Global East. Trubina et al. (2020), for example, show that scholars from the Global East are ever more visible in leading journals and thus challenge the Anglophone hegemony.

The next section analyzes the structural and individual conditions behind the hegemony debate and the gradual rise of English to the status of a lingua franca. Some key themes of this debate are then explored to problematize key questions such as where "relevant knowledge" is produced and who has the right to speak (and be heard) about their own and other contexts. Some illustrations of the persistence of Anglophone dominance in political geography are then presented, and conclusions follow.

Globalizing Science, Neoliberal Rationalities, and the Rise of the Hegemony Debate

Academia is undergoing rapid globalization that reflects many fundamental transformations in social life, such as increased mobility, varieties in capitalist production and information and communications technologies, the spread of rationalism as an important knowledge framework, and the rise of forms of governance that enable new forms of regulation in societies (Scholte 2005). Scientific knowledge is crucial in managing these processes. Likewise, such knowledge is critical in the internationalization, competition, financialization, and regulation of globalization, as well as in building a so-called knowledge-based economy (Slaughter & Leslie 1997; Jessop 2018). At the same time, higher education, academic research, and "society" are becoming increasingly intertwined, as demonstrated by the rise of a new vocabulary and terms such as academic capitalism, entrepreneurial university, McUniversity, or Mode 1/Mode 2.

The common denominator of these trends is neoliberalism. Neoliberalism is not just a monolithic political project associated only with individualism or demands for a "minimalist state," so it is critical to recognize neoliberalism's historical contingency, geographical contextuality, and political complexity (Larner & Le Heron 2005; Jessop 2013). In academia neoliberalism means that knowledge is ever more treated as a commodity whose value is determined by the market. University scholars are also increasingly valued based on their entrepreneurial skills: grants received, patents gained, citations garnered (Sheppard 2013). Recent changes in political, institutional, and funding opportunities have reinforced the power of management and weakened the autonomy of professional researchers, both typical neoliberal conditions behind the growing anxiety in academia (Berg, Huijbens, & Larsen 2016; Coulthard & Keller 2016). Findings from the United Kingdom, for example, show that a neoliberal culture marked by metrication and casualization seriously challenges the traditional academic experience of geography as a vocation, not just a "job" (Mason & Megoran 2022).

Neoliberal rationalities thus accentuate the market-like logic of knowledge production and the new entrepreneurial academic subjectivities, phenomena labeled *academic capitalism* (Slaughter & Leslie 1997). Accordingly, any claim to hegemony must be contextualized within the political economies of globalizing capitalism, the emerging "knowledge economy" (Jessop 2018), and the struggle for symbolic capital (Bourdieu 1975). Cronin (2005: 5, 130) argues that the growing importance of symbolic capital in academic life has led to a "political economy of citation" and an "economics of fame." As competition for attention intensifies, Cronin (2005: 185) contends, tools for "tracking eyeballs" and measuring impact will continue to evolve.

The previous tendencies can be illustrated as a "wheel of power" in which the globalization of academia, political economies, the state and major organizations (like the OECD or the European Union), policy transfer, neoliberal rationalities, academic capitalism, evaluation/indexing/ranking cultures, the clamor to be "international" (publications and conferences as its key manifestations), and the rise of English as a lingua franca come together (Figure 4.1). The Anglophone hegemony debate is therefore related to the processes, events, and institutions identified in the wheel, but also to the struggle over the geopolitics of knowledge, uneven writing spaces, and the operation of the publishing industrial complex.

Figure 4.1 demonstrates the complexity of neoliberal rationality's role in formulating the subjectivities of scholars toward their "entrepreneurial selves" (see Sheppard 2013). Researchers are not just passive victims or compliant subjects under such rationality; they may develop various strategies to operate under the new market-like conditions or to resist them. Academics are to an increasing degree guided to evaluate their activities and to participate in both personal and institutional "competition" over symbolic capital. Yet they are not mere homogenous, individualistic, rational calculators (Bourdieu 2004). Indeed, for some scholars the features displayed in the wheel may provide a stimulus for critical

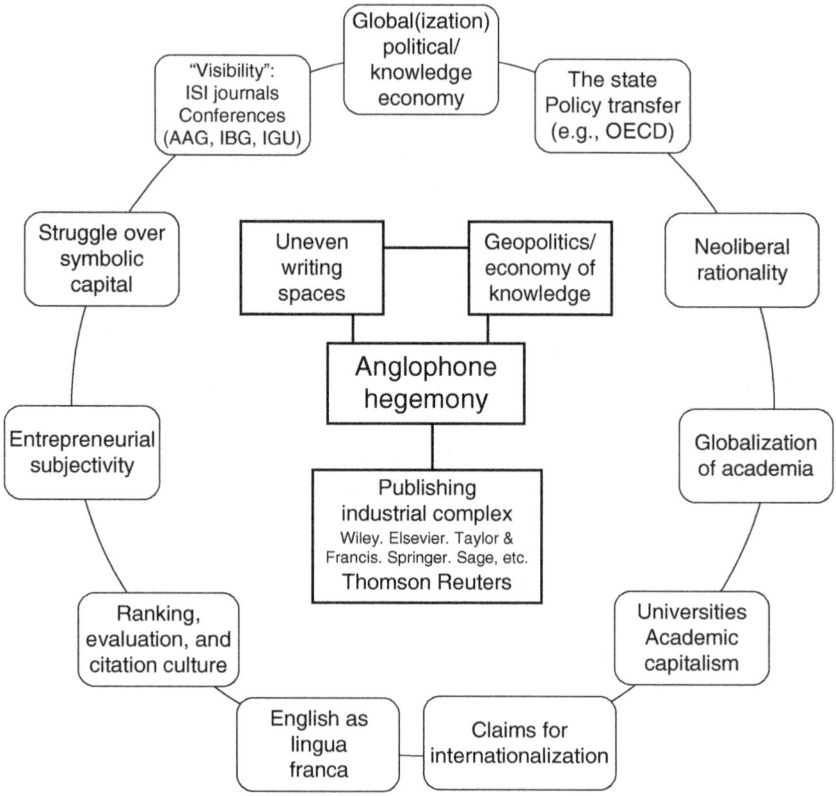

Figure 4.1 Wheel of power: a conceptual framework for understanding the Anglophone hegemony debates.
AAG, American Association of Geographers;
IBG, Institute of British Geographers;
IGU, International Geographical Union;
ISI, International Statistical Institute;
OECD, Organisation for Economic Co-operation and Development.
Source: Paasi 2015a: 512.

assessment, for others they may arouse a desire to contribute to major conferences or publishing in the "best journals," often with the help of people who translate their texts into English. For some "peripheral others" a key strategy may be networking with scholars from the "center" to lower publishing barriers. These features thus position scientific practice in a wider relational space and condition the conduct of researchers, their self-evaluation, and their strategic choices. The wheel thus displays the complicated power relations and struggles embedded in the current neoliberal forms of academic subjectification. Nevertheless, the wheel should not be seen as a "causal chain" or as a set of successive stages. Rather, it distinguishes for analytical purposes some key dimensions that are related to the operation of contemporary academia.

An additional international element of neoliberal rationality and policy transfer is new buzzwords that guide research practices, publication strategies, and ultimately the action of researchers. Expressions such as "world class," "world leading," or "top quality" characterize the new language of performativity. The new calculative message conveyed by these buzzwords is intended to influence academics' practices, motivations, and publishing spaces.

The changes in the use of the citation index are also illustrative of changes in academia. It was developed by Eugene Garfield in the North American Cold War atmosphere in the 1950s. The geopolitical rivalry between the United States and the Soviet Union backed these developments. The competition for control over the flows of knowledge was intensifying. This was accelerated by the space race and the launch of the Soviet Sputnik satellite in 1957 (Risteli 2003; Paasi 2005). At first the recording of publications was seen as crucial in managing the skyrocketing amount of knowledge and in supporting academics to track new publications and information. Gradually the citation index became an end in itself. Garfield seemingly anticipated this when he wrote: "I think it is important not only to be literature minded, but to develop citation consciousness. I'm not sure how you teach this. It requires indoctrination by informed mentors" (cited in Hargittai 1999: 28). Today such "mentoring" is effectively done by ministries of education and university administrators who are accountable for the productivity of universities.

The Key Themes of the Hegemony Debate

Thus, such indoctrination, audit culture, compliance, psychological need for recognition, and the feeling of being marginalized form the multifaceted structural and subjective backgrounds of the hegemony debate. The Gramscian idea of hegemony means the consent and approval of dominant practices and values by the subjugated, rather than a condition where force/violence is used to make them dominated (Williams 1988). Hegemony is thus related to *subjectification*; that is, to various techniques that are mobilized to influence people's thinking, behavior, and actions so that they tend to conform to certain patterns.

What might the Anglophone hegemony in geography mean from these perspectives? At first it was considered a "causal force": "the Anglo-American hegemony forces us [i.e., non-Anglophone scholars] to write articles for UK and US journals" (Aalbers 2004: 321). This kind of view suggests that scholars are passive, and it disregards the power of symbolic capital as both an individual and a structural characteristic in academia, thereby neglecting academic cultures and power structures. "Anglophone hegemony" does not claim that non-Anglophone scholars should publish their work in English in the best Anglophone journals or attend major conferences. Hegemony is structured in more multifaceted ways in academic and societal power relations. There is also the *will* of scholars to participate in the "game," to conform to its rules and the struggle over symbolic capital. Hegemony can thus be understood in part through the clamor toward internationalization, mediated by policy transfer, a new audit culture, and the prestige attached to certain classified journals; partly it expresses the motives of the subjects participating in the game, conditioned by wider institutions (Paasi 2022). As Gramsci noted, by establishing a new cultural hegemony "you can take over people's heads, and their hearts and hands will follow" (Davies and Bansel 2010: 5). Similarly, Brighenti (2021: 325) emphasizes the element of seduction inherent in evaluations and claims that evaluation exploits ambition: "scholars are collaborating with evaluation because of the pleasure to be derived from performing, being recognized and standing out."

Marginalization

The hegemony debate was started by geographers working in the former colonial peripheries of the Anglophone world (Berg & Kearns 1998; Olds 2001) and continental Europe (Minca 2000) around the same time as the term "geopolitics of knowledge" appeared (Mignolo 2002). The debate soon attracted participants from other European states and Singapore. Hungarian scholar Judith Timár (2004) suggested that it is more appropriate to speak of "Western hegemony" than only "Anglo-American." Comments often reflected the experiences of marginalization and even discrimination: "the fact that its [social science knowledge's] production is geographically dominated by the Anglo-American academic world means that knowledge of certain regions and places is likely to be less 'relevant' and less cited, irrespective of the author of this knowledge and their institutional affiliations" (Yeung 2002: 2099).

Yeung raises another issue on the agenda: the growing power of citations and the problems of reviewing that shifted the debate to the hegemony/dominance of Anglophone journals. Some critics expressed frustration with the arrogance of peer reviewers, who often found comments from the non-English-speaking world to be marginal in terms of both geography and topics or unsuited to the "big geographical debates" (Berg & Kearns 1998). Many commentators were disappointed with how Anglophone editors failed to understand how difficult it was for non-Anglophone geographers to access English-language markets and journals. Sometimes critics wrote about their personal experiences of unfair treatment in the editorial process, based on "poor use of English" and being marginalized as linguistic Others (Müller 2021). This criticism is still relevant, although the worst days of the Anglophone hegemony in geography are now apparently behind us (Trubina et al. 2020).

Unequal power relations were based not only on the fact that nonnative speakers are at a language disadvantage in an academic market dominated by the English language, but also on the fact that the most prestigious ranked journals were mainly edited by Anglophone scholars (Garcia-Ramón 2003). The monopolization of "internationalism" by Anglophone journals was thus seen as a complex ethical problem. Many publications promoted (and still promote) themselves as "international flagship journals," but in practice the editors and members of international advisory boards often represented only Anglophone universities (Gutiérrez & López-Nieva 2001). Imhof and Müller (2020) have recently shown that the majority of international geography journals are still deeply anchored in/represent Anglophone countries and scholars.

The experience of marginalization is also an epistemological issue related to the ostensibly inferior quality of knowledge produced in non-Anglophone contexts, thus disparaging such knowledge and implicitly non-Anglophone intellectual subjectivity itself. Critics also suggested that international geography was dominated by US/UK–based theories and that research conducted elsewhere was considered irrelevant, misguided, or at best exotic, local case studies incapable of producing innovative theory or framing the epistemological boundaries of disciplinary practices (Gregson et al. 2003). Simonsen (2002) acknowledged that there was a kind of "center–periphery" assumption at work, where the "center" becomes the referent of philosophical/theoretical reflection and the Anglo-American academic the main agent of theory formation. Many non-Anglophone geographers challenge the divide between the academic core (the Anglophone world) where relevant geographical theory is produced and the periphery (the non-Anglophone world) where receiver-consumers operate.

Several non-Anglophone critics represent theoretical research, so it makes sense that they have been concerned with ignoring/not citing non-Anglophone theoretical work. Unfortunately, the discussion has often included a rather one-dimensional understanding of theory: something that is simply moved from one place to another. A more dynamic view on theory would emphasize conceptualization and the formation of abstractions in concrete contexts (Sayer 1992). Such "conceptual invariances" are also much more likely to move from "peripheries" to "centers" than fixed-order, framework-type theories (Sayer 1992; Paasi 2022). Theories transform too in this process.

While the hegemony debate did not receive much response from the Anglophone community, some scholars were self-critical. Berg (2012) and Kitchin (2005), for example, intended to uncover the forms of hegemony. Outside of the hegemony debate, some political geographers commented on center–periphery thinking, often from a postcolonial perspective. Slater's (1998) proposal that it is crucial to move beyond the image of the Third World as a "conceptually empty space" to be filled with Western knowledge and epistemologies anticipated the postcolonial ideas of the geopolitics of knowledge (Mignolo 2002). Similarly, Sharp (2013) suggested that critical geopolitics has remained a particularly Western body of knowledge that has been less attentive to other traditions. Earlier, Perry (1987) had argued that Anglo-American political geography follows a limited and impoverished version of the discipline that largely ignores the political concerns of four-fifths of humankind. Also, O'Loughlin et al. (2008) had observed that most political geographers in the North American and European core are not very familiar with the evolution of the field in "relative peripheries." Robinson (2003), for her part, warned that political geography is continually subject to parochial forms of theorizing, and to this can be added Berg's (2012) view of Anglophone geography as a predominantly masculine, white enterprise.

It is perhaps an exaggeration to see the "cores" as located only in the West (or in the United Kingdom and the United States) and the "peripheries" elsewhere. In fact, both cores and peripheries are scaled, often in complex ways, vertically within states, universities, and even individual institutions, and they are all situated differently within broader structures of power and the circulation of knowledge.

English as a Lingua Franca

Another important theme focuses on the power of English as a lingua franca and the abduction of the idea of "international" in this linguistic world. The English language acts as an invisible, hegemonic force that powerfully influences the framing and ordering of the global world and its organizations (Tietze & Dick 2009: 120). Hence, English as a hegemonic language ostensibly has a "higher prestige" and validity than other languages (Tietze & Dick 2009). This has strengthened the power of Anglophone publishers and native writers by creating and maintaining asymmetric writing spaces. Similarly, English-language publishers, editors, and reviewers have become significant gatekeepers regulating what is published in English and defining what constitutes the "international," theoretical, or conceptual "cutting edge" (Kitchin 2005; Müller 2021). However, not everyone has criticized the emerging hegemony of the English language. Andres Rodríguez-Pose (2004), a Spanish economic geographer working in the United Kingdom, suggested in the language debate that a very widely shared English language can indeed help preserve the diversity of research among researchers from different countries.

How and when did the hegemony of the English language become institutionalized? Its current supremacy has long roots. It began with voyages to the Americas, Asia, and the Antipodes, and with the growth of the British Empire and its colonies in Africa and the South Pacific in the nineteenth century. Relations between the English language and economic life (trade, business, and commerce) have been close for centuries. Hence the spread of English and linguistic imperialism was associated with the spread of Western capitalism and reinforced by colonialism (Canagarajah 2002; Tietze & Dick 2009).

In Europe, post–World War II political and cultural connections between the United States and Europe were critical to the spread of the English language. The global geopolitical competition between capitalist and socialist camps created new international connections in Western countries. The Marshall Plan, a European recovery program, was designed in the United States after the war to help Western European nations recover and promote the ideals of freedom. The OECD's predecessor, the Organisation for European Economic Co-operation (OEEC), was created in part to implement the Marshall Plan. The OECD, on the other hand, became the most important economic advisor of the capitalist West in the Cold War against the Soviet Union (Kallo 2009). It has been a key international advisory institution behind the demands for increasing the competitiveness of higher education institutions and policy transfer (see Kitchin 2005; Brankovic, Ringel, & Werron 2018). It constantly makes international comparisons of publishing and citations between "nations," fueling an attitude of "national competition," reinforced and coordinated by national higher education systems and individual universities (Paasi 2013).

Another institution promoting connections between European and American researchers is the grant arrangements. For example, the American Fulbright scholarship system was established in 1946 and has been seen as a "cultural variant of the Marshall Plan" (Frijhoff & Spies 2004: 69). Similarly, US philanthropic institutions such as Ford, Carnegie, and the Rockefeller Foundation were significant in helping the global academic elite's networking and generating the United States' scientific dominance (Krige 2006; Parmar 2012). "International science" thus became a vehicle for promoting American values and interests after the war. In fact, it was even more so since the US authorities tried to reshape the European scientific landscape (Krige 2006) and create scientific networks around the world (Parmar 2012). The idea of English as *the* language of science thus spread around the world (O'Neil 2017). Despite criticism, many national science ministries outside the English-speaking world today strongly reinforce and encourage this development (Ammon 2001). Ministries and governments use OECD reports in national science policies and assessments and actually order such reports to justify their decisions (Kallo 2009).

As to the exchange and directions of flows of geographical knowledge across national borders, Anglophone human geography was probably more international before the 1950s–1960s, when scholars seemingly often mastered more languages and paid more attention to what happened in geography in other linguistic contexts. The lists of references in many key Anglo-American geography books, such as Hartshorne (1939) or Wright (1966), display that they not only mastered such languages as German and French but also actively reflected geographical knowledge produced in these contexts when developing their own arguments. Hartshorne (1939) explicitly admitted his debt to continental European geography (Paasi 2015a).

Relatedly, the English language has today become a naturalized part of the neoliberal rationality in the management of scientific practices. Minca (2013) suggests that the step to use English as the only language of so-called international debate and the corresponding tendency of some key sites in the production of geographical knowledge ("best departments in the UK and US") to show less and less interest in what happens in different contexts and in different languages have taken on a new meaning in light of the widespread adoption of neoliberal practices in Europe and beyond. Minca (2013: 8) argues that the ongoing internationalization has not led to academic interaction and exchange of information, but to the dominance of the leading Anglophone journals, where international debate takes place and gains recognition. Some organizations such as the International Geographical Union and the International Critical Geographies Group have tried to promote multilingualism, mainly through conferences, but seemingly not very successfully. These organizations used to be multilingual. Lamego (2020: 115) notes that the International Geographical Conference in Rio de Janeiro in 1956 was a turning point in language diversity toward the consolidation of English as the main, almost exclusive language in international scientific communication in International Geographical Congresses. One way to help scholars with English is translation, but this is costly and translated texts may lose many nuances of the original language (see ACME 2016).

Some scholars in the United States and United Kingdom certainly look at the science produced in continental Europe and elsewhere, but in contrast to the past it is often sociological and philosophical works rather than geographical ones that seem to motivate them – and this often takes place through English translations of these works. References to translations of the works of Foucault, Deleuze, Schmitt, or Agamben, for example, are now popular in the Anglophone political geographical literature. Such circulation of ideas may be understood as part of the neoliberal struggle for symbolic capital and recognition, in which scholars try to profile themselves in geographical and wider social science debates. While such borrowing provides an opportunity for the ever more rapid – mostly unidirectional – adoption and circulation of "new" ideas (that have often been published in their original languages perhaps decades earlier), it is still far removed from a genuine interchange of conceptual ideas in international geography (Müller 2021).

Insiders and Outsiders/Others

Language issues take us to one more theme critically associated with the geopolitics of knowledge and hegemony: What can we say about *other* linguistic-cultural contexts and who has the right to study and speak *about* and *for* them? The question of representation underlies all cases of speaking for another. While representations may have material effects as well as material origins, they are always mediated in complicated ways by location, discourse, and power (Alcoff 1991). This problem has been present in geography and anthropology since the Age of Exploration and the colonial period, where the roots of the expanding hegemonic position of the English language and its ostensible universalism can be situated. This issue, both implying and challenging a sharp state-centric (territorial) inside–outside divide, is crucial in problematizing the production and circulation of knowledge, which are principally relational processes.

Language is not merely a formal category (English, Finnish, etc.) but also a contextual form of reasoning. Accomplishing an understanding of such contextuality is a challenging process and often necessitates "first-hand local knowledge rather than half-baked generalizations culled from second-hand sources" (Agnew 2013: 162). Regardless of this fact, such action has been typical in geopolitics and political geography. Often the maps and texts – both national and international – produced by geopoliticians have been guided by ideological motives, have been distorted, or have divided the world in politically well-motivated ways. Likewise, regional geographers continually divide the earth into metaregions and much world regional geography teaching constantly draws on such divisions and related stereotypes of "us" and "them." Geographers have to be vigilant with this issue, but perhaps they can never completely move beyond it, because it stems from the basic task of geography: to understand the wider world and its places and regions. For critical geographers, this must be an anti-colonial, nonnationalist, and anti-hegemonic enterprise.

The Continuity of the Anglophone Hegemony

Not only do the institutional pressures to publish in major Anglophone journals still continue but the will for international recognition – both individual and institutional – of course seems to be strengthening among non-Anglophone scholars as well. This typically "requires" international publications (see Paasi 2022). As a response to the enduring international critiques of Anglophone hegemony, many publishers have internationalized the editorial boards of geographical journals and in some cases also nominated non-Anglophone editors. The number of indexed geography journals has grown since the millennium and also many new non-Anglophone journals have been indexed. Further, today journals like *ACME* accept submissions in English, French, Italian, and Spanish.

An interesting question is whether these new activities have diversified the authors in international political geography. I analyzed (Paasi 2015b) some time ago the international profile of *Political Geography*, which promotes itself as the flagship journal of political geography. I traced from Web of Science the articles published during two periods, 1992–2002 (N 338) and 2003–2013 (N 408), to map which countries the corresponding author came from. During both periods, most articles came mainly from the United States, the United Kingdom, and other English-speaking countries, with the United States dominating both periods, followed by England. The dominance of English-speaking countries/authors seemed deeply rooted and did not change much between these periods, despite the ongoing hegemony debate. The power of the English language was strengthened by the fact that many other countries also have English-language universities or at least partly English-language curricula (e.g., Singapore, South Africa, Israel). While the number of states from which published articles came expanded from 33 to 38, there was no major tendency toward a wider internationalization of authorship over the two periods. Similar findings have been reached in several other major Anglophone geography journals: more than 70% of articles are written by US or UK authors (Banski & Mariola 2013; see Imhof & Müller 2020).

Internationalization has become a more and more common catchword in perhaps most states. In order to update previous observations, I analyzed the publications in *Political Geography* in the period 2014–2022, this time from a slightly different perspective. I focused on papers that were classified as research articles on the journal's home page. During this period, the journal published a total of 829 such articles with 1055 authors altogether. Compared to the previous periods, there is a strong trend toward internationalization in the list of authors and their affiliated countries. There were authors from 83 countries, practically all over the world. Still, US- and UK-based authors dominate the scene: 37.9% of articles have one or more authors from the United Kingdom (England alone 34%) and 25.3% from the United States, although in both cases coauthors from other countries were common. Other strongly represented countries were Germany (6.8%), Australia (6.5%), and the Netherlands (5.4%). Almost 30 countries were represented by just one author. Internationalization was also exposed by the fact that numerous papers had several coauthors, who wrote together "across state borders." This has been typical in the natural sciences and medicine for decades but is seemingly becoming more common in (political) geography as well. This analysis can be used as a "proxy" to reveal the cooperation of scholars across borders, but the detailed data also indicate the increasing academic mobility or "brain drain" of researchers between countries. This partly challenges the "methodological nationalism" that is often present in country-specific analyses.

Conclusion

This chapter has problematized the globalization of knowledge production, the related unequal power relations, and the neoliberal features of academia. These themes have become highly important in the social sciences and humanities since the turn of the millennium. As an example of these trends in geography, the extensive, continually ongoing debate about the Anglophone hegemony in geography was analyzed in order to reveal the features of the politics of knowledge in this field. Criticism of the hegemony of Anglophone geography and the English language has been maintained primarily by non-Anglophone scholars from continental Europe, Asia, and the Global South and East. Their critique has focused on several topics: the uneven structural features of academic publishing, the regulatory roles of "ranking cultures," the hegemony of the English language, and the implications of neoliberal rationality in general. In contrast to this, only a few Anglophone geographers have found it important to participate in the debate or even to cite references that have been written in other languages.

The difference in interest between these two contexts is based on the fact that publishing in the native language in non-English-speaking countries is discouraged because such publications do not count in (international) "rankings" or in the apparent competition between universities and nations. The contemporary neoliberal business model followed in universities favors native English speakers who can write faster and who probably have a better understanding of the conventions governing English publishing than do nonnative speakers. In fact, the power of English emerges in all key areas of academic work: reading, writing, paper revision, rejection, and conference attendance and presentation, and thus may leave untapped the potential of nonnative speakers in research (Amano et al. 2023).

Interventions focusing on the Anglophone hegemony have been a successful project in the sense that they have made visible the perpetual power of the English language and Anglophone journals, and the privileged position that native English-speaking scholars have in the academic market. Many Anglophone journals have become more international when non-Anglophone members have been appointed to editorial and advisory boards. Whether following from this or not, non-Anglophone authors are now increasingly publishing in most Anglophone journals, pushing through the linguistic barrier. However, recent evidence shows that the authors of major international journals are still largely Anglophone, as are their cited references. Non-Anglophone research work and researchers are often disregarded. Müller (2021) has recently called on geographers around the world to give an adequate hearing to voices and scholarship outside the Anglophone core by reading and also referring to such work. This could create a more equal space for new theoretical approaches, issues, and methodologies and for a real exchange of ideas (Schurr, Müller, & Imhof 2020).

The persistence of hegemony debate reminds us that the social spaces of academic work and writing are not equal, and that knowledge does not flow freely in a "borderless world." The current journal-, impact factor–, and citation-centric culture has gradually become dominant around the world. It has to be resisted in the name of diversity and academic freedom: journals, monographs, and edited collections, as well as novel forms of documenting research (e.g., films), must all have their place in geography and the flow of ideas must be multidirectional. This certainly contributes to the health of the discipline but also to the health of academic scholars.

References

Aalbers, M. (2004). Creative destruction through the Anglo-American hegemony: A non-Anglo-American view on publications, referees and language. *Area 36*(3): 319–322.

ACME (2016). Themed section: Translation as critical practice. *ACME 15*(1): 1–103.

Agnew, J. (2007). Know-where: Geographies of knowledge of world politics. *International Political Sociology 1*(2): 138–148.

Agnew, J. (2009). The impact factor. *AAG Newsletter 44*: 3.

Agnew, J. (2013). Why criticizing grand regional narratives matters. *Dialogues in Human Geography 3*(2): 160–162.

Agnew, J., & Livingstone, D. (2011). Introduction. In Agnew, J., & Livingstone, D. (eds), *The Sage Handbook of Geographical Knowledge*. London: Sage, 1–17.

Alcoff, L. (1991). The problem of speaking for others. *Cultural Critique 20*(Winter 91–92): 5–32.

Amano, T., Ramirez-Castaneda, V., Berdejo-Espinola, V., et al. (2023). The manifold costs of being a non-native English speaker in science. *PLoS Biology 21*(3): 1–27.

Ammon, U. (2001). *The Dominance of English as a Language of Science*. New York: Mouton de Gruyter.

Banski, J., & Mariola, F. (2013). "International" or "Anglo-American" journals in geography. *Geoforum 45*(March): 285–295.

Berg, L. (2012). Geographies of identity I: Geography – (neo)liberalism – white supremacy. *Progress in Human Geography 36*(4): 508–517.

Berg, L., & Kearns, R. (1998). America unlimited. *Environment and Planning D: Society and Space 16*(2): 128–132.

Berg, L., Huijbens, E.H., & Larsen, H.G. (2016). Producing anxiety in the neoliberal university. *Canadian Geographer 60*(2): 168–180.

Bourdieu, P. (1975). The specificity of the scientific field and the social conditions of the progress of reason. *Social Science Information 14*(6): 19–47.

Bourdieu, P. (2004). *Science of Science and Reflexivity*. Cambridge: Polity.

Brankovic, J., Ringel, L., & Werron, T. (2018). How rankings produce competition: The case of global university rankings. *Zeitschrift für Soziologie 47*(4): 270–288.

Brighenti, A.M. (2021). Science evaluation – as it is, as it could be. *Social Science Information 60*(3): 1–5.

Canagarajah, A.S. (2002). *A Geopolitics of Academic Writing*. Pittsburgh, PA: University of Pittsburgh Press.

Coulthard, D., & Keller, S. (2016). Publication anxiety, quality, and journal rankings: Researcher views. *Australasian Journal of Information Systems 20*: 1–22.

Cronin, B. (2005). *The Hand of Science: Academic Writing and Its Rewards*. Lanham, MD: Scarecrow Press.

Davies, B., & Bansel, P. (2010). Governmentality and academic work. Shaping the hearts and minds of academic workers. *Journal of Curriculum Theorizing 26*(3): 5–20.

Dolowitz, D., & Marsh, D. (1996). Who learns what from whom: A review of the policy transfer literature. *Political Studies 44*(2): 343–357.

Frijhoff, W., & Spies, M. (2004). *Dutch Culture in a European Perspective*. Basingstoke: Palgrave Macmillan.

Garcia-Ramon, M.-D. (2003). Globalization and international geography: The questions of languages and scholarly traditions. *Progress in Human Geography 27*(1): 1–5.

Gregson, N., Simonson, K., & Vaiou, D. (2003). Writing (across) Europe: On writing spaces and writing practices. *European Urban and Regional Studies 10*(1): 5–22.

Gutiérrez, J., & López-Nieva, P. (2001). Are international journals of human geography really international? *Progress in Human Geography 25*(1): 53–69.

Hall, S. (2023). A mental-health crisis is gripping science – toxic research culture is to blame. *Nature 617*: 666–668.

Hargittai, I. (1999). Deeds and dreams of Eugene Garfield: Interview by Istvan Hargittai. *The Chemical Intelligencer October*: 26–31.

Hartshorne, R. (1939). *The Nature of Geography*. Lancaster, PE: Association of American Geographers.

Imhof, N., & Müller, M. (2020). How international are geography journals? Not international enough. *Environment and Planning A: Economy and Space 52*(7): 1246–1249.

Jessop, B. (2013). Putting neoliberalism in its time and place: A response to the debate. *Social Anthropology 21*(1): 65–74.

Jessop, B. (2018). On academic capitalism. *Critical Policy Studies 12*(1): 104–109.

Johnston, R. (2005). On journals. *Environment and Planning A 37*(1): 2–8.

Kallo, J. (2009). *OECD Education Policy: A Comparative and Historical Study Focusing on the Thematic Reviews of Tertiary Education*. Jyväskylä: Finnish Educational Research Association.

Kauppi, N., & Erkkilä, T. (2011). The struggle over global higher education: Actors, institutions, and practices. *International Political Sociology 5*(3): 314–326.

Kitchin, R. (2005). Disrupting and destabilizing Anglo-American and English-language hegemony in geography. *Social and Cultural Geography 6*(1): 1–15.

Krige, J. (2006). *American Hegemony and the Postwar Reconstruction of Science in Europe*. Cambridge, MA: MIT Press.

Lamego, M. (2020). How international was the International Geographical Congress in Rio de Janeiro 1956? On location and language politics. In Schelhaas, B., Ferretti, F., Novaes, A.R., & di Friedberg, M.S. (eds), *Decolonising and Internationalising Geography. Essays in the History of Contested Science*. Cham: Springer Nature.

Larner, W., & Le Heron, R. (2005). Neoliberal spaces and subjectivities: Reinventing New Zealand universities. *Organization 12*(6): 843–862.

Mason, O., & Megoran, N. (2022). Geography as a vocation? Becoming a geographer under neoliberalism. *Geografiska Annaler, Series B 105*(1): 17–37.

Mignolo, W. (2002). The geopolitics of knowledge and the colonial difference. *South Atlantic Quarterly 101*(1): 57–96.

Minca, C. (2000). Venetian geographical praxis. *Environment and Planning D: Society and Space 18*(3): 285–289.

Minca, C. (2013). (Im)mobile geographies. *Geographica Helvetica 68*(1): 7–16.

Montgomery, S. (2003). *The Chicago Guide to Communicating Science*. Chicago, IL: University of Chicago Press.

Müller, M. (2021). Worlding geography: From linguistic privilege to decolonial anywhere. *Progress in Human Geography 45*(6): 1440–1466.

O'Loughlin, J., Raento, P., Sharp, J.P., & Sidaway, J.D. (2008). Editorial: Where the tradition meets its challengers. *Political Geography 27*(1): 1–4.

O'Neil, D. (2017). English as the lingua franca of international publishing. *World Englishes 37*(2): 146–165.

Olds, K. (2001). Practices for "process geographies": A view from within and outside the periphery. *Environment and Planning D: Society and Space 19*(2): 127–136.

Paasi, A. (2005). Globalization, academic capitalism and the uneven geographies of international journal publishing spaces. *Environment and Planning A 37*(5): 769–789. doi:10.1068/a3769

Paasi, A. (2013). *Fennia*: Positioning a "peripheral" but international journal under the conditions of academic capitalism. *Fennia 191*(1): 1–13.

Paasi, A. (2015a). "Hot spots, dark-side dots, tin pots"? The uneven internationalism of the global academic market. In Meusburger, P., Gregory, D., & Suarsana, L. (eds), *Geographies of Power and Knowledge*. Dordrecht: Springer, 247–262.

Paasi, A. (2015b). Academic capitalism and the geopolitics of knowledge. In Agnew, J., Secor, A., Mamadouh, V., & Sharp, J. (eds), *The Wiley-Blackwell Companion to Political Geography*. Oxford: Wiley-Blackwell, 509–523.

Paasi, A. (2022). The institutionalization of regions: An autobiographic view on the making of socio-spatial theory in the Nordic periphery. In Jakobsen, P., Jönsson, E., & Larsen,

H.G. (eds), *Socio-Spatial Theory in Nordic Geography: Intellectual Histories and Critical Interventions*. New York: Springer, 273–293.

Parmar, I. (2012). *Foundations of the American Century: The Ford, Carnegie, and Rockefeller Foundations in the Rise of American Power*. New York: Columbia University Press.

Perry, P. (1987). Editorial comment. Political Geography Quarterly: A content (but discounted) review. *Political Geography Quarterly* 6(1): 5–6.

Prince, R. (2012). Policy transfer, consultants and the geographies of governance. *Progress in Human Geography* 36(2): 188–203.

Risteli, L. (2003). Sitaatioindeksin alkuvuosikymmenet ja oppihistorialliset yhteydet [The early decades of citation index and its historical connections, in Finnish]. MA thesis. Oulu: University of Oulu.

Robinson, J. (2003). Postcolonizing geography: Tactics and pitfalls. *Singapore Journal of Tropical Geography* 24(3): 273–289.

Rodríguez-Pose, A. (2004). On English as a vehicle to preserve geographical diversity. *Progress in Human Geography* 28(1): 1–4.

Sayer, A. (1992). *Method in Social Science*. London: Routledge.

Scholte, J. (2005). *Globalization*. New York: Palgrave.

Schurr, C., Müller, M., & Imhof, N. (2020). Who makes geographical knowledge? The gender of geography's gatekeepers. *Professional Geographer* 72(3): 317–331.

Sharp, J. (2013). Geopolitics at the margins? Reconsidering genealogies of critical geopolitics. *Political Geography* 37: 20–29.

Sheppard, E. (2013). Geography and neoliberalizing academy. *AAG Newsletter* (May 31).

Simonsen, K. (2002). Editorial: Global–local ambivalence. *Transactions of the Institute of British Geographers NS*27(4): 391–394.

Slater, D. (1998). Post-colonial questions for global times. *Review of International Political Economy* 5(4): 647–678.

Slaughter, S., & Leslie, L. (1997). *Academic Capitalism: Politics, Policies, and the Entrepreneurial University*. Baltimore, MD: Johns Hopkins University Press.

Tennant, J.P. (2020). Web of science and Scopus are not global databases of knowledge. *European Science Editing* 2020(46): 1–3.

Tietze, S., & Dick, P. (2009). Hegemonic practices and knowledge production in the management academy: An English language perspective. *Scandinavian Journal of Management* 25(1): 119–123.

Timár, J. (2004). More than "Anglo-American", it is "Western": Hegemony in geography from a Hungarian perspective. *Geoforum* 35(5): 533–538.

Trubina, E., Gogishvili, D., Imhof, N., & Müller, M. (2020). A part of the world or apart from the world? The postsocialist Global East in the geopolitics of knowledge. *Eurasian Geography and Economics* 61(6): 363662.

Williams, R. (1988). *Marxismi, kulttuuri ja kirjallisuus* [*Marxism, Culture and Literature*]. Tampere: Vastapaino.

Wright, J.K. (1966). *Human Nature in Geography*. Cambridge, MA: Harvard University Press.

Yeung, H. (2002). Deciphering citations. *Environment and Planning A* 34(12): 2093–2106.

Chapter 5

The Powers of Maps

Karen Culcasi

Department of Geology and Geography, West Virginia University, Morgantown, WV, USA

Maps and the processes of mapping are laden with power and power relations. Political geographers have made many contributions to the study of maps and map-making. There have been substantial analyses of how power and politics are embedded within maps and the mapping process, whether stemming from imperial powers or from social movements working toward equity and justice. Political geographers have also made maps to analyze and visualize political-spatial phenomena. Yet in the pages of major political geography journals – notably *Political Geography* and *Geopolitics* – the making and usage of maps are much less common than the analysis and critiquing of maps. There is, however, incredible potential for political geographers to engage more directly in making maps to analyze and visualize complex relations between political phenomena and space/place.

The most common way in which political geography has engaged with maps is through the intersection of critical geopolitics and critical cartography. Such work has critiqued, deconstructed, and contextualized maps and mapping, typically with a focus on maps produced by imperial or state powers, as well as maps that circulate within popular culture and mass media. Often building from broader poststructural and postcolonial theories, these studies highlight the discursive powers of maps in producing geographical knowledge from the local to the global scale. Researchers working at the intersection of critical geopolitics and critical cartography also investigate how maps are used to control territories and people, revealing that maps enact direct, material power, as well as being part of broader discourses. Counter-mapping is a second, overlapping direction that

The Wiley Blackwell Companion to Political Geography, Second Edition.
Edited by Virginie Mamadouh, Natalie Koch, Chih Yuan Woon, and John Agnew.

political geography has taken toward maps and mapping. This avenue of research examines how marginalized groups create and use maps to counter dominant power structures, drawing from postcolonial and decolonial theories, as well as participatory mapping and geographic information sciences/systems (GIS). Some political geographers have also worked with social and environmental justice movements to make counter-maps.

Except for recent counter-mapping work, most cartographic research within political geography has examined existing maps, primarily approaching them as political texts to deconstruct. In other words, map-making is not central to most political geographical scholarship (Moore & Perdue 2014). Haverluk, Beauchemin, and Mueller (2014: 23) went so far as to refer to the subfield of critical geopolitics as "anti-map." Making maps, however, can expand and deepen political geographical scholarship. Maps and GIS have the power to analyze and visualize spatial data, patterns, issues, phenomena, and politics in ways that may not be evident without maps. As such, mapping can lead to new findings across the many topics that are relevant to political geographers. Critical analysis of existing maps remains necessary and important, but making maps can also "do good" and can help to reveal and correct power inequalities and injustice (Wilson 2017: xi). Or, as Jeremy Crampton (2010) argues, mapping has much to offer *after* critique. For over two decades, work within the areas of critical GIS (Schuurman 1999, 2006), feminist GIS (Kwan 2002; Pavlovskaya & Martin 2007; Elwood & Leszczynski 2018), and critical remote sensing (Bennett 2020) has demonstrated the power of making maps in facilitating critical analyses. Considering the monumental changes in mapping techniques and accessibility over the last couple of decades (Anthamatten 2021: 4), mapping and GIS are well positioned to enrich the work of political geography (Moore & Perdue 2014; Carraro 2023).

In the remainder of this chapter, I explore some of the powers of maps as they relate to political geography. In 1992, Denis Wood published his book *The Power of Maps*. I borrow his book title for this chapter but pluralize the term "power" because I see maps as having multiple powers. First, I focus on research at intersection of critical geopolitics and critical cartography, which concentrates on how maps are part of the production of geographical knowledge, and how maps have been used to assert material power over the Earth and people. Second, I discuss counter-mapping and examples of counter-mapping projects within political geography. Last, I review several examples of political geographers making maps to help reveal spatial-political phenomena on topics that are central to the subdiscipline.

Critical Cartography

Within Anglo-American geography, critiques of maps and mapping processes emerged in the late 1980s. One of the foundational points stemming from these critiques was that maps are powerful because people consider them to be objective, scientific, and accurate representations of the world. Judith Tyner wrote in 1982 that maps are often accepted by map readers as "true and complete representations," that they are "trusted blindly" and accepted "uncritically" (140). J.B. Harley, a scholar who would help launch critical cartography, famously wrote that "much of the power of the map . . . is that it operates behind a mask of seemingly neutral science" (1989: 7). And Chris Perkins (2003: 344) noted that geographers as well as the general populace often view maps as a "mirror of the world." Critical cartography challenges the belief that maps are objective and scientific representations, instead showing how maps and map-making are political, biased, and subjective. It is reasonable to assume that map users (geographers and otherwise) have become more scrutinizing of maps since these early critiques were expounded. However, maps are still commonly perceived as objective representations of the world (Monmonier 2018).

This power of perceived objectivity fuels yet another power of maps: the power to create geographical knowledge. Research on the cartographic production of geographical knowledge has included studies of allegedly apolitical maps, propagandistic ones, as well as ones that fall somewhere in between. Yet the distinction between apolitical and propaganda maps is fuzzy (Pickles 1992). As signaled by Mark Monmonier's famous assertion that "all maps lie," all maps are simplifications and distortions. And all maps are created and used within particular social contexts. But the fuzzy distinction still matters, as seemingly apolitical maps are likely to be perceived as objective and scientific, while propaganda maps often trigger skepticism.

Perhaps the most (in)famous example of a seemingly apolitical map riddled with political and geopolitical biases is Mercator projection maps (Monmonier 1995). This projection was created in 1569 by Gerardus Mercator for the purpose of navigation. However, hundreds of years later in the United States, it is commonly used for world reference maps, including educational material, such as the map in the back of my daughter's Ninth-grade school-issued planner, in Morgantown, West Virginia (Figure 5.1), or as the default base map in the cartographic software ArcGIS and Open Street Map. Unlike most other projections used for world reference maps, the Mercator projection has been critiqued as a Western-centric, imperialist map. The projection has meridians (lines of longitude) and parallels (lines of latitude) intersecting at 90° angles. The uniformity of angles allows a navigator to plot a straight line, which is key for navigation. However, maintaining right angles of a spherical world on a two-dimensional plane means that the parallels must be spaced farther and farther apart from other parallels as they move north and south from the equator. This increased spacing between parallels distorts the size/area of land and water at increasing intensity the farther the parallel is from the equator. In other words, the farther north or south a landmass or body of water is, the greater its size is distorted. (Figure 5.1).

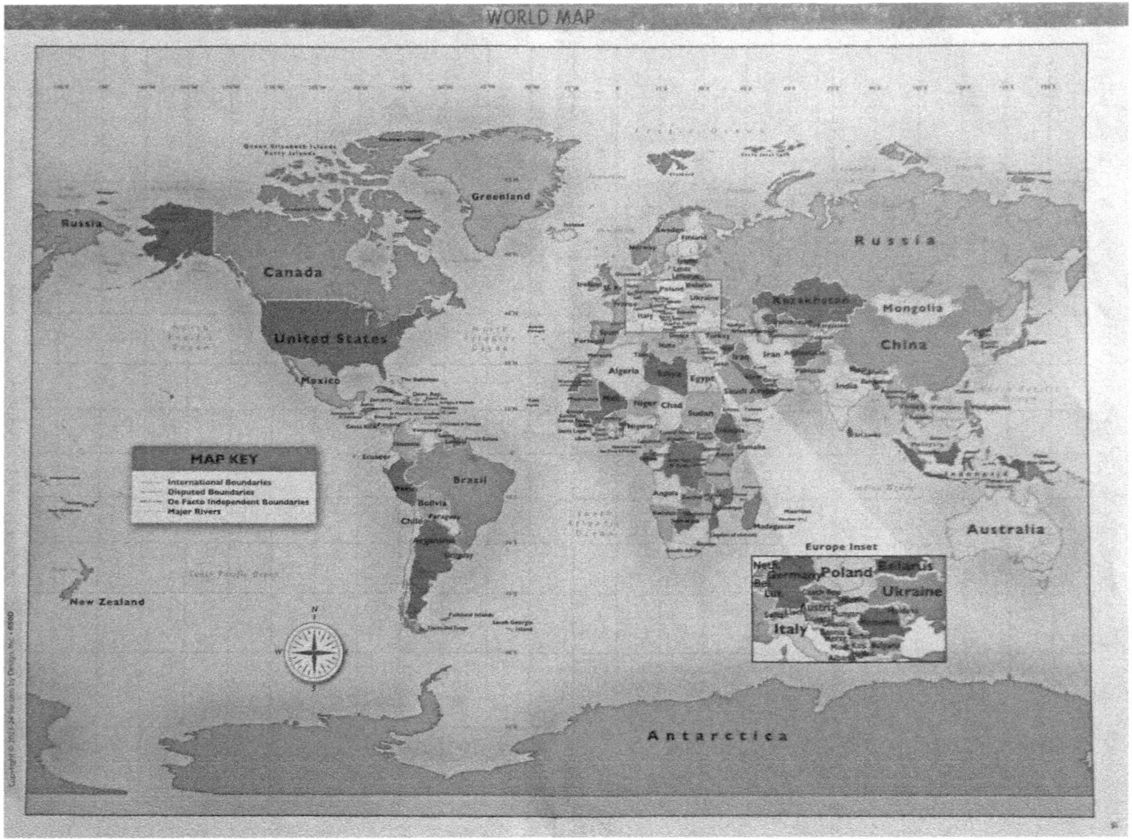

Figure 5.1 World map in the back of a ninth-grade school-issued planner, in Morgantown, West Virginia, USA. *Source:* U.S. DEPARTMENT OF STATE / Public Domain.

For example, the Mercator projection world map shows the area of Greenland as larger than South America and about the same size as Africa. The actual area of South America is eight times larger than Greenland, and Africa is fourteen times large than Greenland. Since Greenland is so far north and both Africa and South America straddle the equator, Greenland's area is exaggerated because its parallels are spaced farther apart than the ones intersecting South America and Africa. Because most of the "Western world," including most of the world's imperialist powers, is within Eurasia and North America, which are far from the equator, its size is exaggerated in relation to land nearer the equator. The territories of less wealthy and formerly colonized places are generally in Africa, South and Central America, the Caribbean, South Asia, Southeast Asia, and Southwest Asia, which are nearer to the equator and thus their size is less exaggerated. The argument then proceeds that a Mercator map is perpetuating Western, imperial superiority and power by falsely increasing its relative size. The Mercator projection debate has made its way into popular culture. For example, in a February 2001 episode of the television series *The West Wing*, the (fictitious) Organization of Cartographers for Social Equality lobbies the West Wing staff to have the president eliminate the use of the Mercator projection in schools because of its "northern bias" and subjugation of less-developed states.

The Mercator debate is quite familiar, but there are many other ways in which seemingly apolitical maps have geopolitical biases. For example, most maps conform to Western cartographic standards. This includes having north on the top of the map and the center of the map along the prime meridian, which passes through Greenwich, England. Most world (and regional) maps also conform to geopolitical standards of dividing and ordering the world into neatly separated territorial states, as opposed to recognizing other territorial divisions, like precolonial regional connections or Indigenous land claims. Representing contested borders and place names is another key issue. For example, the borders and place names used to represent or ignore Israel/Palestine will have political bias, but many map users will not be aware of such contentions. Likewise, the naming and/or borders of Kashmir, Tibet, Taiwan, and the Persian Gulf all have political meaning that likely goes unnoticed by many map users.

Unlike a seeming apolitical world map, many maps are clearly political or propagandistic. For example, both sides of the Israel/Palestine conflict have used maps to cartographically lay claim to territory and to symbolize their homelands (Leuenberger 2013; Quiquivix 2014; Culcasi 2016). Some maps are blatantly propagandistic. For example, a one-page flier and newspaper advertisement published in the early 2000s by the American Jewish Committee (AJC), a lobby group based in New York City (AJC 2004), juxtaposes four maps redrawn from maps published for Arab audiences in Lebanon, Palestine, Syria, and Saudi Arabia. In all four, Palestine is recognized to exist and there is no recognition of Israel. The title of the flier asks: "How can there be peace in the Middle East if Israel isn't even on the map today?" The AJC thus propagates the idea that conflict, or the lack of peace, is elusive because these maps do not recognize Israel. The four maps are offered as "proof" that Arab maps, and by extension Arabs, are the cause of the lack of peace. Of course, the flier does not include maps made in Israel that do not show Palestine or the occupied territories, which are commonplace (Peled-Elhanan & Yellin 2009).

The Mercator and Israel/Palestine examples show that politics are embedded in maps both subtly and blatantly, but many maps fall somewhere in between. Reports on the so-called European refugee crisis of 2014–2015 commonly used maps to show countries of origin, destinations, and numbers of people seeking asylum (refuge). Most of these maps did not seem to be propagandistic to casual observers. Nevertheless, politics and biases are ingrained in the maps. Van Houtum and Bueno Lacy (2020) illustrate this in their analysis of the official 2017 map by Frontex, the European Union's (EU) border enforcement agency, of undocumented migration into the EU (Figure 5.2). The scope of the map includes Europe and Africa from about the equator northward and Southwest Asia eastward to India. Europe is colored in a light blue and has no internal state-territorial borders. The countries (Frontex calls them nationalities) from which the migrants came are in pale yellow and all have borders. The graphic hierarchy of the map is dominated by red and purple arrows that indicate the place of origin and the destination of the migrants. The thickness of the lines indicates numbers of migrants.

While none of these cartographic elements is propagandistic in and of itself, the choice of colors, terms, hierarchy, and size of the arrows reveal a political bias. First, there are two places on the map that refer to "illegal border crossers." Yet migrants who cross a border to seek asylum are not "illegal." Indeed, the right to seek asylum is enshrined in international law to which all EU states are party. The use of large, looming, and "menacing" arrows (Van Houtum & Bueno Lacy 2020: 200) depicts a situation in which huge numbers of migrants seem to be invading Europe. The head of the arrow that originates in Africa is larger than the entire Iberian Peninsula. The color of this largest arrow is red, which is associated with fear and danger, whereas Europe, in blue, connotes tranquility and peace. Again, this map does not at first glance seem propagandistic, yet a critical reading of it shows an anti-migration bias. With the threat so clear on the map, it implies that defense of the EU's territories is justified.

These three examples are all part of the production of geopolitical knowledge. They educate and inform people on geopolitics, whether banal or contentious. There are other maps, however, that are less about knowledge production and more about *enacting* direct, material power over people and territories. The production of knowledge and material enactments are relational and buttress one another. But the role of maps in exerting direct, material power works in different ways, as the cases of European and American imperialism vividly illustrate. It is commonly recognized that the discipline of geography grew in tandem with European imperialism (Cresswell 2013; Muscarà, this volume). The cartographic technologies that European powers developed propelled their ability to navigate

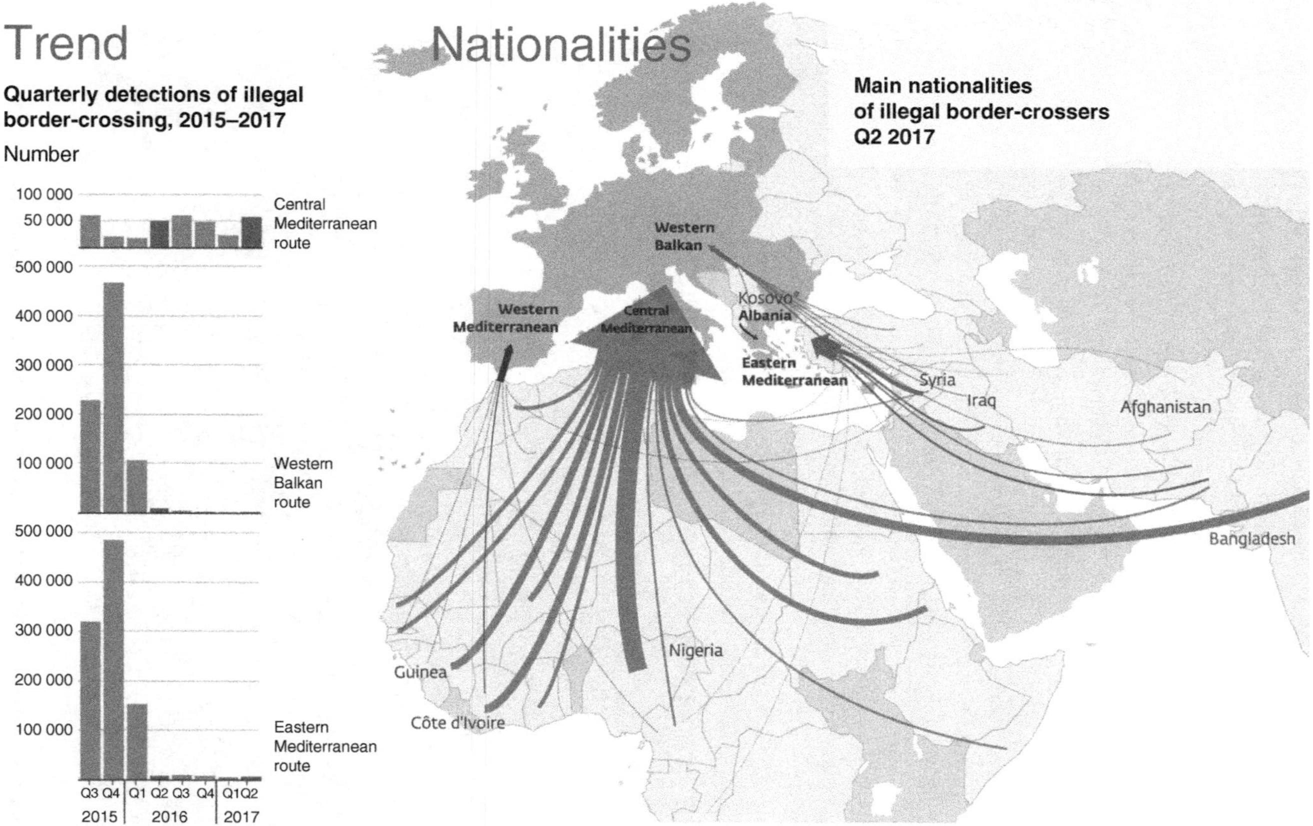

Figure 5.2 The Frontex map, analyzed by Van Houtum and Bueno Lacy (2020). *Source:* https://www.frontex.europa.eu//assets/Publications/Risk_Analysis/FRAN_Q2_2017.pdf. © Frontex, the European Border and Coast Guard Agency. Reproduced by permission.

to lands beyond their own, to cartographically document what they saw, to (theoretically) demonstrate their scientific and intellectual superiority as mapmakers and surveyors, and to divide and control distant territories and peoples (Edney 1997; Elden 2009). In short, geographers and their maps helped to facilitate empire's oppressive and violent practices.

In his study of the colonization of West Africa, Thomas Bassett (1994: 330) notes that "the power to map is synonymous with the power to delimit, to control, and to develop territory." This was exemplified at the Berlin Conference of 1884, during which European leaders (and representatives from the Ottoman Empire and the United States) met to divide control of the African continent. Maps were central to the three-month-long conference, including one gigantic map of Africa that dominated a wall in the main meeting room (Figure 5.3). The maps provided knowledge about Africa that led to divisions of territorial control, which then fueled new colonial practices. For example, by the end of the conference, delegates agreed that King Leopold II would be the sole "owner" of the Congo River Basin. He was then empowered to exploit the Congo's rich natural resources (ivory and rubber) – enacted on the ground through enslaving, torturing, mutilating, and murdering its native inhabitants. It is estimated that

Figure 5.3 Artist's depiction of the European male leaders debating control of different parts of Africa during the Berlin Conference of 1884, Adalbert von Rößler (†1922), published in *Allgemeine Illustrierte Zeitung*. *Source:* Adalbert von Rößler / Wikimedia Commons / Public Domain.

about 10 million people, half of the area's population, were killed under Leopold's reign that lasted until 1908 (Hochschild 1998; Kara 2023). Of course, the maps of Africa were only a part of this brutal past, but they were an important one.

Far from the rooms and halls of power in Europe, maps were also made and used by colonists for exploration, measurement, documentation, and exploitation in North America. In his book *Indigenous Continent*, Pekka Hämäläinen (2022: 267) discusses how maps had a power that aided the taking of land across the continent: "Unable to defeat Native nations in battle, colonial topographic surveys and maps became a means to seize Indigenous land without war." And as Bernard Nietschmann famously quipped, "More Indigenous territory has been claimed by maps than by guns" (as quoted in Wainwright & Bryan 2009: 74). In his examination of the settler colonization of British Columbia, Cole Harris (2004: 175) likewise underscored the central role of maps, writing that maps "enabled newcomers to locate themselves in this space and find their way around." The British colonists made and used maps not only for exploration but to measure, document, and exploit the land and crucially its people. Settlers made maps of Indigenous reserves, properties, and populations, which gave the British the "power to dispossess."

In more contemporary iterations of American empire, maps have figured prominently in military operations and war-making. During the 1991 Persian Gulf War, the US Defense Mapping Agency sent 35 million maps to US troops working to push Iraqi troops out of Kuwait (Harley 1991: 9). The Gulf War was unique because digital maps and GIS were now part of military operations, transforming "the way warfare is fought and staged, and how the public consumes it" (Smith 1992: 257). Almost 10 years later, when the George W. Bush Administration was seeking support for its invasion of Iraq, maps were again integral and had direct, devastating effects. During then Secretary of State Colin Powell's presentation to the UN General Assembly in February 2002, one month before the invasion, he shared a series of remotely sensed photos that had been turned into maps by US State Department officials. With the addition of lines, arrows, polygons, labels, and titles, the maps purportedly identified "chemical munition bunkers," "decontamination vehicles," "missile storage containers," "rocket test facilities," "serum and vaccine institutes," "missile assemblage building," "chemical weapons," and a "chemical complex" (CBS 2003).

According to Powell, the maps provided absolute proof that Iraq had "weapons of mass destruction" – a term he used 17 times during that speech (Mitchell 2023). Powell and other Bush Administration officials bundled these maps with other forms of supposed evidence, like a taped phone conversation between Iraqi military officials, to argue that Iraq was an immediate danger to the United States and that the United Nations must support the US invasion of that sovereign country. Yet, as was quickly discovered after the invasion in early 2003, Iraq had no weapons of mass destruction. Powell's maps demonstrate the multiple powers of maps. They were part of the production of knowledge that Iraq was supposedly a threat to the United States, and thus also a justification for the US invasion.

Counter-mapping

Writing in 1988, J.B. Harley referred to maps as a "language of power, not protest" (1988: 301). In the late 1980s, such a statement was reasonable since mapmaking in the Western world had been controlled by an elite few for hundreds of years. First this included monarchs, ministers, and state governments; later, it was commercial map publishers and the mass media. More recently, however, mapmaking has become more accessible to a broader populace. Indeed, there has been a shift in who controls map production due to online mapping software and social media platforms, concurrent with a greater awareness of social and environmental injustices (Schranz 2021).

The terms counter-map, counter-mapping, and counter-cartographies are now all used to refer to different forms of mapping that protest, challenge, and contest dominant discourses and structures of power, whether sketch maps or digital (Dalton & Mason-Deese 2012). Nancy Peluso (1995: 384) coined the term "counter-map" in an article that examined how local activists in the Indonesian forests created "sketch maps to delineate and formalize claims to forest territories and resources their villages have traditionally managed," territories that had been exploited for decades for timber. As Harris and Hazen (2006: 115) explain, counter-mapping is "any effort that fundamentally questions the assumptions or biases of cartographic conventions, that challenges predominant power effects of mapping, or that engages in mapping in ways that upset power relations." Situated within critical cartography, as well as post- and decolonial theories, Indigenous cartographies, participatory mapping, and qualitative GIS, counter-mapping asserts its own power in highlighting and working toward overturning inequities and injustices.

Critical, alternative, and counter-mapping projects have expanded dramatically in recent years. The publication in 2018 of *This Is Not an Atlas* (https://notanatlas.org) by the kollektiv orangotango is exemplary. This is a collection of mappings from activists and/or academics who share the goal of using maps for the counter-hegemonic goals of resistance and emancipation (Halder & Michel 2018). In 42 short chapters, *This Is Not an Atlas* documents a diversity of counter-mapping projects – from differentiated processes of territorialization in the Amazon to sexual harassment in Egypt – and the potentiality for social justice. The VisionCarto collective likewise makes counter-maps to support a variety of social issues and movements (https://www.visionscarto.net/en). Yet another example is the decolonial atlas (https://decolonialatlas.wordpress.com), which is a digital source showcasing maps that challenge political and cartographic norms and elevate marginalized views and mappings.

Counter-mapping projects place a strong focus on engaging with Indigenous social justice movements (Desbiens, Hirt, & Collignon 2020), in which Indigenous peoples are asserting territorial claims and/or sovereignty (Sparke 1998; Anker 2018; Benson et al. 2023). And just as Nietschmann suggested that maps have claimed more Indigenous territory than guns, he also asserted a corollary that "more indigenous territory can be reclaimed and defended by maps than guns" (quoted in

Wainwright & Bryan 2009: 74). Counter-mapping has been used in movements across the Global South, where Indigenous and other minoritized groups have been marginalized by state leaders and large landowners – especially in South and Central America and the Caribbean (Orlove 1991; Wainwright 2008; Sletto 2009; Wainwright & Bryan 2009; Mollett 2013). For example, Mollett (2013) examined a March 2007 counter-mapping project and conference held in Honduras, which sought to support ancestral land claims of the Miskito and Garifuna populations of Honduras. The counter-mapping project made their land claims legible, implied their future rights over the land, and gave them hope for a better future.

Indigenous counter-mapping projects have also helped to preserve cultural and intellectual heritage, which has been erased or ignored by white colonizers (Gavazzi 2018). In highlighting ways of conceptualizing geography, space, and the environment beyond Western epistemologies and their associated cartographic standards, such counter-cartographies have reinscribed spatial knowledge that had been erased. (Johnson, Louis, & Pramono 2006; Rose-Redwood et al. 2020). Counter-mapping projects are powerful ways to work toward social justice, but there can be contentious dynamics within counter-mapping, including uneven power relations between the marginalized communities and those with the means to map (Oslender 2021). Moreover, counter-mapping projects have the potential to backfire as well, such as when dominant powers like state actors react negatively toward the groups that are counter-mapping (Orlove 1991; Mollett 2013).

Counter-cartographies of migration and border politics are also on the rise. Led by several scholar-activists in Europe, their work demonstrates that making maps can challenge anti-immigration discourses and policies (Mekdjian 2015). For example, Van Houtum and Bueno Lacy (2020) showcase counter-maps that have used creative cartographic techniques (ones that do not conform to Western cartographic standards) to challenge the messaging not only of anti-immigration maps but also from political leaders across Europe. They highlight the counter-cartographies of Philippe Rekacewicz, whose maps refute the message that the EU is overrun by undocumented migrants and instead illustrate how "the EU is part of a system of visas, detention centres and biopolitical border controls that discriminates, dehumanises, illegalises and willingly endangers undocumented migrants" (Van Houtum & Bueno Lacy 2020: 209).

With similar goals, Migreurop is a "Euro-African network of human rights organisations, activists and researchers" who work to "identify, publicise and denounce the consequences of European migration policies on the living conditions and the respect of the rights of people on the move" (https://migreurop.org). In stark contrast to the Frontex map discussed earlier, their maps highlight individual migrant trajectories and the role of EU state policies in migrant tragedies. For example, in 2021 Nicolas Lambert of Migreurop made a map of "dead and missing migrants" who were moving toward or within Europe to seek refuge (Figure 5.4). Using one pink dot to represent one migrant, the clustering of the 3747 dead and missing migrants not only shows important spatial patterns but also clearly focuses on human lives. Drawn over pale blue borderless landmasses

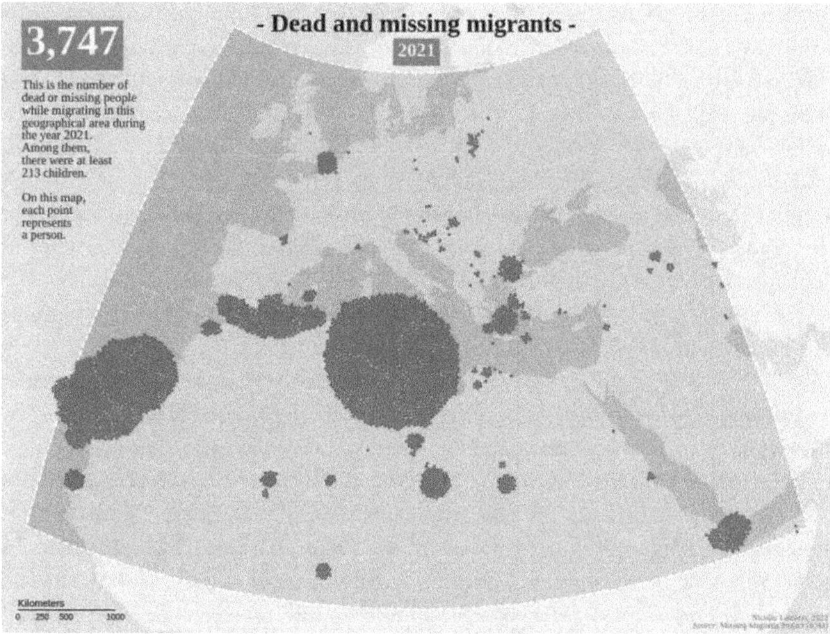

Figure 5.4 *Dead and Missing Migrants*, 2021, by Nicolas Lambert – CNRS/Migreurop. *Source:* https://migreurop.org/article3098.html?lang_article=en&lang_nav=en. Reproduced by permission.

and slightly darker blue water bodies, the map emphasizes humans trying to move, as opposed to European borders and state territories and their supposed vulnerability.

Making Maps in Political Geography

Twenty years ago, in *Progress in Human Geography*, Chris Perkins (2004) observed that theoreticians of the "new" critical cartography were dominating discussions of mapping within human geography. As this new area of scholarship focused on critiques of existing maps, the result was that "mapping and visualizations have become increasingly marginalized from the mainstream of human geographic discourse" (2004: 385). He argued that many geographers prefer to use words, write theory, and deconstruct maps over creating maps or employing critical visualizations. After more than two decades since Perkins leveled this critique, mapping and GIS have proliferated across the discipline of geography, including within human geography. This includes the counter-mapping projects just discussed, but also notable is the work of political ecology and human environment–focused work that regularly leverages mapping and GIS to visualize complex relations between people and the nonhuman world (Hazen & Anthamatten 2004; Harris & Hazen 2009; Carraro 2023).

Concurrently, some political geographers have adopted GIS and other geospatial techniques to make maps to analyze and visualize spatial data. However, the practice of mapping is not widespread within political geography (Moore & Perdue 2014). It is not surprising that many political geographers have not engaged in map-making. The "GIS wars" that proliferated in the 1990s – debates over the ethics of using GIS and geospatial technologies (Schuurman 2000; Kwan 2002; Preston & Wilson 2014) – included strong critiques of GIS. Like more traditional cartography, GIS was considered a tool of power, which could have oppressive and violent consequences. GIS (and remote sensing) was in fact developed and used in great part for military interests, for waging war, and for surveillance (Belcher 2019). As noted earlier, mapping and GIS were integral parts of both the US-led 1991 Gulf War (Smith 1992) and the US-led 2003 invasion of Iraq (Kwan 2002; Gregory 2010). Critiques of GIS also focused on its unequal accessibility: the necessary technical skills can be costly to acquire, while GIS software and geospatial data are expensive and therefore limited to those people and institutions with the means to purchase them. Indeed, for its first several decades, GIS deepened uneven power relations from the macro scales of international war to the micro scales of urban planning.

Yet another concern that slowed political geographers from engaging with GIS was that it evolved from the positivist tradition in geography, during the quantitative revolution of the 1960s. As such, GIS was widely purported to be a science that could achieve objective outcomes. John Pickles (1995) articulated this point in his edited book *Ground Truth*, stating that GIS users are engaged in what is viewed as "value-neutral observation" and a "science as a mirror of reality" (1995: 17–18). Such a view of GIS as an objective science was anathema to theorizing in political geography, which started to take root in the 1980s and considered geographical knowledge production and practices to be subjective and political. Feminist geographies further developed this critique by asserting that GIS was not only as falsely scientific but also as authoritative and masculinist (Pavlovskaya & Martin 2007: 591).

GIS and map-making have evolved substantially since the "GIS wars." Some changes came in response to the critiques articulated in the early debates. For example, there has been notable progress in "democratizing" mapping, meaning that geospatial data and software are more accessible and affordable (Dodge & Perkins 2009). Google Earth, released in 2005, is an exemplar of open access data and mapping tools. In 2012, Esri launched a free, online version of its mapping software. Like counter-cartography projects, making GIS more accessible has facilitated community and participatory GIS projects across the globe (Elwood 2006; Sletto 2009; Oslender 2021; Benson et al. 2023; Preston & Wilson 2014). Concurrently, the specialized areas of interest of GIS and society, critical GIS, and feminist GIS have all burgeoned since the "GIS wars" (O'Sullivan 2006). Rejecting positivist notions of objectivity, such work foregrounds ethics and reflexivity and uses qualitative methods (often alongside quantitative ones) to make maps that are critical of top-down power and work toward social justice (Kwan 2002; McLafferty 2005; Knigge & Cope 2006; Boschmann & Cubbon 2014; Preston & Wilson 2014; Justin 2023).

In addition to these counter-cartography examples, a growing body of political geography research on conflict, displacement, and territorial disputes is now enriched by critical mapping and GIS. In the context of conflict and displacement, mapping projects provide insights into patterns of violence and offer humanitarian suggestions to help alleviate some of the traumas of conflict. Madden and Ross (2009), for example, used satellite imagery of northern Uganda to understand how conflict and displacement have affected millions of people, including the destruction of communities and disruption of agricultural patterns. They paired these data with qualitative methods, mostly personal narratives of those experiencing conflict, to reveal additional details that are not captured through the imagery. Similarly, Sarah A. Son (2022) used remote sensing and GIS to provide new insights into human rights abuses in North Korea. She examined changes to vegetation (which can be discerned through remote sensing) to detect killing and burial sites, in tandem with witness testimonies to reveal and confirm patterns of abuse.

Other studies of conflict that have used remote sensing and GIS include work on the Syrian war (Li & Li 2014), conflicts in South Ossetia and Chechnya (Witmer & O'Loughlin 2011), and conflict in the Rakhine state of Myanmar (Walker 2023). In Ukraine, GIS is helping to locate war-scarred landscapes that can aid in landmine clearance and to identify safe zones for humanitarian supply distribution (Washington Post Creative Group 2022). While many of these studies do not include qualitative research, they nonetheless provide insights into patterns of conflict and destruction, and are doing so in places where getting on-the-ground data can be difficult, dangerous, or impossible.

Data provided by the Armed Conflict Location and Event Data Project (ACLED) and the Uppsala Conflict Data Program Georeferenced Event Dataset (UCDP-GED) also offer a significant resource for political geography research. These two nonprofit organizations collect and share geospatial data on conflicts, violence, military movements, and protests across the globe. The data are free and open source and can be downloaded in multiple file formats (like Excel and Shapefiles), which can then be integrated into GIS software. For example, Zoppolato (forthcoming) has used data from ACLED and UCDP of locations across Pakistan that have experienced violent and nonviolent conflicts since 2016. He then overlaid this spatial data to search for a correlation with sites and corridors of the China Pakistan Economic Corridor (CPEC; part of the controversial Belt and Road Initiative (BRI)). After integrating the geospatial data with field observations, he found a relationship between the locations of conflict and the locations of CPEC projects, as well as the immense securitization infrastructure of CPEC sites. Such findings help challenge the discourse that the BRI is merely an economic project (see also Bennett 2020).

Critical GIS can also help political geographers to visualize contested territorial claims and challenge uneven power relations. Nearly two decades ago, I engaged in a mapping project of various territorial delineations of Kurdistan, including those of Kurdish leaders and organizations and Western powers like the CIA (Culcasi 2006, 2010). Overlaying the different borders highlighted competing territorial claims, the uneven power relations involved in the construction of borders and state territories, and the marginalization of the Kurdish people. More recently, Benson et al. (2023) used Open Street Map and Leaflet (two free online geospatial mapping tools) to create maps to visualize Indigenous oppression in Australia. These maps were part of a larger project aimed at healing the violence of settler colonialism in Australia, as well as challenging Western ideas of sovereignty and states more broadly. Yet other scholars have offered critical mappings of racist US federal lending policies and their long-term effects (Rothstein 2017; Cimons 2020; Wilson 2020; Wang 2024; https://dsl.richmond.edu/panorama/redlining/#loc=10/40.992/-73.302), as well as the politics of environmental and climate issues like sea-level rise (Monmonier 2008), climate change (Fish 2020), and acid rain (McNeil & Culcasi 2015). The power of maps to help visualize political-spatial phenomena is, in short, immense.

Conclusion

In the summer of 2023, controversy arose over a map in the American movie *Barbie*. The cartoonish style "Real World Map" included a line in the South China Sea that is known as the nine-dash line (Diaz 2023). In the mid-twentieth century, demarcation and control of this maritime area became contentious as the Chinese government extended the line on its maps as a cartographic technique to claim a larger maritime extent. The neighboring countries of Vietnam, Malaysia, and the Philippines have rejected Chinese claims in the sea. This issue of control led the Philippines to take China to court in The Hague in 2016. The court ruled in favor of the Philippines and rejected China's demarcation of the nine-dash line. Yet the "Real World Map," which was to guide Barbie from Barbie Land to the real world, includes an eight-dash line on the far right of the map, off a shape that could be read as Asia (Figure 5.5). The Filipino government debated banning the blockbuster film from showing in the Philippines, but ultimately did not although censors requested the dash line to be blurred for distribution in the country. Vietnam, however, did ban it because of the map and its geopolitical significance. In fact, the Vietnamese government had also banned the 2019 DreamWorks' animated movie *Abominable* and Sony's 2022 action movie *Uncharted* for the same reason (Palatino 2023). That this cartoonish map of a disorienting array of landmasses and scribbles could spark such controversy is clearly indicative of the powers of maps.

Maps have many powers – insofar as they are believed to be accurate, as they produce geopolitical knowledge, as they have material effects, and as they can help visualize and analyze spatial phenomena. Critical geopolitics, and political geography more broadly, has been central in making critiques of maps and mapping practices, highlighting uneven power relations and the often oppressive and

Warner Bros

Figure 5.5 The "Real World Map" with an eight-dash line read as China's demarcation of the nine-dash line. *Source:* Barbie (2023) / Direction by Greta Gerwig / Production by Warner Bros. Entertainment Inc.

violent consequences of some maps. This is important and necessary work to continue. Maps are also increasingly being made and used to help elucidate political geographical phenomena, whether part of counter-cartographic projects or for examining conflict and displacement. Yet map-making is not central to most political geographical research. Considering the immense potential that mappings have for political geography topics, the discipline would be richer with even more map-making.

References

American Jewish Committee (2004). How can there be peace in the Middle East if Israel isn't even on the map today?, subject: Refusal of Syria, Palestinian authority, Lebanon Ministry of Tourism and Saudi Arabia to acknowledge the existence of Israel in their maps, *New York Times*, 05/05/04. American Jewish Committee Archives, https://ajcarchives.org/Portal/Default/en-US/RecordView/Index/4480 (accessed November 20, 2024).

Anker, K. (2018). Aboriginal title and alternative cartographies. *Erasmus Law Review 1*: 14–30.

Anthamatten, P. (2021). *How to Make Maps: An Introduction to Theory and Practice of Cartography*. New York: Routledge.

Bassett, T. (1994). Cartography and empire building in nineteenth century west Africa. *Geographical Review 84*(3): 316–335.

Belcher, O. (2019). Sensing, territory, population: Computation, embodied sensors, and hamlet control in the Vietnam War. *Security Dialogue 50*(5): 416–436.

Bennett, M.M. (2020). Is a pixel worth 1000 words? Critical remote sensing and China's Belt and Road Initiative. *Political Geography 78*: 102127.

Benson, E., Brigg, M., Hu, K., et al. (2023). Mapping the spatial politics of Australian settler colonialism. *Political Geography 102*: 102855.

Boschmann, E.E., & Cubbon, E. (2014). Sketch maps and qualitative GIS: Using cartographies of individual spatial narratives in geographic research. *Professional Geographer 66*(2): 236–248.

Carraro, V. (2023). The C-word: How critical cartography, critical GIS and critical data studies can repoliticise disaster-related maps. *Disaster Prevention and Management: An International Journal 32*(6): 1–13.

CBS (2003). Powell's photos. https://www.cbsnews.com/pictures/powells-photos (accessed November 20, 2024).

Cimons, M. (2020). How redlining makes communities of color more at risk of deadly heatwaves. *PBS*, January 23.

Crampton, J. (2010). *Mapping: A Critical Introduction to Cartography and GIS*. Malden, MA: Wiley-Blackwell.

Cresswell, T. (2013). *Geographic Thought: A Critical Introduction*. Malden, MA: Wiley-Blackwell.

Culcasi, K. (2006). Cartographically constructing Kurdistan within geopolitical and orientalist discourses. *Political Geography 25*(6): 680–706.

Culcasi, K. (2010). Locating Kurdistan: Contextualizing the region's ambiguous boundaries. In Diener, A., & Hagen, J. (eds), *Borderlines and Borderlands: Political Oddities at the Edge of the Nation-State*. Lanham, MD: Rowman and Littlefield, 107–120.

Culcasi, K. (2016). Images and imaginings of Palestine: Territory and everyday geopolitics for Palestinian Jordanians. *Geopolitics 21*(1): 69–91.

Dalton, C., & Mason-Deese, L. (2012). Counter (mapping) actions: Mapping as militant research. *ACME 11*(3): 439–466.

Desbiens, C., Hirt, I., & Collignon, B. (2020). Introduction to the special section on Indigenous spatial capital: Incorporating First Peoples' knowledges, places, and relations into mapping processes. *The Canadian Geographer/Le Géographe Canadien 64*(1): 4–9.

Diaz, D. (2023). GOP declares war on . . . Barbie. Politico, https://www.politico.com/news/2023/2007/2007/gop-declares-war-on-barbie (accessed November 20, 2024).

Dodge, M., & Perkins, C. (2009). The "view from nowhere"? Spatial politics and cultural significance of high-resolution satellite imagery. *Geoforum 40*(4): 497–501.

Edney, M.H. (1997). *Mapping an Empire: The Geographical Construction of British India, 1765–1843*. Chicago, IL: University of Chicago Press.

Elden, S. (2009). *Terror and Territory: The Spatial Extent of Sovereignty*. Minneapolis, MN: University of Minnesota Press.

Elwood, S. (2006). Negotiating knowledge production: The everyday inclusions, exclusions, and contradictions of participatory GIS research. *Professional Geographer 58*(2): 197–208.

Elwood, S., & Leszczynski, A. (2018). Feminist digital geographies. *Gender, Place and Culture 25*(5): 629–644.

Fish, C. (2020). Storytelling for making cartographic design decisions for climate change communication in the United States. *Cartographica 55*(2): 69–84.

Gavazzi, R.A. (2018). Indigenous cartography in acre. In Halder, S., & Michel, B. (eds), *This Is Not an Atlas*. Bielefeld: transcript, 111–115.

Gregory, D. (2010). Seeing red: Baghdad and the eventful city. *Political Geography 29*: 266–279.

Group WPC (2022). The hero we've been waiting for is actually a map. kollektiv orangotango+, https://library.oapen.org/handle/20.500.12657/27378 (accessed December 6, 2024).

Halder, S., & Michel, B. (2018). *This Is Not an Atlas*. Bielefeld: transcript.

Hamalainen, P. (2022). *Indigenous Continent: The Epic Contest for North America*. New York: Liveright.

Harley, J.B. (1988). Maps, knowledge and power. In Cosgrove, D., & Daniels, S. (eds), *The Iconography of Landscape: Essays on the Symbolic Representation, Design and Use of Past Environment*. Cambridge: Cambridge University Press, 277–312.

Harley, J.B. (1989). Deconstructing the map. *Cartographica* 26(2): 1–20.

Harley, J.B. (1991). Can there be a cartographic ethics? *Cartographic Perspectives 10*: 9–16.

Harris, C. (2004). How did colonialism dispossess? Comment from an edge of empire. *Annals of the Association of American Geographers 94*(1): 165–182.

Harris, L., & Hazen, H.D. (2006). Power of maps: (Counter) mapping for conservation. *ACME: An International E-Journal for Critical Geographies 4*(1): 99–130.

Harris, L., & Hazen, H. (2009). Rethinking maps from a more than human perspective: Nature-society, mapping and conservation territories. In Dodge, M., Kitchin, R., & Perkins, C. (eds), *Rethinking Maps: New Frontiers in Cartographic Theory*. New York: Routledge, 50–67.

Haverluk, T.W., Beauchemin, K.M., & Mueller, B.A. (2014). The three critical flaws of critical geopolitics: Towards a neo-classical geopolitics. *Geopolitics 19*: 19–39.

Hazen, H.D., & Anthamatten, P.J. (2004). Representation of ecological regions by protected areas at the global scale. *Physical Geography 25*: 499–512.

Hochschild, A. (1998). *King Leopold's Ghost: A Story of Greed, Terror, and Heroism in Colonial Africa*. New York: Houghton Mifflin.

Johnson, J.T., Louis, R.P., & Pramono, A.H. (2006). Facing the future: Encouraging critical cartographic literacies in indigenous communities. *ACME: An International E-Journal for Critical Geographies 4*(1): 80–98.

Justin, J. (2023). Digital cartography and feminist geocriticism: A case study of the Marichjhapi massacre. *Cartographica: The International Journal for Geographic Information and Geovisualization 58*(3). doi:10.3138/cart-2022-00

Kara, S. (2023). *Cobalt Red*. New York: St Martin.

Knigge, L., & Cope, M. (2006). Grounded visualization: Integrating the analysis of qualitative and quantitative data through grounded theory and visualization. *Environment and Planning A: Economy and Space 38*(11): 2021–2037.

Kwan, M.-P. (2002). Feminist visualizations: Re-envisioning GIS as a method in feminist geographic research. *Annals of the Association of American Geographers* 92(4): 645–661.

Leuenberger, C. (2013). Map-making for Palestinian state-making. *Arab World Geographer 16*(1): 54–74.

Li, X., & Li, D. (2014). Can night-time light images play a role in evaluating the Syrian crisis? *International Journal of Remote Sensing 35*(18): 6648–6661.

Madden, M., & Ross, A. (2009). Genocide and GIScience: Integrating personal narratives and geographic information science to study human rights. *The Professional Geographer 61*(4): 508–526.

McLafferty, S. (2005). Women and GIS: Geospatial technologies and feminist geographies. *Cartographica 40*(4): 37–45.
McNeil, B., & Culcasi, K. (2015). Maps on acid: Cartographically constructing the acid rain environmental issue. *Professional Geographer 67*(2): 242–254.
Mekdjian, S. (2015). Mapping mobile borders: Critical cartographies of borders based on migration experiences. In Szary, A.-L.A., & Giraut, F. (eds), *Borderities and the Politics of Contemporary Mobile Borders*. Houndmills: Palgrave Macmillan, 204–224.
Mitchell J. (2023). 20 years ago, the U.S. warned of Iraq's alleged "weapons of mass destruction". NPR, February 3. https://www.npr.org/2023/2002/2003/1151160567/colin-powell-iraq-un-weapons-mass-destruction (accessed November 20, 2024).
Mollett, S. (2013). Mapping deception: The politics of mapping Miskito and Garifuna space in Honduras. *Annals of the Association of American Geographers 103*(5): 1227–1241.
Monmonier, M. (1995). *Drawing the Line: Tales of Maps and Cartocontroversy*. New York: Henry Holt.
Monmonier, M. (2008). *Coastlines: How Mapmakers Frame the World and Chart Environmental Change*. Chicago, IL: University of Chicago.
Monmonier, M. (2018). *How to Lie with Maps*. Chicago, IL: University of Chicago Press.
Moore, A.W., & Perdue, N.A. (2014). Imagining a critical geopolitical cartography. *Geography Compass 8*(12): 892–901.
Orlove, B. (1991). Mapping reeds and reading maps: The politics of representation in Lake Titicaca. *American Ethnologist 18*(1): 3–38.
Oslender, U. (2021). Decolonizing cartography and ontological conflict: Counter-mapping in Colombia and "cartographies otherwise". *Political Geography 89*: 102444.
O'Sullivan, D. (2006). Geographical information science: Critical GIS. *Progress in Human Geography 30*(6): 783–791.
Palatino, M. (2023). Philippines allows "Barbie" film screening amid "9-dash-line" controversy. The Diplomat, July 14. https://thediplomat.com/2023/07/philippines-allows-barbie-film-screening-amid-9-dash-line-controversy (accessed November 20, 2024).
Pavlovskaya, M., & Martin, K.S. (2007). Feminism and geographic information systems: From a missing object to a mapping subject. *Geography Compass 1*: 583–606.
Peled-Elhanan, N., & Yellin, D. (2009). The geography of hostility and exclusion: A multimodal analysis of Israeli schoolbooks. *Journal of Visual Literacy 27*(2): 179–208.
Peluso, N. (1995). Whose woods are these? Counter-mapping forest territories in Kalimantan, Indonesia. *Antipode 27*(4): 383–406.
Perkins, C. (2003). Cartography: Mapping theory. *Progress in Human Geography 27*(3): 341–351.
Perkins, C. (2004). Cartography – Cultures of mapping: Power in practice. *Progress in Human Geography 28*(3): 381–391.
Pickles, J. (1992). Texts, hermeneutics and propaganda maps. In Barnes, T., & Duncan, J. (eds), *Writing Worlds: Discourse, Text and Metaphor in the Representation of Landscape*. London: Routledge, 193–230.
Pickles, J. (1995). *Ground Truth: The Social Implications of Geographic Information Systems*. New York: Guilford Press.
Preston, B., & Wilson, M. (2014). Practicing GIS as mixed methods: Affordances and limitations in an urban gardening study. *Annals of the Association of American Geographers 104*(3): 510–529.
Quiquivix, L. (2014). Art of war, art of resistance: Palestinian counter-cartography on Google Earth. *Annals of the Association of American Geographers 104*(3): 444–459.

Rose-Redwood, R., Barnd, N.B., Lucchesi, A.H., et al. (2020). Decolonizing the map: Recentering indigenous mappings. *Cartographica: The International Journal for Geographic Information and Geovisualization 55*(3): 151–162.

Rothstein, R. (2017). *The Color of Law*. New York: Liveright.

Schranz, C. (2021). *Shifts in Mapping*. Bielefeld: transcript.

Schuurman, N. (1999). Critical GIS: Theorizing and emerging science. *Cartographica: The International Journal for Geographic Information and Geovisualization 36*(4): 53.

Schuurman, N. (2000). Trouble in the heartland: GIS and its critics in the 1990s. *Progress in Human Geography 24*(4): 569–590.

Schuurman, N. (2006). Formalization matters: Critical GIS and ontology in research. *Annals of the Association of American Geographers 96*(4): 726–739.

Sletto, B.I. (2009). "We drew what we imagined": Participatory mapping, performance, and the arts of landscape making. *Current Anthropology 50*(4): 443–476.

Smith, N. (1992). History and philosophy of geography: Real wars, theory wars. *Progress in Human Geography 16*(2): 257–271.

Son, S.A. (2022). Watching North Korea from the sky: Remote sensing and documenting human rights in the Democratic People's Republic of Korea. *Political Geography 92*: 102525.

Sparke, M. (1998). A map that roared and an original atlas: Canada, cartography, and the narration of nation. *Annals of the Association of American Geography 88*: 463–495.

Tyner, J.A. (1982). Persuasive Cartography. *Journal of Geography 81*(4): 140–144.

Van Houtum, H., & Bueno Lacy, R. (2020). The migration map trap. On the invasion arrows in the cartography of migration. *Mobilities 15*(2): 196–219.

Wainwright, J. (2008). Uneven developments: From the *Grundrisse* to *Capital*. *Antipode 40*(5): 879–897.

Wainwright, J., & Bryan, J. (2009). Cartography, territory, property: Postcolonial reflections on indigenous counter-mapping in Nicaragua and Belize. *Cultural Geographies 16*: 153–178.

Walker, J.R. (2023). Rakhine skies: Remote sensing, human rights, and the Rohingya crisis. *Annals of the American Association of Geographers 113*(1): 30–45.

Wang, H. (2024). Party affiliation and the legacy of HOLC redlining maps. *Political Geography 110*: 103075.

Wilson, B. (2020). Urban heat management and the legacy of redlining. *Journal of the American Planning Association 86*(4): 443–457.

Wilson, M. (2017). *New Lines*. Minneapolis, MN: University of Minnesota.

Witmer, F.D.W., & O'Loughlin, J. (2011). Detecting the effects of wars in the Caucasus regions of Russia and Georgia using radiometrically normalized DMSP-OLS nighttime lights imagery. *GIScience & Remote Sensing 48*(4): 478–500.

Wood, D. (1992). *Power of Maps*. New York: Guilford Press.

Zoppolato, D. G. (forthcoming). China's path to (in)securities in the Global South: An analysis of the China-Pakistan economic corridor. Doctoral dissertation, West Virginia University.

(Contested) Key Concepts

Chapter 6

Power

Joe Painter[1] *and Natalie Koch*[2]

[1]*Department of Geography, Durham University, Durham, UK*
[2]*Geography and the Environment Department, Maxwell School of Citizenship & Public Affairs, Syracuse University, Syracuse, NY, USA*

Politics is about power and power lies at the heart of political geography. Indeed, Mark Haugaard and Stewart Clegg (2009: 1) identify power as *the* central concept of all the social sciences. At first glance, power seems straightforward – in everyday language someone is powerful if they can achieve their goals despite opposition. But on closer inspection, power is rather more difficult to grasp. Is it a thing, a substance, a quality, an idea, or a relationship? Can it be won and lost? Does it exist when it is not being used? Is it only a negative force or can it do good? And how does it affect, and how is it affected by, geography?

Power shapes political geographies in profound ways, and yet political geographers did not historically give much attention to theorizing the nature of power itself. Indeed, early political geography largely left power as an unexamined concept, whose meaning has been assumed to be sufficiently obvious not to warrant detailed investigation. But from the 1980s, power became an increasingly important topic for political geographers, and by the early 2000s, it was established as an essential keyword in the subfield. Geographers have since published surveys of power that join established volumes in the social sciences (e.g., Crampton & Elden 2007; Meusburger, Gregory, & Suarsana 2015; Coleman & Agnew 2018). Rather than survey the history of political geography's engagement with power, this chapter highlights some of the main ways in which political geographers have approached the concept in the past several decades.

The Wiley Blackwell Companion to Political Geography, Second Edition.
Edited by Virginie Mamadouh, Natalie Koch, Chih Yuan Woon, and John Agnew.

Taking Power Seriously

It is often said that power is the stuff of politics and that political geography originated in the study of the power-laden conflicts between the most powerful states of the nineteenth and early twentieth centuries. Such talk contains an implicit theory of power; namely, that power is a substance ("stuff"), which is present in political conflicts and of which states can be full or, by implication, empty. Until relatively recently, political geography has been resolutely state centric and power meant state power, and in particular the control of territory. State, power, and territory formed a kind of holy trinity for political geography. The state was understood as an institution, territory as bounded space, and power as control. The state "used" power to control territory and sometimes to expand it, usually at the expense of other states. Power as control; power as tool; power as possession; power as substance; power as power over territory – thus did early political geographers understand one of their central concepts. This understanding was mostly implicit. It seemed self-evident that the "Great Powers" would "wield" power to obtain political objectives, or that the power of a state could increase or decrease over time.

Within this power-politics discourse, untroubled by theoretical introspection, were the seeds of something more complex and even contradictory. The language used by early political geographers implied several different understandings of power, including a distinction between the idea of power as a substance that can be possessed and the idea of power as a relationship of control or domination. It also included instrumental conceptions of power – power as something that can be "used" or "wielded," a tool that can be brought into action at will. And this approach stood in contrast to the idea of power as an attribute of the state – the idea that the state is the state by virtue of its power over territory, for example. We might also contrast the idea of power as control with that of power as capacity, albeit still the capacity of the state to achieve its objectives. Finally, the notion that power is an instrument, by implication held and wielded centrally, is different from the idea that power is spread out over territory – the idea that the power of the state is somehow mysteriously everywhere. These were the seeds of a more nuanced understanding of power in the traditional canon of political geography, but they took a long time to germinate.

Since the early 2000s, but drawing from new theoretical texts published in the 1970s and 1980s, power has joined other key concepts in the lexicon of political geography, including "state" and "territory." Geographical scholarship has sought not only to tease out and clarify differing conceptions of power, but also to develop ideas of power in new directions. This broadening of political geographers' engagement with theories of power has happened for two reasons. First, as the contents of this *Companion* attest, the field has itself become much broader, both empirically and theoretically. Political geography is markedly less state centric than it was and has drawn into its remit a wide range of new topics, from gender relations and sexuality to ecological breakdown, from citizenship to financial crisis, and from health and disease to algorithms and digital surveillance. Second, the traditional core concepts of political geography, such as the state and territory, have themselves been subject to more rigorous critique and conceptualization than was previously the case.

The theoretical and empirical diversification of political geography makes it not only impossible, but also undesirable, to come up with a single concept of power (Flint & Taylor 2011: 28–43). Steven Lukes, a leading theorist of power, argued that power is an "essentially contested" concept (Lukes 1974). In other words, it will never be possible to reconcile competing notions of power with reference to empirical evidence, because different ideas about power reflect their proponents' normative worldviews. There is no consensus about what it is that essentially defines power and little purpose in trying to rule on which definitions are correct (Haugaard & Clegg 2009: 4). Yet some approaches, such as those drawing on the writings of Michel Foucault, have been much more heavily represented in the geographical literature than others.

Moreover, power itself takes many forms and has many modes (Allen 2003). A key distinction is often drawn between power *over* and power *to*. A master has power over a slave, while an orchestra has the power to produce inspiring music. "Power over" speaks of restriction, domination, limitation, and constraint; "power to" of creation, expression, production, and possibility. Power *to* is the older meaning, reflecting the term's origins in the Latin verb *posse* (to be able to). Yet it is not possible to separate entirely power over from power to. The master also wants the slave to be productive, while the orchestra's music arises in part because the conductor has power over the musicians. In both cases, we are talking about power *relations* and opening up *relational* views of power. To understand the value and importance of a relational approach, we need to return briefly to the earlier understanding of power as substance.

Individuals, organizations, and countries are often described as "having" power. For example, the United States is said to have more power than Afghanistan, or the Russian president is imagined as having more power than the Georgian president. Similarly, some political positions are said to have more power than others. Thus, a constitutional monarch has less power than an elected prime minister, and a city mayor has less power than a national president. Political actors and states can also be said to "gain" and "lose" power. China and India are often thought to be gaining power in the global economic and political system. The language of "having," "gaining," and "losing" power suggests that power is some kind of substance that can accumulate or dissipate, be won or lost, and be moved around from one country or leader to another. This can give rise to the assumption that politics is a zero-sum activity – that somehow there is a fixed amount of power in the world and that if the power of one actor increases then the power of another must decline.

Power is not a physical substance; it is not something that we can touch or store. Yet it was precisely this view that defined much of the early political geography scholarship differences in power between states. According to Halford Mackinder's (1861–1947) Heartland thesis, for example, epochal shifts in the balance of power at the world scale were associated with changes in the benefits accruing from being a land-based or a maritime state (Mackinder 1904; see Muscarà, this volume). Similarly, Richard Hartshorne's attempt to systematize political geography in the mid-twentieth century involved identifying geographical factors contributing to "national power," including mineral resources (1954: 175). And in *Geography and Politics in a World Divided* (1963: 8–9), Saul B. Cohen defined six categories in his "power analysis approach" to political geography (the physical environment, movement, raw materials, population, the body politic, and space). In this approach, national power is equated with, or understood to stem from, other characteristics of geographical areas such as population and raw materials and their spatial arrangement.

In fact, in his systematic study of the literature in geopolitics, political geography, and international relations, Karl Höhn (2011) found no fewer than 69 different multivariate formulas and indexes designed specifically to measure national power and to rank countries in terms of their power relative to each other. The earliest of the formulas appeared in Germany in 1741, the other 68 between 1936 and 2010 (Höhn 2011: 258). This large number of attempts to measure national power is testament to the fact that power cannot be quantified directly. Although some early geographers, like Cohen, acknowledged the shortcomings of the power inventory approach, the real problem is more fundamental. Power inventories and indexes of national power are not really measures of power at all. Instead, for the most part, they are measures of different kinds of resources (whether available or potential). Power, then, does not inhere in the resources themselves, but is by implication an attribute or capacity that is generated by them. Furthermore, power inventory approaches seek to express national power on a unitary scale. While they appear to be attentive to multiple *sources* of power (economic, military, cultural, etc.), they distill them to produce a single measure of power. They thus neglect the diverse forms and modes that power can take, a diversity that becomes apparent when power is understood in relational terms. Some scholars have tried to move past these problems in critical research on materialities and state power (e.g., Dalby 2018; Walker & Johnson 2018; Cederlöf & Kingsbury 2019; Paudel & Le Billon 2020; Chandrashekeran 2022; Flint 2024), but a more human-centered, relational approach to power has become the most influential strain of research over the past several decades.

Power as Capacity

The idea of power as capacity or capability is common in the social sciences and appears consistent with the notion of "power to." Specifically, power is often thought of as the capacity to produce effects and transformations, although it is important to remember that the capacity to prevent transformations that would otherwise occur is also an effect of power. The view of power as capacity still leaves us with the sense that power can be held and possessed by one party and used to affect another. However, the notion of power as capacity has been particularly important to feminist understandings of power. For example, the political scientist Nancy Hartsock argues that "women's stress on power not as domination but as capacity, on power as a capacity of the community as a whole, suggests that women's experience of connection and relation have consequences for understandings of power and may hold resources for a more liberatory understanding" (Hartsock 1983: 253; see also Miller 1992). The sociologist John Scott (2001) contrasts the notion of power as a dispositional capacity with the mainstream view of power inaugurated by the early twentieth-century work of Max Weber (1968), in which power is understood as the episodic imposition of the will of one actor over another. In the dispositional view, capacities provide the disposition to act and affect things, even if they are not used.

Geographer John Allen points out that the notion of power as capacity ("power to") is also consistent with thinking of power as control ("power over"): "the idea that individuals or institutions possess specific capacities to secure the compliance of others can be thought about as just so many ways of achieving dominance over others" (2003: 16). Yet the capacity of one party to produce effects depends on the liabilities, susceptibilities, and affordances of the other. A conductor only has the power to produce the effect of music to the extent that the orchestra comprises people who know how to play. A government only has the power to control an international border to the extent that it has sufficient personnel with appropriate training and equipment. These examples hint at some of the complexities of power – it is not simply the exertion of force.

The political thinker Hannah Arendt drew a sharp distinction between power and force. Writing in the middle years of the twentieth century, she produced dynamic and innovative political theory that had relatively limited purchase in geography until the 2000s (Allen 2008; Dikeç 2012a, 2012b, 2013; Menga 2017; Belcher & Schmidt 2020; Tillotson & Mustafa 2021; Koch 2022b), despite the fact that space is integral to her understanding of politics and the possibility of political action (Dikeç 2013; Debarbieux 2017). Arendt's ideas prioritize human action and her theory of power emphasizes human beings acting in concert to "establish relations and create new realities" (Arendt 1958: 200). The word politics derives ultimately from *polis*, the Greek word for city. For Arendt, however, the *polis* "is not the city-state in its physical location; it is the organization of the people as it arises out of acting and speaking together" (1958: 198). The *polis* in this sense is what Arendt describes as a transient "space of appearance," which is potential that arises whenever people gather, but "not necessarily and not forever" (1958: 199). Power, according to Arendt, "is what keeps the public realm, the potential space of appearance between acting and speaking men, in existence." The space of appearance therefore depends on power, since "what first undermines and then kills political communities is loss of power . . .; and power cannot be stored up and kept in reserve for emergencies, like the instruments of violence, but exists on in its actualization" (1958: 200). She thus distinguishes sharply between power and force or strength.

For Arendt, power is not dependent on, or derived from, resources, but comes about from communication between human beings and their development of common purposes. However, this does not mean that people act as a single unit. Arendt emphasizes the plurality of human beings – the fact that while we are all of one species and thus able to communicate and cooperate, we are also all unique individuals. Geographer Mustafa Dikeç argues that this means that, in Arendt's conceptualization of politics, space both brings us into relation but also separates us by revealing our distinctiveness vis-à-vis one another (Dikeç 2012b). Another implication of Arendt's approach is that public space is not a given, but has to be brought into being through political action (Lee 2009; see also McEwan 2005). Equally, as Lee suggests, the fact that it *can* be brought into being when people act in concert undermines arguments that we necessarily are seeing the end of public space through privatization. The popular uprisings associated with the so-called Arab Spring in the early 2010s, which turned many public spaces, such as Cairo's Tahrir Square, into sites of protest, might be seen as examples of Arendtian power (Kallio & Häkli 2017; Lang 2019; Thonhauser 2019). In similar vein, the Black Lives Matter movement has seen widespread public protests in numerous cities around the world, particularly following the murder of George Floyd by a police officer in Minneapolis in 2020 (van Haperen et al. 2020; Barwick & Nayak 2024).

Discipline and Biopower

Arendt's work offers political geographers a positive conception of power that takes space seriously. However, she restricts the exercise of power to the public realm and to positive political action, which she sees as both a precondition and a product of freedom. She excludes violence, strength, and force from her definition of power. Also absent are social relations of the private sphere and influences on human thought and behavior that may be hidden from view – influences that are subconscious or involve deceit, for example (see Kuran 1995; Allen 2020). For a broader understanding of power as something that suffuses all kinds of social relations, whether nominally public or private, and whose operation may not be obvious to those that it affects, many scholars have looked to the work of other political philosophers.

No theorist has had a greater influence on geographers' thinking about power than the French theorist Michel Foucault. While the impact of his ideas was felt first and probably to the greatest extent in social and cultural geography, political geography has also experienced its own Foucauldian turn (Hannah 2007). One of the most important features of Foucault's work is the emphasis that he placed on historical understanding. He showed that power itself has a history – in other words, that the nature of power has changed over time. This approach challenges general definitions of power that, by implication, are intended to be equally applicable in all times and places. Across numerous lectures, essays, and books, Foucault (1975, 1980, 1982, 2004, 2007, 2008) identified three forms of power that have evolved over time. Sovereign power – epitomized by the ritual of public execution – involves submitting to the law of the king or other central authority. From the seventeenth century onward, a new form of power emerges in Europe – disciplinary power – whose techniques

> included all devices that were used to ensure the spatial distribution of individual bodies (their separation, their alignment, their serialization, and their surveillance) and the organization, around those individuals, of a whole field of visibility. They were also techniques that could be used to take control over bodies. Attempts were made to increase their productive force through exercise, drill, and so on. (Foucault 2004: 242)

Unlike the highly visible exercise of sovereign power, disciplinary power operates in subtle and less obvious ways through the inculcation of bodily routines and habits. These are developed through institutions such as schools, prisons, hospitals, factories, and barracks and result in bodies that eventually comply with institutional norms and requirements without constant recourse to force. Disciplinary power thus permeates society, rather than being directed or wielded from a central point (Foucault 1975). The third type of power, biopower, emerges from the eighteenth century onward, and is directed toward populations rather than individual bodies. Biopower is associated more closely with the state than is disciplinary power, since it works through technologies such as statistics, epidemiology, public health programs, eugenics, and, increasingly, environmental interventions and risk management to improve the health, longevity, composition, and size of the population (Foucault 2004, 2007, 2008).

Biopower and its associated political practices (biopolitics) have received considerable attention from geographers. The securitization of territory and borders has been a particular focus of concern (e.g., Hannah 2006; Sparke 2006; Coleman & Grove 2009; Martin 2010; Crampton 2015). But biopower has also been analyzed in relation to war and conflict (Hyndman 2010; Belcher 2018; Hutta 2019; Joronen 2021), the environment and nonhuman nature (e.g., Alatout 2006; Dalby 2013; Clarke-Sather 2019; Andreucci and Zografos 2022; Fleischmann 2023; Varco 2024), colonial administration (Legg 2005, 2008; Usher 2024), migration (Bailey 2013; Wiertz 2021), regional policy (Painter 2013), the governance of emergencies and disasters (Davis and Hayes-Conroy 2018; Anderson et al. 2020), and the Covid-19 pandemic (Jauhiainen 2020), among many other topics.

New Geographies of Power

Moving well beyond the idea of power as control or domination, both Arendt's idea of power as human action in concert and Foucault's theorization of power as operating through a dispersed series of institutions and practices highlight the importance of geography. Arendt's notion of power depends on the space of appearance, while Foucault (1980: 96) proposed that the analysis of power should be concerned with the "points where it becomes capillary" rather than centralized. Yet relatively few geographers have paid attention to the difference that space makes to power. A notable exception is the work of John Allen. In *Lost Geographies of Power*, he emphasizes that power is neither equivalent to resources nor something tangible that can be "wielded or clung on to or flaunted" (2003: 4). Instead, Allen suggests that power is a "relational effect" of social interactions. Moreover, the geographies of those interactions – their situatedness in time and space and whether they involve geographical proximity or reach – are integral to and constitutive of the different forms of power to which they give effect.

Allen (2003) distinguishes between eight modes of power: domination, coercion, manipulation, seduction, authority, inducement, negotiation, and persuasion. Space is integral to all of these modes, but in different ways in each case, and typically in more complex ways than conventional notions of power as either centralized, networked, or immanently everywhere. *Lost Geographies of Power* has gained considerable authority in subsequent geographical research on power – to the extent that it was featured in the "Classics in human geography revisited" section of *Progress in Human Geography* just 10 years after it was published (Agnew 2013; Bulkeley 2013; Haugaard 2013). But Allen has since developed his ideas further, including framing power as inherently contextual and joining theoretical discussions on topologies of power in geography (Allen 2008, 2009, 2011, 2016, 2020; Allen & Cochrane 2010).

The turn toward practice-focused approaches in human geography – whether drawn from Foucault's practice-based methods (e.g., Koch 2014, 2018), nonrepresentational theory (e.g., Anderson & Harrison 2010), or Pierre Bourdieu's social theory (e.g., Hillier & Rooksby 2005) – has begun to have a significant impact in political geography. For example, in calling for a critical geographical perspective on authoritarianism, Koch (2022a, 2022b, 2024) challenges territorially trapped approaches to authoritarian power by instead using a practice-based method to spatialize authoritarianism. Indeed, a growing number of political geographers are beginning to attend to the authoritarian power dynamics that are at play in ostensibly democratic and nondemocratic states alike, and in state and nonstate spaces alike (Tansel 2017; Doshi & Ranganathan 2019; Kantel 2019; Kopack 2019; Conduit 2020; Regilme 2021; Tynen 2021; Hammond 2022; Luger 2022; Fung & Lamb 2023; Murrey 2023).

In addition to the increased attention to the authoritarian nature of state and nonstate power, global political events in the first quarter of the twenty-first century have drawn political geographers back to the topic of sovereign power, as governments have reacted to threats of terrorism and to the impact of the Covid-19 pandemic, nationalist politicians have gained power in many countries, and inter- and intrastate military conflicts have proliferated (Murphy 2022; Culcasi 2023; Landman & Smallman-Raynor 2023; Wolfe et al. 2024). Contemporary events have combined to generate an extensive body of work on the practices of sovereignty, rather than a reassertion of older understandings of sovereign power as simple domination within a clearly demarcated state territory. Indeed, political geographers have long been rethinking the meaning of sovereignty and its geographies beyond conventional imaginings of sovereignty as the ultimate authority in a legal order such as a state (Mountz 2013; Moisio et al. 2020).

In the new geographies of power and sovereignty, the writings of Giorgio Agamben have been particularly influential. Drawing on the ideas of Carl Schmitt, Agamben defines the sovereign as "he who decides the state of exception" (Agamben 1998: 15). In other words, sovereignty is the capacity to suspend the normal rules that apply to all but the sovereign. For Agamben, the paradox of sovereignty is that the sovereign is both the pinnacle of the legal order and yet simultaneously outside it. Political geographers have especially drawn from Agamben's work to identify and investigate "spaces of exception" – enclaves or zones where the usual constitutional or legal limitations on the exercise of state power do not apply. Epitomized in Agamben's writings by the figure of the detention or concentration camp, spaces of exception are populated by those (such as undocumented migrants, foreign terrorist suspects, and so-called enemy combatants) who are beyond the normal legal protections afforded to citizens and lawful residents, and sometimes by those who have had such protections withdrawn.

Geographers have examined a range of spaces of exception, including Nazi concentration camps, the Abu Ghraib prison in Iraq during the US-led occupation between 2003 and 2006, the detention facility at the US naval base in Guantánamo Bay, Cuba, and offshore migrant reception/detention centers (Gregory 2006; Hannah 2006; Minca 2006, 2015; Papastergiadis 2006; Giaccaria & Minca 2011; Hopkins 2019; Coddington, Conlon, & Martin 2020; Martin, Minca, & Katz 2020). Because these spaces are outside the territory of the state, those held there are outside the state's legal jurisdiction but still under its authority. This creates a legal no-man's land, created by the exercise of sovereign power and within which there are no restrictions on sovereign power.

Geographers working on these themes have not treated Agamben's approach uncritically. As Mountz points out, some political geographers have argued that sovereignty increasingly exhibits spatial blurring of "onshore and offshore, internal and external, inside and out" (2013: 833). At the same time, Agambenian notions of sovereign power also involve sharp distinctions. Andrew Neal argues that Agamben sets out the problem of sovereignty in binary terms – either sovereign power involves creating exceptions (by definition limited) to the normal or it exceeds all limits. Neal regards this framing of the problem as "apolitical" and unable to do justice to the "extremely complex principle and practice that is sovereignty" (2004: 375). Grasping that complexity requires a move from seeing space in purely topographical terms to understanding it topologically, with different modes of power involving different forms of proximity and reach. Geographers

Mathew Coleman and Kevin Grove likewise argue that Agamben writes geography out of his account of sovereignty:

> On the one hand, we contend that Agamben's use of biopolitics works . . . to erase the unevenness of political, economic, and social space. We argue that the result is a perverse reconstitution of sovereign (bio)power in a dazzling, all-encompassing, and totalizing spatial form – not unlike the mapping of state power that Foucault found necessary to unpack via his discussion of the biopolitical. (2009: 490–491)

Foucault's writing, they argue, allows for a more polyvalent understanding of the state and power. Indeed, state and sovereign power – even when expressed in warfare – is still a matter of the disaggregated and situated exercise of power in myriad spaces and institutions (Moisio et al. 2020; Mountz 2013).

As John Allen (2016: 147) notes, sovereign borders "no longer map onto the 'inside' and 'outside' of state territory in any meaningful way." The topological character of power involves specific transformations – folding and stretching – of the geographies of the border. This continues to be a theme of rich debate among scholars who think with the writing of Agamben, Foucault, and others, especially in the well-established field of migration and border studies. But political geographers have also explored this stretching and blurring of borders in newer research on the extraterritorial exercise of state power (Collyer and King 2015; Davis 2017; Dalmasso et al. 2018; Conduit 2020; Grant 2020; Kenney-Lazar 2020; Potts 2020; Furstenberg, Lemon, & Heathershaw 2021; Regilme 2021). So to have scholars writing on infrastructure and cyberspace traced the many new spatialities of power that are reflected in and realized through today's material and immaterial networks (Attewell 2021; Barnes 2017; Bevilacqua 2022; Dittmer 2021; Douzet 2020; Oakes 2021; Winter 2021).

Power's Futures

Political geographers' renewed interest in questions of sovereignty and in forms of power as domination is understandable when the governments of major states are, or recently have been, headed by such authoritarian political figures as Donald Trump, Xi Jinping, Vladimir Putin, Narendra Modi, and Jair Bolsonaro, and when global politics has been destabilized by economic shocks, military invasions, resource crises, and intensifying effects of anthropogenic climate change. At first glance, this concern with power stemming from and circulating through state institutions may appear to represent a return to an older style of political geography, in which power is taken to be a substance that states can amass and deploy. However, a careful reading reveals that most contemporary political geographical scholarship is closely attuned to the diverse modalities and complex spatialities of power: the sovereign power of states is not taken to be the unproblematic starting point for a political geography that treats the world as a simple mosaic of discrete territorial units. Instead, sovereign power, and state power more broadly, is viewed as something that itself needs to be unpacked and explained in terms of practices and relations that blur the neat distinction between state and society and stretch across international boundaries.

Political geographers' increasingly nuanced understanding of power stems partly from the subdiscipline's growing conceptual sophistication. But it is also a response to the changing forms of power in contemporary politics. So despite the apparent resurgence of domination and overt forms of sovereign power, "quiet" modes of power – practices that influence political relations and outcomes in hidden and covert ways – are pervasive (Allen 2020). Rapid advances in digital technologies have contributed to a shift in the media and mechanisms through which power is expressed and in its political geographies (Amoore 2018; Douzet 2020; Warf 2021; Adams, this volume). Social media networks, for example, can exert influence through the microtargeting of political messaging, depending less on outright deception and more on dissimulation, insofar as they fail to disclose their intentions and motives. Such practices enable power to have considerable topological reach, so that it is not only the identities and motives of the actors engaged in dissimulation that are veiled, but also their spatial locations and national affiliations (Gentile 2019; Allen 2020; Bevilacqua 2022; Luger 2022).

Political geographers have offered a unique perspective to the conceptualization of power, but it is increasingly difficult to justify a narrowly subdisciplinary focus for such a ubiquitous and indispensable notion. As Coleman and Agnew note, the dominance of political geography in geographical debates about power has "come at a certain conceptual and theoretical cost," and they argue for a move "beyond the power/space couplet as it has been understood by political geographers" (2018: 4). There is certainly much innovative thinking about power and geography beyond the confines of political geography, conventionally defined. Many geographers routinely work across subdisciplines within geography as well as across disciplines beyond geography. Indeed, geographers' interdisciplinary exchanges have helped translate geographical thinking on power to such an extent that nongeographers – mostly political scientists – are increasingly framing their spatially informed studies as "geographies of power" (e.g., Hanieh 2020; Furstenberg, Lemon, & Heathershaw 2021; Schwedler 2022). At the same time, understandings of power in political geography are neither static nor irredeemably state centric. Within or beyond our subfield, plural political geographies can offer rich, and one might even say powerful, resources for future inquiry.

References

Agamben, G. (1998). *Homo Sacer: Sovereign Power and Bare Life*. Stanford, CA: Stanford University Press.

Agnew, J. (2013). Commentary 1: Hidden or lost geographies of power? *Progress in Human Geography* 37(3): 452–454.

Alatout, S. (2006). Towards a bio-territorial conception of power: Territory, population, and environmental narratives in Palestine and Israel. *Political Geography 25*: 601–621.
Allen, J. (2003). *Lost Geographies of Power*. Oxford: Blackwell.
Allen, J. (2008). Pragmatism and power, or the power to make a difference in a radically contingent world. *Geoforum 39*(4): 1613–1624.
Allen, J. (2009). Three spaces of power: Territory, networks, plus a topological twist in the tale of domination and authority. *Journal of Power 2*(2): 197–212.
Allen, J. (2011). Topological twists: Power's shifting geographies. *Dialogues in Human Geography 1*(3): 283–298.
Allen, J. (2016). *Topologies of power: Beyond territory and networks*. London: Routledge.
Allen, J. (2020). Power's quiet reach and why it should exercise us. *Space and Polity 24*(3): 408–413.
Allen, J., & Cochrane, A. (2010). Assemblages of state power: Topological shifts in the organization of government and politics. *Antipode 42*(5): 1071–1089.
Amoore, L. (2018). Cloud geographies: Computing, data, sovereignty. *Progress in Human Geography 42*(1): 4–24.
Anderson, B., Grove, K., Rickards, L., & Kearnes, M. (2020). Slow emergencies: Temporality and the racialized biopolitics of emergency governance. *Progress in Human Geography 44*(4): 621–639.
Anderson, B., & Harrison, P. (eds) (2010). *Taking-Place: Non-representational Theories and Geography*. Farnham: Ashgate.
Andreucci, D., & Zografos, C. (2022). Between improvement and sacrifice: Othering and the (bio)political ecology of climate change. *Political Geography 92*: 102512.
Arendt, H. (1958). *The Human Condition*. Chicago, IL: University of Chicago Press.
Attewell, W. (2021). Just-in-time imperialism: The logistics revolution and the Vietnam War. *Annals of the American Association of Geographers 111*: 1329–1345.
Bailey, A.J. (2013). Migration, recession and an emerging transnational biopolitics across Europe. *Geoforum 44*: 202–210.
Barnes, J. (2017). States of maintenance: Power, politics, and Egypt's irrigation infrastructure. *Environment and Planning D: Society and Space 35*: 146–164.
Barwick, D., & Nayak, A. (2024). The transnationalism of the Black Lives Matter Movement: Decolonization and mapping Black geographies in Sydney, Australia. *Annals of the American Association of Geographers 114*(7): 1587–1603.
Belcher, O. (2018). Anatomy of a village razing: Counterinsurgency, violence, and securing the intimate in Afghanistan. *Political Geography 62*: 94–105.
Belcher, O., & Schmidt, J.J. (2020). Being earthbound: Arendt, process and alienation in the Anthropocene. *Environment and Planning D: Society and Space 39*: 103–120.
Bevilacqua, I. (2022). E-scaping apartheid: Digital ventures of Zionist settler colonialism. *Human Geography 15*: 220–228.
Bulkeley, H. (2013). Commentary 2: Making space for power? *Progress in Human Geography 37*(3): 454–456.
Cederlöf, G., & Kingsbury, D. (2019). On PetroCaribe: Petropolitics, energopower, and post-neoliberal development in the Caribbean energy region. *Political Geography 72*: 124–133.
Chandrashekeran, S. (2022). Energopower, statecraft and political legitimacy. *Environment and Planning E: Nature and Space 5*: 1788–1806.

Clarke-Sather, A. (2019). From the heavens to the markets: Governing agricultural drought under Chinese fragmented authoritarianism. *Annals of the American Association of Geographers 109*: 456–464.

Coddington, K., Conlon, D., & Martin, L.L. (2020). Destitution economies: Circuits of value in asylum, refugee, and migration control. *Annals of the American Association of Geographers 110*: 1425–1444.

Cohen, S.B. (1963). *Geography and Politics in a World Divided*. New York: Random House.

Coleman, M., & Agnew, J. (eds) (2018). *Handbook on the Geographies of Power*. Cheltenham: Edward Elgar.

Coleman, M., & Grove, K. (2009). Biopolitics, biopower, and the return of sovereignty. *Environment and Planning D 27*: 489–507.

Collyer, M., & King, R. (2015). Producing transnational space: International migration and the extra-territorial reach of state power. *Progress in Human Geography 39*: 185–204.

Conduit, D. (2020). Authoritarian power in space, time and exile. *Political Geography 82*: 102239.

Crampton, J. (2015). Collect it all: National security, Big Data and governance. *GeoJournal 80*: 519–531.

Crampton, J., & Elden, S. (2007). *Space, Knowledge and Power: Foucault and Geography*. Burlington, NJ: Ashgate.

Culcasi, K. (2023). *Displacing Territory: Syrian and Palestinian Refugees in Jordan*. Chicago, IL: University of Chicago Press.

Dalby, S. (2013). Biopolitics and climate security in the Anthropocene. *Geoforum 49*: 184–192.

Dalby, S. (2018). Firepower: Geopolitical cultures in the Anthropocene. *Geopolitics 23*: 718–742.

Dalmasso, E., Del Sordi, A., Glasius, M., et al. (2018). Intervention: Extraterritorial authoritarian power. *Political Geography 64*: 95–104.

Davis, S. (2017). Apparatuses of occupation: Translocal social movements, states and the archipelagic spatialities of power. *Transactions of the Institute of British Geographers 42*: 110–122.

Davis, S., & Hayes-Conroy, J. (2018). Invisible radiation reveals who we are as people: Environmental complexity, gendered risk, and biopolitics after the Fukushima nuclear disaster. *Social & Cultural Geography 19*: 720–740.

Debarbieux, B. (2017). Hannah Arendt's spatial thinking: An introduction. *Territory, Politics, Governance 5*(4): 351–367.

Dikeç, M. (2012a). Politics is sublime. *Environment and Planning D: Society and Space 30*(2): 262–279.

Dikeç, M. (2012b). Space as a mode of political thinking. *Geoforum 43*(4): 669–676.

Dikeç, M. (2013). Beginners and equals: Political subjectivity in Arendt and Rancière. *Transactions of the Institute of British Geographers 38*(1): 78–90.

Dittmer, J. (2021). The state, all at sea: Interoperability and the Global Network of Navies. *Environment and Planning C: Politics and Spaces 39*: 1389–1406.

Doshi, S., & Ranganathan, M. (2019). Towards a critical geography of corruption and power in late capitalism. *Progress in Human Geography 43*: 436–457.

Douzet, F. (2020). Cyberspace: The new frontier of state power. In Moisio, S., Koch, N., Jonas, A.E.G., Lizotte, C., & Luukkonen, J. (eds), *Handbook on the Changing Geographies of the State: New Spaces of Geopolitics*. Northampton: Edward Elgar, 325–337.

Fleischmann, L. (2023). More-than-human political geographies: Abjection and sovereign power. *Political Geography 107*: 102949.

Flint, C. (2024). *Near and Far Waters: The Geopolitics of Seapower*. Stanford, CA: Stanford University Press.

Flint, C., & Taylor, P.J. (2011). *Political Geography: World-Economy, Nation-State and Locality*, 6th edn. London: Routledge.

Foucault, M. (1975). *Discipline and Punish: The Birth of the Prison*. New York: Pantheon.

Foucault, M. (1980). *Power/Knowledge: Selected Interviews and Other Writings, 1972–1977*. New York: Pantheon.

Foucault, M. (1982). The subject and power. *Critical Inquiry* 8: 777–795.

Foucault, M. (2004). *Society Must Be Defended: Lectures at the Collège de France, 1975–76*. London: Penguin.

Foucault, M. (2007). *Security, Territory, Population: Lectures at the Collège de France 1977–1978*. New York: Picador.

Foucault, M. (2008). *The Birth of Biopolitics: Lectures at the Collège de France 1978–1979*. New York: Picador.

Fung, Z., & Lamb, V. (2023). Dams, diversions, and development: Slow resistance and authoritarian rule in the Salween River Basin. *Antipode 55*: 1662–1685.

Furstenberg, S., Lemon, E., & Heathershaw, J. (2021). Spatialising state practices through transnational repression. *European Journal of International Security* 6: 358–378.

Gentile, M. (2019). Geopolitical fault-line cities in the world of divided cities. *Political Geography 71*: 126–138.

Giaccaria, P., & Minca, C. (2011). Topographies/topologies of the camp: Auschwitz as a spatial threshold. *Political Geography 30*: 3–12.

Grant, A. (2020). Crossing Khorgos: Soft power, security, and suspect loyalties at the Sino-Kazakh boundary. *Political Geography* 76: 102070.

Gregory, D. (2006). The black flag: Guantánamo Bay and the space of exception. *Geografiska Annaler: Series B, Human Geography 88*: 405–427.

Hammond, T. (2022). Making the millet common: Rethinking authoritarian politics through commemoration following Turkey's July 2016 coup attempt. *ACME: An International Journal for Critical Geographies 21*: 603–622.

Hanieh, A. (2020). New geographies of financial power: Global Islamic finance and the Gulf. *Third World Quarterly 41*: 525–546.

Hannah, M. (2006). Torture and the ticking bomb: The war on terrorism as a geographical imagination of power/knowledge. *Annals of the Association of American Geographers* 96(3): 622–640.

Hannah, M. (2007). Formations of "Foucault" in Ango-American geography: An archeological sketch. In Crampton, J., & Elden, S. (eds), *Space, Knowledge and Power: Foucault and Geography*. Burlington, NJ: Ashgate, 83–106.

Hartshorne, R. (1954). Political geography. In James, P.E., & Jones, D.F. (eds), *American Geography: Inventory and Prospect*. Syracuse, NY: Syracuse University Press, 167–225.

Hartsock, N. (1983). *Money, Sex, and Power: Toward a Feminist Historical Materialism*. Boston, MA: Northeastern University Press.

Haugaard, M. (2013). Commentary 3: The prescience of *Lost Geographies of Power*. *Progress in Human Geography* 37(3): 456–458.

Haugaard, M., & Clegg, S.R. (2009). Introduction: Why power is the central concept of the social sciences. In Clegg, S.R., & Haugaard, M. (eds), *The SAGE Handbook of Power*. London: Sage, 1–24.

Hillier, J., & Rooksby, E. (2005). *Habitus: A Sense of Place*. Farnham: Ashgate.

Höhn, K. H. (2011). Geopolitics and the measurement of national power. PhD thesis, University of Hamburg.

Hopkins, B. (2019). Beyond the Agamben paradigm: The spatial logics of exceptionalism. *Environment and Planning D: Society and Space* 37(6): 953–970.

Hutta, J. (2019). From sovereignty to technologies of dependency: Rethinking the power relations supporting violence in Brazil. *Political Geography* 69: 65–76.

Hyndman, J. (2010). The question of "the political" in critical geopolitics: Querying the "child soldier" in the "war on terror. *Political Geography* 29: 247–255.

Jauhiainen, J.S. (2020). Biogeopolitics of COVID-19: Asylum-related migrants at the European Union borderlands. *Tijdschrift voor Econonische en Sociale Geografie* 111(3): 260–274.

Joronen, M. (2021). To wound life, to prevent its recovery: Enforcing vulnerability in Gaza. In Bissell, D., Rose, M., & Harrison, P. (eds), *Negative Geographies: Exploring the Politics of Limits*. Lincoln, NE: University of Nebraska Press, 206–233.

Kallio, K., & Häkli, J. (2017). Geosocial lives in topological polis: Mohamed Bouazizi as a political agent. *Geopolitics* 22: 91–109.

Kantel, A.J. (2019). Fishing for power: Incursions of the Ugandan authoritarian state. *Annals of the American Association of Geographers* 109: 443–455.

Kenney-Lazar, M. (2020). Relations of sovereignty: The uneven production of transnational plantation territories in Laos. *Transactions of the Institute of British Geographers* 45: 331–344.

Koch, N. (2014). Bordering on the modern: Power, practice and exclusion in Astana. *Transactions of the Institute of British Geographers* 39: 432–443.

Koch, N. (2018). *The Geopolitics of Spectacle: Space, Synecdoche, and the New Capitals of Asia*. Ithaca, NY: Cornell University Press.

Koch, N. (2022a). Authoritarian space-time. *Political Geography* 95: 102628.

Koch, N. (2022b). *Spatializing Authoritarianism*. Syracuse, NY: Syracuse University Press.

Koch, N. (2024). Geographies of authoritarian politics. In Wolf, A. (ed.), *The Oxford Handbook of Authoritarian Politics*. Oxford: Oxford University Press.

Kopack, R. (2019). Rocket wastelands in Kazakhstan: Scientific authoritarianism and the Baikonur cosmodrome. *Annals of the American Association of Geographers* 109: 556–567.

Kuran, T. (1995). *Private Truths, Public Lies: The Social Consequences of Preference Falsification*. Cambridge, MA: Harvard University Press.

Landman, T., & Smallman-Raynor, M. (2023). The politics of COVID-19: Government response in comparative perspective. *Political Geography* 106: 102957.

Lang, A. (2019). Constitutions are the answer! Hannah Arendt and the Egyptian revolution. In Hiruta, K. (ed.), *Arendt on Freedom, Liberation, and Revolution*. Cham: Springer, 225–251.

Lee, N.K. (2009). How is a political public space made? The birth of Tiananmen Square and the May Fourth Movement. *Political Geography* 28(1): 32–43.

Legg, S. (2005). Foucault's population geographies: Classifications, biopolitics and governmental spaces. *Population, Space and Place 11*(3): 137–156.

Legg, S. (2008). Ambivalent improvements: Biography, biopolitics, and colonial Delhi. *Environment and Planning A 40*(1): 37–56.

Luger, J. (2022). Celebrations, exaltations and alpha lands: Everyday geographies of the far-right. *Political Geography 96*: 102604.

Lukes, S. (1974). *Power: A Radical View*. London: Macmillan.

Mackinder, H.J. (1904). The geographical pivot of history. *Geographical Journal 23*: 421–444.

Martin, D., Minca, C., & Katz, I. (2020). Rethinking the camp: On spatial technologies of power and resistance. *Progress in Human Geography 44*(4): 743–768.

Martin, L.L. (2010). Bombs, bodies, and biopolitics: Securitizing the subject at the airport security checkpoint. *Social and Cultural Geography 11*(1): 17–34.

McEwan, C. (2005). New spaces of citizenship? Rethinking gendered participation and empowerment in South Africa. *Political Geography 24*(8): 969–991.

Menga, F. (2017). Hydropolis: Reinterpreting the polis in water politics. *Political Geography 60*: 100–109.

Meusburger, P., Gregory, D., & Suarsana, L. (2015). *Geographies of Knowledge and Power*. New York: Springer.

Miller, J.B. (1992). Women and power. In Wartenberg, T. (ed.), *Rethinking Power*. Albany, NY: State University of New York Press, 240–248.

Minca, C. (2006). Giorgio Agamben and the new biopolitical nomos. *Geografiska Annaler: Series B, Human Geography 88*: 387–403.

Minca, C. (2015). Geographies of the camp. *Political Geography 49*: 74–83.

Moisio, S., Koch, N., Jonas, A.E.G., Lizotte, C., & Luukkonen, J. (2020). *Handbook on the Changing Geographies of the State: New Spaces of Geopolitics*. Northampton: Edward Elgar.

Mountz, A. (2013). Reconfiguring geographies of sovereignty. *Progress in Human Geography 37*(6): 373–398.

Murphy, A.B. (2022). Taking territory seriously in a fluid, topologically varied world: Reflections in the wake of the populist turn and the COVID-19 pandemic. *Geografiska Annaler: Series B, Human Geography 104*: 27–42.

Murrey, A. (2023). A decolonial political geography of resistance and digital infrastructural harm in Cameroon and Ethiopia. *Globalizations 20*: 967–993.

Neal, A.W. (2004). Cutting off the king's head: Foucault's *Society Must Be Defended* and the problem of sovereignty. *Alternatives: Global, Local, Political 29*: 373–398.

Oakes, T. (2021). The Belt and Road as method: Geopolitics, technopolitics and power through an infrastructure lens. *Asia Pacific Viewpoint 62*: 281–285.

Painter, J. (2013). Regional biopolitics. *Regional Studies 47*(8): 1235–1248.

Papastergiadis, N. (2006). The invasion complex: The abject other and spaces of violence. *Geografiska Annaler: Series B, Human Geography 88*: 429–442.

Paudel, D., & Le Billon, P. (2020). Geo-logics of power: Disaster capitalism, Himalayan materialities, and the geopolitical economy of reconstruction in post-earthquake Nepal. *Geopolitics 25*: 838–866.

Potts, S. (2020). Law as geopolitics: Judicial territory, transnational economic governance, and American power. *Annals of the American Association of Geographers 110*: 1192–1207.

Regilme, S.S.F. (2021). Contested spaces of illiberal and authoritarian politics: Human rights and democracy in crisis. *Political Geography 89*: 102427.

Schwedler, J. (2022). *Protesting Jordan: Geographies of Power and Dissent*. Stanford, CA: Stanford University Press.

Scott, J. (2001). *Power*. Cambridge: Polity Press.

Sparke, M.B. (2006). A neoliberal nexus: Economy, security and the biopolitics of citizenship on the border. *Political Geography 25*(2): 151–180.

Tansel, C. (2017). *States of Discipline: Authoritarian Neoliberalism and the Contested Reproduction of Capitalist Order*. New York: Rowman & Littlefield.

Thonhauser, G. (2019). The power of public assemblies: Democratic politics following Butler and Arendt. In Bedorf, T., & Herrmann, S. (eds), *Political Phenomenology: Experience, Ontology, Episteme*. New York: Routledge, 202–219.

Tillotson, M., & Mustafa, D. (2021). An Arendtian geopolitics: Action, power and the deferral of work. *Progress in Human Geography 45*(3): 548–565.

Tynen, S. (2021). "Keep withstanding": Territory in the body, home and market in Xinjiang, China. *Political Geography 84*: 102310.

Usher, M. (2024). Territory, hydraulics, biopolitics: Internal colonization through urban catchment management in Singapore. *Territory, Politics, Governance 12*: 846–865.

van Haperen, S., Uitermark, J., & van der Zeeuw, A. (2020). Mediated interaction rituals: A geography of everyday life and contention in Black Lives Matter. *Mobilization: An International Quarterly 25*: 295–313.

Varco, M. (2024). Volk utopia: Racial futures and ecological politics on the German far-right. *Geoforum 153*: 103823.

Walker, J., & Johnson, M. (2018). On mineral sovereignty: Towards a political theory of geological power. *Energy Research & Social Science 45*: 56–66.

Warf, B. (2021). *Geographies of the Internet*. New York: Routledge.

Weber, M. (1968). *Economy and Society: An Outline of Interpretive Sociology*. New York: Bedminster Press.

Wiertz, T. (2021). Biopolitics of migration: An assemblage approach. *Environment and Planning C: Politics and Spaces 39*(7): 1375–1388.

Winter, T. (2021). Geocultural power: China's Belt and Road Initiative. *Geopolitics 26*: 1376–1399.

Wolfe, S.D., Denysenko, O., Krichker, D., Rebro, O., & Gunko, M. (2024). The intimate and everyday geopolitics of the Russian war against Ukraine. *Geopolitics 29*: 1474–1501.

Chapter 7

The State

Alex Jeffrey

Department of Geography, University of Cambridge, Cambridge, UK

It could be argued that the centrality of the state as a focus of political geographical concern has waned in recent years. In this vein we could explore how scholarly interest has moved to new subdisciplinary concerns and directions, coupled with novel methodological approaches. But much of this work is written "against" or "in tension" with a concern for the state. For example, geographers are at the forefront of challenging the role of the state in masking enduring coloniality, in doing so highlighting plurinational and Indigenous articulations of sovereignty (Mountz 2013; Zaragocin & Caretta, 2021; Radhuber & Radcliffe 2023). This work, in places, is building on feminist political geographies that contest a preoccupation with the "formal" sites of political authority, using participatory and ethnographic methods to trace politics operating at, and in, multiple scales and sites (Hyndman 2019; Kuus 2019; Jackman et al. 2020). Simultaneously, more-than-human geographies seek to confront an anthropocentric imagination of the state as separate to nature in an era of ecological crisis, exploring the material and immaterial entanglements in the Anthropocene (Tsing et al. 2017). Institutional analyses trace alternative geographies of authority that both transgress state boundaries and limit state sovereignty, from "big tech" to credit ratings agencies (Agnew 2020; Bassin 2022). But just as these strands of thought and praxis draw attention to other political geographical entities and events, there is a residual interest in the role of states themselves as shaping political geographies: *inter alia* waging wars (Vorbrugg & Bluwstein 2022), restricting human movement (Coddington, Conlon, & Martin 2020), and regulating ecological life (Loftus 2020; Massé 2022). There is then a paradox, where the concept of the

The Wiley Blackwell Companion to Political Geography, Second Edition.
Edited by Virginie Mamadouh, Natalie Koch, Chih Yuan Woon, and John Agnew.

state may hide many alternative political realities, but the idea of the state as a prominent geopolitical actor endures.

This paradox can be traced back through critical scholarship on the concept of the state. In his landmark paper *Notes on the Difficulty of Studying the State*, Philip Abrams argues that focusing on "the state" grants "spurious concreteness and reality to that which has a merely abstract and formal existence" (1988: 58). Emphasizing this point, later in the paper Abrams seeks to highlight the ideological nature of this political concept, arguing that the state is "not an object akin to a human ear. Nor is it even an object akin to a human marriage. It is a third-order object, an ideological project" (1988: 76). Just as Benedict Anderson (1983) argued regarding *Imagined Communities*, the ideological nature of the state does not render this an illusory entity, but it should orient attention toward how the state is imagined rather than an exploration of its fundamental characteristics. In these terms, we undertake some political labor on behalf of the state by rendering it as a single political geographical apparatus; in doing so we can grant this concept imagined clarity and immutability. It is for this reason that Agnew's (1994) warning of the "territorial trap," where political processes are read off through the lens of state territoriality, remains so prescient. Allied to the concept of "methodological nationalism," this assumption of the state as a backdrop to political life has been a key focus of critique for recent decolonial scholarship and praxis (Koch 2020). One of the first obligations, then, for scholars writing about the state is to assume a dual perspective: on the one hand recognizing and exploring the existence of a large body of work that focuses on the characteristics of "the state," while on the other remaining careful not to reify this apparatus as a timeless ontological truth.

Reflecting this precarious authorial position, a discussion of the state demands a succinct definition of precisely what is meant by this term while also remaining skeptical of its universal application. As Peter Bratsis (2006: 9) asserts, arguments concerning the nature of the state have been undermined by assuming the existence of the state, proceeding "as if" the state "was indeed a universal *a priori* predicate to our social existence rather than the product of our social existence." Such caveats notwithstanding, there exists a set of common reference points around which discussion of the state coheres. The state is often understood as a multidimensional phenomenon, with definitions coalescing around a notion of a relationship between power (sovereignty) and space (territory). This may be expressed in law; for example, Held (1983: 1) suggests the state is an "impersonal and privileged legal or constitutional order with the capability of controlling or administering a given territory." But equally, definitions can focus on the centrality of violence to the maintenance of state authority, exemplified in Weber's (1958 [1919]: 78) definition of the state as a "human community that (successfully) claims the monopoly of legitimate use of physical force within a given territory." As Bratsis (2006) recognizes, the key area of debate within this definition is not necessarily the violence but the centrality of the more ambiguous concept of legitimacy (which is also present in Held's definition through the use of the term "privileged"). As we will see, establishing the legitimacy of rule has both internal and external dimensions: both those within and those beyond the borders of the state regard a particular set of state arrangements as appropriate. Without internal sovereignty, a state struggles to govern its people; without external recognition, a state cannot participate in international affairs.

Both Weber's and (to a lesser extent) Held's definitions introduce another significant element of studies of the state: that this is both a geopolitical and a biopolitical project, where the governance, control, and maintenance of territories and human communities intersect. One common route through this material and embodied geography is to stratify the state into different varieties: for example the medieval state, authoritarian state, imperial state, nation-state, or settler colonial state (Giddens 1995; Inwood & Bonds 2016). The "nation-state" has become the most common shorthand for the fusion between human community and state sovereignty, itself coming under increasing scrutiny as a couplet that protects a particular racialized imagination of the "modern European state" (Mignolo & Bussmann 2023). This term is a helpful reminder that the projection of power is not simply a geopolitical process but is entwined with biopolitical concerns of the management of human populations and the generation of consent for rule. It is for this reason that political scientists from Jean-Jacques Rousseau onward have spoken of the "social contract" that is forged between state and individual, where certain freedoms are surrendered in return for political and social rights. While this has historically been a matter for discussion of scholars of citizenship (Dickinson et al. 2015), political geographers who have studied the material, embodied, and everyday geographies of the state have begun to reflect on the forms of practice that convey the legitimacy and authority of prevailing state systems.

Taking Abrams' longstanding concerns seriously, this chapter seeks to pluralize understandings of the state and trace these through four sections, each overlapping and intertwined. The first explores the imagined foundations of the state as a natural or organic expression of human territoriality. In some of the bolder pronouncements found in this work the state becomes a reflection of natural law, a form of territorialization that is a human quality rather than an expression of power. If the first section outlines how internal sovereignty has been secured, the second looks at the international nature of statehood through a consideration of the state system. Orienting attention to the most influential systemic account, Marxist theories of the state, the section considers how these approaches have embedded a sense of an enduring distinction between state and society. The third section examines how feminist, anthropological, and poststructural approaches to the state have reoriented attention away from formal state institutions and toward the more socially embedded processes through which ideas of the state are reproduced. The final section continues this exploration, examining debates concerning the enduring coloniality of the state and state system. The chapter reflects on the role of political geographers – and academic pursuits more broadly – in confronting the potential role of the state in obscuring the legacies and continuities of colonial power. It should be emphasized that this is not a further theoretical "turn" but a challenge to the very craft and mode of analysis that place the state as the preeminent territorialization of political power.

Tracing the State

When a concept such as the state does so much to defy precise definition, it is no surprise that quests are made for its origin or foundation (see Moisio *et al.* 2020). So pervasive is the state in analysis of the dynamics of contemporary political geography that we may be seduced into thinking that this is a timeless expression of human desires to territorialize. The sense that the state is somehow prepolitical stalks writings on its foundation and characteristics, where the establishment of states reflects the return of humanity to a natural order of things. If we look to historical accounts we can see both premodern and modern expressions of this tendency. For example, Plato's *Republic* considers the ideal state as the *polis* (or Greek city-state). One of the key contentions that organized the *polis* is autochthony, a term that seeks to emphasize the natural origin of its inhabitants, suggesting they "sprang fully formed, born of the earth" (Elden 2013: 22). In his exploration of the emergence of conceptions of territory in Western political thought, Elden turns to Greek myth and tragedy to examine the contention of autochthony, which makes explicit "the close and organic link between the people, the land (*khora*), and the *polis*" (2013: 23). Drawing on the founding myths of Athens, Alexandria, and Thebes, Elden's account traces the interplay between the human body, land, and deity, each playing a part in constructing autochthonous origins. He summarizes the purpose behind such intricate autochthonic myth-making:

> First . . . it provides a unity to the *polis*. Second, the boundaries of the *polis* are set by nature rather than by human agreements. The *polis* is natural, rather than set in opposition to nature. Third, the land is seen to belong to the people by right, by birth. There was no need for conquest and forced movement of previous inhabitants. (Elden 2013: 25–26)

It is perhaps no surprise that those conveying a certain form of rule turn to the materials of nature (be that the body or soil) in order to provide legitimacy for practices of coercion. The era of imperial geopolitics characterized by the work of Ratzel, Kjellén, and Mackinder conveyed a similar sense of the enfolding of natural science and human geography. The work of Ratzel (1896) has risen to prominence for his attempts to provide a neo-Lamarckian (often referred to as Social Darwinist) interpretation of the emergence of states. Just as any consideration of the Greek *polis* cannot be extracted from the poetry and politics of that era, so Ratzel's invocation of the state as a living organism struggling against other states must be inserted into the imperial maneuvers of nineteenth-century Europe. In particular his account provided a justification for the incursion of powerful states on less organized or powerful neighbors. But we must be careful extrapolating from this that Ratzel was merely a scholarly echo chamber for expansionist forms of politics. His work has, of course, cast a long political shadow, famously enrolled within Nazi expansionist politics in 1930s Germany and its anarchistic elements feeding into realist approaches to international relations in the twentieth century (Klinke 2018). But perhaps less frequently addressed are facets of Ratzel's writings that are suggestive of cooperation. On neighboring states, Ratzel was keen to point to the forms of cooperation that are required in the frontier zones between great powers:

> Naturally the efforts of nations are not confined to extension of territory. Even large states come into close contact at length. Neighbouring states share in the advantages of position and natural resources, and hence arises a uniformity of interests and modes of activity. Beside the lines of communication between the Atlantic and Pacific in the United States, Canada has constructed her Canadian Pacific railway, and the navigation of the great lakes is made available by canals on both sides. (Ratzel 1896: 360)

In addition, Ratzel's account may be seen as narrating the nature of conflict; it was also a significant argument for the dynamism of states:

> The territory of a state is no definite area fixed for all time—for a state is a living organism, and therefore cannot be contained within rigid limits—being dependent for its form and greatness on its inhabitants, in whose movements, outwardly exhibited especially in territorial growth or contraction, it participates. (Ratzel 1896: 351)

Thus we must be careful conflating a series of organic approaches to the state (such as autochthonic or neo-Lamarckian) as if each is conveying a similar understanding of state form. Instead, nature exists as a resource deployed to stabilize assertions of rule and challenge the purportedly unnatural characteristics of alternatives. While it is straightforward to recognize the limitations to such monolithic and essentialist ideas of statehood, we need to take seriously the legacy of such perspectives: that the state is imagined as a neutral backdrop against which other forms of political action and spatialization are analyzed.

The State System

Since the state cannot be understood as an immutable characteristic of the human experience, scholarly energies have turned to investigating the emergence of the *state system*. This is both an exercise in understanding the significance of external recognition to the existence of states and also a historicization of the process through which states emerged as the preeminent territorialization of human political life. This entrenches the point that states only exist in the plural, since they require other states to enact and underscore their legitimacy as sovereign actors of specific territories and populations. It also explores how the imagination of horizontal equivalence between states masks enduring inequities and violence in the postcolonial era, a point developed in the final section of the chapter.

The technological, administrative, and ideological processes through which state power may be established and reproduced have long historical roots. In Europe, Charles Tilly (1992) traces these back to the centralization of coercion and violence, as infrastructural and fiscal technologies improve so state authority may be conveyed over larger areas (see also Ogborn 1998). When trying to pinpoint a moment of consolidation for the state system in Europe many look to the Peace of Westphalia, a cluster of diplomatic treaties signed between 1643 and 1648, as the moment that heralded ceasefire arrangements that allowed the establishment of a new diplomatic order of sovereign territories. But this origin-point account does not bear up to scrutiny, and revisionist scholars have been keen to point to the role of these accords in codifying absolutist sovereignty rather than acting as the source of modern state sovereignty, which Teschke (2009), among others, views as a nineteenth-century phenomenon (Elden 2013: 310).

But the state system did not simply summon itself into existence. The state system may be a response to consolidating military and infrastructural technologies, but a larger field of critical scholars have pointed to the role of the rise of capitalism in shaping the emergence of state structures (Harvey 1976; Jessop 1990; Giddens 1995; Clark & Dear 2021). From this perspective (or more properly, these perspectives), the state system exists as a means through which class advantage may be reproduced. Perhaps just as importantly, the capture of state power by a particular social class allows class struggle to be masked behind a system of disinterested domination that appears separate from the realms of economic or corporate life. Etienne Balibar explains the significance of political territories to the capitalist economic system:

> [A] "world-economy" that has developed in the form of a universal market cannot (contrary to liberal myth) form a *homogenous* whole without boundaries; it must be divided into a plurality of *political unities* that allow for a concentration of economic power and the defense of positions of unearned income (*rente*) or "monopoly" by extraeconomic means. There is no market without monopolies, no monopolies without instruments of political (or juridical) compulsion, which in practice means national states. (Balibar 2004: 18, emphasis in the original)

One of the consequences of a Marxist account of the state system is a rather ambiguous position on the relationship between state and society. As a focal point of a transformative politics, Marxist accounts separate the state as a set of institutions that shape social relations in a regressive and counter-revolutionary fashion. But this sits uneasily with a sense of the state as an abstract entity that enters into the reproduction of relations of production through its socially embedded nature (see Abrams 1988: 118).

This paradoxical placement is evident in the Marxist political theory of Nicos Poulantzas (2014). Poulantzas's work crackles with political fomentation, written as it was at a time when the poststructural approaches of Gilles Deleuze and Michel Foucault were rising to scholarly prominence. In a sense, this work stands as a key articulation of the tension between the political theory of the state and the state as the institutionalization of class relations. In a direct challenge to Foucault's invocation of the capillary-like nature of power, Poulantzas (2014: 28) both shuns a view of the state that operates exclusively "through repression, force or 'naked' violence" and instead "calls upon ideology to legitimize violence and contribute to a consensus of those classes and fractions which are dominated from the point of view of political power."

The paradox remains, then, between a critical stance that challenges the ontology of the state as a coherent set of institutions and continued reliance on the distinction between state (as political actor) and society (as subjects of state violence). One of the most powerful attempts to transcend this dichotomy has been provided in the work of Bob Jessop. Jessop has pioneered the translation of "regulation theory" into English-medium debates, work that has traced the role of the state in normalizing and reproducing capitalist relations over time. Through a series of key interventions Jessop has sought to trace the range of social actors that are enrolled within this process in what is termed the strategic-relational approach (Jessop 1990, 2005, 2007; Jessop, Brenner, & Jones 2008). This, he argues, is a process of subject formation and, recalling Althusser's (2001) notion of interpellation, he argues that the "emergence of relatively stable structural ensembles involves not only the conduct of agents and their conditions of action but also the very constitution of agents, identities, interests, and strategies" (Jessop 2005: 53). Despite its theoretical roots, the strategic-relational perspective seeks to historicize and ground the evolution of strategic elements of state practice, emphasizing how the changing nature of the state reflects changes in capitalist systems of production.

The State as Experience

In recent years geographers have sought to reorient attention away from attempts to identify the nature of the state, to look instead at the implications of the *idea* of the state for a variety of political geographies. From this viewpoint, states do not simply exist but are rather accomplishments reified and reformulated through prosaic and quotidian activities (Radcliffe 2001; Painter 2006; Jeffrey 2012). From the 1980s onward – though this approach has a longer scholarly lineage – scholars have sought to understand how ideas of the state reflect uneven landscapes of power, and how in turn "state effects" are distributed through space, place, and society. Capturing this more decentered notion of state scholarship,

Kuus and Agnew (2008: 96) emphasize the need to study states "not as autonomous subjects but as processes of subject-making" and urge "more attention to the spatiality of power beyond that of territoriality." As discussed in the following, feminist and decolonial perspectives have traced in ethnographic detail the everyday geographies of assertions of the state.

While remaining engaged with the formation and reproduction of state power, scholars have moved to an exploration of state effects outside the institutions and spaces that are generally associated with formal state power. As with other areas of political geography, scholars have been drawn to Foucault's concept of governmentality to illuminate the diverse political rationalities of government, its "technologies," and the considerable intellectual labor involved in bringing into being the things, people, and processes to be governed (Mitchell 1991; Painter 2006; Rose 2017). In doing so, "governmentality" has become shorthand for explaining how state power operates to shape the conduct of conduct, as tactics and technologies are deployed to control, subdue, and oppress the citizenry (Huxley 2008). By examining the practice of rule in these terms, the state no longer stands as an unquestioned *source* of power but, rather, its *effect* (Mitchell 1999). For Timothy Mitchell (1991: 81) such effects are diffuse and difficult to grasp, but it is precisely this imprecision that has been the "source of its political strength as a mythic or ideological construct."

In geography the work of Anssi Paasi, and in particular his concept of "spatial socialization," has been at the forefront of work exploring how abstract concepts such as "the state" are made meaningful in everyday lives (see Paasi 2009; Koch 2015). This approach has been particularly prominent in feminist geographical work that has developed empirical perspectives on state effects (Secor 2007; Billo & Mountz 2016). Across a series of landmark empirical and theoretical interventions, feminist scholarship has challenged the simple binary rendered between state and society, exploring instead the embodied and situated moments in everyday life through which ideas of the state are conceived and reproduced. As Ruth Fincher (2004: 49) explains:

> Everyday places like the domestic home, the playground, and the community center exhibit power relations that are differentiated and fractured by relations of gender, ethnicity, age, ability, and class. They are no different from the famously "big-P" political sites of public space, the parliament, the city council and the large unionized workplace.

In order to capture this diverse array of political practices geographers have required new techniques. Using qualitative and anthropological techniques the ambiguities of the state and its embodiment have begun to emerge. Scholars have sought to use ethnographic techniques in order to understand the social and cultural processes through which state ideas are reproduced (Navaro-Yashin 2012; Mountz 2023). The shifting geometry of this form of scholarship is captured in

Corbridge et al.'s (2005) exploration of how the state is apprehended in North India by those who are subject to its rule and decision-making. Entitling this *Seeing the State* is a deliberate retort to James C. Scott's (1998) influential *Seeing Like a State,* a study that traced how state bureaucracies render subjects legible and governable.

There are perhaps two processes under investigation in such political anthropologies of the state. The first examines how subjects negotiate the materials, forces, and affects through which state authority is conveyed. For example, anthropologist Yael Navaro's work builds on earlier ethnographic studies to explore how the state is experienced, lived, and embodied (Harvey 2005; Navaro-Yashin 2012). Through ethnographic fieldwork in both Turkey and the Turkish Republic of Northern Cyprus, Navaro traces what she terms in the later work an "affective geography" of sovereignty, one that commences from a study of the documents, sites, and materials through which sovereign power circulates (Megoran 2006; Reeves 2011). This work is indebted to Latour's actor-network theory, where sovereignty is not a "top-down act of political will or event," it is rather "a worked-on terrain of relationality between human actors, material land and property, and tools or devices of measurement, numeration, and allocation" (Navaro-Yashin 2012: 44). Anna Secor (2007) uses such techniques to trace how the state interpellates – using Althusser's (2001) term – the citizen-subject into particular regimes of power through microsituations (for example, the visit to the hospital or the meeting with state officials in the street). Instead of a stable administrative form, Secor argues that the Turkish state "coheres in the *resonance* (between sites, agents, rationalities, and techniques) that is discursively produced through the circulation and arrest of people, documents, information, money, and influence" (Secor 2007: 49, emphasis in the original).

The second element of anthropological work on the state has centered on the mechanisms through which state authority is reproduced. For example, Ferguson and Gupta (2002) explore the reification of the state across two spatial maneuvers: verticality and encompassment. State verticality refers to the "central and persuasive idea of the state as an institution somehow 'above' civil society, community and the family" (Ferguson & Gupta 2002: 982). This topography of power is reproduced through the rhetoric of "top-down" policies, layers of bureaucracy, or even the notion of the "head" of state. Ferguson and Gupta go on to use the term "encompassment" to refer to the significance of the state idea in an "ever widening series of circles that begins with the family and local community and ends with a system of nation-states" (2002: 982).

Joe Painter (2006) has developed this work through theoretical engagement with the notion of prosaics, in particular examining how certain articulations of the state are made meaningful within everyday life. One of Painter's key conclusions is that we should not understand the state as a static achievement, but rather use the word "statization" to indicate the perpetual process through which the state comes into being. I have tried to explore this ethnographic and quotidian understanding of the state through a consideration of state-building practices after conflict (Jeffrey 2012, 2020). The optic of international intervention provides a useful vantage point for the dynamic and theatrical processes through which a new state secures both internal and external sovereignty (Allen & Cochrane 2010; Toal & Dahlman 2011). This sense of the theatricality of the state, and its entwining with questions of legitimacy, has been developed in McConnell's exploration of the exiled Tibetan state, as it has developed forms of bureaucracy and governance that set it out as a state-in-waiting, even as it currently exists in a deterritorialized form (McConnell 2015). This sense of a "rehearsal state" teases apart elements of internal and external sovereignty, pointing to how unrecognized states continue to seek a social contract, even in the absence of external recognition (Griffiths 2021; McConnell 2015).

Decolonizing the State

There is an intrinsic difficulty – if not impossibility – in an attempt to "decolonize the state," as the concepts and practices of the state system are themselves so entwined with the violence of colonial expansion and consolidation. Consequently, for this chapter to propose a singular decolonial state theory would be to misunderstand the substance of these intellectual and political stakes. There are further challenges and limitations. In the first instance, there is my authorial position. The chapter is written from the perspective of a white man training in a discipline – and positioned within an institution – that has been architect and beneficiary of colonial structures of power. There are no mechanisms for stepping outside these circumstances and their implications flow through the chapter in ways that I cannot control or apprehend. As feminist scholarship has taught us, there is no position of perfect self-knowledge, so the implications of this position are difficult to preempt or offset (Rose 1997). Instead, I acknowledge that this is a partial and situated account that benefits from my learning from – and working with – a range of critical scholars and activists through which the rich and varied character of decolonial praxis and scholarship has come to the fore. This does not erase the vestiges of power, but it does recognize the debt that I have to others in being able to explore these aspects of ideas of the state.

The second – and related – point is the challenge of proposing a singular decolonial state theory. The shift from talking of "postcolonial" to "decolonial" approaches is one of recognizing the continuity of coloniality in the institutions and structures of the contemporary world, including the structures of academic knowledge production. Scholars have argued that the concept of the postcolonial state carries with it an imagination of a severance from colonial structures, a new historical epoch in the wake of colonial violence (Radcliffe 2017). In the shadow of the purported end of European colonization and the independence of successor states, relations and structures have emerged that continue to marginalize, silence, and denigrate voices of the Global South, in particular Indigenous cosmologies and knowledges. In tandem, the economic exploitation of former colonies continues unabated, through exploitative trading relations, debt, environmental harm, and human migration (Yusoff 2018; Thomas 2024). The constitutions of independent states have emulated the political architectures of the Western modern state, in doing so erasing alternative, and locally specific, forms of rule. Decolonial scholars have taught us that understanding the coloniality of geographical knowledge requires a challenge to the imagined universalism of Western theorizing and instead an examination of how racialized logics cast alternatives as marginal or secondary to the supremacy of the European nation-state (Bhandar 2018). It is for this reason that scholars often center their critique around the concept of modernity-coloniality, highlighting the entwined nature of colonial exploitation within the periodization of modernity (Mignolo 2011). So when speaking of the "modern nation-state" we are immediately drawing attention to its roots within colonial structures of power.

Rather than viewing the decolonizing of state theory as an oxymoron, political geographers can explore the scholarly and activist maneuvers required to confront enduring coloniality. There are pathways forward that seek to broaden and challenge the singular narrative of the state while also drawing expertise from activists and thinkers outside the validated institutions of global academia. One route has to been to recast state theory from the Global South, reengaging with work that had been influential in theorizing earlier political geographies of the state (Winichakul 1994). In a more recent example, Eliassi (2021) has sought to theorize the state from the perspective of groups that have been disenfranchised from the modern state system, focusing on the experience of Palestinian and Kurdish groups. This work highlights the exclusionary nature of state sovereignty, where inclusion in regimes of citizenship (understood in both law and practice) is tied to particular ethnic identities. Thus, for Eliassi (2021: 1) decolonizing requires challenging the "politico-cultural supremacy of a single, 'core' ethnicity as the sovereign legislator of the rules and regimes of national belonging and un-belonging." This work builds on the long and growing historical and geographical interest in the role of settler colonial state practices in constituting racial subjectivities. For example, Bhandar's (2018: 8) study of the use of property law as an instrument of settler colonial violence argues that "racial subjects and modern property laws are produced through one another in the colonial context."

Focusing empirically on the practices of "actual existing states" is, in some respects, a necessary extension of critical accounts of the state that have, as has been discussed, centered on the accomplishment of the state as a set of effects as opposed to a stable infrastructure of authority. This extends the focus of state investigations, shifting from a universal imaginary of state sovereignty to the more grounded stories of how political power is secured under the banner of the state (Mountz 2004; Radcliffe & Radhuber 2020; Radhuber & Radcliffe 2023). For example, Radcliffe and Radhuber (2020) have provided a detailed account of the incorporation of *buen vivir* within the constitution of the Ecuadorian state. This work draws attention to the prominence of discourses of "plurinationalism" to emphasize a statehood model "with a multiplicity of politico-administrative territorialities (which are defined by connection to place and racial ethnic diversity)" (Radcliffe & Radhuber 2020: 6). Exploring the case of the Confederation of Indigenous Nationalities of Ecuador (CONAIE), the authors trace the emergence of multiscalar networks of praxis and critique, which underscored a qualitative difference to Western conceptions of devolution or decentralization. This was an alternative territorial imaginary that "prefigures a break with modern-colonial state power as *spatial* and *dehumanising* and in its place asserts Indigenous sovereignties" (Radcliffe & Radhuber 2020: 7, emphasis in the original). A focus on plurinationalism attends to the varied practices through which human communities relate to their territory and seek to organize their political lives. Rather than a model of statehood forged through European colonial encounters, this shift in state practices and imaginaries promotes a decolonized and standardized political geography of the state.

Conclusion

The state has always been an assertion and an accomplishment; it is both fragile in its need for reification and durable through its embedding in social and political life. Rejecting the state as a form of political geography has its seductions. Its constituent concepts (sovereignty, territory, or even legitimacy) seem to carry more precision and avoid either falling into a territorial trap, of universalizing from a Eurocentric view, or reproducing the embedded statism that characterizes affiliated disciplines such as international relations. But this carries both intellectual and political risks. There is a need to trace how state power is secured and reproduced, how sovereignty is legitimized, and the forms of oppression, subjectivity, and exclusion that state effects can foster. Refusing the study of the state risks assuming that states are transhistorical and could overlook how changes in the duties and responsibilities of the state can harm the most vulnerable both within and beyond the territory of a particular state (Jessop et al. 2008; Elden 2013). As shown in McConnell's (2015) work, overlooking the state could also risk ignoring the significance invested in the state as a political project, as a means of securing national survival and making claims to territory. The challenge for political geographers is not to imagine that through abstract theorizing we may come to know the essences of the state, but rather to continue to trace how ideas and discourses of the state are mobilized to legitimize certain transgressions, exclusions, and enclosures. These approaches appear to stand alongside a decolonial perspective that seeks to illuminate alternative political projects and imaginaries of the state, disavowing a unitary understanding of political authority, and instead considering how plurinational practices can challenge and rework spatial accounts of power.

References

Abrams, P. (1988). Notes on the difficulty of studying the state (1977). *Journal of Historical Sociology 1*(1): 58–89. doi:10.1111/j.1467-6443.1988.tb00004.x

Agnew, J. (1994). The territorial trap: The geographical assumptions of international relations theory. *Review of International Political Economy 1*(1): 53–80. doi:10.1080/09692299408434268

Agnew, J. (2020). Geoeconomics and the state. In Moisio, S., Koch, N., Jonas, A.E.G., Lizotte, C., & Luukkonen, J. (eds), *Handbook on the Changing Geographies of the State: New Spaces of Geopolitics*. Cheltenham: Edward Elgar, 161–172.

Allen, J., & Cochrane, A. (2010). Assemblages of state power: Topological shifts in the organization of government and politics. *Antipode 42*(5): 1071–1089.

Althusser, L. (2001). *Lenin and Philosophy and Other Essays*. New York: New York University Press.

Anderson, B. (1983). *Imagined Communities: Reflections on the Origin and Spread of Nationalism*. London: Verso.

Balibar, É. (2004). *We, the People of Europe? Reflections on Transnational Citizenship/Étienne Balibar*. Princeton, NJ: Princeton University Press.

Bassin, M. (2022). "Everything is revealed in maps": The European far right and the legacy of classical geopolitics during the Cold War. *Geopolitics 5*: 1843–1867. doi:10.1080/14650045.2022.2078709

Bhandar, B. (2018). *Colonial Lives of Property: Land, Law, and Racial Regimes of Ownership*. Durham, NC: Duke University Press.

Billo, E., & Mountz, A. (2016). For institutional ethnography: Geographical approaches to institutions and the everyday. *Progress in Human Geography 40*(2): 199–220. doi:10.1177/0309132515572269

Bratsis, P. (2006). *Everyday Life and the State*. Boulder, CO: Paradigm Publishers.

Clark, G.L., & Dear, M. (2021). *State Apparatus: Structures and Language of Legitimacy*. Routledge.

Coddington, K., Conlon, D., & Martin, L.L. (2020). Destitution economies: Circuits of value in asylum, refugee, and migration control. *Annals of the American Association of Geographers 110*(5): 1425–1444. doi:10.1080/24694452.2020.1715196

Corbridge, S., Williams, G., Srivasteva, M., & Véron, R. (2005). *Seeing the State: Governance and Governmentality in India*. Cambridge: Cambridge University Press.

Dickinson, J., Andrucki, M.J., Rawlins, E., Hale, D., & Cook, V. (2015). Introduction: Geographies of everyday citizenship. *ACME: An International Journal for Critical Geographies 7*(2): 100–112.

Elden, S. (2013). *The Birth of Territory*. Chicago, IL: University of Chicago Press.

Eliassi, B. (2021). *Theorizing Statelessness and Stateless Diasporas*. Basingstoke: Palgrave Macmillan.

Ferguson, J., & Gupta, A. (2002). Spatializing states: Toward an ethnography of neoliberal governmentality. *American Ethnologist 29*(4): 981–1002.

Fincher, R. (2004). From dualisms to multiplicities: Gendered political practices. In Staeheli, L., Kofman, E., & Peake, L. (eds), *Mapping Women, Making Politics: Feminist Perspectives on Political Geography*. London: Routledge, 49–70.

Giddens, A. (1995). *A Contemporary Critique of Historical Materialism*, 2nd edn. Houndmills: Palgrave Macmillan.

Griffiths, R.D. (2021). *Secession and the Sovereignty Game: Strategy and Tactics for Aspiring Nations*. Ithaca, NY: Cornell University Press.

Harvey, D. (1976). The Marxian theory of the state. *Antipode* 8(2): 80–89. doi:10.1111/j.1467-8330.1976.tb00641.x

Harvey, P. (2005). The materiality of state-effects: An ethnography of a road in the Peruvian Andes. In Krohn-Hansen, C., & Nudstad, K. (eds), *State Formation: Anthropological Perspectives*. London: Pluto Press, 123–141.

Held, D. (1983). *States and Societies*. Oxford: Robertson in association with the Open University.

Huxley, M. (2008). Space and government: Governmentality and geography. *Geography Compass* 2(5): 1635–1658.

Hyndman, J. (2019). Unsettling feminist geopolitics: Forging feminist political geographies of violence and displacement. *Gender, Place and Culture* 26(1): 3–29. doi:10.1080/0966369X.2018.1561427

Inwood, J., & Bonds, A. (2016). Confronting white supremacy and a militaristic pedagogy in the U.S. Settler Colonial State. *Annals of the American Association of Geographers* 106(3): 521–529. doi:10.1080/24694452.2016.1145510

Jackman, A., Squire, R., Bruun, J., & Thornton, P. (2020). Unearthing feminist territories and terrains. *Political Geography* 80: 102180. doi:10.1016/j.polgeo.2020.102180

Jeffrey, A. (2012). *The Improvised State: Sovereignty, Performance and Agency in Dayton Bosnia*. Oxford: John Wiley & Sons.

Jeffrey, A. (2020). *The Edge of Law: Legal Geographies of a War Crimes Court*. Cambridge: Cambridge University Press.

Jessop, B. (1990). *State Theory: Putting the Capitalist State in Its Place*. University Park, PA: Pennsylvania State University Press.

Jessop, B. (2005). Critical realism and the strategic-relational approach. *New Formations: A Journal of Culture, Theory and Politics* 56: 40–53.

Jessop, B. (2007). *State Power*. Cambridge: Polity.

Jessop, B., Brenner, N., & Jones, M. (2008). Theorizing sociospatial relations. *Environment and Planning D: Society and Space* 26(3): 389–401.

Klinke, I. (2018). Friedrich Ratzel, Lebensraum and the death motif. *Journal of Historical Geography* 61: 97–101.

Koch, N. (2015). "Spatial socialization": Understanding the state effect geographically. *Nordia Geographical Publications* 44(4): 29–35.

Koch, N. (2020). Methodological nationalism. In Kobayashi, A. (ed.), *International Encyclopedia of Human Geography*, vol. 9, 2nd edn. Oxford: Elsevier, 245–248.

Kuus, M. (2019). Political geography I: Agency. *Progress in Human Geography* 43(1): 163–171. doi:10.1177/0309132517734337

Kuus, M., & Agnew, J. (2008). Theorising the state geographically. In Cox, K., Low, M., & Robinson, J. (eds), *Sage Handbook of Political Geography*. London: Sage, 95–106.

Loftus, A. (2020). Political ecology II: Whither the state? *Progress in Human Geography* 44(1): 139–149. doi:10.1177/0309132518803421

Massé, F. (2022). Police power in green: Furthering political ecologies of the state. *Political Geography* 97: 102627. doi:10.1016/j.polgeo.2022.102627

McConnell, F. (2015). *Rehearsing the State: The Political Practices of the Tibetan Government-in-Exile*. Oxford: John Wiley & Sons.

Megoran, N. (2006). For ethnography in political geography: Experiencing and re-imagining Ferghana Valley boundary closures. *Political Geography 25*(6): 622–640.

Mignolo, W. (2011). *The Darker Side of Western Modernity: Global Futures, Decolonial Options*. Durham, NC: Duke University Press.

Mignolo, W., & Bussmann, F.S. (2023). Coloniality and the state: Race, nation and dependency. *Theory, Culture and Society 40*(6): 3–18. doi:10.1177/02632764221151126

Mitchell, T. (1991). The limits of the state: Beyond statist approaches and their critics. *American Political Science Review 85*(1): 77–96.

Mitchell, T. (1999). Society, economy, and the state effect. In Steinmetz, G. (ed.), *State/Culture: State-Formation after the Cultural Turn*. Ithaca, NY: Cornell University Press, 76–97.

Moisio, S., Koch, N., Jonas, A.E.G., Lizotte, C., & Luukkonen, J. (2020). *Handbook on the Changing Geographies of the State: New Spaces of Geopolitics*. Cheltenham: Edward Elgar.

Mountz, A. (2004). Embodying the nation-state: Canada's response to human smuggling. *Political Geography 23*(3): 323–345.

Mountz, A. (2013). Political geography I: Reconfiguring geographies of sovereignty. *Progress in Human Geography 37*(6): 829–841. doi:10.1177/0309132513479076

Mountz, A. (2023). Smoke and mirrors: An ethnography of the state. In Tickell, A., Sheppard, E., Peck, J., & Barnes, T. (eds), *Politics and Practice in Economic Geography*. London: Sage., 38–48. doi:10.4135/9781446212240

Navaro-Yashin, Y. (2012). *The Make-Believe Space: Affective Geography in a Postwar Polity*. Durham, NC: Duke University Press.

Ogborn, M. (1998). The capacities of the state: Charles Davenant and the management of the excise, 1683–1698. *Journal of Historical Geography 24*(3): 289–312. doi:10.1006/jhge.1998.0088

Paasi, A. (2009). Bounded spaces in a "borderless world": Border studies, power and the anatomy of territory. *Journal of Power 2*(2): 213–234.

Painter, J. (2006). Prosaic geographies of stateness. *Political Geography 25*(7): 752–774.

Poulantzas, N.A. (2014). *State, Power, Socialism/Nicos Poulantzas*. London: Verso.

Radcliffe, S.A. (2001). Imagining the state as space: Territoriality and the formation of the state in Ecuador. In Hansen, T.B., & Stepputat, F. (eds), *States of Imagination: Ethnographic Explorations of the Postcolonial State*. Durham, NC: Duke University Press, 122–145.

Radcliffe, S.A. (2017). Decolonising geographical knowledges. *Transactions of the Institute of British Geographers 42*(3): 329–333.

Radcliffe, S.A., & Radhuber, I.M. (2020). The political geographies of D/decolonization: Variegation and decolonial challenges of/in geography. *Political Geography 78*: 102128. doi:10.1016/j.polgeo.2019.102128

Radhuber, I.M., & Radcliffe, S.A. (2023). Contested sovereignties: Indigenous disputes over plurinational resource governance. *Environment and Planning E: Nature and Space 6*(1): 556–577. doi:10.1177/25148486211068476

Ratzel, F. (1896). The territorial growth of states. *Scottish Geographical Magazine 12*(7): 351–361. doi:10.1080/00369229608732897

Reeves, M. (2011). Fixing the border: On the affective life of the state in southern Kyrgyzstan. *Environment and Planning D: Society and Space 29*(5): 905–923.

Rose, G. (1997). Situating knowledges: Positionality, reflexivities and other tactics. *Progress in Human Geography 21*(3): 305–320.

Rose, N. (2017). Still "like birds on the wire"? Freedom after neoliberalism. *Economy and Society 46*(3–4): 303–323.

Scott, J.C. (1998). *Seeing Like a State: How Certain Schemes to Improve the Human Condition Have Failed*. New Haven, CT: Yale University Press.

Secor, A.J. (2007). Between longing and despair: State, space, and subjectivity in Turkey. *Environment and Planning D: Society and Space 25*(1): 33–52.

Teschke, B. (2009). *The Myth of 1648: Class, Geopolitics, and the Making of Modern International Relations/Benno Teschke*. London: Verso.

Thomas, M. (2024). *The End of Empires and a World Remade: A Global History of Decolonization*. Princeton, NJ: Princeton University Press.

Tilly, C. (1992). *Coercion, Capital, and European States, AD 990-1992/Charles Tilly*. Cambridge, MA: Blackwell.

Toal, G., & Dahlman, C.T. (2011). *Bosnia Remade: Ethnic Cleansing and Its Reversal*. Oxford: Oxford University Press.

Tsing, A.L., Swanson, H.A., Gan, E., & Bubandt, N. (eds) (2017). *Arts of Living on a Damaged Planet: Ghosts of the Anthropocene*. Minneapolis, MN: University of Minnesota Press.

Vorbrugg, A., & Bluwstein, J. (2022). Making sense of (the Russian war in) Ukraine: On the politics of knowledge and expertise. *Political Geography 98*: 102700. doi:10.1016/j.polgeo.2022.102700

Weber, M. (1958 [1919]). Politics as a vocation. In Gerth, H., & Wright Mills, C. (eds), *Max Weber: Essays in Sociology*. New York: Oxford University Press, 77–128.

Winichakul, T. (1994). *Siam Mapped: A History of the Geo-body of a Nation*. Honolulu: University of Hawaii Press.

Yusoff, K. (2018). *A Billion Black Anthropocenes or None*. Minneapolis, MN: University of Minnesota Press.

Zaragocin, S., & Caretta, M.A. (2021). Cuerpo-territorio: A decolonial feminist geographical method for the study of embodiment. *Annals of the American Association of Geographers 111*(5): 1503–1518. doi:10.1080/24694452.2020.1812370

Chapter 8

Sovereignty

Joshua Barkan

Department of Geography, University of Georgia, Athens, GA, USA

It is seemingly self-evident that our times and places give us our key concepts. Dictionaries name words of the year, and the reasons that the *Oxford English Dictionary* recently selected terms like "carbon footprint" or "climate emergency" or that *Merriam Webster* highlighted "pandemic" and "vaccine" are clear. One could theorize the phenomenon, as Raymond Williams (1983) attempted to in *Keywords*, discussing how "culture" had come to supplant "literature" on his return to Cambridge after the war. Geographically informed critiques of ideology, such as Gramsci's "conceptions of the world" (Wainwright 2010), provide an avenue not only for understanding the link between concepts and history but for struggling over their meaning, use, and relation to practice. In this regard *sovereignty* can be understood as not just a descriptive term with a settled legal definition but a geographical concept deployed and contested as part of politics. Its dynamic role in contemporary academic and political discourse should attune us to the changing interrelations of space, territory, and authority, to which debates on sovereignty continually refer.

This is certainly true of the early modern European discussions of sovereignty that remain fundamental to the development of modern political thought. Authors such as Jean Bodin (1992) and Thomas Hobbes (1996) articulated their theories of sovereignty in the rhetoric of philosophy, but did so in response to the major political developments of their day. That sovereignty required unified and absolute power was at once a theory for achieving political order, but also an intervention in religious conflicts undermining the French monarchy during the sixteenth-century wars of religion, for Bodin, or the English crown during the

The Wiley Blackwell Companion to Political Geography, Second Edition.
Edited by Virginie Mamadouh, Natalie Koch, Chih Yuan Woon, and John Agnew.

Civil War, for Hobbes. Likewise, mid-twentieth-century discussions of sovereignty can only be understood in relation to the struggles that defined that historical and geopolitical context: the breakup of empires, the Cold War, national liberation movements, and the emergence of postcolonial states.

For this reason, sovereignty is an ambiguous term (see Kalmo & Skinner 2010). Poststructuralism demonstrated that all concepts depend on discursive practices for their stability, but sovereignty seems less stable than most. The way sovereignty is qualified can make it refer to radically different or even incommensurable concepts. The relation of, say, popular to absolute sovereignty is less a different version of a common object than a negation of one by the other. Both practically and theoretically, sovereignty is often discussed as fragmented and failing. In political discourse, discussions of sovereignty are most pronounced when authority and control are violated or in abeyance. Academic interest in sovereignty has coincided with geopolitical moments in which relations between states, territory, and authority are challenged and debated.

In the first edition of this chapter (Barkan 2015), written over a decade ago, I highlighted the US Global War on Terror, economic globalization, and the resulting mass migrations as key phenomena changing the geographies of sovereign power. This was a high point for geographical research on sovereignty, as many expressions of political authority at the beginning of the twenty-first century did not resemble the map of territorially contiguous nation-states. Economic globalization entailed the relaxing of borders for the movement of capital and commodities, while also intensifying surveillance and policing for migrants. Free trade and export processing zones proliferated, with their own regimes of border security, taxation, and tariffs, but so too did carceral spaces including prisons, camps, and detention centers (see Sparke 2006; Mountz 2013). A half century after decolonization seemingly serious scholars were reconsidering empire as part of the political future (most notably Ignatieff 2003). The relations of political authority to state territory appeared out of joint and uneven (see Elden 2009). Sovereignty emerged as an important concept to explain these transformations and their ongoing and contradictory relations to law, state power, and violence.

Today, of course, other concerns are prominent. Geopolitical conflicts, transnational economies, and migrations are still central to the exercise of sovereignty, but current research also focuses on global climate catastrophe and the emergence of authoritarian populisms and economic nationalisms. Contemporary wars in Ukraine and Gaza have revived questions of law and violence while the creation of zones, enclaves, and camps continues apace. As a form of state power and violence, sovereignty is today much more likely to be understood in relation to processes of racial capitalism and colonialism. Many of the most important discussions of sovereign power use the term not only to describe state control over communities and territories, but also in conjunction with counter-hegemonic projects against the dominance of state and capital.

In the chapter that follows, I trace these changing conceptions of sovereignty in political geographical thought and research. I begin by reviewing classic framings of sovereignty that connected the concept to the territorial nation-state. Even mainstream approaches from political science and international relations recognize that full sovereignty has remained elusive for most states most of the time. With the disjuncture between the expression of sovereign power and state sovereignty, geographers were among the scholars most invested in complicating the spatial assumptions of the state–territory–sovereignty nexus. Scholars highlighted the uneven expressions of sovereignty, but also the ways in which unevenness was produced and utilized by states, companies, and individuals to pursue their interests. The second section of the chapter shows how that unevenness was taken seriously in European social thought by engaging accounts that emphasized sovereignty's liminal or exceptional dimensions. Doing so entailed grappling with sovereignty as a form of power over life rather than a property of nation-states. Sovereignty continues to be implicated in the different ways states foster livelihoods but also expose populations to violence. At the same time, sovereignty has also been deployed as part of Indigenous and anti-colonial movements in discourses resisting state violence. The final section explores the tension between these two perspectives and the various attempts to rearticulate sovereignty, along with related concepts of self-determination, self-government, jurisdiction, and authority, as not just a power over life but also a responsibility for care.

State Sovereignty

Intellectual historians suggest that the modern concept of sovereignty emerged in the seventeenth century in tandem with the development of European states, referring to a distinct form of authority over a community and its territory. Quentin Skinner (2010) located diverse sources for this development in the writing of French authors on sovereignty, scholastic discussions of supreme power, and Italian political pamphlets and princely literature, which stressed the importance of a single individual who could act to protect the interest of the realm. By the mid-seventeenth century, this absolute vision of the state was challenged by others that located supreme authority or sovereignty in the population. In early modern social contract theories, including Hobbes's *Leviathan*, the sovereign referred to the head of these political bodies, who was authorized to act in the interest of the realm, including the population and territory. Sovereignty connoted ultimate and exclusive authority over a polity in both social and spatial dimensions.

Today, those early modern conceptions compete with others. International relations theorist Stephen Krasner (1999) noted four types of sovereign power exercised by states. The philosophical tradition of Bodin and Hobbes focused on the organization of government within a state and was thus preoccupied with *domestic sovereignty* as the ultimate authority with a polity. *International legal sovereignty* refers to the mutual recognition of states and state-like entities in the international state system. Although primarily adhering to state governments, legal sovereignty has also been extended to internationally recognized nonstate entities as well as governments that are out of power and cannot exercise effective control over territory or populations. *Westphalian sovereignty* is grounded in the principle that a state is the highest power within its political territory and thus has the ability and right to control the structures of government and authority without external interference. Finally, *interdependent sovereignty* concerns forms of authority associated primarily with globalization and the ability for states to control the cross-border movements of everything from people and capital to pollution and disease. Krasner argues that the leaders of states strategically engage with the norms characterizing the multiple forms of sovereign authority, averring the centrality of some while periodically shirking others, as they pursue their interests within the asymmetric relation of the international state system. Given these multiple definitions and strategic uses, as well as the inability of most states to exercise basic elements of Westphalian sovereignty, Krasner contends that sovereignty functions less as a governing norm or a real fact of world affairs than a kind of myth that elites appeal to in a form of "organized hypocrisy" that structures international relations.

For Krasner, then, sovereignty fails to capture the real ways states and their leaders operate in the international system. Yet a second problem concerns the numerous political actors and processes whose modes of power and authority are not contained within state territory. Whereas Krasner conceptualizes states and state elites as the fundamental agents shaping international politics, political geographers have argued that the exclusive focus on states – whether through the lens of sovereignty or not – gives us a partial view of politics. Geographical thinkers have noted that academic fields such as political science, international relations, political sociology, and law take the territorially organized nation-state as an unproblematic assumption grounding both political analysis and policy recommendations (Agnew 1994). Geographers also have demonstrated that state functions – from the provisioning of welfare to the regulation of capital, labor, national security, and the environment – are increasingly located in new spatial and scalar configurations (Brenner 2004). Only recently have geographers examined the significance of sovereignty to state transformations.

John Agnew's (2017) work is notable in this respect. Engaging debates about the growing separation between effective sovereignty and state territory, Agnew shifts our attention to the multiform ways that sovereign power is exercised geographically. In terms of authority, this implies recognizing the diverse institutions that exercise legitimate forms of control in society besides territorially bounded nation-states. In terms of geography, it means examining the other spatial forms that authority can take besides state territory, including networked forms of political power that flow through specific nodes or locals rather than across bounded terrestrial space. Agnew suggests that sovereignty regimes (and not simply state sovereigns) can be situated within a typology that accounts for both the territorial dynamics of sovereign power and its reliance on central authority. Adopting Michael Mann's (1984) discussion of despotic and infrastructural power, Agnew arrives at a matrix of sovereignty regimes characterized by multiple combinations of strong or weak forms of central state authority, which he links to despotic power, against consolidated or open forms of state territoriality, linked to infrastructural power. The four resulting ideal types of sovereignty regimes – classic (strong state, consolidated territory), globalist (strong state, open territory), integrative (weak state, consolidated territory), and imperialist (weak state, open territory) – characterize the diverse ways authority and territory have been historically and geographically organized. Agnew presents examples ranging from the management of currencies to the regulation of international migration to show how wide variances in policy correspond to different types of regimes. Like Krasner, Agnew questions whether sovereignty was ever a particularly good way of characterizing political power. And though they differ on their understandings of the fundamental forces shaping international politics – with Agnew critiquing Krasner's state-centric descriptions – they both characterize sovereignty as a narrative or fiction that diverse actors appeal to as they pursue their interests in the uneven landscape of international affairs. The actual workings of states and nonstate actors who exercise authority bear little resemblance to state sovereignty as conventionally understood.

These conclusions are supported by Lauren Benton's (2010) historical work on sovereignty and European empires. Benton demonstrates the importance of law and geography to the extension of European colonialism, but also notes its patchworked and negotiated character. Instead of a smoothly spreading imperium that increasingly engulfs areas of the map over time, Benton presents the extension of European empires as a halting process characterized by imperial agents' lack of knowledge of the spaces and people they were attempting to control. Interpreted as such, law served as an "epistemological framework" (2010: 21) in which colonial powers attempted to organize and make sense of the various spaces they encountered, from rivers, oceans, islands, and mountains to the numerous enclaves and imperial zones that were reoccurring features of colonial legal geography. And, of course, this entailed a variety of negotiated relations with extra-European legal structures and people. These relations could entail outright domination, certainly, but also relied on local systems of law, significant instances of resistance, and extensive forms of legal creativity and experimentation that shaped imperial law in novel ways.

Benton's work disrupts a retroactive linkage between sovereignty and territory, which she suggests became intertwined in the nineteenth century and then projected back onto the entire history of European empires. Instead of sovereignty emerging whole cloth in European states and then being exported abroad, Benton identifies variegated expressions of sovereign authority in a dizzying array of spatial forms. In some cases, sovereignty was a "tubulous" power that adhered to European subjects as they attempted to control imperial trade routes, often over watery spaces like rivers and oceans (Benton 2010: 110). In other cases, sovereignty was connected with zones or clusters of imperial authority, but the delegation of sovereignty across national territory could be highly differentiated. For instance, Benton discusses the various classifications of sovereignty and quasi-sovereignty asserted by colonial powers over enclaves such as Native American reservations in the United States, Basutoland in southern Africa, or the princely states in the hill regions of India. In these places, colonists developed a wide array of legal categories to characterize their own authority while also accommodating Indigenous forms of rule that they were unable to subdue or eliminate.

In displacing state territory and emphasizing the uneven geographies of sovereign power, Benton's arguments mirror those of Agnew and Krasner, while also marking a decisive shift. Whereas Agnew and Krasner treat sovereignty as a myth used to secure state authority, Benton examines the *active* aspects of sovereignty as an *assertion* of legal power and a form of legal politics. Sovereignty, on this account, is neither a property of a regime nor simply a narrative deployed by states. Rather, it is an ideological and institutional form that is itself always under construction. In other words, there is no ideal-typical form of state sovereignty that stands behind or outside of the fragmented, incomplete, or "anomalous" legal spaces of imperial sovereignty. Sovereignty, for Benton, does not precede its articulation. Colonialism entailed a continuous process of legal experimentation in which the concept and institutions of sovereignty were created and transformed. As Benton puts it, "the legal and political practices (and, yes, institutions) constituting imperial sovereignty might be seen as shaping discourses about the rule of law, generating a varied set of available narratives about the special virtues and qualities of law in different periods" (Benton 2009: 120). The result is much greater attention to the construction of legal concepts and institutions through the process of legal politics.

From Sovereignty to Sovereign Power

As with the social scientific literature, political and social theorists also approach sovereignty as a mode of legal politics with its own spatiality. Michel Foucault's genealogies of governmental rationality have been notable for challenging the presumed homology between the state and sovereignty. In his later genealogical work, he argued that sovereignty emerged as a power over life and death grounded in the individualistic rule of kings and princes. For Foucault, this power was a "characteristic privilege" (1990: 135) and a "basic attribute" (2003: 240) of the sovereign that derived from the ancient *patria potesta*, a right that allowed the head of a Roman household to "'dispose' of the life of his children and his slaves" (1990: 135). Sovereign power was thus a juridical power (a right) to kill or let live that was subsequently applied to state sovereignty – in however a "diminished form" (1990: 135) – through the early modern period. Its primary characteristic remained the ability to take life, calling on subjects to defend the realm. It also legitimated the appropriation of "a portion of the wealth, a tax of products, goods and services, labor and blood" (1990: 136) in the name of the health, welfare, and security of the state, or the *salus publica*.

One hears echoes of earlier theories in this formulation, and not just that of the *patria potesta*, linking sovereignty to the patriarchal order of ancient Rome. That sovereigns dispose of life or expose life to death in the name of the *salus publica* was also central to theories of sovereignty associated with political theology. Ernst Kantorowicz (1951, 1957) described the transformations in European kingship by which subjects were supposed to sacrifice for the good of the realm, including offering up their own lives for the security of the polity. But the position is most closely (and infamously) linked with Carl Schmitt, the reactionary German legal theorist who argued that "all significant concepts of the modern theory of the state are secularized theological concepts" (Schmitt 1985 [1922]: 36). Concerned about the strength of the liberal constitution during the tumult of the Weimar Republic, Schmitt advocated for an authoritarian sovereign to take decisive action to save the state from existential threats. Opening his 1922 book *Political Theology* with the definition that the "sovereign is he who decides on the exception" (1985 [1922]: 5), Schmitt equated the sovereign with a secular god, who stands beyond the existing legal order yet secures its existence.

Nonetheless, Schmitt and Foucault viewed sovereignty as a waning political force in the contemporary world. For Schmitt this was the tragedy of modern politics. During the Weimar period, Schmitt viewed liberal jurisprudence, with its politics of deliberation, as incapable of sovereign decisionism and thus insufficient to secure the constitutional order in the face of diverse democratic claims. After World War II, he looked with a similarly jaundiced eye at the increasingly global hegemony of US-backed liberal capitalism. Precisely because Schmitt believed that legal systems, and thus sovereignty, required a specific space of application or jurisdictional form, he viewed the tendencies of capital to transgress political boundaries as a threat to the existing spatio-legal order of European public law (Schmitt 2003; see Galli 2010).

For Foucault, however, the decline of sovereignty was accompanied by the rise of disciplinary and governmental powers. Instead of killing or letting live, these new forms of power focused on expanding the capacities of humans "to make live and let die" (2003: 241). Foucault chronicled this transition in his studies of discipline, madness, sexuality, and medicine and would influentially examine the powers aimed at intensifying life under the term "biopolitics." Nevertheless, the relationship between the two forms of power – the "old right" (2003: 241) of sovereignty and the newer forms of discipline, government, and biopolitics – remained complicated. Although Foucault charted a general shift "from a regime dominated by structures of sovereignty to a regime dominated by techniques of government" (2004: 106), he also argued that sovereignty continued to exist as a problem interlaced with newer modes of power.

In terms of understanding sovereignty, the crucial point is less the historical sequence running from sovereignty to biopolitics than the treatment of sovereignty as a form of power over life and death. On this read, sovereignty is a juridical relation incorporated in diverse institutional forms. This shift was important for Foucault's own thinking about the emergence of modern states. Conceptualizing sovereignty not as a property of the state created by contract but a specific right, Foucault was able to situate sovereign power in relation to other modes of biopolitical government. As such, his work directs attention to the multiplicity of forces operative both within and beyond the state and has been critical for redirecting political analysis away from its obsession with central state authority and toward the capillary movement of power in diverse institutional forms.

The centrality of sovereignty in new modes of power was further accentuated by Giorgio Agamben (1998, 2005), whose work has been widely influential across the social sciences and humanities. Agamben drew on the arguments of Schmitt and Foucault to characterize sovereignty as a paradoxical power in which the domain of law is established through its legally authorized suspension. Whereas the exception for Schmitt signified an existential crisis in which the state could legally suspend the law to save itself, Agamben argued that sovereignty was central to the biopolitical government of life that preoccupied Foucault. For Agamben, the legally authorized suspension of law not only created the conditions for the state to save itself, it also created a threshold or zone where some lives are qualified for political existence and others can be unceremoniously discarded as simple, biological beings – what Agamben termed *homo sacer* or bare life – with no place in the political order. Although Agamben traced this biopolitical dimension of sovereign power back to ancient Greece and Rome, he also argued that the relation between sovereignty and life persists into the present, including in democratic regimes. What remains paradoxical about sovereignty for Agamben is that it always includes the potential to expose life to an unceremonious death to promote the salvation of others. Thus he claims that the paradigmatic space of modern law is not the nation-state but the camp, in which bare life is confined and managed such that the political sphere can come into being.

Contemporary Geographies of Sovereignty

Where, then, does this leave geographical research today? Geographers continue to focus on the spatial dimensions of law and regulation as a component of state sovereignty. Discussions of sovereignty have become increasingly central to understanding the legal geographies of the global economy, particularly as it relates to the government of people, capital, and resources that are not territorially or spatially fixed. Places like the oceans (Havice 2018; Campling & Colás 2021) or the melting Arctic (Steinberg et al. 2015) provide fascinating examples in which states, companies, nongovernmental organizations, and Indigenous communities struggle to establish legal frameworks over shifting and dynamic resource landscapes. Like the early modern forms of sovereignty discussed by Benton (2009) or Stern (2012, 2023), state sovereignty emerges in a nexus of relations with other forms of law-making, such as international treatises and property rights, by which groups attempt to claim, control, and channel resources. Against classic divisions between state sovereignty and private property (or its various correlates of politics and economics, regulation and the market), these studies index a long history of exploitation and expropriation that works by shuttling back and forth between the two.

Divisions between sovereignty and property are also central to governing financial flows. Shaina Potts (2020, 2024) has chronicled the role of the United States, and specifically US state courts, in extending jurisdiction over increasingly large swaths of global finance. Focusing on this extension of what she calls "judicial territory," she explains how US courts arrogate to themselves the power to rule on economic policies of sovereign nation-states. Courts engage in "reasoning by dichotomies," thereby expanding their reach by treating foreign economic policies and regulation as private, commercial matters and thereby justiciable in places like New York. For Potts, this is just the latest mode of US commercial imperialism. Yet literature on the "new state capitalism" suggests that many states are attempting to expand geoeconomic power through institutional forms including sovereign wealth funds, development banks, state-owned enterprises, and neomercantilist policies (Alami & Dixon 2020, 2023). It remains an open question what the new state capitalism portends for the sovereignty of host or recipient states, an issue that has received particular attention in the context of China's Belt and Road Initiative and broader investment strategies in developing economies (Lee 2017; Narins & Agnew 2019; Carmody & Wainwright 2022).

Scholars also continue to explore state sovereignty at borders. Whether a sign of the state's power or its fundamental weakness, the rise of nationalist politics has made borders into spectacles of sovereign power through policing, walling, and securitization (Brown 2010; Jones & Johnson 2016). At the same time, the border appears less a clean line of demarcation than an uneven space composed of camps and corridors, by which states, nongovernmental agencies, and new digital technologies manage and direct the flow of people (Jones et al. 2017). As for camps and detention centers, they continued to receive attention from geographers as exceptional political spaces (for overviews see Minca 2015; Martin, Minca, & Katz 2019; Weima & Brankamp 2022). These studies reiterate Agamben's central claim concerning the way state power enables spatial exemptions of law that, in turn, unleash violence directed at some populations in the interest of others. They also challenge key elements of the Agambenian framework. Against Agamben's totalizing narrative of sovereignty, scholars have examined how individuals trapped in spaces or zones of exception continue to struggle, negotiate, and resist state violence (see Martin, Minca, & Katz 2019: 754; Griffiths 2022). Recent scholarship has also stressed the important role that social differences of race, class, gender, sexuality, and nation play in the divisions between lives that are qualified for political recognition and those that are exposed to potential death (Weheliye 2014; Valdez, Coleman, & Akbar 2020).

In this sense, contemporary studies address a conceptual problem long identified in work in the Foucauldian tradition. At least as early as the mid-1990s, scholars identified restricted, underdeveloped, and plain odd discussions of race and colonization in Foucault's writings on modern power (Stoler 1995; James 1996). Scholars also identified a lacuna in the characterization of the camp as the spatial and legal organizational matrix (or *nomos*) of modern political regimes. The concentration camps that preoccupied Agamben were certainly expressions of suspensions of law that enabled the exposure of some lives to death. But scholars also note that the state of exception, as a technology of sovereign power, preceded historically and conceptually the Nazi death camps. Echoing Arendt's (1951) argument that colonization was key to establishing the racial topologies that undergirded the Holocaust, scholars argued that colonial law was replete with exceptions and emergencies that differentiated places and populations. Nasser Hussain's (2003) important work on emergency laws demonstrated the ways legal exemptions structured British rule in South Asia. Achille Mbembe's (2019) concept of necropolitics referenced the sovereign dependence on death that was realized not only in the concentration camp but also in the slave trade, the colony, and the plantation. Adam Sitze (2013) examined legal techniques like indemnity laws as central to racialized violence and apartheid governance in South Africa. Among geographers, Sharad Chari (2008) theorized the implications of sovereignty and biopolitics for racialized systems of spatial control.

Moreover, as Najeeb Jan's brilliant "critical ontology" of the *metacolonial* state in Pakistan shows, Agamben's "sovereigntology" has insights into the predicament of postcolonial power. For Jan, the *meta* in metacolonial indexes a panoply of spatial relations, including "'with,' 'across,' 'after,' 'above,' 'beyond,' and 'behind'" (2019: 236). Conjoined with *colonial*, the metacolonial identifies "a spatio-temporal condition that is both prior to the colonial and ongoing" (2019: 236). Counterintuitively, Jan charts the metacolonial dimensions of Pakistan's postcolonial state formation not historically but through a critical ontology that distinguishes itself as "an incitement towards thinking, an intuitive and creative endeavor that does not rest on discovery – of permanent structures, origins, or facts – but disclosure" (2019: 26). What critical ontology discloses is not the empirical history leading to violent events in the present, but a common problematic or frame of reference that inhabits our concepts. As Lee, Jan, and Wainwright (2014: 655–656) clarify:

> The world is caught up in a series of overlapping relationships of force, all of which accelerate and intensify the threshold of a space of exception and the production of the conditions for bare life. The complex cartographies of power unfolding across the Third World and formerly colonized spaces retain distinctive qualities, while simultaneously participating in an interconnected global catastrophe—a catastrophe that Benjamin identifies as "progress".

Critical ontology attunes us to the dimensions of this catastrophe and the central role of sovereign power and biopolitics in its maintenance and reproduction.

In these studies, colonization and racialized dispossession are intertwined with sovereignty, creating the fabric of the modern world. Current scholarship also examines the reworking of sovereignty as part of anti-colonial struggles. Against the narrative of a steadily spreading liberal, state-centric international order, Adom Getachew (2019) has identified a broad range of political proposals from intellectuals associated with national liberation movements in postcolonial Africa and the Caribbean. Getachew chronicles the development of pan-African and transatlantic projects seeking not just the extension of state sovereignty to postcolonial states, but an international order founded in radical responses to the ravages of empire. These include proposals for the establishment of self-determination as a basic principle of international law, permanent sovereignty over natural resources, and, later, the push to establish the New International Economic Order at the United Nations to work in the interest of Third World states. Getachew, along with historians such as Frederick Cooper (2014), has also charted the variety of federal projects seeking to unite independent nations into confederations as counterweights to the domination of external powers. Although most of these projects depart from their idealized form, they still attest to the possibility of a different content of sovereignty in the international order.

The transformation of state sovereignty also marks Indigenous critiques. Taiaiake Alfred (2005) and Audra Simpson (2020) both demonstrate that, in the Americas, European claims of sovereignty as an absolute and undivided power over territory presuppose the dispossession and disappearance of Native communities. In the 1970s, Native and Indigenous scholars turned to sovereignty, along with concepts of self-determination and self-government, as a means of articulating an alternative path from a liberal, rights-based politics within settler colonial states (Alfred 2005: 39). This makes sovereignty, as Simpson (2020: 687) puts it, "fraught . . . as it is shot through with these hidden and not so hidden experiences of force, displacement, and containment." Nonetheless, Simpson emphasizes that sovereignty is more than the violent right over inclusion/exclusion or life/death decisions connected with colonization and the emergence of nation-states. It also resides in longstanding practices of protection and care. Simpson argues this form of sovereignty is grounded in the ways "Indigenous peoples define themselves through relatedness to each other as families, as clans and nations, in relationship to territory" (2020: 689). As such, it poses a different question: "what does sovereignty mean to people who are conditioned to think of themselves in responsible relation to land, to history, and to each other in this way?" (2020: 689).

Simpson's answer is a practice of refusing to abandon these constitutive relations and a situated political critique of non-Indigenous ways of knowing these forms of life and their responsibilities (see Simpson 2014). As she argues, "'sovereignty,' then, moves through these different contexts as a philosophical and governing system that is carried in the languages and experiences of people who had and *still* have political codes and commitments for life that predate settlement" (Simpson 2020: 691). Geographical dimensions of this form of sovereignty can be seen in ongoing struggles and practices of refusal. For instance, Shiri Pasternak (2017) chronicles the ways the Algonquins of Barriere Lake maintain Indigenous jurisdiction, or "the *authority to have authority*," over their lands in the face of encroachments by both the Canadian state and mining companies (2017: 2, emphasis in the original). Pasternak treats jurisdiction as distinct from sovereignty while noting that the two are interrelated in that jurisdiction is "the apparatus through which sovereignty is rendered meaningful" (2017: 3) and a means to "draw law's boundaries, its subjects, and agents" (2017: 10). Methodologically, Pasternak's approach challenges conventional legal reasoning on conflicts between states, companies, and Indigenous communities by not taking Canadian law, sovereignty, or jurisdiction for granted as the de facto space, either conceptually or territorially, by which conflicts must play out. Showing the primacy of Algonquin jurisdiction, Pasternak indexes a legal regime that precedes Canadian law and persists despite the state's claims. Its warrant to govern "rests in the legal relationship between the political community and the place it inhabits," relationships that are likewise figured in terms of responsibility and care (2017: 3).

Conclusion

In his classic analysis of the relations between ideas and history, Stuart Hall (1986: 43) also drew on Gramsci to probe the "*open horizon*" of theory. Hall argued that ideas are "generated and transformed according to their own laws of development and evolution; though, of course, they are generated *out* of given materials" (1986: 43, emphasis in the original). But he also contended that theory was indeterminate because of the "'openness' of historical developments to practice and struggle" (1986: 43). Today, of course, such a position is not very controversial. We deploy inherited concepts to understand conflicts in the present. They are shaped by their conditions of emergence and their ongoing use. Their significance for the future is limited by the past but also open-ended and unknown.

Sovereignty is one such concept, developed through conflicts and struggles, signifying a range of contradictory and potentially irreconcilable meanings (state power, right over life and death, relational care) and taking a variety of spatial forms (state territory, the camp or colony, Indigenous people's authority over land). This should also remind us that the deployment of sovereignty is, as Hall suggests, without guarantees. In his essay, Hall exposes a similar problematic concerning the term "nation":

> It is perfectly possible for the idea of "the nation" to be given a progressive meaning and connotation, embodying a national-popular collective will, as Gramsci argued. Nevertheless, in a society like Britain, the idea of "nation" has been consistently articulated towards the right. Ideas of "national identity" and "national greatness" are intimately bound up with imperial supremacy, tinged with racist connotations,

> and underpinned by a four-century long history of colonisation, world market supremacy, imperial expansion and global destiny over native peoples. It is therefore much more difficult to give the notion of "Britain" a socially radical or democratic referent. (Hall 1986: 41)

Like the nation, sovereignty is not doomed to be the self-same across time and place, but it does carry with it the history of its development and is thus, as Simpson warns us, not just open-ended but also fraught.

The risk that sovereignty turns toward state violence is clear enough; so too is the risk that the critique of sovereignty confuses the content of legal relations with their form. This risk should be apparent to any reader of Agamben's (2021) Covid-19 writings, which were harshly critical of emergency measures and equated life-saving vaccines to a new form of despotism. To be sure, the pandemic did increase state power in stunning fashion, as well as new configurations of sovereignty relying on multiple levels of government, industries, and nongovernmental institutions. Yet to see the constraints in these acts without realizing their ability to increase the capacity for life (or what Roberto Esposito has referred to as an "affirmative biopolitics") is to be fated by analogy, including analogies incorrectly posed (see Christiaens & De Cauwer 2020). Even with these grave problems, however, Agamben's critique has use in exposing the fraught nature of sovereignty and the risks accompanying its political use. For Agamben, sovereignty named an unthought or unrealized potential for violence within modern political forms. Thus, the key term in Agamben's (1998: 115) outrageous claim that "we are all virtually *homines sacri*" was not *all* but *virtually,* suggesting a *potentiality* of power toward death that he views as ever-present (Neilson 2004). More disturbingly, this potential inhabits places and spaces most closely connected with our desires (Santner 2015).

Today's geographical research on sovereignty sits amid these tensions and impasses. Although the works discussed in this chapter derive from differing philosophical and political traditions, they converge on the deep problems posed by sovereign and sovereign power in the present. Even without consideration of the way sovereignty structures the existential threat of climate change (see Mann & Wainwright 2018), the ontological stakes are made clear in Jan's (2019: 237) calls for critical ontology and "ethical cartography," or Pasternak's (2017: xxvi) moving exhortation to consider "the grounds of authority upon which we rely to make sense of our relationship to the world." Amid such paroxysmal violence, the need to find new forms of law and power is as pressing as ever.

References

Agamben, G. (1998). *Homo Sacer: Sovereign Power and Bare Life* (trans. D. Heller Roazen). Stanford, CA: Stanford University Press.

Agamben, G. (2005). *State of Exception* (trans. K. Attell). Chicago, IL: University of Chicago Press.

Agamben, G. (2021). *Where Are We Now? The Epidemic as Politics*. London: Eris.
Agnew, J. (1994). The territorial trap: The geographical assumptions of international relations theory. *Review of International Political Economy* 1: 53–80. doi:10.1080/09692299408434268
Agnew, J. (2017). *Globalization and Sovereignty: Beyond the Territorial Trap*, 2nd edn. Lanham, MD: Rowman and Littlefield.
Alami, I., & Dixon, A. (2020). The strange geographies of the "new" state capitalism. *Political Geography 82*: 102237. doi:10.1016/j.polgeo.2020.102237
Alami, I., & Dixon, A. (2023). Uneven and combined state capitalism. *EPA: Economy and Space 55*: 72–99. doi:10.1177/0308518X211037688
Alfred, T. (2005). Sovereignty. In Barker, J. (ed.), *Sovereignty Matters: Locations of Contestation and Possibility in Indigenous Struggles for Self-Determination*. Lincoln, NE: University of Nebraska Press, 33–50.
Arendt, H. (1951). *The Origins of Totalitarianism*. New York: Harcourt.
Barkan, J. (2015). Sovereignty. In Agnew, J., Mamadouh, V., Secor, A., & Sharp, J. (eds), *The Wiley Blackwell Companion to Political Geography*. Malden, MA: John Wiley & Sons., 48–60. doi:10.1002/9781118725771.ch5
Benton, L. (2009). Not just a concept: Institutions and the "rule of law". *Journal of Asian Studies 68*: 117–122. doi:10.1017/S0021911809000138
Benton, L. (2010). *A Search for Sovereignty: Law and Geography in European Empires, 1400–1900*. Cambridge: Cambridge University Press.
Bodin, J. (1992). *On Sovereignty: Four Chapters from The Six Books of the Commonwealth* (ed. and trans. J. Franklin). Cambridge: Cambridge University Press.
Brenner, N. (2004). *New State Spaces: Urban Governance and the Rescaling of Statehood*. Oxford: Oxford University Press. doi:10.1093/acprof:oso/9780199270057.001.0001
Brown, W. (2010). *Walled States, Waning Sovereignty*. New York: Zone Books.
Campling, L., & Colás, A. (2021). *Capitalism and the Sea: The Maritime Factor in the Making of the Modern World*. London: Verso.
Carmody, P., & Wainwright, J. (2022). Contradiction and restructuring in the Belt and Road Initiative: Reflections on China's pause in the "Go world". *Third World Quarterly 43*: 2830–2851. doi:10.1080/01436597.2022.2110058
Chari, S. (2008). Critical geographies of racial and spatial control. *Geography Compass 2*: 1907–1921. doi:10.1111/j.1749-8198.2008.00169.x
Christiaens, T., and De Cauwer, S. (2020). The biopolitics of immunity in times of COVID-19: An interview with Roberto Esposito. *Antipode Online*, June 16. antipodeonline.org/2020/06/16/interview-with-roberto-esposito (accessed November 21, 2024).
Cooper, F. (2014). *Africa in the World: Capitalism, Empire, Nation-State*. Cambridge, MA: Harvard University Press.
Elden, S. (2009). *Terror and Territory: The Spatial Extent of Sovereignty*. Minneapolis, MN: University of Minnesota Press.
Foucault, M. (1990). *The History of Sexuality: An Introduction*, vol. 1. New York: Vintage.
Foucault, M. (2003). *"Society Must Be Defended": Lectures at the Collège de France, 1975–1976* (ed. M. Bertani and A. Fontana, trans. D. Macey). New York: Picador.
Foucault, M. (2004). *Security, Territory, Population: Lectures at the Collège de France, 1977–1978* (ed. M. Senellart, trans. G. Burchell.). New York: Palgrave Macmillan.
Galli, C. (2010). *Political Spaces and Global War* (ed. A. Sitze, trans. E. Fay). Minneapolis, MN: University of Minnesota Press.

Getachew, A. (2019). *Worldmaking after Empire: The Rise and Fall of Self-Determination.* Princeton, NJ: Princeton University Press.

Griffiths, M. (2022). Thanato-geographies of Palestine and the possibilities of politics. *EPC: Politics and Space 40*: 1643–1658. doi:10.1177/23996544221099461

Hall, S. (1986). The problem of ideology – Marxism without guarantees. *Journal of Communication Inquiry 10*: 28–44. doi:10.1177/019685998601000203

Havice, E. (2018). Unsettled sovereignty and the sea: Mobilities and more-than-territorial configurations of state power. *Annals of the American Association of Geographers 108*: 1280–1297. doi:10.1080/24694452.2018.1446820

Hobbes, T. (1996). *Leviathan* (ed. R. Tuck). Cambridge: Cambridge University Press.

Hussain, N. (2003). *The Jurisprudence of Emergency: Colonialism and the Rule of Law.* Ann Arbor, MI: University of Michigan Press.

Ignatieff, M. (2003). The American empire; the burden. *New York Times Magazine*, January 5: 22.

James, J. (1996). *Resisting State Violence: Radicalism, Gender, and Race in U.S. Culture.* Minneapolis, MN: University of Minnesota Press.

Jan, N. (2019). *The Metacolonial State: Pakistan, Critical Ontology, and the Biopolitical Horizons of Political Islam.* Hoboken, NJ: John Wiley & Sons.

Jones, R., & Johnson, C. (2016). Border militarization and the re-articulation of sovereignty. *Transactions of the Institute of British Geographers 41*: 187–200. doi:10.1111/tran.12115

Jones, R., Johnson, C., Brown, W., et al. (2017). Interventions on the state of sovereignty at the border. *Political Geography 59*: 1–10. doi:10.1016/j.polgeo.2017.02.006

Kalmo, H., & Skinner, Q. (eds) (2010). *Sovereignty in Fragments: The Past, Present, and Future of a Contested Concept.* Cambridge: Cambridge University Press. doi:10.1017/CBO9780511675928

Kantorowicz, E. (1951). *Pro patri mori* in medieval political thought. *American Historical Review 56*(3): 472–492. doi:10.1086/ahr/56.3.472

Kantorowicz, E. (1957). *The King's Two Bodies: A Study in Medieval Political Theology.* Princeton, NJ: Princeton University Press.

Krasner, S. (1999). *Sovereignty: Organized Hypocrisy.* Princeton, NJ: Princeton University Press.

Lee, C. (2017). *The Specter of Global China: Politics, Labor, and Foreign Investment in Africa.* Berkeley, CA: University of California Press.

Lee, S.-O., Jan, N., & Wainwright, J. (2014). Agamben, postcoloniality, and sovereignty in South Korea. *Antipode 46*: 650–668. doi:10.1111/anti.12070

Mann, G., & Wainwright, J. (2018). *Climate Leviathan: A Political Theory of Our Planetary Future.* London: Verso.

Mann, M. (1984). The autonomous power of the state: Its origins, mechanisms and results. *European Journal of Sociology 25*: 185–213. doi:10.1017/S0003975600004239

Martin, D., Minca, C., & Katz, I. (2019). Rethinking the camp: On spatial technologies of power and resistance. *Progress in Human Geography 44*(4): 743–768. doi:10.1177/0309132519856702

Mbembe, A. (2019). *Necropolitics.* Durham, NC: Duke University Press.

Minca, C. (2015). Geographies of the camp. *Political Geography 49*: 74–83. doi:10.1016/j.polgeo.2014.12.005

Mountz, A. (2013). Political geography I: Reconfiguring geographies of sovereignty. *Progress in Human Geography 37*: 829–841. doi:10.1177/0309132513479076

Narins, T., & Agnew, J. (2019). Missing from the map: Chinese exceptionalism, sovereignty regimes, and the Belt and Road Initiative. *Geopolitics 25*: 809–837. doi:10.1080/14650045.2019.1601082

Neilson, B. (2004). *Potenza nuda*? Sovereignty, biopolitics, capitalism. *Contretemps 5*: 63–78.

Pasternak, S. (2017). *Grounded Authority: The Algonquins of Barriere Lake against the State*. Minneapolis, MN: University of Minnesota Press.

Potts, S. (2020). Law as geopolitics: Judicial territory, transnational economic governance, and American power. *Annals of the American Association of Geographers 110*: 1192–1207. doi:10.1080/24694452.2019.1670041

Potts, S. (2024). *Judicial Territory: Law, Capital, and the Expansion of American Empire*. Durham, NC: Duke University Press.

Santner, E. (2015). *The Weight of All Flesh: On the Subject Matter of Political Economy*. Oxford: Oxford University Press.

Schmitt, C. (1985 [1922]). *Political Theology: Four Chapters on the Concept of Sovereignty* (trans. G. Schwab). Chicago, IL: University of Chicago Press.

Schmitt, C. (2003). *The Nomos of the Earth in the International Law of the Jus Publicum Europaeum* (trans. G. L. Ulmen). New York: Telos Press.

Simpson, A. (2014). *Mohawk Interruptus: Political Life across the Borders of Settler States*. Durham, NC: Duke University Press.

Simpson, A. (2020). The sovereignty of critique. *South Atlantic Quarterly 119*: 685–699. doi:10.1215/00382876-8663591

Sitze, A. (2013). *The Impossible Machine: A Genealogy of South Africa's Truth and Reconciliation Commission*. Ann Arbor, MI: University of Michigan Press.

Skinner, Q. (2010). The sovereign state: A genealogy. In Kelmo, H., & Skinner, Q. (eds), *Sovereignty in Fragments*. Cambridge: Cambridge University Press, 26–46. doi:10.1017/CBO9780511675928.002

Sparke, M. (2006). A neoliberal nexus: Economy, security and the biopolitics of citizenship on the border. *Political Geography 25*: 151–180. doi:10.1016/j.polgeo.2005.10.002

Steinberg, P., Tasch, J., Gerhardt, H., et al. (2015). *Contesting the Arctic: Politics and Imaginaries in the Circumpolar North*. London: I. B. Tauris.

Stern, P. (2012). *The Company-State: Corporate Sovereignty and the Early Modern Foundations of the British Empire in India*. Oxford: Oxford University Press.

Stern, P. (2023). *Empire, Incorporated: The Corporations That Built British Colonialism*. Cambridge, MA: Harvard University Press.

Stoler, A.L. (1995). *Race and the Education of Desire: Foucault's History of Sexuality and the Colonial Order of Things*. Durham, NC: Duke University Press.

Valdez, I., Coleman, M., & Akbar, A. (2020). Law, police violence, and race: Grounding and embodying the state of exception. *Theory & Event 23*: 902–934. doi:10.1353/tae.2020.0056

Wainwright, J. (2010). On Gramsci's "conceptions of the world". *Transactions of the Institute of British Geographers 35*(4): 507–521. doi:10.1111/j.1475-5661.2010.00397.x

Weheliye, A. (2014). *Habeas Viscus: Racializing Assemblages, Biopolitics, and Black Feminist Theories of the Human*. Durham, NC: Duke University Press.

Weima, Y., & Brankamp, H. (2022). Camp methodologies: The "how" of studying camps. *Area 54*(3): 338–346. doi:10.1111/area.12787

Williams, R. (1983). *Keywords: A Vocabulary of Culture and Society*. New York: Oxford University Press.

Chapter 9

Territory

Cristina Del Biaggio[1], *Kirsten Koop*[1], *Carine Pachoud*[2], *and Camille Noûs*[3]

[1]*Pacte, Université Grenoble Alpes, CNRS, Sciences Po Grenoble, Grenoble, France*
[2]*AGIR – INRAE, University of Toulouse, Toulouse, France*
[3]*Laboratoire Cogitamus, www.cogitamus.fr*

The concept of "territory" might well have been swept under the carpet with the postmodernist turn in geography had it been confined to its original ethological meaning, as an appropriated, controlled, and bounded space. But, as Antonsich puts it, the "lesson of the post-modern and post-structural turn" is that concepts "are not fixed, but historically, geographically, and socially specific" (2017: 1) and thus evolve in space and time. Territory is no exception. This is made particularly evident by the increased attention it has received since the 1990s, when geographers sought to understand the processes of globalization, and by its role today in relation to addressing the profound societal and environmental crises facing humanity. The spatial concept of territory has, in fact, turned out to be a powerful lever for thinking about geographies in the making, in both practical and theoretical terms.

In this chapter, we will show how territory is associated with diverse epistemologies grounded in the historical and geographical context to which scholars are bounded. By doing so, we aim to enhance the understanding of what Cox considers "the defining concept" of political geography (2002: 1) and thus contribute to broadening out epistemological knowledge and academic discussions in order to "derive more robust, sophisticated and inclusive theories of territory," as proposed by Woon (2019: 116). The concept of territory has in fact traveled through a variety of sociocultural milieus and has been enriched with different meanings, which have themselves traveled in turn, drawing on new scholarly contributions. In this sense, we believe, with Fall (2012: 174), that "it matters how and where we do things as geographers, where ideas come from and then where

The Wiley Blackwell Companion to Political Geography, Second Edition.
Edited by Virginie Mamadouh, Natalie Koch, Chih Yuan Woon, and John Agnew.

people think about, discuss, and publish them, [and it matters] because they change how we think of the world as a place in which and about which theory is made." Retracing the evolution of the meaning(s) of territory in time and space thus provides insight into varying situated historical and societal contexts, and the changing ways we think about (society in) the world.

We propose to explore the different meanings of territory beginning in different parts of the world during the same period: the 1970s. At this time, in Anglophone geography territory was understood in a largely political – and above all institutional – way, and conceived as the bounded space of state sovereignty (Soja 1971; Gottmann 1973; Cox 1991). Despite some moves toward a more network-based (Jessop, Brenner, & Jones 2008; Painter 2010) and relational understanding (Gottmann 1973), territory has largely remained embedded in political geography, and is still often linked to the state or its (mainly regional) administrative levels. In the Francophone world, in contrast, *territoire* has taken on much more social connotations. The *territoire* is the space of belonging (Frémont 1974; Raffestin 1982; Debarbieux 1995; Di Méo 1998). It is understood as an *espace vécu* (lived space), a space of everyday practices that is strongly linked to the imaginaries and identities of its inhabitants (Frémont 1974) and is embedded in power relations (Raffestin 1982).

From the 1990s onward, the notion of territory found a place within Francophone regional geography, acquiring more economic connotations in the process. Groups of researchers analyzing the growing competition produced by globalization saw the concept of territory as an opportunity to promote the vision of bottom-up development (Campagne & Pecqueur 2014). In South America – particularly in Brazil – and in Italy, the notion of territory was adopted in the same spirit (Abramovay 1999; Saquet 2018; Banini 2019; Guillot & d'Emilio 2020).

It is also in South America, though only since the 2010s, that *territorio* has been mobilized as part of decolonial approaches to studying the struggles for autonomy and self-determination of populations marginalized by modern nation-states. This marks an ontological turn: *territorio* is emancipating itself from the modern Western-centric approach as it is conceived as an anti-hegemonic living space (Haesbaert & Mason-Deese 2020). This conception also goes hand in hand with a relational approach suggesting new ways of analyzing the links between humans and nonhumans (Anthias 2021), thereby moving away from a purely anthropocentric focus.

In the contemporary context of global transformations and crises, territory is now being used to spatialize the dynamics of sustainability transitions, as instigated by both Francophone (Pachoud, Koop, & George 2022) and Italophone geographers (notably Magnaghi 2020). These emerging studies draw on the Francophone *territoire*, with its strong social and (non-state-centered) political roots, and on the Latin American "decolonial territory," thus providing a multidimensional and situated understanding of the dynamics of societal transformations toward sustainability.

Territory as a Political Conception

Anglophone geographers have long favored a rather politico-institutional understanding of territory, related to the state and the notion of sovereignty, thus making it a core concept for political geography (Cox 2002). This modern conception of territory is closely linked to the notion of jurisdiction (Elden 2013b) and was originally defined in Europe in terms of exclusive state sovereignty from the seventeenth century onward. Territory was thus originally conceived within a discipline that considered the world as being divided into vertically nested spaces, each one with a particular function within a hierarchical system for ease of governance and administration (Gottmann 1973). From this perspective, territory was long considered as a "bounded space, a container, under the control of a group of people, nowadays usually a state" (Elden 2010b: 757). These ideas of demarcation and control are also present in the interpretation of territory as a transposition of ethological theories into the social sciences, following Ardrey's argument in his book *The Territorial Imperative* (Ardrey 1970; see also Soja 1971; Sack 1986; Murphy 2012: 159).

Since the 1990s, this conception has been widely criticized. Agnew's well-known article on the "territorial trap" (Agnew 1994) has largely contributed to pushing scholars to reframe "territory" as an open, unbounded, and relational space (Antonsich 2011). The influential contributions from Agnew (1994, 2010, 2013), Elden (2010a, 2013a, 2013b), Jessop et al. (2008), Jonas (2012), Murphy (2010), Paasi (2009), Painter (2009), and others, as well as the creation of the interdisciplinary journal *Territory, Politics, Governance* in 2012 and the 2009 Roundtable on "The Territorial Trap Fifteen Years On" (https://www.tandfonline.com/doi/full/10.1080/14650041003717509), can be seen as efforts to achieve this aim. Analysis of territorial networks has been one of these "escape routes from the 'territorial trap'" identified by Bulkeley (2005: 881). In this vein, Painter (2010), with Jessop et al. (2008), explores the construction of territory in close relation to the notion of "network" (Del Biaggio 2015), showing how territory is in fact the product of networked sociotechnical practices. Another avenue of research has been opened up by Antonsich, who proposes considering territory as the space of political belonging beyond the sociospatial conformity with nation and state (2009).

Yet despite the expanded meanings of territory, and the now extended scope of the political (from the state to the personal and the global), it is through another concept, "place," that the idea of space as a social construct traversed by power relations has been addressed in the Anglophone world (Murphy 2010; Pierce, Martin, & Murphy 2011). Following Massey's influential work (2005), the concepts of "place" and "place-making" offered a way to address the social, cultural, and political dimensions of space, thus indirectly providing a critique of the concept of territory viewed as having rigid delimitations and being under state control (Debarbieux 1999: 42). This led Debarbieux to argue that the meanings given to the term "place" in Anglophone geography recall the innovations in conceptualizing *territoire* in the Francophone discipline (Debarbieux 1999: 42).

Territory as a Social Construct

In Francophone geography, as Antonsich reminds us, *territoire* "has emerged mainly as a social and symbolic concept" (2017: 5). Since the 1980s, as a result of the influential work of Raffestin (1982, 1986a) along with other scholars such as Bonnemaison (1981), territory has been defined in "strictly relational terms" (Klauser 2012: 110) as an entity generated from space, which Raffestin considered as the material support. Territory, in Raffestin's view, is the "result of an action led by a syntagmatic actor (an actor carrying out a program) at whatever geographical scale." By appropriating a space, either concretely or abstractly (through representation, for example), the actor "territorializes the space" (Raffestin 1980: 129). By emphasizing "at whatever scale," Raffestin distances himself from a state-centric vision, and he gives agency to the "syntagmatic actors" by arguing that they "territorialize space."

Although territory is also largely used in French spatial planning as a synonym for local administrative levels, this fundamental conceptual shift has introduced a widely shared constructivist postmodern conception of territory as a social construct, and as the "spatial expression of social change" (Giraut 2008: 62). Behind this idea can be seen, following Lopez and Muchnik, "a space that is socially constructed, culturally marked, and institutionally regulated" (Lopez & Muchnik 1997: 23). Territory is thus seen as the spatial result of the social proximity between actors and their agency. As such, it cannot be confined within defined limits, but is instead blurred and changing (Torre & Beuret 2012) based on the networks the actors build (Giraut 2013). Without ignoring the fact that Anglophone scholars have also increasingly seen territory as a social construct – notably initially via the work of Soja (1971) and Gottmann (1973), and later Antonsich (2009) – it has rarely been used in this way without adopting one of the three types of methodological nationalism identified by Dumitru: state-centrism, groupism, or territorialism (Dumitru 2014).

If territory is defined by these lived spaces and daily practices, it also entails an ideal dimension: the experienced space of social representations (Bonnemaison 1981; Raffestin 1986b; Di Méo & Buléon 2005). This aspect is crucial to the understanding of territory as a human – economic, ideological, political, and social – construction of space (Brunet, Ferras, & Théry 1993: 199) by "groups who give their particular representation of themselves, their history, and their singularity" (Di Méo 1998: 107). Territory thus plays an important role in the production of collective identity, being the minimal identified geographical base that gives identity a material and ideal support, rich in symbols and embedded in landscapes (Di Méo 1998, 2014). Collective identity represents a means of legitimizing a group in a territory that, in turn, contributes to reinforcing the sense of belonging (Di Méo 2014). Territory can therefore be understood as being produced affectively, socially, culturally, and symbolically. Raffestin sums up this immaterial bond in the word *territorialité* (territoriality), which denotes the individual and collective relationships, embedded in power relations, with the territory (Raffestin 1982). This conception of territory, largely shared in French human geography, is thus multidimensional, going beyond the concerns of strictly political geography. While it includes the spatiality of power relations, it stresses their links with the social, ideal, economic, and material facets through which humans know and practice space.

Raffestin's trips back and forth between Geneva and Turin (Minca 2012) led to his ideas crossing the Alps and spreading to Italy. Turco, who has worked closely with Raffestin since the 1980s (Fall 2012; Minca 2012), has developed a "theory of complexity based on the concepts of territory and territoriality" (Banini 2019: 220) that incorporates contributions taken from his research in Africa (Minca 2012: 150). Turco outlines three components in the process of territorialization: denomination, which is what enables the "intellectual control of the territory"; reification, which is the "modification of the physical characteristics of nature" through the activities of production, mobility, and habitation; and structuration, which is the "compartmentalization of the land surface" (Turco 2007). As with Francophone authors, Turco sees *territorializzazione* (territorialization) as involving feelings of belonging, identity, and topophilia (Banini 2019: 223).

Raffestin has influenced other scholars in Italy, notably Dematteis, who proposes a reading of *territorio* as the foundation of socialization. This is particularly evident in Dematteis's statement that "the Earth becomes territory when it is the medium of communication, when it is the means and object of work, production, exchange and cooperation" (1996: 74). Dematteis's contribution, as illustrated by this statement, is to conceive territories as the lever of individual and collective action, since they are the repositories of "things" to which the inhabitants attribute a "value": a "territorial value" (Banini 2019: 224).

Territory as a Bottom-Up Approach to Economic Development and Spatial Planning

In conjunction with the growing concern among European states about widening regional disparities caused by the globalization that has taken place since the 1990s (Torre 2015), Francophone geographers and regional science scholars have forged the model of territorial development, which ties in with the bottom-up approach to local development of the 1980s (Friedmann & Weaver 1979; Stöhr & Taylor 1981). Originally popularized as a strategy for fostering local development in the Global South, this approach has been further developed to conceive endogenous dynamics of economic regeneration for peripheral rural and/or deindustrialized regions in the Global North (Koop 2014). In this context, it has been used to analyze how locally linked actors capitalize on local (material and immaterial) resources to generate territory-specific products and services (Pecqueur 2005, 2008).

The economy-centered conception of *territoire* has largely been shaped by the Grenoble École du territoire (School of Territory; e.g., de Bernardy & Debarbieux 2003; Pecqueur 2006; Gumuchian & Pecqueur 2007; Courlet 2008; Vanier 2009; Giraut 2013) and by the Groupe Européen de Recherche sur les Milieux Innovateurs (GREMI), greatly influenced by the economic and innovation geography developed in Francophone Switzerland (e.g., Crevoisier & Maillat 1991; Kebir & Crevoisier 2007; Crevoisier 2014; Kebir et al. 2017). Here, territorial development is understood as the result of coalitions of actors solving unprecedented problems of production, thereby intervening in the constitution of relational local economic processes (Courlet & Pecqueur 2013). The focus is on the way in which locally networked small and medium enterprises, institutions, and actors from civil society collectively reveal and value material and immaterial resources related to a specific territorial identity, historical heritage, and know-how (Campagne & Pecqueur 2014). Pecqueur's conceptualization is based on Hirschman's work on the discovery of local latent but unexploited resources (Hirschman 1958). He shows that the process of specifying resources – in the sense of making them so specifically linked to their territory that they are irreproducible elsewhere – offers a way to escape international competition (Koop 2014). These ideas, spread through the École du territoire, can be considered the most refined contribution by Francophone academics to thinking about the territorial dimension of economic development (Koop 2014). This economy-centered conception has also been adopted in North African Francophone countries (Koop, Landel, & Pecqueur 2010) and in Latin America since the wave of neoliberal policies of the 1990s. This is especially the case in Brazil (see for example Abramovay 1999; or Schneider 2009), thanks to its longstanding academic links with France (Théry 2018).

Territorial development can thus be understood as an original contribution to conceiving regional development in a globalized world economy alongside other concepts in regional geography, such as Porter's "clusters" (1998) and Aydalot's "innovative milieus" (1986).

Decolonizing Territory

While territory has revealed itself to be a rich concept in Western political and social geography and in regional science over the past few decades, it has also served as a central notion in decolonial, postcolonial, postdevelopment, and Indigenous studies. Scholars in these fields have taken territory out of the modern episteme, or, in decolonial terms, out of the colonial matrix of power (Mignolo 2011). This has occurred in three interconnected ways.

The first avenue by which territory has been decolonized is the challenge to the state-centered political meaning prevailing in the Anglophone world "that remains predominantly rooted in modern, colonial experiences" (Halvorsen 2019: 14). Over the past two decades, the claim for *territorio* has become a strategy of struggles over legal recognition of the right to self-determination by Indigenous communities and social movements in Latin America (Offen 2003; Haesbaert & Mason-Deese 2020). The Zapatistas, for instance, have symbolically created what they call "rebel territories" in southern Mexico. They tend to reject state sovereignty over "their" territory, and offer the possibility of alternative forms of governing it – a "good government" in Zapatista terms (Reyes & Kaufman 2011) – by building horizontal, nonhierarchical power relations (Halvorsen 2019). Some Latin American governments, for example in Ecuador, Colombia, and Bolivia, have even officially recognized Indigenous land as autonomous territories in their constitutions, thus creating overlapping territorialities within the nation-state (Agnew & Oslender 2013).

Scholars studying such processes have analyzed territory as the spatial expression of ethnic/community rights and of autonomous forms of governance, and given it the status of an emancipatory political project (Offen 2003; Castree 2004; Escobar 2008; López Sandoval, Robertsdotter, & Paredes 2017; Oslender 2019). Territory has thus gained an anti-hegemonic political dimension, disentangling it from its "imperialist enclosure" (Jazeel 2017), as evidenced by the rise of expressions such as "dissident territories" (Lopes de Souza 2015) and "territories in resistance" (Zibechi 2012). As Halvorsen observes, these contributions take territory beyond the context of the modern and Eurocentric state to open a dialogue with dominant Western imaginaries and practices of territory (2019).

A second avenue, closely linked to the first, has been opened up by Latin American scholars who have analyzed these territories in resistance in terms of the existence or the reinvention of radically other ways of living, seeing, and knowing the world outside Western epistemologies (e.g., Escobar 2008; Porto-Gonçalves 2009a, 2009b; Zibechi 2012; Halvorsen 2019; Haesbaert 2020). Escobar is the scholar who has explained most explicitly the idea of radical "otherness" from dominant universals. Referring to the writings of Law, Blaser, and de la Cadena, he conceives "otherness" in a much more radical way than purely in cultural terms. In *Territories of Difference*, he shows that, for example, the territory of the Afro-Columbian community reflects the enactment in space of values, identities, imaginaries, practices, and narratives that are considered to be situated "outside" Western ways of knowing and conceiving the world. As such, he follows what is called the ontological turn in anthropology (Escobar 2008).

This conception is illustrated by *buen vivir* (*sumak kawsay*), a principle of the worldview followed by Indigenous communities in the Andes, which has now gained considerable traction due to the international publications by Gudynas (2011). Recognition of the existence of many ways of thinking and practicing the world – rather than solely those from the West – has led to the conception of "a world of many worlds" (de la Cadena 2015) and to what scholars have called "geographies of the pluriverse" (Escobar 2018; Oslender 2019). As a result of its multidimensional conception, "territory" has played a crucial role in this ontological turn. In this understanding, territory is a social and relational construct, a lived place in its ideal, material, economic, political, ecological, and spiritual dimensions. The influence of Francophone geographers can be seen here, and the Brazilian geographer Porto-Gonçalves, a pioneer in conceptualizing territory through the decolonial lens, indeed cites Raffestin in his early work (Porto-Gonçalves 2002).

The third avenue by which territory has been decolonized is through its bio- or ecocentric conception, which challenges a purely anthropocentric understanding of territory. Ways of knowing and conceiving the world "outside" the modern Western ontology play a central role in this shift, along with the conception of a relational territory and the new interpretations of the ties between humans and nonhumans. Indigenous geographies in particular reveal worlds of dense networks of relations between humans and other Earth beings, questioning the dominant human–nature, body–mind, and materiality–spirituality divides of modern ontology (e.g., de la Cadena 2015; Daigle 2016).

The recognition of these territories of difference and their materialization in maps have become central to the work by Indigenous peoples to defend their rights to land and territory, to challenge the dominant geographical imaginary, and to influence public policies relating to territory since the 1990s (Louis, Johnson, & Pramono 2012). Critics, however, argue that communal mapping and land titling perpetuate essentialist conceptions of identity and insert Indigenous territories into broader (state and capitalist) grids of legibility (Anthias 2019). Indigenous claimants are thus asked to simultaneously assert their cultural difference and to conform to a modern conception of territory as "a discrete, sharply bounded and spatially fixed area, premised on spatial exclusivity" (Blomley 2016: 593).

Decolonial scholars from the Global North and Global South have influenced one another through mutual exchanges. Many researchers from the Global South (such as Escobar) have studied or work in Western universities, allowing their ideas to permeate into the Global North. Conversely, some scholars from the Global North, such as Halvorsen, have adopted this new understanding of territory after conducting field research in the Global South.

Territory as a Way of Thinking about Societal Transformation

Scholars have also started using territory as a lens through which to observe and put into practice profound societal transformation, in response to global change and civilizational crises. In this vein, the French philosopher Latour called for radically new ways of "belonging to a territory" in the context of what he termed the "new climatic regime" (2018) – a regime in which the entities of the Earth (in the sense of Gaia) have agency. Inspired by the *cahiers de doléances* ("lists of grievances") drawn up following the French Revolution – the last profound transformation in European history – Latour held a series of workshops across France. During these meetings, symbolically called "*Où atterir?*" (Where to land?), inhabitants were asked to answer this question by focusing on two elements: what they *really* depended on in their daily lives, and what they were *ready* to defend. Through these exercises, they generated alternative descriptions of their territories that "encompass an entire set of animate beings – far away or nearby" (Latour 2018: 95).

Magnaghi, cofounder of the Italian Società dei Territorialisti e delle Territorialiste (Society of Territorialists), also makes territory the focus of interest in his statement that "an effective change of direction, capable of responding strategically to the environmental crisis, is only possible by rebuilding the relationship of inhabitants with the territory in all its complexity, by calling into question all the elements involved in the production of space" (Magnaghi 2022: 16). The territorialist school is based on a critique of the logic of metropolization linked to the neoliberal economic system, and proposes a self-sustaining local development project based on a balance between three objectives: satisfying the fundamental needs of human life, which are not limited to material needs; giving local communities political, economic, and social autonomy; and finally, maintaining biodiversity by improving the quality of the environment (Guillot and d'Emilio 2020: 2).

Magnaghi sees *comunità territoriali* (territorial communities) as a response to the current environmental crisis, and includes them in what he calls the "territorial principle." From this perspective, he stresses the need for a "return to territory" based on "rules, behaviors, cultures, and ecologically sound techniques of living and producing," capable of increasing "awareness of place" and restoring to inhabitants "the capacity to reproduce their own living environments and socio-economic self-government" (Magnaghi 2022: 16). Inhabitants are thus called on not to dominate nature but to "come to terms with it," by "co-producing living neo-ecosystems" (Magnaghi 2022: 21), since territory, in Magnaghi's view, is the "common good par excellence" (2022: 22). He thus urges us to imagine a "regionalist counter-geography," which he defines as a "conflictual reappropriation by local societies . . . of the sovereignty over common goods that has been usurped by the globalized neoliberal economy" (2022: 93–95). Magnaghi sees "common goods" or "territorial heritage" – the latter denoting the way territory is reinterpreted through the mediation of a local society that cares for it – as transforming territory into a resource, within the limits imposed by the reproduction of its existence value (Magnaghi 2022: 121).

More theory oriented, Pachoud, Koop, and George (2022) use the lens of the French multidimensional conception of territory to analyze profound societal transformations toward sustainability. They propose studying transformations in the ideal (encompassing imaginaries, values, and narratives), material (biophysical elements, landscapes, and infrastructure), and political dimensions (norms and rules, and forms of governance). According to these authors, sustainable transformation of territories – which were historically identified by their inhabitants as coherent entities – is triggered by groups enacting new alternative values and narratives in space, assigning value to resources beyond their market exchange value, and developing alternative norms and rules. This is the case for territories in which emerging grassroots communities engage in healthy, just, and resilient futures – such as those of the international Transition Network – via alternative projects (e.g., agroecology, eco-housing, community renewable energy projects, repair cafés, urban gardening, local currency, and a wellbeing economy). Such communities challenge conventional imaginaries, practices, and rules dominating the territory – which creates conflicts over values and the use of resources. Addressing competing imaginaries, identities, and practices as well as power asymmetries though the lens of the multidimensional territory enables researchers and practitioners to explore possible sustainable pathways. Recent work and publications on territorial transition (Armengaud, Degros, & Radulova-Stahmer 2023) and the use of territorial design (e.g., Duhem & Pereira de Moura 2020) to foster socioecological transformations help to address these issues in a practical manner.

Conclusion

This overview of the way in which the meanings and uses of territory have evolved in various academic milieus around the world shows that, in Anglophone geography, the term territory is anchored to political geography and is often still associated with the state and its institutions at various scales. A look at the term territory in Francophone Switzerland, France, Italy, and South America – and the physical journey of geographers between these places – reveals how the concept has been locally appropriated in its iteration and mutually enriched, thus producing a fascinating Romance-language geography. Territory originated in constructivist Francophone human geography, in which it is seen first and foremost as a social construct, a space for everyday life and representations making it possible to address the spatial dimensions of identity, imaginaries, practice, and power relations on equal terms. The concept has also taken on a key role in regional science and planning, particularly with regard to revealing territorial specificities in local bottom-up development processes. The decolonial meaning that has emerged over the past decade, in particular from South America, has led to territory being conceived in terms of community life projects, distancing it from the dominant modern episteme. This has produced a less anthropocentric and more pluriverse conceptualization of territory and territoriality. Last but not least, in these times of systemic crises, territory also serves as a way to address societal transformation in a comprehensive and multidimensional manner.

The numerous conceptual renewals of territory thus largely appear to have taken place outside Anglophone geography. The reason for this may lie in the influence of Raffestin, who has himself traveled between Romance-language worlds, thus transmitting his new conception of territory since the 1980s.

Our semantic journey sketches out a geography of territory marked by a clear divide between Anglophone and Romance-language geography. While the latter has revived the meaning of territory to grasp the complex, multidimensional relationships between societies and space, the former has developed a different strategy, using different terms for analyzing different phenomena. Anglophone geographers tend to use "place" (Del Biaggio 2015) to understand the spatialities of investing the world with meaning (Massey 2005) through social practice, identity, and imaginaries. Place is also the term most often used in Anglophone decolonial and Indigenous geographies (e.g., Daigle 2016: 201; Radcliffe 2017; Barker and Pickerill 2019), as well as in recent attempts to address geographies of sustainability transitions (e.g., Binz et al. 2020). "Region" is the term that continues to dominate in Anglophone economic geography and regional science (e.g., Scott & Storper 2003).

Rather than asking the unproductive question of which concepts best denote what, we prefer to highlight how the meanings that territory takes on – or does not take on – are rooted in historically situated academic contexts. Engaging with the meanings of territory thus requires us to carefully tackle the intellectual and societal context in which it is used. As such, translation of the various meanings from one sphere to another is more a question of transmitting and making comprehensible both the local academic heritage and the societal contexts in which the meanings of the terms have been forged – which goes beyond the matter of language. As Cindi Katz (2001) states, connections between geographically distant milieus play a major role in the transmission of meanings of words. They are partly established through translocal relationships between scholars, and thus due to academic or personal opportunities. What if Raffestin had traveled to North America instead of Brazil? What if his influential book had been translated into English instead of Portuguese in the 1990s? Would this have changed the global geography of the meanings of territory, and thus narrowed the intellectual divide in Anglophone geography between "territory" (for the political) and "place" (for the social)?

Nowadays different factors are in play, and various tools are available for disseminating the different understandings of concepts, such as territory, around the world. Scholars travel easily from one academic context to another and may thus play the role of intermediaries (as with Halvorsen for the Latin American decolonial approach to territory); tools providing quick, cheap, and reliable translations are more widely available; and digital articles can be accessed everywhere via the web. Yet despite this growing global access to scientific output, in practice the process does not occur smoothly. The adoption of new meanings assigned to territory elsewhere is in fact highly dependent on, following Fall (2012: 174), "how and where we do things as geographers, where ideas come from and then where people think about, discuss, and publish them." As she goes on to note, the "place in which and about which theory is made" determines "how we think of the world" (2012: 174) and the societal dynamics and challenges we perceive to be relevant to tackle.

By contributing to the knowledge of the meanings assigned to "territory" in certain non-Anglophone parts of the world, this chapter helps broaden out, even if only partly, the epistemological debate on one of the key concepts in (political) geography. It aims to encourage scholars to translate and contextualize concepts and theories from and in different languages to produce further theoretical hybridizations and renewals. Only by providing such epistemic compasses will geographers feel less "lost in translation."

Note

In 2020, Camille Noûs was born in France, embodying the community's contribution to research work in the form of a collective signature. This co-signature asserts the collaborative and open nature of the production and dissemination of knowledge, under the control of the academic community. Camille Noûs is a member of the multidisciplinary, interdisciplinary, and transdisciplinary Cogitamus research group. By co-authoring this text with Camille Noûs, the authors are endorsing the values promoted by this collective signature. For more information, see: https://www.cogitamus.fr/camilleen.html.

References

Abramovay, R. (1999). Agricultura familiar e desenvolvimento territorial. *Reforma Agrária – Revista Da Associação Brasileira de Reforma Agrária 28*(1).

Agnew, J. (1994). The territorial trap: The geographical assumptions of international relations theory. *Review of International Political Economy 1*(1): 53–80. doi:10.1080/09692299408434268

Agnew, J. (2010). Still trapped in territory? *Geopolitics 15*: 779–784.

Agnew, J.A. (2013). Arguing with regions. *Regional Studies 47*(1): 6–17. doi:10.1080/00343404.2012.676738

Agnew, J., & Oslender, U. (2013). Overlapping territorialities, sovereignty in dispute: Empirical lessons from Latin America. In Miller, B., & Nicholls, W. (eds), *Spaces of Contention. Spatialities and Social Movements*. London: Routledge, 121–140.

Anthias, P. (2019). Beyond territories of resistance. *Human Geography 12*(3): 66–72. doi:10.1177/194277861901200303

Anthias, P. (2021). Rethinking territory and property in indigenous land claims. *Geoforum 119*: 268–278. doi:10.1016/j.geoforum.2019.09.008

Antonsich, M. (2009). On territory, the nation-state and the crisis of the hyphen. *Progress in Human Geography 33*(6): 789–806. doi:10.1177/0309132508104996

Antonsich, M. (2011). Rethinking territory. *Progress in Human Geography 35*(3): 422–425. doi:10.1177/0309132510385619

Antonsich, M. (2017). Territory and territoriality. In *International Encyclopedia of Geography*. Chichester: John Wiley & Sons., 1–9. doi:10.1002/9781118786352.wbieg0022

Ardrey, R. (1970). *The Territorial Imperative: A Personal Inquiry into the Animal Origins of Property and Nations*. London: Collins.

Armengaud, M., Degros, A., & Radulova-Stahmer, R. (eds) (2023). *Towards Territorial Transition*. Zurich: Park Books.

Aydalot, P. (ed.) (1986). *Milieux innovateurs en Europe*. Paris: GREMI.

Banini, T. (2019). *Geografie culturali*. Milan: Franco Angeli.

Barker, A.J., & Pickerill, J. (2019). Doings with the land and sea: Decolonising geographies, indigeneity, and enacting place-agency. *Progress in Human Geography 44*(4): 640–662. doi:10.1177/0309132519839863

Binz, C., Coenen, L., Murphy, J.T., & Truffer, B. (2020). Geographies of transition—from topical concerns to theoretical engagement: A comment on the transitions research agenda. *Environmental Innovation and Societal Transitions 34*: 1–3. doi:10.1016/j.eist.2019.11.002

Blomley, N. (2016). The territory of property. *Progress in Human Geography 40*(5): 593–609. doi:10.1177/0309132515596380

Bonnemaison, J. (1981). Voyage autour du territoire. *Espace Géographique 10*(4): 249–262. doi:10.3406/spgeo.1981.3673

Brunet, Roger, Robert Ferras, and Hervé Théry. 1993. *Les mots de la géographie: Dictionnaire critique*. Paris: Reclus. https://www.librairiecoiffard.fr/livre/654208-les-mots-de-la-geographie-dictionnaire-critiqu--roger-brunet-robert-ferras-herve-thery-reclus (accessed November 22, 2024).

Bulkeley, H. (2005). Reconfiguring environmental governance: Towards a politics of scales and networks. *Political Geography 24*(8): 875–902. doi:10.1016/j.polgeo.2005.07.002

Campagne, P., & Pecqueur, B. (2014). *Le Développement territorial: Une réponse émergente à la mondialisation*. Paris: Charles Léopold Mayer.

Castree, N. (2004). Differential geographies: Place, indigenous rights and "local" resources. *Political Geography 23*(2): 133–167. doi:10.1016/j.polgeo.2003.09.010

Courlet, C. (2008). *L'Économie territoriale. L'économie en plus*. Grenoble: Presses universitaires de Grenoble.

Courlet, C., & Pecqueur, B. (2013). *L'Économie territoriale*. Grenoble: PUG.

Cox, K.R. (1991). Redefining "territory". *Political Geography Quarterly 10*(1): 5–7. doi:10.1016/0260-9827(91)90023-N

Cox, K.R. (2002). *Political Geography: Territory, State, and Society*. Oxford: Blackwell.

Crevoisier, O. (2014). Beyond territorial innovation models: The pertinence of the territorial approach. *Regional Studies 48*(3): 551–561. doi:10.1080/00343404.2011.602629

Crevoisier, O., & Maillat, D. (1991). Milieu, industrial organization and territorial production system – towards a new theory of spatial development. In Camagni, R. (ed.), *Innovation Networks*. London: Belhaven Press, 13–34. https://libra.unine.ch/handle/123456789/9376 (acccessed November 22, 2024)

Daigle, M. (2016). Awawanenitakik: The spatial politics of recognition and relational geographies of indigenous self-determination: Indigenous self-determination. *The Canadian Geographer/Le Géographe Canadien 60*(2): 259–269. doi:10.1111/cag.12260

de Bernardy, M., & Debarbieux, B. (eds) (2003). *Le Territoire en sciences sociales: Approches disciplinaires et pratiques de laboratoire*. Grenoble: Publications de la MSH-Alpes.

de la Cadena, M. (2015). *Earth Beings: Ecologies of Practice across Andean Worlds*. Durham, NC: Duke University Press.

Debarbieux, B. (1995). Le lieu, fragment et symbole du territoire. *Espaces et sociétés 80A*(1): 13–36.

Debarbieux, B. (1999). Le Territoire: Histoires en deux langues. A Bilingual (His-)Story of Territory. In Chivallon, C., Ragouet, P., & Samers, M. (eds), *Discours scientifiques et contextes culturels: Géographies françaises et britanniques à líépreuve postmoderne.* Bordeaux: Maison des sciences de l'homme d'Aquitaine, 33–46.

Del Biaggio, C. (2015). Territory beyond the anglophone tradition. In Agnew, J., Mamadouh, V., Secor, A., & Sharp, J. (eds), *Wiley-Blackwell Companion to Political Geography*, 2nd edn. Wiley-Blackwell, 35–47.

Dematteis, G. (1996). *Le metafore della terra. La geografia umana tra mito e scienza*, 6th edn. Milan: Feltrinelli.

Di Méo, G. (1998). *Géographie sociale et territoires.* Paris: Nathan.

Di Méo, G. (2014). *Introduction à la géographie sociale.* Paris: Armand Colin.

Di Méo, G., & Buléon, P. (eds) (2005). *L'espace social. Lecture géographique des sociétés.* Paris: Armand Colin.

Duhem, L., & de Moura, R.P. (eds) (2020). *Design des territoires: L'enseignement de la Biorégion.* Paris: Editions Etérotopia.

Dumitru, S. (2014). Qu'est-ce que le nationalisme méthodologique? Essai de typologie. *Raisons politiques 54*(2): 9–22. doi:10.3917/rai.054.0009

Elden, S. (2010a). Land, terrain, territory. *Progress in Human Geography 34*(6): 799–817. doi:10.1177/0309132510362603

Elden, S. (2010b). Thinking territory historically. *Geopolitics 15*(4): 757–761. doi:10.1080/14650041003717517

Elden, S. (2013a). *The Birth of Territory.* Chicago, IL: University of Chicago Press.

Elden, S. (2013b). The significance of territory. *Geographica Helvetica 68*: 65–68.

Escobar, A. (2008). *Territories of Difference: Place, Movements, Life, Redes.* Durham, NC: Duke University Press.

Escobar, A. (2018). Farewell to development: An interview with Arturo Escobar. *Local Futures*, March 9, 2018. https://www.localfutures.org/farewell-to-development-an-interview-with-arturo-escobar (accessed November 22, 2024).

Fall, J.J. (2012). Reading Claude Raffestin: Pathways for a critical biography. *Environment and Planning D: Society and Space 30*(1): 173–189. doi:10.1068/d5411

Frémont, A. (1974). Recherches sur l'espace vécu. *L'Espace Géographique 3*(3): 231–238.

Friedmann, J., & Weaver, C. (1979). *Territory and Function: The Evolution of Regional Planning.* Berkeley, CA: University of California Press.

Giraut, F. (2008). Conceptualiser le territoire. *Historiens & Géographes 403*: 57–67.

Giraut, F. (2013). Territoire multisitué, complexité territoriale et postmodernité territoriale: des concepts opératoires pour rendre compte des territorialités contemporaines? *L'Espace géographique 42*(4): 293–305. doi:10.3917/eg.424.0293

Gottmann, J. (1973). *The Significance of Territory.* Charlottesville, VA: University Press of Virginia.

Gudynas, E. (2011). Buen vivir: Today's tomorrow. *Development 54*(4): 441–447. doi:10.1057/dev.2011.86

Guillot, X., & d'Emilio, L. (2020). Le Territoire au cœur de la pensée et de l'action : L'École territorialiste italienne. Le Collectif Paysages de l'Après-Pétrole (PAP). https://www.citego.org/bdf_fiche-document-2063_fr.html (accessed November 22, 2024).

Gumuchian, H., & Pecqueur, B. (2007). *La Ressource territoriale.* Paris: Economica. Anthropos.

Haesbaert, R. (2020). Do corpo-território ao território-corpo (da Terra): Contribuições decoloniais. *GEOgraphia* 22(48). doi:10.22409/GEOgraphia2020.v22i48.a43100

Haesbaert, R., & Mason-Deese, L. (2020). Territory/ies from a Latin American perspective. *Journal of Latin American Geography 19*(1): 258–268. doi:10.1353/lag.2020.0005

Halvorsen, S. (2019). Decolonising territory: Dialogues with Latin American knowledges and grassroots strategies. *Progress in Human Geography 43*(5): 790–814. doi:10.1177/0309132518777623

Hirschman, A.O. (1958). *The Strategy of Economic Development*. New Haven, CT: Yale University Press.

Jazeel, T. (2017). Mainstreaming geography's decolonial imperative. *Transactions of the Institute of British Geographers* 42(3): 334–337. doi:10.1111/tran.12200

Jessop, B., Brenner, N., & Jones, M. (2008). Theorizing sociospatial relations. *Environment and Planning D: Society and Space* 26(3): 389–401. doi:10.1068/d9107

Jonas, A.E.G. (2012). Region and place: Regionalism in question. *Progress in Human Geography 36*(2): 263–272. doi:10.1177/0309132510394118

Katz, C. (2001). On the grounds of globalization: A topography for feminist political engagement. *Globalization and Gender 26*(4): 1213–1234.

Kebir, L., & Crevoisier, O. (2007). Resources development and actors coordination: What role for innovative milieus? *International Journal of Entrepreneurship and Innovation Management* 7(2/3/4/5): 204. doi:10.1504/IJEIM.2007.012882

Kebir, L., Peyrache-Gadeau, V., Crevoisier, O., & Costa, P. (2017). Introduction: Sustainability, innovative milieus and territorial development. In Kebir, L., Crevoisier, O., Costa, P., & Peyrache-Gadeau, V. (eds), *Sustainable Innovation and Regional Development*. Cheltenham: Edward Elgar, 1–24.

Klauser, F.R. (2012). Thinking through territoriality: Introducing Claude Raffestin to anglophone sociospatial theory. *Environment and Planning D: Society and Space 30*(1): 106–120. doi:10.1068/d20711

Koop, K. (2014). Conventional or alternative development? Varying meanings and purposes of territorial rural development as a strategy for the Global South. *Geographica Helvetica* 69(4): 271–280. doi:10.5194/gh-69-271-2014

Koop, K., Landel, P.-A., & Pecqueur, B. (2010). Pourquoi croire au modèle du développement territorial au Maghreb ? Une approche critique. *EchoGéo 13*: 12065. doi:10.4000/echogeo.12065

Latour, B. (2018). *Down to Earth: Politics in the New Climatic Regime*. Cambridge: Polity.

Lopes de Souza, M. (2015). *Dos espaçaos de controle aos rerriorios dissidentes: Escritos de divulgacao cientıfica e analise poltica*. Rio de Janeiro: Consequencia.

Lopez, E., & Muchnik, J. (1997). *Petites entreprises et grands enjeux: Le Développement agroalimentaire local*. Paris: L'Harmattan.

Louis, R.P., Johnson, J.T., & Pramono, A.H. (2012). Introduction: Indigenous cartographies and counter-mapping. *Cartographica* 47(2): 77–79. doi:10.3138/carto.47.2.77

Magnaghi, A. (2020). *Il principio territoriale*. Turin: Bollati Boringhieri.

Magnaghi, A. (2022). *Le principe territoire*. Paris: Eterotopia France.

Massey, D.B. (2005). *For Space*. London: Sage.

Mignolo, W. (2011). *The Darker Side of Western Modernity: Global Futures, Decolonial Options*. Durham, NC: Duke University Press.

Minca, C. (2012). Claude Raffestin's Italian travels. *Environment and Planning D: Society and Space 30*(1): 142–158. doi:10.1068/d5011

Murphy, A.B. (2010). Identity and territory. *Geopolitics 15*: 769–772.

Murphy, A.B. (2012). Entente territorial: Sack and Raffestin on territoriality. *Environment and Planning D: Society and Space 30*(1): 159–172. doi:10.1068/d4911

Offen, K.H. (2003). The territorial turn: Making black territories in Pacific Colombia. *Journal of Latin American Geography 2*(1): 43–73.

Oslender, U. (2019). Geographies of the pluriverse: Decolonial thinking and ontological conflict on Colombia's Pacific coast. *Annals of the American Association of Geographers 109*(6): 1691–1705. doi:10.1080/24694452.2019.1572491

Paasi, A. (2009). Bounded spaces in a "borderless world": Border studies, power and the anatomy of territory. *Journal of Power 2*: 213–234.

Pachoud, C., Koop, K., & George, E. (2022). Societal transformation through the prism of the concept of territoire: A French contribution. *Environmental Innovation and Societal Transitions 45*: 101–113. doi:10.1016/j.eist.2022.10.001

Painter, J. (2009). Territoire et réseau: Une fausse dichotomie? In Vanier, M. (ed.), *Territoires, Territorialité, Territorialisation. Controverses et Perspectives*. Rennes: Presses universitaires de Rennes, 57–66.

Painter, J. (2010). Rethinking territory. *Antipode 42*(5): 1090–1118. doi:10.1111/j.1467-8330.2010.00795.x

Pecqueur, B. (2005). Le Développement territorial, une nouvelle approche des processus du développement pour les économies du sud'. In Antheaume, B., & Giraut, F. (eds), *Le Territoire est mort, vive les territoires ! Une (re)fabrication au nom du développement*. Marseille: IRD Éditions, 295–316.

Pecqueur, B. (2006). Le tournant territorial de l'économie globale. *Espaces et sociétés 124–125*(1–2): 17–32. doi:10.3917/esp.124.0017

Pecqueur, B. (2008). Territorial dynamics: Towards a new model of development facing globalization. In Querejeta, M.J.A., Landart, C.I., & Wilson, J.R. (eds), *Networks, Governance and Economic Development. Bridging Disciplinary Frontiers*. Cheltenham: Edward Elgar, 30–39.

Pierce, J., Martin, D.G., & Murphy, J.T. (2011). Relational place-making: The networked politics of place. *Transactions of the Institute of British Geographers 36*(1): 54–70. doi:10.1111/j.1475-5661.2010.00411.x

Porter, M.E. (1998). Clusters and the new economics of competition. *Harvard Business Review 76*(6): 77–90.

Porto-Gonçalves, C.W. (2002). Da geografia às geo-grafias um mundo em busca de novas territorialidades. *Cuadernos de Trabajo, Instituto de Investigaciones Histórico-Sociales, Universidad Veracruzana 10*: 1–50.

Porto-Gonçalves, C.W. (2009a). De saberes e de territorios: Diversidad y emancipación a partir de la experiencia latino-americana. *Polis 8*(22): 121–136.

Porto-Gonçalves, C.W. (2009b). *Territorialidades y lucha por el territorio en América Latina: geografía de los movimientos sociales en América Latina*. Caracas: Ed. IVIC.

Radcliffe, S.A. (2017). Geography and indigeneity I: Indigeneity, coloniality and knowledge. *Progress in Human Geography 41*(2): 220–229. doi:10.1177/0309132515612952

Raffestin, C. (1980). *Pour une géographie du pouvoir*. Paris: Librairies Techniques.

Raffestin, C. (1982). Remarques sur les notions d'espace, de territoire et de territorialité. *Espaces et sociétés 41*: 167–171.

Raffestin, C. (1986a). Ecogenèse territoriale et territorialité. In Auriac, F., & Brunet, R. (eds), *Espaces, jeux et enjeux*. Paris: Fayard, 173–185.

Raffestin, C. (1986b). Territorialité: Concept ou paradigme de la géographie sociale? *Geographica Helvetica 41*(2): 91–96.

Reyes, A., & Kaufman, M. (2011). Sovereignty, indigeneity, territory: Zapatista autonomy and the new practices of decolonization. *South Atlantic Quarterly 110*(2): 505–525.

Sack, R.D. (1986). *Human Territoriality: Its Theory and History*. Cambridge: CUP Archive.

Sandoval, L., María, F., Robertsdotter, A., & Paredes, M. (2017). Space, power, and locality: The contemporary use of territorio in Latin American geography. *Journal of Latin American Geography 16*(1): 43–67.

Saquet, M.A. (2018). A perspective of counter-hegemonic analysis and territorial transformation. *Geographica Helvetica 73*(4): 347–355. doi:10.5194/gh-73-347-2018

Schneider, S. (2009). Território, ruralidade e desenvolvimento. In Velasquez, F.L., & Medina, J.G.F. (eds), *Las configuraciones de los territorios rurales en el siglo XXI*. Bogotá: Editorial Pontifícia Universidad Javeriana, 67–108.

Scott, A., & Storper, M. (2003). Regions, globalization, development. *Regional Studies 37*(6–7): 579–593. doi:10.1080/0034340032000108697a

Soja, E.W. (1971). *The Political Organization of Space*. Washington, DC: Association of American Geographers.

Stöhr, W.B., & Frazer Taylor, D.R. (1981). *Development from Above or Below? The Dialectics of Regional Planning in Developing Countries*. Chichester: John Wiley & Sons.

Théry, H. (2018). Les sciences humaines et sociales dans les relations scientifiques France-Brésil. *Histoire de la recherche contemporaine VII 2*: 156–167. doi:10.4000/hrc.2362

Torre, A. (2015). Théorie du développement territorial. *Géographie, économie, société 17*(3): 273–288. doi:10.3166/ges.17.273-288

Torre, A., & Beuret, J.-E. (2012). *Proximités territoriales. Economica*. https://hal.science/hal-01197970 (accessed November 22, 2024)

Turco, A. (2007). 'Territorio e territorialità'. In *Enciclopedia Treccani*. Rome: Istituto della Enciclopedia Italiana. https://www.treccani.it/enciclopedia/territorio-e-territorialita_(Enciclopedia-Italiana) (accessed November 22, 2024).

Vanier, M. (ed.) (2009). *Territoires, territorialité, territorialisation: Controverses et perspectives*. Rennes: Presses universitaires de Rennes.

Woon, C.Y. (2019). Translating territory, politics and governance. *Territory, Politics, Governance 7*(2): 115–121. doi:10.1080/21622671.2019.1597558

Zibechi, R. (2012). *Territories in Resistance: A Cartography of Latin American Social Movements*. Oakland, CA: AK Press.

Chapter 10

Scale

Andrew E.G. Jonas

School of Environmental Sciences, University of Hull, Hull, UK

Scale is a powerful construct, which in political geography can be deployed to convey the idea that sociospatial patterns and political processes not only vary from place to place but also are hierarchically structured. At its most basic level, scale is a device used by cartographers to represent the relative distance between real objects and locations on a map. Some maps strive to represent spatial patterns at the global scale, while others seek to represent them at smaller, for example local or regional, scales. In political geography, scale is often used as a shorthand descriptor of political processes that have some sort of spatial reach. Neighborhood activism, insurgent regionalism, civic nationalism, and global climate change politics are examples of how certain political processes of interest to political geographers can assume scalar-specific characteristics.

Many of the societal organizations studied by political geographers, such as supranational institutions, national political parties, city-regions, and regional special-purpose districts, have powers and capacities that correspond more or less to state-defined territorial scales. For its part, the state is organized around a territorial structure comprising administrative powers and fiscal functions, which are assigned to different scales in such a way as to enable the state to manage more effectively spatial interests in society (Cox 1990). Here the language of scale helps to differentiate between the territorial structure of the state *tout court* (i.e., the national state or nation-state) and the allocation of specific powers and resources to the state's respective territorial-administrative scales (national, regional, metropolitan, or local) such that the latter become foci for political lobbying on behalf of different spatial interests in society (Cox & Jonas 1993).

The Wiley Blackwell Companion to Political Geography, Second Edition.
Edited by Virginie Mamadouh, Natalie Koch, Chih Yuan Woon, and John Agnew.

The concept of scale in geography – for example, the local scale of the state – is not directly equivalent to the political science concept of level – as, for example, in the study of multilevel governance processes operating in the European Union (see Marks 1993). Nonetheless, these two concepts are often used interchangeably. Terms such as "politics of scale" (Smith 1992; Swyngedouw 1997) or "scale politics" (Jonas 1994) are frequently used by political geographers to demonstrate that the resulting struggles not only occur around extant state territories and scales but might also generate conditions for the creation of new scales of political organization and governance, many of which do not coincide neatly with the corresponding territorial structures of the state (Cox 2002).

Notwithstanding the multiple locations and scales at which societal interests and political struggles occur, the national scale is often foregrounded in political geographical studies. Yet a more careful analysis can expose hitherto taken-for-granted scalar interdependences underpinning political struggles fought at the national scale. In most democratic states, national elections are conducted under rules that reflect how the national government has internalized taken-for-granted assumptions about how its sovereign territory can, or should, be governed (Jonas 2002). In presidential elections in the United States, for example, the Electoral College serves as a national mechanism for reassigning electoral votes cast for each candidate in proportion to the population of the 50 states in the Union. The winner is usually the candidate who garners the most Electoral College votes rather than the one who wins the popular national vote. Assuming the successful candidate has abided by federal and state election rules, the result of a national election ideally reflects not just the democratic will of the American nation viewed as a unified political territory but also the balance of power between the federal government, the states, and the people as recognized in the US Constitution.

The mantra that "all politics is local" has been attributed to the late Tip O'Neill, who served as the 47th speaker of the US House of Representatives from 1977 to 1987. In using this scalar language, O'Neill was referring to the fact that national lawmakers in the United States are primarily beholden to economic interests in their local constituencies, and therefore are not predisposed to vote in line with national party political interests and ideals. Political scientists counter this claim with the argument that over time US politics has "nationalized" as voting patterns at the state and local levels have become more closely aligned with those at the national level (Hopkins 2018). However, such interpretations are based on the flawed assumption that political issues can be preassigned to the national level, with little or no supporting evidence that place and locality do not make a difference to how voters express their views on such issues (see Agnew 1987).

By way of contrast, the prime minister of the United Kingdom is usually appointed to represent the political party that either wins a majority of *local* seats in the UK parliament or is able to form a winning *national* coalition with another political party. Scotland and Wales are devolved administrative territories within the United Kingdom and have local representatives in the UK parliament. In a national referendum held in Scotland in September 2014, Scottish voters were asked: "Should Scotland be an independent country?" The referendum result was 44.7% in favor of independence and 55.5% not in favor, with a turnout of 84.6% of the eligible voting population. If the outcome of the vote for independence had been a majority in favor, voters in Scotland would in future have voted for representatives in the national parliament of Scotland and not that of the larger-scale UK territory. The referendum outcome exposed significant urban and regional political differences across Scotland: whereas some of the most deprived local authorities in major cities, such as Glasgow and Dundee, returned majority "Yes" votes, "No" votes were concentrated in the largely rural lowlands and border regions of Scotland. In these examples, the national scale is both a material construct and also an ideal, which reflects how the state imagines that its constituent territories (national, local, etc.) should be democratically governed.

The purpose of this chapter is to examine the variety of ways in which scale has been used as a descriptive device, a narrative framework, and, most crucially, a conceptual approach in political geography. In the first section, I examine how political geographers initially sought to describe and map political spaces at different scales. I show how scale became a narrative device for representing the political processes at work shaping territory. I then demonstrate how political geographers have harnessed scale as a theoretical construct. While some political geographers are interested in how and why territorial structures of the state are organized and contested at different scales, others seek to examine nonstate scales and associated forms of contentious politics, many of which challenge received ideas of how state territories and scales are organized. Occasionally there have been disagreements about the intellectual merits of deploying scale as a theoretical lens on sociopolitical processes, and these – along with various attempts at their resolution – will be explored in the later sections. In the conclusion, I suggest that, despite such disagreements, political geographers find scale, alongside concepts of place, territory, and network, to be a useful way of representing and explaining complex and constantly unfolding geographies of political, social, environmental, and economic change.

Cartographic and Quantitative Approaches to Scale in Political Geography

In geography, scale has traditionally conveyed at least two quite distinct ideas about how space is, or can be, represented (for details, see Herod 2011). In the first of these – what Andrew Herod (2011) calls the idealist approach – scale can be a descriptive device or measurement tool deployed, for instance, in cartography to represent real distances on a map. Most maps have a standardized scale that allows the reader to convert map distance into a precise measurement of the distance between actual objects located in space. In this respect, maps try to represent real scales, which in the case of human activities could conceivably range in size from an individual street or neighborhood to the entire planet or globe.

A cursory inspection of some classic political geography textbooks and monographs would reveal that maps have been valuable tools for representing political space at a variety of spatial scales, ranging from the international and national to the regional and urban. For instance, geopolitical maps have sought to represent how the international scale can be divided into "geostrategic regions" (Cohen 2003). Political geographers in addition have mapped political spaces at progressively smaller scales, such as civil divisions inside countries or incorporated and unincorporated parts of a metropolitan area (Glassner & de Blij 1989). Finally, there has been attention paid to the ways in which map scales can be manipulated in order to distort and misrepresent political power (Monmonier 1991). Analysis of these "propaganda maps" or "persuasive maps" can reveal much about the state of geopolitical rivalries at different time periods and in diverse spatial contexts (Boria 2008).

Relatedly, and often linked to the mapping of territories, political geographers have sought to describe how certain political processes tend to function at specific scales. Functional approaches in political geography are premised on the idea that the national scale is the basic geographical entity from which political power emanates. Richard Hartshorne (1950), a key proponent of this approach, divided territory into two sets of political relations, each covering a specific scale of state activity and territorial organization. On the one hand, Hartshorne identified the external relations of the state as those functions that shape international and interregional geopolitical processes and organizations. On the other hand, internal state relations are those functions primarily involved in structuring the territorial organization of regions and political divisions inside the sovereign territory of the state. Hartshorne, therefore, was not just interested in mapping state boundaries and borders; he was also concerned with identifying how different political processes operate around a hierarchy of territorial scales stretching across, and reaching beyond, national state territory.

Over the years, a number of key political geography texts similarly have used scale as a narrative device to represent political processes and problems occurring at discrete scales, such as mapping geographies of welfare distribution within the state against locational conflicts at, respectively, international, national, and urban scales (Cox 1979) or identifying systemic features of geopolitics operating in the world economy, around the nation-state, and within place and locality (Flint & Taylor 2011). A key characteristic of idealist approaches to scale is a tendency to organize knowledge about the world into relatively self-contained scalar categories, thereby obscuring how material processes can in fact span and interconnect the different scales.

The rise of quantitative methods in geography in the 1960s and 1970s – so-called spatial science – inspired further developments in how scale could be used not just to measure and describe spatial patterns but also to explain the spatial dimensions of political processes. This development has, in turn, underpinned the second way of thinking about scale – as a means of theorizing how material processes operate across space (Herod 2011). In the field of electoral geography, for example, there has been considerable interest in why the outcomes of regional and local elections tend to differ from spatial patterns of voting observed at the national level. Pioneering studies conducted by Kevin Cox (1968, 1970) sought to explain such differences by identifying spatial associations between different social and political variables. Cox discovered that regional and local voting outcomes seemed to be influenced by the spatial proximity of voters rather more than by nonspatial factors such as social class or party affiliation, prompting him to propose scalar concepts such as the "neighborhood effect" (Cox 1972) and "neighborhood activism" (Cox & McCarthy 1982) in order to explain different forms of local political behavior.

Cox (1973) further showed that spatial distributional inequalities and their associated political problems, such as the metropolitan fiscal disparity problem in the United States, seem to exhibit scalar-dependent attributes. He argued that metropolitan disparities could be systematically explained by inter alia the presence of exclusionary politics arising from the defense of *local* material interests in and around the urban living place (e.g., local property values). The findings of quantitative political geographical studies such as these suggest that scale can be more than a useful mapping tool. They reveal how space/scale – in this instance, the jurisdictionally fragmented political geography of cities and metropolitan areas – has certain causal attributes, albeit it has taken time for such ideas to become fully embedded within the subfield (see Johnston & Pattie 2012).

From Scale Effects to Scalar Politics

Claims and counter-claims made about the difference that place, space, and scale make to political behavior have prompted occasional critical exchanges between geographers and political scientists, with the latter seemingly unmoved by the former's theoretical assertions regarding "the power of place"; see, for example, discussions in Agnew (1987, 1996) and McAllister (1987). Yet within political geography itself there was surprisingly little discussion about the ways in which the analysis of political processes at different territorial scales might be melded into a unified approach to, or theory of, scale; not, that is, until an important intervention by Peter Taylor, who drew on world systems theory in order to outline what he called a "materialist framework for political geography" (Taylor 1982). In effect, Taylor was making a case for a materialist theory of scale. He argued that hitherto taken-for-granted territorial-*cum*-political scales of analysis – namely, the world economy, the nation-state, and the locality – represent different scalar political constructions of space, which he described as, respectively, the "scale of reality," the "scale of ideology," and the "scale of experience" (Taylor 1982: 25–31). For Taylor, the "national" does not exist at a discrete level between the global and the local. Instead, it is actively brought into being through geopolitical struggles and negotiations such as those that eventually led to the consolidation of smaller territories (kingdoms) into larger nation-states and the Treaty of Westphalia in 1648, which seemingly fixed the idea of sovereignty at the national scale. In Taylor's view, a materialist approach to scale is absolutely central to the development of political geographical theory.

Taylor's materialist political economy of scale inspired other political geographers to be much more explicit in showing how scalar concepts could be used to animate studies of the social and political construction of territory. For example, Neil Smith (1984) showed how the global space of capitalism is internally differentiated into territories that correspond more or less with the international, national, and urban scales. Each scale, in turn, expresses various spatial contradictions and tensions around uneven capitalist development and demands socio-regulatory solutions over the short to medium term in the form of state interventions at each scale. Smith (1993) subsequently refined his argument to address other (nonstate) scales including the human body and attendant struggles around spatial identity and difference.

Since scales are socially produced and politically contested, they reflect how power is distributed across society seemingly in a hierarchical fashion. Many political processes do appear to occupy fixed scalar categories that match the corresponding hierarchical levels of the state, implying that higher scales dominate, or even determine, what happens at lower scales. Harriet Bulkeley (2005) suggests, however, that power does not necessarily cascade downward in a unidirectional manner through the state territorial hierarchy. Instead, political processes can operate in both directions. It is further recognized that in an age of globalization state territory does not coincide neatly with the territorial reach of many societal institutions and political processes, thereby posing challenges to the national state's ability to regulate the economy and society. The resulting "hollowing out" of the state represents a profound socio-regulatory dilemma in the current post-Keynesian era of global capitalism (Jessop 2016). It has engendered a variety of scalar political processes and policy outcomes.

Rescaling the State

The concept of state rescaling describes how state powers, once seemingly fixed at the national scale at the height of Fordism-Keynesianism during the mid-twentieth century, have shifted up, down, and across the state territorial hierarchy and come to reside at scales above and below the national (Brenner 2004). State rescaling, in turn, has empowered nonstate actors operating at and around other (i.e., nonnational state) territorial scales. For example, in many US states business-led "multi-city growth coalitions" (Wachsmuth 2017: 47) have coalesced around new regional spaces of planning and governance in order to attract globally mobile investment. Such spaces occupy larger territorial scales than existing local and metropolitan jurisdictional arrangements.

Notwithstanding their putative global reach, new regionalist scales of planning and economic development at the same time struggle to manage entrenched fiscal disparities and territorial injustices operating within state territories (Jonas & Pincetl 2006). In Europe, for instance, state rescaling has encouraged radical, populist, and redistributive forms of regionalism to flourish alongside, and often in opposition to, the new economic regionalism propagated by supranational organizations including the European Commission (Swyngedouw 2000; Jones & MacLeod 2004). Accordingly, various insurgent regionalist movements, along with claims for citizenship, social justice, and democratic rights, have colonized emerging city-regional spaces, especially in the devolved regions and territories of Europe such as Catalunya and the Basque region of Spain (Calzada 2020).

Given its emphasis on how state powers shift upward and downward between different vertical levels of the state, the state rescaling literature often fails to demonstrate how different spatial interests in society are (re)positioned in relation to such rescaling processes (Cox 2009). Studies in non-Western economies, such as South Korea (Park 2008) and China (Wu 2016), demonstrate not only that the politics of state rescaling are nationally variegated but also that counter-geopolitical tendencies are at work. In China, for example, state rescaling around city-regionalism is not a unidirectional process of "downscaling" state powers in response to globalization and interurban competition. Instead, as Fulong Wu (2016) argues, the Chinese national state has selectively "up-scaled" powers pertaining to territorial administration (e.g., annexation), spatial planning, and collaborative governance to new city-regional arrangements in order to orchestrate the internationalization of national state territory and economy. Extending Wu's analysis, Li and Jonas (2019) argue that the Chinese approach to city-regionalism exposes countervailing geopolitical processes operating both within and beyond the country's national borders.

National differences in the spatial politics underpinning processes of state scalar restructuring prompt John Harrison (2008) to draw a theoretical distinction between "centrally orchestrated" (primarily business-led) regionalism and "regionally orchestrated" (mainly state-led) centralism. Across many territories and regions, business groups, public authorities, and civic society organizations are eschewing national politics and entering into new political alliances and collaborative networks, which do not map neatly onto the extant territories and scales of state and governance. In the process, political struggles in, across, and around state territory have become more spatially diffuse, multiscalar, and even site specific. To put this another way, political struggles around state functions and capacities are no longer – if they ever were – permanently fixed at the national scale.

One concrete manifestation of the spatially diffuse and relational properties of scalar politics and policy-making is the recent rise of "place-based policies" (Neumark & Simpson 2015). Place-based policy-making refers to geographically targeted efforts undertaken by the state to enhance the productivity and economic performance of local areas and encourage local businesses and communities to forge new local area-based collaborative partnerships, which are designed to generate better local job opportunities and higher wages. Such policies have proliferated in countries where local places and people have been "left behind" in an economic sense by uneven development and attendant forces of urban and regional agglomeration.

Viewed at a concrete level, the proliferation of place-based policies is a further sign that the authority of the national state has not so much been hollowed out by globalization as strategically and selectively rescaled (see Jones 1997), giving the appearance that political power has been devolved to local places (and people). When viewed at a more abstract level, however, the fluid and relational properties of places and their constituent networks have prompted some political geographers to question whether scale remains a useful conceptual device for explaining societal forces and political processes that appear to challenge and even undermine received state-centric understandings of territory and spatial policy-making.

Relational Scale: Unbounding Place and Politics

Ontologically, scales are not fixed or pre-given material entities that somehow exist apart from their necessary social relations and spatial contexts (Marston 2000; Herod & Wright 2002). Instead, scales are always socially constructed and politically contested in, around, and through "real-and-imagined" (Soja 1989) spaces, places, and territories. In rejecting overly state-centric renditions of scale and territory, scholars have sought to represent political processes as relational, networked, and topological; that is to say, scale is in an ontological sense "flat" (Marston, Jones, & Woodward 2005). In lieu of scale, Sallie Marston, John Paul Jones III, and Keith Woodward (2005) place the site at the center of their alternative ontology, wherein a site constitutes a material localization of diverse and often quite distant social relations through which one site is connected to others across space (see also Marston et al. 2007). In their view, scalar thinking (as in, e.g., the idea of state rescaling) relies heavily on a ladder-like vertical hierarchy of preregistered scales extending upward from the local to the global and vice versa. Marston et al. (2005) propose instead a site ontology to examine nonbounded places and locales of situated practices (and associated discourses) that stretch and circulate across space through all kinds of more or less spatially extensive networks. They conclude that, so long as scale as material object remains indiscernible from scale as analytical construct in geographical analyses of social and political struggles, human geography would be better off without scale altogether.

Marston et al.'s rejection of scale in preference for a site ontology triggered an extensive debate, as featured in several issues of the international journal *Transactions of the Institute of British Geographers*. Among the contributions to the debate, Helga Leitner and Byron Miller (2007) consider the kinds of social practices underpinning the political construction of scale. For Leitner and Miller, political resistance is always multiscalar in form and composed of many different practices and spatialities – those of territory, spatial connectivity, network, place, space, and scale. They take issue with Marston et al. for dismissing the role of agency in underpinning scalar political struggles and for eliding scale with vertical hierarchies. Leitner and Miller further suggest that many political geographers working with the concept of scale do not assume a priori a world composed exclusively of vertical hierarchies and, moreover, even among those who do believe that society and the state are organized into hierarchies, many do not presume that the global subjugates the local in ways suggested by Marston et al.

In their response, John Paul Jones III, Keith Woodward, and Sallie Marston (2007) accuse Leitner and Miller of conflating site and scale without properly identifying what practices are involved, around which sites, and at what scales or levels. Instead, they find greater affinity with Escobar's (2007) affirmation that site ontology provides an empowering alternative to "established, state-centric, capitilocentric and globalcentric thinking, with their emphasis on 'larger forces', hierarchies, determination and rigid structures" (Escobar 2007: 109, cited in Jones, Woodward, & Marston, 2007: 272). Although Escobar is enthused by the political potential of site ontology, he is reluctant to throw out the notion of scale altogether. In a similar vein, but instead drawing on Michel Foucault's extensive writings on biopolitics and disciplinary power, Nancy Ettlinger (2011) argues that scale-sensitive analysis can illuminate the mobilization of power (especially in regard to its targets) and in so doing decouple knowledge of political practices from preassigned scalar categories pertaining to globalization and national state authority. Revisiting some of the ontological questions raised in the scale debate, Joe Blakey (2021) draws on Jacques Rancière's political philosophy to urge geographers to take existing "common-sense" ideas of scale seriously while remaining sensitive to emergent politics and its diverse spatial forms.

The analysis of actor-network forms of political agency remains a particular feature of topological approaches to spatial politics, where scale – if the term continues to be deployed as such – refers not so much to pre-registered vertical hierarchies – especially state territorial hierarchies – as to emergent horizontal relationships among social actors distributed across the landscape at different sites, be these households, neighborhoods, corporate boardrooms, public buildings and squares, or suchlike (see Marston 2000; Latham 2002). Topological or relational thinking further recognizes that powers exercised at particular sites within a given territory (e.g., city or region) can extend far beyond the formal jurisdictional limits of the territory in question. For example, if many cities and regions have well-defined jurisdictional limits, knowledge of urban and regional political capacities cannot be restricted to mapping conditions *within* such limits – a notable weakness of the now largely discredited traditional regional geography approach to territory (Painter 2008). Instead, regions exist in and through social relations and networks extending well beyond their formal (jurisdictional) boundaries into the global economy at large (Allen & Cochrane 2007; Prytherch 2010). Approaching the region as a discrete analytical site containing disparate practices and/or as a material scale of political organization means not only thinking territorially (e.g., focusing on the regional allocation of powers inside the state) but also relationally (e.g., examining the wider international networks and divisions of labor of which the region is a part) (Jonas 2006).

Scale as a Lens on Territory, Place, and Network

While it is a significant advance on overly state-centric and essentialist renditions of scale, topological thinking about spatiality can nonetheless obscure our view of societal processes and political struggles that can assume multiple spatial forms at the same time. Indeed, Jessop, Brenner, and Jones (2008) go further and argue that researchers often privilege one spatial form (e.g., scale) at the expense of others (e.g., place, networks, and/or territory). Yet,

> [e]ach falls into the trap of conflating a part (territory, place, scale, or networks) with the whole (the totality of sociospatial organization), whether due to conceptual imprecision, an overly narrow analytical focus, or the embrace of an untenable ontological (quasi-)reductionism. This trap is notoriously present in methodological territorialism, which subsumes all aspects of socio-spatial relations under the rubric of territoriality. This is manifested, for instance, in state-centric approaches to globalization studies and in narrowly territorialist understandings of cities, states, and the world economy. (Jessop, Brenner, & Jones 2008: 392)

The accusation of "methodological territorialism" can be directed at not just political geographic studies of the nation-state and nationalism (Koch 2020) but also recent work by economists and regional scientists that privileges globalization as the ultimate scalar lens for analyzing contemporary processes of state hollowing out and rescaling (Moisio et al. 2020). For Sami Moisio (2011), the rise of new scales and territorial formations, such as "global city-regions" and "smart cities," does not so much reflect the emergence of network-like topologies constructed by and around nonstate political actors at the expense of the authority of the state; instead, it discloses – albeit in very subtle ways – how, in a thoroughly globalized knowledge-intensive capitalist economy, the national state is reimagining and reorganizing (its) territory both externally as well as internally in order to achieve wider geopolitical objectives of enhanced international visibility and competitiveness. For Jessop, Brenner, and Jones (2008), the answer to the challenge of methodological territorialism lies not in the outright rejection of state-centric thinking about scale but rather in using scale alongside concepts like territory, network, and place better to represent how and in what form(s) (geo)political processes unfold in and through space.

Indeed, the study of scalar politics has prompted the development of new concepts for showing how place-based political interests and practices map onto scales – including but not limited to state territorial scales – within, beyond, and above the "local." Take, for example, how scale augments our understanding of how different factions of society might mobilize around, or sometimes against, place-based policies implemented by the state. For political geographers, place (or locality) refers to a local territory – be this a neighborhood, a town, or a city – that has both meaning and material significance for local people and businesses (Cox & Mair 1988). At the same time, however, place is relational insofar as local people and businesses have social and material connections and attachments that extend far beyond the jurisdictional limits of the locality. For political geographers, the simultaneously territorial and relational character of place poses a theoretical, if not also a practical, conundrum, insofar as place-based politics and policies can undermine desired societal outcomes pursued at scales beyond the local.

In attempting to resolve this theoretical conundrum, Kevin Cox (1998) draws a conceptual distinction between "spaces of dependence" and "scales of engagement." On the one hand, the concept of "spaces of dependence" is a reference to the various forms of social and material-*cum*-economic attachment to place – be these the local interests of private firms, of local branches of the state, or of people. "Scales of engagement," in comparison, are the territorial structures and institutional arrangements that place-based interests seek to harness in order to reproduce their "spaces of dependence." In positing this distinction, Cox's intention is to show that place-based political struggles often occur around scales that are not necessarily commensurate with policy scales organized or orchestrated by the state. Indeed, where commensurate state scales do not exist, then actors who are materially and socially dependent at those (or, for that matter, other) scales will organize politically to ensure that appropriate scales are constructed. In a response to Cox, Dennis Judd offers a note of caution, suggesting that "[w]hen scales are absent, important political consequences ensue—and political agents will typically find it difficult and sometimes impossible to replace the scales that have not been constructed by the state" (Judd 1998: 30). Nonetheless, Cox's particular contribution to political geographical approaches to scale demonstrates that scalar politics are often fought around the creation of new territorial divisions of responsibility and power inside the state, which can differ from existing territorial structures and scales of the state. Moreover, it shows how locally based political organizations often draw down powers and resources residing outside the local political context as determined and delimited by the state.

A special issue of *Political Geography* edited by David Delaney and Helga Leitner (1997) likewise nudges analysis of the politics of scale in new directions by examining how nonstate political organizations and social movements strategically mobilize around scale and the particular practices and ideas involved in so doing. Notably, Delaney and Leitner (1997) foreground contentious political struggles around, and idealized representations of, scale rather than simply privileging its economic determinants and the formal state spaces in which struggles unfold. Further extending such insights, Hilda Kurtz (2002) argues that the state's use of scale as a regulatory tool can often discipline and exclude certain place-based societal groupings, thereby contributing to environmental and social injustices. Kurtz argues that environmental organizations can challenge such injustices by framing political action at different scales, thereby enabling them to circumnavigate scales of legislative and policy intervention orchestrated by the state. Novel perspectives on the relationships between scale and contentious politics likewise feature in political geographical research on labor struggles (Gough 2010), Indigenous rights (Silvern 1999), peace movements (Miller 1994), community gardening (Smith & Kurtz, 2003), sustainability governance (Kythreotis & Jonas 2012), and climate change (Bulkeley 2005).

Given that political ecologies of nature and society often diverge at urban, regional, national, and planetary scales (Keil 2020), recent analyses of political activism against global climate change take into account how such divergences shape negotiations around international climate policy. On the one hand, climate activists have often adopted site-based protests and organize in and around local-to-global networks in order to challenge national and international climate agreements (North 2011). On the other, nonstate political actors (e.g., business and civic groups as well as climate activists) have worked around and through existing state scales (i.e., with national and subnational state policy-makers) to generate climate adaptation policies at the city-region scale, which have subsequently shaped international climate agreements (Kythreotis et al. 2023). Such scalar-attuned analyses of climate activism and the geopolitics of climate adaptation exemplify a growing appreciation in political geography of how national and international political struggles no longer occur unmediated by political agency operating at and around sites and practices situated at other scales.

Conclusions

This chapter has explored some of the contrasting ways in which political geographers deploy scale in order to animate and analyze political processes. Scale is a way of showing how political struggles in and around territory – especially but not exclusively national state territory – are often organized into spatial hierarchies, ranging from the international and national to the regional, local, and even the household and body. It has also become a useful concept for revealing the spatial dynamics of all sorts of political struggles around and against the state, especially those involving nonstate political actors. In the future we might well see political geographers using concepts like territory, place, network, and scale interchangeably and perhaps even together. Part of this is due to the challenge of adequately representing multifaceted forms of political organization in an increasingly interconnected yet profoundly divided world. It also reflects some challenges and ambiguities associated with methodological territorialism. Notably, the gradual erosion of state power and sovereignty by economic globalization and local citizen protests has demanded new scalar concepts and categories with which to examine new and emerging spaces of politics. The concept of scale certainly helps to cement the idea that such spaces can be differentiated hierarchically.

But the concept itself can be augmented with corresponding notions of place, territory, and network.

Indeed, an intellectual-*cum*-political gap has opened up for political geographers to fill by deploying the language of scale to describe the geographically variegated forms and processes of political action. Strategic use of the language of scale – what Alistair Fraser (2010) describes vividly as "scalecraft" – provides a means by which the politically marginalized and oppressed groups in society can (re)frame collective action around organizational capacities, which empower political actions in, around, and sometimes against an oppressive state as well as the extractive and exploitative practices of capital. When one scale is effectively closed off against political action, another may quickly emerge as a new strategic terrain of struggle and resistance. In these respects, no given scale should be viewed as a fixed arena for political action and activism; instead, scale is always politically (re)constituted in and through the dynamics of societal struggles (Swyngedouw 1997). Thanks to the development of a far more nuanced language of scale, alongside refinements arising from foundational questions of scalar ontology and epistemology (Jones 1998), political geographers today are arguably in a much stronger position than their predecessors when it comes to demonstrating how spatial formations and political processes at one geographical scale are reproduced, come into conflict with, and/or are transformed by those operating at other scales.

Notwithstanding progress toward appreciating the explanatory power of scale, political geographers are still far from achieving a consensus about the difference scale makes to political processes. For some, the concept of scale has arguably superseded that of territory, especially when explaining how nonstate space is politically organized. For others, the close association of scale with state territory seems to create problems of representing the progressive potential of nonstatist political struggles. Nevertheless, scale remains useful for shedding light on the production and contestation of territorial structures both within the state as well as beyond its territorial reach. While this demands of its proponents a high degree of critical reflexivity in how scale is represented, a convincing and sustained case that scale is not a useful concept in political geography has yet to be made.

Acknowledgments

I would like to thank the editors and Andrew Kythreotis for some very helpful comments on an earlier version.

References

Agnew, J.A. (1987). *Place and Politics: The Geographical Mediation of State and Society*. London: Allen and Unwin.

Agnew, J.A. (1996). Mapping politics: How context counts in electoral geography. *Political Geography* 15: 129–146.

Allen, J., & Cochrane, A. (2007). Beyond the territorial fix: Regional assemblages, politics and power. *Regional Studies 41*: 1161–1175.

Blakey, J. (2021). The politics of scale through Rancière. *Progress in Human Geography 45*: 623–640.

Boria, E. (2008). Geopolitical maps: A sketch history of a neglected trend in cartography. *Geopolitics 13*: 278–308.

Brenner, N. (2004). *New State Spaces: Urban Governance and the Rescaling of Statehood.* New York: Oxford University Press.

Bulkeley, H. (2005). Reconfiguring environmental governance: Towards a politics of scales and networks. *Political Geography 24*: 875–902.

Calzada, I. (2020). Emerging citizenship regimes and rescaling (European) nation-states: Algorithmic, liquid, metropolitan and stateless citizenship ideal types. In Moisio, S., Koch, N., Jonas, A.E.G., Lizotte, C., & Luukkonen, J. (eds), *Handbook on Changing Geographies of the State: New Spaces of Geopolitics*. Cheltenham: Edward Elgar, 368–384.

Cohen, S.B. (2003). *Geopolitics of the World System*. Lanham, MD: Rowman and Littlefield.

Cox, K.R. (1968). Suburbia and voting behavior in the London metropolitan area. *Annals of the Association of American Geographers 58*: 111–127.

Cox, K.R. (1970). Geography, social contexts and voting behavior in Wales, 1861–1951. In Allardt, E., & Rokkan, S. (eds), *Mass Politics: Studies in Political Sociology*. New York: Free Press, 117–159.

Cox, K.R. (1972). The neighborhood effect in urban voting response surfaces. In Sweet, D.C. (ed.), *Models of Urban Structure*. Lexington, MA: Lexington Books, 159–176.

Cox, K.R. (1973). *Conflict, Power and Politics in the City: A Geographic View*. New York: McGraw-Hill.

Cox, K.R. (1979). *Location and Public Problems: A Political Geography of the Contemporary World*. Chicago, IL: Maaroufa Press.

Cox, K.R. (1990). Territorial structures of the state: Some conceptual issues. *Tijdschrift voor Economische en Sociale Geografie 81*: 251–266.

Cox, K.R. (1998). Spaces of dependence, scales of engagement and the politics of scale, or: Looking for local politics. *Political Geography 17*: 1–23.

Cox, K.R. (2002). "Globalization", the "regulation approach" and the politics of scale. In Herod, A., & Wright, M. (eds), *Geographies of Power: Placing Scale*. Oxford: Blackwell, 85–114.

Cox, K.R. (2009). "Rescaling the state" in question. *Cambridge Journal of Regions, Economy and Society 2*: 107–121.

Cox, K.R., & Jonas, A.E.G. (1993). Urban development, collective consumption and the politics of metropolitan fragmentation. *Political Geography 12*: 8–37.

Cox, K.R., & Mair, M. (1988). Locality and community in the politics of local economic development. *Annals of the Association of American Geographers 78*: 307–325.

Cox, K.R., & McCarthy, J.J. (1982). Neighbourhood activism as a politics of turf: A critical analysis. In Cox, K.R., & Johnston, R.J. (eds), *Conflict, Politics and the Urban Scene*. London: Longmans, 196–219.

Delaney, D., & Leitner, H. (1997). The political construction of scale. *Political Geography 16*: 93–97.

Escobar, A. (2007). The "ontological turn" in social theory. A commentary on "Human Geography without Scale", by Sallie Marston, John Paul Jones III and Keith Woodward. *Transactions of the Institute of British Geographers 2*: 106–111.

Ettlinger, N. (2011). Governmentality as epistemology. *Annals of the Association of American Geographers 101*: 537–560.

Flint, C., & Taylor, P.J. (2011). *Political Geography: World-Economy, Nation-State and Locality*, 6th edn. Harlow: Pearson Prentice Hall.

Fraser, A. (2010). The craft of scalar practices. *Environment and Planning A 42*: 332–346.

Glassner, M.I., & de Blij, H.J. (1989). *Systematic Political Geography*, 4th edn. New York: John Wiley & Sons.

Gough, J. (2010). Workers' strategies to secure jobs, their uses of scale, and competing economic moralities: Rethinking the "geography of justice". *Political Geography 29*: 130–139.

Harrison, J. (2008). Stating the production of scales: Centrally orchestrated regionalism and regionally orchestrated centralism. *International Journal of Urban and Regional Research 32*: 922–941.

Hartshorne, R. (1950). The functional approach in political geography. *Annals of the Association of American Geographers 40*: 95–130.

Herod, A. (2011). *Scale*. London: Routledge.

Herod, A., & Wright, M. (eds) (2002). *Geographies of Power: Placing Scale*. Oxford: Blackwell.

Hopkins, D.J. (2018). *The Increasingly United States: How and Why American Political Behavior Nationalized*. Chicago, IL: University of Chicago Press.

Jessop, B. (2016). *The State: Past, Present, Future*. Cambridge: Polity.

Jessop, B., Brenner, N., & Jones, M. (2008). Theorizing sociospatial relations. *Environment and Planning D: Society and Space 26*: 389–401.

Johnston, R.J., & Pattie, C.J. (2012). Kevin Cox and electoral geography. In Jonas, A.E.G., & Wood, A. (eds), *Territory, the State and Urban Politics: A Critical Appreciation of the Selected Writings of Kevin R Cox*. Farnham: Ashgate, 23–44.

Jonas, A.E.G. (1994). The scale politics of spatiality. *Environment and Planning D: Society and Space 12*: 257–264.

Jonas, A.E.G. (2002). Local territories of U.S. government: From ideals to politics of place and scale. In Agnew, J.A., & Smith, J.M. (eds), *American Space/American Place: Geographies of the Contemporary United States*. Edinburgh: Edinburgh University Press, 108–149.

Jonas, A.E.G. (2006). Pro scale: Further reflections on the "scale debate" in human geography. *Transactions of the Institute of British Geographers 31*: 399–406.

Jonas, A.E.G., & Pincetl, S. (2006). Rescaling regions in the state: The new regionalism in California. *Political Geography 25*: 482–505.

Jones, J.P., III, Woodward, K., & Marston, S.A. (2007). Situating flatness. *Transactions of the Institute of British Geographers 32*: 264–276.

Jones, K.T. (1998). Scale as epistemology. *Political Geography 17*: 25–28.

Jones, M. (1997). Spatial selectivity of the state? The regulationist enigma and local struggles over economic governance. *Environment and Planning A 29*: 831–864.

Jones, M., & MacLeod, G. (2004). Regional spaces, spaces of regionalism: Territory, insurgent politics and the English question. *Transactions of the Institute of British Geographers 29*: 433–452.

Judd, D.R. (1998). The case of the missing scales: A commentary on Cox. *Political Geography 17*: 29–34.

Keil, R. (2020). The spatialized political ecology of the city: Situated peripheries and the capitalocenic limits of urban affairs. *Journal of Urban Affairs 42*: 1125–1140.

Koch, N. (2020). Nationalism, methodological. In Kobayashi, A. (ed.), *The International Encyclopedia of Human Geography*, vol. 9, 2nd edn. Oxford: Elsevier, 245–248.

Kurtz, H.E. (2002). The politics of environmental justice as a politics of scale. In Herod, A., & Wright, M. (eds), *Geographies of Power: Placing Scale*. Oxford: Blackwell, 249–273.

Kythreotis, A.P., & Jonas, A.E.G. (2012). Scaling sustainable development? How voluntary groups negotiate spaces of sustainability governance in the United Kingdom. *Environment and Planning D: Society and Space 30*: 381–399.

Kythreotis, A., Jonas, A.E.G., Mercer, T., & Marsden, T. (2023). Rethinking urban adaptation as a scalar geopolitics of climate governance: Climate policy in the devolved territories of the UK. *Territory, Politics, Governance 11*: 39–59.

Latham, A. (2002). Retheorizing the scale of globalization: Topologies, actor-networks, and cosmopolitanism. In Herod, A., & Wright, M. (eds), *Geographies of Power: Placing Scale*. Oxford: Blackwell, 115–144.

Leitner, H., & Miller, B. (2007). Scale and the limitations of ontological debate: A commentary on Marston, Jones and Woodward. *Transactions of the Institute of British Geographers 32*: 116–125.

Li, Y., & Jonas, A.E.G. (2019). City-regionalism as countervailing geopolitical processes: The evolution and dynamics of Yangtze River Delta Region, China. *Political Geography 73*: 70–81.

Marks, G. (1993). Structural policy and multi-level governance in the EC. In Cafruny, A.W., & Rosenthal, G.G. (eds), *The State of the European Community. Vol. 2, The Maastricht Debates and Beyond*. Boulder, CO: Lynne Rienner, 391–410.

Marston, S.A. (2000). The social construction of scale. *Progress in Human Geography 24*: 219–242.

Marston, S.A., Jones, J.P., III, & Woodward, K. (2005). Human geography without scale. *Transactions of the Institute of British Geographers 30*: 416–432.

Marston, S.A., Woodward, K., & Jones, J.P., III (2007). Flattening ontologies of globalization: The Nollywood case. *Globalizations 4*: 45–63.

McAllister, I. (1987). Social context, turnout, and the vote: Australian and British comparisons. *Political Geography Quarterly 6*: 17–30.

Miller, B. (1994). Political empowerment, local-central state relations, and geographically shifting political opportunity structures: The Cambridge, Massachusetts, peace movement. *Political Geography 13*: 393–406.

Moisio, S. (2011). Political geographies of the state and scale. *Political Geography 30*: 173–174.

Moisio, S., Jonas, A.E.G., Koch, N., Lizotte, C., & Luukkonen, J. (2020). Changing geographies of the state: Themes, challenges and futures. In Moisio, S., Koch, N., Jonas, A.E.G., Lizotte, C., & Luukkonen, J. (eds), *Handbook on Changing Geographies of the State: New Spaces of Geopolitics*. Cheltenham: Edward Elgar, 1–28.

Monmonier, M. (1991). *How to Lie with Maps*. Chicago, IL: University of Chicago Press.

Neumark, D., & Simpson, H. (2015). Place-based policies. In Duranton, G., Henderson, J.V., & Strange, W.C. (eds), *Handbook of Regional and Urban Economics*, vol. 5. Amsterdam: Elsevier, 1197–1287.

North, P. (2011). The politics of climate activism in the UK: A social movement analysis. *Environment and Planning A: Economy and Space 43*: 1581–1598.

Painter, J. (2008). Cartographic anxiety and the search for regionality. *Environment and Planning A 40*: 342–361.

Park, B.-G. (2008). Uneven development, inter-scalar tensions, and the politics of decentralization in South Korea. *International Journal of Urban and Regional Research 32*: 40–59.

Prytherch, D.L. (2010). "Vertebrating" the region as a networked space of flows: Learning from the spatial grammar of Catalanist territoriality. *Environment and Planning A 42*: 1537–1544.

Silvern, S.E. (1999). Scales of justice: Law, American Indian treaty rights and the political construction of scale. *Political Geography 18*: 639–668.

Smith, C.M., & Kurtz, H. (2003). Community gardens and politics of scale in New York City. *Geographical Review 93*: 193–212.

Smith, N. (1984). *Uneven Development*. New York: Blackwell.

Smith, N. (1992). Geography, difference and the politics of scale. In Doherty, J., Graham, E., & Malek, M. (eds), *Postmodernism and the Social Sciences*. London: Palgrave Macmillan, 57–79.

Smith, N. (1993). Homeless/global: Scaling places. In John Bird, J., Curtis, B., Putnam, T., Robertson, G., & Tickner, L. (eds), *Mapping the Futures: Local Cultures, Global Change*. London: Routledge, 87–119.

Soja, E.W. (1989). *Postmodern Geographies: The Reassertion of Space in Critical Social Theory*. London: Verso.

Swyngedouw, E. (1997). Neither global nor local: "Glocalisation" and the politics of scale. In Cox, K.R. (ed.), *Spaces of Globalisation: Reasserting the Power of the Local*. New York: Guilford Press, 137–166.

Swyngedouw, E. (2000). Authoritarian governance, power, and the politics of rescaling. *Environment and Planning D: Society and Space 18*: 63–76.

Taylor, P.J. (1982). A materialist framework for political geography. *Transactions of the Institute of British Geographers NS7*: 15–34.

Wachsmuth, D. (2017). Competitive multi-city regionalism: Growth politics beyond the growth machine. *Regional Studies 51*: 643–653.

Wu, F. (2016). China's emergent city-region governance: A new form of state spatial selectivity through state-orchestrated rescaling. *International Journal of Urban and Regional Research 40*: 1134–1151.

Chapter 11

Borders

Anne-Laure Amilhat Szary

Université Grenoble Alpes, CNRS, Sciences Po Grenoble – UGA, Pacte, Grenoble, France

---

Boundaries have been a key component of the modern territorial vocabulary. Their origins can be traced to Euclidean thinking, which envisions space as two-dimensional. They also align with Kantian rationality that considers space and time as two symmetric conditions of human experience. They can be considered "a generic term for the linear spatial discontinuity" (Fall 2005: 13). Borders constitute both a restriction and an expansion of the semantic field of boundaries and, in this sense, it is hard to discuss one term without examining the other. "Borders" tends to have a stronger political connotation than "boundaries," having been adopted in modern times to express a specific balance of territorialized powers. This chapter reflects on all of these terms.

Boundaries may appear to be more mundane than borders, as more people have probably had the chance to experience them personally – be it physically, through their presence on urban fringes in "park boundary" signs, or mentally, since history books are replete with references to "boundary treaties" that make boundaries achieve a material consistency for ordinary people. But touching a boundary or a border is hard, even when it is a material barrier or a wall, often because they are designed to restrict access. As delimitations of space, boundaries and borders present a paradox: they constitute *symbolic* components of our environment that *materially* impact our lives. For example, when two Alpine climbers in the early 2010s tried to follow as closely as possible the borders of France and Switzerland through the mountains, they used their own bodies to pose a complex question – whether it is even possible to experience a border or a boundary

The Wiley Blackwell Companion to Political Geography, Second Edition.
Edited by Virginie Mamadouh, Natalie Koch, Chih Yuan Woon, and John Agnew.

as traced on a map (Amilhat Szary 2013). In these mountain fringes, the borders were clearly not meant to be reached.

Language is an important filter to understand the dynamics of bordering. The direct translation of boundary in languages derived from Latin, such as *frontière* in French or *frontera* in Spanish, has its roots in the "front" and attests to the rivalries and battles that took place before linear devices, known as boundaries, were in use. The mythological history of the Western world stained the first tracing of a border with blood: Remus was killed by his twin brother Romulus because of a sacred furrow, plowed to distinguish the soil of Rome from its unknown exterior. Remus mocked the boundary Romulus traced and, as Romulus was lifting the plow to mark a threshold, Remus deliberately walked over the line to denounce its weakness. Romulus responded to Remus's symbolic violence with the ultimate physical act of violence, murdering his brother. In this way, Romulus territorialized his identity through establishing Rome's spatial limit.

Saxon etymology, by contrast, suggests that the boundary or border could be "that which binds," a place of friction or meeting where alterity is negotiated. From this perspective, borders are a space where the relationship with otherness can be developed to allow for identity-building and place-making. This socio-anthropological definition of borders emphasizes the complex relations of spatial divides with distance. In German sociologist Georg Simmel's (1972 [1908]) approach, for example, social links are considered to originate from the connections between that which is close and that which is far, simultaneously fixed and mobile. A border is thus a way of "putting distance in proximity" (Groupe Frontière et al. 2004). The German word for border/boundary, *Grenze*, represents yet another semantic field of reference. It comes from a Polish term for "milestone," a marker once used to demark private property. The word was coined in the context of large borderlands where the line only existed as a mental construction, tied to signs like a point in the landscape. However, in the European context, those different words have evolved "in a similar direction, that of a dematerialization through the metaphorical meaning of the linear limit" (Lask 1995: 68). As this chapter shows, scholars have taken the question of the border's meaning well beyond these linguistic beginnings – and well beyond Europe.

Border Studies Revived

The decades following the end of the Cold War in 1991 saw two opposing movements in territorial thinking. On the one hand, commentators describing this period as the "end of history" assumed that it would entail the symmetric "end of territory" (Badie 1995), as the broadened flows of globalization relied on a decline in barriers. A former director of the World Trade Organization was thus able to advocate for a "world without walls" (Moore 2003) at the very moment when rebordering processes were about to gain new legitimacy, as security became a key issue in international relations following the 9/11 terrorist attacks in the United States in 2001. On the other hand, the same period was marked by an "obsession with borders" (Foucher 2012 [2007]) shared by policy-makers, politicians, and scholars.

Historical knowledge of border demarcation (Lattimore 1988 [1940]) had taken an important step after World War II (Foucher 1986; Sahlins 1996; Nordman 1998), renewing consideration of borders through the research lines set by the Annales school of history (Ancel 1938). After the fall of the Berlin Wall, a few authors led the way (Rumley & Minghi 1991; Donnan & Wilson 1994; Paasi 1996) in analyzing the complex relations between power and space that borders "reified" (Raffestin 1974), thus reactivating the theoretical insights of the previous generation (Guichonnet & Raffestin 1974; Prescott 1978). The input from what was becoming structured as "border studies" was succinctly defined in a widely cited paper by David Newman and Anssi Paasi (1998). The 1990s renewal of scientific paradigms can be characterized by its effort to question the linear dimension of spatial divides and further interrogate their complex spatialities. This breakthrough was made possible by opening up disciplinary boundaries, integrating research in geography, but also political science and international relations, as well as anthropology or sociology; the field of humanities and art was not left out. This bridging of knowledge allowed for a more critical scholarship to emerge. Structurally, the trend was pushed forward by the birth of connecting structures, such as the Association of Borderland Studies (founded in 1976) that organized annual conferences, and of a specialized publication, the *Journal of Borderland Studies* (launched in 1986). These were later bolstered by the Borders in Transition (BRIT) network that started holding conferences in 1994, and other regular border forums such as the Borderscapes conferences (2006–2014).

Since the 1990s, border research has expanded in three directions: one focused on processes of cross-boundary cooperation and sociospatial issues in borderlands, a second on the securitization of border crossings and the renewal of demarcation processes, and a third, more critical one, focused on human mobility as a right preceding the existence of states. Much of this research confronts a longstanding tension between the extent to which borders and boundedness are understood theoretically and empirically. The need for an overarching theory of borders was long sought after, but mostly answered to in terms of modeling (Brunet-Jailly 2005; Hamez et al. 2013; Walther et al. 2023). The concomitant release of two companion books (Wastl-Walter 2012; Wilson & Donnan 2012) marked yet another stage of the maturation of this scientific field. The bridging between critical border studies and migration and refugee research has allowed this theoretical effort to move one step further (Nail 2016). Recent scholarship has drawn on a range of methods to tackle border and boundary together, moving through scales and times. At the same time that the field has undergone methodological diversification, it has also witnessed a shift from the two original geographical cradles of border analysis – that is, the US–Mexico divide and European borders – toward a wider, postcolonial horizon that includes non-Western bordering practices.

Scholars have also been forced to reckon with new spatial practices, which shake the traditional groundings of bordering. Globalization and technological change have challenged well-established spatial notions such as distance and proximity, meaning that boundary-making processes and materialities interact in a technology-dependent context. The projection of bordering practices, both inside and outside national territories, as well as the multiplication of agency in the process, challenges the very notion of linear limits: Is the "everywhere border" still a border? This questioning goes along with the multiplication of exceptional statuses, economically as in extraterritorial financial zones (Nyíri 2009) or politically as with the emergence of various camps and detention centers (Minca 2005, Le Cour Grandmaison, Lhuilier, & Valluy 2007). In fact, borders have always been mobile and dynamic but commonly used static frames of comprehension prevented understanding of this phenomenon. Geographers have long defined borders as palimpsests – those manuscripts that keep the traces of vanished words when new scribes had written on them. For Foucher (1991: 43) they are "time inscribed within space," or, in my own terms, borders represent "space-time that materializes norms" (Amilhat Szary 2020: 196). At the broadest level, these space–time relations are at the heart of political geography scholarship on borders and boundaries.

From Bounded Rationales to Border Thinking

Historically, the invention of physical limits is linked to the process of sedentary settling. The act of materializing compartmentalization into space may date back to the days when the first Sumerian cities were fortified with walls. And even though the borders of territorial states dominate contemporary representations of boundaries, they are a recent phenomenon that goes back only a few centuries. The medieval world was marked by plural forms of bordering, since the feudal system was based on interpersonal allegiances that did not require spatial contiguity, and territorial limits were not the most meaningful. The political order that emerged in modern Europe gave rise to a growing desire to balance political organization and territory. Borders defined as state limits were a solution – commonly said to be "invented" in 1648, with the signing of the Treaties of Westphalia that ended the Thirty Years' War in Europe and the search for a balance of power guaranteed by territorial stability (Foucher 1986). The Westphalian treaties did not usher in an era without war, but they did open up a bounded way of thinking about modernity. The extent of a state's power was thus to be limited by the extent of a rival political domain, resulting in the partitioning of space symbolized by a common separation line that would thereupon be called a border (Albert et al. 2001; Agnew 2009; Longo 2017).

The linear imaginaries underpinning this Westphalian system opened the possibility of an "inter-national" order, negotiated between emerging states. Despite this new state ideal in organizing Europe's political geographies, some regional empires survived, notably Austria-Hungary. But the idea of a European empire was left behind and modern states were consolidated around the notion of territorial sovereignty (Elden 2013). This in turn gave rise to a long-lasting tautology whereby territory is defined by state, state by sovereignty, and sovereignty by territory. The versatility of this equivalence becomes evident when one compares the timeline of border invention with that of colonization: the borders invented in Europe to be stable became the most malleable thing on colonial grounds (Amilhat Szary 2020; Bauder & Mueller 2023). Thinking about borders from a more global perspective imposes a different narrative, notably through a better understanding of their nexus with capitalism, which needs and uses borders rather than merely trying to erase them.

Just as they were drawing their external borders, recently established states reorganized their territories to confirm their rule and govern them, often by introducing new internal divisions and administrative grids, a technique of government analyzed by Foucault (2004), who stressed the power of their mundane materialities. The eighteenth century marked a turning point for governments establishing the administrative grid of their territories, as they built the physical infrastructure needed to develop effective networks such as tax collection or postal distribution. The quest for the "territorial optimum," or the best possible division of space, also dates to this period. For example, the French Revolution included the proposal for a new map that offered to replace all previous territorial (ecclesiastical, judiciary, etc.) subdivisions by a series of geometric figures. Ultimately, however, France was not divided into squares: new administrative boundaries were instead drawn around a newly invented level of administration, the *département* (department). *Département* boundaries followed the dominant spatial rationality of the times – the idea of "natural borders" – and many still bear the names of the rivers that define their outline (Ozouf-Marignier 1989). Indeed, the "natural border" has a long history of political convenience: with naturalized territorial limits, rulers could point to a natural order and deflect questions concerning the justification for their borders (Debarbieux 1997; Fall 2010).

The example of French *départements* highlights the importance of the tool that allows any theory to be enacted. For boundaries and borders, the map is a tool that is much more than illustrative. One could even say that the map has constituted the condition of existence of the political convention that it was meant to represent. If placing a boundary on the map makes it real, it is the existence of the map that makes it possible for the limit to exist as such, as a linear device. The first known European boundary treaties are indeed contemporaneous with the first precise regional maps that made it possible for the decision-makers to provide their governments with tangible evidence of their decisions on the localization of international limits, although technically no maps were attached to them at the time (Branch 2014). In this context, the act of drawing a boundary relied on the elements that *did* figure on those maps, and notably on topographical features (for more, see Culcasi, this volume).

The modern context in which boundaries emerged shows that they are both technical tools and performative creations (Cosgrove 1999). From the moment that boundaries were laid on the vellum of the map, they were given a consistency that retroactively justified the power that had ordered the map; they were born of this visual condition. The Cartesian visual culture of mapping that gave rise to Western boundaries and borders also suggests the difficulty of translating them to places without such mapping cultures. Contemporary African borders, for example, are not more artificial than others, but they reflect an imposed political order that was not traditionally visualized as linearly "mappable" sovereignty. This may explain why, in some regions, collective representations have clashed so much with territorial borders. Rather than considering those situations as mere irregularities to a dominant narrative, decolonial thinkers have favored a positive account of what it meant to think about the border from a peripheral situation. This has led to the drafting of the "border thinking" proposal, as "a specific form of epistemic response from the exteriority" (Tlostanova et al. 2016: 217) and an endeavor to transform "the hegemonic imaginary from the perspectives of people in subaltern positions" (Mignolo 2000: 737).

From Linear to Topological Limits

Boundaries and borders were reinvented during the modern era to define a political order that was conceived as stable, or as part of a process of constant, linear stabilization. Indeed, the inclusion of the word "order" in "b/order" was not a coincidence (Van Houtum, Henk, & Ziefhofer 2004). This reflected an ontological perspective that divided objects into exclusive categories, separated from one another. Few challenged the static ideal of borders in the early history of political geography, but Jean Gottmann (1952) was one of the first to emphasize the importance of flows to understand world organization. Since then, the contemporary global era has repeatedly shown that borders are inherently unstable: they are always undergoing processes of "debordering" and "rebordering," and territorial restructuring is constant in diverse cultural settings (Antheaume & Giraut 2005; Laine 2016; Diener & Hagen 2022a; Paasi 2022).

The globalization of trade and the increase of flows of human beings, goods, capital, and ideas have jointly created new possibilities for opening up and closing down of boundaries around the world, according to "debordering" and "rebordering" processes. Boundaries can actually be under processes of opening and closing at the same time in the same place. This can happen because the forms and functions of borders no longer spatially coincide (Mezzadra & Neilson 2013). Since the 9/11 terrorist attacks on the United States in 2001, many scholars focus on the new regimes of border security and their diverse spatial expressions

(Pickering & Weber 2006; Amilhat Szary & Giraut 2015; Laine et al. 2021; Axelsson 2022). In various situations, authorities are no longer controlling border crossings where a line is crossed but at more diffuse sites like bus and train stations, airports, ports, and logistical centers, through sophisticated means of technological surveillance (Salter 2004; Prokkola 2013; Popescu 2022). The development of artificial intelligence (AI) is currently pushing this logic one step further, making border crossings increasingly dependent on algorithms (Aradau 2023; Amoore 2024; Amoore et al. 2024).

Whether it originates in the proliferation of flows resulting from globalization (Sheller & Urry 2006; Söderström et al. 2013) or in the wearing of mobile lenses to analyze them (Retaillé 2005; Büscher, Urry, & Witchger 2011), the waning of a fixed and integrated border line has led to a renewal of border studies. Scholars have given more attention to the dissociation and crystallization of bordering practices by approaching them as a complex assemblage (Allen & Cochrane 2010; Anderson & McFarlane 2011), a "socio-technical network" (Latour 2005), and a "collective entity associating objects, actors (migrant or border police), places and regulations" (Fourny 2013). This reframing has not erased the politics of border places and bordering processes. Rather, shifting the focus from institutions to practices raises new questions about human agency and the "vernacularisation" of borders (Perkins & Rumford 2013).

If the border has become "a set of discontinuous dots," as Bonditti (2005: 10) describes it, scholars are shifting from privileging place to digital information, from topography to topology. For some, this requires studying the territorial dissemination of border functions, projecting borders inward to and outward from national spaces, as well as the "pixelization" of boundary lines (Bigo & Guild 2005). Pixelization is derived from the sophistication of both technological border control devices and of the networked organizations that manage them. The possible decomposition of any border into sets of data and binary codes that are infinitely malleable could, however, mean the end of the notion of a boundary, because of its obvious dislocation. The mobile lenses required to understand contemporary borders may reveal a necessary step toward the understanding of territorial restructuring: data flows intersecting in ways that produce new grids (Vukov & Sheller 2013; Zhang & Morris 2023).

Scholars considering today's debordering or rebordering processes have also raised the question of whether it is possible to localize the seemingly ubiquitous border. Within Europe's Schengen area, where some customs checkpoints have been transformed into tourist shops or museums and borderland inhabitants seek to preserve their boundary landscapes through public monuments and cultural events (Amilhat Szary 2012), one may witness a nostalgia for the line. The Covid-19 pandemic and closed-border policies adopted by many countries revealed the ongoing power of this linear imaginary (Fall 2020). The materiality of the border enacted by infrastructure, such as walls or barriers, may seem easier to assess on closing borders, even if most of the border control is done through a network of "smart" devices, from video cameras to digital sensors and robots. This technological apparatus is not devoid of its own territoriality, since it is linked to various infrastructures that are mobilized to ensure the continuity of the network (Popescu 2011; Cattaruzza 2019).

New technological infrastructures, including submarine cables and satellite relaying (Starosielski 2015; Bennett & Eiterjord 2023), even impose a three-dimensional vision of border dimensions, if not a more topological approach of boundary analysis. Non-Euclidean perspectives offer innovative interpretations of border complexity (Lévy 2008; Martin & Secor 2014; Li 2024) and what scholars have more recently theorized under the notion of "voluminous sovereignty" (Billé 2020). However, if control technologies have made very clear that the border was no longer "a continuous line demarcating the territory and sovereign authority of the state, enclosing its domain" (Walters 2006: 193), but rather a more complex, reticular, and mobile set of filtering devices, they have not entirely denied its materiality. For example, scholars have investigated novel bordering processes in parts of our ecosphere – the sea and air; spaces that had long escaped territorial management.

Maritime borders have indeed become an active frontier of international law and geopolitics, since GPS and satellite technologies allow surveillance of the legal framework set up by the United Nations Convention on the Law of the Sea (UNCLOS; Glouftsios & Loukinas 2022; see Hung & Lin, this volume). Maritime borders may become the scene for active claims that could be linked to land conflicts, such as with the South China Sea border disputes (Rolf & Agnew 2016; Diener & Hagen 2022b) or the conflict over the waters of the disputed area of Western Sahara (Wilson 2017; Odermatt 2018). In the Mediterranean Sea, the border map is complicated by the sinuous delineation of the often-rocky coastline, the presence of multiple islands, the intense global flows that intersect with military surveillance, and illegal crossings for human and drug trafficking. This is where the nongovernmental organization Watch the Med has set up a framework to monitor the cases of migrant deaths at the borders of the European Union that constitute a violation of human rights supposedly guaranteed by the neighboring states, subsequently engaging in a complex legal discussion over the consequences of territorial sovereignty in a multilayered Europe (see Samers, this volume).

As scholars investigating de/rebordering phenomena have shown, they are not equivalent to processes of de/reterritorialization. Debordering does not necessarily imply a disaffection with territorial stakes. To understand the relations between spatial processes and the possible existence of bounded categories within which they are deployed, reterritorialization appears as an essential hypothesis. The materiality of bordering is such that it can be said that contemporary boundaries evidence a hyper-territorialization process that is not at odds with the existence of multiple networks that endow space with topological metrics. This is also illustrated through the importance of cultural studies for border scholars. As soon as one tries to shift the focus from the institutions to the people who constitute the border by living at and around it, it becomes apparent that the separation line is not essential. It splits up into a multitude of fragments, both functional and ideal. When the Chicana poet and activist Gloria Anzaldúa claimed that she "was" the border (Anzaldúa 1987), she was one of the first to stress the performative dimension of border territorialities and to point out that the embodiment of an institutional convention meant more than the mere designation of a political status. In the Foucauldian sense, the border proves to function as a complex set of apparatuses, a *dispositif* where the political and the physical intersect. More than any other kind of boundary, the border becomes a place where identities are continuously at play and are constantly negotiated through art, poetry, and cultural practice (Brambilla et al. 2015; Schimanski & Wolfe 2017; Antoszek 2022; Wille et al. 2024).

The Mobile Border

The idea of the "mobile border" captures the disjunction between the functions and shapes of borders, especially as it becomes increasingly individualized and flexible (Cuttitta 2006; Weber 2006; Steinberg 2009; Richardson 2013; Jones & Johnson 2014). What kinds of limits can exist between spatial objects if the very notion of division is indeterminate? If the border is everywhere, is it still a border? Answers to these questions are of critical political importance, especially when "borders cease to be purely exterior realities" and "are no longer the shores or the political, but have indeed become . . . *objects* or – let us put it more precisely – *things* within the space of the political itself" (Balibar 2002: 78). This raises key questions about how societies organize collectively, and the foundational premise of territorial sovereignty on which modern democracies have been built.

Early postcolonial approaches to border issues stressed the positive aspects of in-betweenness that borders tend to foster. For example, Price (2000: 105) states that "by its very fluidity, the border constitutes a potential-laden space where old power relations can be reworked and perhaps made less oppressive." But the flexibility of in-betweenness may only be an illusion, as borderlands are regional constructs that surround a linear device. Since the linear dimension of political limits has not totally disappeared, there is still space for liminal conditions of sociality. These can be seen positively as places of multiple exchanges that benefit from the border as a resource without colliding with its separating function. As Raffestin (1980: 153) puts it, "territoriality constitutes the stumbling block of any division." Indeed, border life generally illustrates that linear divides are conventions that allow for multiple interpretations, and that experiences of border crossings are always more than a simple act of moving between two or more points (Plangger 2018; Paasi et al. 2019; Joniak-Lüthi 2020; Meier 2020; Mogiani 2022).

In a sense, the spatial diffusion of contemporary borders can make it impossible to differentiate an inside from an outside, and therefore to separate insiders from outsiders. The personalization of borders through the individualization of boundary experiences and definitions also makes it difficult to build social collectives. When the task of defining identity and alterity is left to the individual alone, they may not be able to endure the violence of the process. Borders, having been invented to materialize state protection over individuals, are becoming the place where the same people now find themselves exposed, without mediation, to the violence of globalization (Jones 2017; Dempsey 2020; Schindel 2022; Ferdoush 2024). In *The Archaeology of Knowledge*, Foucault (1969) writes that there is anxiety in the thinking of the other, because thinking of discontinuity can be fearful. Border culture, if it exists (Konrad & Amilhat Szary 2023), endorses violent components, since the constant exposure to alterity, the experience of *différance* (Derrida 1972), leaves every individual facing a constant need to redefine their identity (or identities) with no certainty about the political outcome of this quest (see Ferdoush, this volume).

Border scholars have thus proposed a new term, "borderities," built on a close reading of Foucault's writings and deriving from governmentality (Amilhat Szary & Giraut 2015). As a specific technology of power, "borderity" refers to any technology of spatial or sociospatial division – the governmentality of territorial limits. It may also appear as a differentiated social and political quality. Examining how political subjects can be disabled and enabled, the borderities approach illuminates the question of how borders can be the site of both power and counterpower. In fact, resistance to today's border regimes cannot be limited to simple acts of transgression through crossing. The only way to pervert this spatial organization, Razac (2013) suggests, would be to inhabit the line. Finding ways to spatialize the mobile border can also be used to denounce its conditions of existence (Zaccaria 2017). This idea is also at the core of a collective of artists' projects: Decolonizing Architecture (DAAR), based in Palestine, produced an installation in 2010 based on what they found within the space traced on the map by the Oslo Accords. Less than 1 mm on the official paper, the space covered by the line traced with a red pen measures more than 5 m in reality, but has no legal status (http://www.decolonizing.ps/site/battir-3). Basing their project on the official line, the collective found exceptional places that the line ran through, including a baroque migrant house, a personal enclave, and the unfinished parliament of Palestine – thus contributing to building a visual denunciation of the partition.

The Covid-19 pandemic also shed new light on the meanings attributed to borders and the multiple roles they play in society, by closing borders (Kenwick & Simmons 2020; Barsrur & Kliem 2021) and implementing new boundaries of all kinds, at all scales (Shin 2021). Previous cross-border cooperation appeared as a useful tool to mitigate severe national policies (Böhm 2023). In many ways, the use of apps to impose and police boundaries, the accelerated dissemination of biometric techniques, and the recourse to AI both illustrated what technology can do at borders and portended their even greater role in future border politics. In many ways, the inequalities of border policies during the pandemic, notably the nationalist-based aid programs and the anti-migrant schemes, stressed the need for more border justice (Casaglia 2021; Heller 2021) and stimulated resistance to restrictive border regimes.

Conclusions

Linear borders have not disappeared, but it is now well understood that linearity represents only one aspect of their complex definition. The diversification of the forms and functions of borders stresses the need for complex geographical analysis of their inherent dynamism. Understanding contemporary borders and the implications of their penetration into many aspects of our social, spatial, and political relations does indeed require a strong epistemological effort. Political geographers have been at the leading edge of such work. The border is a space–time operator of differentiation that materializes social, political, and spatial norms (Amilhat Szary 2020, 2024), which scholars have analyzed through myriad intersectional approaches (Staszak 2017; Schmoll 2020; Freedman & Latouche 2023). The methodological diversity of border studies, combined with the diversity of border practices and materialities themselves, makes it possible to value peripheral positions and their emancipatory potential.

Borders remain to be understood in terms of a dialectic between their secant linearity and the potential for indeterminacy that lies in the in-between situation. From a geographical point of view, their phenomenology has never been so diverse: they take the form of lines, zones, and points linked by networks. The political consequences of this are twofold. First, it has become extremely difficult to cross them, since this means much more than passing from one side of the line to the other. Second, we witness a growing gap between how people can claim rights of residence and citizenship across these divides. In this context, borders have come to operate according to a principle of "differential inclusion" (Mezzadra & Neilson 2010); that is, sorting people according to their economic status, their social capital, and so on. The evolution of the notion of the border is therefore linked to the evolution of the conditions of belonging, now conceived in a less spatialist perspective. It also highlights the violence intrinsic to the mechanisms for sharing the world.

The complex space–time of borders are fertile ground for resistance to various forms of domination. Author and activist Audre Lorde illustrates this in her poem "Berlin Is Hard on Colored Girls," written in 1984, before the fall of the Berlin Wall and assessing the invisible memory of the extermination of the Jews: "I cross her borders at midnight/ the guards confused by a dream" (Lorde 1986). Lorde's intimate views on a historically divided city such as Berlin allow her to express both the multiple layers of order that a border can embody and the role of imagination. By framing alternative visions of spatial divisions, she suggests alternate ways to resist the dominant narrative on divisive boundaries.

The complexity of contemporary borders (Wille, Fellner, & Nossem 2025) imposes that the first boundaries to be crossed to understand them be those of our own academic fields of activity. Border studies is a multidisciplinary field of knowledge, uniting political geography and cognate fields in the social sciences and humanities, as well as artists and activists. Putting border thinking at this crossroads makes borders, more than ever, important laboratories to evidence the evolutions of both spatial thinking and contemporary political transformations (Mezzadra & Neilson 2013).

References

Agnew, J. (2009). *Globalization and Sovereignty*. Lanham, MD: Rowman & Littlefield.

Albert, M., Jacobson, D., & Lapid, Y. (eds) (2001). *Identities, Borders, Orders: Rethinking International Relations Theory*. Minneapolis, MN: University of Minnesota Press.

Allen, J., & Cochrane, A. (2010). Assemblages of state power: Topological shifts in the organization of government and politics. *Antipode 42*: 1071–1089. doi:10.1111/j.1467-8330.2010.00794.x

Amilhat Szary, A.-L. (2012). Border art and the politics of art display. *Journal of Borderlands Studies 27*: 213–228. doi:10.1080/08865655.2012.687216

Amilhat Szary, A.-L. (2013). Towards experimental mountaineering? Perspective on two tours of Alpine borders. *Journal of Alpine Research/Revue de Géographie Alpine 101*. doi:10.4000/rga.2131

Amilhat Szary, A.-L. (2020). *Géopolitique des frontières. Découper la terre, imposer une vision du monde*. Paris: Le Cavalier bleu.

Amilhat Szary, A.-L. (2024). *Qu'est ce qu'une frontière aujourd'hui. . . et demain?* 2nd edn. Paris: PUF.

Amilhat Szary, A.-L., & Giraut, F. (eds) (2015). *Borderities: The Politics of Contemporary Mobile Borders*. Basingstoke: Palgrave Macmillan.

Amoore, L. (2024). The deep border. *Political Geography 109*: 102547. doi:10.1016/j.polgeo.2021.102547

Amoore, L., Campolo, A., Jacobsen, B., & Rella, L. (2024). A world model: On the political logics of generative AI. *Political Geography 113*: 103134. doi:10.1016/j.polgeo.2024.103134

Ancel, J. (1938). *Géographie des frontières*. Paris: Gallimard.

Anderson, B., & McFarlane, C. (2011). Assemblage and geography. *Area 43*: 124–127. doi:10.1111/j.1475-4762.2011.01004.x

Antheaume, B., & Giraut, F. (2005). *Le territoire est mort, vive les territoires*. Paris: IRD.

Antoszek, E. (2022). Challenging the border doxa: Selected examples of border artivism. *Res Rhetorica 9*(2): 121–137. doi:10.29107/rr2022.2.7

Anzaldúa, G. (1987). *Borderlands/La Frontera: The New Mestiza*. San Francisco, CA: Aunt Lute Books.

Aradau, C. (2023). Borders have always been artificial: Migration, data and AI. *International Migration 61*(5): 303–306. doi:10.1111/imig.13186

Axelsson, L. (2022). Border timespaces: Understanding the regulation of international mobility and migration. *Geografiska Annaler: Series B, Human Geography 104*: 59–74. doi:10.1080/04353684.2022.2027260

Badie, B. (1995). *La fin des territoires: Essai sur le désordre international et sur l'utilité sociale du respect*. Paris: Fayard.

Balibar, E. (2002). What is a border? In Balibar, É. (ed.), *Politics and the Other Scene*. London: Verso, 75–86.

Basrur, R., & Kliem, F. (2021). Covid-19 and international cooperation: IR paradigms at odds. *Social Sciences 1*(1): 7. doi:10.1007/s43545-020-00006-4

Bauder, H., & Mueller, R. (2023). Westphalian vs. indigenous sovereignty: Challenging colonial territorial governance. *Geopolitics 28*: 156–173. doi:10.1080/14650045.2021.1920577

Bennett, M.M., & Eiterjord, T. (2023). Remote control? Chinese satellite infrastructure in and above the Arctic global commons. *The Geographical Journal 189*: 398–411. doi:10.1111/geoj.12503

Bigo, D., & Guild, E. (2005). *Controlling Frontiers: Free Movement into and within Europe*. London: Ashgate.

Billé, F. (ed.) (2020). *Voluminous States: Sovereignty, Materiality, and the Territorial Imagination*. Durham, NC: Duke University Press.

Böhm, H. (2023). Five roles of cross-border cooperation against re-bordering. *Journal of Borderlands Studies 38*(3): 487–506. doi:10.1080/08865655.2021.1948900

Bonditti, P. (2005). Biométrie et maîtrise des flux: vers une "géo-technopolis du vivant en mobilité". *Cultures & Conflits 58*: 131–154. 10.4000/conflits.1825

Brambilla, C., Laine, J., Scott, J.W., & Bocci, G. (eds) (2015). *Borderscaping: Imaginations and Practices of Border Making*. Aldershot: Ashgate.

Branch, J. (2014). *The Cartographic State: Maps, Territory, and the Origins of Sovereignty*. Cambridge: Cambridge University Press.

Brunet-Jailly, E. (2005). Theorizing borders: An interdisciplinary perspective. *Geopolitics 10*: 633–645. doi:10.1080/14650040500318449

Büscher, M., Urry, J., & Witchger, K. (2011). Introduction. In Büscher, M., Urry, J., & Witchger, K. (eds), *Mobile Methods*. London: Routledge, 1–19.

Casaglia, A. (2021). Borders and mobility injustice in the context of the Covid-19 pandemic. *Journal of Borderlands Studies 36*(3): 1–9. doi:10.1080/08865655.2021.1918571

Cattaruzza, A. (2019). *Géopolitique des données numériques: Pouvoir et conflits à l'heure du big data*. Paris: Le Cavalier bleu.

Cosgrove, D. (1999). *Mappings*. London: Reaktion Books.

Cuttitta, P. (2006). Points and lines: A topography of borders in the global space ephemera. *Theory & Politics in Organization 6* (1): 27–39. http://www.ephemerajournal.org/sites/default/files/6-1cuttitta.pdf (accessed November 23, 2024).

Debarbieux, B. (1997). La montagne comme figure de la frontière. *Le Globe-Genève 137*: 145–166.

Dempsey, K.E. (2020). Spaces of violence: A typology of the political geography of violence against migrants seeking asylum in the EU. *Political Geography 79*: 102157. doi:10.1016/j.polgeo.2020.102157

Derrida, J. (1972). La différance. In Derrida, J. (ed.), *Marges de la philosophie*. Paris: Editions de Minuit, 1–30.

Diener, A.C., & Joshua Hagen, J. (2022a). *Invisible Borders in a Bordered World: Power, Mobility, and Belonging*. New York: Routledge.

Diener, A.C., & Joshua Hagen, J. (2022b). Bordering the South China Sea: Maritime claims, contested sovereignty, and novel territorialities. In Diener, A.C., & Hagen, J. (eds), *Invisible Borders in a Bordered World: Power, Mobility, and Belonging*. New York: Routledge, 82–104.

Donnan, H., & Wilson, T.M. (1994). An anthropology of frontiers. In Donnan, H., & Wilson, T.M. (eds), *Border Approaches: Anthropological Perspectives on Frontiers*. Lanham, MD: University Press of America.

Elden, S. (2013). *The Birth of Territory*. Chicago, IL: University of Chicago Press.

Fall, J. (2005). *Drawing the Line. Boundaries, Identity and Hybridity in Transboundary Spaces*. Aldershot: Ashgate.

Fall, J. (2010). Artificial states? On the enduring geographical myth of natural borders. *Political Geography 29*: 140–147. doi:10.1016/j.polgeo.2010.02.007

Fall, J. (2020). Fenced in. *Environment and Planning C: Politics and Spaces 38*: 771–794. doi:10.1177/2399654420933900

Ferdoush, M.A. (2024). *Sovereign Atonement: Citizenship, Territory, and the State at the Bangladesh-India Border*. Cambridge: Cambridge University Press.

Foucault, M. (1969). *L'Archéologie du savoir*. Paris: Gallimard.

Foucault, M. (2004). *Sécurité, territoire, Population, cours au Collège de France 1977–78*. Paris: Gallimard-Seuil.

Foucher, M. (1986). *L'invention des frontières*. Paris: Fondation pour les Etudes de Défense Nationale.

Foucher, M. (1991). *Fronts et frontières. Un tour du monde géopolitique*. Paris: Fayard.

Foucher, M. (2012 [2007]). *L'obsession des frontières*. Paris: Perrin.

Fourny, M.-C. (2013). The border as liminal space. A proposal for analyzing the emergence of a concept of the mobile border in the context of the Alps. *Journal of Alpine Research/ Revue de Géographie Alpine 101*. doi:10.4000/rga.2120

Freedman, J., & Latouche, A. (eds) (2023). *The Gender of Borders: Embodied Narratives of Migration, Violence and Agency*. Abingdon: Routledge.

Glouftsios, G., & Loukinas, P. (2022). Perceiving and controlling maritime flows: Technology, kinopolitics, and the governmentalization of vision. *International Political Sociology 16*: olac010. doi:10.1093/ips/olac010

Gottmann, J. (1952). *La politique des Etats et leur géographie*. Paris: Armand Colin.

Groupe Frontière, Arbaret-Schulz, C., Beyer, A. et al. (2004). La frontière, un objet spatial en mutation. EspacesTemps.net, October 29. https://www.literaturadefronteras.cl/wp-content/uploads/2018/03/Groupe-Frontière.-La-frontière-un-objet-spatial-en-mutation.pdf (accessed November 23, 2024).

Guichonnet, P., & Raffestin, C. (1974). *Géographie des frontières*. Paris: PUF.

Hamez, G., Amilhat Szary, A.L., Paris, D., & Reitel, B. (2013). Modelling and benchmarking of borders – modèles de frontières; des frontières modèles? *Belgéo 1*: 10659. doi:10.4000/belgeo.10659

Heller, C. (2021). De-confining borders: Towards a politics of freedom of movement in the time of the pandemic. *Mobilities 16*(1): 113–133. doi:10.1080/17450101.2021.1873563

Houtum, v., Henk, O.K., & Ziefhofer, W. (2004). *B/Ordering Space*. London: Aldershot.

Jones, R. (2017). *Violent Borders: Refugees and the Right to Move*. London: Verso.

Jones, R., & Johnson, C. (2014). *Making the Border in Everyday Life*. London: Ashgate.

Joniak-Lüthi, A. (2020). A road, a disappearing river and fragile connectivity in Sino-Inner Asian borderlands. *Political Geography 78*: 102122. doi:10.1016/j.polgeo.2019.102122

Kenwick, M.R., & Simmons, B.A. (2020). Pandemic response as border politics. *International Organization 74*(1): E36–E58. doi:10.1017/S0020818320000363

Konrad, V.A., & Szary, A.-L.A. (2023). *Border Culture: Theory, Imagination, Geopolitics*. New York: Routledge.

Laine, J., Moyo, I., & Nshimbi, C. (2021). *Expanding Boundaries: Borders, Mobilities and the Future of Europe-Africa Relation*. New York: Routledge.

Laine, J.P. (2016). The multiscalar production of borders. *Geopolitics 3*: 465–482. doi:10.1080/14650045.2016.1195132

Lask, T. (1995). Grenze/frontière: le sens de la frontière. *Quaderni 27*: 65–78.

Latour, B. (2005). *Reassembling the Social: An Introduction to Actor-Network-Theory*. Oxford: Oxford University Press.

Lattimore, O. (1988 [1940]). *Inner Asian Frontiers of China*. Oxford: Oxford University Press.

Le Cour Grandmaison, O., Gilles, L., & Valluy, J. (2007). *Le retour des camps?* Paris: Autrement.

Lévy, J. (2008). Topologie furtive. EspacesTemps.net. http://www.espacestemps.net/articles/topologie-furtive (accessed November 23, 2024).

Li, A.H. (2024). Volumising territorial sovereignty: Atmospheric sciences, climate, and the vertical dimension in 20th century China. *Political Geography 111*: 103106. doi:10.1016/j.polgeo.2024.103106

Longo, M. (2017). From sovereignty to imperium: Borders, frontiers and the specter of neo-imperialism. *Geopolitics 22*(4): 757–771. doi:10.1080/14650045.2017.1355785

Lorde, A. (1986). *Our Dead Behind Us*. New York: W.W. Norton.

Martin, L., & Secor, A. (2014). Towards a post-mathematical topology. *Progress in Human Geography 8*: 420–438. doi:10.1177/0309132513508209

Meier, D. (2020). In-between spaces in the Levant: Conceptual reflections. *Mediterranean Politics 6*: 1–16. doi:10.1080/13629395.2020.1723999

Mezzadra, S., & Neilson, B. (2010). Frontières et inclusion différentielle. *Rue Descartes 67*: 102–108. doi:10.3917/rdes.067.0102

Mezzadra, S., & Neilson, B. (2013). *Border as Method, or, the Multiplication of Labor*. Durham, NC: Duke University Press.

Mignolo, W.D. (2000). The many faces of cosmo-polis: Border thinking and critical cosmopolitanism. *Public Culture 12*(3): 721–748.

Minca, C. (2005). The return of the camp. *Progress in Human Geography 29*: 405–412. doi:10.1191/0309132505ph557xx

Mogiani, M. (2022). Studying borders from the border: Reflections on the concept of borders as meeting points. *Geopolitics*: 1–19. doi:10.1080/14650045.2022.2026329

Moore, M. (2003). *A World without Walls: Freedom, Development, Free Trade and Global Governance*. Cambridge: Cambridge University Press.

Nail, T. (2016). *Theory of the Border*. Oxford: Oxford University Press.

Newman, D., & Paasi, A. (1998). Fences and neighbours in the postmodern world: Boundary narratives in political geography. *Progress in Human Geography 22*: 186–207. doi:10.1191/030913298666039113

Nordman, D. (1998). *Frontières de France*. Paris: Gallimard.

Nyíri, P. (2009). Extraterritoriality. Foreign concessions: The past and future of a form of shared sovereignty. Espacestemps.net, November 19. https://www.espacestemps.net/articles/extraterritoriality-pal-nyiri (accessed November 23, 2024).

Odermatt, J. (2018). Fishing in troubled waters: ECJ 27 February 2018, Case C-266/16, R (on the application of Western Sahara Campaign UK) v Commissioners for her Majesty's Revenue and Customs, Secretary of State for Environment, Food and Rural Affairs. *European Constitutional Law Review 14*(4): 751–766. doi:10.1017/S1574019618000421

Ozouf-Marignier, M.-V. (1989). *La formation des départements. La représentation du territoire français à la fin du XVIIIe siècle.* Paris: Éditions de l'École des hautes études en sciences sociales.

Paasi, A. (1996). *Territories, Boundaries and Consciousness: The Changing Geographies of the Finnish-Russian Border.* Chichester: John Wiley & Sons.

Paasi, A. (2022). Examining the persistence of bounded spaces: Remarks on regions, territories, and the practices of bordering. *Geografiska Annaler: Series B, Human Geography 104*: 9–26. doi:10.1080/04353684.2021.2023320

Paasi, A., Prokkola, E.-K., Saarinen, J., & Zimmerbauer, K. (eds) (2019). *Borderless World for Whom? Ethics, Moralities and Mobilities.* London: Routledge.

Perkins, C., & Rumford, C. (2013). The politics of (un)fixity and the vernacularization of Borders. *Global Society 27*: 267–282. doi:10.1080/13600826.2013.790784

Pickering, S., & Weber, L. (2006). *Borders, Mobility and Technologies of Control.* New York: Springer.

Plangger, M. (2018). De-and re-bordering the alpine space: How cross-border cooperation intertwines spatial and institutional patterns of exclusion and inclusion, subordination and horizontality. *Journal of Borderlands Studies 35*: 443–465. doi:10.1080/08865655.2018.1493943

Popescu, G. (2011). *Bordering and Ordering the Twenty-First Century: Understanding Borders.* Lanham, MD: Rowman and Littlefield.

Popescu, G. (2022). Hiding in plain sight: The power of biometric border technologies. In Diener, A.C., & Hagen, J. (eds), *Invisible Borders in a Bordered World: Power, Mobility, and Belonging.* New York: Routledge, 253–264.

Prescott, J.R.V. (1978). *Boundaries and Frontiers.* London: Croom Helm.

Price, P. (2000). Inscribing the border: Schizophrenia and the aesthetics of Aztlán. *Social and Cultural Geography 1*: 101–116. doi:10.1080/14649369950133511

Prokkola, E.-K. (2013). Technologies of border management: Performances and calculation of Finnish/Schengen border security. *Geopolitics 18*: 77–94. doi:10.1080/14650045.2012.685791

Raffestin, C. (1974). Eléments pour une problématique des régions frontalières. *L'espace géographique 1*: 12–18.

Raffestin, C. (1980). *Pour une géographie du pouvoir.* Paris: Litec.

Razac, O. (2013). La gestion de la perméabilité. *L'espace politique 20.* doi:10.4000/espacepolitique.2711

Retaillé, D. (2005). L'espace mobile. In Anttheaume, B., & Giraut, F. (eds), *Le territoire est mort. Vive les territoires !* Paris: IRD, 175–201.

Richardson, T. (2013). Borders and mobilities: Introduction to the special issue. *Mobilities 8*(1): 1–6.

Rolf, S., & Agnew, J. (2016). Sovereignty regimes in the South China Sea: Assessing contemporary Sino-US relations. *Eurasian Geography and Economics 57*: 249–273. doi:10.1080/15387216.2016.1234403

Rumley, D., & Minghi, J. (1991). The border landscape concept. In Rumley, D., & Minghi, J. (eds), *The Geography of Border Landscapes.* London: Routledge, 1–14.

Sahlins, P. (1996). *Frontières et identités nationales. La France et l'Espagne dans les Pyrénées depuis le XVIIe siècle.* Paris: Belin.

Salter, M.B. (2004). Passports, mobility, and security: How smart can the border be? *International Studies Perspectives 5*: 71–91. doi:10.1111/j.1528-3577.2004.00158.x

Schimanski, J., & Wolfe, S. (eds) (2017). *Border Aesthetics. Concepts and Intersections.* New York: Berghahn Books.

Schindel, E. (2022). Death by "nature": The European border regime and the spatial production of slow violence. *Environment and Planning C: Politics and Spaces 40*: 428–446. doi:10.1177/2399654419884948

Schmoll, C. (2020). *Les damnées de la mer: femmes et frontières en Méditerranée.* Paris: La Découverte.

Sheller, M., & Urry, J. (2006). The new mobilities paradigm. *Environment and Planning A 28*: 207–226. doi:10.1068/a37268

Shin, H.R. (2021). Governing the city through im/mobilities during COVID-19 – technological self-governance. *Eurasian Geography and Economics* 62(4): 507–521.

Simmel, G. (1972 [1908]). In Levine, D.N. (ed.), *Georg Simmel on Individuality and Social Forms.* Chicago, IL: University of Chicago Press.

Söderström, O., Ruedin, D., Randeria, S., D'Amato, G., & Panese, F. (2013). *Critical Mobilities.* Lausanne: EPFL Press.

Starosielski, N. (2015). *The Undersea Network.* Durham, NC: Duke University Press.

Staszak, J.-F. (ed.) (2017). *Frontières en tous genres: Cloisonnement spatial et constructions identitaires.* Rennes: Presses universitaires de Rennes.

Steinberg, P.E. (2009). Sovereignty, territory, and the mapping of mobility: A view from the outside. *Annals of the Association of American Geographers 99*: 467–495. doi:10.1080/00045600902931702

Tlostanova, M., Thapar-Björkert, S., & Koobak, R. (2016). Border thinking and disidentification: Postcolonial and postsocialist feminist dialogues. *Feminist Theory 17*: 211–228. doi:10.1177/1464700116645878

Vukov, T., & Sheller, M. (2013). Border work: Surveillant assemblages, virtual fences, and tactical counter-media. *Social Semiotics 23*: 225–241. doi:10.1080/10350330.2013.777592

Walters, W. (2006). Border/control. *European Journal of Social Theory 9*: 187–203. doi:10.1177/1368431006063332

Walther, O.J., Amilhat Szary, A.L., Brambilla, C., et al. (2023). Border studies at 45. *Political Geography 104*: 102909. doi:10.1016/j.polgeo.2023.102909

Wastl-Walter, D. (2012). *Companion to Border Studies.* Farnham: Ashgate.

Weber, L. (2006). The shifting frontiers of migration control. In Pickering, S., & Weber, L. (eds), *Borders, Mobility and Technologies of Control.* Dordrecht: Springer, 21–44.

Wille, C., Astrid, F., & Eva, N. (eds) (2025). *Bordertextures: A Complexity Approach to Cultural Border Studies.* Bielefeld: Transcript.

Wille, C., Leutloff-Grandits, C., Bretschneider, F., Grimm-Hamen, S., & Wagner, H. (eds) (2024). *Border Complexities and Logics of Disorder.* Baden-Baden: Nomos.

Wilson, A. (2017). Ambivalences of mobility: Rival state authorities and mobile strategies in a Saharan conflict. *American Ethnologist 44*(1): 77–90. doi:10.1111/amet.12427

Wilson, T.M., & Donnan, H. (2012). *A Companion to Border Studies.* New York: Wiley-Blackwell.

Zaccaria, P. (2017). A breach in the wall: Artivist no-border atlases of mobility. *Journal of Mediterranean Studies* 26(1): 37–53. https://muse.jhu.edu/article/685848 (accessed November 23, 2024)

Zhang, C., & Morris, C. (2023). Borders, bordering and sovereignty in digital space. *Territory, Politics, Governance 11*: 1051–1058. doi:10.1080/21622671.2023.2216737

Chapter 12

Citizenship

Patricia Ehrkamp
Department of Geography, University of Kentucky, Lexington, KY, USA

Citizenship is often defined as the formal, legal link between states and their populations, signified by a passport or a birth certificate (Ehrkamp & Leitner 2003). Formal citizenship – in its commonly assumed, liberal democratic form – affords political rights such as the right to vote, freedom of speech, and the right to run for public office (Nyers 2008). Despite originating with the Greek *polis* as an urban concept of governance, alterity, and inclusion (Isin 2002), citizenship has since largely come to be associated with the territory of the state (Bauböck 1994), while the nation remained the dominant political community (Isin 2012a). Liberal democratic citizenship is tied to Western notions of democracy that assume such ideals as equality, liberty, and universalism (Marshall 1950). Citizenship is given meaning through definitions of rights and responsibilities (or duties) that delineate substantive citizenship, and through norms and values that are articulated by various communities and groups (Staeheli et al. 2012). Based on T. H. Marshall's classic 1950 essay, scholarship frequently distinguishes social, civic, and political rights that allow members of the polity to participate in the welfare state, engage in such collective action as demonstrations, and vote or stand in an election (Marshall 1950; see also Shafir 1998 for a summary of debates). A basic assumption of citizenship is that everyone who is part of a polity enjoys the same rights and is held to the same duties.

These universalizing understandings of citizenship are attached to particular notions of modernity and democracy that do not hold for large portions of the globe. Anti-colonial and decolonial scholars insist that the ideals of citizenship such as liberty, equality, and universality obscure the "dark underbelly" of modernity

The Wiley Blackwell Companion to Political Geography, Second Edition.
Edited by Virginie Mamadouh, Natalie Koch, Chih Yuan Woon, and John Agnew.

and democracy that is rooted in imperialism, colonialism, and widespread racial violence that never envisioned the inclusion of colonized people (Wynter 2003). This scholarship also explains that citizenship is racialized and exclusionary. For Mignolo (2008: 318), the "space between the person and the citizen is divided by racism"; that is, citizenship is neither equal nor equally accessible to different racialized groups. These tensions of liberal democratic citizenship as, on the one hand, something one aspires to and, on the other, something one remains excluded from are central to political geographical scholarship on citizenship.

In the remainder of this chapter, I first highlight political geographies of citizenship before turning to questions of rights and political community as they are stretched across space, unevenly distributed, contested, constrained, and occasionally expanded across a variety of struggles and in various spaces and territories.

Political Geographies of Citizenship

Noting that citizenship "is not everywhere the same" (Staeheli et al. 2012: 641), political geographers have made significant contributions to the understanding of citizenship. Among others, they have been examining the relational nature and the uneven spatiality of citizenship – its spatial logics, its application, and the struggles to expand or reduce access to citizenship and rights across space (Ehrkamp & Leitner 2003).

Two central issues stand out in the political geographical debates about citizenship: the question of rights and the question of political community, items whose complex relationship to one another was central to Hannah Arendt's (1948 [1994]) famous observation that "the right to have rights" is contingent on membership in a political community. Emphasizing the spatiality of such a community, for Kesby (2012) membership in a political community is a way of having a place in the world, which points to a key concern of political geographers: that "far from a neutral background to politics, space is political in itself" (McNevin 2011: 37). The relationships between rights and membership in a political community are complex, in large part because of their intricate relationships to space, territory, borders and boundaries, subjectivities, and social practices (Kofman 2002; Ehrkamp & Leitner 2003; Staeheli 2011; Ferdoush 2019).

More broadly, political geographers ask what counts as politics and how the political is to be conceptualized. As an abundance of special issues in scholarly journals shows, geographers have provided lively contributions to the study of citizenship such as the spaces and geographies of citizenship (Painter & Philo 1995; Kurtz & Hankins 2005), expanding conversations across the social sciences about borderwork and citizenship (Rumford 2008), cities and citizenship (Staeheli 2003), and international migration and migrant transnationalism (Ehrkamp & Leitner 2006). Together, these works demonstrate a deep and broad engagement with citizenship as a concept, as a state practice, as a social practice, and as lived reality (Staeheli et al. 2012; see also Kallio, Wood, & Häkli 2020).

While aware of the power of the state, scholars have pointed out the importance of examining lived experiences and the practices of citizenship beyond the state (Benhabib 1999; Ehrkamp & Leitner 2003; Benhabib 2004; Nicholls 2013a) and in various spatialities such as the city (Varsanyi 2006; van Doorn 2013; Mitchell, Attoh, & Staeheli 2015; Summers 2022), a bus (Sziarto & Leitner 2010), across transnational space (Preston et al. 2006; Ho 2010), and in the less tangible spaces of international conservation volunteering (Lorimer 2010).

The Uneven Spatiality of Political Membership

Thinking about these multiple and multiplying spaces of citizenship raises the question of how the community of citizens is defined. Lynn Staeheli (2008: 7) has suggested that "community is where contests are waged over membership and the political subjects and subjectivities that 'belong'." Membership in a political community is often established via birthright citizenship; that is, one is born into citizenship. Citizens acquire citizenship either via the principle of *jus soli* (birth in a territory) or via the principle of *jus sanguinis*, an ethnic definition of citizenship (to whom one is born). Much writing on citizenship takes for granted this birthright citizenship as the entry point into a polity. As Mignolo (2008) observes, the community of birth has replaced the community of faith.

But recent scholarship has pointed out that this birthright affords uneven rights and creates inequalities across the globe by the accident of birth in a polity. Among others, Shachar (2003: 347) sought to unsettle this taken-for-granted idea of the "connection between birth and political membership in any given state" because it affirms certain statuses and rights simply based on place of birth (or birth to a parent who belongs to a political community) without making anyone earn membership. In Shachar's analysis, the birthright citizenship lottery produces and perpetuates global unevenness because the "where" and the "to whom" one is born take precedence over acts of citizenship, enactment of duties, taking on responsibilities, and so forth (Shachar 2009).

Given these realities, numerous scholars have argued for the allocation of rights based on the place where one lives rather than the place where or to whom one is born (Bauböck 1994, 2003; Benhabib 2004; Bauder 2014). Bauböck (2003) proposes expanding formal citizenship for migrants as local/urban citizenship, a call more recently echoed by Bauder (2014), who suggests a move toward *jus domicilus*; that is, residence-based citizenship in an age of human mobility. Bauböck's urban citizenship, like the *jus domicilus* Bauder envisions, affords rights derived from presence within a territory, often spatialities within states such as cities or locales. Much like "the right to the city" (Lefebvre 1996: 158), this scholarship seeks to expand citizenship by adding rights and entitlements for those who are not already included in the (national) polity, but hope to gain access to it through participation and life in cities. Others have sought to move away more clearly from the state as the site of political community and citizenship, arguing that citizenship is increasingly deterritorialized (Soysal 1994), or trying to work toward cosmopolitan ideas of citizenship where national boundaries do not prevent democratic participation in world politics (Archibugi 2008). Soysal (1994) stirred debate with her book *The Limits of Citizenship* over what she called the emergence of postnational citizenship (see also Habermas 2001), arguing that membership in a political community and enjoying citizenship rights are no longer necessarily tied to the same territory because citizenship rights are human rights. This recasting of rights, however, remains problematic as long as the state has to remain the guarantor and enforcer of rights and protections (Russell 2005; Kesby 2012). This work shows that communities of citizens are not necessarily congruent with the territory of states, despite assertions by some that they ought to be (Miller 2008) or that such congruence has been practical and persistent (Calhoun 2007).

In recent years, however, the Covid-19 pandemic has brought into sharp relief the ongoing territoriality of citizenship, made visible in part through border closures and through differential access to healthcare and vaccines (Shachar 2020). Postpandemic citizenship highlights the renewed significance of borders (for some), although a number of scholars have argued that these borders were never fully porous for Black and Brown people (Shachar 2020; Aradau & Tazzioli 2021; Walia 2021).

But even within the space of states, citizenship is unevenly distributed. An important example of this is the *hukou* system in China, which defines certain citizenship rights as defined by citizens' place of residence. The movement of people from rural to urban areas means that they give up certain rights, to health coverage or housing (Fan 2008). This two-class citizenship enables the state to limit and direct rural-to-urban migration in order to regulate its needs for construction labor in the rapidly growing cities (Johnson 2017).

Considering the context of immigration to the United States, Linda Bosniak (2006) usefully illustrates the differentiation of citizenship within a territorially defined polity in her discussion of alienage as a "hybrid legal construct" with regard to immigration and immigrant citizenship. Alienage, on the one hand, refers to legal membership in a national community (Bosniak 2006: 38). On the other hand, alienage defines the "social relationships among territorially present persons" (Bosniak 2006: 38) who may or may not be eligible for equal rights and equal protection under the law, precisely because they are not members of the legal realm that defines membership in the political community. Hence, access to rights and protections varies and is often contradictory in that, for example, undocumented school-aged children may not have the legal right to stay in the United States, but still have a constitutionally guaranteed right to a public-school education. Here, different legal realms and rights compete within the same territorially defined sphere (see also Kanstroom 2012), so aptly demonstrated for African Americans in the United States, making clear that the relationship between political community and citizenship rights remains complex and contradictory.

Political Community

Geographers have also usefully highlighted that conceptions of and membership in a political community rely on a moral community. Drawing on Bell and Binnie (2000), Michael Brown highlights the duties of citizenship and urges us to view citizenship as a social contract that includes "the obligations of the sexual citizen and the notion of membership in a political community or polity" (M. Brown 2006: 875). That is, having rights is conditioned on taking on responsibilities that, in turn, are determined through ideas of morality, norms, and values as the basis of membership in local and national communities.

How does one become a member of such a moral community of citizens, then? For Honig (2009: 122), a "universalist morality" that derives from abstract understandings of cosmopolitanism or human rights as a universally shared good is insufficient to establish community membership. She argues that it "matters whether we relate to those near or distant under the sign of universality or under the sign of the neighbor," positing that the latter might lead to solidarity, while the former may well undermine such solidarity because its abstraction is too far removed from daily life and relationships between people. Similarly, Staeheli et al. (2012) have identified different models of inclusion. One of these models is based on social justice, which is founded on the conviction that people be treated equally and fairly. The other model is based on an ethic of care and explicitly recognizes that not everyone in a polity is equal. In this latter model of an ethic of care, achieving the wellbeing of the citizenry in a democracy requires the "satisfaction of situated and particular needs that arise from the recognition of difference and the practices of accommodation" (Staeheli et al. 2012: 635). These different ethical models, in turn, afford different ways of entry into a moral community, and thereby membership in a polity.

The struggles of undocumented immigrants show that achieving membership in the polity through entering moral communities also becomes possible through establishing deservingness. Unauthorized migrants have, in instances, succeeded at becoming recognized as legitimate political subjects by becoming "good migrants" (and/or "good illegal immigrants") (Chauvin & Garcés-Mascareñas 2012: 249) whose good moral character makes them appear less illegal because it conforms to the dominant moralities and values of the country. Similarly, Nicholls (2013a: 84) argues that in order to be recognized as legitimate, unauthorized migrants "must construct representations of immigrants and their cause in ways that cohere with the core normative and moral values of the nation."

This task is getting more difficult in the globalizing world, however. Mezzadra and Nielsen (2013: 164) observe "overlapping of multiple lines of inclusion and exclusion, blurring the boundary between them and destabilizing the existence of a unified and homogeneous point of reference against which the position of migrants can be ascertained." Recent scholarship shows that the concept of citizenship with its ties to Western-style liberal democracy and to the territory of states needs to be critically examined in the twenty-first century. It is perhaps not surprising, then, that critical research on citizenship has made useful advances by "moving from citizenship as membership to citizenship as political subjectivity" (Isin 2012b: 567). Such a move, according to Isin (2012b), helps to decolonize and deorientalize citizenship by exposing its close ties to European notions of modernity and nationhood. These kinds of conceptual moves are even more important at a time when scholars studying Western democracies are pessimistic about citizenship and its political potential (Balibar 2012).

Struggles over Citizenship Rights

While much Western political theory conceives of citizenship and citizenship rights and duties as universally applicable to members of the same polity, this universality is not reflected in the same ways in the experiences of women, racialized minorities, Indigenous groups, and migrant newcomers (Lister 1997a, 1997b; Chatterjee 2011). The struggles for women's suffrage (Ryan 1992) and of sexual and racial minorities to gain and enact rights (Alexander 2012; van Doorn 2013; Hubbard 2013; Obinna 2022) have shown that the ideals of equality and universality are just that: ideals that are rarely realized without contestations. Accordingly, feminist, Black feminist, as well as anti-colonial, decolonial, and postcolonial scholars (among others) have maintained that difference – rather than equality – is central to the definitions, privileges, and experiences of citizenship.

Black feminist scholars, in particular, argue that racist violence is part and parcel of democracy in the afterlife of slavery that frequently manifests as state violence against Black people in the United States. As Christina Sharpe (2016: 7) so poignantly states, "The ongoing state-sanctioned legal and extralegal murders of Black people are normative and, for this so-called democracy, necessary; it is the ground we walk on." Scholarship in political geography and beyond has thus examined in detail how race, ethnicity, gender, sexuality, or immigration status affects access to membership, rights, and belonging.

From the perspective of marginalized groups, citizenship frequently operates as a technology of exclusion, often through detention or removal. Walters (2002) points to deportation as one such mechanism that aids in the restriction of citizenship. Mass incarceration and the carceral state, which stretches across space beyond prison walls with carceral technologies such as parole and probation, aid in the exclusion of (Black) citizens from citizenship rights. As Wright Rigueur and Beshlian (2019: 272) argue, "The disenfranchisement of incarcerated populations, who are predominantly African Americans and people of color, is a form of citizenship status removal." Despite the achievements of civil rights movements and the more recent Black Lives Matter activist successes in raising awareness, Obinna (2022: 481) concludes that "there are still many miles to go to achieve true citizenship" for African Americans. These insights are not limited to North America or Europe but occur across the globe.

Making racialized exclusions from citizenship visible is at the center of efforts to decolonize (and deorientalize) understandings of citizenship, and to enhance the lives of oppressed groups. As Isin (2023: 424) argues, "Palestinian citizenship in Israel is not merely second-class but it is designed to keep Palestinians in abject domination." For Hawthorne (2021: 718), these efforts include "Globalizing Black geographies and provincializing North America [which] means contending with diverse entanglements of race and nation, as well as a plurality of Black spatial praxes and forms of contestation." Leroy (2016) links these struggles to the dispossession of Indigenous groups under settler colonialism, identifying expressions of solidarity between Black Lives Matter and Palestinian activists.

Lesbian/gay/bisexual/trans/queer/intersex/asexual+ (LGBTQIA+) rights and sexual citizenship continue to be important sites of struggle that illustrate the see-sawing of inclusionary and exclusionary politics as well as the complexity of space in struggles over citizenship. Legal and political moves toward more inclusion, such as the abolition of the Defense of Marriage Act (a federal marriage law positing that marriage was between one man and one woman) in the United States in 2013, eliminated barriers to same-sex marriage (and tax penalties) at the federal level. This leads to such paradoxes as gay marriage being legal in over 30 states of the United States, while officials in states such as Tennessee may refuse the certification of marriage licenses if they personally object to same-sex marriage. In other states, such as Pennsylvania, Utah, and Virginia, same-sex marriage is legal but there are no discrimination protections for gay workers (Egan 2014). For Hubbard (2013: 230), struggles over sexual citizenship highlight that "rights and responsibilities are only partly determined by national laws" (see also Lister 2002; Berlant 2011; van Doorn 2013). Together, this body of scholarship suggests that queering citizenship means to recognize and theorize the intricate connections across private and public space, taking seriously the embodied nature of citizenship and the question of intimacy as deeply political and as determined across multiple spatialities.

Citizenship needs to be understood as a lived experience and a social practice (Ehrkamp & Leitner 2003). While people continue to feel the "prosaic geographies of stateness" (Painter 2006), the deep entanglements of the state with everyday life, geographers and other scholars have usefully expanded conceptions of citizenship beyond state-centered notions and status-based understandings of citizenship, otherwise known as passive citizenship. Rather than simply "being a citizen" by status, Lister (1997b: 36) argues that it is necessary to "act as a citizen" in order to achieve "the full potential" of having citizenship status. A related concept works through the subjectivity of becoming citizens. Isin (2008) argues that it is "acts of citizenship" that define the moments at which subjects establish their political subjectivities as citizens. Hence, this scholarship shows that citizenship is not just a status one holds, but is also enacted by active citizens who express their political subjectivities (Isin 2002).

Focusing on political subjectivities is useful when thinking about political community as well. Geographers have expanded conceptions of citizenship to also include social practices, thereby shifting the focus from the state's influence on people's everyday lives to the ways in which people themselves become political and frequently challenge the state (Ehrkamp & Leitner 2003). Research shows that social practices of citizenship need to be carefully considered when examining the everyday life experiences of ordinary people in a polity. Examples include collective action on behalf of undocumented immigrants (Sziarto & Leitner 2010; Zepeda-Millán 2017), anti-deportation movements (Kocher & Stuesse 2021), and struggles over land justice (Safransky 2017).

Using space to advance political goals gives further meaning to citizenship, as geographers have demonstrated (Varsanyi 2006; Inwood & Bonds 2013; Purcell 2013). For example, struggles over "the right to the city" (Lefebvre 1996: 158) make claims to space by inhabiting it or occupying it. Writing in the context of Buenos Aires, Holston (2008: 48) advances the notion of "insurgent citizenship" to describe everyday practices that disrupt or subvert dominant agendas. He highlights networks of migration, squatter settlements, and migrant labor camps as examples of "sites of insurgence because they introduce . . . new identities and practices that disturb established histories." Geographers have built on such conceptions, for example in analyses of rights claims of immigrants (Leitner & Strunk 2014). Others show how streets and public spaces such as Gezi Park in Istanbul remain important sites for protest, opposition politics, claims to rights, and political expressions more broadly (Bayat 2010; Eken 2014). For Summers (2022), insurgent practices and even fleeting (impermanent) art projects in different cities are useful reminders that vernacular collective practices transform and claim (public) space and challenge anti-Black oppression and police violence.

The Everyday Spaces of Citizenship

In emphasizing the spatial complexity of citizenship, political geographers have also questioned methodological nationalism and challenged such understandings that simply locate citizenship at the level of the state. They show that citizenship "varies across place, across time, and for different people. It is inseparable from the geographies of communities and the networks and relationships that link them, with their attendant inequalities, imperfections, and opportunities" (Staeheli et al. 2012: 641). This scholarship has shown that the sites of citizenship politics often are ordinary, less overt, or even hidden (Staeheli et al. 2012). For example, Jones (2010) examines how rural Muslims in Britain use subterranean spaces of citizenship to establish belonging beyond the formal spaces of politics and beyond the public square. In other instances, places of worship serve as sites where the boundaries of belonging to a political community are being hashed out (Ehrkamp & Nagel 2017). This research highlights the importance of spaces of social reproduction to broader notions of belonging in a polity.

As spaces of sustained encounter, in this case ordinary spaces of citizenship are implicated in the construction of citizenship norms, values, and belonging. Private and intimate spaces are also central to the ways in which individuals may congregate and engage in collective action and political activism or establish norms and values that imbue citizenship with meaning (Ehrkamp & Nagel 2012). The spaces of everyday life such as churches, schools, and community organizations become central sites where norms and values are articulated, and flow into the politics of citizenship and community. Askins's analysis (2015: 476) shows how this process plays out in the "quiet politics" of friendship between long-term residents and asylum seekers in Britain that takes place "in homes, neighbourhoods, cafes, going for walks in the local park, and to local shops."

Relatedly, Ehrkamp and Nagel's work on the precarious hospitality offered to undocumented migrants in churches in the US South illustrates this point (Ehrkamp & Nagel 2014). In these churches, newcomers were welcomed as fellow Christians, racialized as "other," and simultaneously stigmatized as breaking secular law in an era of rampant anti-immigrant legislation. In the contradictions of faith-based welcome and law and order–based exclusion, the importance of churches as sites for the construction of citizenship becomes obvious: their members are engaged in policing the borders not only of church membership but of societal belonging more broadly (Ehrkamp & Nagel 2017).

Schools also serve as important sites for articulating understandings of belonging (Glenn 2011). Walter Nicholls (2013b) describes how the DREAMers movement in California solidified, in part, through networking in social media, in private, and in intimate spaces. The latter are particularly important. As Nicholls (2013b: 68) writes: "Interactions in intimate spaces of daily life . . . have functioned as moments where activists share their fears with one another, celebrate accomplishments, reinforce their belief that their cause is just and right, and express doubts about their situations and concerns over the movement's directions." Intimate spaces, here, are part and parcel of a larger political project because they accommodate emotional and affective interactions that strengthen political convictions and galvanize energy for political organizing and the continued practice of citizenship.

Outlook: Crises of Citizenship, Expansion, and Repair

A number of scholars have commented on the "de-democratization" of public life within what some term a "crisis of democratic citizenship" in Western liberal democracies (Balibar 2012: 437). These developments are brought about in large part by relegating controversial issues into the private realm (W. Brown 2006) and by increasingly "techno-managerial" solutions to political struggles (Swyngedouw 2014: 124). Established Western democracies are suffering from a "pervasive mistrust of elected representatives" and "voter apathy" (Chatterjee 2011: 25). The crisis of citizenship, Balibar and others suggest, is also due to the increased neoliberalization of governance that undermines citizen participation and collective action with their focus on individual responsibility (see also

Mitchell 2006; Naylor & Veron 2021). Indeed, Berlant (2011: 222) argues that "neoliberal interests are well served by the displacement of so many historical forms of social reciprocity onto emotional registers, especially when they dramatize experiences of freedom to come that have no social world for them yet."

By contrast, other scholars consider such crises as integral to the restructuring and transformation of citizenship and democracy in which being political is not limited to one's citizenship status or one's participation in formal and electoral politics (Honig 2009; Staeheli 2011). Rather, being political "occurs in a broad range of relations between people and groups wherein norms, practices, ideas, and ways of organizing material life are challenged, questioned, and potentially reconstructed" (Cowen & Gilbert 2008: 1). Considering the struggles of homeless people for their right to be "sleeping rough" and to challenge city ordinances that criminalize such practices in urban public space, Mitchell, Attoh, and Staeheli (2015: 2637) show how homeless activism goes beyond stakeholder politics as it moves from the streets to the city council to the courtroom. For Honig (2009: 1), this is a form "of popular sovereignty in which action in concert rather than institutional governance is the mark of democratic power and legitimacy." The uprisings in response to the murder of George Floyd and several other Black Americans in 2020, and youth environmental activism such as "Fridays For Future" climate strikes, show powerful ways of such "acting in concert." Youth citizenship can challenge the legacy of colonialism and global inequalities, as young citizens engage in building new worlds in the process (Bowman 2019, 2020; Wood 2022).

Asef Bayat's (2010: 56) analysis of politics in the Middle East shows that more than overt political action, a "quiet encroachment of the ordinary" – that is, "noncollective but prolonged direct action" – serves to enhance the situation of disempowered and poor people through nonovert challenges of the state and the powerful. As a social practice, this quiet encroachment moves beyond direct notions of resistance or overt confrontation to take sometimes protracted action as it aims to fulfill immediate needs. And yet, over time this practice chips away at state and capitalist power to determine people's lives, because they are taking up more and more space and gaining political ground that way. These insights, along with others who identify democratic practices even within authoritarian regimes (Wedeen 2007), are useful in pointing out the political efficacy of practices that do not appear to be formally political or collective.

Similarly, McNevin (2011) argues that new political practices and acts emerge that may not yet be recognizable as political or as enactments of citizenship, especially when they are read exclusively through existing frameworks of citizenship. Hunger strikes by asylum seekers are one such form of political action through which asylum seekers challenge the current system and seek to disrupt the normal order (Conlon 2013). The important aspect here is not the outcome – that is, whether these hunger strikes change the existing order – but the question of how new subjectivities surface in contestations over rights and belonging. This scholarship is interested in how solidarities are being created. Berlant's (2011: 262) conception of "lateral politics" and her focus on the historical present emphasize affective notions of citizenship as she calls for understanding simple acts of solidarity as politics rather than defining them as being outside the political. Lateral politics entails "becoming a political subject whose solidarities and commitments are neither to ends nor to imagining the pragmatics of a consensual community, but to embodied processes of making solidarity itself" (Berlant 2011: 260). Analyzing postcolonial India, Chatterjee (2011: 207) challenges political theory to produce

> a different conceptualization of the subject of political practice—neither as abstract and unencumbered individual selves nor as manipulable objects of governmental policy, but rather as concrete selves necessarily acting within multiple networks of collective obligations and solidarities to work out strategies of coping with, resisting, or using to their advantage the vast array of technologies of power deployed by the modern state.

Political geographers remain well positioned to contribute to new theorizations, to improve on "existing lenses" (McNevin 2011: 155), and to recognize political acts – or "emerging practices" (Chatterjee 2011: 207). These calls for new approaches challenge political geographers to carefully examine our own existing ideas and taken-for-granted notions of rights, political community, and what counts as politics in order to pursue new avenues for thinking about citizenship, starting from the recognition that citizenship is enacted by and through people, across multiple and uneven spatialities, and subject to debate. Hawthorne (2024) remains optimistic "that we are in a time of incredible possibility." Her confidence stems from her analysis of Italian citizenship reform, despite the restrictions this reform brought with it. Instead, she insists that in the current moment of crisis, "Black Italian struggles for racial justice are crucial for understanding how to resist the forces of racial-capitalist-nationalism that are shaping our world today, underscoring the necessity of forging of interconnections among diverse struggles for justice." Hawthorne's work thus highlights some of the many ways in which citizenship can expand through activism and struggles for reform.

Beyond activism and collective actions aiming to expand citizenship and citizenship rights, a number of states have also begun expanding access to formal citizenship. Frost (2022: 456) notes that several European countries have implemented "reparative citizenship laws" that bear the potential to correct some of the injustices inherent in citizenship by "remedying harm that came from exclusion by granting status, rights, and identity to those formerly cast out from the body politic." These historical exclusions are largely related to the atrocities of the Nazi regime and dictatorships, for example in Spain and Portugal (Kandiyoti & Benmayor 2023). For Achiume (2019), reparative citizenship goes further and would be one step in a broader decolonial process that, in their evaluation, would necessarily need to begin with granting citizenship to individuals migrating from former colonies. Writing that migration should be understood as decolonial acts at a personal level, Achiume (2019: 1552) argues "that Third World persons are entitled to a form of First World citizenship as a matter of corrective, distributive justice" because (neo)colonial relations make them political insiders in the (former colonial) polities they might migrate to. Granting formal citizenship would thus affirm existing connectivities, and in the process ameliorate the situation of so-called unauthorized migrants. This is an ambitious proposal. But it is certainly an important one to ponder as questions of repair, reparations, and decolonization occupy political geographers studying citizenship.

In sum, questions of legal conceptions and expansions of citizenship, decolonizing, and repair; struggles over rights; political community; and the lived experiences of citizenship across space remain exciting, timely, and significant topics for contemporary political geographical research.

References

Achiume, E.T. (2019). Migration as decolonization. *Stanford Law Review 71*: 1509.

Alexander, M. (2012). *The New Jim Crow: Mass Incarceration in the Age of Colorblindness*. New York: New Press.

Aradau, C., & Tazzioli, M. (2021). Covid-19 and rebordering the world. *Radical Philosophy* 2(10): 3–10.

Archibugi, D. (2008). *The Global Commonwealth of Citizens: Toward Cosmopolitan Democracy*. Princeton, NJ: Princeton University Press.

Arendt, H. (1948 [1994]). *The Origins of Totalitarianism*. San Diego, CA: Harcourt.

Askins, K. (2015). Being together: Everyday geographies and the quiet politics of belonging. *ACME: An International E-Journal for Critical Geographies 14*(2): 470–478.

Balibar, E. (2012). The "impossible" community of citizens: Past and present problems. *Environment and Planning D: Society and Space 30*(3): 437–449.

Bauböck, R. (1994). *Transnational Citizenship: Membership and Rights in International Migration*. Aldershot: Edward Elgar.

Bauböck, R. (2003). Reinventing urban citizenship. *Citizenship Studies* 7(2): 139–160.

Bauder, H. (2014). Domicile citizenship, human mobility and territoriality. *Progress in Human Geography 38*(1): 91–106.

Bayat, A. (2010). *Life as Politics: How Ordinary People Change the Middle East*. Stanford, CA: Stanford University Press.

Bell, D., & Binnie, J. (2000). *The Sexual Citizen*. London: Polity Press.

Benhabib, S. (1999). Citizens, residents, and aliens in a changing world: Political membership in the global era. *Social Research 66*(3): 709–731.

Benhabib, S. (2004). *The Rights of Others: Aliens, Residents and Citizens*. Cambridge: Cambridge University Press.

Berlant, L. (2011). *Cruel Optimism*. Durham, NC: Duke University Press.

Bosniak, L. (2006). *The Citizen and the Alien: Dilemmas of Contemporary Membership*. Princeton, NJ: Princeton University Press.

Bowman, B. (2019). Imagining future worlds alongside young climate activists: A new framework for research. *Fennia: International Journal of Geography 197*(2): 295–305.

Bowman, B. (2020). "They don't quite understand the importance of what we're doing today": The young people's climate strikes as subaltern activism. *Sustainable Earth 3*(1): 16. doi:10.1186/s42055-020-00038-x

Brown, M. (2006). Sexual citizenship, political obligation and disease ecology in gay Seattle. *Political Geography 25*(8): 874–898.

Brown, W. (2006). *Regulating Aversion: Tolerance in the Age of Identity and Empire.* Princeton, NJ: Princeton University Press.

Calhoun, C. (2007). Nationalism and cultures of democracy. *Public Culture 19*(1): 151–173.

Chatterjee, P. (2011). *Lineages of Political Society: Studies in Postcolonial Democracy, Cultures of History*. New York: Columbia University Press.

Chauvin, S., & Garcés-Mascareñas, B. (2012). Beyond informal citizenship: The new moral economy of migrant illegality. *International Political Sociology* 6: 241–259.

Conlon, D. (2013). Hungering for freedom: Asylum seekers' hunger strikes – rethinking resistance as counter-conduct. In Moran, D., Gill, N., & Conlon, D. (eds), *Carceral Spaces: Mobility and Agency in Imprisonment and Migrant Detention.* Aldershot: Ashgate, 133–147.

Cowen, D., & Gilbert, E. (eds) (2008). *War, Citizenship, and Territory*. New York: Taylor & Francis.

van Doorn, N. (2013). Architectures of the good life: Queer assemblages and the composition of intimate citizenship. *Environment and Planning D: Society and Space 31*(1): 157–173.

Egan, P. J. (2014). More gay people can now get legally married. They can still be legally fired. *Monkey Cage Blog*, October 6. http://www.washingtonpost.com/blogs/monkey-cage/wp/2014/10/06/more-gay-people-can-now-get-legally-married-they-can-still-be-legally-fired (accessed November 23, 2024).

Ehrkamp, P., & Leitner, H. (2003). Beyond national citizenship: Turkish immigrants and the (re)construction of citizenship in Germany. *Urban Geography 24*(2): 127–146.

Ehrkamp, P., & Leitner, H. (2006). Guest Editorial. Rethinking immigration and citizenship: New spaces of migrant transnationalism and belonging. *Environment and Planning A 38*(9): 1591–1597.

Ehrkamp, P., & Nagel, C. (2012). Immigration, places of worship and the politics of citizenship in the US South. *Transactions of the Institute of British Geographers 37*(4): 624–638.

Ehrkamp, P., & Nagel, C. (2014). "Under the radar": Undocumented immigrants, Christian faith communities, and the precarious spaces of welcome in the U.S. South. *Annals of the Association of American Geographers 104*(2): 319–328.

Ehrkamp, P., & Nagel, C. (2017). Policing the borders of church and societal membership: Immigration and faith-based communities in the US South. *Territory, Politics, Governance 5*(3): 318–331.

Eken, B. (2014). The politics of the Gezi Park resistance: Against memory and identity. *South Atlantic Quarterly 113*(2): 427–436.

Fan, C. (2008). *China on the Move: Migration, the State, and the Household.* New York: Routledge.

Ferdoush, M.A. (2019). Acts of belonging: The choice of citizenship in the former border enclaves of Bangladesh and India. *Political Geography 70*: 83–91.

Frost, A. (2022). The rise of reparative citizenship. *Citizenship Studies 26*(4–5): 454–459.

Glenn, E.N. (2011). Constructing citizenship: Exclusion, subordination, and resistance. *American Sociological Review 76*(1): 1–24.

Habermas, J. (2001). The postnational constellation and the future of democracy. In Pensky, M. (ed.), *The Postnational Constellation: Political Essays*. Cambridge: Polity Press, 58–112.

Hawthorne, C. (2021). Making Italy: Afro-Italian entrepreneurs and the racial boundaries of citizenship. *Social & Cultural Geography 22*(5): 704–724.

Hawthorne, C. (2024). Author's conclusion: *Contesting Race and Citizenship* Book Forum. *Society + Space*, February 5. https://www.societyandspace.org/articles/authors-conclusion-contesting-race-and-citizenship-book-forum (accessed November 23, 2024).

Ho, E. (2010). Constituting citizenship through the emotions: Singaporean transmigrants in London. *Annals of the American Association of Geographers 99*(4): 788–804.

Holston, J. (2008). *Insurgent Citizenship: Disjunctions of Democracy and Modernity in Brazil*. Princeton, NJ: Princeton University Press.

Honig, B. (2009). *Emergency Politics: Paradox, Law, Democracy*. Princeton, NJ: Princeton University Press.

Hubbard, P. (2013). Kissing is not a universal right: Sexuality, law and the scales of citizenship. *Geoforum 49*: 224–232.

Inwood, J., & Bonds, A. (2013). On racial difference and revolution. *Antipode 45*(3): 517–520.

Isin, E.F. (2002). *Being Political: Genealogies of Citizenship*. Minneapolis, MN: University of Minnesota Press.

Isin, E.F. (2008). Theorizing acts of citizenship. In Isin, E.F., & Nielson, G.M. (eds), *Acts of Citizenship*. New York: Zed Books, 15–43.

Isin, E.F. (2012a). Citizens without nations. *Environment and Planning D: Society and Space 30*(4): 450–467.

Isin, E.F. (2012b). Citizenship after orientalism: An unfinished project. *Citizenship Studies 16*(5–6): 563–572.

Isin, E.F. (2023). A labour of love (the right to philosophy). *Citizenship Studies 27*(3): 422–426. doi:10.1080/13621025.2023.2171256

Johnson, L. (2017). Bordering Shanghai: China's hukou system and processes of urban bordering. *Geoforum 80*: 93–102.

Jones, R.D. (2010). Islam and the rural landscape: Discourses of absence in west Wales. *Social & Cultural Geography 11*(8): 751–768.

Kallio, K.P., Wood, B.E., & Häkli, J. (2020). Lived citizenship: Conceptualising an emerging field. *Citizenship Studies 24*(6): 713–729. doi:10.1080/13621025.2020.1739227

Kandiyoti, D., & Benmayor, R. (eds) (2023). *Reparative Citizenship for Sephardi Descendants: Returning to the Jewish Past in Spain and Portugal*. New York: Berghahn Books.

Kanstroom, D. (2012). *Aftermath: Deportation Law and the New American Diaspora*. Oxford: Oxford University Press.

Kesby, A. (2012). *The Right to Have Rights: Citizenship, Humanity, and International Law*. Oxford: Oxford University Press.

Kocher, A., & Stuesse, A. (2021). Undocumented activism and minor politics: Inside the cramped political spaces of deportation defense campaigns. *Antipode 53*(2): 331–354.

Kofman, E. (2002). Contemporary European migrations, civic stratification and citizenship. *Political Geography 21*(8): 1035–1054.

Kurtz, H., & Hankins, K. (2005). Guest editorial: Geographies of citizenship. *Space & Polity* 9(1): 1–8.

Lefebvre, H. (1996). *Writings on Cities*, trans. Eleonore Kofmand and Elizabeth Lebas. Malden, MA: Blackwell.

Leitner, H., & Strunk, C. (2014). Assembling insurgent citizenship: Immigrant advocacy struggles in the Washington DC metropolitan area. *Urban Geography* 35(7): 943–964.

Leroy, J. (2016). Black History in occupied territory: On the entanglements of slavery and settler colonialism. *Theory & Event* 19(4).

Lister, R. (1997a). *Citizenship: Feminist Perspectives*. Washington Square, NY: New York University Press.

Lister, R. (1997b). Citizenship: Towards a feminist synthesis. *Feminist Review* 57: 28–48.

Lister, R. (2002). Sexual citizenship. In Isin, E.F., & Turner, B.S. (eds), *Handbook of Citizenship Studies*. Thousand Oaks, CA: Sage, 191–208.

Lorimer, J. (2010). International conservation "volunteering" and the geographies of global environmental citizenship. *Political Geography* 29(6): 311–322.

Marshall, T.H. (1950). *Citizenship and Social Class, and Other Essays*. Cambridge: Cambridge University Press.

McNevin, A. (2011). *Contesting Citizenship: Irregular Migrants and New Frontiers of the Political*. New York: Columbia University Press.

Mezzadra, S., & Neilson, B. (2013). *Border as Method: Or the Multiplication of Labor*. Durham, NC: Duke University Press.

Mignolo, W. (2008). Citizenship, knowledge, and the limits of humanity. *American Literary History* 18(2): 312–331.

Miller, D. (2008). Immigrants, nations, and citizenship. *Journal of Political Philosophy* 16(4): 371–390.

Mitchell, D., Attoh, K., & Staeheli, L. (2015). Whose city? What politics? Contentious and non-contentious spaces on Colorado's Front Range. *Urban Studies* 52(4): 2633–2648. doi:10.1177/0042098014550460

Mitchell, K. (2006). Neoliberal governmentality in the European Union: Education, training, and technologies of citizenship. *Environment and Planning D: Society and Space* 24(3): 389–407.

Naylor, L., & Veron, D. (2021). Geographic education in the Anthropocene: Cultivating citizens at the neoliberal university. *Annals of the American Association of Geographers* 111(3): 958–969.

Nicholls, W.J. (2013a). *The DREAMers: How the Undocumented Youth Movement Transformed the Immigrant Rights Debate*. Stanford, CA: Stanford University Press.

Nicholls, W.J. (2013b). Making undocumented immigrants into a legitimate political subject: Theoretical observations from the United States and France. *Theory, Culture and Society* 30(3): 82–107.

Nyers, P. (2008). No one is illegal between city and nation. In Isin, E.F., & Nielsen, G.M. (eds), *Acts of Citizenship*. New York: Zed Books, 160–181.

Obinna, D.N. (2022). "Warring" ideals: Black citizenship and the struggle for inclusion in the USA. *Journal of African American Studies* 26: 472–484. doi:10.1007/s12111-022-09600-7

Painter, J. (2006). Prosaic geographies of stateness. *Political Geography* 25(7): 752–774.

Painter, J., & Philo, C. (1995). Spaces of citizenship: An introduction. *Political Geography* 14(2): 107–120.

Preston, V., Kobayashi, A., & Man, G. (2006). Transnationalism, gender, and civic participation: Canadian case studies of Hong Kong immigrants. *Environment and Planning A 38*(9): 1633–1651.

Purcell, M. (2013). Possible worlds: Henri Lefebvre and the right to the city. *Journal of Urban Affairs 36*(1): 141–154.

Rumford, C. (2008). Introduction: Citizens and borderwork in Europe. *Space & Polity 12*(1): 1–12.

Russell, J. (2005). Rethinking post-national citizenship: The relationship between state territory and international human rights law. *Space and Polity* 9(1): 29–39.

Ryan, M.P. (1992). *Women in Public: Between Banners and Ballots, 1825–1880*. Baltimore, MD: Johns Hopkins University Press.

Safransky, S. (2017). Rethinking land struggle in the postindustrial city. *Antipode 49*(4): 1079–1100.

Shachar, A. (2003). Children of a lesser state: Sustaining global inequality through citizenship law. In Macedo, S., & Young, I.M. (eds), *Child, Family, and State*. New York: New York University Press, 345–397.

Shachar, A. (2009). *The Birthright Lottery: Citizenship and Global Inequality*. Cambridge, MA: Harvard University Press.

Shachar, A. (2020). Beyond open and closed borders: The grand transformation of citizenship. *Jurisprudence 11*(1): 1–27. doi:10.1080/20403313.2020.1788283

Shafir, G. (1998). *The Citizenship Debates: A Reader*. Minneapolis, MN: University of Minnesota Press.

Sharpe, C. (2016). *In the Wake: On Blackness and Being*. Durham, NC: Duke University Press.

Soysal, Y.N. (1994). *Limits of Citizenship: Migrants and Postnational Membership in Europe*. Chicago, IL: University of Chicago Press.

Staeheli, L.A. (2003). Introduction: Cities and citizenship. *Urban Geography 24*(2): 97–102.

Staeheli, L.A. (2008). Citizenship and the problem of community. *Political Geography 27*: 5–21.

Staeheli, L.A. (2011). Political geography: Where's citizenship? *Progress in Human Geography 35*(3): 393–400.

Staeheli, L.A., Ehrkamp, P., Leitner, H., & Nagel, C.R. (2012). Dreaming the ordinary: Daily life and the complex geographies of citizenship. *Progress in Human Geography 36*(5): 628–644.

Summers, B. (2022). Black insurgent aesthetics and the public imaginary. *Urban Geography 43*(6): 837–847.

Swyngedouw, E. (2014). Where is the political? Insurgent mobilisations and the incipient "return of the political". *Space and Polity 18*(2): 122–136.

Sziarto, K.M., & Leitner, H. (2010). Immigrants riding for justice: Space-time and emotions in the construction of a counterpublic. *Political Geography 29*(7): 381–391.

Varsanyi, M.W. (2006). Interrogating "urban citizenship" vis-à-vis undocumented migration. *Citizenship Studies 10*(2): 229–249.

Walia, H. (2021). *Border and Rule: Global Migration, Capitalism, and the Rise of Racist Nationalism*. London: Haymarket Books.

Walters, W. (2002). Deportation, expulsion, and the international police of aliens. *Citizenship Studies* 6(3): 265–292. doi:10.1080/1362102022000011612

Wedeen, L. (2007). The politics of deliberation: Qāt chews as public spheres in Yemen. *Public Culture 19*(1): 59–84.

Wood, B.E. (2022). Youth citizenship: Expanding conceptions of the young citizen. *Geography Compass 16*(12): e12669. doi:10.1111/gec3.12669

Wright Rigueur, L., & Beshlian, A. (2019). The history and progress of black citizenship. *Du Bois Review: Social Science Research on Race 16*(1): 267–277.

Wynter, S. (2003). Unsettling the coloniality of being/power/truth/freedom: Towards the human, after man, its overrepresentation—an argument. *CR: The New Centennial Review 3*(3): 257–337.

Zepeda-Millán, C. (2017). *Latino Mass Mobilization: Immigration, Racialization, and Activism*. New York: Cambridge University Press.

Chapter 13

Security

Lauren L. Martin
Department of Geography, University of Durham, Durham, UK

What is security? The term can refer to a state of being, psychological stability, postconflict peace, the attenuation of threats to life and personal liberty, stable livelihoods, predictable financial futures, physical safety, enforced territorial borders, or emancipation from oppression, but it is a term through which actors seek to create certain kinds of political possibilities (Neocleous 2008; Ingram & Dodds 2009; Zedner 2009; Philo 2012; Dalby 2022). For state security agencies, achieving security requires surveillance, data analysis, and free rein to inspect, detain, and exclude suspect persons from national territory. For those contesting state security practices, security means defending privacy, ending discriminatory rhetoric, or defining alternative political norms. For those fighting domestic violence, security requires state and community protection and spaces of recovery and autonomy. For those fighting starvation, security means consistent access to affordable sustenance and employment. It is a technical term and a political ideal, a rationale for enforcing social order and the suspension of law; it is the goal of humanitarian aid. Security is, in short, a floating signifier imbued with the power of possibility and mastery of the future.

If the World Trade Center attacks of September 11, 2001, oriented political geographical research on security over the two subsequent decades, the Covid-19 pandemic that began in 2020 shifted popular, political, and academic understandings of security in ways that we have only just begun to conceptualize. The scale of the data collected and transformed into infection rates, hospital bed demands, social distancing recommendations, and population-wide closures was immense, as was that for border closures, quarantines, and negative test requirements

The Wiley Blackwell Companion to Political Geography, Second Edition.
Edited by Virginie Mamadouh, Natalie Koch, Chih Yuan Woon, and John Agnew.

(Dodds et al. 2020; Lyon 2022). The transnational logistics networks enabling global supply chains also required cross-border human mobility as well as in-person production lines. In other words, the virus thrived on political, economic, and social life, on the human and more-than-human geographies of both spectacular and mundane relationships (Andrews et al. 2021). Yet Covid-19 emerged – and flourished – in contexts ridden with existing inequalities: racialized, gendered, and classed access to healthcare, healthy workplaces and housing, clean urban air, furlough schemes, and the ability to work from home (Dodds et al. 2020). States responded with unprecedented claims to emergency powers, invoked through esoteric legal statutes from previous eras that were, nevertheless, difficult to roll back (Toscano 2020; Kipfer & Mohamud 2021; Mamadouh 2022; Torres et al. 2023). We have yet to contend with how the mandate to social distance expanded prophylactic state power, nor with how the ability to isolate itself spatialized various forms of privilege. At its heart, the pandemic was a security crisis: the specter of mass death, empty shop shelves, and spatial isolation. The sociality that gives daily life shape, substance, and rhythm was not only suspended but pathologized.

Other moments have defined security practices as well: spectacular attacks in Bali, Paris, Mumbai, Tunisia, and Manchester; mass shootings in the United States; the Arab Spring in 2011; the 2015 so-called migration crisis in Europe; George Floyd's murder in 2020; and the events beginning on October 7, 2023, in Palestine and Israel have become shorthand for transformative moments in security practices. This list is far from exhaustive but in each event, the security of multiple groups, places, and institutions is at stake. Facing scenes of mass death, whether viral or military, political geographers grapple not only with states' conceptualizations of security and safety but with groups facing – and resisting – annihilation as peoples. Security animates political life, but the political claims over who should be secured, at whose expense, and at what cost remain hotly contested. As anti- and decolonial struggles gain volume and visibility and as climate change alters lifeworlds irrevocably, the persistence or annihilation of people and places invokes geographies of security with renewed urgency. Moreover, the long unfolding of many of these events has provoked interrogation of "long emergencies" (Anderson et al. 2020) and slow violence (Grove et al. 2022).

Political geographers researching security will need theoretical, methodological, and analytical tools to meet these challenging research contexts. This chapter introduces key theories of security in political geography, security studies, and international relations, from realist to poststructuralist to feminist. These theories work from distinct epistemological and ontological perspectives, which, in turn, imply different methodological choices and research questions. As the sections here will show, security research remains diverse, revealing security's power to insert life-and-death urgency into mundane everyday practices and enable new governmental practices.

Theoretical and Methodological Approaches to Security

The relationship between security and state power has a long genealogy and was foundational in Hobbesian and liberal conceptions of the relationship between liberty and security (Foucault 2010). In brief, classical liberal theorists debated the responsibility of governments to create stable futures for their citizens, futures in which citizens could pursue their self-interests in commerce or labor. This concern to provide security to citizens *as populations* came to define nineteenth-century progressive disciplinary institutions seeking to make proper moral subjects, and eventually twentieth-century social security and welfare state institutions. Following World War II, state security efforts focused primarily on *national* security, defined by a bipolar East–West ordering of global power. Cold War security studies understood states as the primary global actors, either as threats to or purveyors of security. Nuclear proliferation and its containment raised the specter of species-wide destruction, and critics of proliferation responded by advocating the creation of an international order, ruled by international laws constraining state actions during conflict. Realist and positivist approaches to security, presuming that states act according to objectively discernible rules of self-interest, benefit maximization, and cost minimization, dominated the field of strategic security studies and international relations (Zedner 2009). During this time, military geography primarily concerned itself with applying geographical methods to military problems, an alliance that met with vehement criticism with the rise of critical geography in the 1970s (Flint 2005; Mamadouh 2005; Woodward 2005) and that continues in relation to recent wars (Belcher 2018; Jones 2020) as well as peace-making (Koopman 2011; Flint & Dempsey 2023). In part responding to the myopia of bipolar Cold War geopolitics and in part to the increasing importance of nonstate actors and ethnonationalism in post–Cold War conflict, mainstream security studies now covers a wide range of actors, problems, and phenomena.

Political geographical research on security has built on critical approaches to geopolitics, international relations, and development. Drawing on the work of Michel Foucault and Jacques Derrida, critical geopolitics extended the critical projects of critical security studies, international relations, and feminist geography to the spatial categories of the international state system (Campbell 1992; Tickner 1992; Dillon 1996; O' Tuathail 1996; O' Tuathail & Dalby 1998; Enloe 2000). For example, the categorization of "rogue" and "failed" states divides the world into zones of order and disorder, enabling particular forms of intervention, such as humanitarian missions or preemptive military operations. "Failed" and "rogue" states are considered incapable of guaranteeing the physical security of citizens and fostering a suitable economy in which to make a livelihood. These maps visualize security threats by dividing and categorizing global space according to neoliberal conventions of good governance that enable the creation of different rules for "us" and "them" (Mitchell 2010). Diagnosing states' failures to achieve particular forms of biopolitical governance and social control, these categorizations are knowledge practices that rely on intertwined neoliberal visions of globalized economic interdependence and security. Thus, neoliberal development discourses (deregulation, privatization, accountability) feed into geopolitical mappings of order and disorder and justify military interventions on humanitarian grounds (Duffield 2007; Elden 2009).

While traditional approaches to security debate the ability of these categories to explain state behavior, for critical political geographers the accuracy of these denominations is not the point. Rather, the task is to ask how these categories produce spatial imaginaries that, in turn, undergird justifications for security interventions. To critique the space–times of these geographical imaginaries, however, we need to understand their impact, power, and circulation, especially their ability to naturalize categories, spatial imaginaries, and interventions. In the following sections I review the theoretical and methodological approaches important to political geographical research on security.

Securitization

Most famously articulated as "securitization theory" by the Copenhagen School (Buzan, Waever, & Wilde 1998; Buzan & Waever 2003), constructivist approaches argue that people, places, or issues are securitized when elite actors make claims about security through illocutionary speech acts and these claims are accepted by public consensus. Ackleson (2005), for example, argues that pre-9/11 US border practices were securitized through elite constructions of border space as chaotic, and then claimed to be potential avenues for terrorists in post-9/11 border policymaking. By coding the US–Mexico border as a disorderly space, border policy elites successfully translated border enforcement into a matter of national security and this framing continues to shape immigration and border policy. Rehashing War on Terror–era discourses from 2003 to 2011, US president Trump's claims that chaotic, uncontrolled borderlands presented an imminent threat to US national security resulted in massive funding infusions, wall-building projects, technological surveillance, and higher staffing, with deadly effect in this first term (Abrego et al. 2017; Boyce et al. 2019).

Performativity

Taking a *performative* approach to securitization, Bialasiewicz et al. (2007) argue that elite claims are more than linguistic speech acts, and that securitization requires more than agreement among elite power networks; certain forms of constructivism – and its critiques – rely on a problematic distinction between linguistic speech acts and material practices. That is, both constructivists who focus on the content of policy texts and critics who argue for on-the-ground analysis of policy impacts rely on a problematic separation between an ideal, discursive realm and a real, material realm. Theories of performativity, however, begin from a different understanding of what language is and how it works. Language is not separate from the material world, useful for describing things, places, or situations that exist independently from representations of them. Rather, the things we take for granted are called into being through discourse, so that ideal and real, discourse and materiality are inextricably bound up with each other. As Bialasiewicz et al. (2007: 407) write, "discourse is then not something that subjects use in order to describe objects; it is that which constitutes both subjects and objects." From this perspective, security discourses produce the worlds they represent, and these discourses gain currency not because certain actors hold power to stipulate policies, but because they reiterate shared mappings of threat and danger, circulate those shared understandings among a variety of institutions, and rework familiar geopolitical tropes. For Jones (2024), these imaginaries of security and threat operate through embodied emotions, producing calculated "emotionscapes" in diplomatic geopolitics. Theorizing language and discourse as performative, political geographers ask how documents and statements circulate, produce authoritative expertise, and produce accepted or "common-sense" geopolitical imaginaries.

Security Assemblages

Similarly, Foucault-inflected approaches ask how imaginations of future catastrophes and dangerous spaces produce particular modes of governing (Dillon 2007; De Goede 2012; Grove 2012; Amoore 2013). Understanding security as a mode of governing, some scholars prefer to focus on an *assemblage* of actors, institutions, embodied practices, texts, narratives, infrastructure, and relations that includes but far exceeds the state. Assemblages are not coherent or closed institutions, but fields of power relations that overlap, conflict, change, and break down. Thinking of security in terms of assemblage rather than institutions or elite actors acknowledges how nonstate actors play key roles. As such, this approach refuses accounts of intentional individual actors making the world according to policy directives. Instead, governmentality approaches focus on a milieu of legal, policy, popular discourses; the production of particular forms of expertise; and how these authorize both formal and informal security practices (Abrahamsen & Williams 2012; Tretter 2023). In turn, political geographers conceptualizing power as dispersed and distributed also trace the continual respatialization of territory, borders, and populations (Burridge et al. 2017).

Security Governmentalities

Working from Foucault and Bourdieu, Didier Bigo's influential work on the securitization of migration argues that the "governmentality of unease" emerges from a field of security professionals with expertise in managing a series of social threats, such as criminals, migrants, terrorists (Bigo 2002). Thus, legitimizing security professionals' expertise requires the active production of insecurity and a generalized sense of "unease" that can be fixed to new threats as they emerge. Generating trust in security experts, the governmentality of unease widens security professionals' discretionary authority, changing the role of policing and surveillance in social life. These approaches require a different explanatory mode from realist and liberal accounts that explain security policy development in terms of intentional acts and individuals.

To summarize, governmentality approaches do not ask whether security discourses accurately represent the world, whether security technologies adequately calculate risks and threats, nor whether there is a gap between policy and practice. Instead, governmentality approaches ask what security interventions, actions, and political rationalities these knowledge practices enable. They ask: What do the geographical imaginations of integration, disconnection, failed states, and gray economies *make possible*? What conditions of possibility do they engender, and how do they work through risk calculations, definitions of safety/threat, alignments of economic governance and national security, visual surveillance technologies, and media representations (De Goede 2012; Amoore 2013)? For Amoore (2013: 5), analyzing security policies across seemingly distinct spheres of governance, foreign policy, and economy reveals "resonances" between different practices, an "infiltration of each one into the other, such that a moving complex emerges." There remains some ambiguity in the literature, however, about how to make sense of these quasi-causal relationships. For some, the discursive production of security and insecurity happens simultaneously; insecurity is the necessary other to security's self-image. For others, insecurity's production emerges from the unintended consequences of security practices and so insecurity cannot be read off textual and visual representations of security. Thus, theoretical and methodological choices about how and where to study security are important to theorizing its spatial politics.

Everyday and Embodied Security

For feminist political geographers, state power is embodied and so, to understand how power operates, researchers should pay close attention to the prosaic and everyday production of state power (Painter 2006; Mountz 2010). Drawing on interdisciplinary feminist theories of embodiment and power, feminist political geographers analyze how power works through bodies, how it inscribes categories onto bodies, and how intimacy, kinship, reproduction, and corporeality are critical sites for the territorialization of state power (Dixon 2015; Smith 2020). For example, securing the homeland invokes feelings of familial love and desire to protect; anxieties over immigration evince "demographic fever dreams" of historically privileged groups being "replaced" by minoritized groups (Gökarıksel et al. 2019). These narratives evoke fears of erasure against internal "others," making the future of the state a matter not of territorial integrity as in the post-9/11 era, but of cultural and social identity. These fears contribute to affective atmospheres of nationalism and terror (Closs Stephens 2022; Fregonese & Laketa 2022).

Focusing on embodied geographies of terror, fear, and insecurity, however, challenges political geographers to turn away from traditional sites of security – state security agencies – and work in the opposite direction: beginning from experiences of terror, fear, and violence in everyday life. For Pain (2014), lived experiences of domestic violence bear striking similarities to the diffuse and totalizing accounts of terrorism, except that intimate, domestic, privatized violence is woefully overlooked by state agencies. Focusing on everyday experiences of marginalization and struggles against it, this work marks out a broader field of political struggle over security. Hiemstra (2019), for example, analyzed the US immigration system by working in Ecuador with families of detained migrants. She connected the emotional, physical, and financial insecurities of Ecuadorian families to specific US immigration detention practices, showing how those practices worked to create insecurity abroad. Thus, when feminist political geographers bring theorizations of embodiment, intimacy, and the everyday to bear on security, they require methodologies of connection (Massaro & Williams 2013), "counter-topographies" (Katz 2001) that question rather than accept mappings of security proffered by states and international organizations.

Ontological Security

Moving security from the state to excluded, marginalized, and abandoned groups raises critical questions: Who is being secured, from what, how, and at whose expense? Recent work on security has, therefore, emphasized conceptualizing marginalized people's *ontological security*, or a stable framework for being and persisting in the world. Ontological security refers to the desire or demand for a stable sense of self over time, a concept drawn from psychoanalysis (see Laing 1960 in Mitzen 2006; Delehanty & Steele 2009; De Leon 2020). In other words, conceptualizing ontological security helps us understand when groups may make policy decisions to maintain identity and morality, rather than being driven solely by physical safety. This analysis can move in different directions. On the one hand, ontological security helps to explain the UK's Brexit vote as shoring up national identity in response to anxieties about changing British culture (Shani 2017; Browning 2018). On the other hand, this approach recognizes that for groups facing genocidal violence, colonial erasure, or permanent displacement, security requires more than surviving violence. Grappling with climate change's challenge to an enduring human presence in island and arctic states, for example, raises questions of how place-based cultures, lifeways, and cosmologies face annihilation (De Leon 2020). Moreover, analyzing the ontological insecurity of colonized and displaced people relocates the problem of security in state practices of marginalization and "group-differentiated vulnerability to premature death" (Gilmore 2017: 225). These concerns refract through geopolitical struggles over resources as well, as different groups and states make competing ontological claims about those resources (e.g., Götz & Middleton 2020). Ontological security thus enables postcolonial scholars to recognize the coexistence of multiple versions of security in their own terms, rather than translating religious, marginalized, or colonized worlds into liberal terms (Shani 2017; Sullivan & Akhtar 2019).

Security connects political geography's core concepts – borders, territory, power, governance, politics – to a wide range of institutions, practices, and places. Debates over the form and function of modern states revolve around security: who provides it, who deserves it, at what scale, and through which individual and collective practices. Next I bring together recent political geography research on security. As you will see, this work cuts across disciplines and engages with security beyond the conceptual frameworks already reviewed.

Securing Volumetric Spaces

Through the research on security, political geographers have grappled with the respatialization of sovereignty, especially how state militaries, police, and border guards engage in "flexible territoriality" (Novak 2011). Security and military tactics of moving through walls and tunnels (Elden 2013) and occupying space through odors and chemical weapons (Joronen 2023) point to the ways in which security practices attempt to weaponize substances, surfaces, containment, movement, and mobility, not to mention disrupted supply chains, energy, and communications (Graham 2011). The 2023 conflict in Gaza, Palestine, for example, exemplifies the use of these tactics. Scholars have therefore argued that theorizing sovereign, state, and securitized spaces requires a theorization of space that goes beyond the two dimensions of cartography (Crampton 2009) to include air (Adey 2014, 2015); oceans and ice flows (Eiterjord 2023); drone policing and warfare (Klauser 2022); gunfire-sensing technologies (Linnemann & Turner 2022); and terraforming (Bennett & Dodds 2024). As Campbell (2019) argues, these efforts to appreciate the changing spatiality of state power resonate with concepts like topological space and Foucault's milieu, which seek to appreciate the complexity and multidimensionality of security (see also vol. 24, no. 1 of *Territory, Politics and Governance*). Scholars have also theorized the respatialization of security – and forms of power expressed as/through securitization – through research on migration and mobility, technology, and environmental politics, which I turn to next.

Security, Data, and Technology

The seeming novelty of visual surveillance and its incorporation into an assemblage of data-based security procedures in the post-9/11 moment was more of a fantasy, which only now sees its partial realization in passport e-gates, passenger prescreening, and imaging technologies. Political geographers have interrogated security technologies along two similar lines: invisible, off-site, remote data-based surveillance of everyday life (Amoore 2006); and the proliferation of increasingly sophisticated visual surveillance technologies (Gregory 2008; Kindervater 2016; Belcher 2018). The first asks how Big Data, algorithms, and machine learning enable both public and private organizations to amass individual and collective data (Aradau & Blanke 2015). The second thread asks what forms of state power emerge through technology's novel "ways of seeing," when "seeing" involves selecting, composing, and visualizing data, such as drone surveillance, remote sensing, and satellite imagery (Kindervater 2017; Gregory 2023; Hall 2024). And a third emerging thread of research interrogates how artificial intelligence and machine learning enable drone and other visual surveillance technologies (Wilcox 2017), raising questions about responsibility for automated decisions in times of war and ethics (Amoore 2013, 2020). Critically, it is not just states using these technologies to monitor security threats, conflict zones, or migration flows: private companies dominate the innovation landscape, while humanitarian organizations, human rights monitoring, and "tech for good" hackathons use similar tools to assist in aid distribution, document human rights abuses, and predict "hotspots" (Burns 2019; Tazzioli 2022). The Covid-19 pandemic threw these strands of research into high relief, as epidemiological surveillance was implemented quickly and en masse (Lyon 2022), then visualized and publicized continuously in dashboards, maps, and charts (Gleeson et al. 2022). For political geographers, this raises important questions about the centrality of the state as the leading actor in security assemblages. Depending on the context, state agencies, regional multilateral security agencies, private companies, and international humanitarian/human rights organizations work with and against each other using these kinds of technologies.

Borders and Mobility

Unauthorized, autonomous cross-border mobilities are paradigmatic objects and subjects of state security politics. From fears of racial mixing and demographic displacement to religious and cultural change to wage depression, migrants are frequent scapegoats for a range of anxieties (Puar 2007; Nevins 2010; Gökarıksel et al. 2019; Jones 2021). The post-9/11 era firmly sedimented the association between foreigners, terrorism, and, particularly, Islam, an association aided by attacks in Paris, Manchester, and Moscow. Border closures, quarantines, and visualizations of invading pathogens/people during the Covid-19 pandemic played on similar associations, with widespread reports of xenophobic violence (Baldacchino 2021; Berescu et al. 2021; Denaro & Boccagni 2022). Thus, the securitization of migration has included the racialization of migrants, creating conditions of permanent foreignness even for multigenerational communities. This move, in turn, has justified the internalization of militarized policing and the externalization of border controls in the Global North (Coleman 2016; Abrego et al. 2017; Hiemstra 2019; Darling 2022). The securitization of migration is part of the wider securitization of mobility in general (Amoore 2006) and the generalization of data collection, database interoperability, and machine learning to identify anomalies in normalized movements (Amoore 2013). Visual surveillance technologies have also been key to sensing imminent arrivals, such as the European Union's surveillance of the Mediterranean and the United States' ground sensors. Thus, technologies in this context are used to obstruct migrants' mobilities (Tazzioli 2023). While states have expanded policing beyond their borders, access to territory to claim asylum or seek safety has shrunk (Mountz 2020). These processes were further intensified during the Covid-19 pandemic, as quarantine enclosures were used alongside detention and encampment to confine migrants (Tazzioli & Stierl 2021; Giacomelli & Walker 2022). Thus, problematizing mobility as a border problem, or borders as a human mobility problem, has enabled a massive expansion of state policing and surveillance functions, which has, in turn, respatialized state power itself (Burridge et al. 2017).

Climate-Changed Futures

From natural resources to terrain to pollution, security has always included consideration of nature, environmental change, and, increasingly, the Earth's climate systems (Dalby 2022). Scholars and policy-makers make the link between environment, climate change, and security in a variety of ways, based on different assumptions about who and what is at risk; these different positions imply very different policy solutions to climate problems as well (e.g., adaptation, mitigation, social change; McDonald 2013). In the last few years, the slow response to climate change has been deeply and widely politicized, so much so that the very future of humanity on planet Earth is at stake at every scale and in every decision, from the banal to the existential (Klein 2015). This is, for Dalby (2022), redefining concepts of security itself and its relationships to the concepts examined earlier. Marketization, insurance, and financial instruments have been especially important for addressing the unpredictability of climate change (Grove 2012; Dalby 2013). The intertwining of ecological and geopolitical concerns has led to calls to conceptualize "geopolitical ecology" (Belcher et al. 2020; Graddy-Lovelace & Ranganathan 2023) and to trace the climate impacts of military operations (Neimark et al. 2023). While conflict- and climate-induced displacements intersect, migration is increasingly presumed to be a climate adaptation strategy. Paired with the demographic anxieties surrounding securitized international migration more broadly, climate migration evokes "racial futurism" and often implicit debates over whose future will survive the Earth's changing climate systems (Baldwin 2022). Research at the intersections of climate change, politics, and security will need to work across subdisciplinary and disciplinary boundaries to track the quick pace of change and the emerging geopolitical imaginaries produced in this quickly changing context.

Securing Logistics

Finally, sitting at the intersection of the themes already discussed are global logistics and supply chains. As Cowen (2014) documents, logistics has become a key knowledge practice enabling global capitalism. "Logistification" transforms ways of organizing resources, people, space, and time around norms of efficiency and speed; the humanitarian sector, for example, operates extensive logistical operations, managed through private-sector technologies and logics (Pascucci 2021). Similarly, military operations require logistical projects, demanding vast resources and, in line with the earlier discussion, producing massive carbon footprints (Neimark et al. 2023). And like the themes already considered, the Covid-19 pandemic highlighted the mutual interdependencies throughout supply chains, particularly for medicine and vaccines (Liu & Bennett 2020). Furthermore, the 2024 Houthi attacks on US Navy and cargo ships highlighted the importance of global shipping to state security (Spender et al. 2024). Maritime shipping requires a vast, flexible machinery of material and labor and is governed by distinct maritime governance regimes, conditions that enable labor exploitation and military interventions (Khalili 2020; Bolton 2024). This research has inspired new analyses of bordering and mobility, as scholars trace the logistics and infrastructure upholding immigration detention and deportation (Walters 2018). Focused on mobility and connectivity, research on logistics, infrastructure, and security draws political economic questions about the operations of capitalism into political geography – and vice versa.

Conclusion

Emerging research on security thereby cuts across and draws together traditional fields of inquiry. Concentrating on security, in particular, focuses analysis on the politics of life and death. Both security and insecurity invoke a sense of existential threat and impending disaster that constricts political deliberation in favor of technocratic solutions. Thus security practices can simultaneously produce political marginalization and close down spaces in which that marginalization can be contested. This exceptional politics has long been a subject of international law and geopolitics, but the imbrication of those spheres with domestic policing and the proliferation of surveillance actively reconfigure the space and time of the political. For minorities, postcolonies, and occupied peoples, these are not new revelations, and the historical geographies of exceptional securitized spaces remain a rich, but understudied, field for political geographers. Much work remains to be done examining security politics beyond liberal democratic states of North America, the European Union, and Australia, as security has been a key political discourse of authoritarian, socialist, and colonial regimes as well. Security mobilizes political norms, produces difference, circulates discourses of threat to human livelihood, and provokes the imagination of multiple futures, threats, and desires. Yet while security discourses may evoke fear, bordering, and exclusion, security also justifies universal healthcare, education, inclusive communities, and solidarity. Political geographers often focus on *insecurity* in the analysis of security practices, but further engagement with feminist approaches to care, transnational solidarity movements, trans and queer practices of kinship, and community- and peace-making offer critically important opportunities for political geographers to analyze how security—and politics—might be reimagined. Our "long emergencies" will not be easily resolved and therefore challenge us to consider how producing political geographical knowledge contributes to these debates.

References

Abrahamsen, R., & Williams, M.C. (2012). *Security Beyond the State: Private Security in International Politics*. Cambridge: Cambridge University Press.

Abrego, L., Coleman, M., Martínez, D.E., Menjívar, C., & Slack, J. (2017). Making immigrants into criminals: Legal processes of criminalization in the post-IIRIRA era. *Journal on Migration and Human Security* 5(3): 694–715. doi:10.1177/233150241700500308

Ackleson, J. (2005). Constructing security on the U.S.–Mexico Border. *Political Geography* 24(2): 165–184. doi:10.1016/j.polgeo.2004.09.017

Adey, P. (2014). Security atmospheres or the crystallisation of worlds. *Environment and Planning D: Society and Space* 32(5): 834–851.

Adey, P. (2015). Air's affinities: Geopolitics, chemical affect and the force of the elemental. *Dialogues in Human Geography* 5(1): 54–75.

Amoore, L. (2006). Biometric borders: Governing mobilities in the war on terror. *Political Geography* 25(3): 336–351. doi:10.1016/j.polgeo.2006.02.001

Amoore, L. (2013). *The Politics of Possibility*. Durham, NC: Duke University Press.

Amoore, L. (2020). *Cloud Ethics*. Durham, NC: Duke University Press.

Anderson, B., Grove, K., Rickards, L., & Kearnes, M. (2020). Slow emergencies: Temporality and the racialized biopolitics of emergency governance. *Progress in Human Geography* 44(4): 621–639. doi:10.1177/0309132519849263

Andrews, G.J., Crooks, V.A., Pearce, J.R., & Messina, J.P. (eds) (2021). *COVID-19 and Similar Futures: Pandemic Geographies*. Cham: Springer International.

Aradau, C., & Blanke, T. (2015). The (Big) Data-security assemblage: Knowledge and critique. *Big Data & Society* 2(2). doi:10.1177/2053951715609066

Baldacchino, G. (2021). Extra-territorial quarantine in pandemic times. *Political Geography* 85: 102302. doi:10.1016/j.polgeo.2020.102302

Baldwin, A. (2022). *The Other of Climate Change: Racial Futurism, Migration, Humanism*. Lanham, MD: Rowman and Littlefield.

Belcher, O. (2018). Anatomy of a village razing: Counterinsurgency, violence, and securing the intimate in Afghanistan. *Political Geography* 62: 94–105. doi:10.1016/j.polgeo.2017.10.006

Belcher, O., Bigger, P., Neimark, B., & Kennelly, C. (2020). Hidden carbon costs of the "everywhere war": Logistics, geopolitical ecology, and the carbon boot-print of the US military. *Transactions of the Institute of British Geographers 45*(1): 65–80. doi:10.1111/tran.12319

Bennett, M., & Dodds, K. (2024). Earthly volumes, voluminous materialities: Working with apprehension. *Territory, Politics, Governance 12*(1): 1–11. doi:10.1080/21622671.2023.2242394

Berescu, C., Alexandrescu, F., & Anghel, I.M. (2021). Vulnerable Roma communities in times of the Covid-19 negative quarantine. *Moravian Geographical Reports 29*(2): 125–136. doi:10.2478/mgr-2021-0011

Bialasiewicz, L., Campbell, D., Elden, S., Graham, S., Jeffrey, A., & Williams, A.J. (2007). Performing security: The imaginative geographies of current US strategy. *Political Geography 26*(4): 405–422. doi:10.1016/j.polgeo.2006.12.002

Bigo, D. (2002). Security and immigration: Toward a critique of the governmentality of unease. *Alternatives: Global, Local, Political 27*: 63–92. doi:10.1177/03043754020270S105

Bolton, J. (2024). Supply nets: The logistics of seafarer abandonment. *Antipode 56*(4): 1172–1190. doi:10.1111/anti.13038

Boyce, G.A., Chambers, S.N., & Launius, S. (2019). Bodily inertia and the weaponization of the Sonoran Desert in US boundary enforcement: A GIS modeling of migration routes through Arizona's Altar Valley. *Journal on Migration and Human Security 7*(1): 25–35.

Browning, C.S. (2018). Brexit, existential anxiety and ontological (in)security. *European Security 27*(3): 336–355. doi:10.1080/09662839.2018.1497982

Burns, R. (2019). New frontiers of philanthro-capitalism: Digital technologies and humanitarianism. *Antipode 51*(4): 1101–1122. doi:10.1111/anti.12534

Burridge, A., Gill, N., Kocher, A., & Martin, L. (2017). Polymorphic borders. *Territory, Politics, Governance 5*(3): 239–251. doi:10.1080/21622671.2017.1297253

Buzan, B., & Wæver, O. (2003). *Regions and Powers: The Structure of International Security*. Cambridge: Cambridge University Press.

Buzan, B., Wæver, O., & de Wilde, J. (1998). *Security: A New Framework for Analysis*. Boulder, CO: Lynne Rienner.

Campbell, D. (1992). *Writing Security: United States Foreign Policy and the Politics of Identity*. Minneapolis, MN: University of Minnesota Press.

Campbell, E. (2019). Three-dimensional security: Layers, spheres, volumes, milieus. *Political Geography 69*: 10–21. doi:10.1016/j.polgeo.2018.11.010

Closs Stephens, A. (2022). *National Affects: The Everyday Atmospheres of Being Political*. New York: Bloomsbury Academic.

Coleman, M. (2016). State power in blue. *Political Geography 51*(1): 76–86. doi:10.1016/j.polgeo.2016.01.008

Cowen, D. (2014). *The Deadly Life of Logistics: Mapping Violence in Global Trade*. Minneapolis, MN: University of Minnesota Press.

Crampton, J.W. (2009). Cartography: Performative, participatory, political. *Progress in Human Geography 33*(6): 840–848. doi:10.1177/0309132508105000

Dalby, S. (2013). The geopolitics of climate change. *Political Geography 37*(1): 38–47. doi:10.1016/j.polgeo.2013.09.004

Dalby, S. (2022). *Rethinking Environmental Security*. Cheltenham: Edward Elgar.

Darling, J. (2022). *Systems of Suffering: Governing Refugee Lives*. London: Pluto Press.

De Goede, M. (2012). *Speculative Security: The Politics of Pursuing Terrorist Monies*. Minneapolis, MN: University of Minnesota Press.

de Leon, J. (2020). Lakota experiences of (in)security: Cosmology and ontological security. *International Feminist Journal of Politics* 22(1): 33–62. doi:10.1080/14616742.2018.1527183

Delehanty, W.K., & Steele, B.J. (2009). Engaging the narrative in ontological (in)security theory: Insights from feminist IR. *Cambridge Review of International Affairs* 22(3): 523–540. doi:10.1080/09557570903104024

Denaro, C., & Boccagni, P. (2022). Migration, borders and "de-bordering" in pandemic times: Voices as interlocution from quarantine ships in Italy. *Journal of Borderlands Studies* 39(2): 351–376. doi:10.1080/08865655.2022.2108110

Dillon, M. (1996). *Politics of Security: Towards a Political Philosophy of Continental Thought*. London: Routledge.

Dillon, M. (2007). Governing through contingency: The security of biopolitical governance. *Political Geography* 26(1): 41–47. doi:10.1016/j.polgeo.2006.08.003

Dixon, D.P. (2015). *Feminist Geopolitics: Material States*. Farnham: Ashgate.

Dodds, K., Broto, V.C., Detterbeck, K., et al. (2020). The COVID-19 pandemic: Territorial, political and governance dimensions of the crisis. *Territory, Politics, Governance* 8(3): 289–298.

Duffield, M. (2007). *Development, Security and Unending War: Governing the World of Peoples*. Cambridge: Polity Press.

Eiterjord, T. (2023). Securitise the volume: Epistemic territorialisation and the geopolitics of China's Arctic research. *Territory, Politics, Governance* 12911: 93–111. doi:10.1080/21622671.2023.2179535

Elden, S. (2009). *Terror and Territory*. Minneapolis, MN: University of Minnesota Press.

Elden, S. (2013). Secure the volume: Vertical geopolitics and the depth of power. *Political Geography* 34: 35–51. doi:10.1016/j.polgeo.2012.12.009

Enloe, C.H. (2000). *Bananas, Beaches and Bases: Making Feminist Sense of International Politics*. Los Angeles, CA: University of California Press.

Flint, C. (2005). *The Geography of War and Peace: From Death Camps to Diplomats*. Oxford: Oxford University Press.

Flint, C., & Dempsey, K.E. (2023). *Making Geographies of Peace and Conflict*. London: Routledge.

Foucault, M. (2010). *The Birth of Biopolitics: Lectures at the Collège de France, 1978–1979*. New York: Palgrave Macmillan.

Fregonese, S., & Laketa, S. (2022). Urban atmospheres of terror. *Political Geography* 96: 102569. doi:10.1016/j.polgeo.2021.102569

Giacomelli, E., & Walker, S. (2022). On board the quarantine-ship as "floating hotspot": Creeping externalization practices in the Mediterranean Sea. *Ethnography*: 146613812211005. doi:10.1177/14661381221100532

Gilmore, R.W. (2017). Abolition geography and the problem of innocence. In Johnson, G.T., & Lubin, A. (eds), *Futures of Black Radicalism*. London: Verso, 225–240.

Gleeson, J., Kitchin, R., & McCarthy, E. (2022). Dashboards and public health: The development, impacts, and lessons from the Irish Government COVID-19 dashboards. *American Journal of Public Health* 112(6): 896–899. doi:10.2105/AJPH.2022.306848

Gökarıksel, B., Neubert, C., & Smith, S. (2019). Demographic fever dreams: Fragile masculinity and population politics in the rise of the global right. *Signs: Journal of Women in Culture and Society 44*(3): 561–587.

Götz, J.M., & Middleton, C. (2020). Ontological politics of hydrosocial territories in the Salween River basin, Myanmar/Burma. *Political Geography 78*: 102115. doi:10.1016/j.polgeo.2019.102115

Graddy-Lovelace, G., & Ranganathan, M. (2023). Geopolitical ecology for our times. *Political Geography 103034*. doi:10.1016/j.polgeo.2023.103034

Graham, S. (2011). *Cities under Siege: The New Military Urbanism*. New York: Verso.

Gregory, D. (2008). 'Rush to the intimate': Counterinsurgency and the cultural turn in late modern warfare. *Radical Philosophy 150*: 8–23.

Gregory, D. (2023). Midnight's victims. *Area Development and Policy 8*(4): 353–386. doi:10.1080/23792949.2023.2237558

Grove, K. (2012). Preempting the next disaster: Catastrophe insurance and the financialization of disaster management. *Security Dialogue 43*(2): 139–155. doi:10.1177/0967010612438434

Grove, K., Rickards, L., Anderson, B., & Kearnes, M. (2022). The uneven distribution of futurity: Slow emergencies and the event of COVID-19. *Geographical Research 60*(1): 6–17. doi:10.1111/1745-5871.12501

Hall, K. (2024). Experimental air power: Early British drone programs and Western violence in the interwar period. *Critical Military Studies 10*(4): 547–561. doi:10.1080/23337486.2023.2300024

Hiemstra, N. (2019). *Detain and Deport: The Chaotic U.S. Immigration Enforcement Regime*. Athens, GA: University of Georgia Press.

Ingram, A., & Klaus Dodds, K. (2009). *Spaces of Security and Insecurity: Geographies of the War on Terror*. Farnham: Ashgate.

Jones, A. (2024). Waging word wars: The "emotionscape" of the United Nations Security Council and the Russian war in Ukraine. *Political Geography 108*: 103032. doi:10.1016/j.polgeo.2023.103032

Jones, C. (2020). *The War Lawyers*. New York: Oxford University Press.

Jones, R. (2021). *White Borders: The History of Race and Immigration in the United States from Chinese Exclusion to the Border Wall*. Boston, MA: Beacon Press.

Joronen, M. (2023). Atmospheric negations: Weaponising breathing, attuning irreducible bodies. *Environment and Planning D: Society and Space 41*(5): 765–783. doi:10.1177/02637758231203061

Katz, C. (2001). On the grounds of globalization: A topography for feminist political engagement. *Signs: Journal of Women in Culture and Society 26*(4): 1213–1234.

Khalili, L. (2020). *Sinews of War and Trade: Shipping and Capitalism in the Arabian Peninsula*. London: Verso.

Kindervater, K.H. (2016). The emergence of lethal surveillance: Watching and killing in the history of drone technology. *Security Dialogue 47*(3): 223–238. doi:10.1177/0967010615616011

Kindervater, K.H. (2017). The technological rationality of the drone strike. *Critical Studies on Security 5*(1): 28–44. doi:10.1080/21624887.2017.1329472

Kipfer, S., & Mohamud, J. (2021). The pandemic as political emergency. *Studies in Political Economy 102*(3): 268–288. doi:10.1080/07078552.2021.2000212

Klauser, F. (2022). Policing with the drone: Towards an aerial geopolitics of security. *Security Dialogue 53*(2): 148–163. doi:10.1177/0967010621992661

Klein, N. (2015). *This Changes Everything: Capitalism vs. the Climate.* London: Penguin.

Koopman, S. (2011). Alter-geopolitics: Other securities are happening. *Geoforum 42*(3): 274–284. doi:10.1016/j.geoforum.2011.01.007

Linnemann, T., & Turner, J. (2022). Three-dimensional policeman: Security, sovereignty and volumetric police power. *Theoretical Criminology 26*(1): 23–39.

Liu, X., & Bennett, M.M. (2020). Viral borders: COVID-19's effects on securitization, surveillance, and identity in Mainland China and Hong Kong. *Dialogues in Human Geography 10*(2): 158–163. doi:10.1177/2043820620933828

Lyon, D. (2022). *Pandemic Surveillance.* Cambridge: Polity Press.

Mamadouh, V. (2005). Geography and war, geographers and peace. In Flint, C. (ed.), *The Geography of War and Peace: From Death Camps to Diplomats.* Oxford: Oxford University Press, 26–60.

Mamadouh, V. (2022). Pandemic geopolitics and the bordering of COVID-19: Academic and lay geographies of the pandemic and policies to contain and mitigate the novel coronavirus. In Brunn, S., & Gilbreath, D. (eds), *COVID-19 and a World of Ad Hoc Geographies.* Cham: Springer International, 65–87.

Massaro, V.A., & Williams, J. (2013). Feminist geopolitics. *Geography Compass 7*(8): 567–577. doi:10.1111/gec3.12054

McDonald, M. (2013). Discourses of climate security. *Political Geography 33*: 42–51. doi:10.1016/j.polgeo.2013.01.002

Mitchell, K. (2010). Ungoverned space: Global security and the geopolitics of broken windows. *Political Geography 29*(5): 289–297. doi:10.1016/j.polgeo.2010.03.004

Mitzen, J. (2006). Ontological security in world politics: State identity and the security dilemma. *European Journal of International Relations 12*(3): 341–370. doi:10.1177/1354066106067346

Mountz, A. (2010). *Seeking Asylum: Human Smuggling and Bureaucracy at the Border.* Minneapolis, MN: University of Minnesota Press.

Mountz, A. (2020). *The Death of Asylum: Hidden Geographies of the Enforcement Archipelago.* Minneapolis, MN: University of Minnesota Press.

Neimark, B., Belcher, O., Ashworth, K., & Larbi, R. (2023). Concrete impacts: Blast walls, wartime emissions, and the US occupation of Iraq. *Antipode 56*(3): 983–1005. doi:10.1111/anti.13006

Neocleous, M. (2008). *Critique of Security.* Edinburgh: Edinburgh University Press.

Nevins, J. (2010). *Operation Gatekeeper and Beyond: The War on "Illegals" and the Remaking of the U.S.–Mexico Boundary.* London: Routledge.

Novak, P. (2011). The flexible territoriality of borders. *Geopolitics 16*(4): 741–767. doi:10.1080/14650045.2010.494190

Ó Tuathail, G. (1996). *Critical Geopolitics.* London: Routledge.

Ó Tuathail, G., & Dalby, S. (1998). *Rethinking Geopolitics.* London: Routledge.

Pain, R. (2014). Everyday terrorism: Connecting domestic violence and global terrorism. *Progress in Human Geography 38*(4): 531–550. doi:10.1177/0309132513512231

Painter, J. (2006). Prosaic geographies of stateness. *Political Geography 25*(7): 752–774.

Pascucci, E. (2021). More logistics, less aid: Humanitarian-business partnerships and sustainability in the refugee camp. *World Development 142*: 105424. doi:10.1016/j.worlddev.2021.105424

Philo, C. (2012). Security of geography/geography of security. *Transactions of the Institute of British Geographers 37*(1): 1–7. doi:10.1111/j.1475-5661.2011.00488.x

Puar, J. (2007). *Terrorist Assemblages: Homonationalism in Queer Times*. Durham, NC: Duke University Press.

Shani, G. (2017). Human security as ontological security: A post-colonial approach. *Postcolonial Studies 20*(3): 275–293. doi:10.1080/13688790.2017.1378062

Smith, S. (2020). *Intimate Geopolitics: Love, Territory, and the Future on India's Northern Threshold*. New Brunswick, NJ: Rutgers University Press.

Spender, T., Cheetham, J., and Gardner, F. (2024). Three killed in Houthi missile attack on cargo ship in Gulf of Aden. *BBC News*, March 7. https://www.bbc.com/news/world-middle-east-68490695 (accessed March 29, 2024).

Sullivan, P., & Akhtar, P. (2019). The effect of territorial stigmatisation processes on ontological security: A case-study of Bradford politics. *Political Geography 68*: 46–54. doi:10.1016/j.polgeo.2018.10.009

Tazzioli, M. (2022). Extract, datafy and disrupt: Refugees' subjectivities between data abundance and data disregard. *Geopolitics 27*(1): 70–88. doi:10.1080/14650045.2020.1822332

Tazzioli, M. (2023). Digital expulsions: Refugees' carcerality and the technological disruptions of asylum. *Environment and Planning C: Politics and Spaces 41*(7): 1301–1316. doi:10.1177/23996544231196680

Tazzioli, M., & Stierl, M. (2021). "We closed the ports to protect refugees." Hygienic borders and deterrence humanitarianism during Covid-19. *International Political Sociology 15*(4): 539–558. doi:10.1093/ips/olab023

Tickner, J.A. (1992). *Gender in International Relations: Feminist Perspectives on Achieving Global Security*. New York: Columbia University Press.

Torres, R.M., Glockner, V., Niño-Vega, N., et al. (2023). Lockdown and the list: Mexican refugees, asylum denial, and the feminist geopolitics of esperar (waiting/hoping). *Environment and Planning C: Politics and Spaces 41*(8): 1503–1520. doi:10.1177/23996544221118906

Toscano, A. (2020). The state of the pandemic. *Historical Materialism 28*(4): 3–23. doi:10.1163/1569206X-12342804

Tretter, M. (2023). Sovereignty in the digital and contact tracing apps. *DISO 2*(1): 2. doi:10.1007/s44206-022-00030-2

Walters, W. (2018). Aviation as deportation infrastructure: Airports, planes, and expulsion. *Journal of Ethnic and Migration Studies 44*(16): 2796–2817. doi:10.1080/1369183X.2017.1401517

Wilcox, L. (2017). Embodying algorithmic war: Gender, race, and the posthuman in drone warfare. *Security Dialogue 48*(1): 11–28. doi:10.1177/0967010616657947

Woodward, R. (2005). From military geography to militarism's geographies: Disciplinary engagements with the geographies of militarism and military activities. *Progress in Human Geography 29*(6): 718–740. doi:10.1191/0309132505ph579oa

Zedner, L. (2009). *Security*. Oxford: Routledge.

Chapter 14

Conflict

Amaël Cattaruzza

French Institute of Geopolitics, Université Paris 8, Saint-Denis, France

The context of the war in Ukraine and the resurgence of the Israeli–Palestinian conflict shed new light on the treatment of conflicts and wars in geopolitics and political geography. Russia's invasion of Ukraine on February 24, 2022, the Hamas attacks on Israel on October 7, 2023, and Israel's massive retaliation in the Gaza Strip have brought types of conflict that were thought to have disappeared back into focus. Tank columns crossing the border, heavy and systematic bombardments of urban targets with no regard for civilian casualties, and all the images we see daily in the press and other media take us back to phenomena we thought we could contain, thanks in part to international institutions and regulatory mechanisms that seemed to have been gradually leading us toward a more peaceful and less violent world. At the beginning of the twenty-first century, some analysts with a Western point of view even promised the end of war itself, based on the factors typically used to evaluate it: a yearly decrease in the number of direct deaths caused by war up to 2012, and therefore a decrease in the number of "wars," statistically defined as armed conflicts with at least 1000 deaths per year. But at the same time, geographers were showing the opposite. In 2009, a special issue of *Annals of the Association of American Geographers* was devoted to "Geographies of Peace and Armed Conflict," covering a number of conflicts and postconflict zones (Afghanistan, Iraq, Israel and Palestine, Sri Lanka, Thailand, Bosnia-Herzegovina, Liberia, etc.). As Audrey Kobayashi (2009: 819) says in her introduction: "A decade into the twenty-first century there seems no utopian end of warfare in sight." Today the situation looks even worse and since 2022 we have been talking about the "return" of war. In fact, the statistical trends

The Wiley Blackwell Companion to Political Geography, Second Edition.
Edited by Virginie Mamadouh, Natalie Koch, Chih Yuan Woon, and John Agnew.

mentioned previously have since reversed, and the number of armed conflicts has increased for almost a decade. These tragic events seem to send observers back to an old-fashioned imaginary of interstate warfare waged by military forces on battlefields and with clearly identified actors.

Does this new context change our understanding of conflict? The concept of conflict is a central theme in political geography and comprises a very broad and diverse literature. But the specter of a "return to war" gives it a new poignancy and geographers have significantly increased their research on conflict in recent decades, expanding the scope of their studies and highlighting the complexity of assessing and analyzing these phenomena. Whether in terms of the actors involved or the territories affected, they show that conflicts are constantly evolving, raising questions about the place of states or the blurred boundaries between civilians and the military and between war and peace. Furthermore, the field itself is constantly evolving depending on social, political, economic, and technical circumstances. The end of the Cold War, the terrorist attacks on the United States of September 11, 2001, the economic crisis of 2008, the Covid-19 pandemic of 2020–2021, or the Russian invasion of Ukraine and the escalation of war in the Middle East have each introduced new approaches to the concrete framing of ongoing conflicts. At the same time, the context of global warming and modern technologies have also changed their forms and their spatial materialization.

In contrast to the idea of a "return to war," which would mean a return to old models of conflict, the aim of this chapter is to relate recent events to contemporary political geography. Indeed, it would be a mistake to use old evaluative factors to quantify and characterize these new conflicts, whose contours today are both highly diverse and far removed from the legal frameworks established since the nineteenth century. Wars – as we have experienced them – are fluid, constantly evolving objects (Flint 2004), whose shape both depends on a global political context (international power relations, influence of international regulators, interstate alliances and rivalries, socioeconomic inequalities worldwide, role and influence of nonstate actors, etc.) and is tied to a sociotechnical context that strongly influences the nature of confrontations (the state of technologies at a given time, the availability of these technologies to different state and nonstate actors, the multiple uses of these technologies in ongoing confrontations), as well as to a socioenvironmental context that sometimes justifies and sometimes disqualifies conflict policies (Dalby 2020; Charbonnier 2024).

This contextual change has been emphasized by the most important research currents in political geography and has influenced the development of the concept of conflict in the field. From critical geography to feminist geography and from the material turn to political ecology, approaches to the concept of conflict have diversified to such an extent that the literature covers a very wide range of topics and methods. I invite the reader to read the comprehensive overview recently published by Virginie Mamadouh (2024). I will attempt here to offer a cross-disciplinary perspective to illustrate the "fluidity" of the concept of conflict and its spatial anchoring – which depends on the academic context (the development of the discipline), on the assessment of the relationships between actors, and on the sociotechnical and socioenvironmental contexts in which it unfolds.

The Notion of Conflict

At first glance, the definition of conflict appears to be rather broad and universal. The term conflict implies a relational situation structured around antagonism, whether due to a power differential, a difference of opinion (divergent values or viewpoints), rivalry (competition for a common goal), or enmity (effects). It thus leads to a specific relationship between actors, groups, or institutions that is driven by incompatible goals. Conflict in this sense is an essential component of all social relationships, as are other forms of relationships such as cooperation and adhesion (Gilmartin 2009). However, this first approach already raises a number of questions that can lead to very different analyses of how to assess conflictuality and antagonism between actors, as conflicts can be latent, without visible manifestations. The concept of "structural violence" developed by Johan Galtung in the late 1960s, which describes the damage suffered by populations living in situations of poverty and exploitation (Galtung 1969), already invites us to reconsider the limits of conflictuality in environments whose structure creates a situation of indirect domination and violence. Since the book *Violent Geographies*, edited by Derek Gregory and Allan Pred (2007), this approach has led to numerous works ranging from research on postconflict – which questions the boundary between "conflict" and "postconflict" in different contexts (Kirsch & Flint 2011; Cattaruzza & Dorier 2015) – to a more intimate geography that examines domestic violence at the intersection of ethnicity, class, and gender (Pain & Staeheli 2014), and a whole range of regional analyses on various topics (geography of fear, memory, etc.). From these different objects of study and methods of analysis, a more fundamental question emerges about the very qualification of conflict and what would distinguish it from a situation of peace, where the one and the other appear deeply intertwined and connected (Koopman 2011; Macaspac & Moore 2024; Cante & Williams, this volume). The fluidity of the concept of conflict thus seems to be linked to the temporal, spatial, and social assessment of the observer as well as to their research posture.

Beyond this first level of questioning, the very methods of political geography have led us to consider other elements of evolution and change in the study of conflicts. Political geographers are interested in how conflicts are rooted and materialized in space, through their various physical, political, social, and cultural manifestations and the representations associated with them – theaters of confrontation, devices reflecting relationships of power and control (borders, checkpoints, etc.), symbolic markings (flags, monuments, etc.), and sociospatial disparities inducing social or political hierarchization, among others (Graham 2011;

Rosière 2021). In this sense, the relationship between geography and conflict is always fluid and shifting, and this at several levels.

First of all, as the French philosopher and political scientist Raymond Aron (1962) reminded us, space is a setting, a stake, and a stage for conflict. A more recent formulation of this idea on the relationship between space and conflict is offered by Clionadh Raleigh in the chapter "Geographies of Conflict" in the previous edition of this volume: "There are three general ways in which political violence research has incorporated geographical variables, aspects, and theory into the larger studies of conflict environments: Geography can be treated as the object of study, its subject, or the context" (Raleigh 2015: 87). Political geography considers space as both a product and a producer of social relations (Flint and Dempsey 2024). In an effort to systematize approaches in the discipline, the French political geographer Stéphane Rosière proposed differentiating between the study of space as a stake, which would be the domain of geopolitics, and the study of space as a framework, which would be the domain of political geography (Rosière 2021). This distinction was deeply influenced by the French school of geopolitics, marked by Yves Lacoste and the journal *Hérodote*, whose scope was particularly interested in the interplay between actors and their spatial representations. In the same vein, the later contributions of critical geopolitics have taken even further the study of actors' strategies and of the performative character of discourses on the evolution of the world. However, more contemporary approaches – feminist geographies and political ecology, among others – are leading to a reevaluation of these analyses. Concepts such as *agencement*, inspired by the philosophy of Deleuze and Guattari, and its English translation "assemblage" are increasingly used by geographers (Müller 2015). They have been able to reintroduce material and environmental dimensions into critical approaches, which, in part, escape the strategies and representations of actors, but also outline a certain hybridity of contemporary conflictuality. They invite us to reassess conflicts, paying particular attention to all the technical, human, and nonhuman interweaving produced at different scales of space and time, and to the way these arrangements modify the sociospatial relations at play.

Evolution in These Approaches

War and conflict are traditional themes in political geography and have been present since the very foundations of the discipline. Indeed, political geography was first developed at the end of the nineteenth century by the German geographer Friedrich Ratzel within a context of colonial domination, in an attempt to define a scientific approach to state power. In his framework of study, marked by the geographical position of states and their relationship with space as the basis for their development, war was seen as a normal and inevitable phenomenon linked to the dynamics of state expansion and intrastate rivalries (Mamadouh 2004). Subsequently, conflicts and power rivalries were at the heart of the reflections of classical geopolitics, from Mahan to Spykman via Mackinder and Haushofer, with a strong strategic dimension to geographical analysis and a partly deterministic approach – war being seen as a natural and technical phenomenon. Technological breakthroughs played a very important role in the thinking of these authors, since they would enable the reconfiguration of the expression of power in space, leading to a whole range of reflections on the projection of power in distinct environments, with expressions such as land power, sea power, air power, and so on (Black 2005; Muscara 2018). The military use of geography, and the ideological dimension this could induce, has since been the subject of much criticism and reflection, from Yves Lacoste in France, and his work *La géographie, ça sert d'abord à faire la guerre* (Lacoste 1976), to the work of critical geopolitics (Agnew & Corbridge 1995; Ó Tuathail 1996) and radical geography (Gregory 2004, 2009).

Nevertheless, the study of conflict and power rivalries is a particularly rich theme for geographers, whether through the study of spaces in conflict – that is, the way in which open conflicts materialize in and through space – or through the study of contested spaces, which reflect, through their planning, relations of force and domination, power rivalries, and violent geographies, despite the absence of open confrontations (Raleigh 2015). This distinction is still highly relevant today. However, it is interesting here to complement this approach by distinguishing a number of geographical fields of study traditionally associated with the study of conflict, whether in the fields of military geography, border studies, urban studies, or migration studies, among others. The geography of conflict is thus part of a very rich field of study that cuts across many fields of human geography, including the study and analysis of actors in conflict, in a wide variety of spatial settings and territories. Additionally, research on postconflict regions has expanded to cover broader subjects such as refugee camps, symbolic territorialization, the gray economy, and so on (Flint 2004).

Finally, one common thread connecting these varied approaches is the study of ongoing actor conflicts, which necessitates incorporating a relational approach into spatial analysis – emphasizing both confrontations on the ground and conflicting actor strategies (see Linke & O'Loughlin, this volume). However, alongside these conflict-related themes, new perspectives within contemporary political geography, and innovative methodologies, have emerged in recent decades, broadening our understanding of conflict phenomena.

Building on critical geopolitics, feminist geopolitics has introduced new themes into international scholarship since the early 2000s. It shifts away from the state- and institution-centric analysis of realist approaches, or from the discourse analysis and performative character of critical studies, to focus on alternative expressions of politics and power. The aim is to highlight more hidden manifestations of discrimination and domination in everyday life and in the private sphere, involving different religious, cultural, and racial minorities in ways that are often unspoken or invisible (Dowler & Sharp 2001; Hyndman 2001). This approach invites the geographer to study conflict as an implicit relationship of domination, impacting communities as well as individuals through a whole range of sociospatial mechanisms that are rarely called into question, and are exercised as much in the public as in the private sphere, even in the intimacy of the home. Scholars such as Rachel Pain (2015) have thus been able to investigate the continuity between domestic violence and war violence. This approach has led to numerous studies, particularly in the analysis of the relationship between violence and space (Springer & Le Billon 2016).

From this perspective, Sara Koopman's proposal to engage in alter-geopolitics broadens the tools of conflict analysis by emphasizing alternative geopolitical practices led by nonviolent actors outside the traditional framework of international competition and power dynamics (Koopman 2011; Macaspac 2023). This perspective extends conflict analysis to peaceful mobilization practices, invoking the making of "spaces of peace," echoing Gandhi's call for peaceful resistance. Movements such as the Gezi Park protests in Istanbul have been studied through the prism of nonviolent action and its influence on participants within an authoritarian context, where parts of the population are excluded from public space (Gulen 2023). More recently, the Ukrainian example was put forward to show how actors could create spaces of peace even within a war zone, creating spaces that defy the logic of conflict or act as counterpoints through local practices and the mobilization of local actors (Koopman 2024). This perspective shows that the study of conflict areas must always be seen in a relational and dynamic context, involving alternative mobilizations and struggles.

Another area of study opened up by political geography concerns the analysis of the technical and sociotechnical contexts in which conflicts occur. The technological resources used by actors in conflict are not merely strategic assets; they are also vectors that reconfigure the conflict space and blur the boundaries of confrontation zones. For instance, the use of aerial drones has led to numerous studies, showing how their increasingly frequent use contributes to a "vertical geopolitics" (Graham 2004), which must rethink power in three dimensions from a volumetric point of view (Elden 2013). Geographer Derek Gregory shows how these technologies have profoundly altered the space and temporality of conflicts. They are enabling remote action over a very wide surveillance zone, thereby modifying the battlefield, which becomes both enlarged and "transparent," but also compressing decision time and the chain of actors involved in the targeting and bombing process (Gregory 2011). Far from being a simple technical solution, the drone changes the very nature of warfare and creates a de facto inequality in the face of death, due to the radical distinction between the space of the observers and that of the observed. The increasing use of these technologies in home surveillance, examined through the prism of feminist geopolitics, raises questions about the consequences of extending their uses into private and intimate spheres (Jackman & Brickell 2022). Such practices introduce surveillance into everyday life, hereby accentuating distinctions of gender, social class, or race in new ways, through phenomena of "drone peeping toms" or through practices of varying degrees of violence, contributing to questioning the boundaries between conflict and peace zones.

The socioenvironmental context is another focus of attention in the geography of conflict. The link between resources and conflict is a widely addressed and traditional theme of political geography (O'Lear 2024). But over the last two decades, scholarship dealing with the consequences of global warming on international security has increased. Indeed, the regional modifications it induces, such as desertification or water scarcity, are factors of tension that can fuel conflicts. Quantitative analyses have established correlations between climate variations and the risk of conflict in nine East African countries (O'Loughlin et al. 2012). Similar studies have tied violence and climate change in Darfur, playing a role in the outbreak of civil war in 2003 (De Juan 2015). The assessment of the relationship between climate change and conflict has been the subject of a rich and detailed production over the last two decades, providing a substantial database for establishing correlations, while challenging deterministic interpretations. Beyond direct conflicts, geographers are also interested in the notion of "slow violence," which stems from environmental degradation resulting from colonial, capitalist, or extractivist exploitation and its impacts on the populations of the South (O'Lear 2024). This notion is drawn from the early work of sociologist Johan Galtung, who identifies environmental destruction and deteriorating living conditions as forms of indirect violence that transcend the conventional distinctions between war and peace.

Critical geopolitics takes a different approach to the relationship between the environment and conflicts, highlighting the weakness of international security discourse in the face of global environmental security issues (Dalby 2022). It also critiques the harmful effects of territorial competition between states on zero-carbon ecological policies, since this actually increases the consumption of carbon resources (Toal 2024), although this last idea is controversial. Paradoxically, French political scientist Pierre Charbonnier recently put forward the notion of "war ecology," according to which energy sobriety policies, in conjunction with sanctions against Russia, were used as a weapon of autonomy and resilience by Europe in the Russian–Ukrainian conflict (Charbonnier 2024). In his view, this "ecology of war" stems from a newfound awareness of Europe's energy dependence and is therefore aimed at increasing the sovereignty of states in the face of the great oil powers, such as Russia, the United States, Saudi Arabia, Qatar, and other petroleum producers in the Middle East.

The contemporary literature on the geography of conflict is thus characterized by a complex approach, articulated around poles of analysis that are partly interdependent of one another and partly autonomous: "traditional" poles studying the interwoven representations and strategies of actors and sociospatial consequences, within a geopolitical context that constantly modifies these relations; and a more "recent" pole that has (re)introduced analyses studying the sociotechnical and socioenvironmental contexts in which these relations are embedded. Nevertheless, the very broad scope of the notion of conflict presented here now seems to be in contradiction with current events characterized by international competition and militarized forms of warfare. These events evoke classical geopolitical analyses, essentially centered on state power, territorial conquests, and the military apparatus. The question we must address at this point is to what extent these major trends in political geography can be useful in understanding the contemporary world, with its return to high-intensity military commitment and events reminiscent of early geopolitical practices – intertwined with military strategies and intended to serve as a political guide for action in the classical game of territorial competition between powers.

Questioning War

How does political geography account for the current global context and can this influence the discipline's scholarly output? It remains clear that examining war and its contemporary manifestations is more than ever a challenge for political geographers. But first, one needs to define what war actually is. How is it more specific than conflict? In fact, finding a consensual definition of the term is complex, given the scope of the phenomenon and the wide range of perspectives. French sociologist and polemicist Gaston Bouthoul's definition is interesting because he provides a broad yet precise framework for the notion of war: "War is a large-scale, organized and bloody confrontation between political groups (sovereign groups in the case of wars between states, internal groups in the case of civil wars)" (Bouthoul 1975: 49). In other words, war implies large-scale implications, lethal violence, and the involvement of political actors with war aims and strategies. His definition remains very close to the statements of Prussian strategist Carl von Clausewitz (1780–1831), for whom war is first and foremost an instrument serving political ends and is a political relationship between actors that manages conflict, just as are cooperation, negotiation, imposition, and so on.

French geographer Stéphane Rosière (2021) points out that war is a convention, governed by rules, principles, and a legal framework, a *jus in bello* and a *jus ad bellum*. These concepts were first introduced by political philosophers such as Hobbes, Grotius, and Rousseau in the seventeenth and eighteenth centuries, and formalized in the late nineteenth and early twentieth centuries. During this period, several international meetings were held to codify and limit warfare, based on conventions that gradually regulated the right to engage in war (*jus ad bellum*) and the rules for conducting warfare (*jus in bello*). This framework established fundamental distinctions that in theory condition the intervention of military groups: the distinction between combatants and noncombatants, between military and civilians. It also set up the protection of victims and prisoners of war. Thus, in principle, military action can only be taken against combatants and noncombatants must be kept out of the fight. The spatial implications of these legal frameworks, their respect or violation, is a recurrent and implicit prism of analysis in political geography. The extension of confrontation zones to civilian spaces enabled by the use of drones on the Afghanistan–Pakistan border, the militarization of the fight against drug cartels on the US–Mexico border, or offensive operations in cyberspace has led Derek Gregory (2011) to question the ambiguity of contemporary combat spaces, which tend to spread ever more widely, resulting in what he calls "the everywhere war." A similar reflection led Stephen Graham (2011) to study "urbicides," the methodical destruction of urban spaces in contemporary conflicts. Similarly, looking at the Israeli–Palestinian conflict, Neve Gordon and Nicolas Perugini (2018) question the notion of a "human shield" in international law, which erodes the legal category of "civilian population" and transforms these populations "into killable subjects."

At first glance, studying war today might therefore seem to require a renewed focus on state politics and military confrontations. But, on closer inspection, interstate conflicts are still the exception on the global stage, as the Uppsala Conflict Data Program (UCDP) database confirms (https://ucdp.uu.se). In fact, while the war in Ukraine, widely regarded in Western countries as a "return to war," has garnered significant media attention, interstate conflicts remain marginal from a statistical point of view, compared to the larger number of conflicts in the world that are mainly intrastate or internationalized intrastate conflicts (see Figure 14.1).

Moreover, the forms of conflict always depend on a specific context, requiring political geographers to constantly adapt their methods and approaches to address situations that are always unique and complex. Between the end of the twentieth century and the 2010s, the manifestations of war have continued to evolve, with the use of military force seemingly playing a supporting role.

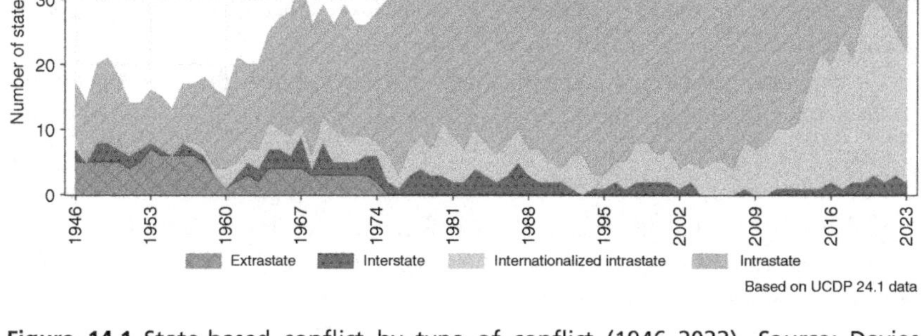

Figure 14.1 State-based conflict by type of conflict (1946–2022). *Source:* Davies et al. 2024. Creative Commons Attribution 4.0 International license (CC BY 4.0).

The conflicts of the 1990s, particularly those in the former Yugoslavia, Rwanda, and the post-Soviet world, led political scientist Mary Kaldor (1998) to introduce the concept of "new wars." These wars were characterized by their occurrence within states, involving private or semi-private groups, and mainly targeting civilian populations. Parallel to these observations, the work of geographers has shown the sociospatial dynamics of these new wars, through the study of ethnic cleansing in Bosnia-Herzegovina, for example (Roux 1992; Toal & Dahlman 2011). On a broader scale, they have studied the multiplication of civil wars in the world and their deadly impact (O'Loughlin 2004). George W. Bush's "War on Terror," following the attacks of September 11, 2001, led to a new wave of asymmetric wars in Afghanistan and Iraq, with a particular focus on the issues of postconflict and peace-building (Kirsch & Flint 2011). Despite significant involvement, these two wars ended in the failure of Western troops and strategies. In both cases, the role of the armed forces was mainly to maintain order, in a context where the use of force remained proportionate and limited, although there were some medium-intensity engagements.

In contrast, the current period seems to be marked by a high-intensity use of military force, at least if we consider the conflicts in Ukraine and the Middle East. But in both cases, the use of military force does not mean a "return" to an old geopolitical order, in which two armies face each other on either side of a front line and in which nonstate actors play only a secondary and negligible role. In fact, despite fighting along front lines in Ukraine, following Russian or Ukrainian advances, the confrontations are in reality punctuated by major technical and strategic innovations on the ground: use of tactical drones, cyberwarfare, informational warfare, and so on. Altogether, they considerably modify the reality of combat, the scales of warfare, and their implications. In the Middle East, belligerents are fighting with highly disproportionate means, using strategies and weapons that defy ethics and international law. Large-scale terrorist practices and the Hamas hostage-taking attack of October 7, 2023, have been followed by the massive bombing of civilians by the Israeli army and the use of artificial intelligence (AI) to rapidly target new locations in Gaza City. In both cases, informational campaigns are launched on different scales by the belligerents to win the support of international opinion, and the conflict is causing strong divisions all over the world. As far as the Israeli–Palestinian conflict is concerned, pro-Palestinian and pro-Israeli demonstrations are multiplying globally. As for the Russian–Ukrainian war, divisions have emerged between nations, opposing Western countries that firmly support Ukraine to countries such as China and Iran that show support for Russia. In this case, the belligerents also use supply-chain disruptions involving such basic necessities as cereals or energy as weapons on the international stage. The scales of the conflict are therefore complex and multiple, extending far beyond the regions concerned.

The term "hybrid warfare" emerged in the late 2000s to describe these new forms of combat, which "came to refer to an approach drawing upon instruments from across the full spectrum, including terrorism, insurgency, criminality, and conventional operations, along with the extensive use of informational operations" (Freedman 2017: 222). Ultimately, hybrid warfare is carried out in a completely new way, involving different scales, different types of actors, and different technologies, with strategies that affect civilian populations to different extents and in different ways, with methods that combine regular and irregular practices, and a very broad spectrum of action combining direct and indirect threats. With this hybridization of warfare, the use by states and nonstate actors of military and nonmilitary means, regular and irregular, economic, political, humanitarian, and cyber, together with the systematic use of new information technologies, thus complicates the spatial dynamics and scales of warfare from the local to the global.

However, critical approaches to the hybrid warfare discourse have also emerged in current writing. Jakub Eberle and Jan Daniel (2022), for example, have shown how the evocation of hybrid warfare has been used in Czech political discourse as a catchword whose function is not to define strategies for facing threats. Instead, it is used as a political lever to bolster a civilizational discourse and diffuse anxiety among the population:

> using the case of Czechia, we have argued that hybrid warfare is a discourse that constantly oscillates between repressing anxiety by geopoliticising the source of danger in East/West terms, and subverting its own constructions by presenting the threats as insidious, invisible, and constantly shifting. Therefore, we contend that the hybrid warfare discourse is structured in a way that cannot achieve its purported ambition to secure populations against "hybrid threats" and instead ends up producing more insecurity and anxiety. (Eberle and Daniel 2022: 8)

The study of war nowadays represents a particular challenge for political geography. Indeed, the current context, far from referring to the approaches of classical geopolitics, calls for the use of all the methods at one's disposal to grasp the full range of issues at stake. The wars in Ukraine and the Middle East, like all other conflicts, need to be analyzed through the lens of assemblage theory. Each of them involves studying entangled actors at different scales and nonhuman agents of varying natural, technical, algorithmic, and energetic kinds. Each war is part of a sociotechnical and socioenvironmental context that influences the strategies of the actors and their representations. Every discourse on war needs to be deconstructed through critical approaches that identify its biases and geopolitical meanings for the actors involved. Feminist perspectives prompt us to examine the implicit power dynamics within society that war both reveals and intensifies, particularly for marginalized populations. Thus, the multiple approaches of political geography provide a robust analytical framework for understanding the complexity of the phenomena and spatialities involved in contemporary conflicts. The study of cyberwarfare offers a good example of the challenges geographers face in addressing the new forms and materialities of contemporary conflicts.

Influence of the Sociotechnical Context: The Example of Cyberwar

The advance of technology and the sociospatial modifications it produces generate new actor relations and new materializations of conflict, as recent studies on cyberwar have shown (Douzet & Géry 2021; Glasze et al. 2023). Indeed, cyberspace, by its networked nature, introduces new forms of spatialization and territorialization that challenge geographers. De facto, it is wrong to imagine that the territorialization of cyberspace would necessarily lead to its fragmentation, as some geographical imaginations suggest (Lambach 2020). The control of digital space by state and nonstate actors does not inevitably involve compartmentalization, but rather multiple territorialization processes (technical, political, legal, economic, etc.) that do not lead to a fractured network. In the context of conflict, cyberspace is becoming both a means of attack and a place of vulnerability whose importance is growing as it becomes increasingly intertwined with aspects of our daily lives (Slesinger 2024). Moreover, the evolution of the sociotechnical framework, of technologies and their uses, is far from being neutral and induces a profound evolution in conflictuality, its manifestations and spatialities, as well as how the world is represented by the various actors involved. Historically, significant scientific revolutions – such as Newtonian physics in the eighteenth century, the development of thermodynamics in the nineteenth century, the rise of computing in the second half of the twentieth century, and today's cybernetics and information paradigms – have redefined how we think about war and conflict. Contemporary perspectives, marked by cybernetic representations, reflect a desire to impose order on the chaos of war through controlled and managed information, based on automated and algorithmic thinking to an extent that eludes human intelligence (Bousquet 2009).

As a result of the transversality of digital space, the idea of a battlefield distinct from zones preserved from combat is gradually fading, in favor of a porosity between war and peace. This reveals different, more or less lethal fronts within physical and digital spaces (Cattaruzza & Taillat 2018). The idea of an enclosed territory of war, closed in on itself, already eroded by terrorist practices, is rendered definitively obsolete in this digitization process. One of the major characteristics of cyberwarfare is the increased fluidity of its contours. This means that the legal framework governing conflicts (*jus in bello* and *jus ad bellum*) must be rethought in a world marked by deep regional interdependence. In fact, technological developments and the spread of digital tools have made information exchanges virtually instantaneous. In so doing, cyberspace has become both an opportunity, in terms of exchanges, and a vulnerability, via the multiplication of cyberattacks (Douzet 2014; Cattaruzza 2019). In this way, cyberwarfare has become a tool in the hands of the powers that be, along with other types of unconventional attack – complementing more traditional offensive operations.

Yet the extension of conflictuality into the cyber domain is also largely transforming the traditional rules of warfare (Stone 2013). Attacks are difficult to attribute. Attacks have a low cost for the attacker, while defense remains complex and expensive for the defender. Attacks are indirect, nonlethal, often below the threshold of an act of war, but can nonetheless be highly destabilizing, affecting all infrastructures, businesses, civil administrations (ministries, hospitals, etc.), as well as all populations (Rid & Buchanan 2014). Also, with the widespread interconnection of all our societies, cyberwarfare, in its many forms (cyberattacks, malware, information operations, etc.), is not limited to a specific battlefield. It blurs the traditional distinctions between space at peace and space at war, between civilian and military, between professional and private spheres. It makes rethinking the different thresholds of war arduous, with rivalries that can play out at different levels of attack, affecting vital infrastructures as much as individuals, and with actors that are harder to identify (criminal actors, political groups, states). Some of them may even act for others due to the difficulty of attributing attacks (Kello 2017; Buchanan 2020).

Furthermore, cyberattacks can be more or less targeted but also spread almost instantaneously within the global space. For example, the WannaCry and NotPetya attacks in May and June 2017, which were ransomware (blocking data and demanding a ransom to unlock it), had spread to around a hundred countries and thousands of infrastructures within a few hours (although Ukraine seems to have been the first target of NotPetya). In 2016, researchers Myriam Dunn-Cavelty and Thierry Balzacq proposed to explore specific forms of malware spatialization by studying the Stuxnet virus (Balzacq & Dunn-Cavelty 2016), which affected Iranian nuclear centrifuges and was discovered in 2010. Using actor-network theory, they looked at the materiality of the digital network affected (with its servers, connections, networks, software, etc.) and the ways in which the virus spread. They thus suggested a triple spatiality of cyberdisruption, acting at once in a topographical space at regional level, with infrastructures affected and precisely located; in a topological space at network level, which finds itself almost instantaneously affected by the same threat; and in a fluid space, corresponding more specifically to the virus's mode of diffusion, which is to spread like a liquid from one machine to another via digital space. This discussion raises questions for geographers about the new spatialities that are occurring alongside conflicts, in light of the contemporary sociotechnical context.

Finally, the cybernetic dimension of conflicts partly extends the battlefield into our daily lives, which are themselves impacted in different ways. The use of the Strava app – a geolocated sports application – by military personnel enabled an Australian student in January 2018 to reveal the location of military bases in theaters of conflict, using the app's data that were accessible at the time as open source. Indeed, the proliferation of open-source data has revolutionized (and in part democratized) contemporary intelligence practices. What is more, the rapid spread of social networks and their increasingly important role in society provide international players with new means of conducting informational operations. As a result, war is increasingly fought on an informational level, and new state actors are emerging to counter these new conflictual practices. In 2015, the European Union, for example, set up the East StratCom Task Force to counter Russian manipulation. In the United States, the fight against disinformation and state and nonstate propaganda is handled by a special unit within the State Department, the Global Engagement Center. In France, following the murder of a teacher, Samuel Paty, in October 2020 by an Islamist terrorist and the disinformation campaign that followed the assassination on social networks, the Defence and National Security Council decided in 2021 to create the VIGINUM agency, whose aim is to detect information manipulation originating abroad.

Beyond cyberwarfare and cyberspace, the issues raised by the digitization of war and conflict are increasingly numerous and represent a real challenge for political geographers. These include the question of autonomous weapons and the use of AI on battlefields, in spaces where decision-making time is increasingly reduced and where the role of human action is increasingly limited (Bousquet 2023). From this perspective, the tactical level would be entirely automated, disappearing in favor of hyper-centralized warfare conducted at the strategic or political level. Meanwhile, the complexity at work on the ground would escape any form of anticipation or prediction. Moreover, the growing use of algorithmic mediation on the battlefield, as in other sectors, raises questions about a potential "depoliticization" of conflicts (Amoore 2013). AI being used to target alleged Hamas combatants in Gaza based on disparate data is a good example. Who is responsible for this automated targeting, for those killed by mistakes or as collateral damage? Is it the AI's designer, the manufacturer, the state that employs it, or the military personnel operating it? Can AI make mistakes? What is the political strategy behind its deployment? How do geolocated data, when combined in such weaponry, create new spaces of insecurity and conflict?

Conclusion

Conflict remains a central notion in political geography. However, it is difficult to define, because it is fluid and multifaceted and depends on multiple interlocking actors and agents, which political geography enables us to apprehend through a number of complementary approaches. The spatialization of phenomena remains a particular challenge given the complexity of contemporary conflicts, which involve a network of human and nonhuman actors and technologies at different scales of space and time. The question of data and their interpretation is one of the challenges of contemporary research. In the age of Big Data and social networks, a profusion of information is accessible on places in conflict, without the candid observer being able to determine their origin, reliability, or political and geopolitical significance. In just a few years, geospatial intelligence (GEOINT) has become an essential tool for military players, while open-source intelligence (OSINT) represents a formidable field of investigation for researchers and journalists interested in these areas.

Nevertheless, the qualitative approaches developed by political geography in recent years – such as critical and feminist approaches – are still essential. They enable us to criticize data – which can be biased or manipulated in times of conflict – and to reveal relations of domination or dependence that are difficult to apprehend through external data. Above all, these factors allow us to question conflicts as complex social phenomena that are deeply intertwined in our daily lives: in the spatial structures that hierarchize, secure, and isolate communities; in the social imaginaries that designate the Other and territorialize the world; in the technologies that govern our daily lives while making us vulnerable; and within a socioenvironmental context increasingly shaped by inequalities and exploitation.

References

Agnew, J.A., & Corbridge, S. (1995). *Mastering Space: Hegemony, Territory and International Political Economy*. London: Routledge.

Amoore, L. (2013). *The Politics of Possibility: Risk and Security beyond Probability*. Durham, NC: Duke University Press.

Aron, R. (1962). *Paix et guerre entre les nations*. Paris: Calmann-Lévy.

Balzacq, T., & Dunn-Cavelty, M. (2016). A theory of actor-network for cyber-security. *European Journal of International Security 1*(2): 176–198.

Black, J. (2005). Geographies of war: The recent historical background. In Flint, C. (ed.), *The Geography of War and Peace*. Oxford: Oxford University Press, 19–25.

Bousquet, A. (2009). *The Scientific Way of Warfare: Order and Chaos on the Battlefields of Modernity*. London: C Hurst.

Bousquet, A. (2023). The persistent appeal of chaoplexic warfare. Towards an autonomous S(war)M machine? In Gruszczak, A., & Kaempf, S. (eds), *Routledge Handbook of the Future of Warfare*. London: Routledge, 227–237.

Bouthoul, G. (1975). *Glossaire polémologique des termes de violence*. Paris: Institut français de polémologie.

Buchanan, B. (2020). *The Hacker and the State: Cyber Attacks and the New Normal of Geopolitics*. Cambridge, MA: Harvard University Press.

Cattaruzza, A. (2019). *Géopolitique des données numériques. Pouvoir et conflits à l'heure du Big Data*. Paris: Le Cavalier Bleu.

Cattaruzza, A., & Dorier, E. (2015). Post-conflit: entre guerre et paix ? *Hérodote 3*(158): 6–15.

Cattaruzza, A., & Taillat, S. (2018). Les enjeux de la numérisation du champ de bataille. *Dynamiques internationales 2018*(3): 1–13.

Charbonnier, P. (2024). *Vers l'écologie de guerre. Une histoire environnementale de la paix*. Paris: La découverte.

Dalby, S. (2020). *Anthropocene Geopolitics: Globalization, Security, Sustainability*. Ottawa: University of Ottawa Press.

Dalby, S. (2022). *Rethinking Environmental Security*. Cheltenham: Edward Elgar.

Davies, S., Engström, G., Pettersson, T., & Öberg, M. (2024). Organized violence 1989–2023, and the prevalence of organized crime groups. *Journal of Peace Research 61*(4): 673–693.

De Juan, A. (2015). Long-term environmental change and geographical patterns of violence in Darfur, 2003–2005. *Political Geography 2015*(45): 22–33.

Douzet, F. (2014). La géopolitique pour comprendre le cyberespace. *Hérodote 152–153*(1): 3–21. doi:10.3917/her.152.0003

Douzet, F., & Géry, A. (2021). Le cyberespace, ça sert, d'abord, à faire la guerre. Prolifération, sécurité et stabilité du cyberespace. *Hérodote 2*(177–178): 329–350.

Dowler, L., & Sharp, J. (2001). A feminist geopolitics? *Space and Polity 5*(3): 165–176.

Eberle, J., & Daniel, J. (2022). Anxiety geopolitics: Hybrid warfare, civilisational geopolitics, and the Janus-faced politics of anxiety. *Political Geography* 92: 102502.

Elden, S. (2013). Secure the volume: Vertical geopolitics and the depth of power. *Political Geography 2013*(34): 35–51.

Flint, C. (2004). Introduction. In Flint, C. (ed.), *The Geography of War and Peace: From Death Camps to Diplomats*. Oxford: Oxford University Press, 3–15.

Flint, C., & Dempsey, K.E. (eds) (2024). *Making Geographies of Peace and Conflict*. London: Routledge.

Freedman, L. (2017). *The Future of War: A History*. Harmondsworth: Penguin.

Galtung, J. (1969). Violence, peace, and peace research. *Journal of Peace Research* 6(3): 167–191.

Gilmartin, M. (2009). Conflict. In Gallaher, C., Dahlman, C.T., Gilmartin, M., et al. (eds), *Key Concepts in Political Geography*. London: Sage, 227–234.

Glasze, G., Cattaruzza, A., Douzet, F., et al. (2023). Contested spatialities of digital sovereignty. *Geopolitics 28*(2): 918–958.

Gordon, N., & Perugini, N. (2018). Human shields and the political geography of international humanitarian law. In Coleman, M., & Agnew, J. (eds), *Handbook on the Geographies of Power*. Cheltenham, Edward: Elgar, 277–296.

Graham, S. (2004). Vertical geopolitics: Baghdad and after. *Antipode 36*(1): 12–23.

Graham, S. (2011). *Cities under Siege: The New Military Urbanism*. London: Verso.

Gregory, D. (2004). *The Colonial Present: Afghanistan, Palestine, Iraq*. Oxford: Wiley Blackwell.

Gregory, D. (2009, 2010). War and peace. *Transactions of the Institute of British Geographers 35*: 154–186.

Gregory, D. (2011). From a view to a kill: Drones and late modern war. *Theory, Culture and Society 28*(7–8): 188–215.

Gregory, D., & Pred, A. (eds) (2007). *Violent Geographies: Fear, Terror, and Political Violence*. London: Routledge.

Gulen, H. (2023). Geopolitics of activism, commoning practices, and survival strategies in Istanbul: Theory of the commons and local politics. Phd thesis, University Paris 8.

Hyndman, J. (2001). Towards a feminist geopolitics. *The Canadian Geographer / Le Géographe canadien 45*(2): 210–222.

Jackman, A., & Brickell, K. (2022). "Everyday droning": Towards a feminist geopolitics of the drone-home. *Progress in Human Geography 46*(1): 156–178.

Kaldor, M. (1998). *New and Old Wars*. London: Polity Press.

Kello, L. (2017). *The Virtual Weapon and International Order*. New Haven, CT: Yale University Press.

Kirsch, S., & Flint, C. (eds) (2011). *Reconstructing Conflict: Integrating War and Post-war Geographies*. London: Routledge.

Kobayashi, A. (2009). Geographies of peace and armed conflict: Introduction. *Annals of the Association of American Geographers* 99(5): 819–826.

Koopman, S. (2011). Alter-geopolitics: Other securities are happening. *Geoforum* 42(3): 274–284.

Koopman, S. (2024). The geography of nonviolent peace activism in the midst of war: From Colombia to Ukraine. In Flint, C., & Dempsey, K.E. (eds), *Making Geographies of Peace and Conflicts*. London: Routledge, 165–182.

Lacoste, Y. (1976). *La géographie, ça sert d'abord à faire la guerre*. Paris: Maspero.

Lambach, D. (2020). The territorialization of cyberspace. *International Studies Review* 22(3): 482–506.

Macaspac, N.V. (2023). Indigenous geopolitics: Creating indigenous spaces of community self-protection and peace amid violent conflict. *Journal of Pacifism and Nonviolence* 1(2): 181–207.

Macaspac, N.V., & Moore, A. (2024). Geographies of peace. In Flint, C., & Dempsey, K.E. (eds), *Making Geographies of Peace and Conflicts*. London: Routledge, 37–53.

Mamadouh, V. (2004). Geography and war, geographers and peace. In Flint, C. (ed.), *The Geography of War and Peace: From Death Camps to Diplomats*. Oxford: Oxford University Press, 26–60.

Mamadouh, V. (2024). Geography and war, geographers and of peace: Expanding research and political agendas. In Flint, C., & Dempsey, K.E. (eds), *Making Geographies of Peace and Conflicts*. London: Routledge, 12–36.

Müller, M. (2015). Assemblages and actor-networks: Rethinking socio-material power, politics and space. *Geography Compass* 9(1): 27–41.

Muscara, L. (2018). Maps, complexity and the uncertainty of power. In Coleman, M., & Agnew, J. (eds), *Handbook on the Geographies of Power*. Cheltenham: Edward Elgar, 362–379.

O'Lear, S. (2024). Geographies of environmental peace and conflicts. In Flint, C., & Dempsey, K.E. (eds), *Making Geographies of Peace and Conflicts*. London: Routledge, 219–236.

O'Loughlin, J. (2004). The political geography of conflict: Civil wars in the hegemonic shadow. In Flint, C. (ed.), *The Geography of War and Peace: From Death Camps to Diplomats*. Oxford: Oxford University Press, 85–110.

O'Loughlin, J., Witmer, F.D.W., Linke, A.M., & Dudhia, J. (2012). Climate variability and conflict risk in East Africa, 1990–2009. *PNAS 109*(45): 18344–18349.

Ó Tuathail, G. (1996). *Critical Geopolitics: The Politics of Writing Global Space*. London: Routledge.

Pain, R. (2015). Intimate war. *Political Geography 44*: 64–73.

Pain, R., & Staeheli, L. (2014). Introduction: Intimacy, geopolitics and violence. *Area* 46(4): 344–347.

Raleigh, C. (2015). Geographies of conflict. In Agnew, J., Mamadouh, V., Secor, A.J., & Sharp, J. (eds), *The Wiley Blackwell Companion to Political Geography*. Oxford: Wiley-Blackwell, 86–99.

Rid, T., & Buchanan, B. (2014). Attributing cyber attacks. *Journal of Strategic Studies* *38*(1–2): 4–37.

Rosière, S. (2021). *Géographie politique et géopolitique*. Paris: Ellipses.

Roux, M. (1992). A propos de la "purification ethnique" en Bosnie Herzégovine. *Hérodote* 67: 49–60.

Slesinger, I. (2024). Hybrid networks: Technology, geopolitics and ontology in digital warfare. In Flint, C., & Dempsey, K.E. (eds), *Making Geographies of Peace and Conflicts*. London: Routledge, 200–218.

Springer, S., & Le Billon, P. (2016). Violence and space: An introduction to the geographies of violence. *Political Geography* 52: 1–3.

Stone, J. (2013). Cyber war will take place! *Journal of Strategic Studies* 36(1): 101–108.

Toal, G. (2024). *Oceans Rise Empires Fall: Why Geopolitics Hastens Climate Catastrophe*. Oxford: Oxford University Press.

Toal, G., & Dahlman, C.T. (2011). *Bosnia Remade: Ethnic Cleansing and Its Reversal*. New York: Oxford University Press.

Chapter 15

Peace

Fabien Cante[1] *and Philippa Williams*[2]
[1]*Department of Geography, University College London, London, UK*
[2]*School of Geography, Queen Mary University of London, London, UK*

The last 15 years have witnessed a transformation in how political geographers engage with the concept and practice of peace. This chapter provides an overview of major thematic approaches to peace by contextualizing their emergence within disciplinary and geopolitical contexts. Although it is now established as a key concept in political geography (see Koopman 2019; Flint & Dempsey 2023), peace has been marginal within the longer history of the (sub)discipline. In part, this is because geography has historically been instrumental to *making war* (see Lacoste 1976) and concerned with furthering state power, imperialism, and military dominance. As Kearns (2008: 1609) put it in his review of Halford Mackinder's influential contribution to the discipline (at least in the United Kingdom, where we write from), "Geography's feet are caked with clay from the battlefield."

To be sure, geographers have critiqued the discipline's warmongering overtones for an equally long time – at least since the propositions of Kropotkin and Reclus (see Springer 2013). As Mamadouh (2023; also 2004) has recently documented, the threat of nuclear conflict after World War II and especially the global "War on Terror" that followed the attacks of September 11, 2001, in the United States spurred political geographers to orient their scholarship and activism toward peace, nonviolence, and nonkilling. The fields of critical and anti-geopolitics, in particular, have worked to expose the "militarist mappings of global space" and to challenge "how contexts are constructed to justify violence" (Dalby 2010: 281; see also Dowler 1998; Ó Tuathail and Dalby 1998; Gregory 2004, 2010, 2011; Dodds 2007).

The Wiley Blackwell Companion to Political Geography, Second Edition.
Edited by Virginie Mamadouh, Natalie Koch, Chih Yuan Woon, and John Agnew.

Yet in its focus on uncovering power asymmetries and critiquing violent socio-spatial relations, political geography has tended to leave peace itself underspecified and undertheorized. Even when geographers did turn their attention to peace (see Pepper & Jenkins 1985; Kliot & Waterman 1991; Flint 2004; Kobayashi 2009), the concept often remained narrowly and negatively interpreted as the absence of violence or war, or alluded to as a goal to be achieved in the aftermath of conflict. Peace, in other words, was given short shrift by a dominant focus on violence.

Tackling this aporia head on, peace geographies, as a research and teaching agenda as well as a realm of political praxis, set out to untangle the complex, at-times ambiguous realities of peace. This chapter begins by unpacking what prompted (political) geographers to take peace seriously. The first section centers a set of influential papers published between 2010 and 2012 that have been key to establishing peace as a focus of critical geographical inquiry. In the next, longer section we review some of the different research strands that have built on these early propositions. We connect developments within political geography to a parallel "spatial turn" in peace studies, while highlighting geographers' specific contributions – in research and in teaching – to analyzing the spatial, material, and temporal politics of peace. We also note cutting-edge research areas in peace geographies that will only gain in importance as "war fever" (Megoran 2023) once again seizes global geopolitics, compounding "slower" (Nixon 2011; Anderson et al. 2021; O'Lear 2023), less visible, yet deeply consequential forms of planetary violence.

Critical Geographies of and for Peace

Roughly 10 years after 9/11 and the publication of many books and articles critically deconstructing the ensuing War on Terror, several geographers in the United Kingdom and North America concurrently published articles that converged around the imperative to make peace more visible within the discipline, as both a concept and a praxis (Megoran 2010, 2011; Inwood & Tyner 2011; Koopman 2011a, 2011b; Williams & McConnell 2011; Loyd 2012). They asked what conceptual frameworks could help understand peace as "real" and critically analyze practices of peace-making within, after, or beyond the spaces and imaginaries of war/violence. Rather than summarize individual contributions to this "agenda-setting" moment (Mamadouh 2023), we highlight four intersecting themes that coalesced at the time that are important for understanding political geographers' engagement with peace and the trajectory of the subdiscipline since then.

First, an explicit geographical focus on peace reaffirmed political geography's conscious commitment to realizing a more just, less violent world. It invited geographers' individual and collective self-reflection on their role as educators and their responsibility for disciplinary socialization within higher education. Writing against the "military-academic-industrial complex," Inwood and Tyner (2011: 42) thus called for political geography to embrace "pro-peace," "pro-justice," and "anti-killing" agendas, and "to build wider disciplinary coalitions that confront a predominant war culture in 21st Century U.S. society" (see also the Center for Global Nonkilling, https://nonkilling.org/center). Drawing inspiration from critical race theory's focus on "the disruption of master narratives that offer neatly (re)packaged versions (and visions) of events which legitimize racism" (Inwood and Tyner 2011: 449), Inwood and Tyner conceived of a pro-peace education as one that "seeks to transform the present human condition by challenging social structures and patterns of thought and by insisting that our current situation is not the only option" (Inwood and Tyner 2011: 449). The call for geographers to mobilize a pro-peace pedagogy was echoed by Nick Megoran (2011: 186), himself recovering Alan Jenkins's (1985) conviction that geography has much to offer peace education, not least by demonstrating frameworks for "collective public engagement" within the discipline.

Second, (political) geography's shift from an "art of war" to an "art of peace" (Gregory 2010: 180–181) required a conceptual effort to reimagine geopolitics beyond state logics, and to assert the possibility of sociospatial relations that are not ultimately underpinned by force. Reflecting on Mackinder's continued influence among so-called realist geopolitics, Kearns (2008: 1601) called on geographers to develop a resolutely "progressive geopolitics" by reorienting inquiry toward "some of the progressive possibilities of the present, non-states and non-violence [that] play a significant role in shaping the modern world order." Taking up Kearns's challenge, Megoran (2010: 383) identified and explored "transformative possibilities for peace" through his study of a Christian Reconciliation Walk. By foregrounding acts of apology as constitutive of new possibilities for interaction and interrelation, Megoran (2010: 385) demonstrated the critical and world-building potential of embracing a "pacific geopolitics" that examines "how actors such as protest movements, governments, think-tanks and film makers frame geopolitical knowledge in ways that contribute towards more harmonious relations between states and other human groups." Likewise, Koopman's (2011a; also 2008) research with international accompaniers in Colombia advanced the idea/concept of "alter-geopolitics" to capture "how geopolitics is *done* differently" (2011a: 280, emphasis in original). Her dialogue with feminist geopolitics and anti-geopolitics scholarship invited political geographers to go beyond critiques of hegemonic power relations by spotlighting and indeed actively nurturing peaceful alternatives, both in published research and "off the page" (2011a: 281). Together, Kearns, Megoran, and Koopman sought to reframe the geopolitical question by asking not (just) why violence and wars persist and how they are justified, but also how (largely nonstate) actors' peace-making practices might already be envisioning a more peaceful world.

Third, political geographers sought to build "critical geographies of peace" (Williams & McConnell 2011) as a unified and dedicated field of study, one that would deploy spatial concepts and modes of analysis established in the discipline to "better conceptualise peace" (Megoran 2011: 187). This agenda entailed giving renewed visibility and coherence to already-existing research that engaged with peaceful concepts, practices, and movements, albeit under different conceptual rubrics (Loyd 2011, 2012; Williams & McConnell 2011). But peace geographers also argued for a specific line of inquiry that would ask: "what is meant by peace, how is it understood within different contexts, who is peace for, and in whose image is it (re)produced?" (Williams & McConnell 2011: 3). By theorizing peace as a sociospatial process, geographers further emphasized its variegated, place-based, and multiscalar nature. Drawing on Dietrich and Sützl (1997), Koopman (2011b) argued for the need to recognize "many peaces" and to take "peace to pieces, not just by place, but also by space" and by scale. Peace, as she put it, "can be experienced as both intimate and global . . . [it can be] created at an individual, family, neighbourhood, community, and other scales, and using the term [peace] can foster seeing these spaces as intertwined and mutually constitutive" (Koopman 2011b: 194). It is worth noting that the interventions brought together under the "critical geographies of peace" umbrella shared an engagement with feminist approaches to space, society, and politics, as well as a commitment to contextually grounded qualitative research.

Lastly, the impetus to center peace in geographical analysis, and to build self-consciously pacific geopolitics, prompted cautionary concern within political geography about the risk of occluding enduring power asymmetries, inequalities, and more structural forms of violence (Loyd 2011, 2012; Ross 2011). As Loyd (2012: 479) highlights, "defining peace and violence is a matter of epistemology, how the terms are conceptualized in relation to each other in specific places and times." Thus, while geographers argued for an engagement with peace in its own right as "process and content" (Williams & McConnell 2011: 928), rather than merely as a referential other to war or violence, they simultaneously underscored that peace and violence were mutually constitutive and part of a "continuum" (Flint 2011: 45; Koopman 2011b). Along with Ross (2011), critical geographers made clear that peace itself could be invoked to enact or reproduce violent inequalities, and that, as a result, critical analysis should never lose sight of "*power* and *equity*" (Ross 2011: 197, emphasis in original). In agreement, Loyd (2012; also 2011) proposed connecting the peace geographies agenda with intersectional feminist conceptualizations of "antiviolence" – understood as an analytic that foregrounds "interrelations of power and oppression" (2012: 481) and differentiated vulnerabilities, as well as a strategic orientation for grassroots struggles, in which peace is indistinguishable from justice and structural transformation.

Geo-graphing Peace: Spatialities, Materialities, Politics, and Pedagogies

This section provides a necessarily partial overview to illustrate how scholarship has extended a geographical understanding of peace since 2012. We begin by highlighting the fact that mapping peace's complex spatialities has mobilized peace studies scholars alongside geographers, although the dialogue could be further consolidated. We then attend to the multiple traditions of social and political critique that have continued to infuse peace geographies and we think ought to play a leading role in animating any global peace agenda in the future. We go on to discuss how an emphasis on spatial entanglements has expanded the geographical study of peace beyond traditional peace-building processes and post/conflict spaces. This conceptual and empirical expansion has been accompanied by deeper reflection on peace's multiple materialities and its often disjunctive, nonlinear temporalities.

Peace Geographies and the "Spatial Turn" in Peace Studies

Macaspac and Moore (2022) highlight that the consolidation of peace as a topic of geographical research has coincided with a "spatial turn" in peace and conflict studies (PCS), most visibly asserted in an edited collection on *Spatialising Peace and Conflict* (Björkdahl & Buckley-Zistel 2016). Björkdahl and Buckley-Zistel's argument (2016: 56) that "space is both the product of and the context through which peace is built and negotiated across scale, articulated through different forms of peace narratives, and informed by uneven geographies of power" strongly resonates with geographical perspectives. The same can be said of Björkdahl's proposed framework for "understanding the constitution of spaces for peace" through "spatial practices," "spatial dynamics," and "space formations" (Björkdahl and Buckley-Zistel 2023: 57). Intrigued, perhaps bemused, by the seeming lack of meaningful dialogue between geographers and PCS scholars on the spatialities of peace, Macaspac and Moore (2022), drawing partly on Gusic (2019), determine that the answer may lie in geography's engagement with peace as a disciplinary critique, in contrast with PCS's engagement with debates around scale (in particular, around the significance of "the local" for peace-building practice – see Mac Ginty & Richmond 2013). Macaspac and Moore (2022: 5) make the incisive observation that the parallel yet divergent trajectories within PCS and geography represent an "odd dynamic whereby geographers often read the spatial turn literature and think 'what's new here?', while spatial turn scholars wonder why political geographers fail to sufficiently theorize the spatiality of peace and conflict." While there remain critical blind spots for each discipline (see Koopman 2020), there have been increasing efforts to bridge disciplinary practices and research at the intersection of peace, space, and scale, as evidenced by more recent special issues (Björkdahl & Buckley-Zistel 2023), conferences (such as the Peace Geographies conference organized at Coventry University in 2019), and edited volumes (Flint & Dempsey 2023) that have proactively brought geography and PCS approaches into exchange.

Many Peaces, Pluriversal Critical Traditions

In addition to dialoguing with PCS, geographers have conversed with various counter-hegemonic traditions of thought and activism to advance and pluralize the theorization of peace. In doing so, they have taken up Koopman's (2011a) and Loyd's (2011, 2012) invitations to connect peace geographies more firmly with social justice movements (such as feminist antiviolence organizing), and built on Koopman's (2011b) insistence that we should conceive of "many peaces" in coexistence.

A prominent strand in this scholarship has been engagement with postcolonial and decolonial critiques of Euro-American hegemony and its associated geopolitical orderings (Daley 2014; Laliberte 2016; Courtheyn 2018, 2023; Feghali et al. 2021). Another, sometimes closely related strand has drawn on Black feminist thought and abolitionist praxis (Loyd 2012; Vasudevan & Smith 2020; Feghali et al. 2021; Piedalue 2022). Together, these critical conversations have helped excavate structural forms of violence (building on Galtung 1969), beyond the immediacies of armed conflict, and to *specify* (Loyd 2012) structural violence's differentiated impact on racialized, classed, and gendered people and spaces. These intersecting forms of violence include state predation, incarceration, and bordering; patriarchal, religious, racial, and caste domination; the devaluation, exploitation, and environmental degradation associated with global capitalism (see Flint 2023); and the dependencies and epistemic erasures that accompany neocolonial and settler colonial humanitarianism.

Mapping such violence in turn helps understand how dominant approaches to peace – especially when they are disconnected from justice – end up reproducing violent inequalities. Writing in an African context, Daley (2014, 2018), Laliberté (2014, 2016), and Feghali et al. (2021) unpack how Western-backed "liberal peacebuilding" and securitization ideologies end up normalizing neocolonial intervention (see also Rutazibwa 2014), reproducing racist tropes, and erasing local agency, even as they profess to empower women in particular. Brickell's (2015) study of initiatives tackling violence against women in Cambodia shows how legal emphases on "harmony" and community leadership can end up reinforcing women's marginalization and unsafety. Courtheyn (2023) discusses how "securing" the Colombian border with Venezuela can in fact reinforce migrants' vulnerabilities, while Ruette-Orihuela et al. (2023) sound the alarm at the violence that continues to target Indigenous leaders and their community after Colombia's Peace Agreement. In the United States, Vaught (2021) illustrates how urban peace – specifically the attempt to address gang violence – continues to be conceived primarily in terms of aggressive policing and draconian criminal sentencing; which is to say, in terms of a "war on crime."

However, radical critical thought has offered more to peace geographies than an expanded understanding of structural violence. As Piedalue notes (2022: 374), writing about the United States and India, "a feeling of futility can accompany the realization of the plurality and immensity of this violence and injustice" – a sense that all peace is merely "putative" (Ross 2011) and a step removed from the scale of transformation that would be required. Yet post/decolonial and Black feminist/abolitionist analyses have *also* contributed to documenting forms of resistance and alternative enactments of peace, to which we turn in the next section.

Everyday Peace, Expanded Politics

Drawing more or less explicitly on the radical traditions already discussed, as well as on a more established strand of scholarship in feminist geopolitics (Koopman 2011a; Williams & McConnell 2011), geographers and PCS scholars have stretched our understanding of where peace takes place and who is involved. Perhaps most prominently, they have highlighted the importance of ordinary spaces of everyday sociality, civic engagement, and expression in making peace and challenging (or not) some of its hegemonic articulations. Williams's (2013, 2015) conceptualization of "everyday peace" in the workplaces, trading circuits, and street-level religious communities of Varanasi, North India, has been influential in this regard. In a similar vein, Woon (2017) investigates how children in the southern Philippines understand and reimagine peace by drawing on their everyday lives, while Lefort (2020) highlights young people's mappings of coexistence in Beirut through encounter and unbounded spatial navigation. More recently, Kynsilheto (2024) has documented the small acts of kindness and care that exist amid everyday peace in contexts of nonwar violence. The notion of everyday peace has also gained prominence in PCS. It has spurred comparative work like Mac Ginty's (2014, 2021) and Firchow's (2018; Firchow & Mac Ginty 2017; also Ware & Ware 2022), who have sought to develop everyday peace indicators and associated evaluative frameworks, as well as ethnographies such as Carabelli's (2018) study of how spontaneous sociabilities and grassroots arts collectives can both uphold and undo ethnonationalist partitioning of space in the Bosnian city of Mostar (see also McMullin 2022).

The last example highlights some of the ambivalences of everyday peace, which is to say its entanglement with power, ideology, and politics – already well demonstrated by Williams's (2013, 2015) study of Varanasi. Kuus's (2007) analysis of "banal militarism" in Europe reminds us that ordinary, gendered subjectivities can internalize hegemonic and deeply problematic geopolitical ideologies through media consumption and reproduce them in everyday discourse. Nonetheless, several scholars have mobilized everyday peace to deliberately spotlight "subaltern peacemaking agency" (Hyndman 2019; Cante 2023a) and demonstrate how marginalized ways of being and doing can inform counter-hegemonic visions of peace. Courtheyn's (2016, 2018, 2022, 2023) work in Colombia has been exemplary in this regard, showing how *campesino* (peasant) and Indigenous ways of life can provide foundations for "radical trans-relational" (Courtheyn 2018) theorizations of peace, rooted in local ecologies and informed by anticolonial perspectives. Courtheyn's scholarship in Latin America is echoed by Macaspac's (2018) theorization of "insurgent peace" based on a careful analysis of the Indigenous-led peace zone of Sagada in the Philippines, as well as by Daley's (2014) and Rutazibwa's (2014) earlier suggestions that African concepts such as *ubuntu* and *agaçiro* might guide more contextually grounded and emancipatory forms of peace-making. In addition, while they do not necessarily use the framework of everyday peace, geographers influenced by Black feminist perspectives have shown how racially marginalized women's practices – from their involvement in nongovernmental organizations for mutual support (Feghali et al. 2021; Piedalue 2022) to more diffuse home-making labor (Vasudevan & Smith 2020) – should be conceptualized as peaceful resistance within and against intersecting structures of (slow) violence. Relatedly, Vaught's (2021) discussion of peace-brokering by incarcerated gang leader Stanley Williams in Los Angeles, and Cante's (2023a) theorization of everyday peace in Abidjan in terms of fugitivity, both point, in different ways, to the productive synergies between peace geographies and abolitionist thought/praxis that Loyd (2012; also Inwood 2009) had begun to map out and remain to be more fully explored. Lastly, Forde et al. (2021) discuss how, in a South African context, acknowledging the structural violence of race and class inequality leads to visions of peace-building as "reparations" and new forms of distributive justice.

Cumulatively, this work on everyday peace has pluralized and extended peacemaking agency beyond traditional spaces of diplomacy and institutional peacebuilding initiatives. In doing so, it has also confirmed peace's fundamentally *political* nature and expanded our understanding of how peace is contested, in practice if not in explicit discourse. Attention to Indigenous and subaltern peacemaking, for example, highlights the importance of community maintenance (Macaspac 2018; Courtheyn 2022), care, and reproductive labor (Vasudevan & Smith 2020), as well as the potentials of refusal (Macaspac 2018; Cante 2023a, 2023b) in crafting visions of peace that, while not being explicitly oppositional, are ultimately incompatible with dominant approaches and the sociospatial orders they seek to uphold. Such examples go some way to answer Bregazzi and Jackson's (2018) call on geographers to approach peace and its politics more ontologically. Bregazzi and Jackson (2018: 74) argue that critical geography's rush to situate everyday peace in an agonistic field of uneven power relations loses sight of its *inherently* political potential – "simply by being always already exemplary." Forms of everyday peace, in other words, are political not only because they are caught up in struggle, but also because they enact positive forms of human (and more-than-human) interrelation, which are valuable and interesting in their own right.

It is worth noting that, by combining a structural approach to violence beyond armed conflict and a focus on peace and/as everyday life, peace geographers and PCS scholars have continued to break down delineations between "peaceful" and "post/conflict" spaces at various scales. In part, this is about attending to spatial variations within a given conflict, showing "capabilities" (Kaldor & Sassen 2020) for resistance and nonengagement (Branch 2013; Jentzsch & Masullo 2022; Mantilla 2023). At a more global level, it is about challenging the idea that peace is only relevant to particular, conflict-affected areas of the world, which are the traditional targets of institutional peace-building initiatives (and are almost all located in the so-called Global South). One of the aims of developing everyday peace indicators in PCS has been to challenge assumptions of "incommensurability"

(Mac Ginty 2014; Firchow & Mac Ginty 2017) between contexts, and thereby to undo exceptionalizing typologies according to which places are separated according to where violence happened, or is susceptible to happening. Challenging contextual incommensurability has also been at the heart of recent work in urban geopolitics (Rokem 2016; Fregonese 2021; Elfversson et al. 2023), which has endeavored to undo the separation between "ordinary" and "divided" cities, pointing to shared configurations of violence as well as opportunities for comparative learning across urban movements for peace.

Materializing Peace

With their attention on the scalar and spatial politics of peace, political geographers have examined the material life of peace, from the body to built infrastructures. Through her research on international accompaniers or "unarmed bodyguards" in Columbia, Koopman (2014) shows how the strategic alignment of differently privileged and networked bodies alongside one another creates space for peace. Meanwhile, the positioning of vulnerable bodies, for example of Ukrainian civilians in front of Russian tanks, represents powerful acts of nonviolence that question the logics of war (see Koopman 2023: 176). Political geographers have emphasized the importance of "peopling peace" (Woon 2015) to recover embodied agency and experience even within institutional initiatives like peace-keeping military missions – or indeed international pacifist conferences (Hodder 2015). This is key to grasping the differentiated experiences of peace in practice, intersected by race, gender, age, and religion (Laliberté 2014, 2016; Williams 2015; Koopman 2020; Forde 2022).

Peace's materialization as a spatial practice is also visible in the production of built form as well as residential and administrative territories (Harrowell 2018). Political geographers have extended a longstanding concern with the bounding and narrating of space to document how places are relationally constituted as "safe cities" and "sanctuary cities" (Hyndman 2003; Darling 2014; Büscher 2020), as "zones of peace" (McConnell et al. 2014; Macaspac 2023), as differentiated and decentralized territories of peace (Cairo et al. 2018; Koopman 2020), or as ethnoreligious enclaves delineated by "peace walls" (Dempsey 2020). The latter example illustrates how the materialization of peace in built form and spatial organization can support the reproduction of "negative" and "violent peace" (Forde 2022; Leonardsson 2023), whereby architectures of separation and/or segregation risk perpetuating hierarchies and animosity rather than transforming relationships. Ruwanpura et al. (2020) show how road-building in postwar Sri Lanka, while appearing to deploy connective infrastructure, arguably ends up materializing peace as the affirmation of Sinhala-Buddhist domination. On the other hand, scholars across geography and PCS have documented "architecture as a framework for living together" (cited in Bădescu 2023: 601) and emphasized the role of built form in the "making of new geographies" (Flint and Dempsey 2023: 8) of peace. Examples include "syncretic place-making" initiatives in Sarajevo (Bădescu 2023), shared civic spaces such as community centers and skate parks (Dempsey 2020; Forde 2022), or the "nonsectarian" reclaiming of public spaces – against both neoliberal amnesia and factional memorialization – through urban social movements in Beirut and Belfast (Nagle 2020).

Alongside the built form of peace, there has been increased attention to its technological infrastructures. Central to this growing agenda is concern among digital geographers and PCS scholars over digital technologies' potential to exacerbate violence and expand the spaces of conflict through cyberattacks, coordinated disinformation, and artificial intelligence-guided military decision-making (Slesinger 2023). In response, a comparatively small but important thread of work has begun to research and theorize how digital technologies might be harnessed explicitly for peace-building (Richmond & Tellidis 2020; Hirblinger et al. 2023) and how they are already used to build peaceful solidarities (Navaro-Yashin 2008; more recently, see Koopman 2023 on the use of WhatsApp). The latter scholarship suggests that digital sociospatial relationships afford generative spaces for knowledge production, communication, and solidarity juxtaposed by a refiguring of the lines between dis/mis/trust, il/legitimacy, and un/certainty. The shifting relationship between peace, spatiality, and data/digital technologies thus bristles with potential for political geography. At the same time, it is important not to lose sight of scholarship on "older" technologies and associated popular geopolitics (Dittmer & Bos 2010), whose insights can continue to inform analyses of technologically mediated peace-making in the digital age. Cante's work in Abidjan, Côte d'Ivoire, for example, shows how local radio as a spatial technology and embedded urban infrastructure crystallizes competing practices and meanings of peace, from government- and donor-backed reconciliation programs (Cante 2020) to the staging of multiethnic encounters by ordinary listeners (Cante 2023a). The focus on radio further highlights how peace gets materialized in spoken discourse, shaping the language used to express shared understandings of peace (see also Georgi 2022 in a Colombian context), authorized modes of postwar public expression and political participation, and ultimately the negotiated contours of what is speakable and what should remain silent (Cante 2020, 2023b). In turn, the materialization of peace in mediated content – be it talk, music, film, public performance (see Till 2012), or the format-blending feed of social media platforms – brings to the fore the *affective* life of peace. How old and new technologies foster and circulate peaceful emotions such as hope and how and at what scale they might sustain atmospheres of peace (however contested, precarious, or ambiguous) remain fruitful questions at the intersection of peace and media/digital geographies (see Laketa 2016; Fregonese 2017 in an urban geopolitics context; also Navaro-Yashin 2012).

Another growing field of study delving into the manifold materialities of peace has been focused on its environmental dimensions. In part, this work emerges from political geography and political ecology's response to "resource war" explanations for conflict, seeking to complicate the scales, relations, and politics that frame nature as a "resource" even before the latter is mobilized as a "cause" of violence (O'Lear 2023). This strand of study has done much to unpick institutional initiatives linking resource control to peace-building, such as mineral tracing in Eastern Congo (Vogel & Raeymaekers 2016) or the creation of peace parks/reservations across contested African borders (Ramuntsindela 2017). More broadly, geographers and PCS scholars have begun theorizing environmental preservation and justice as integral to a "positive ontology of peace" (Barnett 2019: 930) as well as to holistic and sustainable peace-building frameworks (see Ide et al. 2021). Megoran and Dalby (2018: 270) call for a renewed "global ethic" (citing Martin Luther King) and a vision of the Anthropocene that "recontextualizes the human condition as one of interconnection and mutual vulnerabilities that require cooperative efforts to resolve." Similarly, Barnett (2019: 930) conceptualizes a "climate resilient peace" that would account for climate change as "a form of structural violence" and recognize its uneven impacts. Relatedly, scholars in the decolonial tradition evoked earlier have cautioned that "planetary mutuality" (DeBoom 2022) and a global climate-resilient peace should avoid reproducing the inequalities and false universalisms of "climate coloniality" (Sultana 2022), and instead learn from the environmental visions and stewardship of Indigenous communities, echoing Courtheyn's (2022) theorization of peace as "ecological dignity." As Le Billon and Duffy (2018: 241; see also Lederach 2017) summarize it, refiguring peace as a "socio-environmental justice agenda" should aim less to build a unified theory than to draw on the multiplicity of "grassroots movements and inter-governmental appeals to develop more socially just and environmentally sustainable relations across the human and non-human world."

Peace and Pedagogy in Political Geography

As noted in a previous section, the development of peace geographies as a subfield of political geography involved pedagogical concerns as well as an agenda for research (Inwood & Tyner 2011; Megoran 2011). Developing a pro-peace agenda in teaching and education has involved recognizing and pushing back against the "culture of war" that continues to pervade many national public spheres and school systems (see Behr et al. 2017 in the United Kingdom). As Ortiz et al. (2023) discuss, teaching (for) peace requires building syllabi and spaces of collective analysis that begin from students' own lifeworlds to uncover structures of power and violence at different scales, while deconstructing their discursive rationalizations; pointing toward justice and explicitly conceptualizing what positive peace could entail; and, more generally, refusing despair and resignation (Inwood & Tyner 2011; also Megoran 2023).

Beyond lecture content and readings, a key challenge, given the deeply polarized discourses that frame many a geopolitical issue today, is holding space for our diverse students and enabling them to make space for each other – nurturing respect, listening, and "agonistic pluralism" (Askins in Williams et al. 2013: 25) all at once. Colleagues have shared inspiring examples of how to curate engaged and respectful student debates (Gill in Williams et al. 2013) or how to use "nonviolent communication" principles to help students build connection across fundamental differences of opinion (Koopman & Seliga 2021). Nash's (in Williams et al. 2013) example from her Dublin and Belfast field class further demonstrates how in situ learning can foster an emplaced and embodied sense of peace. She describes how collaborating with grassroots peace-making groups to host student activities helps the latter appreciate *both* the structural challenges of peace and the transformative potentials of everyday activism. Building links with such grassroots organizations also makes our own positionality as educators explicit, asking who we learn from and how we ensure that the knowledge we disseminate cultivates an ethical, peaceful orientation to the world. Last but certainly not least, pro-peace pedagogy includes fighting to change the institutions we work for when they are complicit in global militarization (see Corderoy & Stockwell 2023; Lancione 2023).

Conclusion

This chapter has examined the recent trajectory of scholarship within political geography, often engaging allied fields, to unpack peace as a concept and to analyze it critically as a contested sociospatial and material process. We have shown how the consolidation of peace as a key geographical concept emerged within a context of pronounced militarism and state violence post 9/11. In this contemporary period of planetary violence characterized by war fever and slow violence, climate coloniality, new authoritarian alliances, and technological transformations, political geography has a renewed task in conceptualizing and materializing planetary peace(s). Such an approach should be attentive to the trans-scalar assemblages of peace and to refining spatializations of peace, as evidenced by peace studies' debates about the merits of "the local" or "the urban" (Millar 2021; Elfversson et al. 2023) for peace-building practice as well as everyday peace. Beyond scholarship, political geographers have an ongoing role to play in anti-violence and pacifist movements, helping to connect scales and struggles to repair and build the kinds of peaceful (and hopeful) spatial imaginaries that are urgently needed.

References

Anderson, B., Grove, K., Richards, L., & Kearnes, M. (2021). Slow emergencies: Temporality and the racialised biopolitics of emergency governance. *Progress in Human Geography* *44*(4): 621–639.

Bădescu, G. (2023). Urban geopolitics in "ordinary" and "contested" cities: Perspectives from the European South-East. *Geopolitics 28*(5): 1757–1782.

Barnett, J. (2019). Global environmental change I: Climate resilient peace? *Progress in Human Geography 43*(5): 927–936.

Behr, H., Megoran, N., & Carnaffan, J. (2017). Peace education, militarism and neoliberalism: Conceptual reflections with empirical findings from the UK. *Journal of Peace Education 15*(1): 76–96.

Björkdahl, A., & Buckley-Zistel, S. (eds) (2016). *Spatialising Peace and Conflict: Mapping the Production of Places, Sites and Scales of Violence*. Basingstoke: Palgrave Macmillan.

Björkdahl, A., & Buckley-Zistel, S. (2023). Introducing space for peace. *Journal of Intervention and Statebuilding 16*(5): 536–544.

Branch, A. (2013). Gulu in war. . . and peace? The town as camp in Northern Uganda. *Urban Studies 50*(15): 3152–3167.

Bregazzi, H., & Jackson, M. (2018). Agonism, critical geography, and the new geographies of peace. *Progress in Human Geography 42*(1): 72–91.

Brickell, K. (2015). Towards intimate geographies of peace? Local reconciliation of domestic violence in Cambodia. *Transactions of the Institute of British Geographers 40*(3): 321–333.

Büscher, K. (2020). Violent conflict and urbanization in Eastern Democratic Republic of the Congo: The city as a safe haven. In Kaldor, M., & Sassen, S. (eds), *Cities Are War: Global Insecurity and Urban Resistance*. New York: Columbia University Press, 160–183.

Cairo, H., Oslender, U., Suárez, C.E.P., et al. (2018). "Territorial peace": The emergence of a concept in Colombia's peace negotiations. *Geopolitics 23*(2): 464–488.

Cante, F. (2020). Mediating anti-political peace in Abidjan: Radio, place and power. *Political Geography 83*: 1–11.

Cante, F. (2023a). Everyday peace in a "post-conflict" African metropolis: Radio encounters, indeterminacy, fugitivity. *Transactions of the Institute of British Geographers 48*(3): 619–633.

Cante, F. (2023b). Attuning to opacity: Interpreting "post-crisis" refusals on Abidjan's local airwaves. *Anthropological Quarterly 46*(4): 683–710.

Carabelli, G. (2018). *The Divided City and the Grassroots. The (Un)Making of Ethnic Divisions in Mostar*. Basingstoke: Palgrave Macmillan.

Corderoy, J., & Stockwell, B. (2023). UK universities take millions from defence companies arming Israel. *OpenDemocracy*, December 19. https://www.opendemocracy.net/en/dark-money-investigations/universities-millions-defence-companies-israel-gaza (accessed November 24, 2024).

Courtheyn, C. (2016). "Memory is the strength of our resistance": An "other politics" through embodied and material commemoration in the San José peace community, Colombia. *Social and Cultural Geography 17*(7): 933–958.

Courtheyn, C. (2018). Peace geographies: Expanding from modern-liberal peace to radical transrelational peace. *Progress in Human Geography 42*(5): 741–758.

Courtheyn, C. (2022). *Community of Peace: Performing Geographies of Ecological Dignity in Columbia*. Pittsburgh, PA: University of Pittsburgh Press.

Courtheyn, C. (2023). Abandoning the revolution or weaving peace? South-South migration, socialism, and decolonial feminist geopolitics in South America. *Geopolitics*: 1–29. doi:10.1080/14650045.2023.2217750

Dalby, S. (2010). Recontextualising violence, power and nature: The next twenty years of critical geopolitics? *Political Geography 29*(5): 280–288.

Daley, P. (2014). Unearthing the local: Hegemony and peace discourses in Central Africa. In McConnell, F., Megoran, N., & Williams, P. (eds), *Geographies of Peace*. London: I. B. Tauris, 66–86.

Daley, P. (2018). Toward South-South peacebuilding. In Fiddian-Quasmiyeh, E., & Daley, P. (eds), *Routledge Handbook of South-South Relations*. London: Routledge, 320–331.

Darling, J. (2014). Welcome to Sheffield: The less-than-violent geographies of urban asylum. In McConnell, F., Megoran, N., & Williams, P. (eds), *Geographies of Peace*. London: I. B. Tauris, 229–249.

DeBoom, M. (2022). Climate coloniality as atmospheric violence: From necropolitics toward planetary mutuality. *Political Geography 99*: 102786.

Dempsey, K. (2020). Fostering grassroots civic nationalism in an ethno-nationally divided community in Northern Ireland. *Geopolitics 27*(1): 292–308.

Dietrich, W., & Sützl, W. (1997). *A Call for Many Peaces*. Burg Schlaining: Peace Center.

Dittmer, J., & Bos, D. (2010). *Popular Culture, Geopolitics and Identity*. London: Rowman and Littlefield.

Dodds, K. (2007). Steve Bell's eye: Cartoons, geopolitics and the visualization of the "war on terror". *Security Dialogue 38*(2): 157–177.

Dowler, L. (1998). 'And they think I'm just a nice old lady': Women and war in Belfast, Northern Ireland. *Gender, Place & Culture 5*(2): 159–176. doi:10.1080/09663699825269

Elfversson, E., Gusic, I., & Rokem, J. (2023). Peace in cities, peace through cities? Theorising and exploring geographies of peace in violently contested cities. *Peacebuilding 11*(4): 321–337.

Feghali, S., Faria, C., & Jama, F. (2021). "Let us create space": Reclaiming peace and security in a Kenyan Somali community. *Political Geography 90*: 102453.

Firchow, P. (2018). *Reclaiming Everyday Peace: Local Measurement and Evaluation after War*. Cambridge: Cambridge University Press.

Firchow, P., & Mac Ginty, R. (2017). Measuring peace: Comparability, commensurability, and complementarity using bottom-up indicators. *International Studies Review 19*(1): 6–27.

Flint, C. (ed.) (2004). *The Geography of War and Peace: From Death Camps to Diplomats*. Oxford: Oxford University Press.

Flint, C. (2011). Intertwined spaces of peace and war: The perpetual dynamism of geopolitical landscapes. In Kirsh, S., & Flint, C. (eds), *Reconstructing Conflict: Integrating War and Post-war Geographies*. London: Ashgate, 31–48.

Flint, C. (2023). The violence of development and the prospects of peace. In Flint, C., & Dempsey, K. (eds), *Making Geographies of Peace and Conflict*. London: Routledge, 109–127.

Flint, C., & Dempsey, K. (eds) (2023). *Making Geographies of Peace and Conflict*. London: Routledge.

Forde, S. (2022). The violence of space and the spaces of violence: Peace as violence in unequal and divided spaces. *Political Geography 93*: 1–9.

Forde, S., Kappler, S., & Björkdahl, A. (2021). Peacebuilding, structural violence and spatial reparations in post-colonial South Africa. *Journal of Intervention and Statebuilding 15*(3): 327–346.

Fregonese, S. (2017). Affective atmospheres, urban geopolitics and conflict (de)escalation in Beirut. *Political Geography 61*: 1–10.

Fregonese, S. (2021). Shockwaves: Atmospheres beyond the conflict city/ordinary city divide. *Conflict & Society 7*(1): 26–41.

Galtung, J. (1969). Violence, peace and peace research. *Journal of Peace Research 6*(3): 167–191.

Georgi, F.R. (2022). Peace through the lens of human rights: Mapping spaces of peace in the advocacy of Colombian human rights defenders. *Political Geography 99*: 102780.

Gregory, D. (2004). *The Colonial Present: Afghanistan, Palestine, Iraq*. Oxford: Blackwell.

Gregory, D. (2010). War and peace. *Transactions of the Institute of British Geographers 35*(2): 154–186.

Gregory, D. (2011). The everywhere war. *The Geographical Journal 177*: 238–250.

Gusic, I. (2019). The relational spatiality of the postwar condition: A study of the city of Mitrovica. *Political Geography 71*: 47–55.

Harrowell, E. (2018). Towards a spatialised understanding of reconciliation. *Area 50*(2): 240–247.

Hirblinger, A.T.J.M., Hansen, K., Hoelscher, Å.K., Lidén, K., & Martins, B.O. (2023). Digital peacebuilding: A framework for critical–reflexive engagement. *International Studies Perspectives 24*(3): 265–284.

Hodder, J. (2015). Conferencing the international at the World Pacifist Meeting, 1949. *Political Geography 49*: 40–50.

Hyndman, J. (2003). Mind the gap: Bridging feminist and political geography through geopolitics. *Political Geography 23*(3): 307–322.

Hyndman, J. (2019). Unsettling feminist geopolitics: Forging feminist political geographies of violence and displacement. *Gender, Place and Culture 26*(1): 3–29.

Ide, T., Bruch, C., Carius, A., et al. (2021). The past and future(s) of environmental peacebuilding. *International Affairs 97*(1): 1–16.

Inwood, J. (2009). Searching for the promised land: Examining Dr Martin Luther King's beloved community. *Antipode 41*(3): 487–508.

Inwood, J., & Tyner, J.A. (2011). Geography's pro-peace agenda: An unfinished project. *ACME 10*(3): 442–457.

Jenkins, A. (1985). Peace education and the geography curriculum. In Pepper, D., & Jenkins, A. (eds), *The Geography of War and Peace*. Oxford: Basil Blackwell, 202–213.

Jentzsch, C., & Masullo, J. (2022). Violent or non-violent action? Wartime civilian resistance in Colombia and Mozambique. *Political Geography 99*: 102761.

Kaldor, M., & Sassen, S. (eds) (2020). *Cities at War: Global Insecurity and Urban Resistance*. New York: Columbia University Press.

Kearns, G. (2008). Progressive geopolitics. *Geography Compass 2*(5): 1599–1620.

Kliot, N., & Waterman, S. (eds) (1991). *The Political Geography of Conflict and Peace*. London: Belhaven Press.

Kobayashi, A. (2009). Geographies of peace and armed conflict: Introduction. *Annals of the Association of American Geographers 99*(5): 819–826.

Koopman, S. (2008). Imperialism within: Can the master's tools bring down empire? *ACME 7*(2): 283–307.

Koopman, S. (2011a). Alter-geopolitics: Other securities are happening. *Geoforum 42*(3): 274–284.

Koopman, S. (2011b). Let's take peace to pieces. *Political Geography 30*(4): 193–194.

Koopman, S. (2014). Making space for peace: International proactive accompaniment. In McConnell, F., Megoran, N., & Williams, P. (eds), *Geographies of Peace*. London: I. B. Tauris, 109–130.

Koopman, S. (2019). Peace. In Antipode Editorial Collective (ed.),. In *Keywords in Radical Geography: Antipode at 50*. Hoboken, NJ: Wiley Online, 207–211.

Koopman, S. (2020). Building an inclusive peace is an uneven socio-spatial process: Colombia's differential approach. *Political Geography 83*(4): 1–11.

Koopman, S. (2023). The spatialities of nonviolent peace activism in the midst of war: From Colombia to Ukraine. In Flint, C., & Dempsey, K. (eds), *Making Geographies of Peace and Conflict*. London: Routledge, 165–182.

Koopman, S., & Seliga, L. (2021). Teaching peace by using nonviolent communication for difficult conversations in the college classroom. *Peace and Conflict Studies. 27*(3): 1–29.

Kuus, M. (2007). "Love, peace and Nato": Imperial subject-making in Central Europe. *Antipode 39*(2): 269–290.

Kynsilheto, A. (2024). Navigating spaces of everyday peace and violence: Societal protest, care and solidarity in post-18O Chile. *Geopolitics 29*(5): 1583–1603.

Lacoste, Y. (1976). *La géographie, ça sert, d'abord, à faire la guerre*. Paris: François Maspéro.

Laketa, S. (2016). Geopolitics of affect and emotion in a post-conflict city. *Geopolitics 21*(3): 661–685.

Laliberté, N. (2014). Building peaceful geographies in and through systems of violence. In McConnell, F., Megoran, N., & Williams, P. (eds), *Geographies of Peace*. London: I.B. Tauris, 47–65.

Laliberté, N. (2016). "Peace begins at home": Geographic imaginaries of violence and peacebuilding in northern Uganda. *Political Geography 52*: 24–33.

Lancione, M. (2023). *Università e militarizzazione: Il duplice uso della libertà di ricerca*. Turin: Eris.

Le Billon, P., & Duffy, R. (2018). Conflict ecologies: Connecting political ecology and peace and conflict studies. *Journal of Political Ecology 25*(1): 239–260.

Lederach, A.J. (2017). "The campesino was born for the campo": A multispecies approach to territorial peace in Colombia. *American Anthropologist 119*(4): 589–602.

Lefort, B. (2020). Cartographies of encounters: Understanding conflict transformation through a collaborative exploration of youth spaces in Beirut. *Political Geography 76*: 1–12.

Leonardsson, H. (2023). Peace through violence and violence through peace: Peacebuilding practices and a conflictual peace in Lebanon. *Geopolitics 29*(5): 1538–1558. *Online First*

Loyd, J.M. (2011). "Peace is our only shelter": Questioning domesticities of militarization and white privilege. *Antipode 43*(3): 845–873.

Loyd, J.M. (2012). Geographies of peace and anti-violence. *Geography Compass 6*(8): 477–489.

Macaspac, N.V. (2018). Insurgent peace: Community-led peacebuilding of indigenous peoples in Sagada, Philippines. *Geopolitics 24*(4): 839–877.

Macaspac, N.V. (2023). Spatialities of peace zones. *Cooperation and Conflict 58*(2): 175–193.

Macaspac, N.V., & Moore, A. (2022). Peace geographies and the spatial turn in peace and conflict studies: Integrating parallel conversations through spatial practices. *Geography Compass 16*: e12614.

Mac Ginty, R. (2014). Everyday peace: Bottom-up and local agency in conflict-affected societies. *Security Dialogue 45*(6): 548–564.

Mac Ginty, R. (2021). *Everyday Peace: How So-Called Ordinary People Can Disrupt Violent Conflict*. Oxford: Oxford University Press.

Mac Ginty, R., & Richmond, O.P. (2013). The local turn in peace building: A critical agenda for peace. *Third World Quarterly 34*(5): 763–783.

Mamadouh, V. (2004). Geography and war, geographers and peace. In Flint, C. (ed.), *The Geography of War and Peace: From Death Camps to Diplomats*. Oxford: Oxford University Press, 26–60.

Mamadouh, V. (2023). Geography and war, geographers and peace: Expanding research and political agendas. In Flint, C., & Dempsey, K. (eds), *Making Geographies of Peace and Conflict*. London: Routledge, 12–36.

Mantilla, J. (2023). Disruptive hope: The communal repertoires of violence resistance in Cúcuta. *Peacebuilding 11*(4): 358–380.

McConnell, F., Megoran, N., & Williams, P. (2014). *Geographies of Peace*. London: I.B. Tauris.

McMullin, J. (2022). Hustling, cycling, peacebuilding: Narrating postwar reintegration through livelihood in Liberia. *Review of International Studies 48*(1): 67–90.

Megoran, N. (2010). Towards a geography of peace: Pacific geopolitics and evangelical Christian Crusade apologies. *Transactions of the Institute of British Geographers NS35*(3): 382–398.

Megoran, N. (2011). War *and* peace? An agenda for peace research and practice in geography. *Political Geography 30*(4): 178–189.

Megoran, N. (2023). Making sense of the Ukraine war: Geographers should not be afraid of geography. *Transactions of the Institute of British Geographers 48*(4): 821–825.

Megoran, N., & Dalby, S. (2018). Geopolitics and peace: A century of change in the discipline of geography. *Geopolitics 23*(2): 251–276.

Millar, G. (2021). Trans-scalar ethnographic peace research: Understanding the invisible drivers of complex conflict and complex peace. *Journal of Intervention and Statebuilding 15*(3): 289–308.

Nagle, J. (2020). Defying state amnesia and memory wars: Non-sectarian memory activism in Beirut and Belfast city centres. *Social & Cultural Geography 21*(3): 380–401.

Navaro-Yashin, Y. (2008). Pacifist devices: The human/technology interface in the field of conflict resolution. *Cambridge Journal of Anthropology 28*(3): 91–112.

Navaro-Yashin, Y. (2012). *The Make-Believe Space: Affective Geography in a Postwar Polity*. Durham, NC: Duke University Press.

Nixon, R. (2011). *Slow Violence and the Environmentalism of the Poor*. Cambridge, MA: Harvard University Press.

O'Lear, S. (2023). Geographies of environmental peace and conflict. In Flint, C., & Dempsey, K. (eds), *Making Geographies of Peace and Conflict*. London: Routledge, 219–236.

Ortiz, M., Belén Noroña, M., Dowler, L., et al. (2023). Placing peace: The pedagogy of positive peace and environmental justice. In Flint, C., & Dempsey, K. (eds), *Making Geographies of Peace and Conflict*. London: Routledge, 256–271.

Ó Tuathail, G., & Dalby, S. (1998). *Rethinking Geopolitics*. London: Routledge.

Pepper, D., & Jenkins, A. (eds) (1985). *The Geography of War and Peace*. Oxford: Basil Blackwell.

Piedalue, A.D. (2022). Slow nonviolence: Muslim women resisting the everyday violence of dispossession and marginalization. *Environment and Planning C: Politics and Spaces 40*(2): 373–390.

Ramuntsindela, M. (2017). Greening Africa's borderlands: The symbiotic politics of land and borders in peace parks. *Political Geography 56*: 106–113.

Richmond, O.P., & Tellidis, I. (2020). Peace in analogue/digital international relations. *Global Change, Peace & Security 32*(3): 317–336.

Rokem, J. (2016). Beyond incommensurability: Jerusalem and Stockholm from an ordinary cities perspective. *City 20*(3): 472–482.

Ross, A. (2011). Geographies of war and the putative peace. *Political Geography 30*: 197–199.

Ruette-Orihuela, K., Gough, K.V., Vélez-Torres, I., et al. (2023). Necropolitics, peacebuilding and racialized violence: The elimination of indigenous leaders in Colombia. *Political Geography 105*: 1–11.

Rutazibwa, O. (2014). Studying agaciro: Moving beyond Wilsonian interventionist knowledge production on Rwanda. *Journal of Intervention and Peacebuilding 8*(4): 291–302.

Ruwanpura, K.L., Chan, B.B., & Kajotha, V. (2020). Unsettled peace? The territorial politics of roadbuilding in post-war Sri Lanka. *Political Geography 76*: 1–10.

Slesinger, I. (2023). Hybrid networks: Technology, geopolitics and ontology in digital warfare. In Flint, C., & Dempsey, K. (eds), *Making Geographies of Peace and Conflict*. London: Routledge, 200–218.

Springer, S. (2013). Anarchism and geography: A brief genealogy of anarchist geographies. *Geography Compass 7*(1): 46–60.

Sultana, F. (2022). The unbearable heaviness of climate coloniality. *Political Geography 99*: 102638.

Till, K. (2012). Wounded cities: Memory work and a place-based ethic of care. *Political Geography 31*(1): 3–14.

Vasudevan, P., & Smith, S. (2020). The domestic geopolitics of racial capitalism. *Environment and Planning C: Politics and Space 38*(7–8): 1160–1179.

Vaught, S. (2021). No justice, no peace prize: Stanley "Tookie" Williams, the 1992 Bloods-Crips truce, and the limits of racial peacekeeping. *Peace & Change 46*: 353–383.

Vogel, C., and Raeymaekers, T. (2016). Terr(it)or(ies) of peace? The Congolese mining frontier and the fight against "conflict minerals". *Antipode 48*(4): 1102–1121.

Ware, A., & Ware, V.-A. (2022). Everyday peace: Rethinking typologies of social practice and local agency. *Peacebuilding 10*(3): 222–241.

Williams, A., Jeffrey, A., McConnell, F., et al. (2013). Interventions in teaching political geography. *Political Geography 34*: 24–34.

Williams, P. (2013). Reproducing everyday peace in North India: Process, politics, and power. *Annals of the Association of American Geographers 103*(1): 230–250.

Williams, P. (2015). *Everyday Peace? Politics, Citizenship and Muslim Lives in India*. Chichester: Wiley-Blackwell.

Williams, P., & McConnell, F. (2011). Critical geographies of peace. *Antipode 43*(4): 1–5.

Woon, C.Y. (2015). "Peopling" geographies of peace: The role of the military in peacebuilding in the Philippines. *Transactions of the Institute of British Geographers 40*(1): 14–27.

Woon, C.Y. (2017). Children, critical geopolitics, and peace: Mapping and mobilizing children's hopes for peace in the Philippines. *Annals of the American Association of Geographers 107*: 200–217.

Chapter 16

Place and Memory

Karen E. Till[1] *and Joseph S. Robinson*[2]
[1]*Department of Geography, Maynooth University, Maynooth, Ireland*
[2]*Independent scholar*

Halfway through Adania Shibli's (2021) award-winning novel *Minor Detail*, an unnamed woman living in contemporary Ramallah, Palestinian Territories, reads a newspaper article documenting the gang rape and murder of a Bedouin girl by Israeli soldiers that took place in 1949, shortly after what is known to Palestinians as the *Nakba* (catastrophe) and to Israelis as the War of Independence. The woman becomes obsessed with this story and embarks on a quest to recover the girl's memory. In what Gregory (2004) calls "the colonial present," the protagonist navigates the multiple time-spaces of past occupation, settlement, and erasure through what Till (2012) refers to as "memory-work" and Elizabeth Jelin (2003) describes as the "labour of memory." Anxiously transgressing borders and territories, she layers past and present maps to create a palimpsest to "read" the landscape; she questions the heroic narratives of military museums and the absences of archives; and she encounters and follows ghosts in pursuit of the truth. The legacies of (post)colonial occupation and war collide in a state of "diabolical continuity (and inevitability) from an unexpiated past" (Soyinka 1999: 19–20) when Shibli's narrator finds herself in the dunes of a military-restricted zone and is shot for trespass in the very same place where the girl was murdered in 1949.

Artists like Shibli, as well as activist social movements and Indigenous leaders calling for the decolonization of lands, minds, and bodies (Waziyatawin & Yellow-Bird 2012), are engaging in memory-work when they seek "representation and recognition of the violent past" (Hite & Jara 2020: 247) and imagine and create more just worlds. This chapter examines place and memory through

The Wiley Blackwell Companion to Political Geography, Second Edition.
Edited by Virginie Mamadouh, Natalie Koch, Chih Yuan Woon, and John Agnew.

the lens of political geography, with a focus on contested national memories and the significance of intergenerational, spectral, and anticolonial practices of memory-work. We understand the "politics of memory" to include the negotiated processes whereby a group represents the past through place to claim territory, establish social boundaries, justify present political actions, or challenge established social narratives. The spatial and politicized processes and practices of collective memory entwine political imaginations, power relations, and social identities (Johnson 1995; Till 1999, 2003). Places of memory are not passive backdrops that simply reflect a version of history or form of power, but rather are continuously negotiated and reinterpreted by groups seeking to advance their political agendas. Contentious debates often emerge when groups attempt to legitimize their own version of the past as the authoritative one: Whose conception of the past should prevail in a given public realm? How and where are selective narratives depicted and "fixed" through places, including built environments, media spaces, and legal structures? In the next section, we offer an initial overview of the key theoretical works by Maurice Halbwachs and Pierre Nora that have influenced geographers in their understanding of collective memory, followed by a review of more recent relevant literature in the field.

Topographies and Sites of Collective Memory

In the wide-ranging interdisciplinary discussion of collective memory, scholars frequently draw on the writings of French sociologist Maurice Halbwachs and French historian Pierre Nora. Halbwachs rejected Freudian conceptions of memory and recall as processes occurring within the individual psyche. For Halbwachs (1992), memory is not individual experience stored in the unconscious, but the social product of retellings of the past through social environments and relations in the present. Individuals "remember" the past by framing their personal experience in relation to cultural performances of group identity such as through written chronicles, commemorative parades, food traditions, collecting relics, naming places, and observing anniversaries. Social memory and related collective cultural practices thus produce continuities between past and present, temporal connections infused with moral meaning that changes with the needs of the present.

Because stories of the past are always in flux, Halbwachs argues that groups often create "topographies" of memory to make the connection between past and present permanent, tangible, and material. Statues, monuments, and places of symbolic and spiritual significance superimpose the memory of a people onto physical presences in the landscape and thereby communicate permanence. Stories about the past often adapt to those physical presences and/or use their material qualities to communicate group values and meaning. The dense experiential, social, and tangible facets of places and landscapes therefore not only frame or stage social memory, but also situate and spatially constitute group remembrances (Till 2005). Halbwachs has, nonetheless, been criticized for assuming a Durkheimian understanding of the social group and downplaying the complexity and dynamism of intragroup relations, including how collective memories are created and negotiated (Withers 1996).

Pierre Nora's (1997) multivolume *Lieux de mémoire* (sites of memory) project brought Halbwachs's insights to a systematic investigation of the social and collective manifestations of French national memory in everyday settings, including lived environments and sites. Fundamental to his argument is that modern people no longer inhabit lived environments (*milieux*) of memory. For Nora, traditional *milieux* of memory existed in shared "spaces, gestures, images and objects" (1989: 9), whereas modern *lieux de mémoire* are "fundamentally remains," surrogate spaces that "have barely survived in a historical age" (1989: 12). Sites of memory are also opposed to the project of "History," which for Nora is a rationalist, secularist, and scholarly approach to the past that overrides memory's recourse to the sacred (Legg 2005). By positioning memory as resistant to selective narrations of History, Nora's work offers a theoretical framework to understand challenges to national renditions of the past from oppressed and marginalized social groups and memory actors. However, Nora has been criticized for his rigid distinction between a past time of supposedly naturally occurring *milieux de mémoire* and a present wherein memory can only be approached through incomplete *lieux* (Johnson 1999). According to Legg (2005), while Nora's understanding of memory may destabilize dominant historical interpretations of the past, by severing its production from extant power relations, he neglects to consider how sites of memory are subject to complex systems of social relations, including the exclusions in which they are produced.

Official Places of Memory

Scholars drawing on Halbwachs and Nora document and interrogate how state officials (government bureaucrats, leaders, and experts) sanctify places of memory to enact ritually selective narratives of "a people" at prominent locations as part of the process of nation-building. To showcase global status in the nineteenth century, for example, the histories of monuments and statues of military war heroes and high-ranking colonial officials of the British Empire were made and displayed in London and sent to adorn the streets of important colonial centers such as Calcutta (Gosh 2023: 11). The creation of monumental sites and sanctified topographies of memory, moreover, evokes myths of a timeless empire or nation through the adulation of male, heroic bodies in public spaces that naturalize gendered representations of geopolitical power (Sharp 1996; Azaryahu & Kellerman 1999).

In late nineteenth- and early twentieth-century Italy, the Vittoriano Emanuele II Monument in Rome venerated dead male heroes to project a unified nation (Agnew, 1998; Atkinson & Cosgrove 1998). Built in 1878–1882 as a sacred altar honoring a dead king, in 1921 under Mussolini's fascist regime a memorial to the Cult of the Unknown Soldier was added. The grandiose, layered, and ornate designs of this white marble monument – reaching from Earth (symbolized by dead bodies) to the heavens (represented by the Beaux-Arts, neoclassical design) – materialized a mystical understanding of Italy as a transcendental, classical empire; its location, adjacent to the Imperial and Roman Fora, linked new political states to the mythical acropolis of ancient Rome and the ideal of an "eternal city," even as local residents derided it.

As personified in such heroic masculinist monuments and statues, the valor and spirit of "great men" metonymically represent the sacred fraternity of Anglo-European civilization. If female figures are included, they often represent a "motherland" and are depicted as nebulous cultural figures – symbolizing birth, domesticity, family, and tradition – or rendered as passive symbols perpetually in need of male protection (Warner 2001; Cusack 2018). Memorials commemorating war also reinforce gendered understandings of the nation and subnational ethnic groups as a fraternal brotherhood. In Northern Ireland, for example, McDowell (2008) argues that both wartime and postwar forms of commemoration have erased women from the inscriptive landscape, both as active participants and as victims of conflict. Yet embodied female stories of suffering can at once pierce and support nationalist myth-making. In contemporary South Korea, Myadar and Davidson (2021) describe how the stories of "comfort women," or women forced into sexual slavery by Japanese occupiers during World War II, are used to buttress nationalist claims to collective victimization.

The meanings and social identities of "war dead," and of victims, perpetrators, and heroes, change through time depending on different national, local, and geopolitical contexts (Dowler 1998). The "Great War" (World War I), for example, was difficult to commemorate and remember socially because the mass slaughter of that war was unintelligible to citizens and survivors (Winter 1995; Johnson 1999). In contrast, public memory of World War II, especially in the United States and Western Europe, retains a mostly coherent militarist narrative and masculinist aesthetic form, using ideas of "social debt" to the "greatest generation" to provoke uncritical national veneration (Doss 2012). When World War II memory intersects with ongoing issues of colonial erasure, the results may be especially contentious. For example, in Guam, the War in the Pacific National Park has been subject to rigorous debates over how it embeds uncritical narratives of US heroism in the landscape, while eliding Chamorro wartime suffering, sacrifices, and experiences with Japanese and American military conquests (Herman 2008). Thus, the aesthetics of national war *lieux de mémoire* reflect and adapt to changing discussions of how a particular war is publicly remembered at a given time and place.

Places of memory built following times of political transition are often overtly dogmatic projects. Officials in Zimbabwe (Mangena 2020) and Poland (Drozdzewski 2014) seeking to support an emergent political order, for example, renamed streets and urban districts to canonize new versions of the past. In Taiwan, Chinese Nationalists remapped "the geography of the Chinese homeland onto the cityscapes of Taipei" (Leitner & Kang 1999: 221) by renaming streets, schools, theaters, and other public buildings, squares, and spaces in Taipei after 1949. However, opposition movements in Taiwan challenged the Kuomintang's strategy by demanding that these nationalist place names be changed.

Removing and replacing statuary is another strategy used by new governments to claim legitimacy following decolonization movements. The postcolonial Tamil Dravida Munnetra Kazhagam Party, for example, erected new statues, temples, and buildings in Madras after 1949 (Lewandowski 1984), and the postindependence Ceylonese/Sri Lankan nationalist leaders removed and replaced statues of British political figures in Kandy (Duncan 1989). In another example, the Abiy Ahmed regime in contemporary Ethiopia is attempting to navigate a severe legitimation crisis through an ambitious memorialization project that works to position Ahmed as the culmination of an unbroken and glorified Ethiopian great state (Regasser & Emmenegger 2023). In a similar vein, Mongolian leaders have reached far into the past to resurrect the monumental figure of Chinggis Khaan to legitimize their contemporary rule and signal a clear break from nearly a century of Soviet influence (Myadar 2017).

Research about memorials and monuments in authoritarian and formerly authoritarian societies also documents political change and continuity through multifaceted processes of commemoration (Foote, Tóth, & Árvay 2000; Angelucci & Kerschbamer 2017). In National Socialist Germany, for example, an extensive and ambitious program was designed to rewrite the commemorative topography of the nation. Hagen (2020: 166) argues that Nazi commemorative strategies deployed specific places in the German landscape to create a networked archipelago of commemorative sites that "positioned the [Nazi] movement as the path to ultimate redemption in the enduring racial struggle for national survival." Yet such overtly nationalistic memory projects do not necessarily indicate coherent nationalist movements. In the Russian Federation, Forest and Johnson (2002), for example, argue that Russian elites chose to reinterpret rather than erase the Soviet legacy, yet this manipulation was often resisted by popular readings of particular sites. In contemporary Russia, Putin's regime has sought to privilege selective readings of the Soviet past by coopting the popular veneration of places memorializing World War II, including myths of the heroic Soviet soldier and Soviet sacrifice more generally, and the closeness of Russia and former Soviet regimes, especially Ukraine. In the process, the memory of Stalinist terror is displaced and groups and individuals seeking to produce the space for alternative readings are subject to immense risk and pressure (Bernstein 2016; Edele 2017).

Another strategy of authoritarian regimes is the attempt explicitly to break the link between public memories of organized expressions of resistance and dissent formed in and associated with particular places. For example, the Erdoğan regime in contemporary Turkey relentlessly polices transgressive graffiti at Istanbul's Taksim Gezi Park, where multiple and diverse anti-regime demonstrations took place in 2013 (Evered 2018). The regime has also repressed any reproductions of dissident graffiti, textual forms, and imagery, including in social media, from Gezi Park appearing in other sites, thus actively seeking to erase through intimidation all active memory practices tied to places of public dissent. As Hammond (2022) argues, the Erdoğan regime clearly understands that the power of place and transgressive memory is not confined to a physical location per se, but rather that digital and textual recall and recollection can also transform official public spaces into transgressive ones.

Following extreme violence, such as genocidal wars, the creation of places of memory may advance contradictory messages, such as in postgenocide Rwanda, where a dense landscape of more than 250 memorials now exists. After 1994, memorials and museums were established at sites of mass killings, with international support from human rights organizations and governments. However, when Kagame, the former commander of the Rwandan Patriotic Front (RPF), became President in 2000, his regime prohibited any memory projects not clearly conforming to official state historical narratives that celebrate the RPF as heroes (Thomson 2013; Jessee 2017). Nowhere in this official memorial landscape are the Hutu civilians murdered by the RPF remembered; furthermore, female victims are only given space to be remembered as rape victims and mothers (Mannergren Selimovic 2020). Rwanda's memorial landscape thus actively subverts local particularities and experiences of genocide in favor of a "homogenised atrocity narrative" (Lischer 2019: 68). Yet, as Tyner (2017) has argued in Cambodia, even while national heritage landscapes can function to enforce selective recall and enforced forgetting, they can also provide opportunities for transgressive and/or critical remembering due to the fact that the material remnants of genocide remain everywhere in the landscape, and are thereby imbued with latent or potential memorial significance. Kenneth Foote (2003) has demonstrated a similar inability to obliterate or repurpose places and landscapes redolent with difficult pasts in the United States.

Officials of democratic states seeking to acknowledge difficult pasts while politically distancing themselves from those pasts may promote human rights–based discourses to construct official histories that advance linear progressive narratives (Bevernage 2012). In the South African Apartheid Museum in Johannesburg, for example, visitors experience a pervasive sense that the violent racism of the apartheid regime is over and that the period of postapartheid negotiation has resulted in a more benevolent "Rainbow nation" defined by consensus and unity (Teeger & Vinitzky-Seroussi 2007). At the Canadian Human Rights Museum, which opened in 2014 in Winnipeg, "historicized" colonial pasts are represented as aberrant periods of time that no longer reflect Canadian morals and values (Coulthard 2014). Yet in its permanent exhibition, the water used for panoramic reflection pools – intended as a metaphor for universal human rights – is transported through an aqueduct whose construction forcibly displaced members of an Anishinaabe tribe from their traditional homeland (Dean & Failler 2021). Other exhibitions depict Canada's colonial pasts, including the forced assimilation of young Indigenous people through residential schools, as part of a completed past and no longer relevant to the Canadian present, thus supporting the myth of Canadian settler benevolence (Regan 2010) rather than advancing reparative justice for unjust political relations in the present (Failler 2015). As Nelson and Godlewska (2023) argue, such approaches to commemoration-as-reconciliation are a settler-colonial memorial framework that is both inadequate and inappropriate for North American Indigenous understandings of memory and place, a point to which we return later.

The official memory of the Civil Rights movement and struggle, often canonized through the figure of slain Civil Rights leader Martin Luther King, Jr., deploys heroic, timeless, and universalizing frameworks that advance consensual narratives of racial harmony, or that the United States has entered a "post-racial" era (Bruyneel 2014; Hagopian 2020). Such a legacy, however, is not only gendered, it also excludes the multiple legacies and more radical narratives of the Civil Rights struggle, including King's work (Dwyer & Alderman 2008). Research about the fraught topographies of US Civil War memory in the US South, moreover, shows how a particular strand of social memory depicted in Confederate monuments, the so-called Lost Cause narrative, elides the role of slavery in the conflict in favor of a noble, if unsuccessful, struggle for state rights and Southern self-determination (Forest & Johnson 2019). This "Lost Cause" iconography is not traceable to the aftermath of the Civil War, but rather is intimately connected to a white Southern memory politically justifying twentieth-century Jim Crow politics (Hoelscher 2003; Savage 2018). Additionally, this memory landscape includes antebellum Southern museums and heritage tourism sites, such as repurposed plantations that function as places of amusement and forgetting, rather than as places to remember the horror of slavery (Alderman & Modlin 2008).

Richard Schein (2006) argues that all of America's cultural and memorial landscapes are heavily racialized. Founded in 2013, the Black Lives Matter (BLM) movement has called out the foundational racist myths enshrined in the US landscape. Building on their work and the activist memory-work of preceding generations, and following the 2015 mass shooting of innocent people at the oldest Black church in the Southern United States, the 2017 white supremacist "Unite the Right" rally protesting the removal of Confederate monuments that led to violence against counter-protestors, and the 2020 international protests supporting BLM in response to the murder by a white policeman of the unarmed African American George Floyd, in recent years hundreds of Confederate flags have been taken down and statues toppled.

But more work needs to be done to change existing racist power relations (Alderman & Inwood 2016). In 2019, following debates about the statuary along "Monument Lane" in the former Confederate capital of Richmond, artist Kehinde

Wiley was commissioned by the Virginia Museum of Fine Arts to produce a counter-monument. The 27-foot bronze statue *Rumors of War* depicts a young Black man with dreadlocks, wearing a hoodie, ripped jeans, and Nikes, and riding a rearing horse. Randolph (2023) interprets this critical artwork as a "rallying cry" that both reveals white supremacist power structures, as communicated through Confederate monuments and iconography, but also regenerates more just forms of memory. In the next section we consider how acts of memory-work like *Rumors of War* do more than call attention to forgetful official monuments – they also imagine more just futures.

Place-Based Intergenerational, Spectral, and Anticolonial Memory

Activists, community leaders, and artists have used various place-based approaches to memory-work to call attention to state and social injustices and advance reparative, resistant, or rights-based futures. Often led by individuals and groups inheriting the legacies of violent pasts, alternative places and practices of memory may include history workshops, neighborhood tours, creative mappings, historical exhibitions, community centers, public art projects, and occupying public spaces. Unlike official sites of memory that delimit spaces according to linear, progressive timeframes, these entangled and haunted places of memory are better understood according to entangled and blurred "space-time regimes" (Houssey-Holzshuch 2021) connecting generations across geopolitical borders. With the rise of a global media culture, moreover, collective memories of shared pasts may quickly become the object of worldwide mediation (Erll 2011) and lead to creative international forms of solidarity (Featherstone 2012).

Geographers researching places haunted by such past forms of colonial and state violence have contributed to a "spectral turn" in memory studies (Hite & Jara 2020) – a body of research drawing on sociologist Avery Gordon's (1997) concept of "ghostly matters." A ghost, for Gordon, is a felt and known absence whose presence acts on and meddles with taken-for-granted realities in social life (see also Maddrell 2013). Gordon's work focuses on the narrative and material absences in texts, artifacts, and archives that signify larger gaps in authoritative systems of knowledge, including dominant histories. Geographers have paid attention to "spectral traces" in the landscape, those "phantoms, histories, remnants, submerged ways of knowing" that may be given afterlives through activist and artistic place-based memory-work (Jonker & Till 2009: 306; see also Till 2012). Spectral presences haunt particular places and spaces and function as mediators of memory, demanding acknowledgment of past crimes and refusing the censorship of dominant linear understandings of time and official memory. Robinson's (2018, 2022) work with survivor communities in Northern Ireland, for example, argues that the purposive retention of the traumatic past through spatial hauntings in public space functions as resistance against "reconciliatory memory," or the flawed belief that a traumatic past can be harnessed to serve the interests of a transitional future.

Hirsch and Spitzer's (2011) *Ghosts of Home* documents the authors' return to the city of Czernowitz (now Chernivtsi) in western Ukraine from where Hirsch's parents originally came. What does it mean for subsequent generations to return "home" to where National Socialists, supporters, and bystanders attempted to erase all Jewish connections to cultural landscapes and urban places as part of their larger project of the European Holocaust? Once a cosmopolitan, polyglot, and vibrant city, home to one of the major centers of Jewish cultural production in the world, Czernowitz's Jewish material culture was all but destroyed and erased from the visual landscape as a kind of "memoricide" (Pappé 2006). Yet what Hirsch and Spitzer find on their return is not the absence of places of memory, but the vibrant afterlives and "ghosts" of a Jewish past that exist in the places and spaces where people share stories of Czernowitz's past, through preserving and archiving old photographs and guiding visitors to experience and imagine what used to be there. What their analysis powerfully suggests is that no memoricide can ever be completed, even with the erasure of the material attributes of once-inhabited places, if the work of memory continues across generations.

For many survivors, victims, and their families, creating alternative gathering places that are haunted by memories of traumatic loss may be a radical way to destabilize oppressive regimes and governments in transition and advance international calls for justice (Bevernage 2012; Robinson 2018). One such example is the oft-cited case of the human rights groups coalescing under the banner of the Mothers of Plaza de Mayo (hereafter, Madres) in Argentina. Formed in 1977 in response to the forced disappearance of their loved ones by the Argentine *junta*, the Madres regularly occupied a central square in Buenos Aires to publicly demand their relatives' corporeal return, even in the face of deadly repression. Yet even after the fall of the *junta* and the realization that the vast majority of the Disappeared were now dead, some Madres reconfigured their struggle into resisting official attempts to legally mandate institutional forgetting through President Carlos Menem's policy of "National Reconciliation" (Bosco 2004; Crenzel 2020).

The District Six Museum (D6M) in central Cape Town, South Africa, is another significant place of memory challenging official attempts to "move on" from difficult pasts. Between 1966 and 1982, more than 60,000 people were forcibly displaced to the outlying Cape Flats after District Six was declared "Whites only" with the Group Areas Act of 1950. The museum opened in 1994, drawing on the work of anti-apartheid activists and artists who kept alive the memories of the places destroyed and multiracial communities displaced through traces, mappings, and stories (Till 2012; Houssey-Holzshuch 2021). Located in an old Moravian church, the museum's permanent exhibitions and walking tours advance consciousness-raising about the ongoing legacies of the apartheid past for tourists, while curating the stories of former residents as significant in the present day. Addressing unjust legacies of spatial-racial apartheid, the museum works to declare District Six a national heritage site, supports group claims to the "right to return," engages with former residents in a range of translocal cultural memory projects, and is part of African and international networks advancing alternative places of memory known as "historic site museums of conscience."

Such persistent and purposive public acknowledgments of ongoing negative legacies of traumatic pasts may result in a change in national practices of memory, including in societies inheriting the legacies of a "society of perpetrators." In former West Berlin, for example, beneath an overgrown field surrounded by the Berlin Wall, anti-fascist citizen groups uncovered where the former headquarters of the National Socialist terror apparatus, including the Gestapo, SS, and Reich Security Service, was located. After protesting city plans to build a motorway across the field, activists literally and symbolically excavated the landscape to reveal the forgetful nature of official national (and Cold War) memory (Till 2005). Calling this field the "open wound" of the city and nation, activists created the "Topography of Terror," a temporary memorial and exhibition that later became a permanent "memorial museum" and "documentary center" following reunification. It now provides a variety of place-based educational resources for scholars, students, and visitors, advises the national German Memorial Museums Department, coordinates the international Memorial Museum Forum, and collaborates with other national and international memorial museums and initiatives.

Digital forms of co-presence (Kitchin & Dodge 2011; Hoskins 2018) may also contribute to activism and feelings of belonging, memory, and place-making to remember forgotten pasts. Forms of digital place-making, such as Community Facebook groups, can function as a virtual site for self-identified members to share memories of places that are in transition or that no longer exist in order to reconnect with their former homes and create community archives and exhibitions (Ekelund 2022). Sharing traumatic experiences online can function as a powerful means of memorialization and building solidarity, as in the case of survivors of Ireland's Mother and Baby Homes who continue to challenge the government's denial of responsibility for the carceral institutions (Crooke & McDowell 2022), or groups facilitating "the politics of post-terror togetherness" and solidarity following the 2017 bombing of the Manchester Arena (Merrill et al. 2020). Images of transgressive murals and graffiti on state borders, such as Palestinian graffiti on Israeli's segregation walls, may circulate widely through various social media to challenge oppressive everyday geopolitics, assert claims to land and the city, and keep alive Palestinian memories and dreams of homelands (Culcasi 2016; Hasan & Bleibleh 2023).

Relatedly, anticolonial memory-work around the world disrupts the bracketing of the past from the present in official forms of memory by calling attention to colonial legacies in the present day. Anticolonial memory activists understand that dominant narrative productions of space, seemingly cemented in memorial topographies, reify regimes of what Ann-Laura Stoler (2011) refers to as "colonial aphasia," or an occlusion of the continuing relevance of colonialism. In the United States, for example, the hundreds, if not thousands, of monuments, statues, and other reliquaries celebrating the myth of Christopher Columbus's "discovery of America" inculcate a shared, masculine national consciousness that censors the genocide and erasure of Indigenous peoples (Doss 2012). Over time, Columbian commemoration has also come to stand in for the essential contributions of Italian Americans to US society, an ethnic "duty memory" that some groups believe must be protected (Ruberto & Sciorra 2020). Calls to decolonize the symbolism in the landscape and American psyche incorporate significant acts of civil disobedience, including vandalism, defacement, and iconoclasm (Doss 2012). In 1989, Indigenous American activist Russell Means poured blood on a Columbus memorial in Denver, Colorado, triggering calls to replace the colonial national holiday of "Columbus Day" with an "Indigenous Peoples Day." In 2021, President Joe Biden formally replaced this officially sanctioned US holiday. In Lisbon, Portugal, numerous monuments and statues glorifying the imperial past continue to celebrate dominant historical narratives about the "Discoveries" of the former Portuguese Empire while obscuring stories of colonial violence. By exploring the recent *ReMapping Memories* project and the exhibition *Ecos Coloniais*, Lovegrove and Machaqueiro (2024) interpret the public contestations over official sites of memory as contributing to changing dominant forgetful histories by resignifying public spaces and offering alternatives.

These recent Lisbon projects were inspired by other decolonizing and racial justice movements calling for the removal of street and building names, memorials, and monuments honoring colonizers and slave traders in public spaces, most notably the Rhodes Must Fall movement in 2015 in South Africa (Marschall 2017; Chantiluke et al. 2018) and Colston Must Go in 2020 in the United Kingdom (Choksey 2021). The first originated from student demands at the University of Cape Town to remove the centrally located Cecil Rhodes memorial on campus; when these were denied, students covered the statue with feces. Rhodes, the first prime minister of the Cape Colony (1890–1896), was a ruthless mining magnate and white supremacist, but is remembered in the power structures and landscapes of the Commonwealth as a philanthropist, including for his generous endowments to universities and scholarships. In 2020, activists in Bristol, United Kingdom, who were engaged in solidarity protests with BLM following Floyd's murder, pulled down a statue of Edward Colston and threw it into the harbor. Colston was a seventeenth-century slave trader who also used his fortune made by human trafficking to philanthropic ends, resulting in numerous schools, streets, and buildings being named after him (Kearns 2020). The activists involved in defacing and toppling Columbus, Rhodes, Colston, and Confederate statues no doubt understand that a significant part of decolonizing these commemorative sites is to question why such places continue to be socially defined as significant in the present day (Dwyer & Alderman 2008). While symbolic alterations to national commemorative topographies raise consciousness about the past, they cannot substitute for the necessary substantive decolonization of existing structures of power and domination (Post & Rhodes 2022).

Conclusion

As suggested in this chapter, memory-work that is intergenerational, spectral, and anticolonial advances place-based approaches that reimagine contemporary and future political relations as linked to multiple pasts and possibly more just futures. For example, in Jaffa/Yaffa, a city annexed by Israel directly after the 1948 Arab–Israeli War in violation of a 1947 United Nations resolution, Huss (2023) describes how internally displaced activists have developed heritage walking tours that critically locate past and ongoing forms of settler colonialism. One tour guide, speaking at the central mosque, highlights how Israeli heritage tourism allows for such places to remain even as the Israeli state continues its systematic elimination of Palestinian homelands and peoples (Hasan 2019). The walks are a co-performative testimonial exchange with audience members and a chronically traumatized place haunted by past and current violence (Huss 2023).

For Till and Huss (2024), learning from anticolonial activists and community leaders means supporting intergenerational acts of restoration by attending to the wounds of chronic "geotrauma" (after Pain 2021) and caring for places that offer environments for justice, healing, and rebirth. As exemplified by Shibli's unnamed narrator described at the start of this chapter, the work of Palestinian memory will always be led by the generations that challenge regimes of violence and oppression, even as material presences are violently erased from official maps (Pappé 2006). As Gutman (2017: 83) suggests in her study of Palestinian memory activism in Israel–Palestine, translocal memory practices seeking justice and building solidarity, such as Shibli's novel, may indeed be the only "medium for consciousness-raising . . . for Palestinians in Israel that the state cannot disarm." Our scholarly efforts as political geographers should include calling attention to and supporting this important memory-work in the months, years, and decades to come.

References

Agnew, J. (1998). The impossible capital: Monumental Rome under liberal and fascist regimes, 1870–1943. *Geografiska Annaler: Series B, Human Geography 80*(4): 229–240.

Alderman, D.H., & Inwood, J. (2016). Taking down the flag is just a start toward the memory-work of racial reconciliation in white supremacist America. *Southeastern Geographer 56*(1): 9–15.

Alderman, D.H., & Modlin, E.A., Jr. (2008). (In)Visibility of the enslaved within online plantation tourism marketing: A textual analysis of North Carolina websites. *Journal of Travel & Tourism Marketing 25*(3–4): 265–281.

Angelucci, M., & Kerschbamer, S. (2017). One monument, one town, two ideologies: The monument to the victory of Bolzano-Bozen. *Public History Review 24*: 54–75.

Atkinson, D., & Cosgrove, D. (1998). Urban rhetoric and embodied identities: City, nation, and empire at the Vittorio Emanuele II Monument in Rome, 1870–1945. *Annals of the Association of American Geographers 88*: 28–49.

Azaryahu, M., & Kellerman, A. (1999). Symbolic places of national history and revival: A study in Zionist mythical geography. *Transactions of the Institute of British Geographers 24*: 109–123.

Bernstein, S. (2016). Remembering war, remaining Soviet: Digital commemoration of World War II in Putin's Russia. *Memory Studies 9*(4): 422–436.

Bevernage, B. (2012). *History, Memory, and State-Sponsored Violence: Time and Justice*. London: Routledge.

Bosco, F.J. (2004). Human rights politics and scaled performances of memory: Conflicts among the Madres de Plaza de Mayo in Argentina. *Social and Cultural Geography 5*(3): 381–402.

Bruyneel, K. (2014). The king's body: The Martin Luther King, Jr. Memorial and the politics of collective memory. *History and Memory 26*(1): 75–108.

Chantiluke, R., Kwoba, B., & Nkopo, A. (eds) (2018). *Rhodes Must Fall: The Struggle to Decolonize the Racist Heart of Empire*. London: Zed Books.

Choksey, L. (2021). Colston falling. *Journal of Historical Geography 74*: 77–83.

Coulthard, G. (2014). *Red Skins, White Masks: Rejecting the Colonial Politics of Recognition*. Minneapolis, MN: University of Minnesota Press.

Crenzel, E. (2020). The ghostly presence of the Disappeared in Argentina. *Memory Studies 13*(3): 253–266.

Crooke, E., & McDowell, S. (2022). Remembering Mother and Baby Homes online: Digital sharing as means of memorialisation. In Walden, V.G. (ed.), *The Memorial Museum in the Digital Age*. Falmer: Reframe Books, 442–467.

Culcasi, K. (2016). Images and imaginings of Palestine: Territory and everyday geopolitics for Palestinian Jordanians. *Geopolitics 21*(1): 69–91.

Cusack, T. (2018). Introduction: Art, nation, and gender. In Cusak, T., & Bhreathnach-Lynch, S. (eds), *Art, Nation, and Gender: Ethnic Landscapes, Myths, and Mother-Figures*. London: Routledge, 1–9.

Dean, A., & Failler, A. (2021). "An amazing gift"? Memory entrepreneurship, settler colonialism, and the Canadian Museum for Human Rights. *Memory Studies 14*(2): 451–465.

Doss, E. (2012). *Memorial Mania: Public Feeling in America*. Chicago, IL: University of Chicago Press.

Dowler, L. (1998). "And they think I'm just a nice old lady": Women and war in Belfast, Northern Ireland. *Gender, Place and Culture 5*: 159–176.

Drozdzewski, D. (2014). Using history in the streetscape to affirm geopolitics of memory. *Political Geography 42*: 66–78.

Duncan, J. (1989). The power of place in Kandy, Sri Lanka: 1780–1980. In Agnew, J., & Duncan, J. (eds), *The Power of Place: Bringing Together Geographical and Sociological Imaginations*. Boston, MA: Unwin Hyman, 185–201.

Dwyer, O., & Alderman, D.H. (2008). *Civil Rights Memorials and the Geography of Memory*. Chicago, IL: Center for American Places at Columbia College Chicago.

Edele, M. (2017). Fighting Russia's history wars: Vladimir Putin and the codification of World War II. *History and Memory 29*(2): 90–124.

Ekelund, R. (2022). Fascination, nostalgia, and knowledge desire in digital memory culture: Emotions and mood work in retrospective Facebook groups. *Memory Studies 15*(5): 1248–1262.

Erll, A. (2011). Traumatic pasts, literary afterlives, and transcultural memories: New directions of literary and media memory studies. *Journal of Aesthetics and Culture 3*(1): 1–5.

Evered, K.T. (2018). Erasing the place of dissent: Inscriptions and eliminations of Gezi Park graffiti. *Area 51*(1): 155–165.

Failler, A. (2015). Hope without consolation: Prospects for critical learning at the Canadian Museum for Human Rights. *Review of Education, Pedagogy, and Cultural Studies 37*(2–3): 227–250.

Featherstone, D. (2012). *Solidarity: Hidden Histories and Geographies of Internationalism*. London: Zed Books.

Foote, K. (2003). *Shadowed Ground: America's Landscapes of Violence and Tragedy*. Austin, TX: University of Texas Press.

Foote, K., Toth, A., & Arvay, A. (2000). Hungary after 1989: Inscribing a new past on place. *Geographical Review 90*: 301–334.

Forest, B., & Johnson, J. (2002). Unraveling the threads of history: Soviet-era monuments and post-Soviet national identity in Moscow. *Annals of the Association of American Geographers 92*(3): 524–547.

Forest, B., & Johnson, J. (2019). Confederate monuments and the problem of forgetting. *Cultural Geographies 26*(1): 127–131.

Gordon, A.F. (1997). *Ghostly Matters: Haunting and the Sociological Imagination*. Minneapolis, MN: University of Minnesota Press.

Gosh, A. (2023). Moving statues: Monuments to empire from London's Waterloo Place to the Maidan in Calcutta. *Journal of Historical Geography 83*: 10–22.

Gregory, D. (2004). *The Colonial Present: Afghanistan, Palestine, Iraq*. Oxford: Blackwell.

Gutman, Y. (2017). *Memory Activism: Reimagining the Past for the Future in Israel-Palestine*. Nashville, TN: Vanderbilt University Press.

Hagen, J. (2020). Places of memory, historic preservation, and place attachment in Nazi Germany. *Geographical Review 112*(1): 147–170.

Hagopian, P. (2020). The Martin Luther King, Jr. Memorial and the politics of post-racialism. *History and Memory 32*(2): 36–77.

Halbwachs, M. (1992). *On Collective Memory*. Chicago, IL: University of Chicago Press.

Hammond, T. (2022). Making the millet common: Rethinking authoritarian politics through commemoration following Turkey's July 2016 coup attempt. *ACME: An International Journal for Critical Geographies 21*(5): 603–622.

Hasan, D., & Bleibleh, S. (2023). The everyday art of resistance: Interpreting "resistances-capes" against urban violence in Palestine. *Political Geography 101*: 1021833. doi:10.1016/j.polgeo.2023.102833

Hasan, M. (2019). Palestine's absent cities: Gender, memoricide and the silencing of urban Palestinian memory. *Journal of Holy Land and Palestine Studies 18*(1): 1–20.

Herman, R.D.K. (2008). Inscribing empire: Guam and the War in the Pacific National Historical Park. *Political Geography 27*(6): 630–651.

Hirsch, M., & Spitzer, L. (2011). *Ghosts of Home: The Afterlife of Czernowitz in Jewish Memory*. Berkeley, CA: University of California Press.

Hite, D., & Jara, K. (2020). Presenting unwieldy pasts. *Memory Studies 13*(3): 245–252.

Hoelscher, S. (2003). Making place, making race: Performances of whiteness in the Jim Crow South. *Annals of the Association of American Geographers 93*(3): 657–686.

Hoskins, A. (2018). Digital media and the precarity of memory. In Meade, M.L., Harris, C.B., Van Bergen, P., Sutton, J., & Barnier, A.J. (eds), *Collaborative Remembering: Theories, Research, and Applications*. Oxford: Oxford University Press, 371–385.

Houssey-Holzshuch, M. (2021). Keeping you posted: Space-time regimes, metaphors and post-apartheid. *Dialogues in Human Geography 11*(3): 453–473.

Huss, M. (2023). Paths of erasure and trails of resistance in a neo-settler-colonial city. *Antipode 55*(6): 1735–1757.

Jelin, E. (2003). *State Repression and the Labors of Memory*. Minneapolis, MN: University of Minnesota Press.

Jessee, E. (2017). The danger of a single story: Iconic stories in the aftermath of the 1994 Rwandan Genocide. *Memory Studies 10*(2): 144–163.

Johnson, N. (1995). Cast in stone: Monuments, geography, and nationalism. *Environment and Planning D: Society and Space 13*(1): 51–65.

Johnson, N. (1999). The spectacle of memory: Ireland's remembrance of the Great War, 1919. *Journal of Historical Geography 25*(1): 26–56.

Jonker, J., & Till, K.E. (2009). Mapping and excavating spectral traces in post-apartheid Cape Town. *Memory Studies 2*(3): 303–335.

Kearns, G. (2020). Topple the racists 1: Decolonising the space and the institutional memory of the university. *Geography 105*(3): 116–125.

Kitchin, R., & Dodge, M. (2011). *Code/Space: Software and Everyday Life*. Cambridge, MA: MIT Press.

Legg, S. (2005). Contesting and surviving memory: Space, nation, and nostalgia in *Les Lieux de Mémoire*. *Environment and Planning D: Society and Space 23*(4): 481–504.

Leitner, H., & Kang, P. (1999). Contested urban landscapes of nationalism: The case of Taipei. *Ecumene 6*(2): 214–233.

Lewandowski, S. (1984). The built environment and cultural symbolism in post-colonial Madras. In Agnew, J., Mercer, J., & Sopher, D. (eds), *The City in Cultural Context*. London: Allen and Unwin, 236–254.

Lischer, S.K. (2019). Narrating atrocity: Genocide memorials, dark tourism, and the politics of memory. *Review of International Studies 45*(5): 805–827.

Lovegrove, S., & Machaqueiro, R.R. (2024). Contesting monuments, challenging narratives: Divergent approaches to dealing with the colonial past and its legacies in Lisbon, Portugal. *Journal of Historical Geography 83*: 84–95.

Maddrell, A. (2013). Living with the deceased: Presence and absence-presence. *Cultural Geographies 20*(4): 501–522.

Mangena, T. (2020). Rhodes, Mugabe, and the politics of commemorative toponyms in Zimbabwe. *Geopolitics 25*(4): 1015–1036.

Mannergren Selimovic, J. (2020). Gender, narrative and affect: Top-down politics of commemoration in post-Genocide Rwanda. *Memory Studies 13*(2): 131–145.

Marschall, S. (2017). Targeting statues: Monument "vandalism" as an expression of socio-political protest in South Africa. *African Studies Review 60*(3): 203–219.

McDowell, S. (2008). Commemorating dead "men": Gendering the past and present in post-conflict Northern Ireland. *Gender, Place and Culture 15*(4): 335–354.

Merrill, S., Sumartojo, S., Closs Stephens, A., & Coward, M. (2020). Togetherness after terror: The more or less digital commemorative public atmospheres of the Manchester Arena bombing's first anniversary. *Environment and Planning D: Society and Space 38*(3): 546–566.

Myadar, O. (2017). The rebirth of Chinggis Khan: State appropriation of Chingis Khaan in post-socialist Mongolia. *Nationalities Papers 45*(5): 840–855.

Myadar, O., & Davidson, R.A. (2021). Remembering the "comfort women": Geographies of displacement, violence and memory in the Asia-Pacific and beyond. *Gender, Place and Culture* *28*(3): 347–369.

Nelson, E., & Godlewska, A. (2023). Settler ignorance and public memory: Kingston, Ontario. *Geographical Review* *113*(5): 666–684.

Nora, P. (1989). Between memory and history: Les lieux de mémoire. *Representations* *26*: 7–23.

Nora, P. (ed.) (1997). *Realms of Memory: The Construction of the French Past, Volume 1 — Conflicts and Divisions* (trans. A. Goldhammer). New York: Columbia University Press.

Pain, R. (2021). Geotrauma. *Progress in Human Geography* *45*(5): 972–989.

Pappé, I. (2006). *The Ethnic Cleansing of Palestine*. New York: Simon and Schuster.

Post, C.W., & Rhodes, M.A., II (2022). Decolonizing memory work? Textual politics of Settler State historical markers engaging Indigenous peoples in Kansas. *ACME: An International Journal for Critical Geographies* *21*(5): 540–559.

Randolph, N. (2023). *Rumors of War*: Towards the unsettling of the Confederate monumental landscape. *Journal of Historical Geography* *81*: 43–50.

Regan, P. (2010). *Unsettling the Settler Within: Indian Residential Schools, Truth-Telling, and Reconciliation in Canada*. Vancouver: University of British Columbia Press.

Regesser, A.D., & Emmenegger, R. (2023). The emperor, the lion and the peacock: Monuments and contested state sovereignty in contemporary Ethiopia. *Environment and Politics C: Politics and Space* *41*(5): 903–921.

Robinson, J.S. (2018). *Transitional Justice and the Politics of Inscription: Memory, Space, and Narrative in Northern Ireland*. London: Routledge.

Robinson, J.S. (2022). "We have long memories in this area": Ulster Defence Regiment place-memory along the Irish border. *Memory Studies* *15*(5): 995–1010.

Ruberto, L.E., & Sciorra, J. (2020). "Columbus might be dwarfed to obscurity": Italian Americans' engagement with Columbus monuments in a time of decolonization. In Marschall, S. (ed.), *Public Memory in the Context of Transnational Migration and Displacement: Migrants and Monuments*. Cham: Palgrave Macmillan, 61–93.

Savage, K. (2018). *Standing Soldiers, Kneeling Slaves: Race, War and Monument in Nineteenth-Century America*. Princeton, NJ: Princeton University Press.

Schein, R. (ed.) (2006). *Landscape and Race in the United States*. London: Routledge.

Sharp, J.R. (1996). Gendering nationhood. In Duncan, N. (ed.), *Bodyspace: Destabilizing Geographies of Gender and Sexuality*. Oxford: Oxford University Press, 97–108.

Shibli, A. (2021). *Minor Detail*. London: Fitzcarraldo Editions.

Soyinka, W. (1999). *The Burden of Memory: The Muse of Forgiveness*. Oxford: Oxford University Press.

Stoler, A.-L. (2011). Colonial aphasia: Race and disabled histories in France. *Public Culture* *23*(1): 121–156.

Teeger, C., & Vinitzky-Seroussi, V. (2007). Controlling for consensus: Commemorating apartheid in South Africa. *Symbolic Interaction* *30*(1): 57–78.

Thomson, S. (2013). *Whispering Truth to Power: Everyday Resistance to Reconciliation in Postgenocide Rwanda*. Madison, WI: University of Wisconsin Press.

Till, K.E. (1999). Staging the past: Landscape designs, cultural identity, and Erinnerungspolitik at Berlin's Neue Wache. *Ecumene* *6*(3): 251–283.

Till, K.E. (2003). Places of memory. In Agnew, J., Mitchell, K., & Toal, G. (eds), *A Companion to Political Geography*. Oxford: Blackwell, 289–301.

Till, K.E. (2005). *The New Berlin: Memory, Politics, Place*. Minneapolis, MN: University of Minnesota Press.

Till, K.E. (2012). Wounded cities: Memory work and a place-based ethics of care. *Political Geography 31*(1): 3–14.

Till, K.E., & Huss, M. (2024). Anticolonial placemaking. In Courage, C., & McKeown, A. (eds), *Trauma Informed Placemaking*. London: Routledge, 141–152.

Tyner, J.A. (2017). *Landscape, Memory, and Post-Violence in Cambodia*. Lanham, MD: Rowman and Littlefield.

Warner, M. (2001). *Monuments and Maidens: The Allegory of the Female Form*. Berkeley, CA: University of California Press.

Waziyatawin, A.W., & Yellow-Bird, M. (2012). Introduction: Decolonizing our minds and actions. In Waziyatawin, A.W., & Yellow-Bird, M. (eds), *For Indigenous Eyes Only: A Decolonisation Handbook*. Santa Fe, NM: School of American Research, 1–13.

Winter, J. (1995). *Sites of Memory, Sites of Mourning: The Great War in European Cultural History*. Cambridge: Cambridge University Press.

Withers, C.W.J. (1996). Place, memory, monument: Memorializing the past in contemporary Highland Scotland. *Ecumene 3*(3): 325–344.

Approaching Political Geography

Chapter 17

Spatial Analysis

Andrew M. Linke[1] *and John O'Loughlin*[2]

[1]*School of Environment, Society and Sustainability, University of Utah, Salt Lake City, UT, USA*

[2]*Institute of Behavioral Science, University of Colorado, Boulder, CO, USA*

In contemporary spatial analysis of political geographical phenomena, two fundamental and perennial concepts have maintained their significance. Theoretical and empirical understandings and measurements of context (places) and spatial dependencies (autocorrelation) continue to bedevil spatial analysis. Fundamental questions of defining and measuring "neighborhood," the social setting in which political interactions occur, remain only partly answered despite work on these topics since the earliest version of this chapter appeared over two decades ago. Relatedly, key uncertainties around the handling of spatial autocorrelation in geographical data are as important now as in early research within the revival of political geography.

Recent developments in spatial analysis include big improvements in data availability, data analysis, and user-friendly software that have made research more accessible in geography and related social sciences. After reviewing the two primary tenets of spatial analysis and updating on those three recent changes, in this chapter we integrate these discussions into a stylized empirical example of understanding beliefs about violence in the North Caucasus region of Russia, a region with significant localized violence after the collapse of the Soviet Union in the early 1990s. Our geographic information systems (GIS) analysis connects foundational concepts of political geography using both individual- and aggregate-level data.

The Wiley Blackwell Companion to Political Geography, Second Edition.
Edited by Virginie Mamadouh, Natalie Koch, Chih Yuan Woon, and John Agnew.

Timeless Fundamental Concepts

Context and Place

Political geographers, despite their diversity of methodologies and topical foci, share the belief that social settings influence individual- and group-level behaviors and attitudes. Political behaviors such as voting decisions, conflict about territorial control, political boundary demarcation, and public goods provision and allocation result from a constellation of influences that mix factors at scales ranging from the individual to the locality and, ultimately, to the national and international. Unlike some other social sciences where individuals tend to be viewed as autonomous and interactions between them and their social settings underappreciated, political geographers have highlighted the contextual effects of politics for over a century since Vidal de la Blanche (1903) first described the complex "milieux" (geographical settings) of France.

For most political geographers, understanding of place and context has moved away from what O'Loughlin (2000) decried over two decades ago as a geometric or "Cartesian" view; this move now considers relational understandings of the settings in which people live their lives. Relative interactive measures of places, locales, and regions can be defined by links that are related to social identity, political economy, or migration. In this sense, place or context may be defined by group membership or network ties *in addition to* physical location (Massey 1994). While relative and absolute concepts of context are often presented as mutually exclusive, we believe that key concepts such as Agnew's (1987) elements of context (location, locale, and sense of place) can be measured and analyzed in a quantitative spatial manner. Some, for example, have used network analysis to model *relational context* for a host of political issues such as dyads of armed conflict (e.g., Minhas, Hoff, & Ward 2019; Minhas et al. 2019).

In this chapter, we illustrate possible extensions for classical approaches to identifying and measuring the effects of "place" in quantitative political geography, which is conventionally understood to be the "settings and scenes of everyday life" (Agnew 1987: 5). A "structurationist" approach (the two-way interaction between individuals and their milieu that leads to social outcomes) was the root of much foundational research pointing toward the importance of social context in human geography (e.g., Pred 1983). Johnston and Pattie (1992) for example, showed that a "spatial polarization" of voting trends (the link between class and party support) had developed in the British electorate demonstrating the importance of social context in shaping individual choices. Like many other studies of elections, Pattie and Johnston (2000) provided evidence that "people who talk together vote together" and therefore that personal interactions in social settings condition political behaviors.

Relatedly, in political science, Braybeck and Huckfeldt (2002) explicitly identified neighborhood contexts and measured the diffusion of information about political candidates and issues in dispersed and condensed social networks that influence voting behavior. In several experiments with polling station settings (e.g., seemingly innocuous variations in décor and artwork), Bloom, Birenboim, and Hassin (2023) extended this logic to the "(ultra)-micro" scale of the location where votes are cast. Similarly focused on fine spatial resolutions, Maske et al. (2014) developed a novel approach for estimating "micro-level political environments" using observations of political campaign signs and a survey instrument. Such general considerations of contextual effects (see also Enos 2017) have shaped other politically contentious issues, such as the odds that a person in the United States wore a mask during the Covid-19 pandemic (Baxter-King et al. 2022). Some in geography have recently argued that rising support for right-wing politics is linked with the influences of subnational socioeconomic context (Belina 2020; Förtner, Belina, & Naumann, 2021). While place-focused quantitative analysis has mostly concentrated on elections or civic beliefs, it can of course be considered for other political phenomena such as conflict.

The focus of spatial analytical work in political geography is usually the elaboration of the nuanced effects of social forces as expressed in localities at various spatial scales. This theme formed the centerpiece of the spatial analysis chapter in the previous editions of this book. The usual political science study tends to be deductively developed from a formal model (often a rational choice model) and tests observable implications of a theory. Unfortunately, many nongeographers still rely on method choices that "control away" the influence of contexts on human behavior, or, in Gould's (1970: 444) phrase, they are "throwing out the baby with the bathwater." For these researchers, a geographical analysis is a means to an end and not the end itself. There is still a tendency in political science research to narrowly view space and place through proximity and/or contiguity lenses, essentially as geometric measures. Although there have been improvements in spatial analysis outside of geography (we highlight several examples here), issues remain since the first version of this *Companion* chapter and the critique a decade before that by the prominent political scientist Gary King (1996 in response to Agnew 1996). Of course, political geographers are not atheoretical, but, differing from political scientists, they approach hypothesis-testing with a more inductive emphasis on probing the empirics of data.

The contextual theme is also commonly employed in other geographically sensitive research fields. In studies of health outcomes, disease susceptibility, educational achievement, crime rates, and social mobility, analysts often assess the link between contextual qualities (such as neighborhood levels of poverty, racial or ethnic composition, political representation, etc.) and individual characteristics in nested data structures. For example, students learn within classrooms that have different teachers, classrooms are in schools with different allocations of facilities and resources, and those schools are located in school districts that, for example, are differentiated by tax revenue variations. A student's achievement partly depends on their home location, because observable traits like taxes earned from property values and dedicated to education vary widely in American metropolitan areas. Defining the boundary for capturing such contextual effects can influence results greatly.

Following standard econometric practices, most scholars in fact accept that geographical differences can be important and, consequently, use a "fixed-effects" statistical approach. In this method, binary dummy variables are defined for each spatial unit, such as voting districts in a city, and thus can be numerous. If the researcher is probing whether voters' incomes influence electoral participation, they would ideally have an accurate estimate for the effect of poverty after ensuring that other personal and contextual characteristics are not also influencing participation. The accessibility of polling stations at a constituency level could influence participation. If there are no data available for the number of polling station locations, comparing individuals statistically in one constituency only to individuals in that same constituency in a fixed-effects model eliminates the possibility that differences between constituencies are also responsible for the changes in participation rates. The key problem with the dummy variable approach is that while the model controls for baseline differences between spatial units, the results of the analysis cannot tell us *how* the differences between units of observations matter.

A more promising option is to pursue a "geo-nested" approach where a large-N analysis (LNA), typically a quantitative study of many spatially contiguous geographical units, identifies both typical units (close to the key relationship) and outliers, and is then followed by a small-N analysis (SNA), typically a qualitative case-study examination. Building on Lieberman's (2005) early advocacy of the integrated LNA and SNA approach, Harbers and Ingram (2017) use Luc Anselin's (1988) spatial econometric procedures to build spatial lag and spatial error models of county-level homicide distributions and identify "geo-nested" cases for further study of both typical and outlier counties.

Regardless of the chosen statistical model, the delineation of the areal dimensions of the spatial units matters greatly for the conclusions of spatial analysis. The Modifiable Areal Unit Problem (MAUP) is well known but rarely examined methodically. MAUP was outlined by Gehlke and Biehl in 1934 but is most commonly associated with Openshaw (1983). The key MAUP issue is that the size and direction of a statistical association found at one scale of analysis (e.g., census tract) may not hold at others (e.g., county). Related to this inconsistency, changing the boundary shape of an area also changes relationships (also see the later discussion of aggregation bias). For example, if we correlate socioeconomic class and voting for the Republican Party, the coefficient varies across hierarchical scales as a result of the number of data points and the geographical configuration of the districts, and therefore we cannot be sure which coefficient is correct. MAUP dilemmas are different than, but related to, the more conceptual dilemma of the "ecological fallacy." Researchers risk an ecological fallacy when they assume that aggregate-level data represent individual-level processes or phenomena (Robinson 1950). In essence, an ecological fallacy is MAUP at the finest resolution (individual level) and confounds explanations and outcomes across aggregate and individual scales.

As accurate measurement of context is central to spatial analysis in political geography and differing neighborhood bounds can influence results, more flexibility for researchers in defining the relevant neighborhood is warranted. Many empirically derived metrics for selecting neighborhood dimensions exist (Root et al. 2009), but fine-resolution data are required for these analytical tools to be effective. It is impossible to know whether a larger unit of analysis is more appropriate than smaller units if one cannot examine and compare results at different levels. An applied example of aggregation bias impacts is found in Cook and Weidmann (2022), where the correlation among different georeferenced conflict event datasets varies depending on the level of the administrative unit used for aggregation; there is less agreement with smaller regions (e.g., districts) and more with larger areas (e.g., provinces). For analysts who view the world through the lens of spatial scales, decisions about the proper aggregations of information thus become very important. While it is possible to aggregate up from fine resolutions (location-based data) to coarse (large areal unit data), disaggregating from coarse to fine resolutions is impossible (or at least very difficult, requiring estimations of uncertainty). The misalignment of spatial units, either because different variables are measured for different units (precincts, census tracts, counties, etc.) or due to temporal changes in boundaries, often results in different statistical outcomes. Zhukov et al. (2024) clarify how typical GIS agglomeration procedures for mismatched spatial units need additional statistical consideration and have developed open-source software (*SUNGEO*) to make transformation options customizable and intuitive.

If one wants to know how best to understand political violence in Afghanistan, knowing only the provincial-level violence rates limits one's ability effectively to measure the local ebb and flow of conflict as it occurs relative to the border with Pakistan. One would also have no understanding of how violence is geographically distributed within ethnoregional enclaves at localized scales. With location-level data, however, it is possible to aggregate points (and all of the attributes of individual violent events) to a zone defined by any absolute spatial reference (e.g., border) or relative spatial dimension (e.g., ethnic community dominance in a region) (O'Loughlin et al. 2010).

Nonstationary Data and Spatial Autocorrelation

In addition to considerations of context (milieu), another fundamental and timeless quality of spatial analysis in political geography is its nonstationarity quality. Data are nonstationary when the relationship between variables varies across locations within the dataset (spatial heterogeneity) and is not consistent in different regions, such as the explanations of the Nazi party vote across Germany in 1930 (O'Loughlin, Flint, & Anselin, 1994). As Tobler (1970: 236) famously wrote, "everything is related to everything else, but near things are more related than distant things." Spatial dependencies (a term that reflects this distance effect) among units of analysis are important observable artifacts of largely unobserved social processes. Spatial analysts are interested in how, where, and why data are geographically related. If human behavioral patterns cluster in space (and they almost always do), geographers strive to understand how and why, rather than simply controlling for the trend as a nuisance that sullies an otherwise nonspatial model. Spatial terms in models attempt to capture these contextual effects.

Questions about spatially nonstationary data are answered inductively. The calculation of Local Indicators of Spatial Association (LISAs) has been the workhorse of spatial analysts for decades and identifies where a locally specific measure of autocorrelation may not match a global or overall trend (Anselin 1995). LISA statistics are based on a straightforward comparison between observed and simulated (expected uniform or random) spatial distributions. The Getis-Ord Gi* ("hot spot") statistic is one example of a LISA indicator (Ord & Getis 1995). For a study of insurgency in Afghanistan between 2004 and 2009, using daily military reports of "significant activity" (SIGACTS), O'Loughlin et al. (2010), for small 25 km^2 gridcells, were able to identify changes in the locations and sizes of event clusters initiated by both insurgent and coalition forces. As one would expect, some general overlap between the clusters can be observed and they also display differing locational trends. This is an example of using LISA-type statistics inductively, but such procedures can also be employed to identify spatial trends that distort the results of nonspatial econometric analysis. Using a local clustering statistic for the residuals of a regression, for example, can reveal clusters of unexplained variance in a dataset, suggesting that assumptions about the independence of observations (explained later) are violated (see also Harbers & Ingram 2017).

Point pattern process identification, similar to LISA indicators because it compares an observed spatial distribution to an expected distribution, can be used to uncover local nuances and trends in political data. Several advances in this style of analysis were extremely powerful, including the routines implemented in SaTScan, software initially designed to identify space–time clusters for epidemiological and other health data (Kulldorff et al. 2005; Kulldorff and Information Management Services 2009). The benefit of using SaTScan is that the temporal and spatial dimensions of the window that defines clustering are allowed to vary and the output is effectively a cylinder with vertical and horizontal dimensions that can be mapped by location coordinates. For the study of violence in Afghanistan, this method delivers even more detail about the spatiotemporal dimensions of conflict than just a LISA indicator (O'Loughlin et al. 2010: 490).

With geographically dependent data, "knowing one value on the surface provides the observer with a better than random chance of predicting nearby values" (Gould 1970: 444), thus violating the assumption that observation units are independently and identically distributed (IID). IID assumption violations can give researchers faulty assurances that the results of their quantitative analyses are valid because standard errors for testing the significance of a coefficient estimate are artificially small (Anselin 1988). After McCarty's (1954) analysis of geographical patterns of the vote for Wisconsin's right-wing senator Joseph McCarthy – one of the first studies explicitly promoting a spatial approach to statistical analysis of political phenomena – a steady stream of quantitative political geography methods textbooks and articles emerged that provide technical overviews of the confounding influences of spatial dependencies and the many possible solutions (Cliff & Ord 1972; Anselin 1988).

One of the most straightforward treatments of spatial dependency is introducing an autoregressive (also called "spatial lag") term into a regression model. The classical estimator in regression analysis can be represented as $Y_i = \beta X_i + \varepsilon_i$, where for observation *i*, *Y* represents the outcome variable, *X* the main independent variable, and ε the unobserved variation (error) in the relationship. The value of β quantifies the association between *X* and *Y*. More advanced models would include temporal dimensions and control variables that might also contribute to the outcome. A spatial autoregressive model is represented by $Y_i = \rho W_i Y + \beta X_i + \varepsilon_i$, with a spatial weights matrix, *W*, used to define the presence of *Y* within a neighborhood surrounding observation *i*. *W* may be defined by proximity (distance) or contiguity (shared border). The estimate of ρ in the model can be interpreted in substantial terms. Interpreting the autoregressive term of the equation reveals a high level of predictability within a model (e.g., O'Loughlin et al. 2012).

In contrast to lag models that accept the role of geographical affinities in the data and try to use them to understand human behavior, the "spatial error" solution to the problems posed by spatial dependency is an adjustment of the standard error of each coefficient estimate of the independent variables (β in the previous equation) so that their influence on the outcome of interest is not overstated. In effect, a spatial error approach is similar to common adjustments to the error structure of a regression model (e.g., clustered errors at some spatial scale), but is specifically based on geographical distances or contiguities, defining the connections among the units (as is the case for *W* earlier). Because the spatial lag model introduces a new term into the model, consideration of the spatial contiguity effects has been explicitly incorporated into regression models of electoral choices for decades (e.g., O'Loughlin, Flint, & Anselin 1994).

Recent studies illustrate the complex interactions between space and time in modeling geographical and temporal data (Cook et al. 2023); as spillover effects are well documented, we may consider "spillforward" effects too (temporal autocorrelation). As more advanced higher-dimension analyses, there are spatiotemporal autoregressive distributed lag (STADL) models that overcome the biases these dependencies introduce. A relatively new family of integrated nested Laplace approximation (INLA) models similarly account for unmeasured (yet known) space–time dependencies among units in observational data. We have used these in recent applied research (Linke & Shawcroft 2023).

Finally, geographically weighted regression (GWR; Fotheringham, Brunsdon, & Charlton 2002) and the more recent multiscale GWR (MGWR; Fotheringham & Li 2023) are exploratory tools we can use to examine heterogeneity in the relationships that exist in a multivariate model. Such approaches have limitations (due to sampling observations multiple times within a designated bandwidth), yet they would tell a user if, for example, the effect of *X* on *Y* is stronger in one region than in other regions. But as noted already, while patterns and relationships are indicated, the processes underlying them remain unidentified and require the kind of "geo-nested" analysis proposed by Harbers and Ingram (2017) to avoid a fall into the ecological inference fallacy. Recent advances include a localized GWR approach with multilevel models to address aggregation bias issues and arrive at optimal units of analysis (e.g., Feuillet et al. 2024).

Changes in Spatial Analysis

Data Availability

While the foundational concepts remain perennially important, there have been major changes in the landscape of spatial analysis over the last decade. Primary among the changes helping to spur more spatial analysis is a dramatic increase in the availability of diverse types of data that are readily available for political geographical research. On the cusp of important changes in data and analysis for human geography, Johnston et al. (2004: 367) wrote:

> there are many hypotheses regarding neighborhood effects in the geographical and related literatures, but their successful testing has been hampered by the absence of data. In particular, analysts have lacked data on both individuals and their neighborhood milieux, which allow the interactions of different types of people in different types of local context to be explored.

In their article, Johnston et al. (2004) combined survey and electoral outcome data in "bespoke neighborhoods" in a manner that facilitated the creative identification of multiple scales at which social forces operate, thus effectively avoiding criticisms related to MAUP.

Quantitative political geography research had classically relied mostly on administrative unit polygon data, usually in electoral studies. However, quantitative analysis of political geographical subjects is increasingly going beyond only areal unit data to include a variety of formats and types (e.g., surveys in O'Loughlin et al. 2024). Examples of the various formats include network (Radil and Flint 2013), point pattern location (Linke et al. 2012), satellite remote sensing (Henderson et al. 2012), real-time war destruction (Zhukov 2023), and social media (Karrel et al. 2023) data. Some research uses nonuniform polygons at a global scale to capture spatial demographic qualities of territories such as ethnic groupings (e.g., Wucherpfennig et al. 2011).

The push toward sharing data among academics has also picked up pace alongside the increasingly mandated expectation of journals that data for quantitative analysis be made public for replication by other researchers in the field (King 1995). However, concerns regarding the confidential nature of individual data that are anchored to a specific location are growing at a time when social media, governmental, and private information collection has been on the rise since earlier versions of the *Companion* (VanWey et al. 2005).

A major "neogeography" shift in the availability of geospatial data has also seen volunteered private data used for a number of academic and nonacademic applications, in addition to the nonexpert use of GIS to further the aims of a given community. The original rise of such data uses accompanied a revolution in the accessibility of web-based mapping applications, such as Google Maps' API and even the many extensions of the basic Google Maps (and Earth) interface(s). Where nontraditional cartographic tools exist (i.e., those that do not rely on proprietary software like ESRI's ArcMap) and are accompanied by spatial data, the potential for "citizen science" emerges.

Data Use

A second recent trend in spatial analysis relates to the compilation and integration of multiple data formats. In traditional electoral geography, for example, analysis was often conducted with data available at a single scale (vote totals and predictors based on census data). In spatial analysis limited to a single scale of analysis and format of data, important place influences or contextual-level effects may not be captured effectively by distance and contiguity scores. To advance the measurement of possible local-level influences on outcomes of interest in political geography, hybrid units of analysis and mixed data structures are now more commonly used (e.g., the combined use of election outcomes, fieldwork observations, and surveys in Maske et al. 2014).

An analysis of conflict in Iraq can serve as an example of diverse data merging into a single GIS platform, allowing for the discovery of social relationships that are hidden when analysis is bound by a single source, analytical unit, or data format (Linke et al. 2012). Four different data formats and dimensions were merged: socioeconomic status of districts (survey data aggregated into vector polygons), ethnic group distributions (scanned paper map converted to vector polygons), violent event location data (latitude and longitude point coordinates), and satellite nighttime lights to measure urbanization (raster layer). Using a spatial Granger causal effects estimator, a tit-for-tat dynamics of insurgent-regime forces reciprocity emerges where the actions of one side of the combat strongly predicted a timely and local response from the other. These associations varied across different spatiotemporal thresholds and across ethnosectarian, income, and population density regions, a finding that emerged from the compilation of diverse data sources. A similar compilation of survey, population data, and environmental indicators is used in a study of landownership – a highly contentious issue – across dozens of countries of the Global South (Linke & Shawcroft 2023).

This kind of data integration on a specific spatial scale with complementary spatially sensitive methods is developing rapidly. PRIO-GRID (Tollefson et al. 2012) was one example of a major effort to merge multiple freely available data sources for governmental, population, socioeconomic status, terrain, and ecological-climatological data into a ½ degree gridcell unified structure for all world regions. While the basic dataset has a static spatial resolution (thus not automatically resolving MAUP issues), it represents a substantial advance in how conflict inquiry in political science is carried out, with a move away from the "territorial trap" of the nation-state in favor of geographically disaggregated research. More recent similar initiatives include Afro-GRID (Schon & Koren 2022).

Analytical Tools

As well as the important changes in data availability and data use, recent years have also seen a dramatic rise in the number of statistical and graphical tools available for spatial analysts. Free platforms for statistical analysis supporting hundreds of procedures for spatial data management, mapping, and regression can now help researchers in innumerable ways after data are correctly formatted and organized (Bivand et al. 2008; Griffith & Paelinck 2011; Rey et al. 2021). Free standalone software platforms exist for this kind of spatial analysis, including GeoDa (Anselin, Li, & Koschinksy 2023) for exploratory spatial data analysis (ESDA) or LISA calculations (see also SaTScan, noted earlier). There have been recent advances in MGWR that offer exploratory tools for analyzing multivariate relationships through a spatial lens, including open-source software (e.g., SPARC; Oshan et al. 2019). As an alternative to the proprietary ESRI suite of software, QGIS is publicly available and operates as a nearly identical interface for uploading many GIS data formats (raster layer pixels, vector polygons, etc.) and performing analysis, including conventional geoprocessing operations (zonal statistics, adjustments to geometry for boundary delineation, etc.).

Dedicated spatial analysis packages exist in the software platform *R* for managing classes of spatial data (e.g., **sp**), reading and writing shapefiles (e.g., **maptools**), creating and using spatial weights matrices (e.g., **spdep**), point pattern processes (e.g., **spatial**), Bayesian spatiotemporal estimations (e.g., **INLA**), and multilevel modeling (**nlme4**). Python's PySal library (Rey & Anselin 2007) includes numerous spatial analytical tools as well (**esda**, **spreg**, **mgwr**, among others) described succinctly by Rey et al. (2021).

The technical solutions to model estimation challenges (e.g., autocorrelation) and other analysis issues (e.g., varying workarounds for aggregation bias) are ultimately not a perfect solution for what remain conceptual issues. By not deliberating carefully about what data are best, and at what geographical resolution, to answer a research question, complex modeling is putting the cart before the horse. Sometimes, the best data are probably individual-level data, where we can make inferences about behavior most effectively (e.g., voting preferences or other attitudes). In other cases, analyzing regional trends with observational data at an aggregate level is sufficient. After all, those leading developments in these fields make similar qualifications. Even if we often prefer to, these decisions require careful thought: "An adage that runs through this commentary is: *Just because you can model locally, does not mean you should*" (Fotheringham 2023: 191, emphasis in the original).

Effects of Local Violence on Public Opinion in the North Caucasus of Russia

In this section we illustrate some of the principles and methods of spatial analysis in political geography research. Survey data for 2000 individuals from the North Caucasus of Russia were collected in December 2005. The survey was designed as a study of postconflict attitudes toward group reconciliation and prospects for peaceful relations. The North Caucasus conflicts between 1994 and 1996, restarting in 1999, were marked in the later years by guerrilla warfare, terrorist attacks, reprisals by the Russian forces, and a diffusion outward from the original core in Chechnya. Over time, the militancy took on a more Islamist character as its leaders declared a "shariat" (state run by Islamic law) for the Muslim republics across the region. By 2005, the conflict had waned significantly but violent events still occurred on a daily basis in Chechnya and neighboring republics (O'Loughlin & Witmer 2011). Our research question concerns the possible effects of violence in the immediate area on survey respondent attitudes.

The survey of individuals was designed as a geographically stratified one and was conducted in 82 sampling points within the republics of North Ossetia, Dagestan, Karachevo-Cherkassia, Kabardino-Balkaria, and the territory of Stavropol (Figure 17.1). (Chechnya and Ingushetia were excluded from the study because high levels of violence at the time made surveying impossible.) We joined the location of survey respondents with conflict data for the two years before the December 2005 survey. These violent events are coded from newspapers and the data include the exact location, day of the event, the perpetrators and targets, and estimates of casualties. Coders reviewed thousands of LexisNexis stories to identify events with enough information to ensure reliability in geographical precision. The violent event file is precisely georeferenced by UTM Zone X and Y coordinates after the violent event locations are identified, allowing for distance calculations in a projected coordinate system.

Aggregating the Data

With the goal of measuring multiple settings or neighborhoods of violence, we allow the spatial boundaries around survey sampling points to vary in making violent event aggregations. Using a range of dimensions around a location allows us to address the uncertainty of the relevant context for a respondent and to

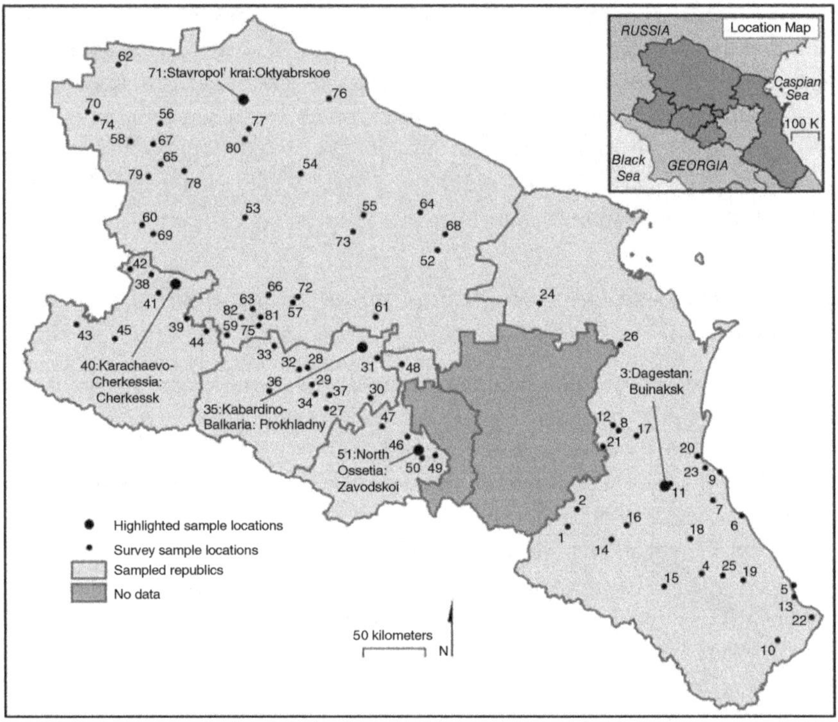

Figure 17.1 The 82 primary sampling locations for the 2005 survey in five republics of the North Caucasus. To facilitate comparison, the sample sites identified in the figure are also highlighted on the graphic in Figure 17.2.

illustrate the MAUP choices. On the map in Figure 17.1, we identify five sampling locations (Oktyabrskoe, Prokhladny, Zavodskoi, Buinaksk, and Cherkessk) that we profile in the tabled event counts of neighborhood violence measures in Figure 17.2. We selected each of these because they have experienced varying levels of violence prior to the survey (some extremely violent and others comparatively peaceful). Additionally, the five locations represent each of the republics within the broader study area. We chose these only for the purposes of illustrating the data aggregation steps and the variation of violence that is captured across space–time dimensions.

For this illustration, we use eight three-month time slices up to a maximum of two years and ten spatial thresholds (10–100 km). Specifically, we first measure the distance between each of the 82 survey locations and every violent event location (total locations of violent incidents number 1367). We use the **pointDistance** function in *R* and count the number of events that has taken place within each space–time buffer around those locations. Each row in Figure 17.2 represents a survey sample point, whose numbers 1–82 correspond to the locations on the map. The thicker black horizontal lines represent borders between republics.

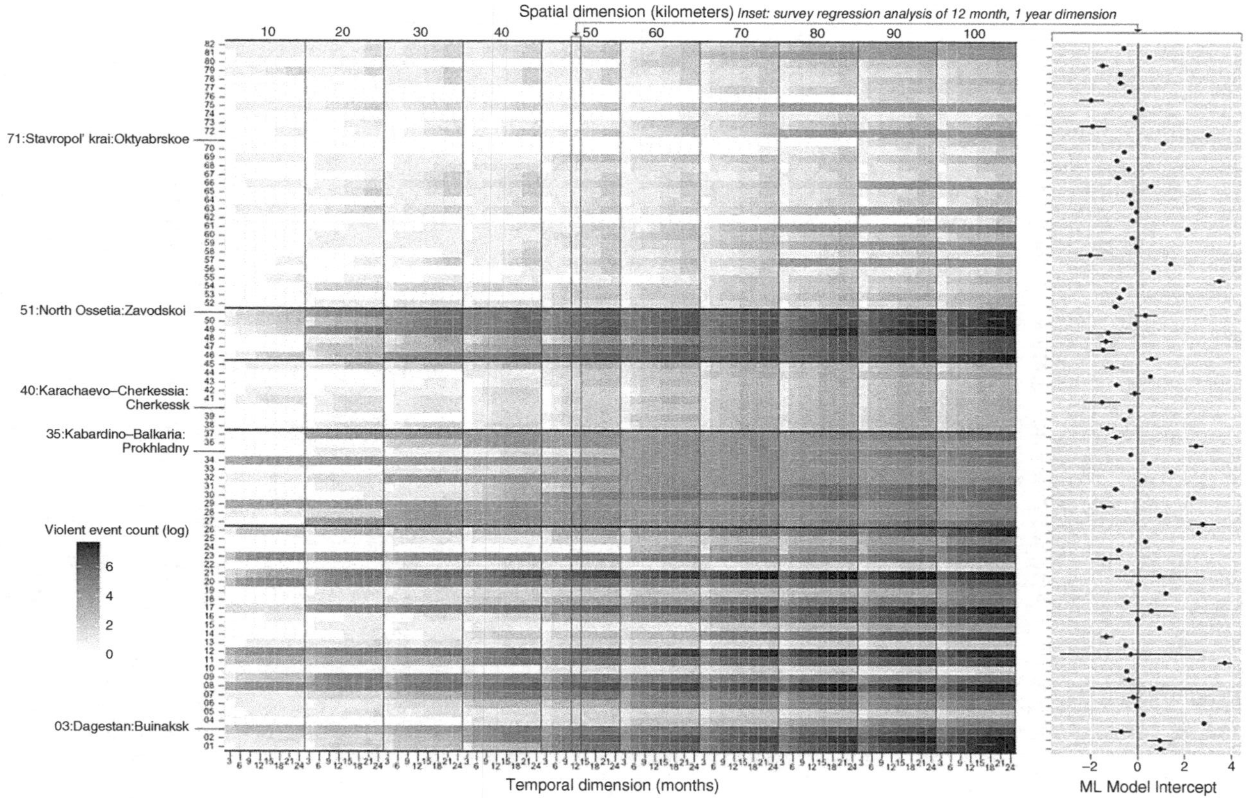

Figure 17.2 The number of violent events aggregated within 80 spatiotemporal dimensions (3-month breaks by 10 km boundaries) for each of 82 survey sample locations in the North Caucasus. Survey locations (horizontal axis) correspond to the map locations in Figure 17.1. Thicker black horizontal lines represent borders of the republics. The smaller graphical figure (right) is an illustration of the multilevel/hierarchical regression model intercept for a single space–time dimension (here, 12 months and 50 km). Similar variation exists for the quantitative modeling of survey responses in each column representing a space–time dimension.

The shade of each cell represents the conflict event count (logged because of some high values) of each space–time buffer. This heatmap presentation of quantitative data is thus very helpful in ESDA. Illustrating a large number of dimensions (here 82 space × 80 time cells = 6560 cells) as a choropleth matrix is more helpful than standard tabular results.

The profiles of all locations show the expected trends. As the spatial buffer becomes very large (100 km) the number of events increases. Many locations have no violence recorded nearby when the 10 km distance is used, but at 100 km almost every location has some violence. Secondarily, as the temporal range used to define our pairing of violence data with locations expands from 3 months to 24 months, a clear trend toward higher conflict event counts is also visible.

Five illustrative locations demonstrate different overall trends in violence. Zavodskoi near Ingushetia (#51) is quite violent, with conflict taking place across nearly all temporal and spatial dimensions (and with very high rates at coarse boundary definitions). In dramatic contrast, Oktyabroskoe (#76) in northern Stavropol' territory is far from the core of violence in Chechnya and thus relatively peaceful. Cherkessk (#40) represents a medium level of violence with no conflict found at very fine temporal and spatial resolutions, but accumulating as distances increase. From this graphical display, we make the straightforward conclusion that the "violence neighborhood" surrounding locations varies dramatically between localities and according to threshold delimitation. The more important consideration relates to our earlier discussion of MAUP and it remains uncertain what the correct range for delimiting the violence context is.

Varying Intercepts in Statistical Models

The multilevel modeling approach is one that allows the evaluation of individual and contextual effects to be measured and compared, since it is now almost axiomatic in political geography that "context matters" in influencing individual choices and behaviors. Multilevel modeling has not achieved prominence in political geography compared to other social sciences, especially epidemiology, criminology, and other public policy research. By allowing the basic relationship between the outcome of interest and a key predictor to vary across settings/contexts – which are known to be qualitatively different in unobserved, unmeasurable, and unknown ways – we allow for the influence of contextual factors to be incorporated into the analysis of survey respondent attitudes as potentially important place influences.

For our illustration, we analyze answers to the survey question "What of the following listed is the most serious danger facing the peoples of the North Caucasus in the next 5 years?" More than crime, ethnic separatism, corruption, and lack of economic development and unemployment, a plurality of people (42%) claimed that "terrorist actions and military conflicts" were the biggest problem. We code those individuals as one and other potential threats as zero for a binary response in a logistic regression model. To take account of alternative explanations of the fear of terrorism, we control for individual-level factors (income and employment status, education, age, and gender).

The relationship we model can be represented as $Y_i = \alpha_0 + \beta X_i + Z_i + \alpha_s + \varepsilon_i$, for observations $i = \{1, 2...2000\}$ and $\alpha_s \sim N(0, \sigma_\alpha^2)$, for survey location level $s = \{1, 2...82\}$, where α_0 is a random effect, Y_i is the outcome, β is the neighborhood violence indicator, and X_i is the vector of individual-level control variables that could confound the association (β) between X_i and Y_i. Unexplained variance is captured in ε_i, the error term. A nonhierarchical generalized linear model (GLM; for the binary outcome) would exclude the α_s term in its empirical estimate, the component of the model that allows the statistical intercept of the relationship under investigation to vary across units at some designated hierarchical scale (survey locations, in our case). In a random effects model, the variance of the intercepts can be informative. Heterogeneity at finer resolutions is likely to be more substantial than at coarse spatial resolutions.

To indicate whether the relationship between individual concerns about terrorism/violence and level of violence in the locality changes by survey sampling point, a plot of the intercepts of the regression model for each of the second-level units (*rayoni*) is helpful. (A plot for the third republic level would be warranted in a more comprehensive research article.) On the right-hand side of Figure 17.2, we present the statistical association between the variable capturing rates of local violence and a respondent's worry about terrorism. As an example, we selected a spatiotemporal dimension (12 months and 50 km) near the middle of the time and distance ranges. Each location identified in the map (Figure 17.1) and choropleth matrix (Figure 17.2, left) is included in this graphical plot.

A key component of our analysis is identifying whether the variation between sampling point intercepts is statistically significant. The intercept values vary from roughly −2 to more than +3.5, indicating a wide divergence in the relationship between violence and respondent worries about terrorism across the study region. The place where a respondent lives, in other words, has a major bearing on what a respondent ranks as an important worry, even after controlling for individual-level influences such as age or education. To see if the random effects estimator (Table 17.1) is statistically significant, analysis of variance (ANOVA) of the log-likelihood for different models is appropriate.

Illustrating Multilevel Model Results

Careful consideration of geographical context and scale can change substantial interpretations of statistical analysis. For the results in Table 17.1, we employed the **lme4** package and the generalized linear model (for binary outcomes) function *glmer*. Using a logistical regression functional form, a statistically significant influence of contextual violence on worries about terrorism emerges (columns b–d). For every additional violent event that took place within 100 km of the survey location in the preceding year, the odds of respondent worry about terrorism increase marginally. Since the ranges of number of events within the space-time

Table 17.1 The influence of violence taking place in various space–time scales on survey respondents' concerns about terrorism and war in their area. Statistical estimator is a random intercept multilevel binary outcome logistical regression model, with intercepts varying by the 82 survey locations. Bold table values are highlighted as the main estimates of interest (effect of violence neighborhood on attitudes).

	(a) Controls only			(b) 20 km			(c) 50 km			(d) 100 km			(e) RE 20 km			(f) RE 50 km			(g) RE 100 km		
	Est	StdEr	OR	Est	StdEr	OR	Est	StdEr	OR	Est	StdEr	OR	Est	StdEr	OR	Est	StdEr	OR	Est	StdEr	OR
(Intercept)	–029	0.10*	0.748	–037	0.10*	0.688	–038	0.10*	0.685	–039	0.10*	0.677	–030	0.19	0.738	–032	0.19*	0.729	–040	0.19*	0.670
12 mo 20 km total				**0.00**	**0.00***	**1.003**							**0.00**	**0.00**	**1.002**						
12 mo 50 km total							**0.00**	**0.00***	**1.002**							**0.00**	**0.00**	**1.001**			
12 mo 100 km total										**0.00**	**0.00***	**1.001**							**0.00**	**0.00***	**1.001**
Education	–009	0.10	0.917	–009	0.10	0.918	–009	0.10	0.911	–010	0.10	0.908	–015	0.11	0.859	–015	0.11	0.860	–015	0.11	0.861
Male	–010	0.09	0.906	–010	0.09	0.904	–010	0.09	0.903	–010	0.09	0.903	–012	0.10	0.885	–012	0.10	0.886	–012	0.10	0.885
Poor	0.31	0.10*	1.357	0.32	0.10*	1.371	0.29	0.10*	1.334	0.28	0.10*	1.328	0.12	0.11	1.130	0.12	0.11	1.132	0.12	0.11	1.129
Rural	–004	0.09	0.961	0.00	0.09	0.997	–002	0.09	0.980	–004	0.09	0.959	0.03	0.20	1.034	0.00	0.21	0.997	0.00	0.20	1.004
Tense interview	–020	0.10*	0.819	–021	0.10*	0.808	–022	0.10*	0.804	–022	0.10*	0.802	–020	0.11*	0.819	–020	0.11*	0.816	–021	0.11*	0.809
AIC	2719.51			2715.52			2711.64			2707.29			2620.86			2622.95			2621.78		
RE significant	N/A			N/A			N/A			N/A			Yes			Yes			Yes		

AIC = Akake information criterion; Est = estimate; OR = odds ratio; RE = Random intercept model; Std Er = standard error; N/A = not applicable; * = Statistically significant at 10%.

buffer are 2–1319, 0–454, and 0–135 for 100 km, 50 km, and 20 km ranges, respectively, this is a noteworthy effect even though the coefficient estimate is small. Drawing conclusions from the basic nonhierarchical model, one would conclude that a context of violence and insecurity influences worries about terrorism and that this relationship is robust across definitions of spatial neighborhood. However, our multilevel model approach shows that the basic relationship between regional violence and worries about terrorism varies widely across survey points, as indicated by the intercepts plotted in Figure 17.2.

The multilevel model results for place-based statistical intercept variations tell a modified but important story (columns e–g of Table 17.1). In contrast to our initial conclusion, the link between regional violence and fear of terrorism is statistically significant only at the 100 km threshold. In other words, a small spatial range (20 km) does not yield significant results because the number of events at such a local level is small. It is the broader, more regional concentration of violence that influences respondent perceptions. In political geography, such place-specific influences are important and consistently appear in different locales and settings, and we should investigate whether our models remain consistent and robust after taking such effects into account. While our example is specific to the violent North Caucasus in 2005 and the distance effect would be different in other settings, we cannot know a priori what the relevant thresholds are until we measure a variety of them and show the variation in the model of contextual effects. Such identifications of the differences allow a "geo-nested" option of further investigation of both "outliers" and "typicals" (Harbers and Ingram's 2017 terms) to understand the processes underlying the patterns.

Conclusion

Political geography is a changing and dynamic discipline, as it should be. However, certain thematic continuities are threaded from the resurgence of the past five decades. Spatial analysis in human (including political) geography should wholeheartedly embrace a nuanced approach to understanding contexts, places, and social settings of human behavior. It must also fully recognize the spatial dependencies measured by different metrics between locales and administrative units, including the linkages among individuals living in those places. While we reiterate the importance of these perennial and timeless issues for political geography – we have referred to early recognition of contextual effects in 1903 and quantitative work from 1934 – there have also been remarkable changes over the past decade in the availability and use of data for spatial social science. The empowering changes can foster robust and valuable improvements to quantitative research, but must be accompanied with thoughtful and meticulous training in the assumptions that underlie statistical methods. The increased availability and accessibility of spatial data do not mean a further avoidance of qualitative approaches.

A firm understanding of theory and concepts in political geography can effectively be merged with cautious and careful quantitative approaches to our discipline; the two are not mutually exclusive and hybrid mixed-methods research agendas will hopefully continue to illustrate this combination in the future.

References

Agnew, J. (1987). *Place and Politics*. London: Allen and Unwin.

Agnew, J. (1996). Mapping politics: How context counts in electoral geography. *Political Geography 15*(1): 129–146.

Anselin, L. (1988). *Spatial Econometrics: Methods and Models*. Dordrecht: Kluwer.

Anselin, L. (1995). Local indicators of spatial association – LISA. *Geographical Analysis 27*(2): 97–115.

Anselin, L., Li, X., & Koschinksy, J. (2023). GeoDa, from the desktop to an ecosystem for exploring spatial data. *Geographical Analysis 54*(3): 439–466.

Baxter-King, R., Brown, J.R., Enos, R.D., Naeim, A., & Vavreck, L. (2022). How local partisan context conditions prosocial behaviors: Mask wearing during Covid-19. *Proceedings of the National Academy of Sciences 119*(21): e2116311119.

Belina, B. (2020). Context and theory. *Political Geography 82*: 102229.

Bivand, R.S., Pebesma, E.J., & Gomez-Rubio, V. (2008). *Applied Spatial Data Analysis in R*. New York: Springer.

Bloom, P.B.-N., Birenboim, A., & Hassin, R.R. (2023). The effect of micro-level context in polling stations on voting. *Political Geography 107*: 102976.

Braybeck, B., & Huckfeldt, R. (2002). Urban contexts, spatially dispersed networks, and the diffusion of political information. *Political Geography 21*(2): 195–220.

Cliff, A.D., & Ord, J. (1972). *Spatial Autocorrelation*. London: Pion.

Cook, S.J., Hays, J.C., & Franzese, R.J., Jr. (2023). STADL Up! The spatiotemporal autoregressive distributed lag model for TSCS data analysis. *American Political Science Review 117*(1): 59–79.

Cook, S.J., & Weidmann, N.B. (2022). Race to the bottom: Spatial aggregation and event data. *International Interactions 48*(3): 471–491.

Enos, R.D. (2017). *The Space Between Us: Social Geography and Politics*. New York: Cambridge University Press.

Feuillet, T., Cossart, E., Charreire, H., et al. (2024). Hybridizing geographically weighted regression and multilevel models: A new approach to capture contextual effects in geographical analysis. *Geographical Analysis 56*(3): 554–572.

Förtner, M., Belina, B., & Naumann, M. (2021). The revenge of the village? The geography of right-wing populist electoral success, anti-politics, and austerity in Germany. *Environment and Planning C: Politics and Spaces 39*(3): 574–596.

Fotheringham, A.S. (2023). A comment on "a route map for successful applications of geographically-weighted regression": The alternative expressway to defensible regression-based local modeling. *Geographical Analysis 55*(1): 191–197.

Fotheringham, A.S., Brunsdon, C., & Charlton, M. (2002). *Geographically Weighted Regression: The Analysis of Spatially Varying Relationships*. New York: John Wiley & Sons.

Fotheringham, A.S., & Li, Z. (2023). Measuring the unmeasurable: Models of geographical context. *Annals of the American Association of Geographers 113*(10): 2269–2286.

Gould, P. (1970). Is statistix inferens the geographical name for a wild goose? *Economic Geography 46*(Supp.): 439–448.

Griffith, D.A., & Paelinck, J.H.P. (2011). *Non-standard Spatial Statistics and Spatial Econometrics*. New York: Springer.

Harbers, I., & Ingram, M.C. (2017). Geo-nested analysis: Mixed-methods research with spatially dependent data. *Political Analysis 25*(3): 289–307.

Henderson, J.V., Storeygard, A., & Weil, D.N. (2012). Measuring economic growth from outer space. *American Economic Review 102*(2): 994–1028.

Johnston, R., Jones, K., Burgess, S., Propper, C., Sarker, R., & Bolster, A. (2004). Scale, factor analyses, and neighborhood effects. *Geographical Analysis 36*(4): 350–368.

Johnston, R., & Pattie, C. (1992). Class dealignment and the regional polarization of voting patterns in Great Britain, 1964–1987. *Political Geography 11*: 73–86.

Karrel, D., Linke, A., Holland, E., & Hendrickson, E. (2023). "Born for a storm": Hard-right social media and civil unrest. *American Sociological Review 88*(2): 322–349.

King, G. (1995). Replication, replication. *PS: Political Science and Politics 28*(1): 443–499.

King, G. (1996). Why context should not count. *Political Geography 15*: 159–164.

Kulldorff, M., Heffernan, R., Hartman, J., et al. (2005). A space-time permutation scan statistic for disease outbreak detection. *PLoS Medicine 2*(3): e59. doi:10.1371/journal.pmed.0020059

Kulldorff, M., and Information Management Services (2009). *SaTScan: Software for the Spatial and Space-Time Scan Statistics, v 9.0.1*, 2009. www.satscan.org.

Lieberman, E.S. (2005). Nested analysis as a mixed-method strategy for comparative research. *American Political Science Review 99*(3): 435–452.

Linke, A.M., & Shawcroft, M. (2023). The effects of environmental stress on global agricultural landownership. *Population and Environment 45*(3): 19.

Linke, A.M., Witmer, F., & O'Loughlin, J. (2012). Space-time Granger analysis of the war in Iraq: A study of coalition and insurgent action-reaction. *International Interactions 38*(4): 402–425.

Maske, T., Minkoff, S.L., & Sokhey, A.E. (2014). Networks, context, and the use of spatially weighted survey metrics. *Political Geography 42*: 79–91.

Massey, D. (1994). *Space, Place, and Gender*. Minneapolis, MN: University of Minnesota Press.

McCarty, H.H. (1954). *McCarty on McCarthy: The Spatial Distribution of the McCarthy Vote 1952*. Iowa City, IA: Department of Geography, University of Iowa.

Minhas, S., Hoff, P., & Ward, M.D. (2019). Inferential approaches for network analysis: Amen for latent factor models. *Political Analysis 27*(2): 2008–2022.

O'Loughlin, J. (2000). Geography as space and geography as place: The divide between political science and political geography continues. *Geopolitics 5*(3): 126–137.

O'Loughlin, J., Flint, C., & Anselin, L. (1994). The geography of the Nazi vote: Context, confession and class in the Reichstag election of 1930. *Annals of the Association of American Geographers 84*(3): 351–380.

O'Loughlin, J., Linke, A., Toal, G., & Bakke, K. (2024). Support for Vladimir Putin in Russia's neighbors. *Political Geography 108*: 103014.

O'Loughlin, J., & Witmer, F. (2011). The localized geographies of violence in the North Caucasus of Russia, 1999–2007. *Annals of the Association of American Geographers 101*(1): 178–201.

O'Loughlin, J., Witmer, F., Linke, A.M., Laing, A., Gettleman, A., & Dudhia, J. (2012). Climate variability and conflict risk in East Africa, 1990–2009. *Proceedings of the National Academy of Sciences 109*: 18344–18394.

O'Loughlin, J., Witmer, F., Linke, A.M., & Thorwardson, N. (2010). Peering into the fog of war: The geography of the Wikileaks Afghanistan war logs 2004–2009. *Eurasian Geography and Economics 51*: 437–471.

Openshaw, S. (1983). *The Modifiable Areal Unit Problem*. Norwich: Geo Books.

Ord, J.K., & Getis, A. (1995). Local spatial autocorrelation statistics: Distributional issues and application. *Geographical Analysis 27*(4): 286–306.

Oshan, T.M., Li, Z., Kang, W., Wolf, L.J., & Fotheringham, A.S. (2019). mgwr: A python implementation of multiscale geographically weighted regression for investigating process spatial heterogeneity and scale. *ISPRS International Journal of Geo-Information 8*(6): 269.

Pattie, C., & Johnston, R. (2000). "People who talk together vote together": An exploration of contextual effects in Great Britain. *Annals of the Association of American Geographers 90*(1): 41–66.

Pred, A. (1983). Structuration and place: On the becoming of sense and place and structure of feeling. *Journal for the Theory of Social Behavior 13*(1): 45–68.

Radil, S., & Flint, C. (2013). Exiles and arms: The territorial practices of state making and war diffusion in post-Cold War Africa. *Territory, Politics, Governance 1*(2): 183–202.

Rey, S.J., & Anselin, L. (2007). PySAL: A Python library of spatial analytical methods. *Review of Regional Studies 37*(1): 5–27.

Rey, S.J., Anselin, L., Amaral, P., et al. (2021). The *PySAL* ecosystem: Philosophy and implementation. *Geographical Analysis 54*(3): 467–487.

Robinson, W.S. (1950). Ecological correlations and the behavior of individuals. *American Sociological Review 15*(3): 351–357.

Root, E.D., Meyer, R., & Emch, M. (2009). Evidence of localized clustering of gastroschisis in North Carolina, 1999–2004. *Social Science and Medicine 68*(8): 1361–1367.

Schon, J., & Koren, O. (2022). Introducing Afro-Grid, a unified framework for environmental conflict research in Africa. *Scientific Data 9*(116) https://doi.org/10.1038/s41597-022-01198-5

Tobler, W. (1970). A computer movie simulating urban growth in the Detroit region. *Economic Geography 46*(Supp.): 234–240.

Tollefson, A.F., Strand, H., & Buhaug, H. (2012). PRIO-GRID: A unified spatial data structure. *Journal of Peace Research 49*(2): 363–374.

VanWey, L.K., Rindfuss, R.R., Gutmann, M.P., Entwisle, B., & Balk, D.L. (2005). Confidentiality and spatially explicit data: Concerns and challenges. *PNAS 102*: 15337–15342.

Vidal de La Blanche, P. (1903). *Tableau de la geographie de la France*. Paris: Hachette.

Wucherpfennig, J., Weidmann, N.B., Girardin, L., Cederman, L.-E., & Wimmer, A. (2011). Politically relevant ethnic groups across space and time: Introducing the geoEPR dataset. *Conflict Management and Peace Science 28*(5): 423–437.

Zhukov, Y.M. (2023). Near-real time analysis of war and economic activity during Russia's invasion of Ukraine. *Journal of Comparative Economics 51*(5): 1232–1243.

Zhukov, Y.M., Byers, J.S., Davidson, M.A., & Kollman, K. (2024). Integrating data across misaligned spatial units. *Political Analysis 32*(1): 17–33.

Chapter 18

Political Economy

Shaina Potts

Department of Geography, University of California Los Angeles, Los Angeles, CA, USA

A mere three decades after the end of the Cold War and the widely perceived triumph of (neoliberal) globalization, we may be on the verge of an epochal shift in state–capital relations. The 2008 financial crisis, the rise of the BRICS (Brazil, Russia, India, China, and South Africa), growing trade wars, Covid-19's effects on supply chains, and more have led to widespread debates about the future of globalization. This has resulted in often simplistic warnings about the return of "geopolitics" to "the economy" in mainstream economic media (e.g., Anon. 2021; Kirk 2024). It has also spurred a more nuanced resurgence of interest in political economy within political geography, and in geopolitics within economic geography (e.g., Lee et al. 2018; Sonn & Hsu 2022; Yeung 2023). While neither the political nor the geopolitical has ever lost its economic significance, these recent changes, alongside the growing crisis of US hegemony, have indeed made the geopolitical dimensions of political economy more apparent than they had been for a long time. This has been compounded by the fact that some countries are growing more openly resistant to the rules of the "liberal international economic order," while those rules themselves arguably no longer benefit even the United States as clearly as they once did. Indeed, if "neoliberalism could well be the most significant political-ideological rationalization for state restructuring over the three past decades across geographical contexts" (Moisio et al. 2020: 12), the question now is whether it is in the process of being replaced by something else.

This makes understanding the complex geographies of state–capital relations more urgent than ever. Of course, this has long been a core concern for political geographers. As Moisio underscores (2020: 158), there is a "close link between

The Wiley Blackwell Companion to Political Geography, Second Edition.
Edited by Virginie Mamadouh, Natalie Koch, Chih Yuan Woon, and John Agnew.

the changing regimes of capital accumulation and states' spatial 'patterning.'" Taking this point seriously means that neither can be fully understood in isolation from the other. At the same time, studying this nexus can be challenging. In their introductory chapter to the *Handbook on the Changing Geographies of the State*, the volume's editors caution against two especially common analytical dangers (Moisio et al. 2020). First, there is the much-discussed issue of methodological nationalism, in which the official political borders of the state are treated as a mere "container" for "national" dynamics, while the state itself is reified as an abstract and often monolithic actor. Second, there is the problem of what the editors call "methodological globalism," in which scholars "take as a pre-given the role of economic globalization as the pre-eminent societal force which moulds, transforms and, increasingly, undermines the state and its constituent national and subnational branches" (Moisio et al. 2020: 14).

Centering capital in analyses of the state can contribute to either danger, whether through reifying the "national economy" or by presenting capital as uniquely mobile in contrast to fixed and bounded states. In this chapter, I suggest that *geographically informed transnational* approaches to (geo)political economy are essential to avoiding both problems. By emphasizing cross-border political economic relations and the heterogeneous articulations of state and capital at and across multiple scales, such work can facilitate the kinds of spatially complex, relational analyses necessary for avoiding either overly economistic or overly state-centric accounts.

There are, of course, multiple ways of conceptualizing the state–capital nexus within a transnational frame. In this chapter I focus on selected approaches to this issue within both political and economic geography. While the line between the two is often ambiguous, these subdisciplines have *tended* to focus on different primary objects of analysis. Where the former foregrounds the state, borders, territory, sovereignty, conflict, nationalism, and citizenship (as reflected, for example, in the chapter titles for the previous edition of this book; Agnew et al. 2015), the latter prioritizes questions of production, distribution, consumption, money and finance, technological innovation, and labor (see, e.g., Barnes & Christophers 2018). Yet political geography has long intersected with political economy and economic geography, and individual scholars and texts regularly blur any clear boundaries between them.

I make no attempt to summarize all the work at this intersection. Rather, I begin with a selective discussion of what I consider to be some of the most useful work on transnational state–capital relations in and at the borders of political geography, including that of Peter Taylor and John Agnew, as well as several scholars developing the concept of "geopolitical economy" and theorizing the relationship between geopolitics and economics. I then turn to two key concepts more closely associated with critical economic geography or, more broadly, "geographical political economy": uneven development and capitalist variegation. Finally, I consider recent work on the question of the "new" state capitalism, located mostly in economic geography, to show how such work makes use of the concepts of uneven development and variegation, while also identifying places in which this conversation already draws on core insights from political geography and places in which it would benefit from further mutual engagement.

Transnational Articulations of State and Capital: Political Geography and Geopolitical Economy Approaches

In and beyond geography, there has been a growing rejection of the "decline of the state" narrative and a refocusing on the ways states both structure and are restructured by globalizing capital (Peck 2010; Agnew 2017; Slobodian 2018; Sparke 2018b). While states are influenced and constrained by transnational capital (perhaps most obviously through its impacts on investment, state budgets, and tax receipts), they themselves also produce and manage transnational flows and markets. How, and how well, any given state does so depends on its domestic capacities and goals as well as its global economic position. But any analysis of state–capital relations that neglects cross-border capital flows altogether is necessarily partial. Conversely, analyzing the intersection of cross-border capital and state power contributes to understanding not only a given state but also how all states are bound together in an interconnected but differentiated capitalist system. That said, there are very different ways of approaching such analyses. In this section, I briefly review two of the most influential scholars theorizing transnational state–capital relations in political geography, before turning to other scholars in political and/or economic geography working with the concept of "geopolitical economy."

Among the earliest and most substantial routes through which political economy has influenced political geography has been Peter Taylor's development of world-systems analysis. Since the early 1980s, Taylor has built on and spatialized the work of Immanuel Wallerstein and other world-systems theorists. As he has explained to multiple generations of political geographers, especially through his influential textbook on *Political Geography* (now in its seventh edition), world-systems analysis is premised on the "single-society assumption" – that is, on the view that we now live not in multiple, neighboring societies but in one interconnected "world-system," which is synonymous with the "capitalist world-economy" (Flint & Taylor 2018: 12). Every society must be understood as a part within this larger whole. A fundamental implication is the rejection of linear theories of (national) development for a focus on interrelated processes of development and underdevelopment.

World-systems analysis also emphasizes a three-tier division of the economy into core, semi-periphery, and periphery, positing the middle tier as the primary driver of global economic dynamism. Similarly, this approach emphasizes three key scales: "a national scale of ideology, a local scale of experience and a global scale of reality," in which the national scale mediates between the other two (Flint & Taylor 2018: 36). Taylor emphasizes the importance of thinking not only about each scale in turn but about interrelations *across* these scales.

Though world-systems analysis has sometimes been critiqued for economism, Taylor rejects this charge. He takes world-systems theory to task for its simplistic conception of bounded national units within the interstate system, insisting instead on the co-constitutive nature of the multiplicity of states, nations, and territories (Taylor 1981, 1995). He also emphasizes the extraeconomic motivations of states, and he has increasingly incorporated both critical geopolitics, with its emphasis on the material power of discourse, and feminist geopolitics, with its emphasis on the scales of the body and the intimate, into his analyses of the world-system.

It is hard to overstate the influence of Taylor and world-systems analysis in political geography. His work has promoted relational analyses of the interstate system within a globally interconnected but uneven capitalist economy. It remains a powerful tool for investigating continued radical inequalities among nation-states. Yet some scholars are wary of such totalizing conceptions of "the" world-system, however nuanced, and eschew neat distinctions between the local, national, and global scales, or between core, periphery, and semi-periphery. Among the most prominent examples in political geography is the work of John Agnew.

Where world-systems analysis aims for a unified theory of the world-system, in which global capital is ultimately the most "real" force, Agnew's work has tended to begin from a critical conceptualization of sovereignty and territory and move from there to the complex relationships between states and globalization. Much of his work has been based on deconstructing powerful ideas about Westphalian or "absolute" territorial sovereignty and binary framings of national sovereignty versus globalization (Agnew 1994, 2010, 2017). In their place, he has developed far more nuanced analyses of the geographies of both state and non-state power. These have included showing that state power is never neatly aligned with strictly demarcated territories but rather operates through spatially diverse "sovereignty regimes"; and that both state and nonstate phenomena can operate through either territorializing or more "networked" forms of power (Agnew 2017).

More recently, Agnew (2022) has developed the concept of "hidden geopolitics" to refer to the many forms of transnational governance not captured by the usual focus on official foreign policy actors and decisions. This term is intended to index a heterogenous array of decisions by state, nonstate, and hybrid state–market actors at multiple scales, involving both economic and extraeconomic calculations, and mobilizing both territorial and networked types of power. If both Taylor and Agnew foreground the intersections of economic and state power within a global context, one might say that Agnew's work inverts world-systems analysis to the extent that, rather than seeing capital as the ultimate domain within which all states operate, he instead seeks to explore "how the geography of power works both globally and nationally to structure and govern the workings of the world's political economy" (Agnew 2022: 3).

Beyond Taylor and Agnew, and in conversation to varying degrees with both, there have been wider conversations at the boundaries of political and economic geography about the intersection of the state and (global) political economy. One important strand of such work has revolved around the concept of "geopolitical economy." Introduced by Agnew and Corbridge (1995), the term has since been further developed by other scholars (Jessop & Sum 2006; Glassman 2011; Lee, Wainwright, & Glassman 2018; Sparke 2018a, 2018b). It has been used in at least two distinct ways. At times it is meant simply as a shorthand for "geographical political economy," while at others it is used to mean "geopolitics plus political economy" (Glassman 2018). While, as Glassman shows, the two meanings are usefully seen as "articulated," I focus in the rest of this section on work that foregrounds geopolitics in the service of theorizing transnational state–capital relations.

In one early piece, for example, Glassman (2011) himself argues for incorporating geopolitical dynamics, including those of war, into analyses of global production networks. He builds on critical geographical theories of geopolitics that reject the Westphalian idea of neatly bounded state space. Such a view, he argues, "neglects the deep interpenetration of states across national boundaries" as well as "the interpenetration of capital and states" (2011: 156, 157). Instead, he sees states (and their geopolitical endeavors) "not as existing external to markets or production networks but rather as being produced and reproduced in the same processes that produce markets and production networks" (2011: 157).

One important focal point of these conversations about geopolitical economy has been the relationship between "geopolitics" and "geoeconomics." With the end of the Cold War and the seeming triumph of globalization in the early 1990s, the idea of a transition from an age of geopolitics, in which territorial competition was central, to one of geoeconomics, in which states have focused on the (supposedly more peaceful) pursuit of economic growth and competition, became prevalent in mainstream academic and foreign policy circles. This temporal division, moreover, has been mapped by such scholars onto physically distinct zones, with "advanced" Western countries being associated with geoeconomics and more "backward" and conflict-ridden ones with geopolitics (Sparke 2018b). Critical geographers, in contrast, have criticized such views as far too conceptually and spatially binary and as based on simplistic definitions of geopolitics (Lee, Wainwright, & Glassman 2018; Sparke 2018a, 2018b; Moisio 2019; Agnew 2020; Mallin & Sidaway 2024). These scholars instead seek to understand the complex relationships between geopolitical and geoeconomic forces in our own as well as in preceding periods. Sparke (2018a), for instance, sees geopolitics and geoeconomics not as distinct policies or domains, but rather as dialectically entwined "geostrategic discourses" that are deployed by both state and nonstate actors for strategic ends (see also Lee, Wainwright, & Glassman 2018; Mallin & Sidaway 2024).

Among scholars emphasizing this mutual entanglement of geopolitics and geoeconomics, productive disagreements about the spatial parameters of this relationship nevertheless remain. Referencing a 1985 essay on geopolitics by David Harvey, Sparke (whose primary object of analysis, it is worth emphasizing again, is geostrategic *discourses*) suggests "that the 'external' dialectic of geopolitics and geoeconomics is best understood as an overdetermined expression of the 'internal' uneven development dialectic in capitalism between spatial fixity and spatial expansion," while "Geopolitical economy can in turn be treated as the analysis of the relays between these internal and external dialectics" (Sparke 2018b: 484). Jessop and Sum (2018: 475) partially overlap in that they define geopolitics as "premised on the territorialization of political power." However, they contrast this with a geopolitical economy/geoeconomics that is not expressly about expansion and deterritorialization, but rather involves "the organization of economic flows . . . [that] can be more-or-less place-based, network-mediated, territorially embedded, multiscalar, and friction-prone" (Jessop & Sum 2018: 475). Lee, Wainwright, and Glassman (2018: 420) go further, rejecting altogether the view that "Geopolitical concerns tend to be equated with territorial integrity, geoeconomic interests with open borders and capital's flow" (see also Moisio 2019). Like Jessop and Sum, they point out that some economic projects are territorial. Conversely, they argue, oversimplified views of the territorial state ignore the significance both of internal territorial fragmentation and of a state's "extra-territorial practices beyond its fixed borders" (Lee, Wainwright, & Glassman 2018: 419). In their view, "territorial logics, either territorial fixity or fluidity, are products of geopolitical and geoeconomic processes which derive from class relevant social struggles" (2018: 421). Such debates about the relative distinctions between and geographies of geopolitical versus geoeconomic logics and strategies provide a useful platform from which to consider the ongoing reconfiguration of state–capital relations, particularly in the face of apparent challenges to US hegemony and the neoliberal consensus, as well as the potential emergence of new forms of state capitalism (as I discuss later).

In sum, while approaching the intersection of political economy and political geography from diverse and sometimes contradictory perspectives, taken together the scholars discussed in this section reject methodologically nationalist accounts of the state and reflect, instead, an effort to theorize states and state power as co-produced with transnational political economic logics and processes. They offer a variety of analytical tools for attempting to do this, from world-systems analysis, to the idea of hidden geopolitics, to the concepts of geopolitical economy and a geopolitics/geoeconomics dialectic. In the following section, I focus on two further analytical tools that are also especially helpful in thinking critically and relationally about the transnational articulation of states and capital, but that have tended to play a more prominent role in economic than in political geography.

Uneven Development and Capitalist Variegation in Economic Geography

The work discussed in the previous section reflects the impossibility of neatly dividing political from economic geography (Sparke and Glassman, for example, are especially difficult to categorize in this way). Nevertheless, the concepts of uneven development and variegated capitalism, to which I turn in this section, have been deployed much less frequently in self-identified political geographical work than in economic geography, or, more broadly, geographical political economy. Here, I follow Sheppard (2011, 2018) in using the latter term to encompass a range of heterodox scholars embracing the views that space and spatial variation are co-constituted with economic activity, and that economic processes "co-evolve" with "biophysical, cultural and social processes" (Sheppard 2011: 321). Scholars in this area also tend to see "capitalism as an unstable economic system, characterized by uneven geographical development" (Sheppard 2011: 320). This emphasis on a dynamic, unstable, and uneven capitalism is fundamental to analyzing state (re) configuration within a globally interconnected capitalist economy today. Here, I suggest that the related concepts of uneven development and variegated capitalism are especially useful for investigating these dynamics while avoiding both methodological nationalism and methodological globalism, and both overly particular and overly homogenizing accounts of state–capital relations.

"Uneven development" is a central concept in Marxian political economy and critical economic geography, but (with the key exception of world-systems analysis) has played a less prominent role in political geography (as evidenced, for example, by minimal references to the term in this book's previous edition; Agnew et al. 2015). Most broadly, theories of uneven development seek to explain why, in contrast to the equalizing predictions of mainstream economics, persistent spatial inequalities remain entrenched within capitalism. More specifically, they emphasize the dynamic relationship between "places of accumulation and sites of devaluation" (Peck, Werner, & Jones 2023: 1396) and see the (re)production of these differences not as a mere byproduct of but rather as central to capitalist accumulation.

In world-systems theory and related traditions, the concept has primarily been used to explain economic differences between a core of "developed" countries and "underdeveloped" peripheral and semi-peripheral countries. Though still powerful in many ways for explaining persistent inequalities between and within North and South, this approach to uneven development has often been too mechanical and (at least in its original formulations) has reified the nation-state as the primary unit of analysis. In part because of this, other ways of conceptualizing uneven development have become more influential in economic geography.

Most famously developed by Neil Smith (1984) and David Harvey (1982), on the one hand, and Doreen Massey (1995), on the other, early geographical debates about uneven development tended to fall into two main camps (see Peck, Werner, & Jones 2023). Smith and Harvey emphasize more abstract and structural accounts of uneven development as driven by capitalism's simultaneous and contradictory tendencies toward spatial equalization and differentiation. These dynamics are located, for example, in the laws of capitalist competition, the problem of fixed capital, the falling rate of profit, and so on. This approach remains influential, as can be seen, for example, in Sparke's work linking the geopolitics/geoeconomics dialectic to the "contradictions of the inner dialectic between spatial fixity and spatial expansion in the uneven development of globalizing capitalism" (Sparke 2018b: 488). In contrast to Smith and Harvey, Massey's approach involves a more situated, contextual, and contingent understanding of uneven development and industrial restructuring, and a stronger emphasis on the changing spatial division of labor.

After the 1980s, the concept of uneven development receded into the background in much work in economic geography, but since 2008 it has been taken up with renewed vigor. This more recent work, however, has tended to steer a middle path between abstract emphases on the "laws" of capitalism and more concretely situated accounts of uneven development (Peck, Werner, & Jones 2023). While still emphasizing capitalism's simultaneous tendencies toward equalization/homogenization and differentiation/fragmentation, it has allowed for more eclectic and contingent accounts of the economic, political, and biophysical processes driving these dynamics, and it has incorporated feminist, postcolonial, and other theoretical approaches (see, e.g., Werner 2015). At the same time, this messier sense of the dynamics of uneven development retains the earlier focus on the centrality of capital to (re)producing difference across space and on the relationships *among* local, regional, and national political economic processes. Some scholars have also called for renewing Trotsky's emphasis on uneven *and combined* development, which arguably better foregrounds not just differentiation but also interconnection and hybridity in the constant emergence of new forms of capitalism (Dunford & Liu 2017; Peck 2019). As Peck argues, an emphasis on "combination problematises questions of interdependence, interconnectivity, situatedness, positionality, and the immanent potential for finding and shaping *new* configurations and pathways" (2019: 51, emphasis in the original).

While both older and newer theories of uneven development foreground the internal contradictions of capitalism, this need not result in economically reductionist analyses. Indeed, among the most important ways in which such theories have evolved is through growing attention to the roles of states, laws, and institutions in (re)producing economic difference. The concept of capitalist "variegation" has become one especially influential way of approaching these dynamics within economic geography (Peck & Theodore 2007; Brenner, Peck, & Theodore 2010), but it has been less widely deployed in political geography (though see Woon 2018; Moisio et al. 2020). While emphasizing the spatiotemporal heterogeneity of the many forms of capitalist governance, variegation is meant as more than a mere description of economic-institutional variation across space (though the word is sometimes used in passing in this way). The concept emerged in large part from the idea that "what we are lacking . . . are institutional theories of the uneven development of capitalism" (Peck & Theodore 2007: 762) or a theory of "regulatory uneven development" (Brenner, Peck, & Theodore 2010: 185). Like the broader concept of uneven development, variegation is meant to help capture the internal drivers of economic-institutional differentiation, while also foregrounding the articulation of state and market, public and private.

This project resonates with "relational" conceptions of the state, associated most famously with Jessop's analyses of the state's role in mediating class struggle and temporarily stabilizing particular "regimes of accumulation" (Jessop & Sum 2006; Jessop 2008; Moisio et al. 2020). Yet variegation – a term Jessop (2014) himself argues for in other contexts – also differs substantially from this framework, including in its attempts to capture a much more spatiotemporally heterogenous set of institutions, rules, and policies. The concept was developed in large part as a corrective to the varieties of capitalism and comparative capitalisms approaches, which categorize economies according to national "types" (e.g., the "American" vs. the "German" model; Peck & Theodore 2007). Variegation, in contrast, aims for spatially differentiated accounts of "economic-geographical difference" at not only the national but also the subnational, regional, and supranational scales (Peck & Theodore 2007: 733). It seeks to avoid methodological nationalism and typologizing and instead to emphasize both particularity and spatiohistorical interconnection – in other words, to "adequately balance national specificity and path-dependency on the one hand with common underlying tendencies in capitalist restructuring on the other" (Peck & Theodore 2007: 731). Work deploying the concept thus not only describes spatial differentiation but also attempts to show how economic governance is shaped by "inherited institutional landscapes" and "mutually recursive policy relays" (Brenner, Peck, & Theodore 2010: 209, 182). Like the concept of uneven development, moreover, variegation emphasizes simultaneous processes of regulatory/institutional equalization and differentiation within and across national borders.

The concept of variegation has been applied perhaps most thoroughly to the neoliberal reconfiguration of states within "an increasingly transnational field of market-oriented regulatory transfer" (Brenner, Peck, & Theodore 2010: 185). It has been taken up in the investigation of the neoliberalization of urban political economy (e.g., Rossi 2016; González et al. 2018; Davidson 2020), as well as to examine the intersections of neoliberalization and financialization (Muellerleile & French 2022) and the sociospatially diverse and contingent geographies of the financialization of housing (Aalbers 2017), firms (Dixon 2011), and investors (Lai 2016). It also intersects with a growing body of work on law in economic geography. Although only a handful of geographers working at the intersection of law and political economy have substantively engaged the concept of variegation so far (e.g., Barkan 2011; Potts 2020, 2023), this area as a whole can be seen precisely as foregrounding the role of law and regulation in constituting spatially differentiated economic geographical processes at multiple scales (e.g., see also Wójcik 2013; Christophers 2015; Blomley 2020).

In sum, by linking relational analyses of uneven economic development to path-dependent variations in institutional and regulatory forms, the concepts of uneven development and variegation provide powerful tools for analyzing ongoing spatial transformations in state–capital relations. Both concepts, moreover, have been important for considering transnational political economic processes, and both are especially useful for efforts to think *across* multiple scales. Important intersections already exist between work on uneven development and variegation, on the one hand, and geographical scholarship on geopolitical economy and geo-economics, on the other. Yet there is room for more engagement, as well as for thinking seriously about how other approaches more closely associated with political geography might be fruitfully combined with this work on uneven development and variegation. In the final section of the chapter, I turn to one area of research on contemporary changes in state–capital relations in order to consider how it does or might intersect with the various approaches discussed so far.

The "New" State Capitalism?

After what turned out to be a somewhat fleeting moment of triumph for the unipolar world in the 1990s, the United States' dominance and the international order it sculpted has been under increasing challenge. In many circles, this has been registered in pervasive hand-wringing about the rise of China and the collapse of the so-called rules-based order. It has also been reflected in widespread debates about the (re)emergence of "state capitalism" in North and South, broadly associated with trade protectionism, new industrial policies, and de-globalization. In mainstream policy and business circles, the term is applied first and foremost to China and other "illiberal" states, as well as to the increasing use of interventionist economic practices from Western governments, which are often cast as *needing* to adopt such tools because of this illiberal threat (see Alami & Dixon 2020b). This discourse is deeply flawed. Yet there have, in fact, been significant changes in state–capital relations over the past two decades, with many states arguably taking on more visible roles in economic activity and management. These practices cannot be easily summarized:

> The label state capitalism indexes an extremely wide constellation of political and institutional forms, including sovereign wealth funds, state enterprises, national development banks, models of postdevelopmentalism, post-neoliberalism and resource-nationalism, capitalism(s) in emerging economies such as China, Russia, Turkey, Indonesia, Singapore or Hungary, the renewal of industrial and neo-mercantilist policies, and many more. (Alami & Dixon 2023: 75)

The question is whether the concept of state capitalism is useful for understanding these diverse phenomena.

This question has recently been taken up by geographers debating the "new" state capitalism, which Whiteside et al. (2023: 63) have dubbed a "focal point" in the "interdisciplinary field concerned with contemporary transformations of the state and capitalism." So far, this conversation has occurred largely within "economic" geography, most prominently in three special issues on the topic in *Environment and Planning A* (see Alami et al. 2023; Dixon et al. 2023; Whiteside et al. 2023). I focus on this conversation here for three reasons. First, the widespread empirical changes sketched earlier should be central to any geographical investigation of state–capital relations in the twenty-first century, whereover one lands on the label of state capitalism itself. Second, the growing importance of these "interventionist" economic moves of various kinds shows more clearly than ever that neither "the state" nor geopolitics can be carved off in any clear way from the operations of the global economy. The simultaneity of these diverse empirical shifts also underscores the impossibility of understanding political economic change in any one country in isolation from either the global economy or the interstate system (or what Taylor 1995 might refer to as the fact of mutual "interstateness"). Third, the current conversation about the new state capitalism in economic geography demonstrates the usefulness of the key concepts of uneven development and variegation for investigating these transformations. At the same time, this conversation would, I will suggest, be further enriched through more engagement with certain ideas from political geography.

The geographical conversation about the new state capitalism is empirically and conceptually rich and characterized by many (dis)agreements, including about how to periodize this era of state capitalism as well as about whether we are seeing anything "new" at all. As a whole, scholars in this debate have attempted (sometimes more successfully than others) to avoid binary readings of state versus capital, while investigating an array of hybrid public/private forms of governance and market activity. I make no attempt to summarize this work here. Rather, I focus on a few selected examples most relevant to the topic of this chapter.

In light of the diversity of state practices considered under the label of the new state capitalism, Alami and Dixon emphasize the importance of avoiding either overly general and homogenizing definitions of the term or overly particular and isolated empirical analyses of particular phenomena. In contrast, they call for attention to the transnational linkages and policy relays shaping seemingly independent state practices, and for analyzing those practices relationally as part of a differentiated but interconnected whole. In their view, the concept of "UCD [uneven and combined development] is essential to grasping the inner nature of the new state capitalism as a variegated, world-historical phenomenon" (Alami & Dixon 2023: 75). More specifically, they identify a set of macroeconomic transformations in global patterns of uneven development itself, which they see as shaping and being shaped by these changing state–capital relations. These include ongoing shifts in economic power from the North Atlantic to the Pacific Rim, the rise of the BRICS and other "emerging economies," technologically driven changes in the international division of labor, and changing patterns of the global distribution of money and finance (Alami & Dixon 2020b, 2023). (Such an emphasis on the rise of China and other emerging economies could be seen as in line with Taylor's focus on the semi-periphery as a key driver of broader political economic transformations.) This macroeconomic view helps make visible part of what is new about this era of state-capitalist practices. Previous forms of state capitalism associated, for instance, with mid-twentieth-century postcolonial governments often involved attempts to disengage to some degree from the global economy. This time around, most state-capitalist practices, from the development of sovereign wealth funds to the renationalization of natural resources, aim to embed states and their national economies *more* thoroughly in circuits of transnational capital (Alami & Dixon 2023).

This conversation about the new state capitalism in geography also aims to avoid monolithic accounts of "the state" and to consider the articulation of capitalist dynamics with an array of sub- and transnationally scaled state institutions, regulations, and policies, and with the various logics and motivations of both state and nonstate actors, though some have called for more attention to the subnational scale (e.g., Paul & Cumbers 2023). In the face of the huge range of hybrid state–market institutional forms and logics that that have been studied under this rubric, Alami and Dixon (2020a: 78) call for "an effective theory of variegation [that] must be able to account for both what is similar (i.e. underlying common processes), and the continuous reproduction of difference, both between state capitalism and other forms of capitalism, and between different varieties of state capitalism." One function of such a theory of variegation is to attend not only to how states are influenced by transformations in global capitalism, but also to how the uneven geography of global capitalism is driven in turn by differentiated regulatory responses. As McGregor and Coe (2023: 716) note, for instance, "the internationalization of state hybrids [e.g. state-owned enterprises and sovereign wealth funds] adds an extraterritorial dimension to the state, which can influence the configuration and governance of global production networks."

The concept of variegation also aims to capture the ways in which state economic practices "develop in inter-referential and cumulative forms across territory, producing further state capitalist modalities" (Alami & Dixon 2023: 72). Endless empirical examples include the synergistic operation of Spanish and Chinese state-owned enterprises to improve the competitiveness of Spanish ports (Jensen 2023), the way the United Kingdom's so-called market-led governance goals have been underwritten by investment from a Qatari sovereign wealth fund (Ward, Brill, & Raco 2023), and mutual competition between China and the United States over harnessing digital platforms for geopolitical–economic ends (Rolf & Schindler 2023). Together, such analyses help demonstrate how it is not only the "internal" dynamics of capitalism but new forms of state–capital engagement that are driving changes in the uneven geography of the global political economy itself.

This work on the new state capitalism should be of immediate interest to political geographers interested in the intersection of political economy and the state. Indeed, while this conversation has taken place primarily in economic geography venues so far, it has already engaged political geography in important ways. For one thing, trying to theorize the new state capitalism requires economic geographers to also have a "firm embedding in state theory" (often based in Jessop's "strategic-relational approach") (Whiteside et al. 2023: 67). This project has also been heavily influenced by critical geopolitics and its emphasis on taking geopolitical discourses seriously as key objects of inquiry in themselves (e.g., Alami et al. 2022).

At the same time, there is significant room for further engagement with political geography, including with deeper theorizations of sovereignty and territoriality. Much of the geographical work on the new state capitalism so far reiterates Agnew's (1994) call to avoid the "territorial trap." Yet there remains much work to do in thinking more deeply about how these changing state-capitalist practices are reconfiguring territoriality and sovereignty themselves (though see Alami & Dixon 2023; Hall 2023; Su & Lim 2023). Agnew's (2017) work on sovereignty regimes provides one potentially fruitful point of connection. The conversation on the new state capitalism would also benefit from engaging with Agnew's concept of hidden geopolitics, which offers an alternate though not necessarily contradictory approach to theorizing the transnational geographies of state and nonstate power. Finally, there has been a surprising lack of direct engagement in the new state capitalism literature with the geopolitics/geoeconomics debate. While that debate emerged at a moment when the term geoeconomics was associated largely with the dominance of *neoliberal* globalization, there is surely much to be gained from asking whether and how the dialectical relationship between these geostrategic logics has shifted in a moment of increasingly open state economic interventions. In short, ongoing changes in state–capital relations in North and South (whether well captured by the concept of a new state capitalism or not) are part of a worldwide and potentially transformative phenomenon that demands analysis via both economic geography and political geography approaches.

Conclusion

Examining the relationships between states and capital, and between political and economic power, has long been a core concern in both political and economic geography. In this chapter I have reviewed key approaches to transnational (geo)political economy within both subfields. What the most powerful approaches in both traditions share is an emphasis on the articulation of state practices with both domestic and transnational economic processes, as well as a growing attention to the proliferation of public and private forms of economic governance and the heterogeneity and mutual interreferentiality of institutional and regulatory practices within and across borders. This project is only becoming more urgent in the context of fundamental shifts in global geopolitical and economic relations, on the one hand, and overly simplistic stories about (de-)globalization, on the other. A key goal of this chapter has been to suggest that further conversation and sharing of concepts between political and economic geographical approaches to the study of transnational state–capital relations would benefit scholars in both traditions. Such synergies would facilitate more nuanced analyses of changing forms of sovereignty, territoriality, and geopolitics, as well as the identification of the hybrid state/nonstate mechanisms driving differentiation within a globally interconnected capitalist economy.

References

Aalbers, M.B. (2017). The variegated financialization of housing. *International Journal of Urban and Regional Research 41*(4): 542–554. doi:10.1111/1468-2427.12522

Agnew, J. (1994). The territorial trap: The geographical assumptions of international relations theory. *Review of International Political Economy 1*(1): 53–80.

Agnew, J. (2010). Still trapped in territory? *Geopolitics 15*(4): 779–784.

Agnew, J. (2017). *Globalization and Sovereignty: Beyond the Territorial Trap*. Lanham, MD: Rowman and Littlefield.

Agnew, J. (2020). Geoeconomics and the state. In Moisio, S., Koch, N., Jonas, A.E., & Lizotte, C. (eds), *Handbook on the Changing Geographies of the State: New Spaces of Geopolitics*. Cheltenham: Edward Elgar, 161–172.

Agnew, J. (2022). *Hidden Geopolitics: Governance in a Globalized World*. Lanham, MD: Rowman and Littlefield.

Agnew, J., & Corbridge, S. (1995). *Mastering Space: Hegemony, Territory and International Political Economy*. London: Routledge.

Agnew, J., Mamadouh, V., Secor, A., & Sharp, J. (eds) (2015). *The Wiley Blackwell Companion to Political Geography*. Chichester: Wiley-Blackwell.

Alami, I., & Dixon, A.D. (2020a). State capitalism(s) redux? Theories, tensions, controversies. *Competition and Change 24*(1): 70–94. doi:10.1177/1024529419881949

Alami, I., & Dixon, A.D. (2020b). The strange geographies of the "new" state capitalism. *Political Geography 82*: 102237.

Alami, I., & Dixon, A.D. (2023). Uneven and combined state capitalism. *Environment and Planning A: Economy and Space 55*(1): 72–99. doi:10.1177/0308518X211037688

Alami, I., Dixon, A.D., Gonzalez-Vicente, R., et al. (2022). Geopolitics and the "new" state capitalism. *Geopolitics 27*(3): 995–1023. doi:10.1080/14650045.2021.1924943

Alami, I., Whiteside, H., Dixon, A.D., & Peck, J. (2023). Making space for the new state capitalism, part II: Relationality, spatiotemporality and uneven development. *Environment and Planning A: Economy and Space 55*(3): 621–635. doi:10.1177/0308518X231156913

Anon. (2021). The new geopolitics of global business: China and America dominate like never before. *The Economist*, June 5. https://www.economist.com/leaders/2021/06/05/the-new-geopolitics-of-global-business (accessed November 25, 2025).

Barkan, J. (2011). Law and the geographic analysis of economic globalization. *Progress in Human Geography 35*: 589–607.

Barnes, T.J., & Christophers, B. (2018). *Economic Geography: A Critical Introduction*. Chichester: John Wiley & Sons.

Blomley, N. (2020). Precarious territory: Property law, housing, and the socio-spatial order. *Antipode 52*(1): 36–57.

Brenner, N., Peck, J., & Theodore, N. (2010). Variegated neoliberalization: Geographies, modalities, pathways. *Global Networks 10*(2): 182–222.

Christophers, B. (2015). The law's markets. *Journal of Cultural Economy 8*(2): 125–143.

Davidson, M. (2020). Post-crash cities: The Great Recession, state restructuring and urban governance. In Moisio, S., Koch, N., Jonas, A.E., & Lizotte, C. (eds), *Handbook on the Changing Geographies of the State: New Spaces of Geopolitics*. Cheltenham: Edward Elgar, 385–398.

Dixon, A.D. (2011). Variegated capitalism and the geography of finance: Towards a common agenda. *Progress in Human Geography 35*(2): 193–210. doi:10.1177/0309132510372006

Dixon, A.D., Peck, J., Alami, I., & Whiteside, H. (2023). Making space for the new state capitalism, part III: Thinking conjuncturally. *Environment and Planning A: Economy and Space 55*(5): 1207–1217. doi:10.1177/0308518X231185587

Dunford, M., & Liu, W. (2017). Uneven and combined development. *Regional Studies 51*(1): 69–85. doi:10.1080/00343404.2016.1262946

Flint, C., & Taylor, P.J. (2018). *Political Geography: World-Economy, Nation-State and Locality*. London: Routledge.

Glassman, J. (2011). The geo-political economy of global production networks: Geo-political economy of global production networks. *Geography Compass 5*(4): 154–164. doi:10.1111/j.1749-8198.2011.00416.x

Glassman, J. (2018). Geopolitical economies of development and democratization in East Asia: Themes, concepts, and geographies. *Environment and Planning A: Economy and Space 50*(2): 407–415. doi:10.1177/0308518X17737170

González, S., Oosterlynck, S., Ribera-Fumaz, R., & Rossi, U. (2018). Locating the global financial crisis: Variegated neoliberalization in four European cities. *Territory, Politics, Governance 6*(4): 468–488. doi:10.1080/21622671.2017.1318713

Hall, S. (2023). Locating state capitalism: Financial centres and the internationalisation of Chinese banks in London. *Environment and Planning A: Economy and Space 55*(5): 1239–1254. doi:10.1177/0308518X221130080

Harvey, D. (1982). *The Limits to Capital*. Chicago, IL: University of Chicago Press.

Jensen, F. (2023). State capitalism and Spanish port development along the maritime silk road. *Environment and Planning A: Economy and Space 55*(3): 636–654. doi:10.1177/0308518X221083683

Jessop, B. (2008). *State Power*. London: Polity Press.

Jessop, B. (2014). Capitalist diversity and variety: Variegation, the world market, compossibility and ecological dominance. *Capital & Class 38*(1): 45–58. doi:10.1177/0309816813513087

Jessop, B., & Sum, N.-L. (2006). *Beyond the Regulation Approach: Putting Capitalist Economies in Their Place*. Cheltenham: Edward Elgar.

Jessop, B., & Sum, N.-L. (2018). Geopolitics: Putting geopolitics in its place in cultural political economy. *Environment and Planning A: Economy and Space 50*(2): 474–478. doi:10.1177/0308518X17731106

Kirk, S. (2024). Should investors care about geopolitics? *Financial Times*, January 19. https://www.ft.com/content/d77cf957-986f-4ec1-860f-bec36b397515 (accessed November 25, 2025).

Lai, K.P.Y. (2016). Financial advisors, financial ecologies and the variegated financialisation of everyday investors. *Transactions of the Institute of British Geographers 41*(1): 27–40. doi:10.1111/tran.12101

Lee, S.-O., Wainwright, J., & Glassman, J. (2018). Geopolitical economy and the production of territory: The case of US–China geopolitical-economic competition in Asia. *Environment and Planning A: Economy and Space 50*(2): 416–436.

Mallin, F., & Sidaway, J.D. (2024). Critical geoeconomics: A genealogy of writing politics, economy and space. *Transactions of the Institute of British Geographers 49*(1): e12600. doi:10.1111/tran.12600

Massey, D.B. (1995). *Spatial Divisions of Labor: Social Structures and the Geography of Production*, 2nd edn. London: Routledge.

McGregor, N., & Coe, N.M. (2023). Hybrid governance and extraterritoriality: Understanding Singapore's state capitalism in the context of oil global production networks. *Environment and Planning A: Economy and Space 55*(3): 716–741. doi:10.1177/0308518X211048197

Moisio, S. (2019). Re-thinking geoeconomics: Towards a political geography of economic geographies. *Geography Compass 13*(10): e12466. doi:10.1111/gec3.12466

Moisio, S. (2020). Introduction: Geographical political economies of the state. In Moisio, S., Koch, N., Jonas, A.E., Lizotte, C., & Luukkonen, J. (eds), *Handbook on the Changing Geographies of the State: New Spaces of Geopolitics.* Cheltenham: Edward Elgar, 158–160.

Moisio, S., Koch, N., Jonas, A.E., Lizotte, C., & Luukkonen, J. (eds) (2020). *Handbook on the Changing Geographies of the State: New Spaces of Geopolitics.* Cheltenham: Edward Elgar.

Muellerleile, C., & French, S. (2022). Variegated intersections of neoliberalism and financialization. *Environment and Planning A: Economy and Space 54*(1): 136–143. doi:10.1177/0308518X211059837

Paul, F.C., & Cumbers, A. (2023). The return of the local state? Failing neoliberalism, remunicipalisation, and the role of the state in advanced capitalism. *Environment and Planning A: Economy and Space 55*(1): 165–183. doi:10.1177/0308518X211050407

Peck, J. (2010). *Constructions of Neoliberal Reason.* Oxford: Oxford University Press.

Peck, J. (2019). Combination. In Antipode Editorial Collective (ed.), *Keywords in Radical Geography: Antipode at 50.* Chichester: John Wiley & Sons., 50–55. doi:10.1002/9781119558071.ch36

Peck, J., & Theodore, N. (2007). Variegated capitalism. *Progress in Human Geography 31*(6): 731–772.

Peck, J., Werner, M., & Jones, M. (2023). A dialogue on uneven development: A distinctly regional problem. *Regional Studies 57*(7): 1392–1403. doi:10.1080/00343404.2022.2116417

Potts, S. (2020). Beyond (de)regulation: Law and the production of financial geographies. In Knox-Hayes, J., & Wójcik, D. (eds), *The Routledge Handbook of Financial Geography.* London: Routledge, 103–121.

Potts, S. (2023). Law's place in economic geography: Time, space, and methods. *Environment and Planning A: Economy and Space 56*(5). doi:10.1177/0308518X231201233

Rolf, S., & Schindler, S. (2023). The US–China rivalry and the emergence of state platform capitalism. *Environment and Planning A: Economy and Space 55*(5): 1255–1280. doi:10.1177/0308518X221146545

Rossi, U. (2016). The variegated economics and the potential politics of the smart city. *Territory, Politics, Governance 4*(3): 337–353. doi:10.1080/21622671.2015.1036913

Sheppard, E. (2011). Geographical political economy. *Journal of Economic Geography 11*(2): 319–331.

Sheppard, E. (2018). Heterodoxy as orthodoxy: Prolegomenon for a geographical political economy. In Clark, G.L., Feldman, M., Gertler, M.S., & Wójcik, D. (eds), *The New Oxford Handbook of Economic Geography.* Oxford: Oxford Academic, 159–178. doi:10.1093/oxfordhb/9780198755609.013.9

Slobodian, Q. (2018). *Globalists: The End of Empire and the Birth of Neoliberalism.* Cambridge, MA: Harvard University Press.

Smith, N. (1984). *Uneven Development: Nature, Capital, and the Production of Space.* Oxford: Blackwell.

Sonn, J.W., & Hsu, L.-F. (2022). Geopolitics and economic geography: A commentary on Andy Pike's geographical political economy. *International Journal of Urban Sciences 26*(1): 36–44. doi:10.1080/12265934.2020.1828148

Sparke, M. (2018a). Geoeconomics, globalisation and the limits of economic strategy in statecraft: A response to Vihma. *Geopolitics* 23(1): 30–37. doi:10.1080/14650045.2017.1326482

Sparke, M. (2018b). Globalizing capitalism and the dialectics of geopolitics and geoeconomics. *Environment and Planning A: Economy and Space* 50(2): 484–489. doi:10.1177/0308518X17735926

Su, X., & Lim, K.F. (2023). Capital accumulation, territoriality, and the reproduction of state sovereignty in China: Is this "new" state capitalism? *Environment and Planning A: Economy and Space* 55(3): 697–715. doi:10.1177/0308518X221093643

Taylor, P.J. (1981). Geographical scales within the world-economy approach. *Review (Fernand Braudel Center)* 5(1): 3–11.

Taylor, P.J. (1995). Beyond containers: Internationality, interstateness, interterritoriality. *Progress in Human Geography* 19(1): 1–15. doi:10.1177/030913259501900101

Ward, C., Brill, F., & Raco, M. (2023). State capitalism, capitalist statism: Sovereign wealth funds and the geopolitics of London's real estate market. *Environment and Planning A: Economy and Space* 55(3): 742–759. doi:10.1177/0308518X221102157

Werner, M. (2015). *Global Displacements: The Making of Uneven Development in the Caribbean*. Chichester: John Wiley & Sons.

Whiteside, H., Alami, I., Dixon, A.D., & Peck, J. (2023). Making space for the new state capitalism, part I: Working with a troublesome category. *Environment and Planning A: Economy and Space* 55(1): 63–71. doi:10.1177/0308518X221140186

Wójcik, D. (2013). Where governance fails advanced business services and the offshore world. *Progress in Human Geography* 37(3): 330–347. doi:10.1177/0309132512460904

Woon, C.Y. (2018). China's contingencies: Critical geopolitics, Chinese exceptionalism and the uses of history. *Geopolitics* 23(1): 67–95. doi:10.1080/14650045.2017.1302429

Yeung, H.W. (2023). Troubling economic geography: New directions in the postpandemic world. *Transactions of the Institute of British Geographers* 48(4): 672–680. doi:10.1111/tran.12633

Chapter 19

Geopolitics and Critical Geopolitics

Sami Moisio

Department of Geosciences and Geography, University of Helsinki, Helsinki, Finland

"Geopolitics" is often used by scholars, journalists, military strategists, and different kinds of management experts both unintentionally and intentionally in multiple ways and for various purposes. It is thus a contested term that has been given different meanings in different spatial and temporal contexts and within different kinds of networks of actors. In short, the word geopolitics "poses a question to us every time it is knowingly evoked and used" (Ó Tuathail 1996: 66).

Over the past few decades, the term geopolitics has, yet again, been expropriated for new usages and it seems pointless to try to capture its "correct" meaning (see Murphy et al. 2004). Two introductory points may be in order, however. First, geopolitics is a scholarly practice, often regarded as a subfield of political geography, political science, and international relations (IR). Geopolitics can also be understood as a constantly remade scholarly field that is articulated in (text) books and articles through certain development trajectories, key scholars, and their ideas (see, e.g., Parker 1998; Agnew 2003; Dittmer & Sharp 2014a; Moisio et al. 2020; Smith 2020; Flint 2022).

Second, geopolitics is a political practice that brings together power, place, "world," and subjects in unique combinations. From a historical perspective, geopolitics must be understood not only as an academic theorizing of politics but also as the political action of all sorts of actors that have sought to mold political spaces. Moreover, geopolitical scholarship has not only informed political practices but has also been impacted by the world political ruptures and emergent and dominant political rationalities of a given time. Because social scientific knowledge production is, albeit often implicitly, tied up with political discourses and

The Wiley Blackwell Companion to Political Geography, Second Edition.
Edited by Virginie Mamadouh, Natalie Koch, Chih Yuan Woon, and John Agnew.

ideologies, it is difficult to draw a blunt line between academic geopolitics and geopolitical action.

This chapter scrutinizes the historical roots (classical geopolitics) and recent developments of geopolitics (foundations, critiques, and research agendas). Throughout, the focus is not restricted to political geographical works that are self-consciously labeled as geopolitical or critical geopolitical – in the title or otherwise. What follows is hence a decidedly subjective mapping of studies that have made a contribution to the field of critical geopolitics, loosely defined, during the past decades.

Historical Roots: Classical Geopolitics

In 1899, the Swedish politician and political scientist Rudolf Kjellén used the concept of geopolitics to denote the territorial or "earthly" aspects of the state. The invention of the term nonetheless "marked only a terminological modification of an existing intellectual agenda, previously labeled 'political geography'" (Heffernan 2000: 28). Kjellén and his German colleague Friedrich Ratzel played a role in developing geopolitics as an organic theory of the state. Within such a frame, the state is not only bound to the Earth but also determined by it. Kjellén and Ratzel understood geopolitics as the territorial dimensions of the state, and they defined the state as a spatial life form. A discernible sociobiological reasoning was not uncommon within European academic circles in the early twentieth century.

The second branch of classical geopolitics is often associated with the British conservative politician and geographer Halford Mackinder and with American naval officer Alfred Thayer Mahan. Although Mackinder never used the term geopolitics, his theory on the global competition between leading states (the so-called Heartland theory) is a telling example of the early tenets of classical geopolitical reasoning (Kearns 2009). Mackinder's political map was constituted through spatial binaries such as sea power/land power and heartland, and cultural distinctions between Western individualist philosophy and Eastern (European) communal traditions (Parker 1982). These binaries constituted a grand theory of world affairs and highlighted the pursuit of supremacy on the part of the most powerful states. In such a view, geography in international affairs refers to a "big picture": timeless laws of nature, political geographical outcomes of technological developments, factors such as proximity or physical geographical landscapes, and the distribution of land and sea. This type of classical "geopolitical gaze," as Painter and Jeffrey (2009: 202) put it, seeks to understand forces emanating from both nature and technology, forces that shape international affairs and purportedly singular national interests. In a classical geopolitical perspective, these forces need to be examined and taken into account in successful strategic statecraft. This kind of geopolitics – which conceives of physical geographical elements, as well as the role of technological development with regard to these elements, as enabling or constraining factors of state strategy – is often dubbed geostrategy.

The classical geopolitical theories were of course products of their own time and reflect the competitive ambitions of European states in the early twentieth century (Ó Tuathail 1996; Parker 1998; Agnew 2003; Dittmer & Sharp 2014b), the anxieties over the balance of power in Europe, as well as the related need to enhance state security in the face of perceived external threats. For instance, Mackinder's attempt to develop the study of geography as an aid to statecraft was explicitly linked with these concerns (Parker 1982).

The two traditions mentioned indicate that the origins of classical geopolitical scholarship in fin-de-siècle Europe, as well as its later developments in the twentieth century, are entangled with changing modes of political action and engagement with the state and state formation, nationalism and nation-building, colonialism and imperialism, as well as with justifications for a particular interstate rivalry (Ó Tuathail 1996; Heffernan 2000; Agnew 2003). As a form of reasoning, classical geopolitics discloses some of the key political characteristics of the twentieth century: command of territory and natural resources were understood as pivotal dimensions of interstate rivalry and fundamental constituents of state status.

After World War II, German geopolitical reasoning (*Geopolitik*), which was based on environmental determinism and thus treated the state as an organic entity, was associated with Adolf Hitler's thinking and politics in particular. Given this historical legacy, geopolitics became a notorious signifier denoting control over space as states vie for power. In the postwar context, geopolitics was a problematic legacy for the discipline of geography. The end of the war marked the demise of geopolitics among academic geographers, to be replaced by a seemingly neutral political geography.

The new political geography that was developed during the Cold War by a handful of individuals understood the spatiality of the world political map as a set of geographical factors, forces, and resulting capacities of state actors that could be analyzed by a trained scholar. This form of geopolitical reasoning, which was possible to conflate with balance-of-power thinking, became prominent particularly in the realist tradition of IR scholarship. Saul Cohen's attempt to divide the world into geopolitical and geostrategic regions is illustrative of the new political geography (see Cohen 1963, 2009). During the Cold War, the legacy of classical geopolitics lived on also in the various branches of strategic studies. Colin Gray's (1988) *The Geopolitics of Super Power* is exemplary in this context, drawing from Mackinder's original conceptual apparatus. Some of the ideas of classical geopolitics had a presence in foreign and security policy practices during the Cold War.

Today, variants of classical geopolitics persist in the ways in which politicians, foreign and security policy experts, military strategists, journalists, and the general public make sense of international affairs (e.g., Megoran 2010). In academic circles, there has been work rehashing (e.g., Owens 1999) and defending various classical geopolitical and geostrategic perspectives (e.g., Wu 2018) to understand the contemporary world. More generally, the recent interest in classical geopolitical perspectives stems from world political developments such as the Belt and Road Initiative, through which China invests almost a thousand billion euros in transport infrastructures and energy production in Asia, Africa, Europe, and South America, and in so doing seeks to gain an economic and political foothold in several countries across the globe (for an analysis, see, e.g., Flint & Zhu 2019). Another important stimulus for the rise of classical geopolitical reasoning has been Russia's invasion of Ukraine in 2022 and what has become one of Europe's biggest land wars since World War II (for an analysis, see Murphy 2022; Dodds et al. 2023). For many representatives of the realist school of IR – such as John Mearsheimer (see Chotiner 2022; also Mearsheimer 2014), the crisis is not about Ukraine per se. Rather, it is about the reconstruction of the global geopolitical order and the related territorial struggle between powerful states – and the values these states represent (Moisio 2022; Ortmann 2023).

Critical Geopolitics: Foundations

Geopolitics experienced a revival in the late 1980s. Critical geopolitics, which challenged most of the tenets of classical geopolitics (Ó Tuathail 1998), was not the only approach that characterized this revival (Mamadouh 1998) but it was the most vocal. At the risk of oversimplifying things slightly, the following pages map out five of the founding characteristics and ideas of critical geopolitics.

First, the emergence of critical geopolitics was influenced by the development of world political developments and the related rise of leftism in Anglo-American academia in the 1970s and 1980s (Agnew 2013). Ó Tuathail writes in retrospect how the

> early critical geopolitics was very critical, and it came from a politicized place. I think, it was a legitimate reaction against the militaristic dangerous politics associated with Cold War geo-politics . . . The first time that Peter Taylor used the term "critical geopolitics" to me – it was in his response to reviews of the manuscript I submitted to *Political Geography* on the US policy to El Salvador – I didn't like the term at all. I protested: "I'm not a geopolitician! I don't want anything to do with geopolitics." I only adopted the term under protest. (Bachmann & Toal 2019: 151)

From the beginning, critical geopolitics was straightforward distancing from earlier modes of geopolitics and political geography that were too much premised on an instrumentalist problem-solving model of science and lacked critical theory (Ó Tuathail 1987: 196–197). The structuralist openings in political geography in the 1970s and early 1980s had already paved the way for critical geopolitics. Structuralism not only provided an example of how geopolitics could be redeployed in a new fashion (Hepple 1986), it also formed a mirror against which agency-centered poststructuralist scholarship could be cultivated. Critical geopolitics emerged in the 1980s and early 1990s as part of the cultural turn in the social sciences and human geography. It thus highlights agency and the constitutive role of language in geopolitical practice. Critical geopolitics conceptualizes geopolitics as culturally embedded spatial practices – both representational and material – of statecraft. Geopolitics is thus understood as a "set of socially constructed, rather than naturally given, practices and ideas through which the international political economy is realized geographically" (Agnew & Corbridge 1995: 4–5).

The origins of critical geopolitics can be traced back to the work of John Agnew, Simon Dalby, and Gearóid Ó Tuathail in particular. There are not only interesting postpositivist intertextualities but also theoretical tensions between the approaches of these scholars – particularly with regard to the central issue of the materiality/textuality of geopolitical discourse and agency. As Agnew (2013: 24) has himself noted in retrospect, this tension "came down to a question of privileging the representational versus emphasizing the socio-economic resources that made some representations more powerful than others." The early tensions in critical geopolitics thus stem from the difference between asking "how questions" and "why questions" in social scientific inquiry.

Second, early critical geopolitics builds on postmodern writings whose main goal is epistemological critique. It reconceptualizes geopolitics through drawing from debates on identity and difference, feminism, positionality, and postcolonialism. In such a view, critical geopolitics demonstrates that geography matters in world politics, but not quite in the classical geopolitical sense. Critical geopolitics is not about the purported natural forces or constraining/enabling geographical factors in politics. Rather, critical geopolitics understands geography in international politics as socially constructed and bound to larger questions regarding identity/difference, for instance. In order to explicate this form of geography, critical geopolitics extends sociospatial theorizing into the study of international affairs. It inquires into the spatial theories and assumptions that guide action in international affairs. One of the founding ideas of critical geopolitics is thus to problematize the seemingly neutral and objective scientific gaze of classical or "conventional" realist geopolitics through deconstruction. It disturbs the putatively scientific "view from nowhere" by arguing that classical geopolitics as a science that emerged in the late nineteenth century is a significant constituent of the modernist perception of international space made visible by geopolitical experts, as "detached viewing subjects" surveying a worldwide stage from above (Ó Tuathail 1996: 29). In so doing, critical geopolitics argues that the spaces of world politics are "produced" in different mapping exercises such as all sorts of strategic plans and related imaginations (Kuus 2010: 683).

Critical geopolitics' emphasis on international affairs is a notable phenomenon. It becomes nonetheless understandable given that the emergence of critical geopolitics cannot be detached from so-called dissident IR. The key dissidents introduced poststructuralism to IR in the 1980s and challenged the foundational ideas of neorealism in particular. In so doing they developed a critique of modernism in

IR theory. They sought to unsettle state-related practices through which an "uncertain" world was constantly territorialized (ordered) by the "sovereign man" employing putatively rational and scientific methods (see Ashley & Walker 1990). For them, statecraft was a specific action "inscribing problems and dangers that can be taken to be exterior to sovereign man and whose exteriority serves to enframe the 'domestic population' in which the state can be recognized as a center and can secure its claims to legitimacy" (Ashley 1989: 302). Critical geopolitics draws on this idea of statecraft. Ó Tuathail and Dalby (1998: 3) propose that "particular cultural mythologies of the state" must be taken into account in critical geopolitical study and that geopolitics is not a distinct school of statecraft, but should rather be understood as the spatial practices (material, representational) of statecraft itself.

The "dissident effect" resulted in a growing awareness of the intertwined nature of foreign/security policy, power, and space. What followed was a set of critical geopolitical investigations of the ways in which the maps of global politics are carved up and simultaneously made public by different kinds of intellectuals of statecraft such as political leaders, pundits, advocates, and advisors (Dalby 1990a; Ó Tuathail 1992). Here geopolitics is not understood as a somewhat contained action by scholars who explicitly speak the language of classical geopolitics. It instead elaborates how international politics is constantly "spatialized" by powerful political actors who depict it as a "world" characterized by specific places, people, and events (Ó Tuathail & Agnew 1992: 192). The representation put together by the powerful influences the ways the world is understood, and these understandings, in turn, have effects on subsequent policy practices (Sharp 2008: 191). The emphasis is thus turned more generally to the politics of spatial specification of politics (Dalby 1991).

Third, critical geopolitics conceptualizes geopolitics as a discursive practice. It is a broad social and cultural phenomenon, so that there is always a particular relationship between geopolitical discourse/text/representation and practice/action (Agnew 2003, 2013: 28). All analyses of international politics make spatial assumptions and are necessarily predicated on dominant geographical assumptions. Critical geopolitics brings these up and puts them under scrutiny by examining discursive operations, the associated "geopolitical truths," as well as the social mechanisms through which geopolitical truth claims and larger geopolitical regimes of truth are produced. Here, critical geopolitics draws on Foucault (1980), who conceptualized discourse as referring to power-driven knowledge, certain accepted truths that are historically constructed and together form larger regimes of truth. The idea is to trace the function of particular geopolitical expertise and related "geopolicy-relevant" knowledge in international affairs.

What follows from the poststructural and deconstructive underpinnings of critical geopolitics is that it does not treat geography as an innocent setting in which international politics is conducted or a structural element that sets the limits for rational policy-making. Critical geopolitics argues that geography and politics are inescapably co-constituted, that space is power, that (global) politics is fundamentally geographical, and that place influences the operation of power. In such a conceptualization, geography must also be understood as earth-writing, active production of "global political space" by institutionally bounded intellectuals (Ó Tuathail 1996: 185; see also Dodds 1994).

Fourth, geopolitics can be understood as a set of traveling theories and imaginations that have been introduced and negotiated in different intellectual, spatial, and temporal contexts (Dodds & Atkinson 2002; An 2020; Suslov 2020) and with different kinds of consequences. Inquiries into different geopolitical traditions/cultures and thinkers typically result from the early critical geopolitical problematizations of classical geopolitical knowledge production. But critical geopolitics has succinctly demonstrated how, as interpretative cultural practices, widely traveled geopolitical theories (such as the Huntingtonian "clash of civilizations") and powerful geopolitical markers (such as East and West) are constituted and played out in an interplay between the universalizing "geopolitical centers" where theories and markers are invented and the margins where they are consumed and appropriated for particular purposes by local intellectuals of statecraft (Kuus 2007).

Fifth, critical geopolitics scrutinizes national identity-building and the associated spatial practices of politico-spatial inclusion and exclusion (see, e.g., Paasi 2016). It points out that political subjectivities, identities, and objects are formed in geopolitical action and within discourses. Following dissident IR, national identities are conceptualized as constituted through powerful narratives of national security, a central constituent of statecraft and sovereign man (see Ashley 1989; Ashley & Walker 1990; Dalby 1990a; Campbell 1992). By following this line of thinking, Dalby (1990b) argues that geopolitics is an inherently ideological process of constructing borders to demarcate the domestic space (here) from the threatening outside of the state, as well as to exclude the threatening "other" (there) and the "otherness" it represents spatially as well as culturally (see, e.g., Jones 2021).

Critical Geopolitics: Critiques

During the past three decades critical geopolitics has become a vibrant scholarly field. As a sign of this vibrancy, some of the foundations of critical geopolitics have been critically discussed. To illustrate, the pioneering text by Ó Tuathail (1996) has been criticized for itself representing a perspective of a detached subject (see Smith 2000), for neglecting the ontological dimension of geopolitics (Agnew 2000), and for being all about textual analysis (see Müller 2008; Müller & Reuber 2008; for a response, see Bachmann & Toal 2019). Conceptual inconsistencies and obscurities in the important theoretical issues of discourse and agency/actor have also been taken up by those researchers critiquing an epistemology-centered critical geopolitics (e.g., Albert, Reuber, & Wolkersdorfer 2014).

Dodds (2001) argued more than two decades ago that critical geopolitics is methodologically disappointing. In particular, he pointed to the lack of ethnographic approaches to how geopolicies emerge in the interaction between institutions, individuals, and textual as well as bodily practices. A number of scholars have taken this critique seriously (e.g., Megoran 2006). Merje Kuus (2014) has developed this perspective in the context of the external relations of the European Union (EU). Her critical geopolitics of "elite occupations" (see Kuus 2021) examines how particular policies get shaped within the EU bureaucracy and how the constitution of these policies involves a particular geopolitical expertise and imaginations that are negotiated among actors who deal with the EU's external affairs (Kuus 2014).

Feminist critical geopolitics emerged as an important critique of early critical geopolitics. Joanne Sharp (2000) suggested that the early critical geopolitics remasculinizes geopolitics and neglects the important issue of subjectification inherent in geopolitical practices (see also Dowler & Sharp 2001). Accordingly, critical geopolitics is overly focused on text, representations, and all kinds of elites (often men) – and pays only scant attention to the embodied practices and materiality of geopolitical discourses. The deconstructive research strategy of the early critical geopolitics was also critiqued as politically powerless in transforming the world. In this sense, feminist critical geopolitics has been premised on critical social theory more than many other "diluted" perspectives in critical geopolitics. Hyndman (2001) argues explicitly that the deconstructive impulses of the early critical geopolitics are insufficient to generate change for building alternative futures.

Since its emergence, feminist critical geopolitics has become a vibrant field that consists of different yet overlapping perspectives. It reconceptualizes geopolitics through drawing from debates on identity and difference, embodiment, positionality, power/knowledge, social reproduction, subjectivity, materiality, postcolonialism, intersectionality, and more-than-human dimensions of the political. In a feminist perspective, space is geopolitical because it is subjective, affectual, experienced, embodied, gendered, racialized, and intimate. To illustrate, scholars have explicated the nexus between feminist geopolitics, intimacies, and domestic spaces (see, e.g., Jackman & Brickell 2022; Sharp 2023).

Feminist critical geopoliticians have attached the concept of geopolitics to people who are frequently absent in state-centered geopolitical analysis. Sharp (2011) argues that quite often critical geopolitical analyses have been rooted in Western or Eurocentric positions and that these examinations reproduce the dominant ways in which the world is usually scripted. She suggests that "subaltern geopolitics" should be developed to understand how political maps are written from the perspective of the dominated (e.g., Woon 2017). This insight has resulted in a number of studies that have decentered the hegemony of Anglo-American geopolitics by grasping the variety and diversity of geopolitical experiences and ways of reasoning in various geographical contexts (see, e.g., Craggs 2018; Feretti 2021).

The conventional notion of the state as the protector of people and their human rights has been challenged and taken up in feminist critical geopolitics (e.g., Martin 2012). As such, the patriarchal territorial state is not a solution but rather a serious problem: it is a territorial social organization that essentially maintains social, spatial, gendered, and racial injustices and related systems of oppression and control. State power, which seeks to control bodies from the perspective of nation, ranges from health policies to conversations about women's clothing (Fluri & Lehr 2017). As such, feminist critical geopolitics has sought to disturb the state-centered view on geopolitics – and its key understandings of war, security, peace, and the nation-state – by bringing the mundane, intimate, affectual, and corporeal aspects of geopolitics into investigation (e.g., Secor 2001; Dixon 2015).

Ongoing Research Agendas in Critical Geopolitics

Today, critical geopolitics is a broad field. Some central themes and features can be singled out, however. The war–environment–security nexus is one of the classical topics of geopolitics and it has also been scrutinized afrom critical perspectives (e.g., Le Billon 2008). This work challenges some of the basic forms of security thinking, which are based on simplistic assumptions about the relationship between environmental change and conflicts (e.g., Chaturvedi & Doyle 2016). Dalby's (2013) plea to radically rethink the basic geopolitical premises in security thinking in light of the insights from Earth system science is exemplary here. His concern is that climate security is too often understood as a matter of environmental change causing political difficulties, while this should be turned upside down. Another example of critical geopolitics of the environment is provided by Natalie Koch (2022), who examines the ways in which Arizona's Sonoran Desert and the Arabian Desert have been linked, and how this connection provides insights into Saudi Arabia's historical encounters with the United States.

The territorial and relational spaces of security and insecurity and their implications for issues such as human rights, public space, and state sovereignty (Coleman 2005; Ingram & Dodds 2009; Elden 2013; Mountz 2013; on sovereignty see also Steinberg, Nyman, & Caraccioli 2012) have been examined conceptually as well as empirically. Scholars of critical geopolitics have also examined the geographies of peace (see Megoran, McConnell, & Williams 2014). A critical examination of diplomacy and self-determination has been another addition to critical geopolitics. Fiona McConnell (2017) has interrogated how political legitimacy is articulated and acted on by marginalized communities, and scrutinized the uneven geographies regarding who has the right to "diplomacy" in the first place (Constantinou & McConnell 2023).

Technology remains an important topic in critical geopolitics. The critical geopolitical work on war touches on new technologies of warfare and the ways in which these technologies open up new social/cultural spaces within the military, form subjectivities and identities, as well as constitute relational battlefield geographies (e.g., Jackman & Brickell 2022). Ian Shaw's (2013) work on the dronified US security strategy and the exercise of biopolitical power in the widening "battlespace" exemplifies the ongoing interest in what Stephen Graham (2004) originally called vertical geopolitics (see also Elden 2013). Here, critical geopolitics inquires into technologies, knowledge, practices, and strategies "arranged around the deployment of drones for targeted killings" (Shaw 2013: 540). Shaw's (2017) research on the constitution of the global power of the United States through robot technologies brought up another important theme that will very likely become a hot topic in critical geopolitics: the role and functioning of rapidly developing artificial intelligence in geopolitical practices related to warfare and security. Gerard Toal (2017: 13) has also paid increasing attention to what he calls the "geopolitical condition": the emergent technological assemblages such as military, transportation, and communication infrastructures, and how these transform the way in which geopolitics is experienced as well as practiced in places.

The elaboration of the state and national identity-building has continued in critical geopolitics (Koch & Paasi 2016; Medby 2018). Studies emphasizing contextuality, relationality, and contingency have sought to denaturalize the nation and state as ontological categories of analysis. Accordingly, both the state and the nation are studied primarily as "a particular way of being and relating" (Ince & Barrera de la Torre 2016: 13). One of the ideas here is to investigate how national identities are narrated and performed, ranging from the upper echelons of state authority to mundane events and related "nationalizing" symbols and rituals. These mundane geographies of stateness have been explored in various studies on everyday state formation (e.g., Ballvé 2012; Neves Alves 2019; Lizotte & Nguyen 2020) and everyday nationalism (e.g., Benwell et al. 2019), as well as on the affective implications and constituents of state and nation (e.g., Woodward 2014).

Popular geopolitics on the circulation of geopolitical narratives and the constitution of geopolitical identities in popular culture has remained central in critical geopolitics. Scholars of popular geopolitics argue that instead of paying attention only to the formal and practical geopolitical reasoning of the intellectuals of statecraft (for these terms, see Ó Tuathail & Dalby 1998), attention must be devoted to visual and nonvisual products such as newspapers and magazines (Sharp 2000), radio shows (Dodds & Pinkerton 2009), films (Carter & Dodds 2011), art (Ingram 2019), comic strips (Dittmer 2013), and cartoons (Dodds 1994). According to popular geopolitics, these are pivotal in enacting and circulating particular popular geopolitics among different "audiences" (Dittmer & Bos 2019; see also Holland 2012; Fluri 2019; dell'Agnese 2021). Focusing on audiences and audience reception is important, for it highlights the dual role of people as both consumers as well as producers of geopolitical discourses (Sharp 2008; Ridanpää 2014; Pierzynska 2023). The work on popular geopolitics has also scrutinized the geopolitical role of social media (e.g., Henry 2021) and the potentially erroneous content that circulates within that domain, and in so doing shapes the ways in which people make sense of geopolitical developments (Suslov 2014; Laketa 2019). This body of research also focuses on the often-hidden emotional spectrum of popular geopolitical scripting of the world. This latter aspect has been highlighted in recent works on the ways in which the geopolitical events and associated military practices are domesticated and commodified in video games, play, and toys (e.g., Woodyer & Carter 2020; Bos 2021).

The research listed reveals that critical scholars have developed new theoretical spaces for geopolitical analysis. In addition, they have examined a more-than-human geopolitics (Raento 2016; Shaw 2017), the affective aspects of geopolitical practices and discourses (Pain 2009), as well as suggested that concepts such as event, assemblage, and complexity would provide new openings for critical geopolitical inquiry (Dittmer 2014; Ingram 2019). By utilizing the notions of assemblage and affect, Jason Dittmer (2017) provides an analysis of foreign policy by tracing the ways in which interactions between people and material "things" shape the decisions and actions of diplomats and policy-makers. This new materialist perspective leads Dittmer to focus on how the geopolitical world operates within the state through the circulation of humans and nonhuman objects.

The critical geopolitical economy approach (Agnew & Corbridge 1995) is alive too. John Agnew has been a central figure in developing this perspective over the past few decades. His Gramscian examination of the spread of the marketplace society as the contemporary form of hegemony – hegemony without a hegemonic state (Agnew 2005) – as well as his recent analysis of "hidden geopolitics" are insightful examinations of the ways in which the contemporary world "is organized on a transversal rather than on a singularly territorial basis with all manner of transnational and global networks and flows drawing distant locations into topological proximity" (Agnew 2023: 165). One of the strengths of such a critical geopolitical economy is that it highlights the inherently geographical nature of (global) politics and underlines the complex geographies of power that include states, transnational ideological movements, supranational polities, international organizations, multinational companies, and regulatory agencies – to mention but a few – in the production of such geography. In sum, one can understand the critical geopolitical economy as a critique of a "negative," state-centered, and narrow conceptualization of geopolitics as if it were all about territorial exclusion, malign nationalism, and anti-liberalism (Moisio 2024).

The critical geopolitical economy approach can also be understood as a critique of mainstream geoeconomics approaches (Sparke 2007; Domosh 2013; Moisio 2018a). In such a view, geopolitics refers primarily to military means and other uses of coercive power, whereas geoeconomics refers to the strategic use of economic means and market mechanisms (Moisio 2019). As a counter-reaction to mainstream scholarship, scholars in critical geopolitical economy have examined the interrelation and entanglement of geoeconomic and geopolitical processes (Bessner 2023; Mallin & Sidaway 2023; Sparke 2024). Morrissey (2017) examines the role of US Central Command (CENTCOM) in the discursive framing of the Gulf region since the 1980s as a geostrategic and geoeconomic geographical space that needs guarding and monitoring by the United States in order to safeguard not only regional business opportunities but also larger structures of the global economy (see also Cowen & Smith 2009). Moreover, the geopolitical expansion of global capitalism has been analyzed by Koch (2014), who examines "liberal" elite university projects in two authoritarian developmental states – Kazakhstan and Saudi Arabia – and how these projects have become part of authoritarian political formations. Moisio (2018b) argues that the global knowledge economy is closely connected to interstate rivalries over territories of technology and global value chains, and that the state has played an important role in this process. This perspective has taken critical geopolitical economy closer to critical urban studies (see Moisio & Rossi 2024).

Conclusion

Critical geopolitics is still a vibrant field that attracts students both within and beyond Anglo-American academia. Critical geopolitics has made a contribution to human geography, and users of the concept of geopolitics no longer have to worry about someone using the subject's past against them within academia (see Hepple 1986). It is easy to agree with Agnew's statement that "[t]here will be a need for critical geopolitics of some sort as long as the established nostrums of how the world is spatialized and divided up for political purposes remain in effect" (Agnew 2013: 30). Whether critical geopolitical thought can respond to contemporary problems and the new possibilities for liberation that arise from changing historical circumstances is another question.

Critical geopolitics brings together scholars who employ different kinds of theoretical and philosophical perspectives and research materials to study the politics and governance of space within and beyond the state. Quite often critical geopolitics is today mentioned only briefly in papers in the context of theoretical framings, and it is not always clear what the word "critical" denotes for scholars. For instance, the relationship of critical geopolitics to classical ideas of critical social science – such as human emancipation, alienation, and reification – has remained loose (Bachmann & Moisio 2020). Also, the contribution of critical geopolitics within academia seems more significant than its impact on the world of policy: scholars of critical geopolitics have been more willing to deconstruct political processes than to imagine and propose alternative strategies (see Dodds 2001: 475; Agnew 2012; however, see Dalby 1990a, 1991).

References

Agnew, J.A. (2000). Global political geography beyond geopolitics. *International Studies Review* 2: 91–99.

Agnew, J.A. (2003). *Geopolitics: Revisioning World Politics*, 2nd edn. London: Routledge.

Agnew, J.A. (2005). *Hegemony: The New Shape of Global Power*. Philadelphia, PA: Temple University Press.

Agnew, J.A. (2012). Of canons and fanons. *Dialogues in Human Geography 2*: 321–323.

Agnew, J.A. (2013). The origins of critical geopolitics. In Dodds, K., Kuus, M., & Sharp, J. (eds), *The Ashgate Research Companion to Critical Geopolitics*. Farnham: Ashgate, 19–32.

Agnew, J.A. (2023). *Hidden Geopolitics*. London: Rowman and Littlefield.

Agnew, J.A., & Corbridge, S. (1995). *Mastering Space*. London: Routledge.

Albert, M., Reuber, P., & Wolkersdorfer, G. (2014). Critical geopolitics. In Schieder, S., & Spindler, M. (eds), *Theories of International Relations*. London: Routledge, 321–336.

An, N. (2020). *Confucian Geopolitics: Chinese Geopolitical Imaginations of the US War on Terror*. Singapore: Springer.

Ashley, R.K. (1989). Living on border lines: Man, postructuralism and war. In Der Derian, J., & Shapiro, M. (eds), *International/Intertextual Relations*. New York: Lexington Books, 259–322.

Ashley, R.K., & Walker, R.B.J. (1990). Speaking the language of exile. Dissident thought in international studies. *International Studies Quarterly 34*: 259–268.

Bachmann, V., & Sami Moisio, S. (2020). Towards a constructive critical geopolitics: Inspirations from the Frankfurt School of critical theory. *Environment and Planning C: Politics and Spaces 38*: 251–268.

Bachmann, V., & Toal, G. (2019). Geopolitics – thick and complex: A conversation with Gerard Toal. *Erdkunde 73*: 143–155.

Ballvé, T. (2012). Everyday state formation: Territory, decentralization, and the narco land-grab in Colombia. *Environment and Planning D: Society and Space 30*: 603–622.

Benwell, M., Núñez, A., & Amigo, C. (2019). Flagging the nations: Citizens' active engagements with everyday nationalism in Patagonia, Chile. *Area 51*: 719–727.

Bessner, D. (2023). Rethinking the relationship between a US-based global security architecture and global capitalism. *Transactions of the Institute of British Geographers 49*(1): e12657. doi:10.1111/tran.12657

Bos, D. (2021). Popular geopolitics "beyond the screen": Bringing modern warfare to the city. *Environment and Planning C 39*: 94–113.

Campbell, D. (1992). *Writing Security*. Minneapolis, MN: University of Minnesota Press.

Carter, S., & Dodds, K. (2011). Hollywood and the "war on terror": Genre-geopolitics and "Jacksonianism" in *The Kingdom*. *Environment and Planning D 29*: 98–113.

Chaturvedi, S., & Doyle, T. (2016). *Climate Terror: A Critical Geopolitics of Climate Change*. Cham: Springer.

Chotiner, I. (2022). Why John Mearsheimer blames the US for the crisis in Ukraine. *The New Yorker*, March 1. https://www.newyorker.com/news/q-and-a/why-john-mearsheimer-blames-the-us-for-the-crisis-in-ukraine (accessed November 25, 2024).

Cohen, S.B. (1963). *Geography and Politics in a World Divided*. New York: Random House.

Cohen, S.B. (2009). *Geopolitics: The Geography of International Relations*, 2nd edn. Lanham, MD: Rowman and Littlefield.

Coleman, M. (2005). US statecraft and the US-Mexico border as security-economy nexus. *Political Geography 24*: 185–209.

Constantinou, C., & McConnell, F. (2023). On the right to diplomacy: Historicizing and theorizing delegation and exclusion at the United Nations. *International Theory 15*: 53–78.

Cowen, D., & Smith, N. (2009). After geopolitics? From the geopolitical social to geoeconomics. *Antipode 41*: 22–48.

Craggs, R. (2018). Subaltern geopolitics and the post-colonial commonwealth, 1965–1990. *Political Geography 65*: 46–56.

Dalby, S. (1990a). *Creating the Second Cold War*. London: Pinter.

Dalby, S. (1990b). American security discourse: The persistence of geopolitics. *Political Geography Quarterly* 9: 171–188.

Dalby, S. (1991). Critical geopolitics: Discourse, difference, and dissent. *Environment and Planning D* 9: 261–283.

Dalby, S. (2013). Rethinking geopolitics: Climate security in the Anthropocene. *Global Policy 5*: 1–9.

dell'Agnese, E. (2021). *Ecocritical Geopolitics: Popular Culture and Environmental Discourse*. London: Routledge.

Dittmer, J. (2013). *Captain America and the Nationalist Superhero*. Philadelphia, PA: Temple University Press.

Dittmer, J. (2014). Geopolitical assemblages and complexity. *Progress in Human Geography 38*: 385–401.

Dittmer, J. (2017). *Diplomatic Material. Affect, Assemblage, and Foreign Policy*. Durham, NC: Duke University Press.

Dittmer, J., & Bos, D. (2019). *Popular Culture, Geopolitics, and Identity*, 2nd edn. Lanham, MD: Rowman and Littlefield.

Dittmer, J., & Sharp, J. (eds) (2014a). *Geopolitics: An Introductory Reader*. Abingdon: Routledge.

Dittmer, J., & Sharp, J. (2014b). General introduction. In Dittmer, J., & Sharp, J. (eds), *Geopolitics: An Introductory Reader*. Abingdon: Routledge, 1–10.

Dixon, D. (2015). *Feminist Geopolitics: Material States*. London: Routledge.

Dodds, K. (1994). Geopolitics and foreign policy: Recent developments in Anglo-American political geography and international relations. *Progress in Human Geography 18*: 186–208.

Dodds, K. (2001). Political geography III: Critical geopolitics after ten years. *Progress in Human Geography 25*: 469–484.

Dodds, K., & Atkinson, D. (2002). Introduction: A century of geopolitical thought. In Dodds, K., & Atkinson, D. (eds), *Geopolitical Traditions: A Century of Geopolitical Thought*. London: Routledge, 1–24.

Dodds, K., & Pinkerton, A. (2009). Radio geopolitics: Broadcasting, listening and the struggle for acoustic spaces. *Progress in Human Geography 33*: 10–27.

Dodds, K., Taylor, Z., Akbari, A., et al. (2023). The Russian invasion of Ukraine: Implications for politics, territory and governance. *Territory, Politics, Governance 11*: 1519–1536.

Domosh, M. (2013). Geoeconomic imaginations and economic geography in the early twentieth century. *Annals of the Association of American Geographers 103*(4): 944–966. doi:10.1080/00045608.2011.653740

Dowler, L., & Sharp, J. (2001). A feminist geopolitics. *Space and Polity 5*: 165–176.

Elden, S. (2013). Secure the volume: Vertical geopolitics and the depth of power. *Political Geography 34*: 35–51.

Feretti, F. (2021). A coffin for malthusianism: Josué De Castro's subaltern geopolitics. *Geopolitics* 26: 589–614.

Flint, C. (2022). *Introduction to Geopolitics*, 4th edn. London: Routledge.

Flint, C., & Zhu, C. (2019). The geopolitics of connectivity, cooperation, and hegemonic competition: The Belt and Road Initiative. *Geoforum* 99: 95–101.

Fluri, J.L. (2019). What's so funny in Afghanistan? Jocular geopolitics and the everyday use of humor in protracted precarity. *Political Geography* 68: 125–130.

Fluri, J., & Lehr, R. (2017). *The Carpetbaggers of Kabul and Other American-Afghan Entanglements: Intimate Development, Geopolitics, and the Currency of Gender and Grief*. Athens, GA: University of Georgia Press.

Foucault, M. (1980). *Power/Knowledge. Selected Interviews and Other Writings 1972–1977 by Michel Foucault*. Brighton: Harvester Press.

Graham, S. (2004). Vertical geopolitics: Baghdad and after. *Antipode* 36: 12–23.

Gray, C. (1988). *The Geopolitics of Super Power*. Lexington, KY: University of Kentucky Press.

Heffernan, M. (2000). Fin de siècle, fin du monde? On the origins of European geopolitics. In Dodds, K., & Atkinson, D. (eds), *Geopolitical Traditions: A Century of Geopolitical Thought*. London: Routledge, 27–51.

Henry, J. (2021). The geopolitics of travel blogging. *Geopolitics* 26: 817–837.

Hepple, L.W. (1986). The revival of geopolitics. *Political Geography Quarterly* 5: 521–536.

Holland, E.C. (2012). "To think and imagine and see differently": Popular geopolitics, graphic narrative, and Joe Sacco's "Chechen War, Chechen Women". *Geopolitics* 17: 105–129.

Hyndman, J. (2001). Towards a feminist geopolitics. *The Canadian Geographer / Le Géographe canadien* 45: 210–222.

Ince, A., & de la Torre, G.B. (2016). For post-statist geographies. *Political Geography* 55: 10–19.

Ingram, A. (2019). *Geopolitics and the Event: Rethinking Britain's Iraq War through Art*. Oxford: John Wiley & Sons.

Ingram, A., & Dodds, K. (eds) (2009). *Spaces of Security and Insecurity: Geographies of the War on Terror*. Farnham: Ashgate.

Jackman, A., & Brickell, K. (2022). Everyday droning": Towards a feminist geopolitics of the drone-home. *Progress in Human Geography* 46: 156–178.

Jones, R. (2021). *White Borders: The History of Race and Immigration in the United States from Chinese Exclusion to the Border Wall*. Boston, MA: Beacon Press.

Kearns, G. (2009). *Geopolitics and Empire: The Legacy of Halford Mackinder*. Oxford: Oxford University Press.

Koch, N. (2014). The shifting geopolitics of higher education: Inter/nationalizing elite universities in Kazakhstan, Saudi Arabia, and beyond. *Geoforum* 56: 46–54.

Koch, N. (2022). *Arid Empire: The Entangled Fates of Arizona and Arabia*. New York: Verso.

Koch, N., & Paasi, A. (2016). Banal nationalism 20 years on: Re-thinking, re-formulating and re-contextualizing the concept. *Political Geography* 54: 1–16.

Kuus, M. (2021). Professions and their expertise: Charting the spaces of "elite" occupations. *Progress in Human Geography* 45: 1339–1355.

Kuus, M. (2007). *Geopolitics Re-framed: Security and Identity in Europe's Eastern Enlargement*. New York: Palgrave Macmillan.

Kuus, M. (2010). Critical geopolitics. In Denemark, R. (ed.), *The International Studies Encyclopedia, Volume II*. Oxford: Blackwell, 683–701.

Kuus, M. (2014). *Geopolitics and Expertise: Knowledge and Authority in European Diplomacy*. Oxford: Wiley-Blackwell.

Laketa, S. (2019). "Even if it didn't happen, it's true": The fantasy of geopolitics in the "post-truth" era. *Emotion, Space and Society 31*: 155–161.

Le Billon, P. (2008). Diamond wars? Conflict diamonds and geographies of resource wars. *Annals of the Association of American Geographers 98*: 345–372.

Lizotte, C., & Nguyen, N. (2020). Schooling from the classroom to the state: Understanding schools as geopolitical sites. *Environment and Planning C: Politics and Spaces 38*: 920–937.

Mallin, F., & Sidaway, J.D. (2023). Critical geoeconomics: A genealogy of writing politics, economy and space. *Transactions of the Institute of British Geographers 49*(1): e12600. doi:10.1111/tran.12600

Mamadouh, V. (1998). Geopolitics in the nineties: One flag, many meanings. *GeoJournal 46*: 237–256.

Martin, L. (2012). "Catch and remove": Detention, deterrence, and discipline in US noncitizen family detention practice. *Geopolitics 17*: 312–334.

McConnell, F. (2017). Liminal geopolitics: The subjectivity and spatiality of diplomacy at the margins. *Transactions of the Institute of British Geographers 42*: 139–152. doi:10.1111/tran.12156

Mearsheimer, J. (2014). Why the Ukraine crisis is the West's fault. The liberal delusions that provoked Putin. *Foreign Affairs 93*(5): 77–89.

Medby, I. (2018). Articulating state identity: "Peopling" the arctic state. *Political Geography 62*: 116–125.

Megoran, N. (2006). For ethnography in political geography: Experiencing and re-imagining Ferghana Valley boundary closures. *Political Geography 25*: 622–640.

Megoran, N. (2010). Neoclassical geopolitics. *Political Geography 29*: 187–189.

Megoran, N., McConnell, F., & Williams, P. (eds) (2014). *The Geographies of Peace*. London: I.B. Tauris.

Moisio, S. (2018a). Towards geopolitical analysis of geoeconomic processes. *Geopolitics 23*: 22–29.

Moisio, S. (2018b). *Geopolitics of the Knowledge-Based Economy*. London: Routledge.

Moisio, S. (2019). Re-thinking geoeconomics: Towards a political geography of economic geographies. *Geography Compass 13*: e12466.

Moisio, S. (2022). Geopolitics of explaining Russia's invasion of Ukraine and the challenge of small states. *Political Geography 97*: 102683.

Moisio, S. (2024). Towards critical geoeconomics? *Transactions of the Institute of British Geographers 49*: e12646.

Moisio, S., Koch, N., Jonas, A.E.G., Lizotte, C., & Luukkonen, J. (2020). *Handbook on the Changing Geographies of the State: New Spaces of Geopolitics*. Cheltenham: Edward Elgar.

Moisio, S., & Rossi, U. (2024). *The Urban Field: Capital and Governmentality in the Age of Techno-monopoly*. Newcastle upon Tyne: Agenda.

Morrissey, J. (2017). *The Long War, CENTCOM, Grand Strategy, and Global Security*. Athens, GA: University of Georgia Press.

Mountz, A. (2013). Political geography I: Reconfiguring geographies of sovereignty. *Progress in Human Geography 37*: 829–841.

Müller, M. (2008). Reconsidering the concept of discourse for the field of critical geopolitics: Towards discourse as language and practice. *Political Geography 27*: 322–338.

Müller, M., & Reuber, P. (2008). Empirical verve, conceptual doubts: Looking from the outside in at critical geopolitics. *Geopolitics 13*: 458–472.

Murphy, A.B. (2022). The modernist territorial order and the Ukraine crisis. *Political Geography 97*: 102667.

Murphy, A.B., Bassin, M., Newman, D., Reuber, P., & Agnew, J. (2004). Is there a politics to geopolitics? *Progress in Human Geography 28*: 619–640.

Neves Alves, S. (2019). Everyday states and water infrastructure: Insights from a small secondary city in Africa, Bafatá in Guinea-Bissau. *Environment and Planning C: Politics and Spaces 39*: 247–264.

Ó Tuathail, G. (1987). Beyond empiricist political geography: A comment on van der Wusten and O'Loughlin. *The Professional Geographer 39*: 196–197.

Ó Tuathail, G. (1992). The Bush administration and the "end" of the cold war: A critical geopolitics of US foreign policy in 1989. *Geoforum 23*: 437–452.

Ó Tuathail, G. (1996). *Critical Geopolitics*. London: Routledge.

Ó Tuathail, G. (1998). Postmodern geopolitics? The modern geopolitical imagination and beyond. In Tuathail, G.Ó., & Dalby, S. (eds), *Rethinking Geopolitics*. London: Routledge, 16–38.

Ó Tuathail, G., & Agnew, J.A. (1992). Geopolitics and discourse. Practical geopolitical reasoning in American foreign policy. *Political Geography 11*: 190–204.

Ó Tuathail, G., & Dalby, S. (1998). Rethinking geopolitics: Towards a critical geopolitics. In Tuathail, G.Ó., & Dalby, S. (eds), *Rethinking Geopolitics*. London: Routledge, 1–15.

Ortmann, S. (2023). Russian spatial imaginaries and the invasion of Ukraine: Geopolitics and nationalist fantasies. *Political Geography 101*: 102784.

Owens, M.T. (1999). In defence of classical geopolitics. *Orbis 59*: 463–478.

Paasi, A. (2016). Dancing on the graves: Independence, hot/banal nationalism and the mobilization of memory. *Political Geography 54*: 21–31.

Pain, R. (2009). Globalized fear? Towards an emotional geopolitics. *Progress in Human Geography 33*: 466–486.

Painter, J., & Jeffrey, A. (2009). *Political Geography: An Introduction to Space and Power*. Thousand Oaks, CA: Sage.

Parker, G. (1998). *Geopolitics: Past, Present, Future*. London: Pinter.

Parker, W.H. (1982). *Mackinder: Geography as an Aid to Statecraft*. Oxford: Clarendon Press.

Pierzynska, J. (2023). "It is Krajina all over again": Geopolitics of spite and geographical analogy in a Serbian newspaper audience. *Territory, Politics, Governance 11*: 1595–1614.

Raento, P. (2016). A geopolitics of the horse in Finland. *Geopolitics 21*: 945–968.

Ridanpää, J. (2014). "Humour is serious" as a geopolitical speech act: IMDb film reviews of Sacha Baron Cohen's *The Dictator*. *Geopolitics 19*: 140–160.

Secor, A. (2001). Towards a feminist counter-geopolitics: Gender, space and Islamist politics in Istanbul. *Space and Polity 5*: 199–219.

Sharp, J. (2000). *Condensing the Cold War: Reader's Digest and American Identity, 1922–1994*. Minneapolis, MN: University of Minnesota Press.

Sharp, J. (2008). Critical geopolitics (1996): Gearóid Ó Tuathail. In Hubbard, P., Kitchin, R., & Valentine, G. (eds), *Key Texts in Human Geography*. Thousand Oaks, CA: Sage, 189–196.

Sharp, J. (2011). A subaltern critical geopolitics of the war of terror: Postcolonial security in Tanzania. *Geoforum 42*: 297–305.

Sharp, J. (2023). Feminist geopolitics and the global-intimacies of pandemic times. *Gender, Place and Culture 30*: 1653–1670.

Shaw, I.G. (2017). Robot wars: US empire and geopolitics in the robotic age. *Security Dialogue 48*: 451–470.

Shaw, I.G.R. (2013). Predator empire: The geopolitics of US drone warfare. *Geopolitics 18*: 536–559.

Smith, N. (2000). Is a critical geopolitics possible? Foucault, class and the vision thing. *Political Geography 19*: 365–371.

Smith, S. (2020). *Political Geography. A Critical Introduction.* Oxford: Wiley-Blackwell.

Sparke, M. (2024). *Defining geoeconomics amid shifts in global hegemony: Critical geographies of new international conjunctures.* Environment and Planning A: *Economy and Space.* doi:10.1177/0308518X241265293

Sparke, M. (2007). Geopolitical fears, geoeconomic hopes, and the responsibilities of geography. *Annals of the Association of American Geographers 97*: 338–349.

Steinberg, P., Nyman, E., & Caraccioli, M.J. (2012). Atlas swam: Freedom, capital and floating sovereignties in the seasteading vision. *Antipode 44*: 1532–1550.

Suslov, M. (2014). "Crimea is ours!" Russian popular geopolitics in the new media age. *Eurasian Geography and Economics 55*: 588–609.

Suslov, M. (2020). *Geopolitical Imagination: Ideology and Utopia in Post-Soviet Russia.* Stuttgart: ibidem Press.

Toal, G. (2017). *Near Abroad: Putin, the West, and the Contest over Ukraine and the Caucasus.* New York: Oxford University Press.

Woodward, K. (2014). Affect, state theory, and the politics of confusion. *Political Geography 41*: 21–31.

Woodyer, T., & Carter, S. (2020). Domesticating the geopolitical: Rethinking popular geopolitics through play. *Geopolitics 25*: 1050–1074.

Woon, C.Y. (2017). Children, critical geopolitics, and peace: Mapping and mobilizing children's hopes for peace in the Philippines. *Annals of the American Association of Geographers 107*: 200–217.

Wu, Z. (2018). Classical geopolitics, realism and the balance of power theory. *Journal of Strategic Studies 41*: 786–823.

Chapter 20

Feminist Political Geography

Jennifer L. Fluri
Department of Geography, University of Colorado, Boulder, CO, USA

Feminist geographers have contributed many radical, insightful, and innovative approaches to the study of political geography. Early interventions identified the lack of women as both researchers and subjects of research, while exposing the gendered hierarchy endemic to various spatial formations of political power (Kofman & Peake 1990; Staeheli & Cope 1994; Staeheli & Lawson 1995). This research challenged existing and conventional approaches and methods for the study of politics spatially, relationally, and at multiple scales. Feminist scholarship moved beyond the inclusion of women and gender as a subject of study and challenged the narrow and problematic limitations of masculinist approaches to political geography (Dowler & Sharp 2001).

Feminist geographers continue to examine political geography and geopolitics across scale, challenging conventional approaches that were limited to national and international scales. Feminist scholars incorporated spaces and places previously thought of as apolitical, such as informal sites, social spaces, private/domestic spaces, and the body. Research on racial, ethnic, religious, and other forms of identity politics illuminates the manifestation of politics through the manipulation of socially constructed gender "norms" and relations (Fenster 1996). Some of this research revealed the tentacles of power through gendered identity politics, such as states attempting to change societal structures through biological and social reproduction (Smith 2013). This research questioned conventional notions of gendered political spaces and made significant contributions to the growing (and now established) field of critical geopolitics.

The Wiley Blackwell Companion to Political Geography, Second Edition.
Edited by Virginie Mamadouh, Natalie Koch, Chih Yuan Woon, and John Agnew.

Many gender analyses draw on intersectionality (Crenshaw 1994; Collins 2019) to examine intersecting identities and multidimensional forms of power hierarchies and power geometries (Massey 1993). Feminist political geographers have expanded their analyses from these foundational approaches. Therefore, this chapter provides an overview of feminist contributions of political geography and critical geopolitics. The following sections explain how traditional political geographical themes (state and nation) have been gendered, how new themes (body politics and domestic spaces) were explored, and how this contributes to new analyses of traditional political geographical themes (conflicts, borders).

Methodologies and Research Methods

Feminist political geographers have used and continue to use research methodologies that incorporate fieldwork and qualitative methods, from interviewing research participants to conducting intensive ethnographies. These qualitative approaches have contributed empirically sound and theoretically rich analyses. Feminist geographers' contributions have challenged research objectivity by incorporating multiple perspectives and the concept of intersubjectivity between researchers and research populations (Cope & Hay 2021). This includes acknowledging research participants as experts with specific epistemologies of the political engagements and entanglements associated with their own experiences. Many feminist political geographers incorporate the concept of situated knowledges (Haraway 1988) to illustrate that the production of knowledge and understanding of politics must be contextualized across space at multiple scales: "Situated knowledges, knowledges that are explicit about their positioning, sensitive to the structures of power that construct these multiple positions and committed to making visible the claims of the less powerful" (McDowell 1992: 413; also see Rose 1993).

Other theoretical interventions include counter-topographies, a term coined by Cyndi Katz (2001), which provides a metaphorical elaboration of the contour line to track the lineage of theory and practice in the production of knowledge (also see Wright 2009). Feminist geographers have also incorporated topology to address the ways in which power can become endless twisted formations of abuse and oppression (Koopman 2011). In addition to these methodological interventions, feminist contributions to political geography incorporate fieldwork to provide empirically rich linkages between theory and praxis. Most feminist political geographers conduct fieldwork and view this as integral to political analyses and as providing a rich and nuanced understanding of small-scale politics (Sharp 2005; Dixon & Marston 2011).

Gender, State, Nation

Feminist political geographers examine how gender intersects with discursive national ideologies and state power. Feminist interventions have challenged the dichotomous representations of the state as male/masculine and the nation as female/feminine, which both physically and socially positioned women as biological and social reproducers rather than social producers and active political subjects or participants (Sharp 1996; Mayer 2000). Gendered political analyses critique and confront the material and ideological struggles involved in the making of nations as well as counter or revolutionary groups poised to challenge the state (Cowen & Gilbert 2008). Research that counters masculinized representations of the state calls attention to feminist nationalist political movements and activism that challenge heteronormative nationalisms and dichotomous gender roles and relations (Mayer 2000; Fluri 2009; Puar 2017). Examining gendered embodied experiences integral to place formations and power relations remains central for a robust understanding of the dynamic affiliations between gender, state, and nation (Mayer 2000).

Feminist political geographers have also explored women's participation and agency within the mechanisms of statecraft (Gilmartin & Kofman 2004). Other forms of critical scholarship analyze the use of gender as a political tool, such as the weaponization of gendered tropes that situate both the internal and distant "other" as inferior, which remains integral to international aid and development geopolitics (Strüver 2007; Bhungalia 2023).

Critical examinations of national politics also draw on feminist decolonial geographies, toward disrupting and challenging conventional, Anglophone, and North Atlantic geographies of knowledge, power, and subjectivity (De Leeuw & Hunt 2018; Naylor et al. 2018; Radcliffe & Radhuber 2020). Examinations of decoloniality through the lens of political geography and geopolitics are exemplified by various scholars. For example, Alqaisiya (2023) places Palestine as a contemporary site of settler colonialism and incorporates decolonization as a method for examining the "cultural, and geo-political condition enabling the continuity of the [Israel's] settler colonial present" (2023: 103). In a different way, Courtheyn (2022) examines community-based peace-building by drawing on decolonial politics toward community autonomy and ecological dignity. In addition to these interventions, feminist geographers' contributions also pay attention to the corporeal as a fundamental scale of analysis.

Corporeal Geographies and Body Politics

By focusing on the body as a site of analysis, feminist geographers have challenged conventional political analysis that centers the state and macro-scale processes as *the* expected scale of analysis. For example, Dowler's (2002) analysis of Ground Zero (after September 11, 2001) in New York City identified how gender roles were reinforced politically through the visualization of the Twin Towers as the body of a male firefighter and a male police officer. The ways in which bodies carry representations of citizenship and its associated privileges elucidate important linkages among citizenship, identity, transnationalism, and mobility studies.

Transnational solidarity groups in Guatemala and Colombia put the bodies of privileged volunteers to work accompanying threatened activists to provide secured mobility. This form of "proxy citizenship" attempts to alter conflict dynamics through corporealized protection (Henderson 2009). Koopman (2011) coined the term "alter-geopolitics" as an analytical entry point for incorporating on-the-ground approaches to peace-building and broadening our understanding of security, peace-building geopolitics, and community (also see Courtheyn 2022).

Other conceptualizations of bodies as proxies include research on sex workers and brothels as sites for legitimate male sexual aggression, particularly by soldiers and related personnel during times of political conflict or war (Enloe 1993, 2000). For example, during the 2001–2021 US occupation of Afghanistan, sex workers were identified as providing a proxy corporeal space of protection for mostly white privileged international female aid/development workers, meaning that predominantly male international workers could "release" their sexual aggression via sex workers rather than harass or rape female international aid/development workers or local Afghan women (Fluri 2009). The illegitimacy and illegality of sex workers place their corporeality as a proxy for the release of male sexual aggression and have been mythologized by some men as a method to "protect" certain privileged or geopolitically designated women from this aggression. Sex work, as argued by Wright (2004), became a discursive strategy for questioning the morality of female murder victims in Ciudad Juárez, Mexico, as a discursive method of victim-blaming rather than providing protections for women or properly investigating these murders. These analyses question conventional assumptions that excuse the sexual aggression of men through victim-blaming or viewing sex workers as sexual proxies to protect *other more privileged* women from this aggression.

Feminist researchers illustrate the ways in which women's bodies become flexible nodes for intimately aggressive geopolitics that diverts attention away from male actors/perpetrators and the structures of aid/development and urban governance, along with the unwillingness of these institutionalized structures to address or curtail these abuses.

Intimate/Intimacy Geopolitics

Intimate linkages between political conflict and daily life are further explored in Sara Smith's work (2009, 2011, 2012, 2013, 2020). She examines the complicated ways in which the rise of Hindu nationalism radically changed the formerly accepted practice of intermarriage between Buddhists and Muslims in Ladakh, India. As she ardently argues, "when population becomes part of a territorial struggle, the body itself becomes a geopolitical site" (Smith 2011:456–457). In many respects, bodies signify and approximate the fragile scale at which processes and promises of geopolitics and political identities resonate (Dixon & Marston 2011). Situating bodies at the center of research on political conflict is necessary to ensure that the bodies of individuals and collectives count rather than being reduced to aggregate sums associated with the collateral excesses of war (Hyndman 2007). Intimacy geopolitics places corporeality at the center of analyses and examines the intimate and visceral geographies of political power (Pain & Staeheli 2014). Gender-based violence is another area of research where feminist geographers draw on intimacy geopolitics and corporeal geographies to address the role of the state in mitigating domestic violence and the linkages between geopolitical and intimate partner abuse (Cuomo 2013; Pain 2015; Pain & Cahill 2022). For example, Brickell (2014) further examines intimate geopolitics through the forced evictions of Cambodia women's activism and how it vacillates across scales.

Research focusing on the interlinkages between the intimate and the global have disrupted grand narratives that dominated hegemonic and imperial political reasoning to chart a path of research inquiry and analyses outside conventional or traditional frameworks (Wright 2009; Koopman 2011; Pratt & Rosner 2012). As part of feminist political geographers' engagements with scale and calls for examining private, social, and apolitical spaces and places, the body increasingly resonates within feminist study as both a site for political geographical analysis and a methodology for destabilizing the disembodied macro-scale study of politics and space. Feminist political geographers' analyses of the corporeal have significantly altered the geographical understanding of space by emphasizing the importance of embodied practices, place, and intimate knowledge as integral to research (Dowler & Sharp 2001; Mountz & Hyndman 2006; Sharp 2007; Clark 2017).

Gender Aesthetics and Geopolitics

Gendered aesthetics of violence, war, and militarism materialize through hierarchical representations of soldiers, enemies, and vulnerable bodies. Many film industry tropes visualize the "villainous other" as "ugly" by way of disfigurement, disability, or the grotesque, in stark contrast to the gender-specific beauty of the heroine/hero (Kirby 2020). Other corporeal war representations rely on truncated beauty regimes that draw on and reinforce existing tropes of masculinity and femininity (Goldstein 2001; Dowler 2002; Hasso 2005).

The aesthetics of gendered bodies and geopolitics, as discussed by feminist political geographers, examine the representative ways in which idealized forms of womanhood become representative of nationalism and citizenship through beauty pageants and related symbolic representations of femininity (Sharp 1996; Oza 2006; Fluri 2009; Faria 2010, 2013; Elledge & Faria 2020). National pageants operationalize women's bodies to represent ideal femininity that strengthens national norms and seductively erases difference, while profoundly cementing racist notions, stereotypes, and ideologies. As Faria and Fluri (2022) argue, the geopolitics of beauty situate aesthetic hierarchies of aliveness. The designation of beauty in some examples confers access and resources on the aesthetically pleasing, while denying the same access to the nonaesthetically pleasing other. Conversely, beauty can be a site of exploitation, such as beauty regimes or expectations that cause harm, and the role of corporeal aesthetics in sex and human trafficking.

Sartorial Geopolitics

Additional examinations focus on politics that influence the ways in which states, institutions, communities, and families restrict or place limits on gendered corporeal representations and mobility in public spaces. For example, the geopolitics of the hijab remains a political tool of exclusionary citizenship, spatial marginalization or acceptance, and security in various spaces (Hancock & Mobillion 2019; Shaban 2021; Schenk et al. 2022). Hijab geopolitics originated during Western European colonialism, where women from Muslim-majority countries were narrowly conceptualized and falsely represented as a singular aggregated and objectified group. This representation of Muslim women was significantly revived in association with the gendered geopolitics and discursive framing of Muslim women by various Western countries during the United States' so-called Global War on Terror (GWOT). Contemporary experiences of sartorial geopolitics by women wearing the hijab in Europe include everyday forms of microaggressions and policing of Muslim women's bodies in non-Muslim-majority spaces (see Hancock & Mobillion 2019; Najib & Hopkins 2019).

The geopolitics of fashion, explored by feminist geographers and other scholars through pious or veiling fashion, represents a counter to Western misperception/misunderstanding of veiling (Gökarıksel & Secor 2009; Tarlo 2010). Gökarıksel and Secor's (2009, 2010, 2012) work on veiling fashion underscores the contradictions of hijab politics as both a symbol of piety and religious devotion while simultaneously functioning as a motif of capitalist production and consumption. The politics of the veil and its very intentionality are therefore disrupted, while becoming entrenched within the equally oppressive and problematic mechanisms of neoliberalism (also see Secor 2002; Gökarıksel & Secor 2010, 2012). This scholarship offers a poignant and complicated view that challenges popular preconceptions in mostly Western, non-Muslim-majority spaces about gender and Islam. The narrow representations of Muslim women in popular media and journalistic accounts are further challenged by examinations that focus on the diversity of women's experiences, agency, and methods of negotiating sartorial practices and navigating through public spaces (Falah & Nagel 2005; Fluri & Lehr 2017; Bagheri 2019).

Cuerpo-Territorio/***Body-Territory***

Dress is inherently linked to bodies, but territory has not been linked to bodies in conventional political geography analyses. Scholarship on corporeal geopolitics and territory explores the social, economic, and political hierarchies of infrastructure and the relationships between body *and* territory and bodies *as* territory. Truelove and Ruszczyk (2022) identify the intersectional relationship between class, caste, and gender in Delhi, India, for households without access to piped water. Women's bodies operate as human infrastructure, physically moving water from its public source to the household (Truelove 2021). Indigenous Latin American scholarship focuses on bodies as territory through the concept of *cuerpo-territorio* (body-territory), defined as a distinct geographical, "decolonial feminist method grounded in the ontological unity between bodies and territories" (Zaragocin & Caretta 2021: 1503). Body mapping provides a qualitative research method that draws from *cuerpo-territorio* as a decolonizing feminist approach centering the body to highlight the epistemologies of Indigenous and other marginalized individuals and groups. This approach provides data collection and analytical methods for tackling the relationship between embodiment, territory, and various forms of resistance against land dispossession (Glockner et al. 2023).

Bodily Autonomy

The politics of gendered bodily autonomy is of increasing importance among feminist political geographers, due to the growing political challenges that seek to limit individual and collective autonomy over the body and body-territory. The spatial and temporal politics of reproductive rights is one area of focus. Geographers have framed fertility through a spatial and temporal lens to identify and explicate the links between land, body, and bodily tissue in settler colonial contexts, in association with citizenship, border management, and population control (Mansfield 2012; Murphy 2017; Hiemstra 2021). For example, Calkin (2019: 69) calls for feminist scholarship to reconceptualize abortion as a "multiscalar process of contestation and resistance that implicates a variety of legal, medical, and social domains, rather than as a right permitted or denied by the nation-state."

Limitations on gendered bodily autonomy can be used as a weapon of state control, to limit individual freedoms or discursively provision these freedoms through violent violations of sovereignty, war, and related political conflicts (Fluri 2009; Greenburg 2023). Feminist geographers highlight the gendered justifications used to garner public support for military interventions or invasions under the guise of protecting or saving women from their own national or community-based forms of patriarchy (Fluri & Lehr 2017; Shaban 2021), while, simultaneously, military interventions are regularly positioned within the structures of heteronormative patriotism (Puar & Rai 2002).

Intimate/intimacy geopolitics further addresses the gendered politics of reproduction and corporeal safety and security (Smith 2020; Pain & Staeheli 2014). For example, US government policies that negatively impact and restrict children's bodily autonomy range from foster care systems to family separation at the US–Mexican border (Torres et al. 2022b). Political and social control over bodies remains situated within hierarchical power relations that mitigate autonomy based on race, class, ethnicity, gender, and sex/sexuality. For example, feminist scholars highlight the linkages between white supremacy and controls over biological reproduction, particularly for individuals who are the targets of political marginalization and oppression (Hooks 1987; Eaves 2017; Smith & Vasudevan 2017). This includes control over Black bodies, including the economic value of reproductive bodies, rape, and child separation during slavery (McKittrick 2006). Indigenous feminist geographers further underscore the relationship between bodily and community autonomy in relationship to land theft, displacement, and continued and contemporary experiences of settler colonialism and white supremacy.

A decolonial feminist approach remains attentive to non-Anglophone ways of conceptualizing bodily autonomy. Centering community in analyses of feminist political agency, the body is framed as connected to land in relation to settler colonialism, extractive industries, territorial conflicts, and agricultural production (González-Hidalgo et al. 2022). Alternative epistemologies and ontologies based on body–land and communities have been introduced to critical geopolitics to enrich a decolonized feminist geopolitics (Zaragocin 2019). Recent calls for reframing the concept of autonomy in political geography articulate autonomy as an embodied practice of community rather than of separate individuals/individuality (Naylor 2017). In addition to centering the body as a site of analysis, feminist political geographers examine spaces and places formerly considered apolitical or informal.

Domestic Spaces and Informal Politics

Feminist geographers challenge the implicit political division between public (political) and private (apolitical), which has been foundational to conventional political analyses (Hyndman 2004). Feminist research on the public and private and the ways in which they influence, enact, and reproduce each other elucidates the interlinkages between these places, rather than viewing them as a binary (Secor 2001, 2002). Radcliffe and Westwood's (1996) book focuses on place, identity, and politics in Latin America, and examines several ways in which gendered expectations of private lives and behavior were used to confront politics in public spaces. Communities marked by racial, class, and gendered marginalization have worked to gain political legitimacy by appropriating public space for political action, such as political demonstrations (Clark 1994).

Feminist scholarship argues that political participation includes the involvement of individuals operating in both formal and informal political spaces (Cope 1997; see also Ali & Hopkins 2012). Counter-geopolitics challenges the historical representations of politics as operating among states by examining "alternative and multiple loci of power" (Secor 2001: 193). Thus, what constitutes power, space, and scale remains fluid, geometric, and multifarious rather than static, hierarchical, or dichotomous. Some forms of political action have effectively operationalized conventional gender norms toward nonconventional political action. For example, some women activists have opportunistically manipulated the predisposition of the home as an apolitical space to plan or execute covert political actions against the state (Dowler 1998). Female activists have incorporated representations of the home, motherhood, and other gendered identities as an act of spatial transgression (Cresswell 1996; Bosco 2001; Blunt & Dowling 2006; Nelson 2006).

Feminist geographical approaches also decenter the state as the primary site of analysis through examinations of the everyday and social reproduction (Pearson & Crane 2017). These studies illustrate disparate forms of gendered spaces that challenge the existing dichotomous demarcations of public and private spaces and the use of gendered norms or tropes to counter and challenge the actions of the state and its military. In many spaces the home remains a gendered site, and in some places a site of women's seclusion, immobility, or abuse, while conversely (and sometimes simultaneously) being a site of comfort, care, and retreat from abuse. For example, within some domiciles both conflict and its mitigation occur in tandem, which challenges the "expected" and "effective" role of outside forces to mitigate domestic conflicts or violence (Fluri 2011b; Cuomo 2013).

As feminist scholars have shown, the home is a complex and complicated space and one that offers important insights for social, cultural, and political epistemologies (Blunt & Dowling 2006). Feminist geographers' examinations of domestic labor provide a nuanced understanding of the fluidity and relational and intersectional linkages between public and private spaces. For example, paid domestic work has historically been seen as less important and outside state regulation, which rendered domestic workers more vulnerable to abuse by employers (Pratt & Yeoh 2003; Pratt 2004, 2012). Women working as domestics abroad became integrated into Peruvian nationhood because these predominantly female laborers exemplified their "proper" place within the national imaginary through their embodied representations of gender, ethnicity, and class (Radcliffe 1990). Research on Filipina migrant domestic workers in Canada illustrates a complicated mix of intersecting and at times conflicting social and political stipulations from the state, domestic workers, and their employers (Pratt 2004). In other examples, the state has been actively involved in planning and regulating remittances from its migratory labor force (Silvey 2004). This research demonstrates the interweaving of global geopolitical maneuvers with the embodiment of everyday experiences, emotions, and actions (Sharp 2009). Gendered national identities are regularly reshaped or reinforced during times of heightened nationalism, such as in situations of war and related violence and questions of national security. Feminist geographers explore the complex, complicated, and at times contradictory relationships between violence and security across interrelated spatial scales.

Conflict and Security

Static and disembodied analyses of political conflict and war have been challenged by examining the complex and multilayered analyses of the body as a geographical site, corporeal experiences of conflict, and bodies as a geopolitical tool or weapon (Hyndman & De Alwis 2004; Mountz 2004, 2017; Fluri 2011a, 2011b; Koopman 2011). For example, throughout the GWOT the United States relied on gendered tropes of Muslim women's vulnerability and docility to reinforce or solidify the "need" for military occupations and development interventions in an effort to "save or protect" women in locations such as Afghanistan and Iraq (Riley, Mohanty, & Pratt 2008; Fluri 2011a, 2011b; Fluri & Lehr 2017).

Feminist geographers' study of conflict and security includes analyses of security at macro scales (international, regional, national) as they intersect with local, community, domestic, and corporeal scales (Enloe 1993, 2000; Hyndman 2007). Within locations embroiled in political conflict or war, security is distinctly spatialized and considerably contentious, leading feminist researchers to examine relational experiences of security and question "whose security is of concern and to whom" (Hyndman & De Alwis 2004: 535; Fluri 2011a; Clark 2013).

Additionally, war and political conflict garner more media, political, and economic attention than other forms of gender-based violence. In response, Pain (2014) labels domestic violence "everyday terrorism" to call attention to the imbalance of resources spent on geopolitical terrorism in comparison to mitigating or assisting victims and survivors of domestic violence. Pain (2014) reconceptualizes domestic violence as terrorism to understand how these differential forms of violence are interrelated and extensively mediated by dominant constructions of gender, race, class, and nationality.

Feminist geographers have also contributed significant, insightful, and empirically rich analysis of geopolitical conflicts that incorporate the everyday experiences of individuals living in, resisting, or fleeing from conflict. For example, Ehrkamp, Loyd, and Secor examine the geopolitics of trauma, displacement, and ablenationalism for refugees and asylum seekers who are embroiled in series of displacements, replacements, profound uncertainty, and security regimes that ultimately produce trauma, reinforce spatial exclusions, entrench existing stereotypes, and lead to racialized assumptions that frame vulnerable populations with suspicion and fear-based othering (Loyd, Ehrkamp, & Secor 2018; Ehrkamp, Loyd, & Secor 2022; Secor, Ehrkamp, & Loyd 2022; Loyd, Secor, & Ehrkamp 2023). This scholarship is also connected with research on migration, border and mobility management, and security.

Borders, Mobility, and Security

Mobility has in some respects become a "barometer" for the study of gender, class, and other sociopolitical relations (Hyndman 2004). Feminist scholars examine the intersectionality between gender and other forms of identity that spatially situate persons as central to, associated with, or at the margins of political machinations and power relations (Fenster 1998). Research on borders, migration, and mobility both questions and reveals the ways in which the gendering of states or nations resonates as part of a particularly and often politically crafted spatial imaginary.

Research on migrant family detention explicates the ways in which the state determines what is both "secure" and in the "best interests of the child" when incarcerating families that are seeking refuge (Martin 2011). State geopolitics resituates childhood vulnerability by framing migratory laborers traveling with children as criminals rather than caregivers (Torres et al. 2022a). Research on prisons and borders challenges the conventional belief that these practices create social order and security by illustrating how they act as forms of coercive mobility or oppression against poor and displaced persons (Loyd, Mitchelson, & Burridge 2012; Torres et al. 2022b).

Feminist political geographers have significantly shaped the study of mobility as a political process (Silvey 2004). Embodied analysis of mobility and asylum-seeking offers a nuanced understanding of geopolitical and economic modes of security and insecurity. Research on border, mobility, and security includes multi-scalar understandings of power relations, particularly from the perspectives and experiences of the marginalized and disenfranchised (Loyd 2020).

Borders are spaces that incorporate an array of social, economic, and political activity such as human mobility, extensive waiting, detention, incarceration, and political economy. The experiences of many individuals attempting to cross borders include limitations on their mobility. Structural violence at the border remains evident through the militarization of these spaces by armed border guards, and practices of detention and incarceration of refugees and asylum seekers (Mountz 2011). The criminalization of bodies and mobilities at borders is manipulated through biased classifications of race, gender, class, and sexuality (Papada 2021; Hawthorne 2022). The management of migratory bodies reveals a crisis of geopolitical care, particularly for individuals fleeing political conflict or environmental disaster, and the subsequent gendered trauma and retraumatization of refugees and asylum-seekers within the morass of institutionalized management (Ehrkamp, Loyd, & Secor 2022; Torres et al. 2022a).

The linkages between immigration laws for seeking asylum and bodies reveals how these laws restrict the free movement of women, children, and lesbian/gay/bisexual/trans/queer/intersex/asexual+ (LGBTQIA+) individuals and groups (Grungras et al. 2009; Shakhsari 2014). While borders have been spaces of extensive bureaucratic management, feminist inquiries draw attention to the embodied experiences of migrants and their agentic methods for survival (Mountz 2011; Borrelli 2022; Dempsey 2022; Myadar 2022). In other cases, border control agents have seized on the gendered vulnerability of migrants by indulging in pornographic reenactments of sexual abuse against female migrants by border patrol officers (Casaglia 2022). These analyses remind us that migrant and refugee vulnerability exists alongside individual and collective agency. Survival and resiliency through various creative methods to subvert institutional control and seek dignity and some forms of autonomy are key areas of agency for migrants and refugees (Dempsey 2022).

Conclusion and Feminist Political Geography Futures

Feminist geographers have made significant and lasting contributions to political geography and critical geopolitics by incorporating gender as an intersectional category of analysis. This has included examining the body as a political space and centering the experiences of individuals and communities over national or international scales. Feminist geographers' engagement with corporeal geopolitics takes an intersectional approach to analyses and incorporates race, ethnicity, sex/sexuality, class, and other markers of politicized difference. While feminist geographers have made significant contributions, there remains room for more incorporation of and collaboration with other forms of critical scholarship, such as (but not limited to) decolonial and anti-colonial theories and Indigenous, Black, Latinx, and Asian geographies. The politics of gender, racial, ethnic, sexual, and class hierarchies that mitigate individuals' and communities' ability to meet their basic needs and access resources is another area of growth for feminist political geography. There remains an increased focus on bodily and community autonomy as an entry point for examining political corporeal control through the prism of gendered intersectional power geometries. This is particularly timely with the rise of various populous and conservative movements across the globe, and the legal gendered limits being placed on reproductive and sexual freedoms.

Feminist geographers should continue to expand critical geopolitical inquiry to further understand the ways in which gender and gendered bodies are politicized through various expressions of nationalism, sovereignty, and supranational and non-state actor involvement in security, conflict, border control, and mobility. Providing more discussion and analyses of the complexities, complications, challenges, and contradictions of the divergent feminist approaches or feminism*s* remains a vital area for additional examination by feminist political geographers. This includes grappling with gender in new ways, such as learning from new and emerging scholarship from but not limited to queer, trans, Black, and Indigenous geographies.

Feminist political geography/geopolitics has spent the past 20 plus years contributing to and pushing critical geopolitics and political geography by identifying and filling the epistemological gaps of spatial political analyses. Since feminist geography has moved from a marginal subfield of human geography to a mainstream form of academic inquiry, it is time to reflect on these contributions to pause and reflect on what we can learn from other forms of progressive analysis and intervention. This includes continuing to engage, think, and remain open to questions, conversations, and debates about the ways in which new positions, epistemologies, and ontologies can inform feminist contributions in the future.

References

Ali, Rahielah, and Peter Hopkins. 2012. Everyday making and civic engagement among Muslim women in Scotland. In *Muslims in Britain: Making Social and Political Space* (ed. W. I. U. Ahmad and Z. Sardar). Abingdon: Routledge, 141–155.

Alqaisiya, W. (2023). Beyond the contours of Zionist sovereignty: Decolonisation in Palestine's Unity Intifada. *Political Geography 103*: 102844. doi:10.1016/j.polgeo.2023.102844

Bagheri, N. (2019). Tehran's subway: Gender, mobility, and the adaptation of the "proper" Muslim woman. *Social & Cultural Geography 20*(3): 304–322. doi:10.1080/14649365.2017.1356364

Bhungalia, L. (2023). *Elastic Empire: Refashioning War through Aid in Palestine*. Redwood City, CA: Stanford University Press.

Blunt, A., & Dowling, R. (2006). *Home*. New York: Routledge.

Borrelli, L.M. (2022). Should (s)he stay or should (s)he go? – Street-level suspicion and the construction of the "(un)deserving migrant". *Geopolitics 27*(4): 1093–1116. doi:10.1080/14650045.2020.1814257

Bosco, F.J. (2001). Place, space, networks, and the sustainability of collective action: The Madres de Plaza de Mayo. *Global Networks 1*(4): 307–329.

Brickell, K. (2014). "The whole world is watching": Intimate geopolitics of forced eviction and women's activism in Cambodia. *Annals of the Association of American Geographers 104*(6): 1256–1272.

Calkin, S. (2019). Towards a political geography of abortion. *Political Geography 69*: 22–29. doi:10.1016/j.polgeo.2018.11.006

Casaglia, A. (2022). Pornography at the border: Ethnosexual borderscapes, gendered violence, and embodied control. *Geopolitics 27*(1): 185–205. doi:10.1080/14650045.2020.1755266

Clark, H. (1994). Taking up space: Redefining political legitimacy in New York City. *Environment and Planning A 26*(6): 937–955.

Clark, J.H. (2013). "My life is like a novel": Embodied geographies of security in Southeast Turkey. *Geopolitics 18*(4): 835–855.

Clark, J.H. (2017). Feminist geopolitics and the Middle East: Refuge, belief, and peace. *Geography Compass 11*(2): e12304. doi:10.1111/gec3.12304

Collins, P.H. (2019). *Intersectionality as Critical Social Theory*. Durham, NC: Duke University Press.

Cope, M., & Hay, I. (2021). Where are we now? Qualitative research in human geography. In Copy, M., & Hay, I. (eds), *Qualitative Research Methods in Human Geography*. Oxford: Oxford University Press, 3–17.

Cope, M. (1997). Participation, power, and policy: Developing a gender-sensitive political geography. *Journal of Geography 96*(2): 91–97.

Courtheyn, C. (2022). *Community of Peace: Performing Geographies of Ecological Dignity in Columbia*. Pittsburgh, PA: University of Pittsburgh Press.

Cowen, D., & Gilbert, E. (2008). *War, Citizenship, Territory*. New York: Routledge.

Crenshaw, K. (1994). Mapping the margins: Intersectionality, identity politics, and violence against women of color. *Stanford Law Review 43*(6): 93–118.

Cresswell, T. (1996). *In Place/Out of Place: Geography, Ideology, and Transgression*. Minneapolis, MN: University of Minnesota Press.

Cuomo, D. (2013). Security and fear: The geopolitics of intimate partner violence policing. *Geopolitics 18*(4): 856–874.

De Leeuw, S., & Hunt, S. (2018). Unsettling decolonizing geographies. *Geography Compass 12*(7): e12376.

Dempsey, K.E. (2022). Migrant agency and counter-hegemonic efforts among asylum seekers in the Netherlands in response to geopolitical control and exclusion. *Geopolitics 27*(2): 402–423. doi:10.1080/14650045.2021.1912019

Dixon, D.P., & Marston, S.A. (2011). Introduction: Feminist engagements with geopolitics. *Gender, Place and Culture 18*(4): 445–453.

Dowler, L. (1998). "And they think I'm just a nice old lady": Women and war in Belfast, Northern Ireland. *Gender, Place and Culture: A Journal of Feminist Geography 5*(2): 159–176.

Dowler, L. (2002). Women on the frontlines: Rethinking war narratives post 9/11. *GeoJournal 58*(2–3): 159–165.

Dowler, L., & Sharp, J. (2001). A feminist geopolitics? *Space and Polity 5*(3): 165–176.

Eaves, L.E. (2017). Black geographic possibilities: On a queer Black South. *Southeastern Geographer 57*(1): 80–95. doi:10.1353/sgo.2017.0007

Ehrkamp, P., Loyd, J.M., & Secor, A.J. (2022). Trauma as displacement: Observations from refugee resettlement. *Annals of the American Association of Geographers 112*(3): 715–722. doi:10.1080/24694452.2021.1956296

Elledge, A.M., & Faria, C. (2020). "I want to . . . let my country shine": Nationalism, development, and the geographies of beauty. *Environment and Planning D: Society and Space 38*(5): 829–848. doi:10.1177/0263775820911953

Enloe, C. (2000). *Maneuvers: The International Politics of Militarizing Women's Lives*. Berkeley, CA: University of California Press.

Enloe, C.H. (1993). *The Morning After: Sexual Politics at the End of the Cold War*. Berkeley, CA: University of California Press.

Falah, G.-W., & Nagel, C.R. (2005). *Geographies of Muslim Women: Gender, Religion, and Space*. New York: Guilford Press.

Faria, C. (2010). Contesting Miss South Sudan: Gender and nation building in diasporic discourse. *International Feminist Journal of Politics 12*(2): 222–243.

Faria, C. (2013). Styling the nation: Fear and desire in the South Sudanese beauty trade. *Transactions of the Institute of British Geographers 39*: 318 330.

Faria, C., & Fluri, J.L. (2022). Allure and the spatialities of nationalism, war and development: Towards a geography of beauty. *Geography Compass 16*(9): e12652. doi:10.1111/gec3.12652

Fenster, T. (1996). Ethnicity and citizen identity in planning and development for minority groups. *Political Geography 15*(5): 405–418.

Fenster, T. (1998). Ethnicity, citizenship, planning and gender: The case of Ethiopian immigrant women in Israel. *Gender, Place and Culture: A Journal of Feminist Geography 5*(2): 177–189.

Fluri, J.L. (2009). "Foreign passports only": Geographies of (post) conflict work in Kabul, Afghanistan. *Annals of the Association of American Geographers 99*(5): 986–994.

Fluri, J.L. (2011a). Armored peacocks and proxy bodies: Gender geopolitics in aid/development spaces of Afghanistan. *Gender, Place and Culture 18*(4): 519–536.

Fluri, J.L. (2011b). Bodies, bombs and barricades: Geographies of conflict and civilian (in) security. *Transactions of the Institute of British Geographers 36*(2): 280–296.

Fluri, J.L., & Lehr, R. (2017). *The Carpetbaggers of Kabul and Other American-Afghan Entanglements: Intimate Development and the Currency of Gender and Grief*. Athens, GA: University of Georgia Press.

Gilmartin, M., & Kofman, E. (2004). Critically feminist geopolitics. In Staeheli, L., Kofman, E., & Peak, L. (eds), *Mapping Women, Making Politics*. London, Routledge, 113–126.

Glockner, V., Borzacchiello, E., Torres, R.M., et al. (2023). The cuerpo-territorio of displacement: A decolonial feminist geopolitics of re-existencia. *Geopolitics 29*(4): 1220–1244. doi:10.1080/14650045.2023.2213639

Gökarıksel, B., & Secor, A. (2010). Islamic-ness in the life of a commodity: Veiling-fashion in Turkey. *Transactions of the Institute of British Geographers 35*(3): 313–333.

Gökarıksel, B., & Secor, A. (2012). "Even I was tempted": The moral ambivalence and ethical practice of veiling-fashion in Turkey. *Annals of the Association of American Geographers 102*(4): 847–862.

Gökarıksel, B., & Secor, A.J. (2009). New transnational geographies of Islamism, capitalism and subjectivity: The veiling-fashion industry in Turkey. *Area 41*(1): 6–18. doi:10.1111/j.1475-4762.2008.00849.x

Goldstein, J. (2001). *War and Gender: How Gender Shapes the War System*. Cambridge: Cambridge University Press.

González-Hidalgo, M., Del Bene, D., Iniesta-Arandia, I., & Piñeiro, C. (2022). Emotional healing as part of environmental and climate justice processes: Frameworks and community-based experiences in times of environmental suffering. *Political Geography 98*: 102721. doi:10.1016/j.polgeo.2022.102721

Greenburg, J. (2023). *At War with Women: Military Humanitarianism and Imperial Feminism in an Era of Permanent War*. Ithaca, NY: Cornell University Press.

Grungras, N., Levitan, R., & Slotek, A. (2009). Unsafe haven: Security challenges facing LGBT asylum seekers and refugees in Turkey. *PRAXIS: The Fletcher Journal of Human Security 24*: 41–61.

Hancock, C., & Mobillion, V. (2019). "I want to tell them, I'm just wearing a veil, not carrying a gun!" Muslim women negotiating borders in femonationalist Paris. *Political Geography 69*: 1–9. doi:10.1016/j.polgeo.2018.11.007

Haraway, D. (1988). Situated knowledges: The science question in feminism and the privilege of partial perspective. *Feminist Studies 14*(3): 575–599.

Hasso, F.S. (2005). Discursive and political deployments by/of the 2002 Palestinian women suicide bombers/martyrs. *Feminist Review 81*(1): 23–51. doi:10.1057/palgrave.fr.9400257

Hawthorne, C. (2022). Black Mediterranean geographies: Translation and the mattering of black life in Italy. *Gender, Place and Culture 30*(3): 484–507.

Henderson, V.L. (2009). Citizenship in the line of fire: Protective accompaniment, proxy citizenship, and pathways for transnational solidarity in Guatemala. *Annals of the Association of American Geographers 99*(5): 969–976.

Hiemstra, N. (2021). Mothers, babies, and abortion at the border: Contradictory U.S. policies, or targeting fertility? *Environment and Planning C: Politics and Spaces 39*(8): 1692–1710. doi:10.1177/2399654421998368

hooks, b. (1987). *Ain't I a Woman: Black Women and Feminism*. New York: Pluto Press.

Hyndman, J. (2004). Mind the gap: Bridging feminist and political geography through geopolitics. *Political Geography 23*(3): 307–322.

Hyndman, J. (2007). The securitization of fear in post-Tsunami Sri Lanka. *Annals of the Association of American Geographers* 97(2): 361–372.

Hyndman, J., & De Alwis, M. (2004). Bodies, shrines, and roads: Violence, (im)mobility and displacement in Sri Lanka. *Gender, Place and Culture 11*(4): 535–557.

Katz, C. (2001). On the grounds of globalization: A topography for feminist political engagement. *Signs 26*(4): 1213–1234.

Kirby, P. (2020). Battle scars: *Wonder Woman*, aesthetic geopolitics and disfigurement in Hollywood film. *Geopolitics 25*(4): 916–936. doi:10.1080/14650045.2018.1510389

Kofman, E., & Peake, L. (1990). Into the 1990s: A gendered agenda for political geography. *Political Geography Quarterly* 9(4): 313–336.

Koopman, S. (2011). Alter-geopolitics: Other securities are happening. *Geoforum 42*(3): 274–284.

Loyd, J. M. (2020). Boundaries of policing. *Society and Space*, October 1. https://www.societyandspace.org/articles/boundaries-of-policing (accessed November 27, 2024).

Loyd, J.M., Ehrkamp, P., & Secor, A.J. (2018). A geopolitics of trauma: Refugee administration and protracted uncertainty in Turkey. *Transactions of the Institute of British Geographers 43*(3): 377–389. doi:10.1111/tran.12234

Loyd, J.M., Secor, A.J., & Ehrkamp, P. (2023). Geopolitics of disability and the ablenationalism of refuge. *Geopolitics 1–28*. doi:10.1080/14650045.2023.2185139

Loyd, J., Mitchelson, M., & Burridge, A. (eds) (2012). *Beyond Walls and Cages: Prisons, Borders, and Global Crisis*. Athens, GA: University of Georgia Press.

Mansfield, B. (2012). Environmental health as biosecurity: "Seafood choices," risk, and the pregnant woman as threshold. *Annals of the Association of American Geographers 102*(5): 969–976. doi:10.1080/00045608.2012.657496

Martin, L. (2011). The geopolitics of vulnerability: Children's legal subjectivity, immigrant family detention and US immigration law and enforcement policy. *Gender, Place and Culture 18*(4): 477–498.

Massey, D. (1993). Power-geometry and a progressive sense of place. In *Mapping the Futures* (ed. J. Bird, B. Curtis, T. Putnam et al.). London: Routledge, 59–69.

Mayer, T. (2000). *Gender Ironies of Nationalism: Sexing the Nation*. London: Psychology Press.

McDowell, L. (1992). Doing gender: Feminism, feminists and research methods in human geography. *Transactions of the Institute of British Geographers 17*(4): 399–416. doi:10.2307/622707

McKittrick, K. (2006). *Demonic Grounds : Black Women and the Cartographies of Struggle*. Minneapolis, MN: University of Minnesota Press.

Mountz, A. (2004). Embodying the nation-state: Canada's response to human smuggling. *Political Geography 23*(3): 323–345.

Mountz, A. (2011). Where asylum-seekers wait: Feminist counter-topographies of sites between states. *Gender, Place and Culture 18*(3): 381–399. doi:10.1080/0966369X.2011.566370

Mountz, A. (2017). Political geography III: Bodies. *Progress in Human Geography 42*(5): 759–769.

Mountz, A., & Hyndman, J. (2006). Feminist approaches to the global intimate. *Women's Studies Quarterly 34*(1/2): 446–463.

Murphy, M. (2017). *The Economization of Life*. Durham, NC: Duke University Press.

Myadar, O. (2022). Place, displacement and belonging: The story of Abdi. *Geopolitics 27*(2): 462–477. doi:10.1080/14650045.2020.1837115

Najib, K., & Hopkins, P. (2019). Veiled Muslim women's strategies in response to islamophobia in Paris. *Political Geography 73*: 103–111. doi:10.1016/j.polgeo.2019.05.005

Naylor, L. (2017). Reframing autonomy in political geography: A feminist geopolitics of autonomous resistance. *Political Geography 58*: 24–35. doi:10.1016/j.polgeo.2017.01.001

Naylor, L., Daigle, M., Zaragocin, S., Ramírez, M.M., & Gilmartin, M. (2018). Interventions: Bringing the decolonial to political geography. *Political Geography 66*: 199–209. doi:10.1016/j.polgeo.2017.11.002

Nelson, L.K. (2006). Geographies of state power, protest, and women's political identity formation in Michoacán, Mexico. *Annals of the Association of American Geographers 96*(2): 366–389.

Oza, R. (2006). *The Making of Neoliberal India: Nationalism, Gender, and the Paradoxes of Globalization*. Boca Raton, FL: CRC Press.

Pain, R. (2014). Everyday terrorism: Connecting domestic violence and global terrorism. *Progress in Human Geography 38*(4): 531–550.

Pain, R. (2015). Intimate war. *Political Geography 44*: 64–73.

Pain, R., & Cahill, C. (2022). Critical political geographies of slow violence and resistance. *Environment and Planning C: Politics and Spaces 40*(2): 359–372. doi:10.1177/23996544221085753

Pain, R., & Staeheli, L. (2014). Introduction: Intimacy-geopolitics and violence: Introduction: intimacy-geopolitics and violence. *Area 46*(4): 344–347. doi:10.1111/area.12138

Papada, E. (2021). Engaging the geopolitics of asylum seeking: The care/control function of vulnerability assessments in the context of the EU–Turkey agreement. *Geopolitics 28*(1): 257–281. doi:10.1080/14650045.2021.1884548

Pearson, Z., & Crane, N.J. (2017). The challenge of feminist political geography to state-centrism in Latin American geography. *Journal of Latin American Geography 16*(1): 185–193. doi:10.1353/lag.2017.0011

Pratt, G. (2004). *Working Feminism*. Philadelphia, PA: Temple University Press.

Pratt, G. (2012). *Families Apart: Migrant Mothers and the Conflicts of Labor and Love*. Minneapolis, MN: University of Minnesota Press.

Pratt, G., & Rosner, V. (2012). *The Global and the Intimate: Feminism in Our Time*. New York: Columbia University Press.

Pratt, G., & Yeoh, B. (2003). Transnational (counter) topographies. *Gender, Place and Culture 10*(2): 159–166. doi:10.1080/0966369032000079541

Puar, J.K. (2017). *Terrorist Assemblages: Homonationalism in Queer Times*. Durham, NC: Duke University Press.

Puar, J.K., & Rai, A. (2002). Monster, terrorist, fag: The war on terrorism and the production of docile patriots. *Social Text 20*(3): 117–148.

Radcliffe, S. (1990). Ethnicity, patriarchy, and incorporation into the nation: Female migrants as domestic servants in Peru. *Environment and Planning D: Society and Space 8*(4): 379–393.

Radcliffe, S.A., & Radhuber, I.M. (2020). The political geographies of D/decolonization: Variegation and decolonial challenges of/in geography. *Political Geography 78*: 102128. doi:10.1016/j.polgeo.2019.102128

Radcliffe, S., & Westwood, S. (1996). *Remaking the Nation: Place, Identity and Politics in Latin America*. London: Routledge.

Riley, R.L., Mohanty, C.T., & Pratt, M.B. (2008). *Feminism and War: Confronting US Imperialism*. London: Zed Books.

Rose, G. (1993). *Feminism and Geography: The Limits of Geographical Knowledge*. Minneapolis, MN: University of Minnesota Press.

Schenk, C.G., Gökarıksel, B., & Behzadi, N.E. (2022). Security, violence, and mobility: The embodied and everyday politics of negotiating Muslim femininities. *Political Geography 94*: 1–4.

Secor, A.J. (2001). Toward a feminist counter-geopolitics: Gender, space and Islamist politics in Istanbul. *Space and Polity 5*(3): 191–211.

Secor, A.J. (2002). The veil and urban space in Istanbul: Women's dress, mobility and Islamic knowledge. *Gender, Place and Culture: A Journal of Feminist Geography 9*(1): 5–22.

Secor, A.J., Ehrkamp, P., & Loyd, J.M. (2022). The problem of the future in the spacetime of resettlement: Iraqi refugees in the U.S. *Environment and Planning D: Society and Space 40*(3): 508–527. doi:10.1177/02637758221088865

Shaban, N. (2021). The influence of geopolitics on the resistance movement against compulsory hijab in Iran. Master's thesis, University of Colorado at Boulder.

Shakhsari, S. (2014). The queer time of death: Temporality, geopolitics, and refugee rights. *Sexualities 17*(8): 998–1015. doi:10.1177/1363460714552261

Sharp, J. (2005). Geography and gender: Feminist methodologies in collaboration and in the Field. *Progress in Human Geography 29*(3): 304–309.

Sharp, J. (2007). Geography and gender: Finding feminist political geographies. *Progress in Human Geography 31*(3): 381–387.

Sharp, J. (2009). Geography and gender: What belongs to feminist geography? Emotion, power and change. *Progress in Human Geography 33*(1): 74–80.

Sharp, J.R. (1996). Gendering nationhood: A feminist engagement with national identity. In Duncan, N. (ed.), *Bodyspace: Destabilising Geographies of Gender and Sexuality*. London: Routledge, 97–108.

Silvey, R. (2004). Transnational domestication: State power and Indonesian migrant women in Saudi Arabia. *Political Geography 23*(3): 245–264.

Smith, S.H. (2009). The domestication of geopolitics: Buddhist-Muslim conflict and the policing of marriage and the body in Ladakh, India. *Geopolitics 14*(2): 197–218.

Smith, S. (2011). "She says herself, 'I have no future'": Love, fate and territory in Leh District, India. *Gender, Place and Culture 18*(4): 455–476.

Smith, S. (2012). Intimate geopolitics: Religion, marriage, and reproductive bodies in Leh, Ladakh. *Annals of the Association of American Geographers 102*(6): 1511–1528. doi: 10.1080/00045608.2012.660391

Smith, S. (2013). "In the past, we ate from one plate": Memory and the border in Leh, Ladakh. *Political Geography 35*: 47–59.

Smith, S. (2020). *Intimate Geopolitics: Love, Territory, and the Future on India's Northern Threshold*. New Brunswick, NJ: Rutgers University Press.

Smith, S., & Vasudevan, P. (2017). Race, biopolitics, and the future: Introduction to the special section. *Environment and Planning D: Society and Space 35*(2): 210–221. doi:10.1177/0263775817699494

Staeheli, L.A., & Cope, M.S. (1994). Empowering women's citizenship. *Political Geography 13*(5): 443–460.

Staeheli, L.A., & Lawson, V.A. (1995). Feminism, praxis, and human geography. *Geographical Analysis 27*(4): 321–338.

Strüver, A. (2007). The production of geopolitical and gendered images through global aid organisations. *Geopolitics 12*(4): 680–703. doi:10.1080/14650040701546111

Tarlo, E. (2010). *Visibly Muslim: Fashion, Politics, Faith*. Oxford: Berg.

Torres, R.M., Glockner, V., Niño-Vega, N., et al. (2022a). Lockdown and the list: Mexican refugees, asylum denial, and the feminist geopolitics of esperar (waiting/hoping). *Environment and Planning C: Politics and Spaces 41*(8): 23996544221118906.

Torres, R.M., Swanson, K., Faria, C., Herrera, T.S., & Blue, S. (2022b). Bordering through care and control: Policing and sheltering central American migrant youth in Mexico. *Political Geography 98*: 102719. doi:10.1016/j.polgeo.2022.102719

Truelove, Y. (2021). Gendered infrastructure and liminal space in Delhi's unauthorized colonies. *Environment and Planning D: Society and Space 39*(6): 1009–1025. doi:10.1177/02637758211055483

Truelove, Y., & Ruszczyk, H.A. (2022). Bodies as urban infrastructure: Gender, intimate infrastructures and slow infrastructural violence. *Political Geography* 92: 102492. doi:10.1016/j.polgeo.2021.102492

Wright, M.W. (2004). From protests to politics: Sex work, women's worth, and Ciudad Juarez modernity. *Annals of the Association of American Geographers 94*(2): 369–386.

Wright, M.W. (2009). Gender and geography: Knowledge and activism across the intimately global. *Progress in Human Geography 33*(3): 379–386.

Zaragocin, S. (2019). Gendered geographies of elimination: Decolonial feminist geographies in Latin American settler contexts. *Antipode 51*(1): 373–392. doi:10.1111/anti.12454

Zaragocin, S., & Caretta, M.A. (2021). *Cuerpo-territorio*: A decolonial feminist geographical method for the study of embodiment. *Annals of the American Association of Geographers 111*(5): 1503–1518. doi:10.1080/24694452.2020.1812370

Chapter 21

Postcolonial and Decolonial Political Geographies

Chih Yuan Woon

Department of Geography, National University of Singapore, Singapore

---

In 2017, the journal *Transactions of the Institute of British Geographers* published a special section with the provocative title "Decolonialising Geographical Knowledges." This themed intervention was released in conjunction with the 2017 RGS-IBG Annual Conference, which was publicized under the exact same label, albeit with the subtitle "Opening Up Geography to the World." In the introductory essay to the set, Radcliffe (2017: 332) asserts that featuring this theme at the conference attests to the "colonial modern resources and authority that continue to pervade British public life—including some of its most respected institutions" (see also Legg 2017). Yet it simultaneously represents a call for a *disciplinary-wide engagement* to open up geography to Southern knowledges, to "re-think the world *from* Latin America, from Africa, *from* Indigenous places and *from* the marginalized academia in the Global South" (Legg 2017: 331, emphasis in the original). Indeed, as Radcliffe (and colleague) argue elsewhere, a decolonization injunction involves decentering and allowing for diversity, a multiplicity of perspectives, heterogenous political actions, and other ways of knowing and being (Radcliffe & Radhuber 2020; see also Naylor et al. 2018).

Following this high-profile intervention, there has been no lack of voices claiming that geography has undergone a "decolonial turn," with decolonial perspectives and praxis increasingly debated in critical geographical scholarship in the so-called Global North (see, e.g., Stanek 2019; Radcliffe 2022; Kamal & Courtheyn 2024). Indeed, geographers have gone to great lengths to foreground decolonial geographies' emancipatory potential while also explicating the

The Wiley Blackwell Companion to Political Geography, Second Edition.
Edited by Virginie Mamadouh, Natalie Koch, Chih Yuan Woon, and John Agnew.

contradictions and paradoxes between epistemic decolonization and anticolonial practice (Davies 2019; Muller 2021; Zaragocin & Caretta 2021). Despite this wider disciplinary development, this chapter insists on placing postcolonial and decolonial debates alongside each other in order to foreground their complementarities and synergies in critiquing and questioning the longstanding legacies of colonialism and arrive at alternative, more cosmopolitan ways of imagining and characterizing the world.

Without trying to draw a strict dichotomy between the "postcolonial" and the "decolonial," I concur with Bhambra (2014: 115) that, despite their disciplinary affiliations and geographical origins and remit, there are "connected sociologies" between these two fields of study and both are equally acknowledged for their "radical potential in unsettling and reconstituting standard processes of knowledge production." I am aware of the dividing line that is often drawn between the two areas of scholarship – as some decolonial writers have it, a decolonial shift constitutes a project of delinking, whereas postcolonial theory is a project of scholarly transformations within the academy (Asher 2013). However, it is my contention that postcolonial theory does not simply belong to the realm of academia. Rather, in adapting Jazeel's (2017) observation, I uphold that some political geographers' forays into the "postcolonial" have already started to look toward the academy's "outside," to engage with Indigenous, Southern, Black, and feminist thinkers for epistemic perspectives and cosmologies heretofore obscured by Eurocentric rationality. As will be shown in the later parts of the chapter, similar efforts have also been initiated under the rubric of the "decolonial," which further attests to how postcolonial and decolonial perspectives are united by their careful attention to exposing colonialism's lingering legacies and the ways in which they have been brought into/manifested in the present in different and novel ways (see Young 2012).

Building on these developments, this chapter documents and assesses the contributions of political geography to understanding various contemporary (re)configurations of the postcolonial/decolonial. Specifically, I showcase that political geographers' emerging research forays into these new postcolonial and decolonial formations not only help to locate and expose the hidden rhizomes of colonialism's historical reach (of what remains invisible, unseen, silent, or unspoken), but simultaneously go beyond critiques to open up possibilities for projects of empowerment – to reconstruct Western knowledge formations, reorient ethical norms, turn the power structures of the world upside down, and refashion the world from below. The aims of the chapter are thus threefold. First, I map out the genealogy and development of postcolonial and decolonial theories, in order to elucidate their impacts on political geography. This will then provide the basis for me to illuminate and account for novel and emerging research agendas of postcolonial and decolonial political geography. The final part of the chapter will reflect on the methodological question of "how" to do postcolonial and decolonial political geographies in order to chart out future possibilities for the field.

Origins and Developments of Postcolonial and Decolonial Theories

In a paper that attempts to foster productive dialogues between postcolonial and decolonial scholars, Bhambra (2014) highlights both the differences and connections between these two schools of thought. According to Bhambra, postcolonialism emerged as a consequence of the work of diasporic scholars from the Middle East and South Asia and, for the most part, refers back to those locations and their imperial interlocutors (Europe and the West). Decoloniality similarly emerged from the work of diasporic scholars (albeit those from South America, including Fernando Coronil and Walter Mignolo) that addresses the critical intersections between coloniality and modernity over a longer time frame. Indeed, whereas postcolonialism refers mainly to the nineteenth and twentieth centuries, decoloniality starts with the earlier European incursions into the lands that came to be known as the Americas from the fifteenth century onward. But as Bhambra is quick to remind us, despite their differences in spatial origins and exact foci, postcolonial and decolonial arguments have been most successful in pointing to the partiality (and insularity) of historical narratives and historiographical traditions emanating from Europe. This has been particularly so in the context of "demonstrating the parochial character of arguments about the endogenous European origins of modernity in favour of arguments that suggest the necessity of considering the emergence of the modern world in the broader histories of colonialism, empire, and enslavement" (Bhambra 2014: 115).

In reflecting more deeply on the theoretical impetuses for postcolonial/decolonial scholars' emphasis on criticizing the material and discursive legacies of colonialism, Bhambra asserts that the works of three writers – namely, Edward Said, Homi Bhabha, and Gayatri Chakravorty Spivak – have been particularly influential. This is of course a view that has been echoed by geographers such as Jazeel (2019) and Sharp (2003). Said is best known for his book *Orientalism* (1978), which is often considered to be the founding text of contemporary postcolonial and decolonial theory. Orientalism, according to Said, offers a critical pathway to conceptualize the history of relations between what we might commonsensically call the "West" and the "East" or the Occident and the Orient. Rather than accept the term as one that designates an area of neutral scholarly expertise (be it Oriental languages, literatures, or customs), Said argues that Orientalism was and continues to function as a discourse in which the West's knowledges about the Orient are inextricably bound up with its domination over it. Evoking Michel Foucault's proposition that all forms of knowledge are productive of power, Said assesses the implications of the Western construction of the Orient as an object of knowledge during colonial expansion. Because he refuses to accept the innocence of knowledges about and representations of the Orient, Said is able to consider how Orientalism's classification of the East as different and inferior has performative effects insofar as it legitimized Western interventions and rule.

This relates to the process of "Othering" and how power functions to place the European experience at the center while positing other cultures/peoples in other places as inferior and abnormal.

As with Said, Bhabha is concerned with how Western negative imag(in)ings of postcolonial countries have hinged on classifying the latter group as a homogenous category. But Bhabha presents a profound challenge to the geography of Orientalism, insofar as he refutes the binary logic inherent in Orientalism by redefining the postcolonial/decolonial positively in terms of uncertainty, ambivalence, a third space of mixing, and multiple cultural borders. Indeed, in *Location of Culture*, Bhabha (1994) introduces concepts such as mimicry and hybridity to critically question the colonial production of binary oppositions (center/margin, civilized/savage, superior/inferior) so as to underscore that cultures interact, transgress, and transform each other in much more complex and nuanced ways than dichotomous thinking allows. Hence, hybridity has the potential to intervene and dislocate processes of domination through reinterpreting and redeploying the dominant discourse – the spaces where differences meet become important, which "ultimately heralds the end of colonialism because it is not possible to keep Occidentalism pure and separate from Orientalism" (Sharp 2009: 121). In Bhaba's conceptualization, then, the hybrid is a form of resistant politics that does not simply redraw boundaries but actively seeks to subvert them.

Spivak's ethico-political project, on the other hand, is concerned with the least powerful (the impoverished, women, and other marginalized people), whom she collectively refers to as the subaltern. In her widely cited essay "Can the subaltern speak?" Spivak (1988) argues that Western ways of knowing have always been deemed the *only* legitimate way of knowing, thereby downplaying or even negating other (non-Western or Indigenous) forms of knowledge. Spivak is thus engaging in "a critique of the self-assured, scientific approach to studying other cultures that has been characteristic of Western scholarship, which makes the Western self the subject of history and the nonwestern other its object" (Sharp 2009: 111). As a result, in order for their voices to be heard and taken seriously, the subaltern have to adopt Western thought, language, and reasoning. In this sense, Spivak questions whether the subaltern can truly speak for themselves or if their views necessarily have to be represented (sometimes even in a distorted manner). The issue here, according to Spivak, does not lie with the inability of the subaltern to speak but with the unwillingness of the dominant groups to listen. As such, the subaltern must always be "caught in translation, never truly expressing herself, but always already interpreted" (Briggs & Sharp 2004: 665).

Crystallizing the theoretical insights from Said, Bhabha, and Spivak, it can be argued that postcolonial/decolonial perspectives possess two critical agendas. First, they address issues such as identity, race, ethnicity, and gender, challenging the ways in which they are intimately bound up with the power/knowledge nexus, both in terms of how colonial powers produce and use the knowledges of colonized peoples for their own agendas and in how these knowledges continue to structure inequitable relations between the colonizers and those formerly colonized. However, a second concern goes beyond critique to reignite interest in the agency of the marginalized/subaltern groups in articulating, reclaiming, and celebrating their cultural identities, subjectivities, and knowledges. Here, the goal is to decenter the hegemonic Western knowledge regime, thereby constructing and legitimizing other knowledge systems by exploring alternative epistemologies, ontologies, and cosmologies.

Despite the significant uptake of postcolonial and decolonial theories in a number of disciplines within the social sciences and the humanities, there have been ongoing contemplations pertaining to their actual impacts and continued relevance. There have been growing concerns that the intellectual agenda of decolonizing knowledges has yet to be fulfilled. Although there has been more than a decade of talk about decolonizing knowledges, it has been argued that postcolonialism and decoloniality represent the interests of Western-based intellectual elites (Bhaba, Said, and Spivak were employed at universities in Europe and/or the United States when they made their most influential interventions, for example) who speak the language of the Western academy, thereby perpetuating the exclusion of the colonized and oppressed. Furthermore, there is the sentiment that postcolonial and decolonial perspectives seem to have lost some vigor, especially in the context of various disciplines' fast-changing theoretical predilections and that certain empirical settings may feel as if past the postcolonial/decolonial (e.g., Europe) (Sidaway et al. 2014). However, critics have pointed out that postcolonial and decolonial perspectives remain vital and relevant in many other places and that scholars should not lose sight of temporal and geographical variations with respect to postcolonial and decolonial theories and their relevance (see Jazeel 2019; Sharp 2023).

In reflecting on these concerns, Young (2012: 22) argues that thinking through the postcolonial (and indeed, I would argue the same for the decolonial) still has yet-to-be-realized potentials. In particular, he suggests that the postcolonial remains insofar that it "lives on, ceaselessly transformed in the present into new social and political configurations." As such, Young is essentially imploring for contemporary postcolonial analyses to take on the challenge of understanding how the postcolonial itself has been changed in response to the historical transformations that have been occurring in the last decades. In other words, what conditions and situations have risen to new visibility on the radar of postcolonialism? But even if such questions are increasingly important, Young cautions that to truly decolonize knowledge, postcolonial research must avoid the tendency to employ the language of the "Other" while trying to deconstruct this exact same process of "Othering" in Western discourses. If not, postcolonialism will simply be "accept[ing] the discriminatory gesture of social and political othering that it appears to contest" (2012: 37). Hence, for Young, the critical task for postcolonial (and decolonial) researchers is to delineate possible pathways to stop majority groups from othering minorities altogether, "at which point minorities will be able to represent themselves as they are, in their specific forms of difference, rather than as they are othered" (2012: 37).

The rest of the chapter will critically discuss political geographers' ventures into the postcolonial and decolonial. Apart from documenting how the critical agendas of postcolonialism/decoloniality have shaped the developmental trajectory of political geography, I will also offer insights into how political geographers have, in line with Young's critical expositions, avoided the language of the "Other" to engage with the novel geopolitical, economic, and cultural logics that imperial and neocolonial geographies proliferate as well as the resistances to them.

Postcolonial and Decolonial Political Geographies

Up till the 1990s, there was little dialogue between postcolonialism/decoloniality and political geography, reflecting differences in disciplinary traditions, politics, and divergences in the languages and concepts used to articulate core issues. However, the possibility of producing a truly postcolonial decolonized knowledge in political geography became a subject of keen interest and considerable debate during the early 1990s. Postcolonial and decolonial perspectives are posing challenges to dominant ways of apprehending North–South relations and have begun to pose significant questions for political geography, particularly the material effects of discourse, the (geo)politics of knowledge, and notions of agency and power.

In the previous iteration of this chapter (Woon 2015), I attempted to highlight how political geographers have deployed a postcolonial lens to critically interrogate the rise of "new empires" (e.g., China) and the radical possibilities offered by subaltern ideas, knowledges, and practices, by drawing particularly on important interventions by Sharp (2011, 2013) and Sidaway (2012). These research directions remain extremely productive in defining and shaping the scope of postcolonial and decolonial (political) geographies, especially when writings on subaltern geopolitics have been gaining a significant following and traction among political geographers (e.g., Craggs 2018; Ferretti 2021). Crucially, there have been some interesting geographical writings that broach the interface of decolonial, subaltern, and liminal debates to articulate standpoints that are not completely other, resistant, or alternative to dominant geopolitics, but adopt an ambiguous position of marginality (McConnell 2017; Ferretti 2024). Moreover, in recent years political geography textbooks have emerged that take into serious consideration how decolonial debates and theories can expand the conceptual and empirical imaginations of the subdiscipline (Smith 2020; Squire & Jackman 2023).

But as the following discussions reveal, there are new(er) "real-world" and theoretical developments that political geographers have exhorted should have urgent postcolonial and decolonial interventions. I will highlight two recent research pathways in which postcolonial and decolonial inflections of political geography have allowed for the decentering of "Western" knowledges and practices and the initiating of culturally sensitive scholarship across the North–South divide at this (geo)political juncture.

Critiquing the "Colonial Present": The Wars in Gaza and Ukraine

In his widely acclaimed monograph *The Colonial Present*, Derek Gregory (2004) makes the persuasive argument that colonialism continues to manifest itself in the present through the construction of imaginative geographies or "constructions that fold distance into difference through a series of spatializations" (2004: 17). In this formulation, Gregory maintains that the absence of traditional colonies does not imply the absence of colonialism. Specifically, he extends Said's idea of Orientalism to "real" spaces (i.e., Afghanistan, Iraq, and Palestine) to show how the United States–initiated global "war on terror" was framed in cultural terms and reminiscent of the colonial past. The ultimate goal of *The Colonial Present*, then, is to highlight the physical and corporeal manifestations of contemporary colonialism. In doing so, Gregory exposes the brutality of violence that accompanies imperial (territorial) control in the Middle East that a complacent, if not complicit, American government and media have tried to shield from the public.

Twenty years later, the Israeli military's ongoing war in Gaza has once again reminded us of the veracity of Gregory's observations – that the "colonial present" persistently haunts and engulfs Palestine. These military interventions by Israel are arguably in response to the October 7, 2023, chain of events where Hamas, an Islamist organization with goals that extend well beyond establishing a Palestinian state, and several Palestinian militant groups not only launched over 3000 rockets against Israel but, more significantly, simultaneously undertook massive incursions into Israel, a large-scale massacre of civilian communities, as well as the kidnapping of Israeli hostages (including soldiers and children) and taking them into Gaza. While Hamas's actions were immediately labeled acts of terror by world organizations/leaders, Israel's destructive retaliatory actions invited widespread criticisms as well. The UN (2023) has already described these devastating assaults on one of the most vulnerable populations on the planet as a "true human catastrophe" and critiques abound that Israel's incursions into Gaza are a "textbook case of genocide" (e.g., Segal 2023). Going beyond the veneer of these intensified events, it has been emphasized that October 7, 2023, cannot offer a starting point for narrating the situation in Palestine. Rather, this contemporary war needs to be seen as a continuation of a long history of European colonialism in the region, influenced by British colonial interest and its facilitation of the Zionist movement's desire to establish a Jewish homeland in Palestine through various means (Thompson 2019; Khalidi 2021). As such, a settler colonialism framework has been mobilized by a multiplicity of scholars to enable a reading of these critical historical continuities while also making visible the differing power relations and politics that are playing out in Gaza and the West Bank (Velednitsky et al. 2020; Nabulsi 2023). John Agnew (2023b) further reminds us, however, that the Gaza situation is not a unique case and that the story of settler colonialism and ethnic displacement is more pervasive worldwide. He argues that "the expropriation of Palestinian territory by Israeli settlers on the West Bank has a prehistory in the way in which British settlers incorporated parts of Ireland and the British governments depopulated the Scottish Highlands." Thus, in Agnew's words (2023b: 7), "the Zionist state is a perfect replication of the idealized European form projected into the rest of the world" and the war in Gaza represents the wider tragedy of the (ethno-)nation-state that has exclusionary territorialized logic as its basis. From this perspective, a Palestinian state engaged in removing all Israelis would similarly not be decolonial but would be conducting exactly the same process inherited from the European "ideal" of an ethnically homogenous nation-state. This corresponds to Yiftachel's (2024) critical reminder to make a clear distinction between the "counter-colonial" and the "decolonial" – that the Palestinian and Islamic counter-colonial visions and the political and terrorist mobilizations these visions encapsulate must also form part of the story around the destructive events in Gaza.

Political geographers have tended to miss this complexity and argue that the settler colonial past and present make the question of Palestine an inherently one-sided geographical one. From this viewpoint, the most recent rendition of violence in Gaza is only serving to further accentuate the territorial strategies and the un/making of space and place that have resulted in a longstanding landscape of spatial segregation that defines the everyday geographies and realities of ordinary Palestinians. Crucially, it has been pointed out that Palestinians are severely restricted in movement (Rijke & Minca 2019; Griffiths and Repo 2020), facing forced displacements and demolition of their homes (Harker 2011; Feldman 2015), subjected to military practices and urban land-grabbing (Jones 2023; Porter & Yiftachel 2017), and suppressed from making any political and cultural expressions, thereby leading to their diminished hopes for the future (Hassouna 2023; Jarvi 2023). In proposing a "decolonial geography" in looking at the Gaza situation, Agha et al. (2024) contend that all these aforementioned geographical inquiries into the day-to-day debilitating experiences of Palestinians can help to contribute to a decolonial politics. However, such academic acts of rethinking the world from marginalized perspectives "should not supplant structures, institutions and praxis as the basis of decolonial politics" (2024: 4). Given this approach, they simultaneously call on (political) geographers to engage in scholarly activism with the Palestinians themselves. This may involve listening to and supporting Palestinian colleagues and students who are often the vanguard of Palestinian solidarity movements on university campuses; building links with Palestinian institutions and academics; and working with national and institutional restrictions to make ethical choices on research partnerships and in pursuing questions around university investments. These concrete actions, according to Agha et al. (2024: 6), would help to develop an academic praxis/ethos that "works meaningfully toward an end to the siege of Gaza, towards a reckoning with the political alliances that prop up Israel's settler colonial regime and normalise apartheid policies across Palestine." Whether this must ultimately require an exclusively Palestinian state in place of an Israeli one is left unclear.

While the colonial histories underpinning the Gaza conflicts have been well documented, the same cannot be said for another war that has preoccupied public and scholarly imagination in recent times – that is, Russia's invasion of Ukraine in 2022 (see, e.g., Dodds et al. 2023). (There are longer historical roots to this war, but Russia's annexation of Crimea in 2014 is widely seen as the precursor to the 2022 conflict; see Wood et al. 2016.) Indeed, as Moisio (2022) pointedly notes, realist geopolitical perspectives have more often than not dominated discussions about this incident. On the one hand, Western narratives routinely frame the crisis as stemming from Russian political leaders' everlasting obsession with occupying and mastering territories and uniting the Slavic peoples (Smirnova 2023). On the other hand, there are discourses around Russia engaging in a preemptive war because of the fact that its geopolitical standing has become increasingly compromised/vulnerable due to NATO's expansion (see also Megoran 2022; Klinke 2023). Political geographers have produced some fascinating works deconstructing the inherent weaknesses within such dualistic reasoning; see the virtual forum in *Political Geography* (2022) for related interventions into this event. Although constituting a minority, a handful of them have started to reflect on the conceptual utility in deploying a postcolonial/decolonial lens to interrogate the war in Ukraine. They are however quick to remind us that postcolonial/decolonial thought stemming from contexts with a Western (settler) colonial history "does not offer blueprints that could be simply applied to Eastern Europe and Central Asia" (Vorbrugg & Bluwstein 2022: 2).

Indeed, as Bonnett (2023) emphasizes, the invasion of Ukraine reveals that colonialism does not come in one flavor. Russian's model of colonialism (as signified by its treatment of Ukraine) is done through absorption rather than the conquest of distant lands. Although a case can be made for Russian's expansion being less destructive (as compared to traditional "Western" forms of colonialism), Bonnett's (2023) criticism is that the "consequences of absorption is a lack of recognition that conquest has taken place and, hence, an almost impermeable sense of entitlement." The outcome is a firm belief that contiguous lands are not real countries but mere annexes of Russia. This point comes through even more clearly in Agnew's (2023a) piece where he calls for a focus on "how" the war is conducted (rather than explaining it through various universal theories), which can in turn highlight what purpose(s) the war is supposed to serve. Critically, Agnew demonstrates in a nuanced way how Putin's military intervention into Ukraine has fallen into a "territorial trap." Part of his argument resonates with Bonnett's – he reveals Putin's (flawed) view of states as static storage containers of people, resources, and sentiments, so that when understood from Putin's perspective, Ukraine simply has no meaning as a separate state or society because of its historical subordination to Russia, as part of the all-Russian triune nation – although he stresses that most of the territory of post-Soviet Ukraine was colonized and developed by the Russian empire after conquest over the Ottomans.

Alternatively, Dubrova (2022) and Vorbrugg and Bluwstein (2022) agree that realist international relations accounts have drowned out and downplayed the colonial nature of Russia–Ukraine relations. Moreover, there is the perpetuation of the Orientalist myth that Ukraine is simply a buffer zone that demarcates the nomadic "East" and the enlightened, settled "West." In so doing, these two papers urge postcolonial/decolonial perspectives to decenter knowledge production, so as to allow for learnings from and engagements with scholars and activists with profound knowledge of or ties to Ukraine and the region. Similar calls for the decolonializing of knowledge production about Ukraine are also emerging in other disciplinary scholarship (Burlyuk & Musliu 2023; Byford et al. 2024). This can simultaneously spur productive inquiries into the persistence of violence frontiers in Ukraine and demand explorations of where Ukrainian borders are and for whom these borders are for.

All in all, a decolonial/postcolonial analysis allows us to go beyond the nationalist framings that have often accompanied Israel's and Russia's justifications for war in Gaza and Ukraine, respectively (e.g., claims of a historical homeland in pre-Islamic and pre-Christian Jerusalem and via the evocation of Kievan Rus, respectively). Indeed, it draws attention to the ways in which these violent endeavors are underpinned by colonial logics, frameworks, and legacies. This includes, of course, both Palestinian and Ukrainian aspirations to a European-style nationhood.

"New" Empires? The Belt and Road Initiative and Chinese Geopolitics

Recent years have witnessed significant interest in the economic ascendancy of countries outside of the "West" that contribute to a shifting of paradigms of geo-economic and geopolitical power that has certainly modified the sensibility of colonial dependency. One area that has arguably attracted widespread attention is "Asia," as attested by the buzz around the discourse of "Emerging Asia" (or what some term "Rising Asia"; see Roy 2016; Sheppard 2016). Increasingly, one could argue that China has come to serve as the perfect embodiment of the Emerging Asia discourse, especially given the Chinese government's ambitious Belt and Road Initiative (BRI), which has been touted to have profound implications for governance and development regimes/frameworks at the global level (see Liu & Dunford 2016). As such, it is my contention that the BRI and the ways it positions China on the international stage require us to critically reexamine the logics of (Western) imperialism and empire.

Indeed, framed by Chinese political elites as constituting a set of infrastructural projects that are open to any interested party, the BRI's "contemporary connectivity projects . . . have generated heated discussion about their potential to transform the global geopolitical landscape" (Blanchard & Flint 2017: 223). China's reliance on roads, railways, ports, and pipelines to facilitate economic partnerships with multiple state and nonstate actors has led some scholars to theorize the BRI as initiating "infrastructure diplomacy" (Jia & Bennett 2019). Despite the seemingly "win–win" and "inclusive" nature of the BRI, criticisms have emerged that question the BRI's close associations with debt-trap diplomacy (Cooley & Heathershaw 2017). Many critics have argued that China seeks to deliberately entrap various BRI-participating countries (especially in Africa and South Asia) in webs of debt in order to secure certain strategic and/or material advantages (Gopaldas 2018). This, in turn, raises the question of whether the BRI is merely reproducing hierarchical and unequal relations in international politics, thereby allowing China to displace "Western" powers and regimes to become a new imperial power and global hegemon (Beeson & Crawford 2023). Additionally, China's emphasis on constructing an international system based on Chinese visions and ideas can create suspicions about its true intent in formulating and implementing the BRI. The close associations drawn between the BRI and the Chinese world vision fall into the trap of what Callahan (2016) terms "Sino-speak." According to Callahan, Sino-speak is the emerging dialect for articulating the interface between a new Orientalism and Chinese exceptionalism, whereby the uniqueness of Chinese values is celebrated as the moral panacea to a US/Western-centered geopolitical world that is constructed as decadent and chaotic. Consequently, in a bid to leverage the BRI to foreground Chinese exceptionalism for the ushering in of a distinct, nonhierarchical model of international relations, China has, according to some scholars, paradoxically manufactured its own discursive legitimacy for Sinocentric hegemony in the twenty-first century (see Woon 2018 for a summary of these views).

Political geographers have not remained detached from such debates around the BRI. One of their most important contributions has been to investigate how the BRI has been constructed and implemented as a political and economic project in a variety of geographical settings (Flint & Zhu 2019; Murton 2023; Poon et al. 2024). This research agenda appears to stem from a fundamental desire to eschew seeing the BRI as a monolithic program that is designed in Beijing and imposed on others for geopolitical and imperial ends. As Oliveira et al. (2020: 81) explicate in a special issue in *Political Geography* titled "China's Belt and Road Initiative: Views from the ground," it is imperative to "invert analysis of the BRI from a top-down coherent strategy to a relational, contested process that occurs in specific places." In other words, in order to understand how the BRI is reshaping global development, it must be examined empirically "from the ground." What this essentially means is that there is no clear-cut and straightforward answer as to whether the BRI is enabling a new "Chinese empire" to emerge on the global stage. Rather than subscribing to a binary viewpoint on whether China is embarking on the route to empire, political geographers are paying attention to the ways in which the BRI (and China's power and influence) is being negotiated, interpreted, or even resisted by a multiplicity of actors (Chinese or otherwise) in their everyday life (see also Cheng & Apostolopoulou 2023).

Beyond empirical expositions, some political geographers have also seen the BRI as an opportunity to "postcolonialize/decolonize" the subdiscipline. Specifically, the Chinese philosophies and ideas that have been drawn on by Chinese elites to justify the BRI have been given careful reflection. It is important to highlight that these political geographical works have neither sought to reproduce the fallacies of Chinese exceptionalism (as pointed out earlier) nor attempted to advocate the relative merits and superiority of a Chinese model of globalization/governance. In contrast, their main aims are to provide more situated accounts and explanations of Chinese foreign policy agendas and to contemplate how these Chinese theories and ideas can be put into productive conversations with other worldviews so as to foster more cosmopolitan and accurate understandings of the (geo)political environments in which we live. For example, An et al. (2021) postulate that existing Western geopolitical frameworks are insufficient for grasping the complex theories and historical legacies that underpin China's foreign policy. This, in turn, leads them to propose a close appreciation of Confucian philosophies and their uptake in Chinese state theory, geographical imaginations, and geopolitical scripts related to the BRI (see also An 2020).

Alternatively, Woon (2021) focuses on how China has counterposed the "win–win" and mutually beneficial nature of the BRI with a "Western" geopolitical order that is zero-sum and hierarchical in nature. These positive values undergirding the BRI draw on an assemblage of Chinese traditions (notably Confucianism and Daoist dialectic) to foreground a relational ontology that is premised on egalitarian and inclusive "Self–Other" geopolitical interactions. This Chinese-inflected narrative of the BRI, according to Woon, is thus instructive in offering new concepts that are contextually grounded, thereby contributing to the broader movement of stimulating dialogues between Sinophone and Anglophone scholarship for a global relational theorizing of world politics.

Conclusion: A Pluriverse of Political Geographies

This chapter has foregrounded some key topical areas that have yielded postcolonial interventions on the part of political geographers. In closing, I want to reflect on the future of postcolonial political geography and am reminded of Jazeel's (2017) caution that a truly postcolonial/decolonial imperative for geography does not only involve a pluralization of voices within our disciplinary community but also demands an ethical commitment to "abide by" the places where we work and study. How can we then attend to those "elsewheres" as well as intervene in issues even when they do not matter to the Anglophone world and community? To answer these questions, I am increasingly drawn to the epistemological, methodological, and ontological sensibilities that are opened up by the idea of the pluriverse.

Pluriverse, as a term, is not entirely foreign to political geographers. The subdiscipline's experimentations with the idea are connected by the acknowledgment that there are multiple worlds out there that have historically been suppressed and marginalized by a dominating "Western" cosmology. This understanding resonates strongly with postcolonial and decolonial impetuses, which has consequently culminated in calls for the coexistence of many worlds and the entanglements of diverse cosmologies (of which "Western" universalism is but one). In illuminating how the notion can be put into operation, Oslender (2019), in understanding the political organization of Afro-Colombian communities, advocates engagement with the on-the-ground experiences of what it means to live in a pluriverse. But, as Ferretti (2023) cautions, this is not a simple endorsement of the act of documenting more case studies in/of the so-called Global South. Rather, in citing Italian writer Franco Cassano's explorations of "Southern Thought" and reflecting on his own research in Latin America, Federico (2023: 8) is adamant that the pluriverse signifies "a posture of learning and translation that is not about going on a sightseeing tour inside the culture of the other but about taking it seriously and learning something from it." Recent exchanges on Muslim geographies in *Dialogues in Human Geography*, I contend, can further enrich debates on the pluriverse (and the decolonial), especially in light of Sidaway's exhortation to "explor[e] *premodern and non-European* imperialism, [thereby] bringing many other geographies and destinations into view" (2023: 358, emphasis mine).

Hence, if we take into account the possibilities of the pluriverse in the "doing" of postcolonial political geographies, the methodological implications are significant. (Postcolonial) political geographers should not reduce the language and communities of the Southern Hemisphere to mere objects of study; rather, they must take up the position of a learning subject inclined to "learn from below" (Spivak 2003: 15). This type of learning is not about speaking for an individual or group in order to produce something like a notionally more cosmopolitan geographical knowledge. Instead, it seeks to develop new positions through interactions between researchers and people in disparate locations. This hinges on a greater sensitivity to the relationship of power, authority, positionality, and knowledge, and an ethic that demands consistent critical reflections in an effort that seeks to learn between contexts and constituencies (Jazeel & McFarlane 2010). It requires continual questioning and reflexivity about one's relative position (e.g., nationality, class, and gender) and how that can potentially have negative effects on attempts to create knowledge and to represent others. A good example of this responsible learning process is embodied through the methodological commitment to praxis that refocuses on resistance, agency, and action – there are already ongoing efforts by political geographers to move out of the ivory tower and collaborate with marginalized and underrepresented communities and social movements. This can involve employing field techniques that broach a collaborative ethos, such as those offered by participatory action research, so that the dangers of misrepresenting different subaltern groups can be avoided through culturally informed understandings of the agency of research participants in the (re)making of their individual lifeworlds. Thus, the key lesson here for political geographers is that they must always be prepared to work with the uncertainties inherent in a postcolonial/decolonial approach to learning and come to accept that their knowledges, ideas, or theories may be altered, transformed, or even rejected through responsible engagements with different constituencies.

References

Agha, Z., Esson, J., Griffiths, M., & Joronen, M. (2024). Gaza: A decolonial geography. *Transactions of the Institute of British Geographers 49*: e12675.

Agnew, J. (2023a). Vladimir Putin's territorial trap: What the invasion of Ukraine reveals about the contemporary war-sovereignty nexus. In Daho, G., & Richard, Y. (eds), *War, State and Sovereignty: Interdisciplinary Challenges and Perspectives for the Social Sciences*. London: Palgrave Macmillan, 43–70.

Agnew, J. (2023b). The language of intractability and the Gaza war: Conflating anti-Semitism and anti-Zionism is historically problematic and misses how much contemporary Israel HAS become a role model for ethno-nationalists worldwide. *Human Geography 17*(2). doi:10.1177/19427786231220046

An, N. (2020). *Confucian Geopolitics: Chinese Geopolitical Imaginations of the US War on Terror*. Singapore: Springer.

An, N., Sharp, J., & Shaw, I. (2021). Towards a Confucian geopolitics. *Dialogues in Human Geography 11*(2): 218–235.

Asher, K. (2013). Latin American decolonial thought, or making the subaltern speak. *Geography Compass 7*(12): 832–842.

Beeson, M., & Crawford, C. (2023). Putting the BRI in perspective: History, hegemony and geoeconomics. *Chinese Political Science Review 8*: 45–62.

Bhabha, H. (1994). *The Location of Culture*. London: Routledge.

Bhambra, G. (2014). Postcolonial and decolonial dialogues. *Postcolonial Studies 17*(2): 115–121.

Blanchard, J.M.F., & Flint, C. (2017). The geopolitics of China's maritime silk road initiative. *Geopolitics 22*(2): 223–245.

Bonnett, A. (2023). Comment: The invasion of Ukraine points to a colonial history. Newcastle University, March 22. https://www.ncl.ac.uk/press/articles/archive/2022/03/commentthecolonialhistoryofukraine (accessed February 27, 2024).

Briggs, J., & Sharp, J. (2004). Indigenous knowledges and development: A postcolonial caution. *Third World Quarterly 25*: 661–676.

Burlyuk, O., & Musliu, V. (2023). The responsibility to remain silent? On the politics of knowledge production, expertise and (self-)reflection in Russia's war against Ukraine. *Journal of International Relations and Development 26*(4): 605–618.

Byford, A., Doak, C., & Hutchings, S. (2024). Decolonizing the transnational, transnationalizing the decolonial: Russian Studies at the crossroads. *Forum for Modern Language Studies 60*(3): 339–357.

Callahan, W. (2016). Sino-speak: Chinese exceptionalism and the politics of history. *Journal of Asian Studies 71*(1): 33–55.

Cheng, H., & Apostolopoulou, E. (2023). Locating the Belt and Road Initiative's spatial trilectics. *Geography Compass 17*(4): e12683.

Cooley, A., & Heathershaw, J. (2017). *Dictator without Borders: Power and Money in Central Asia.* New Haven, CT: Yale University Press.

Craggs, R. (2018). Subaltern geopolitics and the post-colonial commonwealth, 1965–1990. *Political Geography 65*: 45–56.

Davies, A. (2019). *Geographies of Anticolonialism: Political Networks across India, c. 1900–1930.* London: Wiley-Blackwell.

Dodds, K., Taylor, Z., Akbari, A., et al. (2023). The Russian invasion of Ukraine: Implications for politics, territory and governance. *Territory, Politics, Governance 11*(8): 1519–1536.

Dubrova, M. (2022). Decolonial paths through borderlands: Perspectives from the Ukrainian heartland. *Political Geography 99*: 102747.

Feldman, I. (2015). What is a camp? Legitimate refugee lives in spaces of long-term displacement. *Geoforum 66*: 244–252.

Ferretti, F. (2021). Geopolitics of decolonisation: The subaltern diplomacies of Lusophone Africa (1961–1974). *Political Geography 85*: 102326.

Ferretti, F. (2023). Geography, pluriverse and "southern thought": Engaging with decoloniality from the Mediterranean. *Political Geography 107*: 102990.

Ferretti, F. (2024). Conciliating subaltern and liminal geopolitics: Reggio Emilia's city diplomacy and the geographies of Lusophone Africa's decolonisation. *Geopolitics*: 1–28. doi:10.1080/14650045.2023.2291067

Flint, C., & Zhu, C.P. (2019). The geopolitics of connectivity, cooperation, and hegemonic competition: The Belt and Road Initiative. *Geoforum 99*: 95–101.

Gopaldas, R. (2018). Lessons from Sri Lanka on China's "debt-trap diplomacy". Institute for Security Studies, February 21. https://issafrica.org/iss-today/lessons-from-sri-lanka-on-chinas-debt-trap-diplomacy (accessed February 28, 2024).

Gregory, D. (2004). *The Colonial Present: Afghanistan, Palestine, Iraq.* Malden, MA: Blackwell.

Griffiths, M., & Repo, J. (2020). Women's lives beyond the checkpoint in Palestine. *Antipode 52*(4): 1104–1121.

Harker, C. (2011). Geopolitics and family in Palestine. *Geoforum 42*(3): 306–315.

Hassouna, S. (2023). Cultivating biodiverse futures at the (postcolonial) botanical garden. *Transactions of the Institute of British Geographers 49*: e12639.

Jarvi, T. (2023). Expectations to fulfil: Anticipating the familial future in Palestine refugee camps. In Griffiths, M., & Joronen, M. (eds), *Encountering Palestine: Un/making Spaces of Colonial Violence.* Lincoln, NE: University of Nebraska Press, 176–199.

Jazeel, T. (2017). Mainstreaming geography's decolonial imperative. *Transactions of the Institute of British Geographers 42*(3): 334–337.

Jazeel, T. (2019). *Postcolonialism.* London: Routledge.

Jazeel, T., & McFarlane, C. (2010). The limits of responsibility: A postcolonial politics of academic knowledge production. *Transactions of the Institute of British Geographers NS35*: 109–124.

Jia, F.Q., & Bennett, M. (2019). Chinese infrastructure diplomacy in Russia: The geopolitics of project type, location, and scale. *Eurasian Geography and Economics 59*(3–4): 340–377.

Jones, C. (2023). Gaza and the great march of return. *Transactions of the Institute of British Geographers 48*(2): 249–262.
Kamal, A., & Courtheyn, C. (2024). Where is the North? Southern ruptures in decolonizing theory. *Political Geography 108*: 103029.
Khalidi, R. (2021). *The Hundred Years' War on Palestine*. London: Picador Books.
Klinke, I. (2023). Of tanks and tankies: What's "left" for geography after the invasion of Ukraine. *Transactions of the Institute of British Geographers 48*(4): 811–815.
Legg, S. (2017). Decolonialism. *Transactions of the Institute of British Geographers 42*(3): 345–348.
Liu, W.D., & Dunford, M. (2016). Inclusive globalization: Unpacking China's Belt and Road Initiative. *Area Development and Policy 1*(3): 323–340.
McConnell, F. (2017). Liminal geopolitics: The subjectivity and spatiality of diplomacy at the margins. *Transactions of the Institute of British Geographers 42*(1): 139–152.
Megoran, N. (2022). "Russian troops out! No to NATO expansion!" A Pacific geopolitics for a new Europe. *Political Geography 98*: 102699.
Moisio, S. (2022). Geopolitics of explaining Russia's invasion of Ukraine and the challenge of small states. *Political Geography 97*: 102683.
Muller, M. (2021). Worlding geography: from linguistic privilege to decolonial anywheres. *Progress in Human Geography 45*(6): 1440–1446.
Murton, G. (2023). Beyond the BRI: The volumetric presence of China in Nepal. *Territory, Politics, Governance 12*(1): 72–92.
Nabulsi, J. (2023). "To stop the earthquake": Palestine and the settler colonial logic of fragmentation. *Antipode 56*(1): 187–205.
Naylor, L., Daigle, M., Zaragocin, S., Ramirez, M., & Gilmartin, M. (2018). Interventions: Bringing the decolonial to political geography. *Political Geography* 66: 199–209.
Oliveira, G., Murton, G., Rippa, A., Harlan, T., & Yang, Y. (2020). China's Belt and Road Initiative: Views from the ground. *Political Geography 82*: 102225.
Oslender, U. (2019). Geographies of the pluriverse: Decolonial thinking and ontological conflict on Colombia's Pacific coast. *Annals of the American Association of Geographers 109*(6): 1691–1705.
Political Geography (2022). Virtual Forum: War in Ukraine. *Political Geography 109*.
Poon, J., Peng, P., Qian, X.Y., & Yang, Y. (2024). End of the Asia-Pacific: Geoeconomics, Belt and Road and the Indo-Pacific. *Political Geography 108*: 103026.
Porter, L., & Yiftachel, O. (2017). Urbanizing settler colonial studies. *Settler Colonial Studies 9*(2): 177–186.
Radcliffe, S. (2017). Decolonising geographical knowledge. *Transactions of the Institute of British Geographers 42*(3): 329–333.
Radcliffe, S. (2022). *Decolonializing Geography: An Introduction*. London: Polity Press.
Radcliffe, S., & Radhuber, I. (2020). The political geographies of D/decolonization: Variegation and decolonial challenges in/of geography. *Political Geography 78*: 102128.
Rijke, A., & Minca, C. (2019). 'Inside checkpoint 300: Checkpoint regimes as spatial political technologies in the occupied Palestinian territories. *Antipode 51*(3): 968–988.
Roy, A. (2016). When is Asia? *The Professional Geographer 68*(2): 313–321.
Said, E. (1978). *Orientalism*. New York: Random House.
Segal, R. (2023). A textbook case of genocide. *Jewish Currents*, October 13. https://jewishcurrents.org/a-textbook-case-of-genocide (accessed February 27, 2024).

Sharp, J. (2003). Feminist and postcolonial engagements. In Agnew, J., Mitchell, K., & Toal, G. (eds), *A Companion to Political Geography*. London: Blackwell, 59–74.
Sharp, J. (2009). *Geographies of Postcolonialism*. London: Sage.
Sharp, J. (2011). Subaltern geopolitics: Introduction. *Geoforum 42*(3): 271–273.
Sharp, J. (2013). Geopolitics at the margins? Reconsidering genealogies of critical geopolitics. *Political Geography 37*: 20–29.
Sharp, J. (2023). *Geographies of Postcolonialism: Spaces of Power and Representation*. London: Sage.
Sheppard, E. (2016). Emerging Asias: An introduction. *The Professional Geographer 68*(2): 309–312.
Sidaway, J. (2012). Subaltern geopolitics: Libya in the mirror of Europe. *The Geographical Journal 178*(4): 296–301.
Sidaway, J. (2023). Muslim geographies: Beyond the decolonial. *Dialogues in Human Geography 13*(3): 337–358.
Sidaway, J., Woon, C.Y., & Jacobs, J. (2014). Planetary postcolonialism. *Singapore Journal of Tropical Geography 35*(1): 4–21.
Smirnova, V. (2023). Russia's spatial anxiety and the construction of geographical knowledge. *Transactions of the Institute of British Geographers 48*(4): 816–820.
Smith, S. (2020). *Political Geography: A Critical Introduction*. Hoboken, NJ: Wiley-Blackwell.
Spivak, G.C. (1988). Can the subaltern speak? In Nelson, C., & Grossberg, L. (eds), *Marxism and the Interpretation of Culture*. Urbana, IL: University of Illinois Press, 271–313.
Spivak, G.C. (2003). *Death of a Discipline*. New York: Columbia University Press.
Squire, R., & Jackman, A. (2023). *Political Geography: Approaches, Concepts, Futures*. London: Sage.
Stanek, M. (2019). Decolonial education and geography: Beyond the 2017 Royal Geographical Society with the Institute of British Geographers annual conference. *Geography Compass 13*(12): e12472.
Thompson, G. (2019). *Legacy of Empire: Britain, Zionism and the Creation of Israel*. London: Saqi Books.
UN (2023). Settlement expansion in occupied Palestinian territory violates international law, must cease, many delegates tell Security Council. United Nations, September 23. https://press.un.org/en/2023/sc15424.doc.htm (accessed February 27, 2024).
Velednitsky, S., Hughes, S., & Machold, R. (2020). Political geographical perspectives on settler colonialism. *Geography Compass 14*: e12490.
Vorbrugg, A., & Bluwstein, J. (2022). Making sense of (the Russian War in) Ukraine: On the politics of knowledge and expertise. *Political Geography 98*: 102700.
Wood, E., Pomeranz, W., Merry, W., & Trudolyubov, M. (2016). *Roots of Russia's War in Ukraine*. New York: Columbia University Press.
Woon, C.Y. (2015). Postcolonialism. In Agnew, J., Mamadouh, V., Secor, A., & Sharp, J. (eds), *The Wiley Blackwell Companion to Political Geography*. London: Wiley-Blackwell, 248–264.
Woon, C.Y. (2018). China's contingencies: Critical geopolitics, Chinese exceptionalism and the uses of history. *Geopolitics 23*(1): 67–95.
Woon, C.Y. (2021). "Provincialising" the Belt and Road Initiative: Theorising with Chinese narratives of the "digital silk road" (数字丝绸之路). *Asia Pacific Viewpoint 62*(3): 286–290.

Yiftachel, O. (2024). 'Colonial—and counter-colonial: The Israel/Gaza war through multiple critical perspectives. *Palestine/Israel Review 1*(1): 228–236.

Young, R. (2012). Postcolonial remains. *New Literary History 43*(1): 19–42.

Zaragocin, S., & Caretta, M. (2021). Cuerpo-Territorio: A decolonial feminist geographical method for the study of embodiment. *Annals of the American Association of Geographers 111*(5): 1503–1518.

Doing Political Geography

Chapter 22

Electoral Geographies

Shaun Johnson and Barney Warf
Department of Geography, University of Kansas, Lawrence, KS, USA

Elections are the *sine qua non* of democracy, a key foundational institution in democratic societies, expressions of the popular political will, and the means by which the citizenry, at least in theory, chooses its public officials. Elections, when conducted in a manner that accurately reflects the choices of the voters, are the will of the electorate made real (Johnson 2022). Elections are the preferred means by which the peaceful transfer of power is accomplished. Elections are not perfect processes, however, and are subject to innumerable influences from both within and outside the political system. Moreover, elections do not guarantee democracy, as sham elections in totalitarian societies provide an ostensibly legitimate cover for manifestly undemocratic forces retaining power.

Electoral geography is, broadly, the analysis of how and why people participate in electoral systems and how that process varies over space, in different places, at different scales, and when done by different groups. In a sense, the election is the perfect social phenomenon on which to conduct research, being a scheduled event with publicly published results tied to spatial coordinates. The study of relationships between various phenomena and election returns has been prominent since the broader discipline of geography underwent the "quantitative revolution" of the 1950s and 1960s, with elections being described as a "geographer's delight, providing researchers with large amounts of data that can be mapped and spatially analysed" (Johnston & Pattie 2009, quoted in Temple 2023: 3).

Along with examinations of issues like these, electoral geographers drill down into the multitude of inherently geographical influences on voting such as gerrymandering, malapportionment, class, race, and gender, and regional, local,

The Wiley Blackwell Companion to Political Geography, Second Edition.
Edited by Virginie Mamadouh, Natalie Koch, Chih Yuan Woon, and John Agnew.

and neighborhood effects on voting. Elections, like other social processes, are inherently tied to space, and any attempt to examine them without an accounting for geographical influences will fall short of a comprehensive understanding.

This chapter summarizes the history, topics of analysis, and emerging areas of study in electoral geography. It discusses some of the varying subjects of analysis in electoral geography, covering issues in the spatial-analytic tradition, the historical materialist tradition, the feminist tradition, and the sociocultural tradition. Finally, this chapter summarizes recent topics of study in electoral geography, in particular digital electoral geographies, voter suppression, and expanding electoral geographical study outside of the traditional Anglo-American context.

A Brief History of Electoral Geography

The earliest studies of elections were conducted well before the advent of the computer and typically focused on simple empirical questions. Are there spatial patterns of voting? What could be the cause of such patterns? Forest (2018) identifies André Siegfried's 1913 work *Tableau politique de la France de l'ouest sous la troisième république* as the first work of early electoral geography. Siegfried sought to understand relationships between the physical characteristics of voters and their voting behavior, and compared election returns to deposits of minerals at chosen locations throughout France (Siegfried 1913). Electoral geography today is grounded in the work of J. R. Prescott, notably "The function and methods of electoral geography" (1959). Though writing during the early stages of the "quantitative revolution," Prescott anchored his electoral geography firmly in the regional geography of Richard Hartshorne that was quickly falling out of favor. This early form of electoral geography never gained widespread traction, although later works by J. Clark Archer (1981) and Peter Taylor (1978) reworked the concept of the region and adapted it to describe larger political identities such as the "Solid South."

Electoral geography evolved in the context of the quantitative revolution and the rise of positivism, becoming a field consumed by quantitative analyses and electoral returns. The spatial approach to understanding elections emerged as a way of both making elections manageable topics of study and analyzing voting patterns as a function of other voter characteristics such as class, income, and cultural status. These contributions relied heavily on the assumptions that had become characteristic of positivist political science, namely that voters always act in a way that maximizes their political utility (Johnston, Shelley, & Taylor 2014). Through the works of these geographers, electoral geography began to develop a reputation as a subdiscipline obsessed with empirical patterns and atheoretical approaches, an aspersion that often persists to this day.

The empiricist obsession did not pertain to the entire field, with geographers like Kevin Cox (1969) arguing that voters were enmeshed in their spatial contexts, seeing voters not as isolated decision-makers but as nodes in a larger community network. The idea that voters were influenced by their neighborhood and local contexts was expanded by sociologists like Paul Lazarsfeld and Bernard Berelson, who utilized social surveys to map out patterns of political influence in small and middle-sized towns across the United States (Berelson, Lazarsfeld, & McPhee 1986; Pattie & Johnston 2009).

Works by later electoral geographers like Johnston and Pattie took the contextual arguments developed earlier and fused them with ideas like structuration theory (Johnston & Pattie 2006). A place-based understanding of voting like that espoused by Johnston and Pattie argued that political contexts develop in their locality and with distinctive political cultures, and that there was no clear way to separate voters from their contexts. At the same time as the context and composition debates were ongoing in electoral geography, other forms of research in the subfield were developing, the first of these being electoral cartography. Cartography had been at the center of electoral geography since its inception in the early twentieth century, but it had split from being the focus of electoral study around the time of the "quantitative revolution" of the 1960s. In-depth analyses of spatial patterns of voting would be pioneered by geographers like Archer (1981), who analyzed electoral maps and drew conclusions about larger geographical trends in voting. As Forest (2018) points out, the work of these academic cartographers was not to produce maps for a larger audience; rather, their job was to draw conclusions from the data they were mapping, and to that effect complicated statistical analyses were employed to analyze the spatial data (Archer, Shelley, & White 1986).

While some areas of electoral geography, such as electoral cartography, persisted in popularity, the field of electoral geography in general declined over the course of the late 1980s and well into the 2000s. Warf and Leib (2011) argue that the decline of the subfield was rooted in the unrepentant positivism of its supporters at a time when human geography was moving decisively into more conceptual and theoretical realms of inquiry. Further criticism focused on the heavy dominance of Anglo-American electoral geographies, with almost no electoral work having been done on any democracies in the Global South. Electoral geography acquired at this time the aspersion that had once been levied on geography in general, that of the "moribund backwater," and accordingly the number of publications in the subfield fell to an average of just 12 per year (Leib & Quinton 2011).

Recent events have led to a minor resurgence of interest in electoral geographies, with attempts to theorize electoral study or collect works in electoral geography into edited collections of work (e.g., Leib & Quinton 2011; Shin 2015). Some recent work has highlighted the increasingly large number of neogeographical sources of electoral maps, with Shin (2009) noting how the diversification of producers of electoral data has, aptly, democratized electoral geography.

Distorting the Outcomes of the Popular Will

Even in democratic political systems, several mechanisms exist that distort electoral outcomes and may lead to candidates winning who have lost the popular vote. This section examines several such forces, including gerrymandering, malapportionment, and voter suppression.

Gerrymandering

Perhaps no topic in electoral geography has been as well studied as gerrymandering, the process of redrawing electoral districts to unfairly benefit a given political party or candidate. Redrawing districts is a routine part of the electoral process in most democracies, necessitated by the differential growth of populations across space and the desire (or mandate) to create districts that equally represent different communities and groups that are members of the electorate. When districts are drawn, what requirements are placed on them, and who draws their boundaries are typically matters established by the law of the state in question.

The term "gerrymander" famously originates from the early history of the United States, when in 1812 Massachusetts governor (and later fifth vice president of the United States) Elbridge Gerry signed into effect a new map of electoral districts that were so oddly shaped the then editor of the *Boston Gazette* referred to them as "A new species of *Monster* appearing in the Essex South District" (Martis 2008).

The process of gerrymandering typically has its greatest effects in plurality or "first-past-the-post" electoral systems, where a candidate can win election by simply attaining the greatest number of the votes cast (though notably not an absolute majority of votes). In the United States and the United Kingdom this manner of voting translates to single-candidate districts where candidates compete for a simple majority on election day, with voters choosing only one candidate. In practice this leaves a large portion of "wasted" votes or votes that are cast that have no impact on the electoral outcome, and these wasted votes are a large part of why the principal methods of gerrymandering, "packing" and "cracking," are effective (Johnston 2002). "Cracking" refers to the process of splitting up communities that have been determined to be electorally opposed to the political party drawing the districts. This process also takes the form of splitting up majority-minority areas into several larger districts that dilute their power as a voting bloc, ensuring they cannot elect a representative of their community. "Packing" refers to the opposite of gerrymandering, where a map-maker will instead cluster together voting groups of an opposing party into one or a few districts that overwhelmingly support that party, thereby leaving the remaining districts relatively uncompetitive. This method of gerrymandering works by creating the greatest number of wasted votes possible, leading to districts that vote overwhelmingly for candidates of opposing political parties and a greater number of outlying "safe" districts.

While redistricting in other countries often revolves around explicit discrimination against voters of other political parties, in the United States the process also encounters a history of racial bias, a legacy that continues in the present. Throughout the Jim Crow era in the American South and continuing well into the 1960s, electoral districts were often drawn with the explicit intention of discriminating against Black Americans, culminating in *Gomillion v Lightfoot*, a Supreme Court case that held that racial discrimination in district drawing is illegal (O'Loughlin 1982; Palandrani & Watson 2020). Despite this ruling, and the later Voting Rights Act of 1965, racial discrimination in district drawing has persisted, with Justice Elena Kagan of the Supreme Court noting in its 2016 invalidation of North Carolina's congressional district maps that the Republican Party had used racial geographical data to construct its districts and deprive African Americans of an impactful vote "with almost surgical precision" (Lang & Hebert 2017: 780).

Forest (2004) notes that in many cases concerning redistricting in the 1990s the American Supreme Court often struck down district maps based on their shapes being "irregular" or noncompact. This raises an important question: What criteria define a "good" or "regular" district? In the United States the requirements are few, with the only strictly enforced standard being the necessity of equal, or as equal can be attained, populations among districts. Though the focus on district shape has been emphasized by the US Supreme Court a number of times, Forest (2018) holds that even smooth or compact districts contain no guarantee against gerrymandering.

One solution that has been presented to resolve the issue of gerrymandering is independent redistricting commissions (IRCs), which have been adopted in several countries and American states. While independent redistricting aims to thread the needle of fairness in drawing districts, Rossiter, Johnston, and Pattie (1998) note that even when districts in Northern Ireland were drawn by ostensibly neutral commissions, the results were heavily challenged and still had a partisan effect. In England, Wales, and Scotland, Boundary Commissions are utilized as a form of IRC, though Johnston (2002) notes that these commissions led to advantages and disadvantages for political parties even though they were largely nonpartisan, with the Labour Party having a lower percentage of effective votes than the Conservative Party from 1957 to 1997 despite receiving similar vote shares (Taylor & Gudgin 1976; Johnston 2002).

In the United States the effect of IRCs has been similarly murky, as a patchwork implementation in different states has been met with mixed results. Henderson, Hamel, and Goldzimer (2018) found that politically "neutral" IRCs did not author districts that promoted competition among candidates, often producing the same number of safe seats for each party that had existed under the previous system. Edwards et al. (2017) found that IRCs tended to draw districts that were more compact and were more likely to respect local political subdivisions such as city boundaries.

Malapportionment: Counting Regions Not Voters

A related but separate issue that has been studied by electoral geographers is malapportionment, or the assignment of electoral seats in a manner that misrepresents the number or population in each area. An example of malapportionment is the US Senate. While each member of the House of Representatives represents roughly 750,000 voters, the voting base for US Senators varies wildly: the two senators from Wyoming represent around 290,000 people each, whereas each of the two from California represents over 19 million (The Learning Network 2022; Johnson & Miller 2023).

This power imbalance is not confined to upper-house legislative chambers, however, as misallocations of power occur in lower-house elections as well. Again, the United States is the best example of this phenomenon because of the manner of assigning representatives laid out in its Constitution, although malapportionment is not exclusively an American problem.

How, then, does this process correspond to power imbalances? The House of Representatives last expanded its number in 1911 after the 1910 census. That census showed a decisive shift of population away from rural areas and into cities, and this trend continued in the 1920 census (US Census Bureau 2021). Rural politicians were fearful that the growing urbanization across the United States would gradually sap their political power as more seats were added to represent urban areas, and as a result representatives for rural districts stalled Congress for nearly a decade, allowing no new seats to be added or districts redrawn from 1920 to 1929 (US Congress, House of Representatives 1929). By the time districts were reassigned in the Reapportionment Act of 1929, the disparity had grown so large that while a single member of Congress from rural Missouri represented roughly 200,000 constituents, a single member in Detroit, Michigan, represented over 1.3 million (Johnston 2002; Desilver 2018). The 1929 Act permanently solidified the situation by capping the number of representatives at 435, leading to ever-widening disparities in representation: while a single representative from 1910 to 1920 had around 200,000 constituents, the number today has grown to one member for around every 750,000 constituents.

Malapportionment is not only an American phenomenon and can be found in nearly every representative democracy across the globe. The United Kingdom famously gave rise to some of the most egregious examples of this phenomenon in "rotten" or "pocket" boroughs, electoral districts that were either nearly or entirely empty of voters and could be used to easily elect members to the House of Commons with only a handful of votes. Johnston (2002) notes that despite parliamentary reforms enacted in 1832 ridding the UK of rotten boroughs, malapportionment still plays a role in deciding elections due to the parliamentary approval necessary to approve redistricting; if the party in control of the Commons would lose seats with new district maps it will be unlikely to approve said maps. Similar examples of party interference in the addition or subtraction of electoral districts have been highlighted in countries like Brazil (Hiroi & Neiva 2013), Romania (Giugăl 2015), and Ghana (Resnick 2017).

Finally, a somewhat different form of malapportionment may be observed in the dynamics of the US Electoral College (Warf 2009), an anachronistic political body if there ever was one. Inserted in the Constitution during the Constitutional Convention of 1787, it formed a compromise between those who favored a popular vote and those who wanted Congress to select the president. The institution allocates electoral votes to states based on their allotted congressional representation, with a minimum of three: one elector for each House of Representatives seat and one for each Senate seat. This process allows for presidential candidates to win the election without winning the popular, or majority, vote, with examples including the victories of George W. Bush in 2000 and Donald Trump in 2016. It is widely believed that the College gives disproportionate influence to smaller, rural, and conservative states at the expense of larger, more urbanized, and liberal ones.

Voter Suppression

The Republican Party (or GOP), which has won the popular vote in only two presidential elections (2004 and 2024) since 1988, has resorted to increasingly authoritarian measures to cling to power. A central tactic in this regard is voter suppression. Republican attempts to shrink the electorate and make voting more difficult have taken various forms, including curtailing of early voting and mail-in ballots, reducing the number of polling places, and prohibiting ballot collections. Informal voter suppression included the intimidation of election officials, including death threats. GOP politicians in Georgia stripped local election officials of their power by implementing new identification requirements, banned mobile voting centers, made it illegal to mail absentee ballot applications to all voters, and made it a misdemeanor for volunteers to give food or water to people waiting in long lines to vote (Corasaniti & Epstein 2021). These efforts were replicated all over the country as the GOP introduced hundreds of bills in dozens of states to curtail voting rights.

Unsurprisingly, these measures were aimed like a laser at Democratic voters, particularly Black ones, in large cities, essentially heralding a return of the old Jim Crow era. Thus, the current GOP effort is part of a long history of voter suppression designed to protect white supremacy. By making it more difficult to vote, the Republican Party seeks to depress turnout, particularly among minorities, an attempt to "stack the deck" in its favor.

Digital Electoral Geographies

The growth of the internet has had important repercussions for electoral geography in what Temple (2023) calls the field's "digital turn." Political campaigns have deployed websites and used social media since the popularization of the internet. From the time the first campaign websites were deployed nationally for the Clinton versus Dole election in 1996, digital campaigning has become something of an arms race among electoral campaigns, with each round bringing newer and more expensive tools to bear. The Trump campaign famous utilized a company called Cambridge Analytica to harvest the data of over 87 million Facebook users and target them with precision advertising during the 2016 election, an effort that bore immense fruit for Trump (Hu 2020). Far removed from the early campaign websites of 1996, recent experiments with artificial intelligence (AI) technology were utilized by campaigns for the 2024 presidential election, with the Democratic primary challenger Dean Phillips creating an AI bot to answer questions on his behalf (it was quickly shut down; Chatterjee & Fernandez 2024). Another AI-backed operation mimicked President Joe Biden's voice and auto-called voters, telling them not to head to the polls for the New Hampshire primary elections (Seitz-Wald & Memoli 2024). Though these are just the latest phase of the digital electoral war, they are not likely to be the last.

More recent, and contentious, is internet voting, which although in many respects it is an extension of voting by mail that has a long tradition in many countries, raises unique issues of security. Estonia is the world's leader in experimenting with internet voting, having allowed the practice for decades (Ehin et al. 2022). France allows its citizens living abroad to use the method, and Switzerland and Norway have adopted it (Dandoy & Kernalegenn 2021). In the United States, Oregon and notably Arizona have experimented with it (Solop 2001).

The Spatial-Analytic Tradition

Gerrymandering and malapportionment are examples of what Temple (2023) calls the spatial-analytic tradition of electoral geography. This mode of thought focuses heavily on spatial approaches to examining elections, relying on the abundance of electoral data to inform quantitative analyses of voting patterns.

Other areas of study important to the spatial-analytic tradition concern a wide range of different topics that can be studied through the lens of electoral returns. Research has considered how campaigns can most wisely spend funds in different places, or how directed campaign communication strategies can influence neighborhood opinion on candidates (Denver & Hands 2004). Other work has emphasized the economic effects on voting, considering topics such as socioeconomic class, labor relations, the state of national economic sentiments, and local macroeconomic effects on voting patterns (Estrada-Correa & Johnson 2012). The spatial-analytic tradition has explored a number of different effects on voting, but all of them have in common a focus on spatial and local voting effects grounded in quantitative research.

The spatial analysis tradition has been put to good use in revealing the subtleties of voting behavior. Using European case studies, several authors have studied the distance-decay effects on political competition among parties, including the advantages of campaigning in home regions, "friends and neighbors" effects, how localism shapes voter perspectives, and how travel times affect voter turnout (Evans et al. 2017). This line of thought has been extended to include intraparty competition (Put, Schoultz, & Isotalo 2020).

Using techniques such as multiscalar geographically weighted regression, Fotheringham, Li, and Wolf (2021) were able to differentiate the impacts of local context from other forces shaping voting outcomes. Similarly, Fotheringham and Li (2023) compared three types of models that incorporate geographical context in voting, including different forms of geographically weighted regression, spatial error models, and Monte Carlo simulations. Such works have aptly demonstrated the profound significance of local context: electoral opposition to immigration, for example, is most evidence in places exhibiting a significant influx of migrants.

In short, the spatial-analytic tradition, despite the epistemological problems concerning its positivist roots, continues to be an important part of electoral geography. Increasingly sophisticated quantitative techniques have shed light on the manifold subtleties that accompany voting patterns over space.

Historical Materialist and Political Economic Electoral Geographies

Around the time when geography as a wider discipline was engaging with dialectical and historical materialism and reorienting itself away from positivism, there emerged a new perspective in political geography, one that recognized the need for change. Political geography in general had been on something of a downward trend during this period, with the largest subdivision of the field at the time, electoral geography, seeing the number of publications slump considerably (Leib & Quinton 2011). A large part of the decline of political geography during the late 1970s and early 1980s was the atheoretical nature of the discipline at a point when historical and dialectical social theory was providing huge insights into the nature of society (Agnew 1990). Electoral geography was not immune to this atheoretical basis, as noted by Reynolds (1990: 22): "If political geography lacked a general theoretical coherence ... electoral geography certainly did too."

Responding to the theoretical dearth in electoral geography were historical materialist geographers, led by Peter Taylor. Taylor was well established in electoral geography, with his 1978 summary "Political geography," a call for the development of a more coherent structure to positivist electoral geography, being one of the most cited pieces in political geography around the time of publication. Taylor homed in on many of the assumptions that underline electoral geography, utilizing an historical materialist critique to pick away at the classical norms that characterize political geographical analyses. Why does electoral geographical research see democratic government as responsive to its citizens, and why does research in this field commonly discard historical and political influences on voting (Osei-Kwame & Taylor 1984; Taylor 1985)? Why are elections that are not conducted under the conditions of what is considered liberal democracy often discarded as not being worth studying (Taylor 1990)? These questions, while profound, would inspire few answers, with Leib and Quinton (2011) noting that the world-systems and political economic approaches generated few publications. Following this period, interest in electoral geography began to wane, with the number of publications in the subfield declining to around 12 per year from 1990 to 2007 (Leib & Quinton 2011).

Sociocultural and Poststructuralist Electoral Geographies

At the end of the twentieth century, a small number of poststructuralist and sociocultural electoral geographers sought to bring electoral geography back from the senescence in which it had found itself by utilizing new perspectives. Perhaps the best example of these is John Agnew, who has written extensively on sociocultural and place-making relationships in regard to voting in the context of Italy (see Agnew 2002, 2007). Agnew combines several methods to make the argument that while previous research in electoral geography often treated elections as one-time proceedings influenced by national characteristics or economic outlooks, elections are not singular and isolated events. Electoral geography, instead of looking at the broadly based census characteristics of individual voters, should focus instead on the history of places and political parties and how they impact voting decisions, in order to provide a clearer picture of the contextually informed voting espoused by spatial-analytic electoral geographies (Agnew 1990).

Other work in this vein has focused on changing sociocultural and economic patterns and the deeper impacts on voting trends that follow. Warf (2011) highlights the changing nature of support for political parties across class lines in the context of the 2008 US presidential election. This decoupling of class from party was only reinforced in the 2016 election, where Donald Trump won two-thirds of noncollege-educated white voters, and 56% of nonvoters reported incomes under $30,000 annually (Pew Research Center 2018; Hull 2020).

Electoral geographical research has also pointed to shifting geographies of racial alignment with political parties in the American context. The Democratic Party, which built a coalition of college-educated, middle-class, and minority voters, saw declines in support among Black and Latinx voters in the 2016 and 2020 election cycles when Donald Trump was on the ballot (Towler & Parker 2018; Sommer & Franco 2023). While this phenomenon may have occurred for a variety of reasons, the growing number of minority voters supporting Trump provides a pathway for the Republican Party to build support in the coming decades, as their traditionally white base declines as a share of the voting population.

Feminist Electoral Geography

Research in electoral geography has been overwhelmingly published by men, with Leib and Quinton (2011) noting that only five articles could be found published by women in the subdiscipline between 1990 and 2007. Electoral geography has since had many calls for the inclusion of both female-authored work and feminist perspectives on elections, with Secor (2004) giving first voice to the development of a feminist electoral geography. Despite this call for feminist perspectives, McGing (2014) notes that even for a male-dominated field like geography, the inclusion of female voices in electoral geography has been markedly slow.

Many feminist contributions to electoral geography take the form of a critique of the spatial-analytic vein and the generalizations on which it relies. Electoral geographers have infrequently considered issues of gender, or if gender has been addressed it is often in comparison to voting patterns that exist among men. Some work has been done in revealing the gender gap of elected officials, with Webster (2000) noting that while women represented a greater percentage of voters than men in the United States during the 1999–2001 electoral period, only 12% of the representatives elected to the Congress were women. Other research by Teele (2023) examined the political choices of women in Sweden, concluding that women in densely populated municipalities were far more likely to support left-leaning politics than were their rural counterparts.

Other work that could be grouped under feminist electoral geographies pertains to emotional and performative electoral geographies. Schurr (2013) argues for the adoption of an electoral geography that recognizes the performative and emotional nature of elections, highlighting the case of the 2009 electoral campaigns in Ecuador as a case study of populist movements channeling voters' emotions. Schurr develops a framework of thought that focuses on how emotional campaign practices create political communities of voters and calls for further study in this area. Other work in this vein has focused on voting as a performative act done in a performatively public space but meant to be private, an expression of a voter's authentic political will that often amounts to a disempowering result (Staeheli 2008; Hollweg 2015). Voting, seen in this way, is not just a one-time event undertaken by a citizen, but an embodied and affectual social practice that is mediated by a voter's gender, ethnicity, race, history, and other defining characteristics (Hollweg 2015).

A smaller but related area of study concerns the intersection of queer identity and electoral politics. Chapman (2011) noted how the banning of gay marriage in Florida by ballot referendum in 2010 reinforced the electoral divides already known to exist throughout the state, with those in more rural areas voting more heavily in favor of the ban. Thompson and Turnbull-Dugarte (2024) highlight a growing rate of queer participation in elections and political participation in urban areas, accounting for the difference by close urban landscapes and intra-group contact.

Indigenous Electoral Geography

Another small but growing area of study in electoral geography focuses on Indigenous electoral geographies. Indigenous political geographies have focused on how Indigenous political power has been repressed in settler-colonial states, relegated to the underside of politics by imperial/colonial states that, although efforts have been made to internally decolonize them, maintain models of political power that confine Indigenous peoples to the position of the political "other" (Coombes et al. 2013; Naylor et al. 2018).

Some work has examined Indigenous political power through the lens of electoral geographies. Dabin, Daoust, and Papillon (2018) discuss patterns of Indigenous voting among communities of Canada in the context of the settler-colonial state, noting that Indigenous voters are more likely to participate in electoral processes when Indigenous candidates are on the ballot. Work by Templeman (2018) has noted how the creation of reserved seats for Indigenous representatives in the Taiwanese legislature did not often lead to positive outcomes for Indigenous communities, with socioeconomic, factional, partisan, and tribal divisions getting in the way of collective Indigenous action (Templeman 2018).

Analyses of electoral returns following the 2020 American presidential election noted how Indigenous voting may have tipped the scales in Arizona for the Biden campaign, with the Navajo community reportedly voting 97% for the then Democratic candidate. The case in Wisconsin also reflects this reality, with Menominee county, home of the Menominee Tribe reservation, voting 82% for Biden in a state that had a margin of only 0.63% difference between the candidates (Chisholm 2020). It remains to be seen whether the participatory Indigenous voting bloc remains constant in future electoral contests, as Herrick et al. (2022) found no evidence that Indigenous voters participated more significantly than other groups in elections from 2006 to 2016.

Electoral Geography Outside of the Anglo-American Sphere

Electoral geography has long been dominated by analyses of elections in the United States. In an article titled "A politics of failure: The political geography of Ghanaian elections, 1954–1979," Taylor outlines a criticism that is a common remark on electoral geography even today: that the focus of the field was far too Anglo-American centered (Osei-Kwame & Taylor 1984; Taylor 1985). Recognizing the need to broaden the field of electoral geography beyond the United Kingdom and the United States, Taylor (1990) argued for the development of a new electoral theory of elections that draws on world-systems theory to develop a theoretical framework by which non-Anglo-American elections could be analyzed.

More recently, studies of elections in other countries have appeared, widening the spatial scope of the field and illustrating how elections vary in their structure, procedures, and consequences. A raft of works has explored the spatiality of elections in Britain (Furlong 2019). Brexit, Britain's decision to leave the European Union, also exhibited uneven electoral support (Harris & Charlton 2016), as did Scotland's referendum on independence (Curtice 2018).

In a long series of works, Agnew (2007), often with Shin (Shin & Agnew 2007), explicated the electoral dynamics of Italy, notably how voting patterns are wrapped up in dynamic place configurations and the lack of electoral bipolarity. More notably, the country's longstanding north–south divide has waxed and waned in significance among various groups and parties. Recent extensions of this work uncovered the spatiality of Italian populism (Agnew & Shin 2017).

Elsewhere in Europe, electoral geography has received scrutiny in Poland (Górecki, Bartnicki, & Alimowski 2022), the Czech Republic (Maškarinec 2019), Hungary (Kovalcsik & Bódi 2023), Bosnia (Reményi et al. 2023), and Georgia (Sichinava 2020). In Germany, the success of right-wing parties such as Alternative for Germany prompted geographical studies of the bases of their support (Förtner, Belina, & Naumann 2021). The surge of right-wing nationalism in Spain likewise motivated study of the distribution of its supporters (Iglesias-Pascual, Paloma, & Benitez 2021).

In other parts of the world, the geographies of elections have been analyzed in the African context (Boone et al. 2022). Given the highly authoritarian nature of many African governments and political divisions that often run along tribal lines, violence before and after elections has been an ongoing concern (Wahman 2023). Postapartheid South Africa also exhibits profound electoral cleavages (Rulea 2018).

Asian electoral geography remains largely unexplored. In India, examples include the Lok Sabha (India's lower house of parliament) elections in Uttar Pradesh (Biswas et al. 2023). Other examples from Asia include papers on Mongolia (Maškarinec 2023: 20), Japan (Osipov & Elatskov 2020), and Iran (Heidarian, Ahmaipour, & Roumina 2018). Latin America has received very little attention, although Schurr (2013) has offered a compelling analysis of affect during electoral campaigns there.

Conclusion

Elections are the primary expression and exercise of political power in democratic societies. Although in principle democracies ensure that all members have equal political status and a voice in governance, the reality is quite different. In theory, elections give equal voice to all concerned parties; in reality, elections reflect a constellation of forces that shape voter opinions, political campaigns, and outcomes. All these processes play out unevenly over space, giving rise to myriad spatialities.

Electoral geography has surprisingly deep roots in the discipline, with origins that can be traced back as far as the early twentieth century. Over time, the subdiscipline has evolved and adapted in response to the changing political circumstances and intellectual shifts of geography itself. The subdiscipline has also exhibited a studied concern for how political authorities and parties manipulate electoral systems to their advantage, in the process undermining the democratic claim to equal representation.

From its empiricist origins, electoral geography shifted into positivist accounts, with an obsession for mathematical analyses that was enabled by the large volumes of electoral data readily available. This move came at a price: the neglect of theory and the dismissal of social context. For most of the twentieth century the field was paralyzed by this stance, relegating it to the margins of human geography.

More recently, under the impetus offered by social theory, electoral geography turned its attention to pressing matters such as class, gender, ethnicity, and sexuality. The field witnessed blossoming literatures in feminist and Indigenous electoral studies. In the process, it also began to loosen its obsession with Anglo-American elections to study elections in a wider variety of social and political environments.

References

Agnew, J. (1990). From political methodology to geographical social theory? A critical review of electoral geography, 1960–87. In Johnston, R.J., Shelley, F., & Taylor, P.J. (eds), *Developments in Electoral Geography*. New York: Routledge, 15–21.

Agnew, J. (2002). *Place and Politics in Modern Italy*. Chicago, IL: University of Chicago Press.

Agnew, J. (2007). Remaking Italy? Place configurations and Italian electoral politics under the "second republic". *Modern Italy 12*(1): 17–38.

Agnew, J., & Shin, M. (2017). Spatializing populism: Taking politics to the people in Italy. *Annals of the American Association of Geographers 107*(4): 915–933.

Archer, J. (1981). *Section and Party: A Political Geography of American Presidential Elections, from Andrew Jackson to Ronald Reagan*. Baldock: Research Studies Press.

Archer, J., Shelley, F., & White, E. (1986). *American Electoral Mosaics*. Washington, DC: Association of American Geographers.

Berelson, B., Lazarsfeld, P., & McPhee, W. (1986). *Voting: A Study of Opinion Formation in a Presidential Campaign*. Chicago, IL: University of Chicago Press.

Biswas, F., Khan, N., Ahamed, M., & Rahaman, M. (2023). Dynamic of electoral behaviour in Uttar Pradesh: A study of Lok Sabha elections from 2009 to 2019. *GeoJournal 88*(2): 1317–1340.

Boone, C., Wahman, M., Kyburz, S., & Linke, A. (2022). Regional cleavages in African politics: Persistent electoral blocs and territorial oppositions. *Political Geography 99*: 102741.

Chapman, T. (2011). Sexual politics and the sunshine state: Electoral geographies of Florida's marriage protection amendment. In Barney, W., & Leib, J. (eds), *Revitalizing Electoral Geographies*. Farnham: Ashgate, 219–232.

Chatterjee, M., & Fernandez, M. (2024). "An arms race forever" as AI outpaces election law. *POLITICO*, February 7. https://www.politico.com/news/2024/02/07/ai-2024-elections-00140005 (accessed November 28, 2024).

Chisholm, N. (2020). Native American voters helped swing crucial states in 2020 presidential election. *Colorlines*, November 9. https://colorlines.com/article/native-american-voters-helped-swing-crucial-states-2020-presidential-election (accessed November 28, 2024).

Coombes, B., Johnson, J., & Howitt, R. (2013). Indigenous geographies II: The aspirational spaces in postcolonial politics – reconciliation, belonging and social provision. *Progress in Human Geography* 37(5): 691–700.

Corasaniti, N., & Epstein, R. (2021). How Republican states are expanding their power over elections. *The New York Times*, June 19. https://www.nytimes.com/2021/06/19/us/politics/republican-states.html (accessed November 28, 2024).

Cox, K. (1969). The voting decision in a spatial context. *Progress in Geography* 1: 81–117.

Curtice, J. (2018). The 2017 election: Scotland re-enters British politics? *Scottish Geographical Journal* 134(1–2): 39–44.

Dabin, S., Daoust, J., & Papillon, M. (2018). Indigenous peoples and affinity voting in Canada. *Canadian Journal of Political Science / Revue canadienne de science politique* 52(1): 39–53.

Dandoy, R., & Kernalegenn, T. (2021). Internet voting from abroad: Exploring turnout in the 2014 French consular elections. *French Politics* 19(4): 421–439.

Denver, D., & Hands, G. (2004). Labour's targeted constituency campaigning: Nationally directed or locally produced? *Electoral Studies* 23(4): 709–726.

Desilver, D. (2018). U.S. population keeps growing, but House of Representatives is same size as in Taft era. *Pew Research Center*, May 31. https://www.pewresearch.org/short-reads/2018/05/31/u-s-population-keeps-growing-but-house-of-representatives-is-same-size-as-in-taft-era (accessed October 30, 2023).

Edwards, B., Crespin, M., Williamson, R., & Palmer, M. (2017). Institutional control of redistricting and the geography of representation. *Journal of Politics* 79(2): 722–726.

Ehin, P., Solvak, M., Willemson, J., & Vinkel, P. (2022). Internet voting in Estonia 2005–2019: Evidence from eleven elections. *Government Information Quarterly* 39(4): 101718.

Estrada-Correa, V., & Johnson, M. (2012). Foreclosure depresses voter turnout: Neighborhood disruption and the 2008 presidential election in California. *Social Science Quarterly* 93(3): 559–576.

Evans, J., Arzheimer, K., Campbell, R., & Cowley, P. (2017). Candidate localness and voter choice in the 2015 general election in England. *Political Geography* 59: 61–71.

Forest, B. (2004). The legal (de)construction of geography: Race and political community in Supreme Court redistricting decisions. *Social & Cultural Geography* 5(1): 55–73.

Forest, B. (2018). Electoral geography: From mapping votes to representing power. *Geography Compass* 12(1): e12352.

Förtner, M., Belina, B., & Naumann, M. (2021). The revenge of the village? The geography of right-wing populist electoral success, anti-politics, and austerity in Germany. *Environment and Planning C: Politics and Spaces* 39(3): 574–596.

Fotheringham, A.S., & Li, Z. (2023). Measuring the unmeasurable: Models of geographical context. *Annals of the American Association of Geographers 113*(10): 2269–2286.

Fotheringham, A.S., Li, Z., & Wolf, L. (2021). Scale, context, and heterogeneity: A spatial analytical perspective on the 2016 US presidential election. *Annals of the American Association of Geographers 111*(6): 1602–1621.

Furlong, J. (2019). The changing electoral geography of England and Wales: Varieties of "left-behindedness". *Political Geography 75*: 102061.

Giugăl, A. (2015). Geography of electoral districting or gerrymandering and malapportionment, Romanian-style. *Sfera Politicii XXIII*(3.185): 43–54.

Górecki, M., Bartnicki, S., & Alimowski, M. (2022). Local voting at local elections revisited: "Friends and neighbors voting" at mayoral elections in rural Poland. *Political Geography 94*: 102559.

Harris, R., & Charlton, M. (2016). Voting out of the European Union: Exploring the geography of leave. *Environment and Planning A 48*(11): 2116–2128.

Heidarian, M., Ahmaipour, Z., & Roumina, E. (2018). Analysis of the spatial pattern of voting; case study: Tenth parliamentary elections in northern electoral district of Ilam Province. *Geopolitics Quarterly 13*(48): 53–75.

Henderson, J., Hamel, B., & Goldzimer, A. (2018). Gerrymandering incumbency: Does nonpartisan redistricting increase electoral competition? *Journal of Politics 80*(3).

Herrick, R., Davis, J., & Pryor, B. (2022). Are Indigenous Americans unique in their voting in US national elections? *Politics, Groups, and Identities 10*(2): 276–294.

Hiroi, T., & Neiva, P. (2013). Malapportionment and geographical bases of electoral support in the Brazilian Senate. *Journal of Politics in Latin America 5*(1): 127–150.

Hollweg, B. (2015). Political sensibilities, affect, and the performative space of voting. *Contemporary Theatre Review 25*(2): 177–189.

Hu, M. (2020). Cambridge Analytica's black box. *Big Data & Society* 7(2): 2053951720938091.

Hull, K. (2020). Lost and found: Trump, Biden, and white working-class voters. *Atlantisch Perspektief 44*(5): 11–16.

Iglesias-Pascual, R., Paloma, V., & Benitez, I. (2021). The role of contextual factors in the electoral resurgence of extreme right-wing forces in Spain: The case of Andalusia. *Political Geography 86*: 102356.

Johnson, R., & Miller, L. (2023). The conservative policy bias of US Senate malapportionment. *PS: Political Science & Politics 56*(1): 10–17.

Johnson, S. (2022). An ailing republic: How Covid-19 affected the 2020 presidential election. MA thesis, University of Kansas.

Johnston, R. (2002). Manipulating maps and winning elections: Measuring the impact of malapportionment and gerrymandering. *Political Geography 21*(1): 1–31.

Johnston, R., & Pattie, C. (2006). *Putting Voters in Their Place: Geography and Elections in Great Britain.* Oxford: Oxford University Press.

Johnston, R., & Pattie, C. (2009). Geography: The key to recent British elections. *Geography Compass 3*(5): 1865–1880.

Johnston, R., Shelley, F., & Taylor, P. (eds) (2014). *Developments in Electoral Geography.* London: Routledge.

Kovalcsik, T., & Bódi, M. (2023). Geographical realignment of the Hungarian voting behaviour between 2014 and 2022. *Modern Geográfia 18*(1): 59–77.

Lang, D., & Hebert, J. (2017). A post-Shelby strategy: Exposing discriminatory intent in voting rights litigation collection: Five years after *Shelby County v. Holder*: The ongoing fight for voting rights. *Yale Law Journal Forum 127*: 779–798.

Leib, J., & Quinton, N. (2011). On the shores of the "moribund backwater"? Trends in electoral geography research since 1990. In Warf, B., & Leib, J. (eds), *Revitalizing Electoral Geography*. Farnham: Ashgate, 9–27.

Martis, K. (2008). The original gerrymander. *Political Geography 27*(8): 833–839.

Maškarinec, P. (2019). The rise of new populist political parties in Czech parliamentary elections between 2010 and 2017: The geography of party replacement. *Eurasian Geography and Economics 60*(5): 511–547.

Maškarinec, P. (2023). Malapportionment in Mongolian elections: Does institutional structure matter? *Cogent Social Science 9*(2): 2261232.

McGing, C. (2014). Towards a feminist electoral geography. SSRN Scholarly Paper 2457175. Rochester, NY: Social Science Research Network.

Naylor, L., Daigle, M., Zaragocin, S., Ramírez, M., & Gilmartin, M. (2018). Interventions: Bringing the decolonial to political geography. *Political Geography 66*: 199–209.

O'Loughlin, J. (1982). The identification and evaluation of racial gerrymandering. *Annals of the Association of American Geographers 72*(2): 165–184.

Osei-Kwame, P., & Taylor, P. (1984). A politics of failure: The political geography of Ghanaian elections, 1954–1979. *Annals of the Association of American Geographers 74*(4): 574–589.

Osipov, K., & Elatskov, A. (2020). Japan's electoral landscape in the 21st century. *Pskov Journal of Regional Studies 44*: 3–15.

Palandrani, J., & Watson, D. (2020). Racial gerrymandering, the For the People Act, and *Brnovich*: Systemic racism and voting rights in 2021. *Fordham Law Review Online 89*: 124–139.

Pattie, C., & Johnston, R. (2009). Electoral geography. In Kitchin, R., & Thrift, N. (eds), *International Encyclopedia of Human Geography*. Amsterdam: Elsevier, 405–422.

Pew Research Center (2018). An examination of the 2016 electorate, based on validated voters. *Pew Research Center*, August 9. https://www.pewresearch.org/politics/2018/08/09/an-examination-of-the-2016-electorate-based-on-validated-voters (accessed November 28, 2024).

Prescott, J. (1959). The function and methods of electoral geography. *Annals of the Association of American Geographers 49*(3): 296–304.

Put, G., Schoultz, Å., & Isotalo, V. (2020). Fighting over friends and neighbors: The effect of inter-candidate geographic distance on intra-party competition. *Political Geography 81*: 1–12.

Reményi, P., Gekić, H., Bidžan-Gekić, A., & Sümeghy, D. (2023). Contextual factors influencing election results in Bosnia and Herzegovina. *Quality & Quantity 57*(2): 1–20.

Resnick, D. (2017). Democracy, decentralization, and district proliferation: The case of Ghana. *Political Geography 59*: 47–60.

Reynolds, D. (1990). Whither electoral geography? A critique. In Johnston, R., Shelley, F., & Taylor, P. (eds), *Developments in Electoral Geography*. New York: Routledge, 22–35.

Rossiter, D., Johnston, R., & Pattie, C. (1998). The partisan impacts of non-partisan redistricting: Northern Ireland 1993–95. *Transactions of the Institute of British Geographers 23*(4): 455–480.

Rulea, S. (2018). Geography and voting: The growth of Urban opposition in South Africa two decades after democratisation. *South African Geographical Journal 100*(2): 141–161.

Schurr, C. (2013). Towards an emotional electoral geography: The performativity of emotions in electoral campaigning in ecuador. *Geoforum 49*: 114–126.

Secor, A. (2004). Feminizing electoral geography. In Staeheli, L., Kofman, E., & Peake, L. (eds), *Mapping Women, Making Politics*. London: Routledge, 261–272.

Seitz-Wald, A., & Memoli, M. (2024). Fake Joe Biden robocall tells New Hampshire Democrats not to vote Tuesday. *NBC News*, January 22. https://www.nbcnews.com/politics/2024-election/fake-joe-biden-robocall-tells-new-hampshire-democrats-not-vote-tuesday-rcna134984 (accessed November 28, 2024).

Shin, M. (2009). Democratizing electoral geography: Visualizing votes and political neogeography. *Political Geography 28*(3): 149–152.

Shin, M. (2015). Electoral geography in the twenty-first century. In Agnew, J., Mamadouh, V., Secor, A., & Sharp, J. (eds), *The Wiley Blackwell Companion to Political Geography*. Chichester: John Wiley & Sons, 279–296.

Shin, M., & Agnew, J. (2007). The geographical dynamics of Italian electoral change, 1987–2001. *Electoral Studies 26*(2): 287–302.

Sichinava, D. (2020). Cleavages, electoral geography, and the territorialization of political parties in the Republic of Georgia. In O'Loughlin, J., & Clem, R. (eds), *Political Geographies of the Post-Soviet Union*. London: Routledge, 202–222.

Siegfried, A. (1913). *Tableau politique de la France de l'ouest sous la troisième république*. Paris: A. Colin.

Solop, F. (2001). Digital democracy comes of age: Internet voting and the 2000 Arizona democratic primary election. *PS: Political Science & Politics 34*(2): 289–293.

Sommer, U., & Franco, I. (2023). Trump's African Americans? Racial resentment and Black support for Trump in the 2020 elections. *Politics, Groups and Identities 12*(4): 921–947.

Staeheli, L. (2008). Publicity, privacy, and women's political action. In Agnew, J., & Mamadouh, V. (eds), *Politics: Critical Essays in Human Geography*. London: Routledge, 407–425.

Taylor, P. (1978). Political geography. *Progress in Human Geography 2*(1): 153–162.

Taylor, P. (1985). All organization is bias: The political geography of electoral reform. *Geographical Journal 151*(3): 339–343.

Taylor, P. (1990). Extending the world of electoral geography. In Johnston, R., Shelley, F., & Taylor, P. (eds), *Developments in Electoral Geography*. London: Routledge, 257–271.

Taylor, P., & Gudgin, G. (1976). The myth of non-partisan cartography: A study of electoral biases in the English Boundary Commission's redistribution for 1955–1970. *Urban Studies 13*(1): 13–25.

Teele, D. (2023). The political geography of the gender gap. *Journal of Politics 86*(2): 428–442.

Temple, L. (2023). Reckoning with the digital turn in electoral geography. *Progress in Human Geography 47*(4): 555–574.

Templeman, K. (2018). When do electoral quotas advance indigenous representation? Evidence from the Taiwanese Legislature. *Ethnopolitics 17*(5): 461–484.

The Learning Network (2022). What's going on in this graph? Senate representation by state. *The New York Times*, October 27. https://www.nytimes.com/2022/10/27/learning/whats-going-on-in-this-graph-nov-9-2022.html (accessed November 28, 2024).

Thompson, J., and Turnbull-Dugarte, S. (2024). Sexuality and the city: The geography of LGBT political participation in the United States. Unpublished manuscript. https://osf.io/preprints/osf/zv3hp (accessed November 28, 2024).

Towler, C., & Parker, C. (2018). Between anger and engagement: Donald Trump and black America. *Journal of Race, Ethnicity, and Politics 3*(1): 219–253.

US Census Bureau (2021). About the 1920 Decennial Census. https://www.census.gov/programs-surveys/decennial-census/decade/1920/about-1920.html (accessed December 6, 2024).

US Congress, House of Representatives (1929). *Apportionment of Representatives in Congress. June 10, 1929*. Serial Set No. 9190. Washington, DC: US Congress.

Wahman, M. (2023). *Controlling Territory, Controlling Voters: The Electoral Geography of African Campaign Violence*. Oxford: Oxford University Press.

Warf, B. (2009). The U.S. electoral college and spatial biases in voter power. *Annals of the Association of American Geographers 99*(1): 184–204.

Warf, B. (2011). Class, ethnicity, religion and place in the 2008 US presidential election. In Warf, B., & Leib, J. (eds), *Revitalizing Electoral Geography*. New York: Routledge, 133–155.

Warf, B., & Leib, J. (2011). *Revitalizing Electoral Geography*. New York: Routledge.

Webster, G. (2000). Women, politics, elections, and citizenship. *Journal of Geography 99*(1): 1–10.

Chapter 23

Nationalism and Nation-Building

Benjamin Forest

Departments of Geography and Political Science, McGill University, Montreal, Canada

At their root, nationalism, nation, nationalist, national identity, and related terms refer to a specific idea of political community that fully emerged in nineteenth-century Europe and came to dominate the global political order in the twentieth century. The idea of nationalism is closely related to the modern (Westphalian) state, but the two concepts are distinct. A nation in the modern sense is a population characterized as having fundamental political solidarity based on common culture, common allegiance, mutual self-recognition, and/or ancestry that claims the right of self-determination. The right to self-determination generally means control of a state, or significant autonomy over a territorial unit within a state. Nationalism can thus be defined (1) as a political movement that seeks to create or control a state for a national group; and (2) more generally as a project that seeks to create, cultivate, or reinforce a sense of national identity to further a political goal. A variety of groups may undertake the latter endeavor, including existing states, political parties, and nonstate political elites.

It is common, even in scholarly work, to conflate nation and state, although their modern conceptions are closely related. A state is a political institution that, in Weber's (1958: 78) terms, holds a monopoly over the use of legitimate violence within a specific territory. A nation, in contrast, is a social, cultural, and political group, typically rooted in a real or imagined homeland, and may or may not have institutions associated with it. The first definition of nationalism mentioned assumes the existence of (modern) states, and states commonly engage in the practice of nation-building. Arguably, the global political order rests on the idea of the nation-state, where the boundaries of a nation coincide with the boundaries

The Wiley Blackwell Companion to Political Geography, Second Edition.
Edited by Virginie Mamadouh, Natalie Koch, Chih Yuan Woon, and John Agnew.

of a state, and where "national self-determination" is the customary basis for the legitimacy of a state. Nonetheless, a nation may exist without a corresponding state and not all states claim to represent nations.

Williams (1983: 213–214) notes that the word "nation" – rooted in the Latin word *nāscī* ("to be born") – emerged in English around the thirteenth century, connoted a group of people related by common ancestry, and was largely synonymous with the term "race." By the seventeenth century, nation began to take on a political meaning, designating the people in a territory, but the term "nationalism" did not come into common use until the eighteenth or nineteenth century (Hobsbawm 1992: 3). The separation of meaning was not linear or complete, as late nineteenth-century nationalist movements turned to emerging racial theories to boost claims of ethnic nationalism (Hobsbawm 1992: 107–109). In any case, both the terms and concepts of race and nation are often conflated in both language and practice (e.g., Figure 23.1). In this chapter, I focus on the transformation of "nation" into an explicitly political grouping in Europe in the nineteenth century and its use and meaning globally in the twentieth and twenty-first centuries.

Definitions and Characteristics of Nationalism

Nationalist movements define membership in their nation in a variety of ways that vary over time, making it difficult to identify a precise definition with broad application. Indeed, while the "fuzziness" of membership criteria presents a problem from a scholarly perspective, it is what makes nationhood "unusually convenient for propagandist and programmatic" uses (Hobsbawm 1992: 6). Consequently, defining the criteria of membership in a particular nation is distinct from the more scholarly project of describing the criteria used by nationalist movements. While a nation is a social, cultural, and political group, it is also more than that.

The Modernist School

The Modernist or Constructivist perspective dominates scholarship on nationalism, and asserts that national identities as we know them today were essentially invented in the last 200 years (Hobsbawm & Ranger 1992). Many scholars in the Modernist tradition sought to explain the puzzle that has troubled Marxist thinkers since World War I: Why does solidarity based on national identity tend to override class solidarity?

Perhaps the best known among the Modernist definitions of nation is Benedict Anderson's (1991: 6) evocative phrase "an imagined political community . . . both inherently limited and sovereign." A nation can only be imagined rather than experienced because unlike small-scale communities of propinquity, national populations are simply too large to meet in person. Nations are "limited" because their membership necessarily excludes members of other nations. Unlike messianic

Figure 23.1 Ethnic/national diversity in the Austro-Hungarian Empire 1911. *Source:* Shepherd 1911: 168. https://maps.lib.utexas.edu/maps/historical/shepherd_1911/shepherd-c-168.jpg, public domain.

religions, nationalist movements do not seek to include all humans within their fold but to draw distinctions among them. Nations are imagined to be sovereign political entities with the right of self-determination, a capacity that can be ultimately fulfilled only in the form of a nation-state. Finally, nations are communities because "the nation is always conceived as a deep, horizontal comradeship," even in the face of substantial intranational inequalities (Anderson 1991: 7). Communications technology plays an important role in cultivating national consciousness; the production of mass-circulation newspapers permits both the common experience and the awareness of common experience vital to cultivating national identity. More generally, the technology of "mechanical reproduction" (see Benjamin 2008) enables national symbols (both image and text) to be disseminated widely among a population and aids the creation of uniform national histories and school curriculums.

Anthropologist Gellner (1997: 3–4) emphasizes the role of culture as the foundation of national solidarity: "Whatever principles of authority may exist between people depend for their legitimacy on the fact that the members of the group concerned are of the same culture (or, in the nationalist idiom, of the same 'nation')." Nationalist movements base membership on a shared culture and the mutual recognition among members of that commonality. Such membership is both exclusionary and exclusive; only those sharing the culture are considered part of the nation, and all members of the culture group must belong to the nation. For Gellner, culture includes language, belief systems, ways of behaving, and styles of communication (clothing, food, and the like). He notes that cultural traits can both change rapidly or persist over time. The key distinction between a mere culture group and nation, however, is the belief – necessarily cultivated by nationalist leaders and institutions – that such common cultural traits engender mutual rights and duties among members.

The association of cultural similarity with mutual obligation is perhaps the key conceptual leap of nationalism because it provides political meaning to ostensibly nonpolitical, everyday practices. Hobsbawm (1992: 9) pushes this observation further by observing that nationalist movements assert that membership engenders a specifically political duty to both co-nationals and to the institution(s) that represent(s) the nation; that is, the state. These duties override all other obligations in the public sphere (civil society). In routine circumstances, for example, one is expected to put the interests of the nation above the interests of one's labor union, professional organization, or religious institution. In extreme circumstances (i.e., war), duty to the nation overrides both public and private obligations, justifying not only the sacrifice of one's life over familial or personal responsibilities but the deliberate killing of others (Billig 1995: 1–4).

The association between national membership and duty to the state thus creates a powerful tool and leads to at least two possible strategies of nationalism, corresponding to the two definitions given in the introduction to this chapter. First, elites can cultivate national consciousness among a culture group and use that to make a claim for a new state, typically carved out of a multinational or multiethnic one. Second, states can shape, encourage, or force cultural similarity and national solidarity among their populations to lay claim to the obligations of national members.

While both definitions and paths of nationalism emphasize top-down nation-building, a critical measure of the "success" of a nationalist movement is the embrace of national identity by a substantial number of people; that is, the degree of national consciousness. "The existence of a nation . . . is an everyday plebiscite," notes Renan (1994 [1882]: 17). That is, to exist as a nation, members of the nation must believe that they are members, so that national identity is never simply an elite project nor a matter of spontaneous mass consciousness (Connor 1990). Rather, nationalist movements – whether led by a state or a group of nonstate elites – face the difficult task of persuading a large group of people that they share certain cultural (or hereditary) similarities, that these similarities constitute a nation, and that national membership is the exclusive legitimate source of political solidarity. National identity may face many competing claims along lines of religion, class, status, ethnicity, and the like, so the success of a nationalist movement is not preordained. In short, the existence of a national project is not the same as the existence of a nation.

Primordialist Definitions

The Modernist perspective is typically set against so-called primordial conceptions of national identity, a position generally taken by nationalist movements themselves. Such political primordialism holds that nations have ancient roots and identity and that a particular nation has always existed (although it may not always have had its own state). Members of the same nation have similar cultural, linguistic, and physical characteristics because the nation has existed as a discrete and identifiable population since prehistoric times. In this sense, nations exist outside of history because the national population is carried as a discrete unit through time. Such political primordial accounts tend to emphasize heredity and physical appearance as the key criteria of national membership, and thus tend toward exclusivity.

Nonetheless, there is a scholarly form of primordialism that challenges the universality of the Modernists' claims. Smith (1998) in particular argues that some ethnic populations self-identified as distinct groups before the rise of nationalist movements. In doing so, Smith makes two important contributions. First, he emphasizes the distinction between nationalism and nation. While conceding that the former is modern (nineteenth century and later), he argues that the latter is not, or at least that modern national groups cannot be clearly distinguished from much older "*ethnie* [ethnic communities]. . . a named community of shared origin myths, memories and one or more element(s) of common culture, including an association with a specific territory" (Smith 2002: 15). Second, Smith decries the

Eurocentrism of the Modernist school, arguing that a focus on nationalism excludes consideration not only of pre-nineteenth-century history but also of non-European cases. Ethnic states existed in ancient China, Sumer, and Egypt, while dynastic states existed in medieval England, France, Spain, and Sweden. Neither type of state was synonymous with nations, but they had "a collective proper name, a fund of ethnic myths and shared memories, the wider dissemination of a public culture to adult males, the growth of a standard law code and ritual practice for all members, and a powerful collective attachment to an ancestral homeland with clear (if in practice shifting) borders" (Smith 2002: 27). In short, numerous historical populations have the same characteristics used to define modern nations in terms of cultural homogeneity and territorial rootedness.

Nonetheless, membership in such historical nations was not a key element of political solidarity in their respective (premodern) states. That is, an ancient or medieval state might have been ethnically homogenous, but political unity would have been based on religious grounds (e.g., divine right of the ruler), loyalty to the sovereign, or other principles, rather than on the co-nationality of the population. Scholarly primordialism is a useful corrective to Modernists' narrow historical and geographical focus and provides a potential pathway (albeit a deeply historical one) to discuss non-European forms of national identity. Yet as Smith himself also notes, scholarly primordialism must guard against cooption of its projects by political movements that would use historical evidence in the service of exclusionary nationalism.

"Classic" Nationalism: Nineteenth-Century Europe

The modern conception of nation, and indeed much of the Modernist school of nationalism, is rooted in the struggle between nationalism and empire in nineteenth-century Europe. In the wake of the Napoleonic Wars in 1815, the four victors (British Empire, Austrian Empire, Russian Empire, and Kingdom of Prussia) drew new state boundaries for Europe at the Congress of Vienna. Among other things, this was an attempt by multiethnic, multinational empires to roll back the threat, introduced by the French Revolution, of national identity as the basis of political solidarity. The next hundred years saw two important developments. First, nationalist movements within these empires, most notably the Austro-Hungarian, cultivated national identities first through cultural activities (developing "national" languages, literatures, and arts) and subsequently through political activities, including the failed revolutions of 1848. Second, the various German and Italian states were consolidated into titular nation-states by the early 1870s.

At the end of the "long" nineteenth century – the start of World War I – nationalism had served as a successful unifying force (in Germany and Italy) and as a failed centrifugal force (in the Austro-Hungarian Empire, among others). Some states, principally in Western Europe, had relatively homogenous "national" populations that spoke the same language, practiced the same culture, and thought of themselves as part of a national group represented by their state. In contrast, eastern Europe, dominated by the Austro-Hungarian Empire, included diverse, geographically dispersed ethnic/national groups – see Figure 23.1 from 1911. (Note, as well, the use of the term "race" to refer to the ethnic/national groupings shown on the map.) The map even understates the diversity of the region by representing all groups in the Balkans as a single category of "Croats and Serbians," a bitterly ironic cartographic choice in light of the interethnic wars of the 1990s (Kaufman 2001). Political solidarity – the principle that held the Austro-Hungarian state together – rested on loyalty to the emperor, rather than bonds of mutual rights and obligation defined by membership in a nation.

World War I saw the destruction of three multinational empires – Russian, Ottoman, and Austro-Hungarian – with the former disintegrating and reforming as the Soviet Union and the latter two being broken up by the war's victors. In doing so, the victors elevated the principle of national self-determination as the legitimizing principle of statehood, and produced new nation-states and expanded others (e.g., Romania). This principle was used unevenly and cynically as it was applied only within Europe to the losing side of the war. Moreover, some of the new states (i.e., Czechoslovakia and Yugoslavia) were composed of several "national" groups. Nonetheless, this reconfiguration of European states, along with the founding of the League of Nations, cemented the tie between nation and political legitimacy. In principle, the path of a population to autonomy, self-determination, and power lay in recognition as a nation. The Russian Empire's successor state, the Soviet Union, ostensibly offered a nonnational model of sovereignty, but one that ultimately resembled the old imperial model, albeit with the selective manipulation of national identities (Martin 2001).

If nationalism acted as a powerful anti-imperial principle within Europe in the nineteenth century, it became perhaps an even more powerful anticolonial principle outside Europe in the twentieth. As Anderson notes, "since World War II every successful revolution has defined itself in *national* terms" (1991: 2, emphasis in the original). While not disputing the power of nationalism, Getachew (2019) argues that anticolonial thinkers did not simply adopt European notions of nationalism, but sought even more fundamental revolutionary change in the global state system. Rather than seeking to build nations simply to join the existing state system, revolutionary thinkers engaged in "world-building" to create a system of global governance founded on egalitarianism and mutual aid rather than hierarchy and competition. This project was largely unsuccessful, however, and nationally infused anticolonialism is often coopted by reactionary movements (Zhang 2023).

Nation-Building

If nations do not (generally) consist of distinct primordial populations, then both the idea of the nation and the adoption of "national consciousness" must be invented, cultivated, and promoted. In short, nations require nation-building. For revolutionary nationalist movements – those using national identity as the basis of a new state – nation-building is typically limited to cultural rather than explicitly political activities, as in the nineteenth-century Austro-Hungarian Empire. In contrast, existing states generally have the full range of options available to them, and nation-building is often high on the agenda of new states. Indeed, for postcolonial states in the postwar era, cultivating a sense of national solidarity and building state institutions were closely intertwined.

Anderson (1991: 163–185) identifies three activities/institutions as key to nation-building, particularly in postcolonial states: the census, maps, and museums. Although all three were essential to the operation of colonial administrators, they were taken over by newly independent states. While Anderson discusses these three institutions in the context of nation-building in postcolonial states, it is important to note that all states classify, manage, and in a sense create their populations and territories (e.g., Foucault 1991). In the pursuit of greater control – for good or ill – states seek to render their populations and territories "legible" through processes of simplification (Scott 1998). It behooves nation-states to both imagine and create homogenous national populations.

Late nineteenth-century colonial governments created census population categories that typically classified populations by "race" and with little regard for their subjects' self-identification. Moreover, census categories were both exclusive and exhaustive: "The fiction of the census is that everyone is in it, and that everyone has one – and only one – extremely clear place" (Anderson 1991: 166). Colonial officials then used census classifications to administer and control their populations, seeking to reify essentially fictional categories. At least in Southeast Asia, however, interactions with Indigenous religious, educational, and judicial institutions remained guided by self-identification, and thus became centers of resistance to colonial rule. Religious identity in particular cut across "racial" colonial categories. Nonetheless, postcolonial states ultimately adopted the logic of exclusive and exhaustive census classifications to define national categories, even where "national" identity acted as a proxy for religious affiliation.

The Thai case illustrates the role of the map in cultivating national identity (Anderson 1991: 171–178; Thongchai 1994). Indigenous maps represented either cosmological relationships or very local areas from a largely horizontal perspective rather than territories. Borders were marked inconsistently and were not necessarily congruent with the markers set by other states. Sovereignty, in terms of both the state jurisdiction and the power of the sovereign, varied by the distance from the capital or other markers of authority. The introduction of Western cartography in the late nineteenth century offered a fundamentally different way of representing the state's territory, as a bounded space with clearly marked borders representing a sphere of uniform sovereignty. In a fundamental way, making a map created a national territory or, as Thongchai (1994) terms it, Siam's "geo-body." Being Thai no longer meant simply being a subject of the Thai monarch, but a member of the geo-body, qua nation (Thongchai 1994: 135). In other instances – Anderson (1991) cites examples from Indonesia – territorial maps served as logos and symbols for burgeoning nationalist movements who came to claim disparate populations and cultures as their own.

Scholars have long recognized that museums, both within and outside Europe, are profoundly political and serve as potent instruments of state power. In particular, nineteenth-century European imperial and national museums distinguished between "civilized" and "uncivilized" people and presented progress as a national achievement, but they have been used to create national narratives throughout the world (Knell 2011; Koch 2024).

As with censuses and maps, colonial administrators had their own reasons for uncovering and documenting the local histories of their jurisdictions. Such activities included not only the creation of orthodox museums, but also crafting archeological heritage sites and reproducing their images in local textbooks, on stamps, postcards, money, and the like (Anderson 1991: 181–182). Such sites and images gave postcolonial states a means to represent "national" histories that provided a sense of primordial origins free of the colonial experience and/or to construct a narrative of national liberation (Lepawsky 2008; Gillen 2014; Cawood & Fisher 2022).

The effort of states to shape national identities by constructing their pasts is not limited to museums. Indeed, for Nora (1989), creating places of memory (*lieux de mémoire*) embodying national identity is a central function of modern states (see Legg 2005; Till 2005). National monuments, memorials, and statues are especially powerful tools for doing so because they encourage "passionate and consensual understanding of nationhood" (Doss 2010: 20). That is, monuments and accompanying commemorative rituals not only make factual claims about a nation's history (as with a school textbook), but also facilitate an affective attachment, alchemically transforming national memories into personal ones (Misztal 2005). Consequently, new states and new regimes – and sometimes their publics – seek to rapidly transform their national symbolic landscapes at moments of change (Rose-Redwood, Alderman, & Azaryahu 2010; Mamvura 2022). These transformations can be as dramatic as the destruction of old monuments, but often involve the selective use of existing elements to help maintain continuity with prior expressions of nationhood (Forest & Johnson 2002; Muzaini 2014).

Moments of revolution and commemorative activities associated with national origins are dramatic events that can create strong affective attachments to the nation. Nonetheless, such events may not be the most effective or common form of nation-building. Billig (1995: 8) argues that "banal nationalism" is a more powerful and pervasive practice: "The metonymic image of banal nationalism is not a flag which is being consciously waved with fervent passion; it is the flag hanging unnoticed on the public building." That is, everyday state practices that are woven into the daily lives of citizens cultivate a sense of national belonging without dramatic emotional force.

Banal nationalism has been embraced enthusiastically both within and beyond political geography, although sometimes in ways that do not always align with its original meaning (Koch & Paasi 2016; Duchesne 2018). For example, Militz and Schurr (2016) draw attention to embodied and emotional practices ("affective nationalism") of activities like dance to extend Billig's original idea, although in many ways this is closer to a Durkheimian emphasis on the power of emotional excitement to generate group consciousness (Misztal 2003; Antonsich & Skey 2017). More generally, scholars have argued for renewed attention to the emotional dimension of nationalism (Antonsich et al. 2020). Similarly, geographers have suggested that Billig overstates both the power of banal practices to generate uniform understandings of national identity and the distinction between "hot" and banal nationalism (Antonsich 2016; Culcasi 2016; Paasi 2016).

Ethnic and Civic Nationalism

By definition nationalism is exclusionary, but the us/them distinction may be drawn in different ways. Ethnic and civic nationalism represent two ideal types; nationalist movements typically fall on a spectrum between the two (Brubaker 1992). While civic nationalism seems to offer an inclusive, open form of identity, ethnic national appeals have greater emotive force and are easier for elites to shape into political action (Geertz 1963; Kaufman 2001; Ram 2008).

Ethnic nationalism bases membership on common descent and is essentially synonymous with biological notions of race, albeit with more narrow categories than traditional "racial" divisions. Brubaker (1992) characterizes this as "German-style" nationalism. As membership can only be inherited, ethnic nationalism is necessarily exclusionary because there is no way to join a nation; one must be born into it. There is a strong affinity between ethnic nationalism and primordialist narratives because both rely on the idea of a population moving unchanged through time. Admitting nonmembers or marriage outside the national group is a threat to the nation because descendants would not trace their ancestry back to the original population. Consequently, ethnic nationalist movements tend to emphasize the question of "purity" in determining membership. Cultural practices – language, religion, and the like – are often conflated with biological ones, so that national members are seen to carry both cultural and bodily markers.

In contrast, civic nationalism (what Brubaker describes as "French-style" nationalism) defines membership in terms of common territory or common values, particularly professed commitment to a set of laws, political principles, or constitution. In the extreme, all people residing within the boundaries of a territory would be considered members of the nation and/or a nation is defined by a common commitment to revolutionary/democratic ideals. In theory, this leads to an inclusive national principle, open to the integration of nonmembers, and reflects the late eighteenth-century revolutionary-democratic tradition of the French Revolution.

The distinction between civic and ethnic nationalism is useful analytically but has also been subject to criticisms. The first is the assumption that there is a developmental or even teleological progression from ethnic to civic nationalism, with increasing civic nationalism accompanying economic development (Brubaker 2004). Similarly, the goals of nationalist movements – particularly between those seeking greater autonomy within an existing state and those seeking their own state (secession) – influence the nature of nationalist claims (Lecours 2000). Civic nationalism is more amenable to autonomy-seeking movements while ethnic nationalism is more effective for secessionists. Finally, Tamir (2019) argues that the distinction between ethnic and civic "nationless" nationalism serves an ideological purpose. Claiming to practice a civic nationalism sharply distinguished from its ethnic form permits Western democracies to both ignore inequalities and oppressions within their own societies while also criticizing exclusions in non-Western ones. The final section further explores the complexities of the distinction between ethnic and civic nationalism.

Inclusions and Exclusions

While nationalist movements and nation-states may be relatively more ethnic or civic in principle, the distinction is typically blurred in practice. For example, while civic nationalism holds that ascribed characteristics based on ancestry are irrelevant to membership, as in France, there is often significant discrimination along lines of religion, origin, and the like (Najib 2021). Indeed, Agnew (2017) describes the nation-state itself as a flawed and "tragic" political form due in part to the exclusions it requires. The complex mix of ethnic and civic nationalism and their accompanying inclusions and exclusions in practice can be illustrated through the relationships of language, religion, citizenship, and gender to national identity.

Language

Language has long been used for both marking and generating national identity, yet is a nonexclusive, achieved characteristic; people may learn new languages and may speak more than one (see Berdichevsky 2004). The revival (and in some cases invention) of formalized Slavic languages was critical to the nineteenth-century nationalist movements in the Austro-Hungarian Empire (Kamusella 2009; Holubec & Mocicková 2023). Indeed, many of the classic nationalist movements in East and Central Europe started as efforts to codify vernacular languages and to develop a literary and theatrical tradition using them. In these cases, a distinct "national" language was central to the strategy of nationalist movements.

Similarly, the sovereigntist movement in Quebec has defined Québécois nationhood around language since the mid-twentieth century (Kaplan 1994; Peker 2024). Yet support of sovereignty is linked more closely to anxiety over the status of French in Quebec rather than to knowledge of French itself, and the movement has sometimes sought to project a territorial rather than linguistic definition of the nation (Desbiens 2000; Medeiros 2017). Contemporary debates over the status of French in Quebec center on measures of linguistic competence. Populist-sovereigntists focus on the relative decline in the number of "mother-tongue" Francophones (those whose first language is French), a measure that necessarily excludes a large proportion of immigrants and treats language as an ascribed and exclusionary characteristic. Opponents point out that most non-Francophone Quebec residents have learned French as a second language, treating language as an acquired, and thus civil, characteristic.

It is typically challenging, however, for nationalist movements to use a colonial language as a basis for identity. First, the language may be used by many different national groups and states. French, for example, is used as a marker of cosmopolitan internationalism and to create links among national groups within the Francophonie (Glasze 2007). A colonial language, such as English, may be the only common tongue spoken by members of different national or ethnic groups within a state (e.g., Dua 1993; Oliveira 2024). Indeed, the current Indian government's promotion of both Hinduism and Hindi as markers of Indian identity may mean that the use of English (rather than Hindi) is associated with greater tolerance toward minorities (Singh & Tir 2024). While Hindu nationalism would fundamentally redefine the multiethnic, multireligious, and multilingual state of India, the emotive appeal of English as a uniting force is weak because it is a colonial legacy.

In short, nationalist movements may use language in both exclusionary and inclusionary ways, as markers of distinctiveness or of cosmopolitanism, as carriers of culture, or simply as a tool of communication.

Religion

The relationship between religion and nationalism is complicated because religion can be considered as either an ascribed or an achieved characteristic. Nearly all religions have mechanisms for incorporating new members, and messianic ones explicitly seek to expand by conversion. Insofar as the definition of a national identity includes religion, conversion would operate as a civic principle of openness. The French principle of *laïcité* turns this logic on its head and holds that the exclusion of religion from the public sphere is a fundamental, and distinctively French, civic principle (Lizotte 2020, 2024). If the practice of religion involves expression or forms of embodiment, however, application of *laïcité* can exclude some members of the public. In particular, the general principle of secularism, which might require a ban on religiously symbolic clothing, can be easily transformed into a tool of Islamophobia when bans specifically target hijabs or veils (Hancock & Mobillion 2019). This can require Muslim women to adopt careful spatial strategies to simply navigate public spaces, although it also provides an opportunity for resistance and protest (Najib & Hopkins 2019). Similarly, turning away from a religious basis of identity was also fundamental to constructing the identity of a modern Turkish nation after the end of the Ottoman Empire (Grigoriadis 2009). As in France, however, the norm of "secularism" falls heavily on observant women, who must manage their religious expression by negotiating the boundaries of public and private spaces (Gökariksel & Secor 2015).

Other Western democracies, notably Quebec and Switzerland, have followed similar strategies of secularism, typically targeting minority religions as "religions" while classifying dominant religious traditions (i.e., Christianity) as "culture" or "heritage" (Barras 2021). In contrast, both religious neutrality and religious expression are core constitutional principles in the United States, which leads to a greater variety of identity claims and outcomes (Howe 2009; Airriess, Hawkins, & Vaughan 2012).

Finally, religion shared by multiple national groups presents a similar issue to shared language: it may demonstrate commonality rather than distinctiveness and provide a competing principle of political solidarity across nation-states. In short, a nationalism that includes a particular religion is not necessarily exclusionary in principle, just as secularism is not necessarily inclusive in practice.

More commonly, however, religious identity and belief are treated as nonrational, deeply held, and functionally immutable characteristics. Consequently, nationalist movements that incorporate religion tend to be exclusionary. This may involve absolute exclusion, where the state or other political actors (re) define national identity to exclude some segment of a state's population from full membership in the political community (Bacchetta 2010; Perry, Schnabel, & Grubbs 2022), or more subtle exclusion, where religion is invoked as a marker of national identity (Moser 2010; Reeves 2015). For example, the Malaysian state constructed a new capital city Putrajaya, using a fantasy "High Islam" style to assert the religious character of the nation while ignoring actual Indigenous architectural traditions (Moser 2012). In some instances, nevertheless, "exclusion" through religious distinctiveness can be a form of resistance, as in the case of Tibet (Houston & Wright 2003).

Citizenship and Migration

Simple birthright citizenship (*jus soli*) is common among states in the Americas, but rare elsewhere in the world (Safran 1997; Kashiwazaki 1998). Even France, for example, limits birthright citizenship based on parentage and/or length of residency, and extended *jus soli* in the late nineteenth century for complex political reasons rather than in a burst of revolutionary fervor (Brubaker 1993). At the same time, Germany adopted significant reforms in 1999 that introduced some recognition of place of birth as relevant to citizenship (Green 2012). While it is important to keep in mind that national membership and state citizenship are not synonymous (e.g., Koch 2016), both are markers of membership in the political community of a nation-state. Birthright citizenship challenges national membership based on ethnicity or ancestry and can therefore be thought of as a relatively civic principle. It provided a means to bring former slave populations and immigrants into the polity but is also currently a source of anxiety for political movements seeking to restrict access to citizenship in Canada and the United States (Hiemstra 2021; Freckelton 2022; Harden 2023). Conversely, distributing citizenship by place of birth is arguably no less arbitrary than determining national membership by ancestry; both are immutable but neither is an act of volition, blurring the distinction between an ethnic and a civic principle.

The possibility (or not) of immigrants obtaining citizenship in their new location – or merely the presence of immigrants – often provokes anxiety in both receiver societies and migrant communities about the nature of their identity (Ehrkamp 2019). Ehrkamp (2006), for example, shows how "Turks" and "Germans" construct identities in opposition to each other, often centered on specific geographical locations. At the same time, migrants maintain relatively stable, multiple "national" identities rather than adopting new exclusive national, nonnational, or transnational ones (Leitner & Ehrkamp 2006; Ralph & Staeheli 2011).

Gender and Sexuality

Gender is fundamentally implicated in all nationalisms; the interaction of gender and religious practice discussed earlier illustrates just one way in which this is the case. Nonetheless, the relationship is perhaps clearest in its ethnic forms because insofar as national identity is a matter of ancestry, national reproduction literally depends on the biological reproduction of national members. More generally, women are typically characterized as the symbolic carriers of national identity; that is, as "mothers of the nation." Women or feminine images are used to represent the nation as weak, threatened, or domestic, while masculine ones are used to present protection, strength, and wealth (Koch 2011; Edenborg 2020; Wang & Chen 2023). This symbolic role can lead not only to gendered inequalities within nations but to the erasure of the actual women involved in nationalist movements (Kearns 2004). Maintaining the "purity" of the nation involves policing women's sexual relations and hence ethnic nationalist movements are generally patriarchal (Rahbari, Longman, & Coene 2019). Rape is used as a weapon of war because it is not simply an act of personal or gendered violence, but a way to destroy the "purity" of an enemy nation (Mayer 2004; Burds 2009; Christian, Dowler, & Cuomo 2016; Trenholm et al. 2016). Ethnic nationalism also tends to be highly heteronormative because queer identities both challenge the moral "purity" of the nation and do not contribute to its (biological) reproduction (Suchland 2018; Oswin 2019). As Puar (2007) argues, however, nationalism is a flexible ideology, such that twenty-first-century Western nationalisms have embraced queer identities – or at least tolerance of queer identities – in a kind of renewed

Orientalism ("homonationalism") to define themselves against an intolerant, terroristic Islamic "other." Geographers have both applied this insight beyond the US case and critiqued scholars' oversimplified embrace of homonationalism (Kulpa 2014; Ritchie 2015).

Conclusion

Studies of nationalism and nation-building remain a robust area of research in contemporary political geography, with thousands of articles published in the last five years using these concepts. The attention to nationalism would have been difficult to anticipate at the start of the twenty-first century as both scholarly and popular discourse focused on globalization and the (alleged) coterminous decline of national distinctions and nation-states.

While it is hazardous to predict the future, it seems several areas are likely to attract special interest over the next decade. These include the rise of populist movements and their use of various forms of nationalism (Lamour 2023; Beveridge, Naumann, & Rudolph 2024; Lee 2024); the ongoing struggle over Israel/Palestine (Abreek-Zubiedat & Nitzan-Shiftan 2021; Mason 2023; Agnew 2024); and the upsurge of climate-induced migration that will place increased pressure on state borders as divisions not simply among national groups but between livable and unlivable environments (Woodworth 2024).

References

Abreek-Zubiedat, F., & Nitzan-Shiftan, A. (2021). The right to an urban history: The Gaza Master Plan, 1975–1982. *Environment and Planning D: Society and Space 39*: 254–270.

Agnew, J. (2017). The tragedy of the nation-state. *Territory, Politics, Governance 5*: 347–350.

Agnew, J. (2024). The language of intractability and the Gaza War: Conflating anti-Semitism and anti-Zionism is historically problematic and misses how much contemporary Israel has become a role model for ethno-nationalists worldwide. *Human Geography 17*: 203–209.

Airriess, C., Hawkins, M., & Vaughan, E. (2012). Situating banal nationalism, the culture wars, and civil religion: Governing localized geographies of national identity in Indiana. *Social & Cultural Geography 13*: 49–67.

Anderson, B.R. (1991). *Imagined Communities: Reflections on the Origin and Spread of Nationalism*. London: Verso.

Antonsich, M. (2016). The "everyday" of banal nationalism: Ordinary people's views on Italy and Italian. *Political Geography 54*: 32–42.

Antonsich, M., & Skey, M. (2017). Affective nationalism: Issues of power, agency and method. *Progress in Human Geography 41*: 843–845.

Antonsich, M., Skey, M., Sumartojo, S., et al. (2020). The spaces and politics of affective nationalism. *Environment and Planning C: Politics and Spaces 38*: 579–598.

Bacchetta, P. (2010). The (failed) production of Hindu nationalized space in Ahmedabad, Gujarat. *Gender, Place and Culture 17*: 551–572.

Barras, A. (2021). Formalizing secularism as a regime of restrictions and protections: The case of Quebec (Canada) and Geneva (Switzerland). *Canadian Journal of Law and Society 36*: 283–302.

Benjamin, W. (2008). *The Work of Art in the Age of Mechanical Reproduction*. London: Penguin.

Berdichevsky, N. (2004). *Nations, Language, and Citizenship*. Jefferson, NC: McFarland.

Beveridge, R., Naumann, M., & Rudolph, D. (2024). The rise of "infrastructural populism": Urban infrastructure and right-wing politics. *Geography Compass 18*(2): e12738.

Billig, M. (1995). *Banal Nationalism*. London: Sage.

Brubaker, R. (1992). *Citizenship and Nationhood in France and Germany*. Cambridge, MA: Harvard University Press.

Brubaker, R. (1993). Migrants into citizens: The imposition of jus-solis in France during the late 1800s. *Actes de la recherche en sciences sociales 99*: 3–25.

Brubaker, R. (2004). The Manichean myth: Rethinking the distinction between "civic" and "ethnic" nationalism. In Kriesi, H. (ed.), *Nation and National Identity: The European Experience in Perspective*. West Lafayette, IN: Purdue University Press, 55–71.

Burds, J. (2009). Sexual violence in Europe in World War II, 1939–1945. *Politics and Society 37*: 35–73.

Cawood, S., & Fisher, J. (2022). "It should be a constant reminder": Space, meaning and power in post-liberation Africa. *Political Geography 99*: 102782.

Christian, J., Dowler, L., & Cuomo, D. (2016). Fear, feminist geopolitics and the hot and banal. *Political Geography 54*: 64–72.

Connor, W. (1990). When is a nation? *Ethnic and Racial Studies 13*: 92–103.

Culcasi, K. (2016). Warm nationalism: Mapping and imagining the Jordanian nation. *Political Geography 54*: 7–20.

Desbiens, C. (2000). Something straight in our landscapes: Looking at the "Lemieux effect" in Quebec nationalism. *Ecumene 7*: 211–232.

Doss, E. (2010). *Memorial Mania: Public Feeling in America*. Chicago, IL: University of Chicago Press.

Dua, H.R. (1993). The national language and the ex-colonial language as rivals: The case of India. *International Political Science Review 14*: 293–308.

Duchesne, S. (2018). Who's afraid of banal nationalism? *Nations and Nationalism 24*: 841–856.

Edenborg, E. (2020). Saving women and bordering Europe: narratives of "migrants' sexual violence" and geopolitical imaginaries in Russia and Sweden. *Geopolitics 25*: 780–801.

Ehrkamp, P. (2006). "We Turks are no Germans": Assimilation discourses and the dialectical construction of identities in Germany. *Environment and Planning A: Economy and Space 38*: 1673–1692.

Ehrkamp, P. (2019). Geographies of migration II: The racial-spatial politics of immigration. *Progress in Human Geography 43*: 363–375.

Forest, B., & Johnson, J. (2002). Unraveling the threads of history: Soviet-era monuments and post-Soviet national identity in Moscow. *Annals of the Association of American Geographers 92*: 524–547.

Foucault, M. (1991). Governmentality. In *The Foucault Effect: Studies in Governmentality*. Chicago, IL: University of Chicago Press, 87–104.

Freckelton, A. (2022). Could Canada abolish birthright citizenship? *Journal of International Migration and Integration* 23: 701–731.

Geertz, C. (1963). The integrative revolution: Primordial settlements and civil politics in the new states. In Geertz, C. (ed.), *Old Societies and New States: The Quest for Modernity in Asia and Africa*. New York: Free Press, 105–157.

Gellner, E. (1997). *Nationalism*. Washington Square, NY: New York University Press.

Getachew, A. (2019). *Worldmaking after Empire: The Rise and Fall of Self-Determination*. Princeton, NJ: Princeton University Press.

Gillen, J. (2014). Tourism and nation building at the war remnants museum in Ho Chi Minh City, Vietnam. *Annals of the Association of American Geographers 104*: 1307–1321.

Glasze, G. (2007). The discursive constitution of a world-spanning region and the role of empty signifiers: The case of Francophonia. *Geopolitics 12*: 656–679.

Gökariksel, B., & Secor, A. (2015). Post-secular geographies and the problem of pluralism: Religion and everyday life in Istanbul, Turkey. *Political Geography 46*: 21–30.

Green, S. (2012). Much ado about not-very-much? Assessing ten years of German citizenship reform. *Citizenship Studies 16*: 173–188.

Grigoriadis, I.N. (2009). Islam and democratization in Turkey: Secularism and trust in a divided society. *Democratization 16*: 1194–1213.

Hancock, C., & Mobillion, V. (2019). "I want to tell them, I'm just wearing a veil, not carrying a gun!" Muslim women negotiating borders in femonationalist Paris. *Political Geography 69*: 1–9.

Harden, M. (2023). Unmade in America: The gendered politics of birth tourism in the United States. *Political Geography 107*.

Hiemstra, N. (2021). Mothers, babies, and abortion at the border: Contradictory US policies, or targeting fertility? *Environment and Planning C: Politics and Spaces 39*: 1692–1710.

Hobsbawm, E.J. (1992). *Nations and Nationalism since 1780: Programme, Myth, Reality*. Cambridge: Cambridge University Press.

Hobsbawm, E.J., & Ranger, T.O. (1992). *The Invention of Tradition*. Cambridge: Cambridge University Press.

Holubec, S., & Mocicková, J. (2023). Ethnic mapping in Central Europe, 1810–1945: The case of the Czech-German language border. *Imago Mundi 75*: 1–27.

Houston, S., & Wright, R. (2003). Making and remaking Tibetan diasporic identities. *Social & Cultural Geography 4*: 217–232.

Howe, N. (2009). Secular iconoclasm: Purifying, privatizing, and profaning public faith. *Social & Cultural Geography 10*: 639–656.

Kamusella, T. (2009). *The Politics of Language and Nationalism in Modern Central Europe*. Basingstoke: Palgrave Macmillan.

Kaplan, D.H. (1994). Two nations in search of a state: Canada ambivalent spatial identities. *Annals of the Association of American Geographers 84*: 585–606.

Kashiwazaki, C. (1998). *Jus sanguinis* in Japan: The origin of citizenship in a comparative perspective. *International Journal of Comparative Sociology 39*: 278–300.

Kaufman, S.J. (2001). *Modern Hatreds: The Symbolic Politics of Ethnic War*. New York: Cornell University Press.

Kearns, G. (2004). Mother Ireland and the revolutionary sisters. *Cultural Geographies 11*: 443–467.

Knell, S.J. (2011). *National Museums: New Studies from Around the World*. London: Routledge.
Koch, N. (2011). Security and gendered national identity in Uzbekistan. *Gender, Place and Culture 18*: 499–518.
Koch, N. (2016). Is nationalism just for nationals? Civic nationalism for noncitizens and celebrating National Day in Qatar and the UAE. *Political Geography 54*: 43–53.
Koch, N. (2024). Scientific nationalism and museums of the future in Germany and the UAE. *Political Geography 113*: 103144.
Koch, N., & Paasi, A. (2016). *Banal Nationalism* 20 years on: Re-thinking, re-formulating and re-contextualizing the concept. *Political Geography 54*: 1–6.
Kulpa, R. (2014). Western *leveraged pedagogy* of Central and Eastern Europe: Discourses of homophobia, tolerance, and nationhood. *Gender, Place and Culture 21*: 431–448.
Lamour, C. (2023). Orban placed in Europe: Ukraine, Russia and the radical-right populist heartland. *Geopolitics 29*: 1297–1323.
Lecours, A. (2000). Ethnic and civic nationalism: Towards a new dimension. *Space and Polity 4*: 153–166.
Lee, J.Y. (2024). Left-wing nationalist-populist movement and identification: A psycho-political analysis of the "comfort woman" statue movements in South Korea. *Political Geography 108*: 103039.
Legg, S. (2005). Contesting and surviving memory: Space, nation, and nostalgia in *Les Lieux de Memoire*. *Environment and Planning D: Society and Space 23*: 481–504.
Leitner, H., & Ehrkamp, P. (2006). Transnationalism and migrants' imaginings of citizenship. *Environment and Planning A: Economy and Space 38*: 1615–1632.
Lepawsky, J. (2008). A museum, the city, and a nation. *Cultural Geographies 15*: 119–142.
Lizotte, C. (2020). Laïcité as assimilation, laïcité as negotiation: Political geographies of secularism in the French public school. *Political Geography 77*: 102121.
Lizotte, C. (2024). The prosaic stateness of secularism: Diversity, incoherence and divergence in the application of laïcité. *Geopolitics 29*: 962–986.
Mamvura, Z. (2022). Reconstituting the cultural geography in Zimbabwe: Place renaming in Zimbabwe's "new dispensation". *Geopolitics 27*: 972–994.
Martin, T. (2001). *The Affirmative Action Empire: Nations and Nationalism in the Soviet Union, 1923–1939*. Ithaca, NY: Cornell University Press.
Mason, O. (2023). The geographies of colonial infrastructures: Mobility, im/materiality, and politics on walking trails in the Middle East. *Transactions of the Institute of British Geographers 48*: 506–524.
Mayer, T. (2004). Embodied nationalism. In Staeheli, L.A., Kofman, E., & Peake, L. (eds), *Mapping Women, Making Politics: Feminist Perspectives on Political Geography*. New York: Routledge, 153–167.
Medeiros, M. (2017). Refining the influence of language on national attachment: Exploring linguistic threat perceptions in Quebec. *Nationalism and Ethnic Politics 23*: 375–390.
Militz, E., & Schurr, C. (2016). Affective nationalism: Banalities of belonging in Azerbaijan. *Political Geography 54*: 54–63.
Misztal, B.A. (2003). Durkheim on collective memory. *Journal of Classical Sociology 3*: 123–143.
Misztal, B.A. (2005). Memory and democracy. *American Behavioral Scientist 48*: 1320–1338.

Moser, S. (2010). Putrajaya: Malaysia's new federal administrative capital. *Cities 27*: 285–297.

Moser, S. (2012). Circulating visions of "High Islam": The adoption of fantasy Middle Eastern architecture in constructing Malaysian national identity. *Urban Studies 49*: 2913–2935.

Muzaini, H. (2014). The afterlives and memory politics of the Ipoh Cenotaph in Perak, Malaysia. *Geoforum 54*: 142–150.

Najib, K. (2021). Spaces of Islamophobia and spaces of inequality in Greater Paris. *Environment and Planning C: Politics and Spaces 39*: 606–625.

Najib, K., & Hopkins, P. (2019). Veiled Muslim women's strategies in response to Islamophobia in Paris. *Political Geography 73*: 103–111.

Nora, P. (1989). Between memory and history: *Les lieux de mémoire*. *Representations 26*: 7–24.

Oliveira, J.P. (2024). The progressive pilgrim: Real and mythical Indian geography in contemporary retellings of the *Rāmāyana*. *Nations and Nationalism 30*: 719–734.

Oswin, N. (2019). *Global City Futures: Desire and Development in Singapore*. Athens, GA: University of Georgia Press.

Paasi, A. (2016). Dancing on the graves: Independence, hot/banal nationalism and the mobilization of memory. *Political Geography 54*: 21–31.

Peker, E. (2024). Postcolonial imaginaries in the West: Secular state-building and cultural defence in Quebec since the quiet revolution. *Politics Religion & Ideology 25*: 218–242.

Perry, S.L., Schnabel, L., & Grubbs, J.B. (2022). Christian nationalism, perceived anti-Christian discrimination, and prioritising "religious freedom" in the 2020 presidential election. *Nations and Nationalism 28*: 714–725.

Puar, J.K. (2007). *Terrorist Assemblages: Homonationalism in Queer Times*. Durham, NC: Duke University Press.

Rahbari, L., Longman, C., & Coene, G. (2019). The female body as the bearer of national identity in Iran: A critical discourse analysis of the representation of women's bodies in official online outlets. *Gender, Place and Culture 26*: 1417–1437.

Ralph, D., & Staeheli, L.A. (2011). Home and migration: Mobilities, belongings and identities. *Geography Compass 5*: 517–530.

Ram, U. (2008). Why secularism fails? Secular nationalism and religious revivalism in Israel. *International Journal of Politics, Culture, and Society 21*: 57–73.

Reeves, D. (2015). Religious themes in Central European postage stamps, 2006–2010. *Journal of Cultural Geography 32*: 169–194.

Renan, E. (1994 [1882]). Qu'est-ce qu'une nation? [What is a nation?]. In Hutchinson, J., & Smith, A.D. (eds), *Nationalism*. New York: Oxford University Press, 17–18.

Ritchie, J. (2015). Pinkwashing, homonationalism, and Israel-Palestine: The conceits of queer theory and the politics of the ordinary. *Antipode 47*: 616–634.

Rose-Redwood, R., Alderman, D., & Azaryahu, M. (2010). Geographies of toponymic inscription: New directions in critical place-name studies. *Progress in Human Geography 34*: 453–470.

Safran, W. (1997). Citizenship and nationality in democratic systems: Approaches to defining and acquiring membership in the political community. *International Political Science Review 18*: 313–335.

Scott, J.C. (1998). *Seeing Like a State: How Certain Schemes to Improve the Human Condition Have Failed*. New Haven, CT: Yale University Press.

Shepherd, W.R. (1911). *Historical Atlas*. New York: Henry Holt.

Singh, S.P., & Tir, J. (2024). Less human than human: Threat, language, and relative dehumanization. *British Journal of Political Science 54*: 612–628.

Smith, A.D. (1998). *The Ethnic Origins of Nations*. Malden, MA: Blackwell.

Smith, A.D. (2002). When is a nation? *Geopolitics 7*: 5–32.

Suchland, J. (2018). The LGBT specter in Russia: Refusing queerness, claiming "whiteness". *Gender, Place and Culture 25*: 1073–1088.

Tamir, Y. (2019). Not so civic: Is there a difference between ethnic and civic nationalism? *Annual Review of Political Science 2019*: 419–434.

Thongchai, W. (1994). *Siam Mapped: A History of the Geo-Body of a Nation*. Honolulu: University of Hawaii Press.

Till, K.E. (2005). *The New Berlin: Memory, Politics, Place*. Minneapolis, MN: University of Minnesota Press.

Trenholm, J., Olsson, P., Blomqvist, M., & Ahlberg, B.M. (2016). The global, the ethnic and the gendered war: Women and rape in eastern Democratic Republic of Congo. *Gender, Place and Culture 23*: 484–502.

Wang, C.Y., & Chen, Z.F. (2023). From "motherland" to "daddy state": A genealogical analysis of the gender undertone in China's nationalist discourses. *Nations and Nationalism 29*: 751–767.

Weber, M. (1958). Politics as a vocation. In Herth, H.H., & Wright Mills, C. (eds), *From Max Weber: Essays in Sociology*. New York: Oxford University Press, 77–128.

Williams, R. (1983). *Keywords: A Vocabulary of Culture and Society*. New York: Oxford University Press.

Woodworth, F. (2024). Exclusion of climate migrants from the global compact on refugees. *Geopolitics 29*: 118–147.

Zhang, C.C. (2023). Postcolonial nationalism and the global right. *Geoforum 144*: 103824.

Chapter 24

Multilevel Governance

Herman van der Wusten
Human Geography, Planning and International Development, University of Amsterdam, Amsterdam, The Netherlands

Multilevel governance (MLG) started in 1993 as part of the title of a paper on a new governing practice in one policy sector within the just emerging European Union (EU). The approach in the paper became more frequently used and seemed to acquire a reputation as a paradigm in the way Kuhn (1962) had prescribed; others called it a mantra (Bache 2008: 29). Its scope increased, covering not only a set of policies but also the accompanying politics and polity. Its name was born within academia in political science and policy analysis (Marks 1993). Soon, the idea was also picked up in political geography (MacLeod 1999; Boyle 2000). In the course of time MLG was used in research inside and outside the EU. Within the EU it became temporarily a slogan in policy debates and political declarations.

In the rest of the chapter I first outline the ways in which MLG has evolved, taking in perspectives that had been developed earlier (notably federalism), and then how it flourished, notably with a series of major publications comprising datasets and theoretical interpretations, and at the same time was somewhat overtaken by reality. I concentrate on the work of Liesbet Hooghe and Gary Marks, central authors in the development of MLG, often accompanied by a series of other authors, who together formed a working group successively in different locations in the United States and Europe. In the next section I indicate a few brief encounters of MLG with political geography and the potential for ramification between the two in metropolitan regions across the world, notably in China. The final section closes the storyline of MLG provisionally.

The Wiley Blackwell Companion to Political Geography, Second Edition.
Edited by Virginie Mamadouh, Natalie Koch, Chih Yuan Woon, and John Agnew.

How Multilevel Governance Has Evolved

Mature MLG encompasses the whole political trilogy of polity (an identifiable political entity of people on a territory), politics (authorities and those directly connected in political practice), and policy (political plans in preparation or action with ends, means, and a timeline). It is perhaps best characterized as the counterpart of an ideal-type political order of nation-states internally ruled by a single center without any interest in international cooperation. It opens up questions on internal cultural homogeneity and a centralized government with complete territorial sovereignty over such states. Apart from central state authorities, the ML in MLG leaves room for several distinctive authorities/jurisdictions in parts of each state's realm and also for international authorities transcending these realms with at least some freedom of maneuver. The last letter in MLG stands for governance. While in governing there used to be a stronger separation between the individuals who govern and the others (they are "the society"), in governance some of the others may well take an active part in the steering of that society.

MLG was first picked up and studied in the EU policy sector of disadvantaged area issues (named after its aim: cohesion) where public and private/civic governing bodies at supranational, national, and subnational levels mutually negotiated and then jointly administered programs based on mixed funding (Marks 1993). Cohesion policy had a precursor in "structural policy" financed with a structural fund. In 1988 that became considerably larger as the Single Market was agreed and the new southern member states acceded. In cohesion policy small areas were determined to execute programs on low incomes, industrial decline, high unemployment, occupational integration for young people, agricultural improvement, and community-building. Partnership (different authorities, public and private participants), additionality (combining structural funds and member state finance), programming, and (territorial) concentration formed ideally the basis for continuous negotiations in nonhierarchical networks resulting in joint policy results (Schmitter 1996; Bache 2008: 39–53).

Eventually it turned out that this kind of networking also took shape in various other policy sectors. In the EU realm polity structures seemed to emerge not merely as a multiple of member states but also as a new EU polity. Hooghe and Marks further elaborated their model and did more research, in the process systematically extending their interest beyond the EU with other international governmental organizations (IGOs), states, and governmental authorities within states, collecting data that would later be meticulously published. In their research they concentrated on two aspects of government construction. First, they provided a typology by distinguishing between authorities with multipurpose decision-making powers and others in charge of single public goods. Second, they used the literature on federal systems and designed new indicators and indexes to come up with an operationalization to measure the self-rule and shared rule of individual regional authorities vis-à-vis each one's national authority, and in a similar way the autonomous authority of IGOs vis-à-vis their member states.

In a frame already broader than that of the EU, Hooghe and Marks (2003: 235, 241) thus produced a binary typology of MLG meant to encompass all concepts of governing that are "radical departures from unitary government" (polities ruled in all policy sectors from the same single center). Type I had a restricted number of levels, with larger and smaller numbers of units per level operating at respectively smaller and larger territories. They all possessed multipurpose decision-making powers and their areas of jurisdiction did not intersect at any level. Units of smaller partitioned levels fitted seamlessly into larger units of the next degree of partition. Type II was formed to provide single public goods (education, utilities, but also trade regulations and defense in international organizations performing at the suprastate level). The single goods of this functional differentiation were provided in larger or smaller territories dependent on efficiency considerations irrespective of overlap with other jurisdictions. It turned out that both types could well exist within one polity. While Type I systems tended to be sticky and difficult to change, Type II systems tended to be nimble and flexible. Type I units therefore tended to form the basic structure of state government architecture, while Type II units were often more loosely coupled to the rest of that framework, either substate or indeed suprastate. In both systems there was polycentricity and there could be mere governing or more broad-based governance. In later stages Hooghe and Marks concentrated their attention on Type I cases that they indicated as the basic architecture of government, at the same time stressing the flexibility of that architecture as they found it. It was perhaps not so sticky as had once been thought, just like real architecture.

Federalist governments are so far the clearest counterpart of the unitary governments that Hooghe and Marks wanted to exclude completely from their typological project. Unitary and federalism ideal types have origins in the works of Jean Bodin (1992, first published in 1576) for singular state sovereignty in the interest of the king of France and Johannes Althusius – first published using contemporary terms in 1603 – for sovereign self-rule in a diversity of communities within the shared rule commonwealth in the Holy Roman Empire of the German nation, and also in support of the Dutch confederal secession of the Spanish Empire (Carney 1964). In the current world there are many intermediate cases with various amounts of decentralization within the unitary government structures and a variety of intergovernmental relations between federal and state levels within federalist governing structures.

Despite considerable overlap of unitary and federalist governance structures, much relevant literature is still focused on one of these original ideal types. This applies in particular to federalism, although in the entire collection of states only about 15% call themselves or are considered federations. Of course, the largest states in terms of land surface nearly all have federal governments, with China as the most significant exception. But this emphasis on federations has certainly also to do with the fact that the United States, still the most discussed country on Earth, is a federation. And domestically, its federalist structure is incessantly discussed as an essential part of its polity. Agnew (2022) considers the shift in federalist governance in the United States since the 1980s from "polyphonic" to "dualist" as a major factor in its chaotic response to the Covid-19 pandemic (2020–2022), resulting in very negative consequences for public health compared to other federalist countries like Germany, Canada, and Australia. This dualist understanding of federalism harks back to the original United States of pre–Civil War times. States now once more hardly collaborate and the central state now plays only a weak part in coordinated efforts for the public good, contrary to its role in the New Deal starting in the 1930s that basically continued to the 1970s and resulted in a polyphonic type of federalism. Ideological changes and changes in the support patterns of the Republican Party provide parts of an explanation for the shifts since the 1980s.

In a paper in *Publius*, a journal originally dedicated to the study of federalism, Hooghe and Marks (2013) showed how meaningful it could be to replace a dichotomous split between states in unitary and federalist structures by a more gliding scale of governing capacity at the substate level(s), by means of self-rule by the individual authorities at the substate tiers and by means of control of the central state government by these substate authorities through shared rule. In the end they constructed two indexes with five items (in a previous version there had been four) for each that in fact covered the essence of the federalist ideal type of self-rule and shared rule on a scale picturing the unitary–federalist continuum. They then tried to find out how the size of either area or population would further the chances for federal-oriented government. Population size was more strongly related to more federalist institutions and therefore probably had a greater role in establishing such institutions. In this period commonalities in the research on federalism and MLG came more frequently to the fore.

MLG started to appear also in larger texts. Piattoni (2010) wrote a stimulating study empirically based on the already mentioned cohesion policy and then also on the environment policy, which already had a longer prehistory in European cooperation from before the EU, and on the beginnings of policy-making on higher education in the same generation of EU policy-making as that on cohesion. MLG posits connections without a "prime mover" between the blurring of center–periphery divides, the trespassing of state–society boundaries, and overcoming domestic–international distinctions.

Multilevel Governance in Bloom and in Upcoming Crosscurrents

MLG was finally itself baptized as a postfunctionalist theory (Piattoni 2010; Hooghe, Lenz, & Marks 2019) that was also relevant for cooperation processes from substate to state to suprastate beyond the EU. Meanwhile, Hooghe and Marks had collected a group of like-minded researchers working in the same vein and organizing symposia to consult about the progress of the project as a small republic of letters (and numbers), first centered in Chapel Hill, NC, and then with additional temporary seats at different venues in Europe and the United States, powerfully supported by permanent connectivity through digitalization across the world.

In 2020 Hooghe and Marks (2020: 820–826) published "A postfunctionalist theory of MLG" in the *British Journal of Politics and International Relations*, an excellent place for this missionary purpose. In short:

> Multilevel governance describes the diffusion of authority away from the central state . . . Postfunctionalism has three foundations. The first – the functionalist premise – is that multilevel governance is a functional adaptation to the benefits of scale diversity in the provision of collective goods. The second – the community premise – is that the form governance takes at any level depends on the sociality of the participants (willingness to invest in maintaining long-term cooperative relations, agreed mutual constraint through shared norms). The third – the premise of politicisation – is that both functionality and sociality are constructed in political conflict. (2020: 820, 822–824)

In general terms they wrote that the architecture of government had greatly changed since 1945. Authority was in the early period predominantly concentrated in national states, but it has later increasingly shifted in the vertical dimension in two directions: downward and higher up.

In the vertical dimension there has been a shift to lower-level, particularly regional, governments. They may cover different tiers between the local and national levels of their states. Stronger regional governments signifying primarily self-rule often also imply shared rule on the national level (co-decision-making under a federalist label). Above the national state, international organizations have on the whole also got increased authority in recent decades, notably those connecting states in a macroregional combination. By far the most significant is the European international organization that became the EU. In an index of authority it now scores the highest position as the only IGO playing a significant role in all policy sectors in a portfolio of 25 sectors; its important supranational institutions (European Court of Justice, European Commission, European Central Bank) have been empowered in a sequence of reforms.

National central governments have played a pivotal role in all these territorial transformations, sometimes taking a leading role in reform and at other times a reactive one. MLG may encompass authorities from the village to the globe. All these authorities are in some way interconnected and they should be considered as one single phenomenon. The recent extensions and growths of subnational and supranational governance, together with the already longer-enduring governance at the national level, provide new systemic puzzles: MLG questions.

In the 2016–2019 period, Hooghe and Marks' group of researchers produced four volumes under a common subtitle: *A Postfunctionalist Theory of Governance*. The series consisted of two pairs on (intrastate) regional authorities (Hooghe et al. 2016; Hooghe & Marks 2016)and IGOs (Hooghe et al. 2017; Hooghe,

Lenz, & Marks 2019). The first volume of each pair presented the datasets, the second concentrated on the variations that turned up and their interpretation. They also offered occasional examples of the effects of both types of authorities. Numbers, strengths, and scope of both types increased. This does not necessarily mean that the authority of states decreased absolutely, but in any case state authorities had to deal with ever more meaningful others. Clearly, the double focus of MLG on substate and suprastate authorities developing at some distance from the central states was again obvious.

The first pair of volumes is organized around the 3465 regional authority annual double indexes (RAIs) in the 1950–2010 period for individual regions in 81 countries. They show different profiles in strengths and portfolios of competences in self-rule and shared rule. Far from permanent stability, they show quite a few changes over time. After sharp political changes in a state center, there often follows a new preference for the "ladder of government" across the state territory. A new map with portfolios per tier can then be crafted. The classic example is Napoleon's map of the new *départements* installed during the French Revolution. All the later adaptations and changes in newly determined tiers of the ladder notwithstanding, the *départements* are still there.

There are quite different trajectories where other actors than the state architects of the ladder of government are also active in its construction. Apart from the scale question from the standpoint of efficiency of public goods provision, the question in this perspective is notably: What is the collectivity that wants to/should be in charge of a certain level of self-rule and/or a certain level of shared rule in the center? In fact these imagined collectivities will in the end be expressed in authorities that show up in territorial jurisdictions. And this entire arrangement can over time change again. In these center–periphery contestations, three peripheral assets are indicated as conditions for building up communities that might survive as part of a more encompassing state. Hooghe and Marks (2016) designate these different forms of peripheral self-expression in new state formations as Rokkanian in honor of Stein Rokkan, who opened this research road in an earlier period (after his premature death in 1979 his complete work was reconstructed by Peter Flora in Rokkan 2000). These assets are being very distant and difficult to reach, a living language, and earlier statehood. It turns out that the mere distance that is generally accompanied by small populations does either result in authorities totally dependent on the central state without pronounced self-rule (Svalbard, Rokkan's place of origin) or an active bilateral relationship with the central state and pretty extensive self-rule (examples Aceh, Greenland). More centrally located, often larger communities conditioned by a living language and/or a statehood past tend to have considerable self-rule and shared rule. They tend to start with pressure for self-rule and may eventually end with high levels of self-rule and shared rule after political mobilization in political parties and social movements, as well as multilateral negotiations with the central state and other regional authorities in the same tier (Catalonia, or Russia under Yeltsin with Tatarstan in particular). Political mobilization gets easily emulated in the process. This may result in slightly different solutions in authorities of the same tier. All in all, the territorial structure of authority gets adapted for individual regions with some special need or circumstance. Particularly in the last type there is MLG through general tension in territorial rule on account of community diversity and parties and movements as prime drivers.

The second pair of volumes on IGOs is organized around 67 of these organizations with their member states from different parts of the world. They are analyzed in a mirror-like perspective, with processes of delegation toward IGOs as the central measuring concept instead of self-rule and shared rule in the first pair. An IGO comprises at least three states and has a distinct location, a formal structure, at least 30 permanent staff, a written constitution, and a decision-making body meeting at least once per year. Autonomy in political and judicial practices, independence of bodies within an IGO of member state control, and a role in the IGO decision-making body all count for an annual summary in a delegation score during 1950–2010. During that period there was an uptick of IGOs, a broadening of portfolios, and a strengthening of delegation, with acceleration particularly in the 1950s and from the 1990s. A follow-up paper (Lenz et al. 2023) shows how the increase in delegation was tailing off from about 2010.

One result in Hooghe, Lenz, and Marks (2019) is a provisional answer to the question of why states do in fact pool authority. Although there may of course be many public goods to be captured by cooperation (the functional perspective), strong desires at state level for self-rule remain and the drive for delegation of authority will be small. In general transnational community looks thin. But in fact in the empirical results some indicators for cultural conformity and open contracts are much stronger than trade interdependence, power asymmetry, democracy, and affluence or economic diversity to explain delegation and the growth of IGO portfolios. As in the regional pair of volumes, scale and community turn out to be important, with community gain as nearly indispensable. That is the dominant meaning of the nametag of postfunctionalism.

In the four volumes that form *A Postfunctional Theory of Governance*, the winds of change in the empirical world are already tangible. The Brexit referendum is mentioned in passing. In finishing the series in 2019 there was a renewed effort to specify how far postfunctionalism could reach and how long it could endure (referring back to earlier passages about massively supported nationalist populisms in the United States and in Europe):

> (T)his is a study of a single case, the liberal world order following World War II. This world order has been sustained by factors that lie outside our theory, including American material and ideational dominance and an extended period devoid of world war. It is possible that the functional and cultural variables in our theory are time-bound in ways that are not yet evident. As Postfunctionalism argues, the chief constraint on international governance today is nationalism. Whether this will remain

> so, we cannot yet tell . . . . (A)n era of relatively benign transnationalisms based on a permissive consensus seems to be drawing to a close . . . . The most direct challenge seems to come from within the liberal core. (Hooghe, Lenz, & Marks 2019: 6–7)

It seems that in this way the powerful resistance of authoritarian regimes against the postwar embedded liberalism in other parts of the world has been left too much out of the limelight.

In 2021 Tatham, Hooghe, and Marks (2021) provided a pretty comprehensive stock-taking of research results on the model and a view of its current stance. Its title – "The territorial architecture of government" – suggests a worldwide view. But in fact its content refers to contemporary Western societies, also called democracies, from around the world. This summarizing paper (together with three supporting papers) was first presented at the University of Bergen, Norway, where Stein Rokkan worked in the 1960s and 1970s. One of the highlights of Rokkan's famous program on state formation and democracy was an edited book (Lipset & Rokkan 1967). Gary Marks wrote his PhD some years later with Lipset. Tatham, the first author of the summarizing article, is now one of the professors in Rokkan's Department of Comparative Politics.

MLG has continued to operate under the strains of the scale diversity of functional benefits resulting from optimal attribution of public goods and also of strains arising from the multitude of created collective identities. These two types of strain are connected, because the capacity for providing public goods is enhanced where individuals are part of a community. In connection with MLG this effect of community is in fact double-edged. The norms that facilitate governance within communities simultaneously constrain governance among them. The acceptance of such norms is socially constructed and can be politically contested. They can be strengthened and weakened according to circumstances and by political intent. From a functionalist perspective, MLG should consist of a limited number of tiers reflecting economies of scale. From a community perspective it should consist of patterns of collective identities with resolved political struggles on how competing identities could be compromised in forms of self-rule. These functional and identitarian puzzles should be solved under the various strains of each one, with in every provisional "final" resolution a new version of the ultimate political triptych of polities, politics, and policies.

Regarding policy, based on one of the other papers (Peters 2021), shifts in both directions of the vertical dimension are studied in their effect on the citizen–policy link, which is in this case the relation between citizen preferences and public spending on social policy per state. The link is stronger in polities with more important regional authorities and in particular in those with more pronounced shared rule. There, citizens have an extra channel to make their preferences known. In the other direction of the vertical dimension, greater linkage to the external world directly erodes the citizen–policy link, but simultaneously it indirectly reinforces it through its positive effect on regional authority (and its feedback through shared rule). By provoking increases in regional authority, external openings seem to generate a positive response to their own shortcomings, hampering the citizen–policy link.

The Governance of Metropolitan Regions

In an extensive discussion on regional authorities in MLG (Eaton et al. 2019), one researcher in that field (Harbers 2019: 274–278) makes a strong plea for a spatial perspective renouncing the normal route of research in comparative politics through comparisons of nation-states as bounded and more or less independent units. In a section called "Interdependency in overlapping systems," the possibilities of geographic information systems (GIS) and geocoding of all these regional authorities are opened up in a clear awareness of a fruitful terrain for overlapping disciplines. In contributions from the field of political geography, MLG has mostly been referred to so far in early publications like Marks (1993) and Hooghe and Marks (2001). I now indicate in some more depth how political geography has further perceived and assessed MLG, and thereafter focus on recent work on upcoming metropolitan regions. This is where researchers of MLG, and political geographers in particular, labor on the same terrain as more or less competing colleagues.

Political geography commentary on MLG has repeatedly used the term "level" as a way to distinguish its own approach from others. In relation to level, a hierarchy of authorities and the spatial size of jurisdictions is observed but considered unclear or too one-sided, and jurisdiction should apparently not be the preferred term in this context. For instance, Jonas (2015: 26–35, notably 27, 30) marks the concept of level as more or less similar to scale. In his view political scientists' level and MLG are at least in one sense inferior to the use of scale in political geography. Political geographers acknowledge "not only that political processes reflect how power is distributed hierarchically through the different levels of the state but also that struggles within or across territory can produce new scales of organization and governance, which do not map directly onto the corresponding levels of the state" (2015: 27).

In a similar vein, Agnew (2016) declares territory to be completely tied to the state in current literature – he denotes the combination as a fetish. To develop a richer spatial epistemology as recommended by Wendel (2015), he follows that author's argumentation. This starts from "spaces of encounter, where individuals make demands on each other and the spaces where these encounters occur locate, represent, or form political subjectivity, engagement and power" (Wendel 2015: 3). This suggests a strong bottom-up process of political mobilization where existing or emergent authorities do not count. However, particularly in governance problems, this seems only rarely a fruitful start as governance rarely begins from scratch. In political governance (not governance of single businesses or associations), territory is still a defining part of the authorities and jurisdictions that are involved. Such authorities are not necessarily Westphalian states and are thus not necessarily defined in terms of a (more less magical) sovereignty. In a spatial methodology concerning political governance, territory should play its own role and not be part of a fetish. In MLG territorially marked authority at different scales/levels should be retained as fruitful parts of a still highly relevant set of problems.

Under the title "Polymorphic political geographies," Jones (2016) has introduced and summarized a set of papers that in different ways relate to MLG. At least two papers in particular continue the preceding intervention by Jonas. Alcantara et al. (2016) narrow the concept of MLG compared to the way it is used by Hooghe and Marks by limiting it to incidences and excluding it from longer-term structures. In an instructive case study of the joint regulatory history of water management on both sides of the American–Canadian border, incidences of MLG versus intergovernmental relations (slightly differently defined) are shown within a longer-standing structure as a cooperative venture of two federalist states. This seems to be a question of conceptualizing preferences. While Alcantara et al. make a tiny place for MLG within the traditional way of dealing with federalism and intergovernmental relations, Hooghe and Marks (2016: 12) use the basic idea of federation formation for the construction of a new "ladder of governing" by introducing systematic variations in the size and sorts of self-rule and shared rule.

A second paper by Jessop (2016) proposes a further critical analysis of MLG and at the end a new term, multispatial metagovernance, as an overarching alternative (in short, from MLG to MSMG). In Jessop's version states are now no longer only functioning as dominant primary kernels of governance but foremost as arbiters and coaches of all kinds of actual political governing acts by others. They have the governance of governance in hand. In MLG the state is less centrally positioned in studying the acts of political governing. Complementary authorities play their additional roles, possibly in the longer term changing the distribution of roles. In this case the political change is in the withholding of final authority in the hands of states, in the constant uncertainty and negotiation about where "final" authority rests, and possibly in sharing. Both MLG and MSMG perceive the art of governing as very different from the traditional image of Westphalian statehood: single-center final authority on bounded territory with population, in principle completely autonomous. While MLG points in the direction of heterarchical structures and political heritage and innovations as potential engines for political fortunes, MSMG is more oriented to restructuring state theory following in the footsteps of Gramsci and Foucault.

During the same period in which MLG started to flourish, Brenner (2004) discussed shifts of governance, notably in Western Europe, as in the case of MLG. As an urban theorist and geographer, he called it state rescaling in light of a political economy perspective. The prime mover here is the development of capitalism, whereas MLG is more at home in a world seen through a lens of political sociology without prime movers. While many analyses of globalization and the changing state focused on the construction of new supranational political regimes, such as the EU, Brenner stressed that also at subnational scales, particularly those of major urban regions, strategic institutional arenas turned up in which far-reaching transformations of state spatiality were unfolding. Processes of urban governance, he argued, represented a key mechanism for the rescaling of state space. Clearly, here was the potential for some ramification of MLG and state rescaling. Much work was afterward done on these assumed new institutional arenas, often demarcated as metropolitan scales.

In a series of three papers, an international group of researchers (with geographers and planners, plus political sociologists and political scientists) reported their work on the definition and measurement of a metropolitan scale and on possible determinants, notably globalization and possibly others as well. Their research program was instigated by Brenner's idea. It had attracted a wide audience of theorists and practitioners, but according to this collectivity it still lacked much empirical analysis in order to better link abstract concepts to specific contexts. The papers were all based on combinations of descriptions of projects and policies apparently aimed at or demonstrating the emergence of a metropolitan scale in case studies in six urban regions, three in Europe (Berlin, Paris, Rome) and three elsewhere (Delhi, New York, Shenzhen). They made use of a number of additional perspectives often mentioned in conjunction with Brenner's state rescaling.

In their introductory overview, d'Albergo and Lefèvre (2018) consider the emerging metropolitan scale – the product of rescaling – as a three-dimensional concept (economic, spatial, political), with each dimension developing only partially connected with the others. The economic dimension refers to the external connectivities and relationships of urban agglomerations, the spatial dimension to the internal functional and spatial polarization of the expanding agglomeration, and the political dimension to the state's relevant scalar changes in its territorial organization, including governance and policy formation. To track down the determinations of the emerging metropolitan scale, they look at globalization as an external context of market forces and cognitive pressures of economic ideologies and beliefs of international organizations, the form of policy networks, and the results of material actions and discursive representations. The three following papers take care of several of these determinants.

In the six cases studied, signs of an emergent metropolitan scale are present in the economic and spatial dimensions, but evidence of "new political spaces" is very weak. There is no direct relation between the urban challenges brought about by globalization and the construction of metropolitan scales as the prevailing political solution. If solutions for such challenges eventually turn up, political aims and forms do not refer to globalization but are somehow related to local and national factors. Businesses in general are only rarely instigators of efforts to build up a political metropolitan scale. Big transnational firms in particular perceive their space of dependence (Cox 1998) in terms of those factors of production that enhance their global competitiveness, such as connectivity and innovation. This also enhances the chances for support and endorsement of policy processes aimed at metropolitan scale building. The case of Shenzhen showed examples. Other businesses, and particularly those organized around real estate and land rent, have a different kind of dependence on place and space. They do not aim to assist in building a metropolitan scale in either economic, spatial, or political terms. This was illustrated in the case of Rome.

National states and the policies formulated and ratified at that level are still central for metropolitan scale development. However "Europeanized," such policies are still mainly institutionalized in a national context. Path dependencies are often crucial. By way of example, the neocorporatist traditions detected in the European case studies (Berlin, Paris, Rome) did not show up in the other ones (New York, Delhi, Shenzhen). This helped to interpret the convergence of European state–business relations and their absence elsewhere in North America and Asia in efforts to pursue the construction of a metropolitan scale. Businesses and national public actors express their preferences through symbolic–cognitive representations of space. Such representations are also conveyed by prominent international political organizations, which in turn often rely on representations and language produced by academics or thinktanks. For a while a mainstream vision may emerge, whereby developing metropolitan space and scale is perceived as a condition for the global competitiveness of cities, while metropolitan governance provides the form of collective action that enables metropolitan regions to act as stakeholders in global networks. Nevertheless, when "applied" to local contexts, these patterns risk not matching with the specific features of places in terms of economic interests, political institutions, and local cultures.

Clearly, the jurisdictions that score on the RAI in MLG are somehow akin to the strategic institutional arenas that Brenner saw emerging in new state rescaling, but by no means largely overlapping. The RAI entities are defined by way of the geographical demarcation of their competences and described in organizational terms. In governance these authorities can still play different roles, as "the society" may also acquire more or less prominent governing parts. The arenas probably refer foremost to the metropolitan spaces and places that strategic actors have in mind, but that collection of actors and the form of their networks and decision circuits are still open. The form of a specific, metropolitan authority foremost in charge of the metropolitan scale is rare according to the cases just discussed.

During China's formidable urbanization – from 20% to 60% in 40 years (1980–2020) – existing cities exploded, major new settlements with urban densities were built, and the functional profiles of agglomerations experienced revolutionary changes. A few of the very greatest urban concentrations across the country were assigned as *chengshi qun* (city clusters or city-regions) in the New Urbanization Plan (NUP) launched in 2014. In fact each consisted of a number of adjacent administrative entities, with major cities of at least five million officially registered inhabitants as centers. Three of them were the growth centers and their surroundings were close to the coast: the Pearl Diver Delta with Shenzhen as one major center, the Yangtze River Delta with Shanghai as the major center, and Beijing accompanied by the port of Tjanjiu and the province of Hebei. More in the interior part of China were Chendu-Chongqing, another cluster in the Middle Reaches along the Yangtze, and finally one on the Central Plain, all on domestic transport nodes but also important in the planning of the global Belt and Road Initiative in the direction of Europe. Chu (2020) has collected some data on these six new entities. He gives the impression that there may eventually be room for additional clusters. After all, there was still a sizable flow of rural–urban migrants at the time. It is also interesting that the clusters were apparently not presented as one specific category in the plan. Their diverse nametags in the tables of the paper were separately presented as references to singular regional development plans.

Globalization and the incorporation of China in the world economy had been a subject taken up in five-year plans from the 1990s. The beginning of the strategy for regional coordinated development (RCD) in urban regions had already been politically determined in 1993. Sassen's global city was discussed and Taylor's methodology to measure advanced producer service networks to assess the positions of cities within the globalized economy became used in Chinese academia. The NUP seemed to be an effort to prepare the Chinese economy for this future by way of a Chinese road to a metropolitan scale. Wang and Zheng (2024) see China as a state that is part of the south, with an authoritarian political regime and an economy following a neoliberal course. The new urbanization was an effort to rescale the state to allow China to benefit from a prominent position in the globalized economy. The overall aim of NUP for the urban clusters was first to further concentrate advanced producer services in the coastal clusters and encourage more industrial growth in the interior, also through transfers from the coastal areas. As the interior clusters might profit from the Belt and Road Initiative, they could in the longer term also alter their industrial profile to that of the globalized city.

Wang and Zheng (2024) stipulate that RCD should be read against the background of the longstanding, vertically organized, rigid administrative division system (ADS). Cities have therefore so far been governed in the context of what these authors call Foucauldian structural discursive practices, moving downward from province through prefecture, county, and township. The different authorities through which the discourse passed could only marginally deviate by interpretations on the basis of single interests without deliberation with other parties. The ongoing construction of city-regions has not yet been institutionalized into the ADS. Instead, it has been constructed as a temporarily localized administrative division by the RCD national strategy. The term city-region is used quite flexibly and vaguely, referring here particularly to cross-border interaction among cities. Accordingly, discursive practices at the city-regional scale are in this context more contingent and flexible. Higher authorities leave more space. Lower-tier authorities see more chances to cross jurisdictional borders to gain development opportunities in the name of the RCD national strategy. All this paves the way for more selective (thus nonstructural) discursive practices at city-regional scale.

Two of these selective practices are reported in detail in Wang and Zheng's (2024) paper. Both are concerned with development zones instigated by two cities, each under the RCD. One is within the Pearl Delta, involving Shenzhen and another less fortunate place, the other within the Beijing–Tjanjiu–Hebei city-region, with a major new town at the edge of Beijing as large as the instigator of the zone plus another less well-situated place. The analysis is focused on two sequences of narratives. Starting from the basic structural practice following the higher levels of the administration in proclaiming general improvement of economic development through cooperations within the RCD, in the Shenzhen case a narrative of pairing assistance is developed, in which massive regional economic disparities between unequal partners will be reduced. After this had failed, an innovative reform narrative followed (how to organize new business activity including practices) that was in the end successful for both parties and notably was accompanied by a certain shift in the administrative structure, although it remained outside the ADS. In the case of Beijing's new town the win–win structural practice was initially given substance by an industrial integration narrative (providing new land in order to shift and expand industries), which ended in contestation between the partners and was followed by a deterritorialization narrative (which left action in the hands of a powerless management committee); the project ended without a result.

If all this shows small steps with uncertain success in the direction of a metropolitan scale, it is again interesting to see how difficult it is to break out of path dependencies, in this case out of the ADS that in principle prevents possibilities of horizontal and bottom-up interaction. The territorial politics within this authoritarian polity are now under temporary RCD rules without a time line. They play out between political authorities of different tiers within these enormous city clusters. There is a certain similarity with the MLG model as proposed by Hooghe and Marks, with a shared emphasis on political games but considerable differences in context concerning the rules of those games. Nonetheless, it suggests potential room for mutual ramifications and inflection points in further MLG and state rescaling studies.

Provisional Completion

From the 1990s onward MLG has evolved as a promising road for the study of political governance, focused on its necessarily multilevel and forever contentious character. MLG was first geared to at least three levels or scales, encompassing the EU, its member states, and an intrastate regional level. Efforts of cooperation and negotiation by actors representing these three types of jurisdictions would give rise to changes in the authority distribution of institutions, with increases for the EU and the regional level and diminution for the member states. Notably, cooperation between the EU and the regional level could result in such outcomes. All these efforts were political processes. No end result could be entirely determined beforehand. The scales were not merely functionally determined, there were also communities linked with these jurisdictions.

While MLG was exercised across the territory of the EU, the EU extended from 12 to 28 member states and then lost one toward the end, thus finishing with 27. Simultaneously, MLG studies were carried out that also took in other parts of the world or simply single countries. In those studies all the three types of jurisdictions and their interrelations were not always particularly predominant. It turned out that insufficient data were available to get a full overview of the three relations necessary to come to a final conclusion concerning a trend of authority distributions across scales of jurisdictions.

In fact, for intrastate regional jurisdictions quite a few states had not one but a variety of such jurisdictions; and for suprastate political organizations similar comments can be made. In the end it turned out that for regional jurisdictions as well as for international organizations, special datasets had to be created in order to get some grip on their authority over time and across the units during a single year. These datasets have been published with much attention on the construction of both measurements through a widening of federalist principles and a limitation of authority to the sphere of legal modernity. They have been analyzed with a light touch to find new openings for turning on scale and community, not to speak of historical and geographical contexts as important basics for specific forms of multilevel political governance. In these datasets a series of annual measurements for the period 1950–2010 were entered.

Only at the end was there again an effort to connect the three types of jurisdictions. But it was now less a question of shifts in the distribution between the three types and more a way in which the combination of intrastate regional types and international organizations in mutual coherence would together push up or restrict the social expenditure of the various member states. This time it was more a question of extra consequences through the design of secondary parts of the member states than one of shifts in the final design of the political order.

References

Agnew, J. (2016). Spatializing politics. *Territory, Politics, Governance* 4(3): 265–268.

Agnew, J. (2022). Failing federalism? US dualist federalism and the 2020–22 pandemic. *Regional Studies, Regional Science* 9(1): 149–171.

Alcantara, C., Bröschek, J., & Nelles, J. (2016). Rethinking multilevel governance as an instance of multilevel politics: A conceptual strategy. *Territory, Politics, Governance* 4(1): 33–51.

Bache, I. (2008). *Europeanization and Multilevel Governance: Cohesion Policy in the European Union and Britain*. Lanham, MD: Rowman and Littlefield.

Bodin, J. (1992). In Franklin, J.H. (ed.), *On Sovereignty*. Cambridge: Cambridge University Press.

Boyle, M. (2000). Euro-regionalism and struggles over scales of governance. *Political Geography 19*(6): 737–769.

Brenner, N. (2004). Urban governance and the production of new state spaces in western Europe, 1960–2000. *Review of International Political Economy 11*(3): 447–488.

Carney, F.S. (ed.) (1964). *The Politics of Johannes Althusius: Abridged Translation of Althusius' Main Text of 1614*. Boston, MA: Beacon Press.

Chu, Y.-w. (2020). China's new urbanization plan: Progress and structural constraints. *Cities 103*: 1–10.

Cox, K.R. (1998). Spaces of dependence, scales of engagement and the politics of scales. *Political Geography 17*(1): 1–23.

D'Albergo, E., & Lefèvre, C. (2018). Constructing metropolitan scales: Economic, political and discursive determinants. *Territory, Politics, Governance 6*(2): 147–158.

Eaton, K., Faguet, J.-P., Harbers, I., Schakel, A.H., Hooghe, L., & Marks, G. (2019). Symposium measuring and theorizing regional governance. *Territory, Politics, Governance 7*(2): 265–283.

Harbers, I.M. (2019). Contributions of the postfunctionalist theory of governance to subnational research in comparative politics. *Territory, Politics, Governance 7*(2): 274–278.

Hooghe, L., Lenz, T., & Marks, G. (2019). *A Theory of International Organization: A Postfunctionalist Theory of Governance*, vol. *IV*. Oxford: Oxford University Press.

Hooghe, L., & Marks, G. (2001). *Multi-Level Governance and European Integration*. Lanham, MD: Rowman and Littlefield.

Hooghe, L., & Marks, G. (2003). Unraveling the central state, but how? Types of multi-level governance. *American Political Science Review 97*(2): 233–243.

Hooghe, L., & Marks, G. (2013). Beyond federalism: Estimating and explaining the territorial structure of government. *Publius: The Journal of Federalism 43*(2): 179–204.

Hooghe, L., & Marks, G. (2016). *Community, Scale, and Regional Governance: A Postfunctionalist Theory of Governance*, vol. *II*. Oxford: Oxford University Press.

Hooghe, L., & Marks, G. (2020). A postfunctionalist theory of multilevel governance. *British Journal of Politics and International Relations 22*(4): 820–826.

Hooghe, L., Marks, G., Lenz, T., Bezuijen, J., Ceka, B., & Derderyan, S. (2017). *Measuring International Authority: A Postfunctionalist Theory of Governance*, vol. *III*. Oxford: Oxford University Press.

Hooghe, L., Marks, G., Schakel, A.H., Niedzwiecki, S., Chapman Osterkatz, S., & Shair-Rosenfield, S. (2016). *Measuring Regional Authority: A Postfunctionalist Theory of Governance*, vol. *I*. Oxford: Oxford University Press.

Jessop, B. (2016). Territory, politics, governance and multispatial metagovernance. *Territory, Politics, Governance 4*(1): 8–32.

Jonas, A.E.G. (2015). Scale. In Agnew, J., Mamadouh, V., Secor, A.J., & Sharp, J. (eds), *The Wiley Blackwell Companion to Political Geography*. Chichester: Wiley-Blackwell, 26–34.

Jones, M. (2016). Polymorphic political geographies. *Territory, Politics, Governance 4*(1): 1–7.

Kuhn, T.S. (1962). *The Structure of Scientific Revolutions*. Chicago, IL: University of Chicago Press.

Lenz, T., Ceka, B., Hooghe, L., Marks, G., & Burlikov, A. (2023). Discovering cooperation: Endogenous change in international organizations. *Review of International Organizations 18*: 631–666.

Lipset, S.M., & Rokkan, S. (1967). *Party Systems and Voter Alignments: Cross-National Perspectives*. New York: Free Press.

MacLeod, G. (1999). Place, politics and "scale dependence": Exploring the structuration of Euro-regionalism. *European Urban and Regional Studies* 6(3): 231–253.

Marks, G. (1993). Structural policy and multilevel governance in the EC. In Cafruny, A., & Rosenthal, G. (eds), *The State of the European Community*. New York: Lynne Rienner, 391–410.

Peters, Y. (2021). Social policy responsiveness in multilevel contexts: How vertical diffusion of competences affects the opinion-policy link. *Governance 34*(3): 687–705.

Piattoni, S. (2010). *The Theory of Multi-level Governance. Conceptual, Empirical, and Normative Challenges*. Oxford: Oxford University Press.

Rokkan, S. (2000). *Staat, Nation und Demokratie in Europa*. Frankfurt am Main: Suhrkamp.

Schmitter, P.C. (1996). Imagining the future of the Euro-polity with the help of new concepts. In Marks, G., Scharpf, F.W., Schmitter, P.C., & Streeck, W. (eds), *Governance in the European Union*. London: Sage, 121–150.

Tatham, M., Hooghe, L., & Marks, G. (2021). The territorial architecture of government. *Governance 34*(3): 607–620.

Wang, X., & Zheng, G. (2024). Narrating regional coordinated development: Discursive practices in China's city-regions. *Cities 146*(4): 104768.

Wendel, D.D.B. (2015). Introduction: Toward a spatial epistemology of politics. In Wendel, D.D.B., & Aidoo, F.S. (eds), *Spatializing Politics: Essays on Power and Place*. Cambridge, MA: Harvard University Graduate School of Design, 2–13.

Chapter 25

Social Movements

Sara Koopman

School of Peace and Conflict Studies, Kent State University, Kent, OH, USA

Political geography has traditionally focused on states and formal politics in and across them. It was not until recently that political geography textbooks even mentioned social movements, and then usually very briefly, in an add spice-and-stir sort of way (e.g., Agnew 2002; Cox 2002), though Painter (1995) is a notable exception, now in the second edition as Painter and Jeffrey (2009), as is Smith (2020). But there has been a broadening of what counts as "doing politics" such that it is not only big P Politics (in and with the state, e.g., politicians and voting) but now includes other forms of political engagement by social movements as part of a reshaping of the discipline and reworking of understandings of the relationship between space and power, the political and the spatial (Cox, Murray, & Robinson 2008; Dodds, Kuus, & Sharp 2013).

These changes are in part because feminists have for years pushed for the integration of "small p" politics (of the everyday, e.g., sexual harassment) and a recognition of how public and private spheres are intertwined and processes happen across, and through, both the global and the intimate (Rosner & Pratt 2006; Sharp 2007). Some social movements focus more on the small p, some on influencing large P Politics, and many on both, at times using one or the other. Many think of their work as social or economic more than political. Likewise, some geographers researching movements identify not as political geographers but more as social, economic, urban, development, or feminist geographers. I include their work here as part of the shift in the field to a broader understanding of what is political.

The Wiley Blackwell Companion to Political Geography, Second Edition.
Edited by Virginie Mamadouh, Natalie Koch, Chih Yuan Woon, and John Agnew.

Much of the work done by geographers on social movements has actually used categories other than movement, such as contentious politics, collective action, connective action, or network. In this chapter I begin by reviewing these issues of definition. I then review some of the major contributions that geographers have made to understanding the work of social movements, particularly how various spatialities shape and are shaped by movements, and the spatialities of solidarity. I briefly review territoriality as one spatiality of movements that is emphasized by Latin American geographers. I then discuss relationships between geographers and social movements, and point to some of the collaborative ways in which geographers have gone about thinking with and for movements. I argue that geographers have made and can make important contributions to building the power of movements to make a better world.

What Is a Movement, or "Who Moves What How?"

Social movement is not a clearly defined or well-agreed upon for collective action and organizing for social change (Pile & Keith 1997; Routledge & Cumbers 2009; McCarthy 2011; Chatterton, Featherstone, & Routledge 2013). What counts as social movements, how to categorize them, and even what they should be called have been contested by both academics and activists – sometimes for theoretical or methodological reasons, sometimes for political or ethical ones. Issues have included: How many groups does it take to count as a movement? Do they have to be connected and/or coordinated? Do they have to be effective at creating social change to count? Can groups have paid staff? Can they be involved in electoral campaigns? Can they take up arms? Engage in property destruction? What if groups have no office or no formal structure? What if they receive aid or training from the state or, more likely, the city? What if it is from a foreign state?

In recent years it has become more and more possible for groups in many places to self-organize, informally and transnationally. Information and communication technologies (ICTs) have played a key role in this new era of looser-linked organizing with much broader frames (like "We are the 99%") that can encompass many different issues and identities (Bennett 2005; see also Adams 2025, this volume). ICTs help make this sort of organizing possible, but it is also born of generational and societal shifts, and the breakdown in all sorts of group membership and institutional loyalties in late modernity.

Bennett and Segerberg (2012) call this new wave "connective action," because the quick resonance and spread of these movements, like Occupy for example, happened through a back and forth between online and offline activisms that feed and shape each other. The changing online tools available, such as the recent ubiquity of both social media and smartphones, also change what is possible on the streets, and those actions continue to change how the online tools are used. This is quite different than what used to be one-way communication from an organization to supporters; it is much more than the internal listservs that were widely used for organizing in the 1990s and 2000s, and has grown beyond the indymedia aggregation sites that were early harbingers of this. This is about more than just the low-input clicktivism of signing a petition or sending a form letter. This online communication goes in many directions, in personalized ways, through action sites where people post pictures, videos, stories, and analysis; comment on posts; share posters for wheatpasting up onto telegraph poles and such; add calendar events, say to an interactive map; or do other participatory geovisualization, like open street mapping of Occupy sites. It is about actions off the main action site, like posting clickthrough action links and widgets but also, importantly, radicalizing social media with personalized social media posts, perhaps geolocated, and using hashtags. But again, the change is not just in virtual space but is entangled with physical space in ways that Gregory (2013) argues depend on, and create, a particular performance of space.

New ICT tools are yet again changing the look and feel of organizing and tactics and what it means to "do" politics. Burgeoning smartphone use has made the internet available to many more people in the Global South, but widespread autonomous horizontal organizing also has a longer history on the other side of the digital divide, notably in the El Alto uprising in Bolivia in 2005 and the "Que se vayan todos" ("Throw them all out") uprising in Argentina in 2001. But these were not entirely without ICTs of a sort, for pirate and community radio stations have played an important role in organizing throughout Latin America.

How ICT tools are used for organizing has changed since Pickerill and colleagues conducted an in-depth study of how the anti-war movement was shaped by ICTs from 2005 to 2007 (Gillan, Pickerill, & Webster 2008). McLean (2020) offers a more recent and wide-ranging exploration of how various movements shape and are shaped by digital geographies. This is a rich area for more geographers to explore, but findings from studies like that by van Haperen, Uitermark, and Nicholls (2023) about how the #blacklivesmatter hashtag had different geographical connections when used by supporters than by the alt-right (on what was then Twitter) point to how quickly social media contexts change, both socially and geographically. (For more on the political geographies of social media see Adams 2025, this volume.) Evolving digital tools continue to shape the spatiality of protest in new ways, both in how protesters use them and in how the state uses them to suppress protest. Stokols (2023) does a powerful analysis of this and how these two shaped each other in Hong Kong pro-democracy protests, where protesters used Telegram channels for crowdsourced sousveillance (surveilling those doing the surveillance) to track repression and shift gatherings away from it. Artificial intelligence (AI) makes it easier for repressors to sift through massive datafiles, say using facial recognition technology, so how movements are responding, and how AI thus shapes the spatiality of movements, is another area ripe for research.

Older forms of collective action still coexist with connective action – and often we see hybrid actions where traditional brick-and-mortar membership organizations also participate in these loose multi-issue networks, say around particular events or days of action (Bennett & Segerberg 2012). Yet much of the wave of protests of the 2010s, notably most Occupy sites and the Indignados movement, rejected the participation of formally organized groups as such, though some accepted material support. For example, the port-a-potties at Vancouver Occupy were provided by a labor union. Some activists bemoan professionalization, and say that once a group has paid staff and nonprofit status it has bureaucratized and is no longer in a movement. It is a double-edged sword, as Katz (2005) details, but some nongovernmental organizations (NGOs) do function as part of, rely on, and even seek to build a larger movement, even when their grassroots grow into "large sequoias," as Brown (2008) puts it. Certainly there are some that are primarily handmaidens to neoliberal states, but NGOs should not be written out of the social movement story (Townsend, Porter, & Mawdsley 2004; Jeffrey 2013), particularly given the growing trend of cities funding and nurturing organizing, sometimes called new municipalism (Thompson 2021). Geographical thinking about social movements could be strengthened by more conversation across these literatures.

The Spatialities of Social Movements

Whether they use the term movement, network, NGO, contentious, collective, or connective action, what geographers have brought to understanding the work of movements is a focus on space. Social movement studies had until recently tended to write about movements as if they happened "on the head of a pin" (Nicholls 2007). Geographers have asked: How do movements use space, how does space shape movements, and how do movements shape space? We have used different spatial concepts to get at this: place, networks, and scale in particular, though also sometimes territory, city, and region. Nicholls, Miller, and Beaumont (2013) argue that in the 1930s–1950s it tended to be region, 1960–1970s space, 1980s place, 1990s–2000s scale, and in the 2010s networks and mobility. They offer a helpful review of how each has been taken up for movement studies.

Each of these concepts exists in relation both to movements and to each other. This argument was first made by Leitner, Sheppard, and Sziarto (2008), who have been widely cited for arguing that too often, as they call it, a favorite "master spatiality" has been plopped down on a movement rather than recognizing that these spatialities are, as they term it, co-implicated. Scale and territory, for example, not only shape each other but also, as Neil Smith puts it, "the scale of struggle and the struggle over scale are two sides of the same coin" (Smith 1992: 74, cited in Martin 2013: 332).

Miller argues in the conclusion of *Spaces of Contention: Spatialities and Social Movements* (Nicholls, Miller, & Beaumont 2013) that this is also true for other spatialities of social movements. That is, movements work to shape these spatialities, and they also shape movements. Though this will always be contextual, he offers as a guide a complex grid that, in relation to movements, grafts the Jessop et al. framework for the relationship between territory, place, scale, and network (TPSN) onto the Lefebvrian triad of material space, conceptual space, and lived space, and points to the different sociospatial power relations created in each (Jessop, Brenner, & Jones (2008), as cited in Miller 2013: 334).

Assemblage (*agencement* in Deleuze's French original) is a more fluid take on these relations, which in political geography of social movements was the wave of the 2010s. McFarlane (2009) argues for shifting from networks to assemblages to understand the spatiality of movements. Davies (2012) uses the practices of the Tibet Support Group to make a strong argument for using assemblage to convey how movements use both territorial and relational spatialities. Davis (2017) looks at how it is that translocal activist assemblages are built, with a focus on movements against foreign military bases. Halvorsen (2017), however, warns that the approach can fail to see and explain the inherent contradictions of social movements, and particularly of their spatialities. He proposes instead an elegant use of spatial dialectics, showing how the contradictions in spatial strategies push activists to shift the spatiality they prioritize.

In work on assemblage there is often slippage between it and actor-network theory (ANT; Müller & Schurr 2016). Rutland (2013) used ANT to great effect in detailing how the movement for a new Portland energy policy was made possible and constituted through a concatenation of human and nonhuman elements associated across time and space. But, he warns, the approach ignores processes of *dis*association and exclusion, and attends to minute details at the cost of obscuring broad systemic conditions, like capitalism and patriarchy, under which associations are formed (2013: 255).

Urban movements like the one in Portland are more likely to be discussed in urban than in political geography journals. Likewise, geographies of social movements of all sorts are being done in development, feminist, cultural, anarchist, peace, economic, and labor geography. Castree, Featherstone, and Herod's review (2008) compares labor geographies to geographies of transnational organizing, and argues that the former tend to emphasize scale, as opposed to the network focus of the latter. But community unionism is all about networks, and Wills (2009) in particular has made important contributions to thinking about these. Herod (2017) reviews how labor movement geographies have engaged with space, place, and scale. Again, there is no one dominant, or right, spatiality for understanding movements, nor is there one right subfield for geographers interested in movements.

Within political geography some work has even been done within the subfield of critical geopolitics, looking at how social movements try to influence, and themselves engage in, geopolitics (Routledge 2003; Koopman 2011). Critical geopolitics has been associated with textual analysis, but from its early conception Dalby argued for it to critically engage with social movements as a way to reconceptualize security and push for peace, and over the years he has taken seriously the discourses of movements (Dalby 1991, 1993, 1999; Mackenzie & Dalby 2003). More recently feminist geopolitics has pushed the field to more engagements with tensions across the material and the discursive, and this has led to work that looks at both the discourses and actions of movements, such as that by Casolo and Doshi (2013), Williams and Boyce (2013), and the in-depth work of Loyd (2014). In peace geographies work has been done on nonviolent action as a spatial mechanism (Hodder 2021; Koopman 2023).

Work by geographers on the spatialities of social movements has been broad and varied. What is clear is that space is a key part of how social movements create change, whether the best way to think of that in the case of a given movement is scale, territory, network, place, space, or assemblage – or some combination of these (Blank 2016; Routledge 2017).

Spatialities of Solidarity

Through their work on, with, and across social movements, geographers have also contributed to broader debates about the spatialities of solidarity. Much of political geography is about the relationship between here and there, us and them. Some of what political geographers have offered, as I will review here, is a perspective on what can shift and bring those categories together, and how movements build connections across distance and difference. This has ranged from analyses of commonalities, to how difference is reworked, to the emotional work involved.

Neil Smith's (1992) argument that movements can be strengthened when they "jump scale" was highly influential. Yet it is not always obvious how to expand resistance in this way. Castree, Featherstone, and Herod (2008: 318) argue that the most useful contribution geographers could make to movements is to "draw lessons" as to "the most effective basis for otherwise place-based constituencies to unite"; that is, common identities, enemies, or goals. Routledge (2007) urges geographers to draw comparisons and account for why social movements emerge *where* they do. He has long written about place as being able to tell us why movements happen where they do and how, and the importance of looking at how movements can take the uneven ways in which state power and economic power intersect with people's everyday lives in a place and use them to create "terrains of resistance" – both literally (barricades) and metaphorically (imaginaries) (Routledge 1994). Yet grand overarching patterns about what sort of places inspire what sort of movements, or what always works best for building solidarity across them, seem difficult to find.

However, geographers are well "placed" to do analyses of who is organizing where, how, and for, or against, what – and how different whos are organizing against similar whats, or how organizing here is or is not like organizing there, and how each is facing similar or related dynamics of not only capitalism but other systems of oppression. In so doing geographers have drawn what Katz (2001) has called countertopographical lines that could help movements learn from and connect with each other. Some topics seem particularly ripe for geographers, such as resistance to road expansion (Perz 2012) or land grabs (Baletti et al. 2008). Miller (2000) could be seen as drawing short lines between four places in metropolitan Boston with different demographics and ties to the defense industry that shaped their different anti-nuclear organizing. Many of these lines have been short in the sense that they were drawn within parts of a broader movement, for example local alternative currencies (North 2007) or Occupy sites (Uitermark & Nicholls 2012), or between fairly similar movements, such as racialized radical activists in Los Angeles (Pulido 2005). But other lines are much longer. Hart (2006), for example, looks at how struggles around land dispossession in South Africa and East Asia (particularly Taiwan) are constituted in relation to each other. Mackenzie and Dalby (2003) connect two struggles, in Scotland and Canada, around siting superquarries.

It is those surprising connections, à la the countertopographical lines that Katz draws from New York City to Sudan, that are a particularly useful contribution that geographers can offer to movements. Dalby (1999) warns that it is common wishful thinking to assume compatibility between diverse resistance in different places, but that is precisely why it is worth doing careful work to draw specific connections. Casolo and Doshi (2013: 827) draw such lines between two gendered movements in rural Guatemala and urban India, from which they gain insights into how practices of social reproduction can be key to both geoeconomic processes of accumulation by dispossession and to their resistance. Routledge (2017) uses a framework of different types of "sites of intervention," ranging from sites of production to destruction, consumption, circulation, and so on, as a way to highlight connections across movements, though they themselves may not (yet) be in active alliance.

Massey (2008) suggests that movements that are working to build solidarity across distance and inequality do so by stressing the ways in which those places are connected, either through similarities between the places, or their political and economic relations, or a common enemy such as the World Trade Organization. Brown and Yaffe (2014) argue that these are often used in combination and that solidarity can and often does flow in more than one direction, not just back and forth but laterally as well (in their example, between groups in various countries working against South African apartheid). As they put it, political solidarity campaigns try to change "the connections and flows between places, refusing to participate in the reproduction of inequalities and oppression" (Brown & Yaffe 2014: 39). Unfortunately, even as movements work against uneven power relations, it is easy to unintentionally reproduce them. But then again, those with more privilege (particularly geopolitical, economic, and racial) can sometimes leverage that position to work in solidarity with those with less, without necessarily reinforcing the structures that create those inequalities (Koopman 2011).

Featherstone (2012: 5) considers solidarity to be a "transformative relation," one that is "forged through political struggle which seeks to challenge forms of oppression" and works to create "new ways of relating." He writes about the various practices of solidarity used by movements across time and place to affect and change uneven power relations between here and there, and the many ways in which imaginaries of us and them have been reworked. Rather than focus on likeness, Featherstone emphasizes the new relationships that are in constant creation through solidarity practices. Brown and Yaffe (2014) extend his arguments by emphasizing that the way solidarity is framed cannot be separated from how it is performed, and that micropractices at key sites matter for what sort of alternatives are created. They give moving examples from the four-year-long Non-Stop Picket in front of the South African embassy in London in the late 1980s. They present the daily life on the picket as an assemblage of practices through which solidarity was performed. These included learning and singing South African songs banned in that country loudly enough to disrupt embassy business, as well as singing to cohere the group in the face of police pressure. In so doing the group both disrupted and reconfigured spatial relationships. Brown (2012) also writes about how these acts of standing in solidarity released powerful emotions.

Geographers have contributed insights into how the geographies of emotion shape and are shaped by social movements. That singing on the picket, for example, helped overcome a sense of isolation and made it enjoyable to be "unruly" in public space (Brown 2012). In an influential piece, Brown and Pickerill (2009) argue that examining the spaces of activism is a useful way to get at the relationship between emotions and activism, and in particular to improve emotional reflexivity in movements and make activism more sustainable. (That paper is part of a special issue that they edited for *Emotion, Space and Society* on emotions and movements. For a fuller review of literature in and around geography on emotion and protest, see Wright 2010.) Bosco (2007) has argued that the way social movements grow and maintain their networks is shaped by the strategic mobilization of emotions and activists' emotional labor to create shared feelings of outrage, love, and solidarity, even across distance.

Solidarity, these geographers have argued, is more than sameness. Indeed, solidarity builds connections across difference and distance. The practices, emotions, and imaginaries involved in those connections work to change uneven relationships between here and there, us and them. Hopefully geographers will continue to offer analyses to strengthen those ties, build new ones, and be part of the process of building coherence between movements.

Movements' Territoriality

Geographers also need to build more connections within our discipline, not only between subfields but also across linguistic and national borders. Latin American geographers' work on movements tends to emphasize territorial praxis. Bernardo Mançano Fernandes (2000) makes a distinction between what he calls sociospatial movements and socioterritorial movements, the latter being those who use territory as a *trunfo* (trump card), like the Movimento dos Trabalhadores Rurais Sem Terra (MST), the landless rural workers' movement with which he has long collaborated. They have gaining and keeping territory as a central aim; that territory shapes their identity and is a site of socialization, and eventually institutionalization (Halvorsen, Fernandes, & Torres 2019). Territory can mean many things (see Del Biaggio et al., 2025, this volume). As do many Latin American geographers and activists, Fernandes uses territory differently than most work in English, which tends to focus on the modern sovereign state (Halvorsen, Fernandes, & Torres 2019).

Carlos Porto-Gonçalves died in 2023 and was widely hailed as a leading light in decolonial geography. He wrote (2001, 2009) about movements establishing "new territorialities" through a process of symbolic-material appropriation that he describes as "making the earth" or geo-graphing. His own work was with the *seringueiros*, or rubber tappers, and he was particularly interested in movements' new ways of understanding and enacting territoriality as a way of building sustainability.

The most widely recognized writer in Spanish (and the most translated into English) on the spatial dynamics of social movements is not a geographer but a journalist, Raúl Zibechi (2012), although he often slides into romanticizing resistance (Sparke 2008). Circling back to the terms used for movements, he argues that they are so different in Latin America today that they should be called "societies in movement." Zibechi highlights how movements are "territorialized" – by which he means rooted physically, culturally, emotionally, and even spiritually in spaces that they have either recuperated or retained through struggle. This non-state take on territory goes beyond Indigenous and Afro-Colombian territories legally recognized by the state (Agnew & Oslender 2013), or even Colombian small farmers resisting state displacement in peace communities (Koopman 2011) or plantation takeovers by the landless. He includes factory takeovers by workers and land takeovers by new urban dwellers that become marginalized urban communities ("slums"), as do Halvorsen, Fernandes, and Torres (2019). Latin American feminists go further and widely argue that our bodies are our first territory and that territory is necessarily always embodied, using the term *cuerpo-territorio* ("body-territory") together to convey this (Hernández & Jiménez 2019).

There is a geopolitics to whose knowledge counts, as Sharp (2013) has argued, and language issues are part of it. We can move toward decolonizing our discipline by taking seriously and translating more of the voices and ideas of thinkers in the Global South, or writing collectively across our linguistic strengths, as Halvorsen, Fernandes, and Torres did (2019). We can also foster more pieces in English that review and dialogue with work in other languages, such as Halvorsen (2017) did for work on territory, or the pieces in the revived Geography in the World section of *Transactions* (https://tinyurl.com/2s3s8n9e). Both work in Spanish and French on territory and social movements have recently had increasing impact on work in English. Notable is Moradi's (2023) work on the productions of Kurdish territory and Clare, Habermehl, and Mason-Deese (2018) on the dialectic of movements organizing power *to* territory rather than power *over* (*potencia* vs. *poder*).

How Geographers Move with Movements

Geographers working on social movements are often motivated to do so because we feel strongly about the issues the movement works on, as Askins (2009) compellingly describes. Sometimes this means we work *with* movements, but of course there are also all sorts of reasons you might *not* want to collaborate with the movement you are researching (Gillan & Pickerill 2012). For example, Williams and Boyce (2013) work to understand the vigilante anti-immigrant movement in the US–Mexican borderlands as a way to support immigrant rights.

It is becoming increasingly common for geographers to collaborate with movements that they support, and are even a part of, as engaged scholars or activist academics, committed to contributing to the strength of the movement. Sometimes this means analyzing the processes that movements are opposing or working to nurture. Other times geographers help movements strengthen their conjunctural analysis. Often the research is about the movement itself and how it organizes.

The process of doing the research itself can also be a contribution to the movement. Geographers' discussions with activists about how they understand and use various spatial strategies and tactics for social change and the sociospatial relations they create (Martin 2013) may themselves offer insights to activists that change their strategies. Research can also help movements to be more accountable for what they do and how they do it by fostering conversations about internal dynamics (Halvorsen 2015). Movements operate in a world full of systems of oppression, so of course forms of domination show up inside resistance (Sharp et al. 1999).

Sometimes this work is called people's geography, sometimes community, participatory, or public geography (Koopman 2024). There are different forms and levels of collaboration, but for this work to be more able to contribute to the work of movements themselves, Castree, Featherstone, and Herod (2008) point to the importance of writing accessibly and in venues beyond books and journals. Many geographers do regularly present their research to movements, both in person and in movement newsletters and similar venues. This can be done as a way to keep the conversation going more than as simply one-way dissemination. There now many online ways to do this and be not only more accessible but also more *accountable* to movements (Fuller & Askins 2010).

Many geographers are having these sorts of conversations and learning from, thinking with, and acting with activists. There are all sorts of creative ways to do this, from theater to discussions that also serve as blockades (Mason & Askins 2012). For more on ways to think, and act, together see Kindon, Pain, and Kesby (2007). Three guiding lights, as geographers who have done ongoing collaborative research and writing with grassroots groups, are Geraldine Pratt, Linda Peake, and Richa Nagar. Pratt has collaborated for over 20 years with the Philippine Women Centre of British Columbia in Canada on issues involved in temporary labor migration, including family separation (Pratt 2002). They have used theater, in both Canada and the Philippines, as a way to share this research and stimulate community debate. Peake has collaborated for three decades with the Guyanese women's organization Red Thread on research on issues such as the impact of structural adjustment, domestic violence, and reproductive health, and written about the doing of transnational activist-academic praxis (Peake & de Souza 2010). Nagar writes with the Sangatin collective of seven Indian grassroots activists about how NGOs both foster and stifle solidarity (Nagar, Ali, & Sangatin Women's Collective 2003).

Unfortunately, this sort of collaboration, and open discussion in and among movements, is sometimes hampered by the potential that such research conversations could be used against a movement. In an era of high military surveillance, exemplified by Edward Snowden's National Security Agency revelations, and now that the US military has expressly shifted its focus from anthropology to geography as a tool for controlling the so-called human terrain (Medina 2016), it is quite possible that research by geographers could also be weaponized; that is, actually used to harm or even kill activists or those they work with (Koopman 2020). Of course, the researcher could also be put at risk by the research, even with citizenship and other status privilege, as was horribly clear with the murder of Italian Cambridge PhD student Giulio Regeni in 2016, while he was researching independent trade unions in Egypt. The risk is of course particularly great in authoritarian contexts and in conflict zones, but as police forces around the world are increasingly militarized and respond more violently to mobilizations, it is all the more important for geographers working with movements to be cautious about how their work might be destructively misused.

Geographers have collaborated with social movements in all sorts of ways, from going undercover to research the target of a movement's campaign (Routledge 2002), to investigating the counter-movement, to fostering conversations inside a movement about how systems of domination are shaping their resistance. The challenge now is for us to support each other in being creative about how to do this work in ways that are safe and ethical in the current context of high surveillance.

Conclusion

There are various ways in which geographers contribute to the power of social movements. More and more identify as scholar activists, but not all are holding collaborative research conversations in the trenches. Some geographers focus on the scale of movements, others on how they function as networks, and recently the assemblage approach has become increasingly popular. Not only have geographers looked at how spatialities shape movements, but also how movements shape spatialities.

Geographers have analyzed the spatialities of solidarity in various ways. Rather than solidarity around sameness, this work has tended to focus on connections across difference and distance and how movements can change the relationships between here and there, us and them. Geographers have contributed to growing connections between movements by drawing countertopographical lines between not only capitalist violences but also how movements resist and create alternatives; that is, between the pieces of a greater peace with justice that is being built around the world. Geographers in Latin America have particularly focused on the territoriality of movements, by which they mean not just physical rootedness but also cultural, emotional, and spiritual connections to territory gained through struggle.

Some geographers use the term movement, others prefer network, grassroots groups, NGOs, contentious, collective, or connective politics. Some consider themselves political geographers, many do not. I have argued here for more solidarity in geography and for making more connections: across geographies that use different terms for social movements, across subdisciplines of geography, across languages of research, and, most of all, with and across various movements. I particularly hope that geographers will pay increasingly more attention to the collaborative thinking being done through and with the action of movements around the world. Let us honor all of these ways of doing this research and continue to contribute to these movements in our various ways – so that together we can build a better world for us all.

References

Adams, P.C. (2025). Digital media. In: Mamadouh, V., Koch, N., Woon, C.H. and J. Agnew (Eds) *The Wiley Black Companion to Political Geography*. Wiley Blackwell. 468–485.

Agnew, J.A. (2002). *Making Political Geography*. Lanham, MD: Rowman and Littlefield.

Agnew, J., & Oslender, U. (2013). Overlapping territorialities, sovereignty in dispute: Empirical lessons from Latin America. In Beaumont, J., Miller, B., & Nicholls, W. (eds), *Spaces of Contention: Spatialities and Social Movements*. Farnham: Ashgate, 121–140.

Askins, K. (2009). "That's just what I do": Placing emotion in academic activism. *Emotion, Space and Society* 2(1): 4–13. doi:10.1016/j.emospa.2009.03.005

Baletti, B., Johnson, T.M., & Wolford, W. (2008). "Late mobilization": Transnational peasant networks and grassroots organizing in Brazil and South Africa. *Journal of Agrarian Change* 8(2–3): 290–314. doi:10.1111/j.1471-0366.2008.00171.x

Bennett, W.L. (2005). Social movements beyond borders: Understanding two eras of transnational activism. In Porta, D.D., & Tarrow, S.G. (eds), *Transnational Protest and Global Activism*. Lanham, MD: Rowman and Littlefield, 203–226.

Bennett, W.L., & Segerberg, A. (2012). The logic of connective action. *Information, Communication & Society* 15(5): 739–768. doi:10.1080/1369118X.2012.670661

Blank, M. (2016). De-fetishizing the analysis of spatial movement strategies: Polymorphy and *trabajo territorial* in Argentina. *Political Geography* 50: 1–9. doi:10.1016/j.polgeo.2015.09.002

Bosco, F.J. (2007). Emotions that build networks: Geographies of human rights movements in Argentina and beyond. *Tijdschrift voor Economische en Sociale Geografie 98*(5): 545–563.

Brown, M. (2008). Working political geography through social movement theory: The case of gay and lesbian Seattle. In Cox, K.R., Low, M., & Robinson, J. (eds), *The SAGE Handbook of Political Geography*. London: Sage, 353–377.

Brown, G. (2012). Unruly bodies (standing against apartheid). In Cameron, A., Dickinson, J., & Smith, N. (eds), *Body-States*. Aldershot: Ashgate, 145–157.

Brown, G., & Pickerill, J. (2009). Space for emotion in the spaces of activism. *Emotion, Space and Society 2*(1): 24–35. doi:10.1016/j.emospa.2009.03.004

Brown, G., & Yaffe, H. (2014). Practices of solidarity: Opposing apartheid in the centre of London. *Antipode 46*(1): 34–52. doi:10.1111/anti.12037

Casolo, J., & Doshi, S. (2013). Domesticated dispossessions? Towards a transnational feminist geopolitics of development. *Geopolitics 18*(4): 800–834. doi:10.1080/14650045.2013.811644

Castree, N., Featherstone, D., & Herod, A. (2008). Contrapuntal geographies: The politics of organizing across sociospatial difference. In Cox, K.R., Low, M., & Robinson, J. (eds), *The SAGE Handbook of Political Geography*. London: Sage, 305–321.

Chatterton, P., Featherstone, D., & Routledge, P. (2013). Articulating climate justice in Copenhagen: Antagonism, the commons, and solidarity. *Antipode 45*(3): 602–620.

Clare, N., Habermehl, V., & Mason-Deese, L. (2018). Territories in contestation: Relational power in Latin America. *Territory, Politics, Governance* 6(3): 302–321. doi:10.1080/21622671.2017.1294989

Cox, K. (2002). *Political Geography: Territory, State and Society*. Chichester: John Wiley & Sons.

Cox, K. R., Murray, L., & Robinson, J. (Eds.). (2008). SAGE Handbook of political geography. SAGE.

Dalby, S. (1991). Dealignment discourse: Thinking beyond the blocs. *Current Research in Peace and Violence 13*(3): 140–155.

Dalby, S. (1993). The "Kiwi disease": Geopolitical discourse in Aotearoa/New Zealand and the South Pacific. *Political Geography 12*(5): 437–456. doi:10.1016/0962-6298(93)90012-V

Dalby, S. (1999). Against "globalization from above": Critical geopolitics and the World Order Models project. *Environment and Planning D: Society and Space* 17(2): 181–200. doi:10.1068/d170181

Davies, A.D. (2012). Assemblage and social movements: Tibet support groups and the spatialities of political organisation. *Transactions of the Institute of British Geographers* 37(2): 273–286. doi:10.1111/j.1475-5661.2011.00462.x

Davis, S. (2017). Sharing the struggle: Constructing transnational solidarity in global social movements. *Space and Polity 21*(2): 158–172. doi:10.1080/13562576.2017.1324255

Del Biaggio, C., Koop, K., Pachoud, C., & C. Noûs. (2025). Territory. In: Mamadouh, V., Koch, N., Woon, C.H. and J. Agnew (Eds) *The Wiley Black Companion to Political Geography*. Wiley Blackwell. 144–160.

Dodds, K., Kuus, M., & Sharp, J. (eds) (2013). *The Ashgate Research Companion to Critical Geopolitics*. Farnham: Ashgate.

Featherstone, D. (2012). *Solidarity: Hidden Histories and Geographies of Internationalism*. London: Zed Books.

Fuller, D., & Askins, K. (2010). Public geographies II: Being organic. *Progress in Human Geography 34*(5): 654–667. doi:10.1177/0309132509356612

Gillan, K., & Pickerill, J. (2012). The difficult and hopeful ethics of research on, and with, social movements. *Social Movement Studies 11*(2): 133–143. doi:10.1080/14742837.2012.664890

Gillan, K., Pickerill, J., & Webster, F. (2008). *Anti-War Activism: New Media and Protest in the Information Age*. Basingstoke: Palgrave Macmillan.

Gregory, D. (2013). Tahrir: Politics, publics and performances of space. *Middle East Critique 22*(3): 235–246. doi:10.1080/19436149.2013.814944

Halvorsen, S. (2015). Taking space: Moments of rupture and everyday life in Occupy London. *Antipode 47*(2): 401–417. doi:10.1111/anti.12116

Halvorsen, S. (2017). Spatial dialectics and the geography of social movements: The case of Occupy London. *Transactions of the Institute of British Geographers 43*(3): 445–457. doi:10.1111/tran.12179

Halvorsen, S., Fernandes, B.M., & Torres, F.V. (2019). Mobilizing territory: Socioterritorial movements in comparative perspective. *Annals of the American Association of Geographers 109*(5): 1454–1470. doi:10.1080/24694452.2018.1549973

van Haperen, S., Uitermark, J., & Nicholls, W. (2023). The swarm versus the grassroots: Places and networks of supporters and opponents of Black Lives Matter on Twitter. *Social Movement Studies 22*(2): 171–189. doi:10.1080/14742837.2022.2031954

Hart, G. (2006). Denaturalizing dispossession: Critical ethnography in the age of resurgent imperialism. *Antipode 38*(5): 977–1004.

Hernández, D.T.C., & Jiménez, M.B. (2019). *Cuerpos, territorios y feminismos: Complilación latinoamericana de teorías, metodologías y prácticas políticas*. Mexico City: Editorial Abya-Yala.

Herod, A. (2017). Labor unions and economic geography. In Barnes, T.J., & Sheppard, E. (eds), *A Companion to Economic Geography*. Chichester: John Wiley & Sons., 341–358. doi:10.1002/9781405166430.ch21

Hodder, J. (2021). Casting a black Gandhi: Martin Luther King Jr., American pacifists and the global dynamics of race. *Journal of American Studies 55*(1): 48–74. doi:10.1017/S0021875819000033

Jeffrey, A. (2013). Non-governmental organizations. In Dodds, K., Kuus, M., & Sharp, J.P. (eds), *The Ashgate Research Companion to Critical Geopolitics*. Farnham: Ashgate, 387–403.

Jessop, B., Brenner, N., & Jones, M. (2008). Theorizing sociospatial relations. *Environment and Planning D: Society and Space 26*(3): 389–401. doi:10.1068/d9107

Katz, C. (2001). Vagabond capitalism and the necessity of social reproduction. *Antipode 33*(4): 709–728.

Katz, C. (2005). Partners in crime? Neoliberalism and the production of new political subjectivities. *Antipode 37*(3): 623–631. doi:10.1111/j.0066-4812.2005.00516.x

Kindon, S.L., Pain, R., & Kesby, M. (eds) (2007). *Participatory Action Research Approaches and Methods: Connecting People, Participation and Place*. Hove: Psychology Press.

Koopman, S. (2011). Alter-geopolitics: Other securities are happening. *Geoforum 42*(3): 274–284.

Koopman, S. (2020). Avoiding the weaponization of your research. In Wilson, H.F., & Darling, J. (eds), *Research Ethics for Human Geography: A Handbook for Students*. London: Sage, 97–104.

Koopman, S. (2023). The spatialities of nonviolent peace activism in the midst of war: From Colombia to Ukraine. In Flint, C., & Dempsey, K.E. (eds), *Making Geographies of Peace and Conflict*. London: Routledge, 165–182.

Koopman, S. (2024). Will the real people's geography please stand up? Community, public, and participatory geographies in conversation. *Geography Compass 18*(4): e12745.

Leitner, H., Sheppard, E., & Sziarto, K.M. (2008). The spatialities of contentious politics. *Transactions of the Institute of British Geographers 33*(2): 157–172. doi:10.1111/j.1475-5661.2008.00293.x

Loyd, J.M. (2014). *Health Rights Are Civil Rights: Peace and Justice Activism in Los Angeles, 1963–1978*. Minneapolis, MN: University of Minnesota Press.

Mackenzie, A.F.D., & Dalby, S. (2003). Moving mountains: Community and resistance in the Isle of Harris, Scotland, and Cape Breton, Canada. *Antipode 35*(2): 309–333. doi:10.1111/1467-8330.00325

Mancano Fernandes, B. (2000). Movimento social como categoria geográfica. *Terra Livre 15*: 59–85.

Martin, D.G. (2013). Place frames: Analyzing practice and production of place in contentious politics. In Nicholls, W., Beaumont, J., & Miller, B. (eds), *Spaces of Contention: Spatialities and Social Movements*. Farnham: Ashgate, 85–102.

Mason, K., & Askins, K. (2012). "Us and us": Faslane 30 and academic direct action. *Medicine, Conflict, and Survival 28*(4): 282–288. doi:10.1080/13623699.2012.739045

Massey, D. (2008). Geographies of solidarities. In Clark, N., Massey, D., & Sarre, P. (eds), *Material Geographies: A World in the Making*. London: Sage, 311–362.

McCarthy, J. (2011). Social movements. In Gregory, D., Johnston, R., Pratt, G., et al. (eds), *The Dictionary of Human Geography*. Malden, MA: John Wiley & Sons, p. 695.

McFarlane, C. (2009). Translocal assemblages: Space, power and social movements. *Geoforum 40*(4): 561–567. doi:10.1016/j.geoforum.2009.05.003

McLean, J. (2020). *Changing Digital Geographies: Technologies, Environments and People*. Amsterdam: Springer International. doi:10.1007/978-3-030-28307-0

Medina, R.M. (2016). From anthropology to human geography: Human terrain and the evolution of operational sociocultural understanding. *Intelligence and National Security 31*(2): 137–153. doi:10.1080/02684527.2014.945348

Miller, B.A. (2000). *Geography and Social Movements: Comparing Antinuclear Activism in the Boston Area*. Minneapolis, MN: University of Minnesota Press.

Miller, B. (2013). Spatialities of mobilization: Building and breaking relationships. In Nicholls, W., Miller, B., & Beaumont, J. (eds), *Spaces of Contention: Spatialities and Social Movements*. Farnham: Ashgate, 285–298.

Moradi, S. (2023). Producing territory, resisting the state: Embodiment, discourse, and symbolism in street demonstrations in Iranian Kurdistan. *Geopolitics 29*(1): 261–287. doi:10.1080/14650045.2022.2160711

Müller, M., & Schurr, C. (2016). Assemblage thinking and actor-network theory: Conjunctions, disjunctions, cross-fertilisations. *Transactions of the Institute of British Geographers 41*(3): 217–229. doi:10.1111/tran.12117

Nagar, R., Ali, F., & Sangatin Women's Collective (2003). Collaboration across borders: Moving beyond positionality. *Singapore Journal of Tropical Geography 24*(3): 356–372. doi:10.1111/1467-9493.00164

Nicholls, W.J. (2007). The geographies of social movements. *Geography Compass 1*(3): 607–622. doi:10.1111/j.1749-8198.2007.00014.x

Nicholls, W., Miller, B., & Beaumont, J. (eds) (2013). *Spaces of Contention: Spatialities and Social Movements*. Farnham: Ashgate.

North, P. (2007). *Money and Liberation: The Micropolitics of Alternative Currency Movements*. Minneapolis, MN: University of Minnesota Press.

Painter, J. (1995). *Politics, Geography and Political Geography: A Critical Perspective*. London: Hodder Arnold.

Painter, J., & Jeffrey, A. (2009). *Political Geography: An Introduction to Space and Power*, 2nd edn. London: Sage.

Peake, L., & de Souza, K. (2010). Feminist academic and activist praxis in service of the transnational. In Nagar, R., & Lock Swarr, A. (eds), *Critical Transnational Feminist Praxis*. Albany, NY: SUNY Press, 105–123.

Perz, S. (2012). Social mobilization in protest of trans-boundary highway projects: Explaining contrasting implementation outcomes. *Development and Change* 43(3): 797–821. doi:10.1111/j.1467-7660.2012.01773.x

Pile, S., & Keith, M. (1997). Introduction. In Keith, M., & Pile, S. (eds), *Geographies of Resistance*. London: Routledge, 1–32.

Porto-Gonçalves, C.W. (2001). *Geo-grafías: Movimientos sociales, nuevas territorialidades y sustentabilidad*. Mexico City: Siglo XXI Editores.

Porto-Gonçalves, C.W. (2009). De Saberes y de Territorios – diversidad y emancipación a partir de la experiencia latino-americana. *Polis: Revista Latinoamericana* 22(April 8). http://journals.openedition.org/polis/2636 (accessed December 21, 2024)

Pratt, G. (2002). Collaborating across our differences. *Gender, Place and Culture* 9(2): 195–200.

Pulido, L. (2005). *Black, Brown, Yellow, and Left: Radical Activism in Los Angeles*. Oakland, CA: University of California Press.

Rosner, V., & Pratt, G. (2006). The Global and the Intimate. *WSQ* special issue *34*(1–2).

Routledge, P. (1994). Backstreets, barricades, and blackouts: Urban terrains of resistance in Nepal. *Environment and Planning D: Society and Space* 12(5): 559–578. doi:10.1068/d120559

Routledge, P. (2002). Travelling east as Walter Kurtz: Identity, performance, and collaboration in Goa, India. *Environment and Planning D* 20(4): 477–498.

Routledge, P. (2003). Anti-geopolitics. In Agnew, J., Mitchell, K., & Tuathail, G.Ó. (eds), *A Companion to Political Geography*. Malden, MA: Blackwell, 236–248.

Routledge, P. (2007). Transnational political movements. In Cox, K.R., Murray, L., & Robinson, J. (eds), *SAGE Handbook of Political Geography*. London: Sage, 335–349.

Routledge, P. (2017). *Space Invaders: Radical Geographies of Protest*. London: Pluto Press.

Routledge, P., & Cumbers, A. (2009). *Global Justice Networks: Geographies of Transnational Solidarity*. Manchester: Manchester University Press.

Rutland, T. (2013). Energizing environmental concern in Portland. In Nicholls, W., Miller, B., & Beaumont, J. (eds), *Spaces of Contention: Spatialities and Social Movements*. Farnham: Ashgate, 250–271.

Sharp, J.P. (2007). Geography and gender: Finding feminist political geographies. *Progress in Human Geography* 31(3): 381–387. doi:10.1177/0309132507077091

Sharp, J.P. (2013). Geopolitics at the margins? Reconsidering genealogies of critical geopolitics. *Political Geography* 37: 20–29. doi:10.1016/j.polgeo.2013.04.006

Sharp, J.P., Paddeson, R., Philo, C., et al. (1999). *Entanglements of Power: Geographies of Domination/Resistance*. London: Routledge.

Smith, N. (1992). Geography, difference and the politics of scale. In Doherty, J., Graham, E., & Malek, M. (eds), *Postmodernism and the Social Sciences*. Basingstoke: Palgrave Macmillan, 57–79.

Smith, S. (2020). *Political Geography: A Critical Introduction*. Chichester: John Wiley & Sons.

Sparke, M. (2008). Political geography – political geographies of globalization III: Resistance. *Progress in Human Geography 32*(3): 423–440. doi:10.1177/0309132507086878

Stokols, A. (2023). From the square to the shopping mall: New social media, state surveillance, and the evolving geographies of urban protest. *Urban Geography 44*(6): 1166–1191. doi:10.1080/02723638.2022.2056366

Thompson, M. (2021). What's so new about New Municipalism? *Progress in Human Geography 45*(2): 317–342. doi:10.1177/0309132520909480

Townsend, J.G., Porter, G., & Mawdsley, E. (2004). Creating spaces of resistance: Development NGOs and their clients in Ghana, India and Mexico. *Antipode 36*(5): 871–889. doi:10.1111/j.1467-8330.2004.00459.x

Uitermark, J., & Nicholls, W. (2012). How local networks shape a global movement: Comparing Occupy in Amsterdam and Los Angeles. *Social Movement Studies 11*(3–4): 295–301. doi:10.1080/14742837.2012.704181

Williams, J., & Boyce, G.A. (2013). Fear, loathing and the everyday geopolitics of encounter in the Arizona Borderlands. *Geopolitics 18*(4): 895–916. doi:10.1080/14650045.2013.780035

Wills, J., Datta, K., Evans, Y., Herbert, J., May, J., & McIlwaine, C. (2009). Religion at work: The role of faith-based organizations in the London living wage campaign. *Cambridge Journal of Regions, Economy and Society 2*(3): 443–461. doi:10.1093/cjres/rsp016

Wright, M.W. (2010). Geography and gender: Feminism and a feeling of justice. *Progress in Human Geography 34*(6): 818–827. doi:10.1177/0309132510362931

Zibechi, R. (2012). *Territories in Resistance: A Cartography of Latin American Social Movements*, trans. R. Ryan. Chico, CA: AK Press.

Chapter 26

Identity Politics

Md Azmeary Ferdoush

Department of Geography and Environment,
Loughborough University, Loughborough, UK

Identity politics can be located virtually anywhere and everywhere in society where categories are variously privileged. This is because identity itself is not a neutral term: it comes with political baggage and throws identity, rights, justice, "race," religion, exploitation, privilege, belonging, and politics into sharp relief. As an academic term and a topic of inquiry, identity politics is relatively young. According to Mary Bernstein (2005), it was in 1979 that the phrase was first used in an academic setting by Renee Anspach (1979). Anspach defined identity politics in connection to the political activism of physically challenged people as "societal movements which seek to alter the self-conceptions and societal conceptions of their participants" (1979: 765). However, since the range of political activism and movements that "may be described as identity politics is broad . . . there is no straightforward criterion that makes a political struggle into an example of 'identity politics,'" suggests Cressida Heyes (2020: para. 3). Rather, according to Heyes (2020), the term signifies a "loose collection of political projects" undertaken by a representative mass in response to being neglected, erased, or suppressed. It is often why many designate identity politics to be too slippery to define (Lichterman 1999).

Although numerous American movements in the 1960s, for instance Black Power, Women's Liberation, and Gay Liberation, later came to be defined under the umbrella of identity politics, it was not until the 1970s and 1980s that identity politics started to gain major public attention, especially in settler societies such as Canada, the United States, and Australia (Nicholson 2008; Heyes 2020; Kobayashi 2020). Canadian geographer Audrey Kobayashi (2020) divides these

The Wiley Blackwell Companion to Political Geography, Second Edition.
Edited by Virginie Mamadouh, Natalie Koch, Chih Yuan Woon, and John Agnew.

movements into two broader categories – one that aimed at establishing an alternative form of nationalism and another that aimed at establishing alternative meanings of multiculturalism. As the scope of identity politics is broad, so is its theoretical basis. It is common to credit feminist and Black scholars for providing a solid theoretical foundation on which many social movements in the 1960s built their stronghold. Liz Bondi, nonetheless, argues that the theoretical foundation for a politics of identity occurs earlier than the 1960s. Bondi credits Karl Marx and Sigmund Freud for providing a solid theoretical basis for a "politics in which a counter-hegemonic construction of the human subject is marshalled to challenge dominant groups and to organize the subordinated, a politics in which an oppositional redefinition and assertion of identity is central" (1993: 84). Identity politics however has not always been viewed through a positive lens. The phrase is often synonymized with terms such as "extremist" and, even worse, used as a derogatory synonym for anti-racism, feminism, and anti-heterosexism (Fraser 1997; Nicholson 2008). Thus, as identity politics is celebrated, it is also shunned by many.

Trained geographers started expressing interest in identity politics around the 1960s with the development of "radical" geography. Since then geographers have played the crucial role of teasing out the spatial dimension of identity by a concrete positioning of space and place in the scholarship (Keith & Pile 1993). But their interest was not limited to an academic setting only. In 1968, the geographer William Bunge took radical geography from papers to the streets with the founding of the Detroit Geographical Expedition and Institute (DGEI) in Detroit, Michigan. Bunge collaborated with colleagues and local residents to conduct and publish research in racial justice with participatory mapping in response to the nationwide struggle for civil rights (Garrido 2021). Although identity politics enjoyed relatively less attention during the late 1970s and the 1980s, it went through a regenerative phase in the early 1990s as poststructuralist scholarship debunked the pseudoscientific basis of "race" and discredited primordial definitions of culture (Kobayashi 2009b). Further, feminist political geographers played an instrumental role in center-staging identity politics by drawing attention to the micro and macro aspects of politics, specifically by revealing the masculinist character of the field. By shifting attention to the "locations for politics," they demonstrate how power, politics, and difference are not just produced at the grand stages of nation and the state but also on bodies within domestic arenas (Mitchell 2004; Staeheli, Kofman, & Peake 2004: 5; Ahmed 2017). Eventually, identity politics took center stage within geographical scholarship, made evident through regular encyclopedic entries (Kobayashi 2009b, 2020) and papers in flagship geography journals such as *Political Geography* (Drozdzewski & Matusz 2021), *Geopolitics* (Aalto, Dalby, & Harle 2003; Mansbach & Rhodes 2007), and *Progress in Human Geography* (Mitchell 2004; Dowling 2009, 2010; Browne 2021; Alexander 2023).

The most widely studied category in political geography is perhaps national identity. The question of national identity and the diverse politics around it have garnered continuous attention from geographers, resulting in an impressive body of work. Because national identity fundamentally rests on the question of belonging, the debate of "us" versus "them" dominates the scholarship. However, who are defined as "us" and who as "them" touch on numerous aspects such as territory, "race," ethnicity, language, religion, culture, time, and the banal (Gellner 1983; Billing 1995; Anderson 1996; Koch & Paasi 2016; Koch & Vora 2020; Dempsey 2022; Ferdoush & Väätänen 2022; Koch 2023). Recent decades, nonetheless, have witnessed a renewed interest in the examination of religion and sexuality in relation to national identity and identity politics, especially focusing on their intersectionality; that is, how one such category is intimately related to other categories in defining one's identity (Hopkins 2019; Oswin 2019; March 2021; Sidaway 2023). Continuing the dialogue in this chapter, I shed light on identity politics specifically through religion and sexuality. That is because, although increasing in recent years, these two have traditionally received marginal attention from geographers compared to topics such as territorial belonging, sense of place, space, and scales, and because other chapters dedicate detailed attention to such issues in the current volume. With this in mind, in the discussion that follows, I first shed light on some of the major attributes of identity politics. I then examine identity politics through religion and sexuality. Finally, I conclude with a note on the untapped potential and underrecognized risk of reading and doing identity politics.

Attributes of Identity Politics

Identity is an elusive term and identity politics is ubiquitous across space and place (Tilly 1996; Brubaker & Cooper 2000). Identity and politics around identity are diverse, complex, multilayered, and multiscalar. It is therefore not a simple task to draw up an exhaustive list of attributes that characterize identity politics. However, for the same reason – that is, the ubiquity of identity politics – it is possible to pin down a few attributes that can be located across contexts and scales. In this regard, I draw and build on James Tully's seminal work on the most common and identifiable characteristics of identity politics (Tully 1995, 2008). According to Tully, the first attribute of identity politics is its heterogeneity. Identity politics is not a straightforward struggle of a group of unified people against another clearly defined group or institution. Neither is it a politics of separate and uniformly organized minorities seeking a common end. But rather, in most cases, the demands are expressed around overlapping and often competing interests that crisscross multiple allegiances such as religion, nationality, culture, region, ethnicity, sexuality, and more. Therefore, the struggle of one group to obtain their demands ends up being a struggle against not just a dominant group but other competing groups too. In this sense, identity politics often runs the risk of a struggle for (in the sense that it is against the dominant group) and against (concurrently against other less dominant groups and their demands) diversity (Connolly 1995). For example, one Indigenous group's demand for self-determination is likely to find itself against not just the dominant national elite but other competing Indigenous groups, minorities, and ethnic groups seeking similar goals. This is because, as Edward Said (1994: xxv) has it, "the ideological concern over identity is understandably entangled with the interests and agendas of various groups . . . that wish to set priorities reflecting these interests."

The second characteristic of identity politics stems from the first one: it is by design exclusionary. Identity politics is essentially about prioritizing one identity over other(s). The priority granted to one identity, the spokespeople who represent an identity-related group, and their accommodation overpower others (Tully 2008). Consequently, identity politics consists of three different processes of negotiation that exist in an entangled fashion to determine the outcome of a given identity's degree of priority (Tully 1995, 2008). The first step of the process starts with the members of the group who are seeking to establish an identity for themselves. To do so, the group members first must come to an understanding on the nature of the identity that they all agree to identify with, what it means, what it includes, and therefore what it excludes. Once such an agreement has been reached, the second process of negotiation starts. In this phase, the struggle is taken outside of the group to where its members seeking to establish their identity find themselves against those to whom the demand is being made; that is, the dominant group (often the political elite). And the third process of negotiation involves other competing groups whose identities get thrown into question as the members of the identity-seeking group demand their recognition. This is simply because, as mentioned earlier, no identity has a clear boundary but overlaps with others. Therefore, in every instance when one group seeks to establish an identity for itself, it inevitably involves others with crisscrossing interests.

The third and perhaps the fuzziest feature of identity politics is the concept of identity itself. Rogers Brubaker and Frederick Cooper (2000: 1) eloquently capture the elusive nature of identity: "'Identity,' we argue, tends to mean too much (when understood in a strong sense), too little (when understood in a weak sense), or nothing at all (because of its sheer ambiguity)." For example, an identity document makes an individual unique but at the same time groups them under a certain category that makes them disappear within a broader group. Regardless, identity is a fundamental element of everyday life and politics across time and space (Béland 2017). Thus, when it comes to identity politics, it is essential that we conceive of identity in one way or another. In this regard, James Fearon's simple yet effective way of distinguishing two different meanings of identity remains helpful (Fearon 1999, quoted in Béland 2017). One is identity as a social category, which is defined by membership rules, attributes, and expected conduct. The other is identity as a socially disguising feature that an individual takes pride in connecting with. There is no doubt that the meaning of identity can be classified in many other ways, for example personal/collective, psychological/social, latent/manifest, stigmatized/privileged, and more. However, as the focus of this chapter is on identity politics and not on the complexity of the concept itself, I limit my discussion to identity as a social category and its spatial dimensions.

Categories provide the basis for difference in the production and sustenance of inequality across societies. However, they do not themselves produce inequalities: the way they are positioned provides the source for inequality. In this regard, Rogers Brubaker's (2015) thought experiment with categories offers useful insight. He argues that categories may produce difference, not inequality. They become mechanisms for producing inequality only when positioned vertically on top of each other. It is not just how categories are placed but also how their boundaries are drawn that play a crucial role in the politics of categories (Jones 2008a). As boundaries are drawn between and among categories, new power relations are imposed and knowledge is legitimized that silences other alternatives (Foucault 1971). Those boundaries are not always clear but, as mentioned earlier, are crisscrossed. As a result, the categories in question often overlap and hence must be viewed from diverse axes rather than a single point of difference. In other words, they must be approached from an intersectional perspective, frequently overlooked in much of social and political geography, suggests Peter Hopkins (2019). Concurrently, how these categories are manifested across space and place and, similarly, how space and place influence the materialization of the categories must also be considered seriously. To put it in another way, identity politics through social categories must be spatialized as well, because identity is intimately connected to place and place remains shaped by the politics of identity (Staeheli et al. 2004). This is the key, I suggest, to a critical reading of identity politics – to identify the category and categorical boundaries, as well as their positionality (horizontal and vertical), intersectionality, and spatiality. Therefore, the discussion that follows is based on the assumption that identity politics is essentially a politics of categories, at once social and spatial. With this in mind, I now focus on two such categories.

Religion

In response to the September 11, 2001, attacks on the United States, the world order was significantly reimagined and reshaped, especially in the name of the "war on terror" led by the United States with its allies. The war on terror did not just operate on a broader geopolitical scale but also diffused throughout the everyday lives of millions, especially under the disguise of religion. Based on religious allegiance, the world became divided to such an extreme extent that it often came across as a divide between "those with a God (the Christian West) and those with the wrong God (the Islamic Middle East)" (Strum 2015: 352). Islamophobia and anti-Arab racism saw a sharp rise not just in the United States but around the world (Salaita 2006). Religious allegiance (assumed or assigned) and identity politics with religion garnered renewed attention, giving rise to new tensions. The renewed primacy of religion in international geopolitics was called the future of geopolitical research by David Newman (quoted in Strum 2015). Religion and religious identity politics also garnered attention in local geopolitics, especially in the Global North, where public expression of Muslim identity through certain female clothes became a topic of heated debate. As Audrey Kobayashi (2009a: 284) has it, "Nowhere is this tension more pronounced than in contemporary debates over the expression of Muslim identity . . . secular multiculturalism has provided the ironic and somewhat disingenuous rationale for curtailing certain public Islamic expressions, such as female dress." This offers an illuminating example of the intersectionality of two different types of identities in focus. While the debate is primarily about the right to express religious identity through a particular way of dressing, it is simultaneously about gender. The dress in question concerns only females and thus sparks debate around lopsided gender rights in the name of religion. Furthermore, it reveals the spatial aspect of identity politics, as the question of space and place (in particular, public space and place) is thrown into sharp relief with religion and gender identities. Referring to similar debates about sexual violence, migrants, and gender value in Sweden, Emil Edenborg (2020) suggests that such identity politics gain traction within society because they tap into the local geopolitical narrative of othering. With this in mind, I shed light on religion and identity politics in this section. In so doing, I offer the cases of Bangladesh and India to show how religious identity and politics play out in the lives of ordinary people on a daily basis, and at the same time, how they shape bilateral relations and regional geopolitics, with a transition through a brief note on religious identity itself.

There are a multitude of ways to define and debate religious identity. For the purpose of this discussion, religious identity refers to one's sense of belonging based on a religious belief and cultural traits commonly held with a larger group of people. The group, hence, refers to a religious group. When it comes to the political relevance of religious identity in everyday lives and in national settings, there are three different theories that concurrently complement and compete with each other. These are primordialism, instrumentalism, and constructivism (Brass 1974; Robinson 1974; Hussain 2013; Fox 2018). The primordial explanation suggests that religious allegiance has been an essential aspect of cultural and political identities since antiquity. In other words, religion has remained an essential aspect of political identity for such a long time that the origin of their entanglement remains difficult to trace. In such a context, Fox (2018: 34) suggests, "religious identity is relevant today because it was relevant yesterday." Because religion, culture, and identity remain entangled for generations, religious conflict also tends to last for a long period of time in such settings, for example the Arab–Israeli conflict. The instrumentalist path for religion to become relevant in politics and thus make its way into identity politics refers to a comparatively recent period. Such cases are found in societies where religious identity existed for a while without particular importance until someone sought to activate it for political gain. Political entrepreneurs, numerous governments, and political parties often tend to do so. As religion is activated within politics and identities in such a fashion, it gets frequently staged on minorities who end up being the scapegoats. They are excluded as "them" while the religious majority are included as "us." Constructivism shares similarities with instrumentalism in the sense that at some point, political entrepreneurs activate a new identity to gain power and influence. However, unlike instrumentalists who activate an existing identity, constructivists engineer a new one. Such instances are most commonly found within the narratives that are created to justify a new national identity. The cases of Bangladesh and India in this sense are examples of societies where religion was introduced in a constructivist and an instrumental fashion, respectively, to which I now turn.

India, as a sovereign nation-state, started its journey in 1947 after the partition of colonial India. Until the 1940s, the Indian National Movement was essentially a movement for independent India driven by the underlying discourse of "unity in diversity," which aimed to accommodate religious, cultural, and linguistic diversities within its national body (Chatruvedi 2002). However, the Two-Nation Theory, popularized by Mohammad Ali Jinnah from the 1940s, was a major blow to the Indian National Movement. It claimed that Hindus and Muslims in India were two radically different peoples, incapable of living peacefully and forming a coherent nation. Thus two separate nation-states were necessary for them (Chatruvedi 2002; Van Schendel 2005). The Two-Nation Theory materialized in 1947, scissoring colonial India in an awkward fashion. The state of Pakistan was created for the Muslims but was divided into two wings – West (present-day Pakistan) and East (present-day Bangladesh). India was created for the non-Muslims sitting between the two wings of Pakistan.

Besides the perplexing cartographic divide, the inherent problem with such a partition was twofold. One, India hosted more Muslims than Pakistan within its boundaries at the time of partition; and two, religion was introduced as the "glue" for binding these two nations only very recently. As the imposed national binding factor – that is, religion – started to gain primacy in the nation-making projects in these two states, Muslims in India were imagined to have potential extraterritorial allegiance (Chatruvedi 2002: 154). Muslims were othered and excluded from the nation's geo-body. Although Indian secularism thrived in the early decades of independence, political elites, especially led by the Bharatiya Janata Party (BJP), were instrumental in introducing and eventually championing *Hindutva* (conservative Hindu nationalism) in Indian politics (Kothari 1998; Chatruvedi 2002). As *Hindutva* gained further ground in an instrumental fashion, Muslims who had started as the "other" became further othered and further excluded. Religion, however, was not just limited to the everyday politics of identity inside the nation-state. It played a crucial role in India's geopolitical framing and bilateral relations with its neighboring Muslim-majority states of Bangladesh and Pakistan. India followed the United States in the war on terror and assumed an active role in depicting the Muslim-majority state of Bangladesh as the "barbarian other" beyond its borders that allegedly posed a threat to its internal peace (Jones 2012). Religion diffused with geopolitics – which Lari Nyroos (2001) calls "religeopolitics" – became one of the pillars of Indian geopolitics.

The case of Bangladesh offers a mirroring story. Bangladesh (East Pakistan) was part of the Muslim successor state Pakistan from 1947 until 1971. It gained independence in 1971 and started its journey as a secular state, with secularism as one of its constitutional pillars. Bengali nationalism too was not based on a religious identity but on shared cultural and linguistic ties (Jones 2008b; Ferdoush 2019). However, the country started to go through a process of Islamization from 1975 after the assassination of its first prime minister, Sheikh Mujibur Rahman. Subsequent governments backed up by the state army brought fundamental changes to its constitution, discourse of nationalism, and geopolitical orientation. The most significant were (1) a change in the constitution, replacing secularism with *Bismillah-ar-Rahman-ar-Rahim* ("in the name of Allah, the beneficent, the merciful"); (2) changing secular Bengali nationalism to religiously charged Bangladeshi nationalism; (3) reorienting the country's foreign policy ties from the India–Soviet bloc to the Middle East Islamic bloc; and (4) eventually the formal introduction of Islam as the state religion in 1988 (for details see Huda 2004; Jones 2008b; Riaz 2016; Ferdoush 2019). Such a constructivist introduction of religion into national politics came with fateful consequences. Within a very short period of time, Bangladesh saw the rise of fundamental Islamist groups, sharp religious divides, exclusions of minorities especially Hindus from national discourses, and an acute sense of religious identity in everyday practices. Simple statistics demonstrate this effectively. The proportion of Bangladesh's Hindu population in 2022 had come down to half what it was in 1974 (Bangladesh Bureau of Statistics 2023). Although the state's rhetoric justifying this decrease was to blame the higher fertility rates of Muslims compared to Hindus, it is well known that Hindus have been forced to leave the country for decades (Guhathakurta 2013). Again, similar to India, the geopolitical orientation of Bangladesh has been on the one hand, to depict India as a hostile Hindu state, especially from the late 1970s, and on the other hand, to reorient its international allegiance to the Islamic world (Ferdoush 2024).

The cases of Bangladesh and India are telling in numerous ways when it comes to identity politics around religion. Although India was created as a nation for non-Muslims, religion did not gain primacy in forming a national identity until it was instrumentally injected by conservative political groups. One of the primary factors for Bangladesh's separation from Pakistan was that Bengali nationalism was never about religion but about shared language and culture. Yet as a new Bangladeshi nationalism based on religio-territorial identity was engineered by its politicians, religion became an essential part of everyday identity politics. Religion now defines who belongs and who does not within these two nation-states.

Furthermore, their bilateral relations saw sharp changes too with a strong religious nationalism in place – from being friendly to often hostile neighbors. Religious identity and politics carry significant spatial and scalar consequences that, as mentioned earlier, get relatively little attention, especially from political geographers. Political geography therefore would do better with a wider accommodation of geographies of religion within its research agenda.

Queering Identity Politics

Identity politics around sexuality starts with a binary positioning of every individual in a biological frame with a prescribed behavior of masculinity or femininity and expecting that they would experience sexual intimacy only with opposite-sex partners (Strum 2015). Such a binary positioning of males and females is based on an unproblematic male view of the world. Such a view is known as "heteronormative" and functions within a heterosexual understanding of the world to normalize certain sexual behaviors. Consequently, heteronormativity excludes others who do not fit into the binary of heterosexual man and woman; that is, lesbian, gay, bisexual, transgender, queer, and intersex (LGBTQI) people. And as such, LGBTQI rights and recognition have become an important site of struggle for identity politics (Brown 2012; Ehrkamp & Jacobsen 2015; Browne 2021). Heteronormativity is not merely limited to the politics of sexual identity, it is rather over-encompassing and spatially distributed, suggests Natalie Oswin (2019). In studying the heteronormative characteristics of the Singapore state, Oswin shows that as a state, Singapore allows only heteronormativity to be reproduced. While the LGBTQI community is the obvious group on which the state's heteronormativity is most visibly expressed, Oswin proposes, heteronormativity reaches far beyond the question of gender and sexual identity in Singapore. The state is spatially organized in such a way that both the private space of the home and public spaces such as a park are governed to reproduce heteronormativity. The Singapore case remains a strong demonstration that sexuality is not just social and psychological but equally spatial; as Oswin (2019: 105) has it, "The sexual is spatial and the spatial is sexual."

Notwithstanding this, from the 1990s there has been a significant rise in awareness and support for LGBTQI rights, especially across the Global North. Struggles and rising levels of awareness about diverse sexual orientations have led to the recognition of a select set of sexual identities, often labeled homonormativity. Yet this does not necessarily imply that every sexual orientation and identity have been "normalized." Homonormativity, contra heteronormativity, refers to the way a select group of lesbians and gay men are recognized for performing a "proper" way of being homosexual as they participate in same-sex marriage, engage in monogamous relationships, and exhibit "gender-appropriate" roles that reproduce the dominant social order, including "white, middle-class aspirations and aesthetics" (Strum 2015: 367). Homonormativity, understood in this fashion, has a dual role for the countries that recognize it (typically the Global North countries). On the one hand, it enables these states to project an image of a "progressive" nation that does not discriminate based on sexual orientation. On the other hand, homonormativity facilitates a state in performing homonationalism, suggests Jasbir Puar (2007, 2013). Puar contends that homonationalism is used as a tool for evaluating national sovereignty, civility, democracy, and progress by Global North states, allowing them to label others as "backward" and "uncivilized." Furthermore, this positioning is often used by Global North countries to justify violence, migrant humiliation, and vilification of certain nations and peoples because they do not support LBGTQI rights (Strum 2015).

The use of sexual identity in developing and fostering an image of a superior nation and culture is not only limited to the Global North. Many Global South nation-states resort to sexual identity in an effort to distinguish themselves from the Global North (or the West) – a process that has been understood as "occidentalism" by many (Carrier 1992; Chen 1995; Zhang 2015). Occidentalism may be conceived as the flip side of Edward Said's (1979) famous thesis of orientalism, in which the East appropriates the West in creating an image of the "other" to achieve numerous cultural and political ends. In this regard, upholding an image of a heteronormative society at the expense of homonormativity has proved to be a popular technique. While many African nation-states resort to occidentalism by depicting homosexuality as "un-African" and an import from the West, Uganda stands out in its effort to exclude and punish homosexuality (Dicklitch, Yost, & Dougan 2012; DeJong & Long 2014; Fumanti 2017). Ugandan political elites have been at the forefront of not just depicting homosexuality as foreign to African culture but also introducing bills and passing laws with severe punishments, including life imprisonment. The Anti-Homosexuality Bill proposed in 2009 and then reintroduced in 2012 by David Bahati, an Ugandan member of parliament, came to be infamously known as the "Kill the Gay Bill" since it proposed the death penalty for homosexual acts (Dicklitch, Yost, & Dougan 2012; DeJong & Long 2014). The bill gained traction among powerful political actors in Uganda, including President Yoweri Museveni when he signed it into law in 2014, although it replaced the death penalty with a life sentence (Al Jazeera 2014). Labeling homosexuality as "un-African" and a colonial imposition thus provides a powerful tool for Ugandan state elites, since it allows them to draw a boundary between the precolonial and postcolonial Ugandan identities.

Similar instances of occidentalism can be traced in the Russian context too, where exclusion of the LGBTQI population from the nation's geo-body gained unprecedented attention after Vladimir Putin returned to the presidency in 2012. As Russia started to reimagine its geopolitical position and narrative in relation to the European Union and the West, identity politics with sexuality gained center stage (Foxall 2019). It was precisely at this juncture that the phrase "Gayropa" started to appear in Russian (social) media and was made popular within a very short period of time with prominent political figures including Putin. Gayropa is commonly used to make a clear distinction between Russia and Europe based on sexual identity and role, where Russia is positioned as the "civilized" nation upholding "natural" family traditions and values, while Europe and the West are depicted as the bearers of uncivilized and uncouth values that allow same-sex marriage and families (Foxall 2019). Commenting on the importance of Gayropa in Russian domestic geopolitics, Riabov and Riabova suggest that "[b]y putting a label on European civilization, the concept of *Gayropa* helps to support the collective identity of Russians" (2014: 3, emphasis in the original).

The politics behind homo/heteronormativity demonstrates that identities around sexuality are not just limited to individuals' sexual orientation but span other sociospatial axes such as nation, national territory, and citizenship. Citizenship is not asexual but rather the opposite, it is sexualized, and thus citizens' exclusion and inclusion within the nation's geo-body also vary depending on their sexual orientations (Bell & Binnie 2000). Every nation-state provides a set of criteria for determining its citizens' rights and entitlements. To avail themselves of those, citizens must fit into the set criteria. States thus expect certain sexual behaviors, whether hetero- or homonormative, from their citizens to be able to determine their entitlements (Richardson 2000). The scope of sexual citizenship therefore does not remain limited to the private space of home and domestic life but is acutely concerned with the public space. Citizens must be able to publicly "express and publicize" their sexual identities and orientation and, simultaneously "to have them acknowledged as legitimate" to be able to display them in the public sphere (Hubbard 2013: 225). This is where the intersection of state, citizens, and sexual politics becomes visible. In a heteronormative setting, the struggle is usually between groups that do not fit into the binary category of heterosexual man and woman and those who refuse to recognize diversity beyond such categorization, as shown in the contexts of Uganda and Russia. In the context of the homonormative United States, where LGBTQI identities have been established, the struggle is not anymore about being legally recognized but about abolishing discrimination on grounds linked to sexual orientation, for example not getting fired for being openly gay (Tilcsik 2011). Politics around sexual identities as such assumes different forms across space and scale; it remains in flux. Sexual identity and politics therefore offer fertile ground for further geographical research in close relation to core political geography concepts such as citizenship, belonging, nation, and the state, yet they remain underrecognized.

Conclusion

The scope of identity politics is not simply limited to questions of who we are and where we belong. It is equally about citizenship, belonging, religion, sexuality, mobility, nationalism, state, and sovereignty. Furthermore, identity politics is also crucial in naming and recognizing power relations, fighting hegemony and oppression, and working collectively for social change (Ahmed 2017). It spans time, scale, and space. But what remains crucial in an examination of identity politics, to repeat myself, is a closer look at categories – categories that are artificially created and reified so that they appear "natural" to us. Seeing through and beyond categorical boundaries hence remains an important task for geographers and other social scientists. When it comes to political geography in particular, scholarship on identity politics is in abundance. However, some categories such as citizenship, borders, "race," and nation attract much of the attention, while others, for example religion and sexuality, garner less, although they are not less important. Consequently, they exist as rich soil for future research agendas. One way to address this uneven attention is to tease out the intersectionality between the more and the less obvious categories, for instance between the categories of national identity and sexuality or those of citizenship and religion. Thus the social and spatial dimensions of these categories must be examined not as stand-alone phenomenon but as "mutually constitutive" (Hopkins 2019: 937). Another way of attaining such a "balance" would be through seriously considering calls for decolonizing the production of geographical knowledge through a commitment to "pluriversity" – that is, not just acknowledging the Euro-centric worldview but also being committed to the participatory production of knowledge (Van Klinken 2020). Especially when it comes to geographies of religion, many argue that geographical research on Muslim communities in the United States and Europe largely operates on the imaginary that Muslims are a given minority in society who form a distinct group, and thus the vast diversity and spatial difference among Muslim communities remain unnoticed (Mills & Gökarıksel 2014). But contrary to this, research carried out in predominantly Muslim societies about Islam powerfully demonstrates the fluidity, diversity, and cultural dynamics within many groups that often question the category of "Muslim" itself (Sidaway 2023). However, such scholarship is often positioned as "discourses outside the discipline" (Sidaway 2023: 342) and therefore its potential to inform the disciplinary debates remains untapped.

An important dimension of identity politics is the possibility of easily connecting theories, practice, and practitioners. In this regard, William Bunge's DGEI serves as a powerful example and a tool. In addressing the nationwide racial politics of segregation in the 1960s and 1970s, Bunge directly involved residents by training them to recognize numerous planning and spatial issues in their neighborhoods. This allowed them to directly influence city planning and thus racial segregation in the Detroit neighborhood of Fitzgerald (Garrido 2021). The DGEI demonstrates that when planned and executed collaboratively with local people, geographical tools and training could be used effectively in addressing both identity politics and politics of identity. Juxtaposing identity politics and politics of identity hence presents a great avenue for a direct bridging of theory, policy, and practice – a difficult terrain that academics find challenging to navigate.

Finally, we must keep in mind that new categories of identity are always at risk of ending up being the same tool that they opposed, at the first place, to come into being. Therefore, whether reading or doing identity politics, one would do well to take Judith Butler's (1999: XXVI) caution into serious consideration: "The mobilization of identity categories for the purposes of politicization always remain threatened by the prospect of identity becoming an instrument of the power one opposes."

References

Aalto, P., Dalby, S., & Harle, V. (2003). The critical geopolitics of Northern Europe: Identity politics unlimited. *Geopolitics 8*(1): 1–19. doi:10.1080/14650040412331307682b

Ahmed, S. (2017). *Living a Feminist Life*. Durham, NC: Duke University Press.

Al Jazeera (2014). Ugandan president signs anti-gay law, February 25. https://www.aljazeera.com/news/2014/2/25/ugandan-president-signs-anti-gay-law (accessed June 4, 2024).

Alexander, D. (2023). Decolonizing energy justice: Identity politics, attachment to land, and warfare. *Progress in Human Geography 47*(2): 348–350. doi:10.1177/03091325221149553

Anderson, B. (1996). *Imagined Communities: Reflections on the Origin and Spread of Nationalism*. London: Verso.

Anspach, R.R. (1979). From stigma to identity politics: Political activism among the physically disabled and former mental patients. *Social Science & Medicine Part A: Medical Psychology & Medical Sociology 13*: 765–773. doi:10.1016/0271-7123(79)90123-8

Bangladesh Bureau of Statistics (2023). *Population and Housing Census*, vol. *1*. Dhaka: Ministry of Planning. https://tinyurl.com/yc8bshww (accessed November 29, 2024)

Béland, D. (2017). Identity, politics, and public policy. *Critical Policy Studies 11*(1): 1–18. doi:10.1080/19460171.2016.1159140

Bell, D., & Binnie, J. (2000). *The Sexual Citizen: Queer Politics and Beyond*. Cambridge: Wiley-Blackwell.

Bernstein, M. (2005). Identity politics. *Annual Review of Sociology 31*: 47–74.

Billing, M. (1995). *Banal Nationalism*. London: Sage.

Bondi, L. (1993). Locating identity politics. In Keith, M., & Pile, S. (eds), *Place and the Politics of Identity*. New York: Routledge, 82–99.

Brass, P. (1974). *Language, Religion, and Politics in North India*. Cambridge: Cambridge University Press.

Brown, M. (2012). Gender and sexuality I: Intersectional anxieties. *Progress in Human Geography 36*(4): 541–550. doi:10.1177/0309132511420973

Browne, K. (2021). Geographies of sexuality I: Making room for lesbianism. *Progress in Human Geography 45*(2): 362–370. doi:10.1177/0309132520944494

Brubaker, R. (2015). *Grounds for Difference*. Cambridge, MA: Harvard University Press.

Brubaker, R., & Cooper, F. (2000). Beyond "identity". *Theory and Society 29*(1): 1–47. doi:10.1023/A:1007068714468

Butler, J. (1999). *Gender Trouble: Feminism and the Subversion of Identity*. London: Routledge.

Carrier, J.G. (1992). Occidentalism: The world turned upside-down. *American Ethnologist 19*(2): 195–212.

Chatruvedi, S. (2002). Process of othering in the case of India and Pakistan. *Tijdschrift voor Economische en Sociale Geografie 93*(2): 149–159.

Chen, X. (1995). *Occidentalism: A Theory of Counter-discourse in Post-Mao China*. Oxford: Oxford University Press.

Connolly, W.E. (1995). *The Ethos of Pluralization*. Minneapolis, MN: University of Minnesota Press.

DeJong, C., & Long, E. (2014). The death penalty as genocide: The persecution of "homosexuals" in Uganda. In Peterson, D., & Panfil, V.R. (eds), *Handbook of LGBT Communities, Crime, and Justice*. New York: Springer, 339–362.

Dempsey, K.E. (2022). Fostering grassroots civic nationalism in an ethno-nationally divided community in Northern Ireland. *Geopolitics 27*(1): 292–308. doi:10.1080/14650045.2020.1727449

Dicklitch, S., Yost, B., & Dougan, B.M. (2012). Building a barometer of gay rights (BGR): A case study of Uganda and the persecution of homosexuals. *Human Rights Quarterly 34*(2): 448–471.

Dowling, R. (2009). Geographies of identity: Landscapes of class. *Progress in Human Geography 33*(6): 833–839. doi:10.1177/0309132508104998

Dowling, R. (2010). Geographies of identity: Climate change, governmentality and activism. *Progress in Human Geography 34*(4): 488–495. doi:10.1177/0309132509348427

Drozdzewski, D., & Matusz, P. (2021). Operationalising memory and identity politics to influence public opinion of refugees: A snapshot from Poland. *Political Geography 86*: 102366. doi:10.1016/j.polgeo.2021.102366

Edenborg, E. (2020). Saving women and bordering Europe: Narratives of "migrants' sexual violence" and geopolitical imaginaries in Russia and Sweden. *Geopolitics 25*(3): 780–801. doi:10.1080/14650045.2018.1465045

Ehrkamp, P., & Jacobsen, M.H. (2015). Citizenship. In Agnew, J., Mamadouh, V., Secor, A., & Sharp, J.P. (eds), *The Wiley Blackwell Companion to Political Geography*. Chichester: John Wiley & Sons., 152–164. doi:10.1002/9781118725771.ch13

Fearon, J. D. (1999). What is identity (as we now use the word)? Unpublished manuscript, Stanford University. https://web.stanford.edu/group/fearon-research/cgi-bin/wordpress/wp-content/uploads/2013/10/What-is-Identity-as-we-now-use-the-word-.pdf (accessed November 29, 2024).

Ferdoush, M.A. (2019). Symbolic spaces: Nationalism and compromise in the former border enclaves of Bangladesh and India. *Area 51*(4): 763–770. doi:10.1111/area.12539

Ferdoush, M.A. (2024). *Sovereign Atonement: Citizenship, Territory and the State at the Bangladesh-India Border*. Cambridge: Cambridge University Press.

Ferdoush, M.A., & Väätänen, V. (2022). Anticipatory state identity: Understanding the Finnish state's approach to the Arctic. *Area 54*(4): 618–626. doi:10.1111/area.12802

Foucault, M. (1971). *The Order of Things: An Archeology of the Human Sciences*. New York: Pantheon.

Fox, J. (2018). *An Introduction to Religion and Politics: Theory and Practice*. London: Routledge.

Foxall, A. (2019). From Evropa to Gayropa: A critical geopolitics of the European Union as seen from Russia. *Geopolitics 24*(1): 174–193. doi:10.1080/14650045.2017.1417848

Fraser, N. (1997). *Justice Interruptus: Critical Reflections on the "Postsocialist" Condition*. New York: Routledge.

Fumanti, M. (2017). The politics of homosexuality in Africa. *Critical African Studies 9*(1): 1–8. doi:10.1080/21681392.2017.1282724

Garrido, G.J.L. (2021). Radical geography and advocacy mapping: The case of the Detroit Geographical Expedition and Institute (1968–1972). *Journal of Planning History 20*(4): 291–307. doi:10.1177/1538513220988673

Gellner, E. (1983). *Nationas and Nationalism.* Oxford: Blackwell.

Guhathakurta, M. (2013). Amidst the winds of change: The Hindu minority in Bangladesh. In Fazal, T. (ed.), *Minority Nationalism in South Asia.* London: Routledge, 126–139.

Heyes, C. (2020). Identity politics. In Zalta, E.N. (ed.), *The Stanford Encyclopedia of Philosophy.* Stanford, CA: Stanford University Press. https://plato.stanford.edu/archives/fall2020/entries/identity-politics (accessed November 29, 2024)

Hopkins, P. (2019). Social geography I: Intersectionality. *Progress in Human Geography 43*(5): 937–947. doi:10.1177/0309132517743677

Hubbard, P. (2013). Kissing is not a universal right: Sexuality, law and the scales of citizenship. *Geoforum 49*: 224–232. doi:10.1016/j.geoforum.2012.08.002

Huda, Z. (2004). Problem of national identity of the middle class in Bangladesh and state-satellite television. PhD thesis, University of Warwick.

Hussain, D. (2013). *Boundaries Undermined: The Ruins of Progress on the Bangladesh/India Border.* London: Hurst Publishers.

Jones, R. (2008a). Bounding categories, fencing borders: Exclusionary narrative and practices in the borderlands of India and Bangladesh PhD dissertation, University of Wisconsin, Madison.

Jones, R. (2008b). Searching for the greatest Bengali: The BBC and shifting identity categories in South Asia. *National Identities 10*(2): 149–165.

Jones, R. (2012). *Border Walls: Security and the War on Terror in the United States, India, and Israel.* London: Zed Books.

Keith, M., & Pile, S. (eds) (1993). *Place and the Politics of Identity.* London: Routledge.

Kobayashi, A. (2009a). Geographies of peace and armed conflict: Introduction. *Annals of the Association of American Geographers 99*(5): 819–826. doi:10.1080/00045600903279358

Kobayashi, A. (2009b). Identity politics. In Kitchin, R., & Thrift, N. (eds), *International Encyclopedia of Human Geography*, vol. 5. London: Elsevier, 282–286.

Kobayashi, A. (2020). Identity politics. In Kobayashi, A. (ed.), *International Encylopedia of Human Geography*, 2nd edn. London: Elsevier, 151–155.

Koch, N. (2023). Geographies of nationalism. *Human Geography 16*(2): 200–211. doi:10.1177/19427786231157712

Koch, N., & Paasi, A. (2016). *Banal Nationalism* 20 years on: Re-thinking, re-formulating and re-contextualizing the concept. *Political Geography 54*: 1–6. doi:10.1016/j.polgeo.2016.06.002

Koch, N., & Vora, N. (2020). Islamophobia and the uneven legal geographies of ethnonationalism. *Political Geography 83*: 102187. doi:10.1016/j.polgeo.2020.102187

Kothari, R. (1998). *Communalism in Indian politics.* Ahmedabad: Rainbow Publishers.

Lichterman, P. (1999). Talking identity in the public sphere: Broad visions and small spaces in sexual identity politics. *Theory and Society 28*(1): 101–141. doi:10.1023/A:1006954611027

Mansbach, R., & Rhodes, E. (2007). The national state and identity politics: State institutionalisation and "markers" of national identity. *Geopolitics 12*(3): 426–458. doi:10.1080/14650040701305633

March, L. (2021). Queer and trans* geographies of liminality: A literature review. *Progress in Human Geography 45*(3): 455–471. doi:10.1177/0309132520913111

Mills, A., & Gökarıksel, B. (2014). Provincializing geographies of religion: Muslim identities beyond the "West". *Geography Compass 8*(12): 902–914. doi:10.1111/gec3.12188

Mitchell, K. (2004). Geographies of identity: Multiculturalism unplugged. *Progress in Human Geography 28*(5): 641–651. doi:10.1191/0309132504ph510pr

Nicholson, L. (2008). *Identity before Identity Politics*. Cambridge: Cambridge University Press. doi:10.1017/CBO9780511551956

Nyroos, L. (2001). Religeopolitics: Dissident geopolitics and the "fundamentalism" of Hamas and Kach. *Geopolitics 6*(3): 135–157. doi:10.1080/14650040108407732

Oswin, N. (2019). *Gloabl City Futures: Desire and Development in Singapore*. Athens, GA: University of Georgia Press.

Puar, J. (2007). *Terrorist Assemblages: Homonationalism in Queer Times*. Durham, NC: Duke University Press.

Puar, J. (2013). Rethinking homonationalism. *International Journal of Middle East Studies 45*(2): 336–339. doi:10.1017/S002074381300007X

Riabov, O., & Riabova, T. (2014). The decline of Gayropa? *Eurozine*, February 5. https://www.eurozine.com/the-decline-of-gayropa/?pdf (accessed November 29, 2024).

Riaz, A. (2016). *Bangladesh: A Political History since Independence*. New York: I.B. Tauris.

Richardson, D. (2000). Claiming citizenship? Sexuality, citizenship and lesbian/feminist theory. *Sexualities 3*(2): 255–272. doi:10.1177/136346000003002009

Robinson, F. (1974). *Separatism among Indian Muslims: The Politics of the United Provinces*. Cambridge: Cambridge University Press.

Said, E. (1979). *Orientalism*. New York: Vintage.

Said, E. (1994). *Culture and Imperialism*. New York: Vintage Books.

Salaita, S. (2006). Beyond orientalism and islamophobia: 9/11, anti-Arab racism, and the mythos of national pride. *New Centennial Review 6*(2): 245–266.

Sidaway, J.D. (2023). Beyond the decolonial: Critical Muslim geographies. *Dialogues in Human Geography 13*(3): 337–358. doi:10.1177/20438206221121651

Staeheli, L.A., Kofman, E., & Peake, L.J. (eds) (2004). *Mapping Women, Mapping Politics: Feminist Perspectives on Political Geography*. London: Routledge.

Strum, T. (2015). Religious movements. In Agnew, J., Mamadouh, V., Secor, A.J., & Sharp, J.P. (eds), *The Wiley Blackwell Companion to Political Geography*. Chichester: John Wiley & Sons, 352–365. doi:10.1002/9781118725771.ch27

Tilcsik, A. (2011). Pride and prejudice: Employment discrimination against openly gay men in the United States. *American Journal of Sociology 117*(2): 586–626. doi:10.1086/661653

Tilly, C. (1996). Citizenship, identity and social history. In Tilly, C. (ed.), *Citizenship, Identity and Social History*. Cambridge: Cambridge University Press, 1–18.

Tully, J. (1995). *Strange Multiplicity: Constitutionalism in an Age of Diversity*. Cambridge: Cambridge University Press.

Tully, J. (2008). Identity politics. In Ball, T., & Bellamy, R. (eds), *Cambridge History of Twentieth-Century Political Thought*. Cambridge: Cambridge University Press, 517–533.

Van Klinken, A. (2020). Studying religion in the pluriversity: Decolonial perspectives. *Religion 50*(1): 148–155. doi:10.1080/0048721X.2019.1681108

Van Schendel, W. (2005). *The Bengal Borderland: Beyond State and Nation in South Asia*. London: Anthem Press.

Zhang, Q. (2015). Sexuality and the official construction of Occidentalism in Maoist and early post-Mao China. *European Journal of Cultural Studies 18*(1): 86–107. doi:10.1177/1367549414557807

Chapter 27

Digital Media

Paul C. Adams

Department of Geography and the Environment,
University of Texas at Austin, Austin, TX, USA

In the introduction to *Political Matter*, Braun and Whatmore (2010: x) argue: "From cell phones to stem cells, stuff of all kinds increasingly makes us what we are." This argument helpfully draws attention to the material side of political geography, but it is crucial to remember that our relationship with the material world is not deterministic, as if controlled by matter itself, any more than humans are able to drive the entire people–matter interaction. The relationship is reciprocal: as we act on the world with things in our hands, over our eyes, and on our ears, we constantly rediscover ourselves as agents. There is no one in charge of this process except for an always already hybrid agency.

Viewed historically, the progression of artifacts from sharpened stones to touchscreens is, at the same time, a progression of actor-networks. Media must be understood within the perspective of this ongoing human encounter with the material world, as not merely technological artifacts or systems but rather as structures of opportunity. For some scholars, more-than-human agency in political geography includes the efficacy of things as diverse as cast-off bottle caps, abandoned gloves, and dead rats caught on storm drains, described as having an "energetic vitality" (Bennett 2010: 39). My analysis sticks closer to conventional notions of political power but joins other attempts to appreciate the material-political by including media within the broader assumption that there "is no moment at which humanity comes to be contaminated by technical objects and practices – no fall into a world of things – because there can be no human without them" (Braun & Whatmore 2010: xix).

The Wiley Blackwell Companion to Political Geography, Second Edition.
Edited by Virginie Mamadouh, Natalie Koch, Chih Yuan Woon, and John Agnew.

In addition, digital media are not merely constitutive of material-political interdependencies and hybridities; they also carry discourses. As Glasze et al. (2023: 921) argue: "the question of how the digital transformation is politically produced and discursively embedded, and which spaces are thereby newly articulated, seems to us to be central to the debates, demands, and strategies of a 'digital sovereignty' in various geographical contexts." So the political geography of media needs to include how media are discursively constructed and how media convey political discourses, and both, in turn, point to the key concept of sovereignty, along with associated concepts of power and boundaries.

The chapter starts with media affordances – the ways in which particular media facilitate certain kinds of agency. Next it turns to a discussion of sovereignty – a concept that is troubled and assisted by digital geographies in contrasting ways depending on whether one considers personal sovereignty or state sovereignty. Sovereignty involves the maintenance of security in both cases, but what is rendered secure is a matter of overt and covert contestation at a range of geographical scales. Likewise, what is afforded agency differs depending on how media are deployed and appropriated, including the means and measures designed to protect state or personal sovereignty. We will consider state sovereignty and personal sovereignty in turn, but first we must look more closely at media affordances.

Media Affordances

In the words of Biekart and Fowler (2013: 529), technological changes are not a direct cause of political changes, "but they have certainly opened up innovative avenues for people to challenge existing configurations of power." In media studies, any opening up of new ways to achieve something through a medium implies an affordance. Affordances are opportunities associated with a medium, in the sense that "tools do not determine action, they afford it" (Nielsen 2013: 174). Just as it would be ludicrous to argue that hammers do not install nails, people do (a paraphrase of an even more unhelpful comment about guns), it makes no sense to say that phones do not call people, people do. The action of making a phone call is afforded by the phone when someone recognizes its affordances as a tool and temporarily forms a hybrid agent including both person and tool. Just as what hammers in nails is a person-with-a-hammer, what makes a phone call is a person-with-a-phone. The implication of this is less obvious: what posts an incendiary comment on social media is not just a person, or just a computer or an online platform, but rather a hybrid agent made up of a person recognizing certain affordances of an online platform, appropriating the computer and the platform, and using them in ways that cannot be reduced to autonomous actions by either the tool or the person.

Media affordances facilitate new ways of organizing actions in space and time, and therefore enable evolutions, and occasionally revolutions, in political action. Hybrid actors – humans acting in conjunction with technology – demonstrate a kind of agency that has been called *technicity* (Simondon 1992; Mackenzie 2002). Technicity pushes us to think beyond the politics of who has media access and who lacks access, though it is always important to keep in mind the political geography of haves and have-nots (Warf 2019). Media politics are a complex and multidimensional set of relations between the human and the material. On this account, appropriations of media can be understood abstractly as "an exteriority that is necessarily also an interiority" (Braun & Whatmore 2010: xix): our technologically afforded actions reach out into the world having political effects, but also, by doing so, the experience of a particular hybridization reaches into the agent and reworks its *sense of political agency*. What changes through tool use is not just the external world but also the human component's (or actant's) perceived and actual autonomy, and subsequently the hybrid agent's ability and inclination to make things happen.

If we set aside the habitual division of the world into humans and nonhumans and view the political agent as a heterogenous network and an outcome of technicity, then our inhabitation of media can be seen as enabling us to constantly evolve as spatially extensible beings and political actors (Latour 2005). As humans have extended themselves through space and time via media – for example through the use of parchment, paper, Instagram, or Twitter – each corresponding form of technicity has shaped the ways in which we can (and cannot) become political, engaging in media politics that are hegemonic, compliant, creative, or resistant, to give just a few examples (Mumford 1934; Braun & Whatmore 2010; Kitchin & Dodge 2011). There is no finite list of digital media affordances, but a few of the more significant ones for our purposes include the following:

- *Replicability*: there is no appreciable difference between the time and cost to produce 1 copy and 100 copies of a given creative work.
- *Instantaneity*: flows of digital media products are virtually instantaneous compared to previous media, notwithstanding technical issues like "latency."
- *Virality*: as a consequence of their replicability and instantaneity, digital media products are easily shared with a wide audience in a short period of time, leading to extremely rapid diffusion.
- *Translatability*: a digital product can generally be accessed via phone, computer, or tablet through a range of different platforms and applications (apps).
- *Ephemerality*: individual copies of digital media products are culturally and economically treated as disposable and are deleted or forgotten rather quickly.
- *Durability*: despite ephemerality, a digital media text can technically be preserved and retrieved indefinitely without deterioration of any sort, and replicability (see earlier) makes it virtually impossible to ensure the complete deletion of any particular piece of information.
- *Modifiability*: the "final" digital media product is easy to modify, thus rendering it a work in progress, open to reinterpretation and the proliferation of countless altered versions.

- *Ubiquity:* the adoption of digital technologies in all domains of life including work, leisure, transportation, and commerce, wherein various devices and systems can "talk to each other."

More abstractly, we can think of the moral and ethical implications of this evolution toward hybrid, digital-political agents in terms of "a human 'gesture' upon the space suggested by newer media" (Papacharissi 2010: 9). To gesture is to move in a meaningful way, using the body to send a signal, or be a signal, or both. All of the affordances described above depend not just on technology but also on particular human gestures, both literally and figuratively speaking.

Whether we think in terms of the articulation of grievances, discovery of opportunities, leveraging of resources, or formation of identities, all political processes in a mediatized world depend on the appropriation of media affordances in ways that both accept and contest existing boundaries, and on political "gestures" that are complicit in dominant scalings of power yet also trouble those scalings. This generally means acquiescing in top-down power but also contesting, negotiating, and redefining that power. For example, digital media enable inversion of the surveillant gaze, recognition of marginalized groups, crowdsourcing of geospatial information, and grassroots mobilization (Elwood & Leszczynski 2013), but at the same time digitalization leads to a logic of enclosure whereby communications are subsumed within a logic of extraction, commodification, and "rendering" that facilitates political and economic manipulation (Andrejevic 2007; Zuboff 2019). A user cannot choose one or the other. Rather, any use of digital tools opens one up to being used via those tools. Any gesture of resistance becomes, simultaneously, a gesture of acquiescence. These are the flip sides of a hybrid political agency that incorporates multiple and diverse media affordances. This duality is often captured in references to state sovereignty and personal sovereignty.

Digital Media and Scales of Sovereignty

References to sovereignty in connection with digital technology have increased severalfold over the past two decades, both in academic writing and in public discourse (Couture & Toupin 2019). Included in this trend are references to "data sovereignty," "technological sovereignty," and "digital sovereignty." The trend is less evident in the United States for a geopolitical reason: much of the digital sovereignty discourse signals efforts to circumvent the existing US dominance of global digital infrastructure (Cattaruzza et al. 2016; Couture & Toupin 2019: 2310). A geopolitical dynamic emerges around digital media because "past social contingencies and political choices are materially sedimented" in digital infrastructure (de Goede & Westermeier 2022: 2), and the biases and tendencies of a global digital infrastructure have been shaped in significant ways by the United States America. The existing global financial exchange networks, data management software, telecommunications services, computer operating systems, smartphone operating systems, and cutting-edge computing all manifest dominance by the United States. This American hegemony plays out in the realms of foreign policy, domestic policy, monetary policy, the military, companies, consumers, legal systems, and technology research and development (Agnew 2005b). From outside the United States, these diverse aspects of hegemony tend to be conflated as a unified American threat, although these impulses are in no way unified.

Sovereignty is an extremely flexible concept, with applications to state autonomy as well as individual, group, local, or city autonomy (Lynch 2020; Galis & Makrygianni 2022) and it locates strands of both top-down and bottom-up power. However, it is also a problematic concept that needs to be theoretically interrogated. Digital sovereignty includes ideas of state control over cross-border digital flows as well as grassroots efforts to appropriate digital media and platforms as tools for local autonomy. In democratic countries, state sovereignty depends on the sovereign subject, an agent who carves out spaces of autonomous action within the physical and institutional spaces of the state (Glasze et al. 2023: 922). Elsewhere, state sovereignty may apprehend most forms of digital autonomy as a threat to state sovereignty. For the purpose of analysis, then, it helps to separate state sovereignty from personal sovereignty and discuss them in turn.

State Sovereignty

Questions about state sovereignty often arise in connection with digital media, whether in the context of international hacking at the behest of a state, like the actions of the Russian-backed Fancy Bear, or the disruption of sovereignty by an individual using a publishing platform like WikiLeaks (Geller 2016), or because of viral cross-border flows of ideology, strategy, and affect, like the Arab Spring (Lynch 2006; Howard & Hussain 2013), or, as is most often the case, in conjunction with complaints about the domination of the global digital media system by certain countries or groups of states (Cattaruzza et al. 2016; Couture & Toupin 2019: 2310; de Goede & Westermeier 2022: 2; Kumar & Thussu 2023; Lin & de Kloet 2023). While the ontological status of state sovereignty is highly uncertain (Agnew 2005b), the state's *assertion* of sovereign control is what comes into play in these debates, and that assertion is itself a major feature of contemporary geopolitics. Digital communication flows cross state borders and percolate up from within the state, appearing to challenge state sovereignty both externally and internally.

One of the main challenges to discussing sovereignty is the assumption that states are sovereign *because of* their borders and *within* their borders. As Agnew argues, this "territorialization of sovereignty" is contradicted by everything from military cooperation to flows of pollution, from denationalized currencies to migration. Such phenomena and processes lead to the "attenuation of territoriality as sovereignty's primary mode of geographical organization" because de facto sovereignty has always fallen short of de jure sovereignty (Agnew 2005a: 438). De facto sovereignty exhibits patterns quite different from the cellular space of state borders, concentrating in particular locations and connecting to spatial networks that do not fill the boundaries of the state uniformly, while invariably reaching beyond the boundaries of the state. Therefore, as we counterpose digital media to state sovereignty here, it is not because they represent a new and radically different challenge to sovereignty but rather because they extend the familiar challenges to state sovereignty in particular ways.

Geographical approaches to state sovereignty are quite varied. Those approaches include familiar Westphalian notions of state autonomy; however, they also increasingly acknowledge the *interdependent sovereignty* emerging through neoliberal relations between countries; *insular sovereignty* in the form of "zones of confinement" around bodies, islands, exclaves, and military bases; and *ambiguous sovereignty* marked by simultaneous abandonment and entrapment of people within "'gray', exceptional, and asymmetrical expressions of power" (Mountz 2013: 833–834). Digital affordances including replicability, instantaneity, durability, and modifiability are essential to these transformations of sovereignty, whether we consider viral TikTok videos, leaked government documents, cross-border meddling in political campaigns, or the use of biometric population management technologies.

We have established that state sovereignty is certainly never fully realized, but is a principle, an objective, or an ideal. However, principles, objectives, and ideals all have the power to affect how specific political actors achieve or protect things, which makes for real effects. The rhetoric of state sovereignty prompts very real efforts to forge alliances and affect international organizations, as well as the countervailing efforts to impose closure on flows in and out of territories (Agnew 2005a; Glasze et al. 2023: 921). Sovereignty discourse, although effective in these ways, is not coherent. Western contributions to state sovereignty discourses assume that the sovereign political subject, a free and autonomous, enfranchised individual, is the foundation of state sovereignty. Elsewhere, states are conceptualized as having power over individuals, in some cases employing a rhetoric of sovereignty to justify cracking down on digitally mediated dissent and unrest (Pétiniaud et al. 2023). This makes state sovereignty not just a vague term but one that does service for contrary political agendas with regard to personal sovereignty. Regardless of ideological support for personal sovereignty, states of all sorts inscribe borders on human bodies through databases, making the insider/outsider distinction indelible and persistent (Amoore 2006).

Research on state sovereignty can trace how various actors try to achieve *sovereignty over* the digital as well as how they work to attain *sovereignty through* the digital (Bellanova et al. 2022). Digital infrastructure challenges state sovereignty – most notably through its instantaneity, replicability, and modifiability – because despite being international it is hegemonically dominated by US private and governmental actors (de Goede & Westermeier 2022) and by private and governmental actors from a handful of other powerful states (Kumar & Thussu 2023; Lin & de Kloet 2023). Paradoxically, even the task of securing borders against flows of undocumented immigrants draws on software and algorithms that are imported and exported (Aradau 2023); thus the project of sealing a border often depends on cross-border flows of bordering products.

Insofar as much global discourse around state-level digital sovereignty emerges in reaction to the multifaceted digital dominance by the United States, calls for "digital sovereignty" serve as "leverage for political, economic, social and strategic power" by other states (most conspicuously Russia) and nonstate actors (Glasze et al. 2023: 920). China has been singled out, as well, and bans on the video-sharing app TikTok show how a digital entertainment platform may emerge as a symbol of state espionage (Lin & de Kloet 2023) and of "moral and social degradation" (New Straits Times 2021). The "sovereignty" that states try to protect ranges from absolute control over all internal and incoming digital flows, as attempted by China and Russia, to a more measured stance that attempts to retain a measure of control over certain flows, as in the European Union (EU) (Cattaruzza et al. 2016; Bellanova et al. 2022). Digital affordances of replicability, instantaneity, translatability, and modifiability all challenge such containment efforts.

In some cases, an individual holds the reins of a powerful company or corporation that has become enmeshed in the machinery of state sovereignty, and the digital affordances challenging sovereignty have much to do with the privatization of global infrastructure. In September 2022, SpaceX founder Elon Musk received a request from Ukraine to extend Starlink communication satellite connectivity to Sevastopol. He denied the request, thereby disabling Ukrainian drone submarines that had been advancing on the Russian naval fleet in the Black Sea in response to Russia's control of eastern Ukraine (Coppa 2023; Dress 2023). Musk explained that the "obvious intent [of Ukraine was] to sink most of the Russian fleet at anchor. If I had agreed to their request, then SpaceX would be explicitly complicit in a major act of war and conflict escalation" (Dress 2023). But the decision not to extend Starlink coverage was a political decision as well, negatively affecting Ukraine's ability to maintain control over territory. Private company control over communication infrastructure can hinge state sovereignty on the whims of a powerful private-sector actor.

Another individual implicated in discussions of state sovereignty has been WikiLeaks' founder Julian Assange. The Australian publisher/hacker/free-speech activist demonstrates another way in which digital affordances affect state sovereignty. In 2016, at the height of the US presidential campaign, WikiLeaks released tens of thousands of emails acquired through a spear phishing attack from the Gmail account of Hillary Clinton's campaign manager, John Podesta. The breach was traced to Fancy Bear, a hacking organization associated with Russia's GRU intelligence agency (Geller 2016). Assange (who was at this time living and working from the geographical limbo of the Ecuadorian Embassy in London) refused to elucidate his political motives, disparaging both presidential candidates and claiming he was not playing favorites but simply had no information to leak about Donald Trump or his campaign. In this case, we see a familiar list of digital affordances. The ability to quickly access, copy, and publish large amounts of politically damaging information depends on the instantaneity, durability, and replicability of digital data.

In a published discussion forum dedicated to the earlier activities of WikiLeaks, Springer et al. (2012) associated the platform with an "epistemic shift" in sovereignty. Wikileaks received broad support from these geographers as "progressive," "counterhegemonic" (Warf in Springer et al. 2012: 699), "democratic" (Crampton in Springer et al. 2012: 690), and useful for destabilizing "the hegemony of state power" (Springer and Chi in Springer et al. 2012: 683) and challenging the "internal sovereignty" of the state (McConnell in Springer et al. 2012: 692). McConnell sees WikiLeaks as engaging a "new frontier of sovereignty – between states and informal web-based networks," which raises the question of whether WikiLeaks is perhaps even "emerging as a new sovereign authority in and of itself" (McConnell in Springer et al. 2012: 693). However, the subsequent activities of the platform, including its ties to Russian intelligence operations and partisan interference in the 2016 election, did not so much subvert state sovereignty in general as amplify the political biases of a particular state (Russia) and a particular individual (Assange).

A crucial year for the shift among European countries to strengthen digital sovereignty was 2013, after Edward Snowden (a former US intelligence contractor) leaked thousands of documents obtained from the US National Security Agency (NSA). This material, also released on WikiLeaks, exposed a digital surveillance operation that was global in scale. In response, the German minister of the interior, Thomas de Maizière, and the French minister of culture, Catherine Morin-Desailly, called for a collective effort on the part of the EU to promote digital sovereignty (Bellanova et al. 2022: 347). Further complicating the picture was economic anxiety regarding "market dominance by large foreign corporations" and the negative influence this had on local technological innovation in European countries. So cooperation at the supranational level (the EU) was intended, in part, to protect state economic interests as well as state sovereignty (Bellanova et al. 2022: 346). And yet, in addition to counterespionage and economic interests, the European call for digital sovereignty reflected the maintenance of personal sovereignty as an objective.

The EU Court of Justice determined in 2015 that the Safe Harbor Agreement with the United States should be annulled after an EU citizen filed a complaint that his personal data, transferred from the servers of Facebook's Irish subsidiary to America, failed to adhere to EU data protection standards, and that the United States was not a "safe harbor" for the personal data of European citizens (Cattaruzza et al. 2016; Bellanova et al. 2022). Concerns with personal data management spurred calls within the EU not simply to limit personal data flows to the United States, but also to develop infrastructure for secure data storage and management within the EU. Notable about this case was the way in which control over personal data (a facet of personal sovereignty) led to a retraction of an international data management agreement (with America), and the outcome was the call to develop a different international data management framework (situating greater data sovereignty within the EU as a whole). This exposure of the United States as an "unreliable partner for data protection" (Coyne 2019: 73) powerfully influenced the shape of the EU's General Data Protection Regulation (GDPR), which was adopted in 2016 and went into effect in 2018. The situation is complicated because what we could call "macroregional sovereignty," implemented at the supranational level, safeguarded aspects of personal privacy and hence personal sovereignty.

These examples highlight that the state is porous in particular ways vis-à-vis digital data flows, and that state sovereignty is impacted by the affordances of digital media. They demonstrate interlinkages of state sovereignty and personal sovereignty. We turn now to personal sovereignty issues more directly.

Personal Sovereignty

The micropolitics of digital media use engage with digital citizenship, arising from and giving rise to power dynamics at scales from the urban to the transnational (Hearn 2006; Basu 2019; Juzefovičs & Vihalemm 2020; Luger 2022; Simpson & Sheller 2022). Authors have identified new political identities internalized by citizens in the form of behavioral patterns, linking digital governmentality to particular forms of subjectification (e.g., Dammann, Eichenmüller, & Glasze 2022). This approach recalls the finding by Kasapoglu, Masso, and Calzati (2021) that people recognize the challenge to personal sovereignty presented by the algorithms in digital media, and that such recognition can prompt contestation of the digital spaces of state power. The pursuit of personal autonomy has been complicated by the emergence and proliferation of hybrid code-spaces mingling corporate and governmental power (Kitchin & Dodge 2011; Benhabib 2016; O'Hara & Hall 2018).

It is well known that digital spaces reflect and perpetuate social and spatial unevenness (Warf 2019). The dynamic is more complicated than advantaging the digital haves and disadvantaging the have-nots, because to have digital access is, at the same time, in subtle ways, *to be accessed*. Profits are generated for third parties when people can be induced to live in a "bubble" of mutually confirming digital content that seems diverse in format and source but in fact strategically reduces their encounters with unfamiliar ideas, maximizing attention capture and user obsession (Pariser 2011; Zuboff 2019), leading to what Dammann, Eichenmüller, and Glasze (2022) call "platform-based governing." Clicking "agree" to proceed to a website or online service is a small, mundane act that often escapes attention, but it is not without political implications. This act allows one's subsequent actions to be included within regimes of power based less on visible governmental apparatuses than on hybrid spaces of digital governance and governmentality (Andrejevic 2007; Leszczynski 2016; Kasapoglu, Masso, & Calzati 2021; Dammann, Eichenmüller, & Glasze 2022). In addition, as Louise Amoore (2018: 5) points out, "the visualization of a figurative cloud stands in for the complexity of the internet" in public discourse and corporate promotional material, and yet data that are "on the cloud" are in fact located within geographical territories and subject to geopolitical power dynamics.

Personal sovereignty is therefore severely challenged by the political dynamics of digital attention capture and surveillance. On the one hand, it is profitable to capture attention online, whether this does or does not entail surveillance (Davenport & Beck 2001). On the other hand, digital surveillance produces valuable data regardless of whether people pay attention (Zuboff 2019). Bringing together attention capture and digital surveillance facilitates microscopically monitored and strategically *streamed* activity (Jansson 2022). This streaming of activity is at the heart of what could be called the attention-surveillance economy, using distraction to coax people to voluntarily surrender their privacy (Zuboff 2019). This economy involves a novel convergence of the affordances translatability, ephemerality, and ubiquity.

These reworkings of personal sovereignty have variously been called *algorithmic governance*, *algorithmic governmentality* (Kitchin & Dodge 2011; Leszczynski 2016; Kasapoglu, Masso, & Calzati 2021), *digital governmentality* (Dammann, Eichenmüller, & Glasze 2022), *infrastructural politics* (Bellanova & Glouftsios 2022), and *logistical media* (Jansson 2022). This terminology points to the complex and powerful yet opaque decision-making processes built into digital platforms. People are manipulated by proprietary algorithms embedded in platforms wielded by unknown gatekeepers who profile users then filter people's worldviews with little or no transparency (Zook & Graham 2007). This manipulation increasingly impacts daily life, steering people in both physical and virtual spaces (Kasapoglu, Masso, & Calzati 2021). Digital governmentality merges with ubiquitous computing in "hybrid, sensing and adaptive environments" that track bodies through digital and physical space so as to both control populations and increase consumption (Dammann, Eichenmüller, & Glasze 2022: 2).

The corrosion of personal sovereignty is replicated in the spheres of politics and policy. Close ties between corporate power and political power, between advertising campaigns and political campaigns, and also between consumer identities and political identities have all led to what Ash, Kitchin, and Leszczynski (2018: 38) describe as "a strong neoliberal ethos underpinning" various appropriations of data and a logic of "technological solutionism." People are given access to certain kinds of data, but in using data they in turn acquiesce to data capture and sociospatial manipulation.

Reclaiming personal sovereignty requires thinking through both directions in every digital communication flow. For example, digital media facilitate asymmetric types of transparency in the "smart city" (Basu 2019), where users of digital services receive information about urban amenities, utilities, events, and the like, all the while creating reciprocal data flows that are almost entirely opaque with regard to their nature, extent, and use. Ubiquity and translatability work together to embed people in an asymmetric communication flow as "humans are implored to welcome the new era of urban innovation at precisely the moment when ubiquitous, tele-mobile communication, Big Data, and artificial intelligence seem to render humans obsolete" (Wiig & Wyly 2016: 487). This is not to say that humans are rendered valueless in their own eyes, but rather they are treated as means to an end within administrative and commercial systems, so their inherent value becomes an afterthought as their uniqueness and autonomy are treated instrumentally. Such processes and routines sidestep conventional modes of deliberative governance and undermine the foundations of the sovereign individual.

We now turn to people's efforts to evade, resist, reclaim, and appropriate digital media and how such efforts cultivate personal sovereignty. The ubiquity of digital technologies in various aspects of daily life spurs individual efforts to "raise and defend boundaries" (Fast, Lindell, & Jansson 2021: 81) against the intrusive and disruptive aspects of digital technology. Some of the most privileged users of digital media make substantial, coordinated, and deliberate efforts to avoid surveillance and monopolization of their attention. The micropolitics of daily digital life among more privileged users carves out a space of personal control within the hybrid code-spaces of digital media (Fast, Lindell, & Jansson 2021: 81) and this project often involves the effort to establish physical or virtual boundaries in support of "post-digital territoriality" (Adams & Jansson 2023). This kind of personal sovereignty does not seek complete elimination of digital incursions, but rather establishes a degree of separation from the attention-surveillance economy. In doing so, it works against several affordances at once, particularly ubiquity, instantaneity, and virality.

In a complementary gesture, digital media can help people create interstitial spaces for "expressive performances of citizenship" (Juzefovičs & Vihalemm 2020: 9) and temporary evasion of the state's sovereign power (Simpson & Sheller 2022). Evasive and resistant uses of digital media can loosen the bonds between political agents and particular locations, via virtual networking or "swarming" (Luger 2022). Digital media can also facilitate reappropriation of physical spaces such as urban squares through the organizational affordances of media (virality, instantaneity, translatability, and modifiability) building affective energies into political resistance, for example by circulating images that arouse anger toward authority figures, coordinating mass gatherings, and creating a feedback loop between "actions in the street and actions online" (Ziter 2013: 117; see also Amin 2009; Schejter & Tirosh 2012: 307; Gana 2013: 8–9; Tufekci 2013).

There are also ways in which digital affordances support reclamation of personal sovereignty from the state, for example through the "digital recreation of space that spreads beyond disciplinary and taxonomy mechanisms" (Galis & Makrygianni 2022: 8). For example, when state bureaucracies identify, control, and regulate immigrants, "migrants' engagement with Information and

Communications Technologies (ICTs) reveals a wide spectrum of practices that create ruptures in the 'sovereignty scapes' and borderlands as well as generating new spaces of autonomy" (Galis & Makrygianni 2022: 8). Urban citizens may similarly undertake initiatives "to build alternative economies and institute new forms of decentralized democratic control over processes of technological change" (Lynch 2020: 2). Such efforts point to the power of "an informal network of community groups, cooperatives, and activists" to reassert local control over communications and social services, and resist "dominant technocapitalist subjectivities" (Lynch 2020: 7). The translatability of digital media can also be employed as a means of foregrounding neglected urban stakeholders through hybrid digital-physical artwork revealing "new emancipatory contact zones" (Datta & Thomas 2022: 234). Here modifiability and translatability are key affordances for appropriating digital media.

More mundane, but quite wide-ranging, are everyday appropriations of digital media that foster personal sovereignty, as indicated by all sorts of everyday activities, including information gathering, entertainment, social networking, personal expression, labor, and political mobilization, and all of these are permeated at this point with digital practices that resonate with political collectives and bodies. Leszczynski (2020: 1194) draws attention to the "digital mundane," highlighting the routines and habits through which "digital technologies are implicated in and by the lives of ordinary people." Banal nationalism (Billig 1995) is both implicated in these routines and challenged by them. The everyday may be "the locus where people creatively and self-consciously mobilize nationhood in their social interactions" (Antonsich 2016: 33), but everyday digital media use may trouble orthodox nationalist narratives. Such was the case when a meme temporarily circulated on Chinese digital media labeled *Ruguanxue* (Veg 2022). This was a reinterpretation of Chinese geopolitics that provocatively cast China in the role of the "barbarian" Manchu who had invaded and established a new dynasty in the seventeenth century, and the United States was equated with the Ming Dynasty, indicating its vulnerability to invasion. Though pro-China, this counternarrative was out of alignment with the orthodox geopolitical narratives of the Chinese Communist Party (CCP) and the user's accounts were taken offline by the CCP.

People sense their own agency when appropriating social media, even if their autonomy remains compromised. Other examples are the use of virtual private networks (VPNs) to circumvent repressive internet blackouts in Cameroon (Murrey 2023) and the appropriation of digital media to support networks of advocacy, mutual assistance, and care among Indigenous people in Aotearoa New Zealand and Australia (Frazer, Carlson, & Farrelly 2022; Cupples & Glynn 2024). Such uses of social media intervene in state narratives constituting Indigenous peoples "as passive 'receivers' of care, particularly from the welfare state, rather than deeply committed and capable providers of care" (Frazer, Carlson, & Farrelly 2022: 8). Personal sovereignty is supported through such community appropriation of digital media.

People welcome digital media affordances that augment aspects of their autonomy, in a gesture that goes beyond mere appropriation to include reclamation. For example, Kasapoglu, Masso, and Calzati (2021) find that immigrants into Europe distinguish between what they see as *technologies of power* and *technologies of the self*. Technologies of power include the government algorithms used to guide the relocation of refugees. Technologies of self drive social media recommendations and online ads. Users critiqued the technologies of power for neglecting contingencies of place and their individuality as persons, but they perceived technologies of self (which arguably did much the same thing) as less threatening. This selective media appropriation supports strategies for reclaiming personal sovereignty.

The advent of artificial intelligence (AI) has prompted policy-makers at state and regional levels to envision a sort of race (Braun & Hummel 2023). Not exactly an arms race, the race to dominate AI is more subtle and conceptually ambiguous, since the potential applications of AI include governmental initiatives around defense and offense in the international security arena, but also internal administration and assertion of governmental legitimacy, which may infringe on personal sovereignty. AI can vastly accelerate the speed at which cross-border processes occur, such as cyberattacks and responses to such attacks (Timmers 2024: 583), and it can identify patterns in complex data (such as social media use) that elude (unaided) human observer-analysts. These changes extend the affordances of previous digital media, particularly ubiquity and instantaneity, while adding affordances of autonomous pattern recognition and data synthesis. In place of modifiability, AI introduces self-modifiability, the ability of machines to adapt and "learn" on their own.

These observations can be summed up with the observation that AI creates countless, unprecedented political challenges. Responses include the objective of developing "human-centric, sustainable, secure, inclusive and trustworthy AI," as articulated by the European Commission in 2021. Of course this sidesteps international disagreement regarding the values and ethics implied by "human-centric" (Braun & Hummel 2023: 933), which is itself a political gesture. Another example is the governmental regulation of social media platforms such as TikTok. The perceived threat involves the collection of data about users that is rich enough to support AI-based cross-border manipulation of people's interests, outlooks, and worldviews, and hence their "perception of who 'we' are" (Timmers 2024: 582, 583). AI presents a range of difficult, pressing choices that reconfigure the dynamics between personal and state sovereignty in new and troubling ways.

Conclusion

Digital media are caught up in both state and personal sovereignty. To understand the stakes of digital media in political processes, we must see them not merely as devices, infrastructures, and material systems, but as fragile sociotechnical achievements. Digital media include texts and discourses, codes and conventions, practices and norms that have evolved in tandem with particular technologies and the hybrid agents enabled and constrained by the affordances of particular media. This chapter has shown how the concept of sovereignty is central to discussions about state power and individual power, and their interactions, as well as the various ways in which hybrid agents achieve their objectives. A range of digital media affordances have been identified at the heart of political processes.

Future research should develop these insights by building greater theoretical coherence and further articulating the complex interplay of sovereignty (both personal and state) in terms of digital/physical assemblages (Clark 2021) or actor-networks (Thielmann 2010; see also Müller & Schurr 2016; Machold 2018: 93). Rather than trying to settle on the best theoretical framework, or the most complete list of affordances, ongoing research should think through how digital media, as sociotechnical systems, offer distinct political affordances, and how these, in turn, constitute novel, multilevel political terrains.

References

Adams, P.C., & Jansson, A. (2023). Postdigital territoriality: Disentangling from digital media as a return to place. *Annals of the American Association of Geographers 113*(3): 658–674.

Agnew, J. (2005a). Sovereignty regimes: Territoriality and state authority in contemporary world politics. *Annals of the Association of American Geographers 95*(2): 437–461.

Agnew, J. (2005b). *Hegemony: The New Shape of Global Power*. Philadelphia, PA: Temple University Press.

Amin, R. (2009). The empire strikes back: Social media uprisings and the future of cyber activism. *Kennedy School Review 10*: 64–66.

Amoore, L. (2006). Biometric borders: Governing mobilities in the war on terror. *Political Geography 25*(3): 336–351.

Amoore, L. (2018). Cloud geographies: Computing, data, sovereignty. *Progress in Human Geography 42*(1): 4–24. doi:10.1177/0309132516662147

Andrejevic, M. (2007). *iSpy: Surveillance and Power in the Interactive Era*. Lawrence, KS: University Press of Kansas.

Antonsich, M. (2016). The "everyday" of banal nationalism – ordinary people's views on Italy and Italian. *Political Geography 54*: 32–42.

Aradau, C. (2023). Borders have always been artificial: Migration, data and AI. *International Migration 61*(5): 303–306.

Ash, J., Kitchin, R., & Leszczynski, A. (2018). Digital turn, digital geographies? *Progress in Human Geography 42*(1): 25–43.

Basu, I. (2019). Elite discourse coalitions and the governance of "smart spaces": Politics, power and privilege in India's smart cities mission. *Political Geography 68*: 77–85.

Bellanova, R., Carrapico, H., & Duez, D. (2022). Digital/sovereignty and European security integration: An introduction. *European Security 31*(3): 337–355. doi:10.1080/09662839.2022.2101887

Bellanova, R., & Glouftsios, G. (2022). Controlling the Schengen Information System (SIS II): The infrastructural politics of fragility and maintenance. *Geopolitics 27*(1): 160–184.

Benhabib, S. (2016). The new sovereigntism and transnational law: Legal utopianism, democratic scepticism and statist realism. *Global Constitutionalism 5*(1): 109. [ii]

Bennett, J. (2010). Thing-power. In Braun, B., & Whatmore, S.J. (eds), *Political Matter: Technoscience, Democracy, and Public Life*. Minneapolis, MN: University of Minnesota Press, 35–62.

Biekart, K., & Fowler, A. (2013). Transforming activisms 2010+: Exploring ways and waves. *Development and Change 44*(3): 527–546. doi:10.1111/dech.12032

Billig, M. (1995). *Banal Nationalism*. London: Sage.

Braun, B., & Whatmore, S.J. (2010). The stuff of politics: An introduction. In Braun, B., & Whatmore, S.J. (eds), *Political Matter: Technoscience, Democracy, and Public Life*. Minneapolis, MN: University of Minnesota Press, ix–xl.

Braun, M., & Hummel, P. (2023). Sovereign power: Artificial intelligence and Europe's digital sovereignty. *Geopolitics 28*(2): 932–937.

Cattaruzza, A., Danet, D., Taillat, S., & Laudrain, A. (2016). Sovereignty in cyberspace: Balkanization or democratization. In *2016 International Conference on Cyber Conflict (CyCon US)*. Washington, DC: IEEE, 1–9.

Clark, J. (2021). "The Darroch affair": Assemblage lines, components, and transformations. *Political Geography 84*: 102304. doi:10.1016/j.polgeo.2020.102304

Coppa, T. (2023). Elon Musk's refusal to provide Starlink support for Ukraine attack in Crimea raises questions for Pentagon. *PBS*, September 11. https://www.pbs.org/newshour/economy/elon-musks-refusal-to-provide-starlink-support-for-ukraine-attack-in-crimea-raises-questions-for-pentagon (accessed November 30, 2024).

Couture, S., & Toupin, S. (2019). What does the notion of "sovereignty" mean when referring to the digital? *New Media & Society 21*(10): 2305–2322.

Coyne, H. (2019). The untold story of Edward Snowden's impact on the GDPR. *Cyber Defense Review 4*(2): 65–80.

Cupples, J., & Glynn, K. (2024). *Transmedia Geographies: Decoloniality, Democratization, Cultural Citizenship, and Media Convergence*. New Brunswick, NJ: Rutgers University Press.

Dammann, F., Eichenmüller, C., & Glasze, G. (2022). Geographies of "digital governmentality": Platform-based governing through adaptive environments. *Digital Geography and Society 3*: 100034.

Datta, A., & Thomas, A. (2022). Curating #AanaJaana [#ComingGoing]: gendered authorship in the "contact zone" of Delhi's digital and urban margins. *Cultural Geographies 29*(2): 233–252.

Davenport, T.H., & Beck, J.C. (2001). *The Attention Economy: Understanding the New Currency of Business*. Boston, MA: Harvard Business School Press.

de Goede, M., & Westermeier, C. (2022). Infrastructural geopolitics. *International Studies Quarterly 66*(3): sqac033.

Dress, B. (2023). Musk acknowledges he turned off Starlink internet access last year during Ukraine attack on Russia military. *The Hill*, September 8. https://thehill.com/policy/defense/4193788-musk-acknowledges-he-turned-off-starlink-internet-access-last-year-during-ukraine-attack-on-russia-military (accessed November 30, 2024).

Elwood, S., & Leszczynski, A. (2013). New spatial media, new knowledge politics. *Transactions of the Institute of British Geographers 38*(4): 544–559.

European Commission (2021). *Europe fit for the Digital Age: Commission proposes new rules and actions for excellence and trust in artificial intelligence.* Brussels: European Commission. https://ec.europa.eu/commission/presscorner/detail/en/ip_21_1682 (accessed December 20, 2024)

Fast, K., Lindell, J., & Jansson, A. (2021). Disconnection as distinction: A Bourdieusian study of where people withdraw from digital media. In Jansson, A., & Adams, P. (eds), *Disentangling: The Geographies of Digital Disconnection*. New York: Oxford University Press, 61–90.

Frazer, R., Carlson, B., & Farrelly, T. (2022). Indigenous articulations of social media and digital assemblages of care. *Digital Geography and Society 3*: 100038.

Galis, V., & Makrygianni, V. (2022). Analog flows in digital worlds: "Migration multiples" and digital heterotopias in Greek territory. *Political Geography 95*: 102599.

Gana, N. (2013). Tunisia. In Amar, P., & Prashad, V. (eds), *Dispatches from the Arab Spring: Understanding the New Middle East*. Minneapolis, MN: University of Minnesota Press, 1–23.

Geller, E. (2016). Russian hackers infiltrated Podesta's email, security firm says. *POLITICO*, October 20. https://www.politico.com/story/2016/10/russia-responsible-podesta-wikileaks-hack-230095 (accessed November 30, 2024).

Glasze, G., Cattaruzza, A., Douzet, F., et al. (2023). Contested spatialities of digital sovereignty. *Geopolitics 28*(2): 919–958.

Hearn, J. (2006). The implications of information and communication technologies for sexualities and sexualised violences: Contradictions of sexual citizenships. *Political Geography 25*(8): 944–963.

Howard, P.N., & Hussain, M.M. (2013). *Democracy's Fourth Wave? Digital Media and the Arab Spring*. Oxford: Oxford University Press.

Jansson, A. (2022). *Rethinking Communication Geographies: Geomedia, Digital Logistics and the Human Condition*. Cheltenham: Edward Elgar.

Juzefovičs, J., & Vihalemm, T. (2020). Digital humor against essentialization: Strategies of Baltic Russian-speaking social media users. *Political Geography 81*: 102204.

Kasapoglu, T., Masso, A., & Calzati, S. (2021). Unpacking algorithms as technologies of power: Syrian refugees and data experts on algorithmic governance. *Digital Geography and Society 2*: 100016.

Kitchin, R., & Dodge, M. (2011). *Code/Space: Software and Everyday Life*. Cambridge, MA: MIT Press.

Kumar, A., & Thussu, D. (2023). Media, digital sovereignty and geopolitics: The case of the TikTok ban in India. *Media, Culture and Society 45*(8): 1583–1599.

Latour, B. (2005). *Reassembling the Social: An Introduction to Actor-Network-Theory*. Oxford: Oxford University Press.

Leszczynski, A. (2016). Speculative futures: Cities, data, and governance beyond smart urbanism. *Environment and Planning A: Economy and Space 48*(9): 1691–1708. doi:10.1177/0308518X16651445

Leszczynski, A. (2020). Digital methods III: The digital mundane. *Progress in Human Geography 44*(6): 1194–1201.

Lin, J., & de Kloet, J. (2023). TikTok and the platformisation from China: Geopolitical anxieties, repetitive creativities and future imaginaries. *Media, Culture and Society 45*(8): 1525–1533.

Luger, J. (2022). Celebrations, exaltations and alpha lands: Everyday geographies of the far-right. *Political Geography 96*: 102604.

Lynch, C.R. (2020). Unruly digital subjects: Social entanglements, identity, and the politics of technological expertise. *Digital Geography and Society 1*: 100001.

Lynch, M. (2006). *Voices of the New Arab Public: Iraq, Al-Jazeera, and Middle East Politics Today*. New York: Columbia University Press.

Machold, R. (2018). Reconsidering the laboratory thesis: Palestine/Israel and the geopolitics of representation. *Political Geography 65*: 88–97.

Mackenzie, A. (2002). *Transductions: Bodies and Machines at Speed*. London: Continuum.

Mountz, A. (2013). Political geography I: Reconfiguring geographies of sovereignty. *Progress in Human Geography 37*(6): 829–841.

Müller, M., & Schurr, C. (2016). Assemblage thinking and actor-network theory: Conjunctions, disjunctions, cross-fertilisations. *Transactions of the Institute of British Geographers 41*(3): 217–229.

Mumford, L. (1934). *Technics and Civilization*. New York: Harcourt, Brace and World.

Murrey, A. (2023). A decolonial political geography of resistance and digital infrastructural harm in Cameroon and Ethiopia. *Globalizations 20*(6): 967–993. doi:10.1080/14747731.2022.2149162

New Straits Times (2021). Bangladesh court orders ban on TikTok, PUBG, Free Fire to "save children". *New Straits Times*, August 16. https://www.nst.com.my/world/region/2021/08/718633/bangladesh-court-orders-ban-tiktok-pubg-free-fire-save-children (accessed November 30, 2024).

Nielsen, R.K. (2013). Mundane internet tools, the risk of exclusion, and reflexive movements – Occupy Wall Street and political uses of digital networked technologies. *Sociological Quarterly 54*(2): 173–177. doi:10.1177/1461444810380863

O'Hara, K., & Hall, W. (2018). Four internets: The geopolitics of digital governance. CIGI Papers No. 206. Waterloo: Centre for International Governance Innovation.

Papacharissi, Z.A. (2010). *A Private Sphere: Democracy in a Digital Age*. Malden, MA: Polity Press.

Pariser, E. (2011). *The Filter Bubble: How the New Personalized Web Is Changing What We Read and How We Think*. London: Penguin.

Pétiniaud, L., Limonier, K., & Bertran, M.-G. (2023). Russia's pursuit of "digital sovereignty": Political, industrial and foreign policy implications and limits. *Geopolitics 28*(2): 924–928. doi:10.1080/14650045.2022.2050070

Schejter, A.M., & Tirosh, N. (2012). Social media new and old in the Al-'Arakeeb conflict: A case study. *The Information Society 28*(5): 304–315.

Simondon, G. (1992). The genesis of the individual. In Crary, J., & Kwinter, S. (eds), *Zone 6: Incorporations*. Cambridge, MA: Zone Books, 296–319.

Simpson, I., & Sheller, M. (2022). Islands as interstitial encrypted geographies: Making (and failing) cryptosecessionist exits. *Political Geography 99*: 102744.

Springer, S., Chi, H., Crampton, J., et al. (2012). Leaky geopolitics: The ruptures and transgressions of WikiLeaks. *Geopolitics 17*(3): 681–711.

Thielmann, T. (2010). Locative media and mediated localities: An introduction to media geography. *Aether v*(A): 1–17.

Timmers, P. (2024). Sovereignty in the digital age. In Werthner, H., Ghezzi, C., Kramer, J., et al. (eds), *Introduction to Digital Humanism*. Cham: Springer, 571–592.

Tufekci, Z. (2013). "Not this one" social movements, the attention economy, and microcelebrity networked activism. *American Behavioral Scientist 57*(7): 848–870.

Veg, S. (2022). The grassroots intellectuals. *Dissent*, Spring. https://www.dissentmagazine.org/article/the-grassroots-intellectuals (accessed November 30, 2024).

Warf, B. (2019). Teaching digital divides. *Journal of Geography 118*(2): 77–87. doi:10.1080/00221341.2018.1518990

Wiig, A., & Wyly, E. (2016). Introduction: Thinking through the politics of the smart city. *Urban Geography 37*(4): 485–493. doi:10.1080/02723638.2016.1178479

Ziter, E. (2013). The image of the martyr in Syrian performance and web activism. *Drama Review 57*(1): 116–136.

Zook, M.A., & Graham, M. (2007). Mapping DigiPlace: Geocoded Internet data and the representation of place. *Environment and Planning. B, Planning & Design 34*(3): 466.

Zuboff, S. (2019). *The Age of Surveillance Capitalism: The Fight for a Human Future at the New Frontier of Power*. New York: Public Affairs.

Chapter 28

Migration

Michael Samers

Department of Geography, University of Kentucky, Lexington, KY, USA

This chapter concerns the political geography of *international* migration. In the discussion that follows, I examine mainly the control and facilitation of migration (entry into national territories), including asylum-seeking and refugee movements (sometimes called "forced migration"), and to a lesser extent issues of citizenship (settlement within national territories). Migration and citizenship raise pivotal questions about the dynamic relationship between human mobility, space, state, nonstate, and quasi-state policies and practices, as well as immigrant mobilization and political participation.

While political geographers may sometimes tackle questions of space with more explicit intent, they hardly have a monopoly on using spatial metaphors to explain migration-related issues. This presents an immediate challenge for articulating a "political geography of international migration," since inter alia anthropologists, legal scholars, political scientists, and sociologists have all produced seemingly endless research on migration. Other social scientists may not interrogate spatial metaphors with the same fervor, but the relevant work of a broad range of social scientists needs to be considered. This chapter offers a brief survey of such research, principally over the last decade, and is therefore a selective cut through developments in the cavernous migration/immigration-oriented literature (for an earlier survey, see Samers 2015a).

Twenty-five years ago, political scientists observed a potential shift in the governance of migration in Europe from the "national" to the "supranational," from the "national" to the "local," and from state-based to privatized governance (Guiraudon & Lahav 2000). For many political geographers, this would

The Wiley Blackwell Companion to Political Geography, Second Edition.
Edited by Virginie Mamadouh, Natalie Koch, Chih Yuan Woon, and John Agnew.

eventually be captured in the concept of "multiscalar" and, given the apparent growth of private actors in migration, the neoliberalization of governance. On the other hand, political scientists especially extended Guiraudon and Lahav's insights by using the concept of "multilevel governance" (MLG) in the context of migration (Caponio & Jones-Correa 2018; Scholten, Caponio, & Zapata-Barrero 2019; Bazurli, Caponio, & de Graauw 2021; de Graauw 2022; see also van der Wusten, this volume). In this approach, the emphasis is on the "level" of government as a starting point for analyzing the effects on society and not necessarily the other way around.

We can use this idea of multiple "levels" first to discuss – albeit reservedly – the relationship between migration and municipal, metropolitan, and other subnational forms of governance (US states, German *Länder*, Canadian provinces, and so forth), then to address the relationship with national states, and finally to consider the relationship between migration and interstate relations, supranational governance (e.g., the European Union), or "global governance." Yet there are limitations to a state-centric approach to the governance of migration, and thus in the midst of this threefold analysis I broach the privatization of migration or the prevalence of nonstate actors, which is also unavoidably spatial.

In reviewing the relevant scholarship, this chapter assumes three aspects of such spatial metaphors: that privileging a particular metaphor over another is not necessarily fruitful (Leitner et al. 2008); that metaphors are constructed through *discourses* and *imaginations* as well as policies and practices; and that the invocation of spatial metaphors is intimately related to political strategies of research (e.g., Dikeç 2012). Spatialities (like those of territoriality) do not simply regulate or even constitute migration and migrants, but rather migrants shape these spatialities. And as Feldman (2012: 8) writes, "like any other subject-position, the 'migrant' does not exist ontologically but rather as someone who is enticed into a social role and negotiates it in a historically constituted field of political, economic, and social relations."

Migration and Local Territorial States

"Local" political geographies of migration encompass "local territorial states," understood as municipal, urban, or metropolitan discourses, policies, and practices. By the 2010s, social scientists began to speak of a "local turn" in the politics of migration *and* in the *study* of the politics of migration itself (e.g., Kreichauf & Glorius 2021; Ambrosini 2024). City governments are argued to shape or even develop their own migration policies and practices (such as in Sweden), rather than passively receiving them from national governments (Kerr 2019; Meer et al. 2021), and they may even contribute to the global governance of migration (Stürner-Siovitz 2023). Watson (2019) and Hillmann and Samers (2024) distinguish between urban governments that are more reluctant, restrictive, or "reactive" in terms of governing or recruiting immigrants and those that are more proactive or "welcoming" with respect to recruiting and governing migrants. This is not to be understood as a dichotomy but rather as a spectrum, and might also include smaller towns and rural areas (insofar as each of these is distinguishable and can be defined at all). Concerning cities or towns that are more "reactive," scholars have, over the last two decades, examined what might be called the "local exclusionary politics and policies" of municipal governments, nongovernmental organizations (NGOs), and civil society groups (e.g., Varsanyi 2011; Winders 2012; Togral-Koca 2019; Mescoli & Roblain 2021).

"(Re)bordering" and "border internalization" are similar terms for such politics, policies, and practices of subnational exclusion and border enforcement. For example, during the Covid-19 pandemic, national governments established subnational borders (checkpoints) or other internal mobility restrictions to attenuate the propagation of the virus, which also restricted the mobility of legal and undocumented immigrants alike (e.g., Bissonnette & Vallet 2022; Lan 2022). However, this has hardly been limited to the moment of the pandemic, and such bordering can involve anything from controlling access to healthcare and education to obtaining a driver's license. As Coleman observed almost two decades ago in the United States, "border enforcement – is increasingly *everywhere*" (2007: 64, emphasis in original).

Nonetheless, research in the United States has mainly highlighted the discord between local immigration politics on the one hand, and often more oppressive federal and substate politics and policies on the other (e.g., Arizona's 2010 SB1070 law required immigrants to carry proof of legal status, and required police officers to determine an immigrant's status if they were stopped legally, detained, or arrested) (Coleman 2009, 2012; Wong 2012; Wallace 2014). Varsanyi et al. (2012) refer to this as a "multi-layered jurisdictional patchwork," which appears to be a more legalistic version of MLG. Yet it is not simply a matter of law: there is a gap between legislation and actual enforcement practices, whether in the United States (Coleman 2012) or elsewhere (see, e.g., Yüksel 2023 on Turkey). In America, Coleman (2012) argues that site-specific immigration practices ("site" seems to be a synonym for "local") are intervening in the spaces of reproduction (e.g., traveling to work, shopping, etc.) by nonfederal immigration officers asking for proof of citizenship. This means that immigrants are as exposed in the spaces of *reproduction* as they are in the spaces of *production* (e.g., working informally).

Beyond this bordering literature, there are numerous studies of resistance to the settlement of immigrants by municipal governments, NGOs, or citizen publics in cities and towns in other countries, such as Italy (Ambrosini 2024), Serbia (Obradovic-Wochnik 2018), and South Africa (Hammerstad 2012). In this respect, Ambrosini (2024:27) conceptualizes the local governance of asylum as a "battleground," in which many local authorities and citizens see themselves as "victims" of "an invasion" by "them." These "local exclusionary politics and policies" such as in Saronno, Italy, can be seen as "institutionalized forms of intolerance," and are manifested for example in the resistance to the geographical placement and operation of asylum detention centers (Dikeç 2009; Coddington et al. 2020). In Serbia, Obradovic-Wochnik (2018) examines the contestation over public space between refugees, developers, and the state in the informalized, "wild capitalism" of international property development in Belgrade. Using the lens of the "biopolitical rationalities of neoliberalism" (2018: 65), the author shows how displacement of refugees from their city-center squats is supported by European policies and NGOs that inadvertently facilitate private development and assist the state in providing support for moving them to "camps" or refugee settlement centers (on the camp in political geography, see Minca 2015).

In contrast to state policies and practices that are imposed on migrants, we can cite the work on the political mobilization, formal political participation, and informal or formal political protests of immigrants in favor of legalization and against poor housing and working conditions, deportations, the police/immigration authorities, racism, and racialized violence (e.g., Boyce, Launius, & Aguirre 2017). In the Global North, then, cities have become "sites of refuge and resistance" (Mayer 2018). There is a long history of such protests, especially in cities such as Chicago and Los Angeles (Nicholls & Uitermark 2017; Gomberg-Munoz & Wences 2024), and some writers argue that the 2000s and 2010s witnessed heightened protest among immigrants in the United States (Tyler & Marciniak 2013; Kocher & Stuesse 2021).

For example, Leitner and Strunk (2014) have explored activism among undocumented immigrants in Washington, DC. Swerts engages with what Goffmann called the "front-stage" (2017: 394) and "back-stage" (390) to understand the struggles and "urban interstices" (380) of undocumented life (the spaces between formality and informality; between illegality and legality, and visibility and invisibility) and the "liminal politics" (380) of undocumented immigrants through a comparison of Chicago and Brussels. Kocher and Stuesse detail the tactics used in deportation defense campaigns, including "online petitions targeted at regional ICE officers or elected officials, media reports that humanize immigrants' stories, direct action such as rallies and sit-ins to pressure members of Congress, and legal advocacy to create a lawful path out of removal proceedings" (2021: 2).

Elsewhere in the world, scholars have discussed the mobilization of Zimbabweans in detention in the United Kingdom (McGregor 2011); "A Day Without Us" marches and strikes in France, Greece, Italy, and Spain in 2011; the emergence of the international European NoBorders Network or No One Is Illegal movement (Tyler & Marciniak 2013; Nicholls & Uitermark 2017); and pro-refugee resistance in the Athens neighborhood of Exarcheia (Turam 2024). These various forms of mobilization all demonstrate the significance of either putatively localized, national, or trans-state/transnational networked migrant protests against their political, economic, and social conditions (for one review, see Mora et al. 2018).

Local governance is not necessarily "reactive" or "restrictive," however. In fact, some municipal or subnational governments in the Global North have also encouraged migration to cities and towns, such as in Atlantic Canada (Pottie-Sherman & Graham 2024), or if they do not encourage it, then urban governments and NGOs either "tolerate" or actively provide help with housing, job search assistance, language classes, protection from deportation, and so forth. Some such efforts have ossified around the idea of "sanctuary or solidarity cities," which have gained wide coverage since the late 2000s, especially in the Global North (e.g., Bauder 2017, 2020; Darling & Bauder 2019; Roy 2019; Darling 2020; Jeffries & Ridgeley 2020; De Graauw 2022). However, several authors have insisted that sanctuary cities are limited in their ability to protect immigrants, given that they are interwoven with urban policing systems (Roy 2019; Davis 2020; Gomberg-Munoz & Wences 2024). For Roy (2019: 767), "the local refusal to enforce federal immigration law rests on local authority, notably the authority of the police." Sanctuary cities are therefore "a paradigm of liberal inclusion, extending rather than limiting state violence" (2019: 767).

Migration and National Territorial States

To understand the relationship between national territorial states and migration, one might distinguish between the research objectives of political scientists and political sociologists on the one hand, and those of political geographers on the other. The former tend to be more concerned with *explaining the politics of* migration and immigration policies ("outputs") or their consequences for migration ("outcomes") by assuming the continual salience of the state (e.g., Ellerman 2021; Hollifield et al. 2022). Political geographers, on the other hand, are typically more preoccupied with *critiquing national state effects*, or even the ontological status of the state (e.g., Gill 2010).

With respect to political science, Hollifield's (2004) "emerging migration state" (or more simply now the "migration state") set off a debate across migration studies. Hollifield suggested that states must balance openness in terms of trade, investment, and people with the closure of "security" (what he called the "liberal paradox"), or what Hampshire (2013) terms a "politics of closure" and "a politics of openness." While the "migration state" concept might reify or "essentialize" the state (Favell 2023; Nagel 2023; Triadafilopoulos & Taylor 2023), be Eurocentric, or involve a "typological approach of 'adjectives+migration states'" (Adamson, Chung, & Hollifield 2023: 569; Chung et al. 2024), it seems reasonable to argue that national governments practice contradictory forms of "migration management." This might include highly skilled recruitment, temporary work programs, the malign or strategic neglect of undocumented workers or asylum-seekers/refugees, and humanitarian operations on the one hand, and militarized borders and fluctuating interdiction and deportations on the other (e.g., Geiger & Pécoud 2010; Hampshire 2013; Işleyen 2018).

In some cases, these incongruous migration management strategies are combined simultaneously, as in the "military-humanitarian" operations of the

Italian navy in the Mediterranean (Heller & Pezzani 2018). In other cases, such as Tunisia, the government has followed a "laissez-faire" approach toward Libyan refugees who are tolerated and even praised, but not allowed to officially settle or claim citizenship (Müller-Funk & Natter 2024). In Japan there are similar paradoxes, where highly skilled migration is prioritized and other forms of migration (e.g., less-skilled temporary work programs or asylum) continue to be highly restricted regardless of rhetoric (e.g., Wakisaka 2023; Higuchi 2024). It is not surprising then that Natter (2024) does not find any stark differences between so-called liberal and autocratic regimes with respect to migration reforms, what she calls the "illiberal paradox."

However, explaining why particular national states have certain immigration policies and not others is a tricky affair. In this respect, Ellerman (2021) reviews the last 30 years of thinking in political science on immigration policy in the Global North, from practitioners representing interest groups, courts and judges, political parties, and bureaucracies to broader social concepts (e.g., national identities, and domestic and global norms such as "human rights"). She argues that three types of "insulation" – "popular insulation from the demands of an unorganized public, interest group insulation from the claims of lobbies, and diplomatic insulation from the lobbying of foreign governments" (2021: 19) – explain why some national governments choose a path between the liberalization of migration and restrictionism. Ellerman's analysis does not necessarily extend to the Global South and, for example, Müller-Funk, Fröhlich, and Bank's (2023) explanation of the governance of "forced migration" in Lebanon and Syria offers different "constellations of drivers" (see also the various chapters on the state and refugee movements in Fiddian-Qasmiyeh et al. 2014).

Although some authors argue that migration policies have become more liberal since 1945 (de Haas, Natter, & Vezzoli 2018), critical or activist scholarship among political geographers and others has put a spotlight on critiquing the violent, damaging, or exclusionary effects of such liberal state discourses, policies, and practices over the last two decades. This includes how they are classed, gendered, racialized, sexualized, criminalized, illegalized/irregularized, faced with Islamophobia, militarized, neoliberalized, and securitized in the Global North and South (e.g., Jones 2016, 2021; Peterson 2017; Hyndman 2019; Mitchell, Jones, & Fluri 2019; De Genova & Roy 2020; McMahon & Sigona 2021; Squire 2021; Feldman 2022; Tudor 2022). A further question is whether these processes are increasingly inflected by nationalist or populist discourses, parties, policies, and practices (Dennison & Geddes 2019), often in conjunction with (rather than counter to) neoliberalism (Joppke 2024; Samers & Rydgren 2024).

Contributions lamenting the neglect of gender in analyses of state institutions, policies, and practices concerning migration governance have attracted particular attention (Kofman & Raghuram 2015; Elias 2018; Bilecen et al. 2019; Cuban 2022; Joseph et al. 2022; Walton-Roberts, Rajan, & Joseph 2022; Foley 2023). For example, Cuban (2022) details the gaps and problems in social policies facing migrant women in Chile that are exacerbated by national and international politics; Walton-Roberts, Rajan, and Joseph (2022) and Joseph et al. (2022) examine the gendered emigration restrictions on Indian and Sri Lankan women working in the Gulf States, particularly nurses; and Foley (2023) scrutinizes international norms and institutions of migration in the context of Malaysia. In general, Bilecen et al. (2019) deplore the continual production of gender-blind state theory with respect to migration. Some scholars combine the inattention to gender with other themes, including the law (Jacobsen 2022) or the role of nonstate or quasi-state actors. To do so, many (but not all) reference the work of French theorist Michel Foucault and his concept of biopolitics (for reviews, see Kuusisto-Arponen & Gilmartin 2015; Samers & Collyer 2017; Hyndman 2019; Tazioli 2019). Perhaps indirectly in the same Foucauldian vein, other authors stress how calculations and statistics figure in the governance of migration (Heller & Pécoud 2020; Savatic et al. 2024).

More influenced by Mbembe and thanatopolitics than Foucault and biopolitics, McMahon and Sigona (2021) examine the governance of migrant deaths in the Mediterranean by different national governments and various European initiatives such as Triton, which combined both refoulement and rescue. From the point of view of abolitionist critique, Squire (2021) ties together the "prison-industrial complex" with the oppressive politics of asylum and other forms of migration, and Feldman (2022) moves beyond the liberal paradox and migration management to speak of "militarized migration management" within the history of global capitalism.

From yet another critical perspective, Naylor et al. (2018) urge the decolonization of political geography or a postborder epistemology that is equally critical of all racial and ethnic hierarchies, and the often-naturalized state categories that limit individuals and groups to certain roles (Collins 2022), or even the very category of "migrant" (Raghuram 2021). Indeed, there are now numerous calls for the decolonization or "decentering" of state theory with respect to migration and migration studies more generally (e.g., Triandafyllidou 2022; Adamson, Chung, & Hollifield 2023; Bashi 2023; Kibria 2023; Chung et al. 2024). This is echoed in a number of interventions that seek to decenter the European Union in the process of "border externalization" (El Qadim 2014; Cobarrubias et al. 2023; İşleyen & El Qadim 2023) or, for example, critique the lack of democratic decision-making and "moralizing" on immigration by Western European states with respect to Eastern European migration policies (Kovacs, Ramakrishnan, & Thieme 2023).

Migration, Interstate Relations, and Supranational and Global Governance

Three decades ago, Hollifield (1992) asked why there is no "global liberal order" for migration as there is for trade or finance. In the 1990s, political scientists and sociologists addressed Hollifield's question indirectly through a migration-related literature from the perspective of Marxism, realist (or neorealist) international relations (IR), and international (or global) political economy (IPE/GPE) (for a review, see Betts 2012). In hindsight, IR- and IPE-related approaches could be easily accused of methodological nationalism, the "territorial trap," embedded nation-statism, state rationality, or state essentialism, but newer research does not necessarily avoid these conceptual "folk devils" either.

In any case, IR- or IPE-related studies of migration are not necessarily methodologically nationalist. Diverse authors build on neorealism (which assumes states to be situated within a complex interstate or regional system that involves asymmetries of power) by combining it with a political sociological approach (e.g., Delano 2009; Samers 2015b), or investigating how politics are shaped by diasporic or transnational political movements or (transnational) governmentalities (e.g., Lafleur 2013; Pearlman 2014). In a sophisticated modification of IR/IPE analyses, Adamson (2024) develops the idea of "entangled states" in the context of France and Algeria and shows how migration figured in the processes of state-building, rather than being a simple consequence of state policy.

A second body of research examines questions of cooperation between states with respect to migration. This has given rise to discussions of the regional governance of migration, especially in Europe and the Mediterranean (e.g., Geddes et al. 2019), as well as the global governance of migration or "global mobility regimes" (e.g., Betts 2012; Pécoud & Thiollet 2023). Betts defines global governance as the "norms, rules, principles and decision-making procedures that regulate the behavior of states (and other transnational actors)" (2012: 4). In practice, the "global" governance of migration is a combination of bilateral, multilateral, regional, and interregional agreements. Thus, while he acknowledges that there is no explicit and binding global regime of migration governance, this does not mean that "there is no global governance" (2012: 8).

Indeed, Betts argues that there is both a "thin multilateralism" (rather than a dominant binding set of global institutions and policies) that regulates the behavior of states with respect to labor migration, international travel, or refugee protection, and a "facilitative multilateralism" that does not entail "formal multilateral cooperation but is designed to encourage states to become involved in dialogue and the sharing of information" (2012: 12). The global governance of migration is in fact a mixture of formal and informal state, extrastate, or nonstate institutions that operate through different actors and "levels" of governance, and different institutions or agreements regulate the various forms of migration through not only direct but indirect policies such as those of the International Organization for Migration (IOM), International Labour Organization, United Nations Development Program, UN Refugee Agency UNHCR, and the UN Global Compact for Safe, Orderly and Regular Migration (ratified in 2018) (Betts 2012; Triandafyllidou 2022; Pécoud & Thiollet 2023).

In contrast to the "liberal," institution-oriented global governance literature, a third body of research entails a more critical and theoretically eclectic range of work that embraces a mixture of critical cartographies of migration (Walters 2010), critical border studies (e.g., Bauder 2011; Szary & Giraut 2015), critical feminist geography or geopolitics (e.g., Basok & Piper 2012; Hyndman 2019), critical security studies, postcolonial critique or the urgency of a decolonial approach (see discussion in the previous section), poststructuralism (especially Foucault and governmentality), a concern with the privatization of governance, and transnationalism. In this wide-ranging literature, the critique of methodological nationalism, Global North centrism, the naturalization of borders, state centrism, and often state essentialism is either implicit or explicit. Yet it also reflects a special concern for the classist, racist, and gendered discourses of governance, as much as a concern for space.

A fourth constellation of studies examine the relationship between gender and global governance, sometimes with surprisingly positive assessments. For example, Mahon (2020) examines the IOM's emergent and potentially more positive policies with respect to gender and migration. Similarly, Cleton and Meier (2023) explore how the analytic of "intersectionality" shapes knowledge production concerning the governance of migration in Europe, and how therefore it might contest the EU's and member states' "hegemonic classification power" (2023: 3028). In contrast, Hennebry and Petrozziello (2019) critically examine the relationship between gender and governance in the Global Compact for Migration.

A fourth body of scholarship focused on global governance has placed greater emphasis on nonstate or quasi-state actors in the interstate/supranational or global governance of migration, including airport/airline officials, ferry companies, and other carriers; money transfer firms; immigration lawyers; international human rights organizations; labor brokers and other labor intermediaries; security firms; smuggling cartels; and social media (e.g., Xiang & Lindquist 2014, 2018; Yeoh 2021; Pécoud & Thiollet 2023). These actors have (sometimes with the collusion of states) served to reinforce border control well beyond the boundaries of nation-states, which today can be described as the actions of a "transnational state" (Lacroix 2022) or perhaps more commonly as "border externalization" (e.g., Bialasiewicz 2012; Casas-Cortes, Cobarrubias, & Pickles 2013; Cobarrubias et al. 2023; Álvarez Velasco 2024). Ultimately, this worldwide "remote control" of the refoulement and incarceration of asylum-seekers and refugees may reflect the quasi "death of asylum as an ethical ideal" (Mountz 2020).

Conclusion

Since the early 2000s, the political geography of migration has been concerned with the *effects* of governance, and political geographers (alongside a range of other social scientists) have provided vigorous critiques of the colonial, intersectionalized, biopolitical (class, racialized, gendered, etc.), nationalist, neoliberal, and securitarian governance of migration. In spatial terms, much of this research by political geographers over the last two decades seems to have three substantive, critical, or theoretical emphases: (1) how urban or localized politics, policies, and practices of state governance (or the politics and practices of migrants themselves) are entangled with other levels, territories, or scales, notwithstanding the problems of these spatial metaphors (see Hillmann & Samers 2024); (2) border externalization processes, including the heinous policies and practices of detention; (3) employing decolonial or abolitionist epistemologies (including engaging more with scholars and migrants from the Global South in the theorization of migration) to question a Global North–centered understanding of the governance of migration and/or to subvert the exercise of state power. Generally speaking, there seems to be a distaste for methodological nationalism.

Ironically though, given geographers' apparent predilection for "spatial correctness," political geographers have not interrogated spatial metaphors with (I would argue) the necessary energy. This unlikely negligence concerns, for example, problematic references to the vague and confused concept of "scale/multiscalar" or to "transnationalism," or a failure to specify exactly what "urban" or "local" means. After all, spatial metaphors may be more than simply a "mirror of nature" but rather a "mode of political reasoning" (Dikeç 2012). Whether or not one accepts this point, more scrutiny of the spatial metaphors that we employ in understanding migration appears to be an elusive but necessary objective.

References

Adamson, F. (2024). Entangled migration states: Mobility and state-building in France and Algeria. *Journal of Ethnic and Migration Studies 50*(3): 597–616.

Adamson, F.B., Chung, E.A., & Hollifield, J.F. (2023). Rethinking the migration state: Historicising, decolonising, and disaggregating. *Journal of Ethnic and Migration Studies 50*(3): 559–577.

Álvarez Velasco, S. (2024). Between "*Trochas*", orphans and mourning: Migrant mobilities and the effects of US "soft" remote control in Ecuador. *Geopolitics*: 1–27.

Ambrosini, M. (2024). The urban governance of asylum as a "battleground": Policies of exclusions. In Hillmann, F., & Samers, M. (eds), *Cities, Migration and Governance: Beyond Scales and Levels*. London: Routledge, 21–39.

Bashi, V. (2023). The practice of decolonizing migration studies. *Sociological Forum 38*(3): 886–895.

Basok, T., & Piper, N. (2012). Management versus rights: Women's migration and global governance in Latin America and the Caribbean. *Feminist Economics 18*(2): 35–61.

Bauder, H. (2011). Toward a critical geography of the border: Engaging the dialectic of practice and meaning. *Annals of the Association of American Geographers 101*(5): 1126–1139.

Bauder, H. (2017). Sanctuary cities: Policies and practices in international perspective. *International Migration 55*: 174–187.

Bauder, H. (2020). Migrant solidarities and the politics of place. *Progress in Human Geography 44*(6): 1066–1080.

Bazurli, R., Caponio, T., & de Graauw, E. (2021). Between a rock and a hard place: Mayors, migration challenges and multilevel political dynamics. *Territory, Politics, Governance 10*(3): 297–305.

Betts, A. (ed.) (2012). *Global Migration Governance.* Oxford: Oxford University Press.

Bialasiewicz, L. (2012). Off-shoring and out-sourcing the borders of Europe: Libya and EU border work in the Mediterranean. *Geopolitics 17*(4): 843–866.

Bilecen, B., Barglowski, K., Faist, T., & Kofman, E. (2019). Gendered dynamics of transnational social protection. *Comparative Migration Studies 7*(47): 1–14.

Bissonnette, A., & Vallet, E. (2022). Internalized borders and checkpoints: How immigration controls became normalized tools for COVID-19 responses in North America. *Journal of Borderlands Studies 37*(4): 679–697.

Boyce, G., Launius, S., & Aguirre, A. (2017). Drawing the line: Spatial strategies of community and resistance in post-SB1070 Arizona. *ACME 18*(1): 187–216.

Caponio, T., & Jones-Correa, M. (2018). Theorising migration policy in multilevel states: The multilevel governance perspective. *Journal of Ethnic and Migration Studies 44*(12): 1995–2010.

Casas-Cortes, M., Cobarrubias, S., & Pickles, J. (2013). Re-bordering the neighbourhood: Europe's emerging geographies of non-accession integration. *European Urban and Regional Studies 20*(1): 37–58.

Chung, E.A., Draudt, D., & Yunchen, T. (2024). The developmental migration state. *Journal of Ethnic and Migration Studies 50*(3): 637–656.

Cleton, L., & Meier, P. (2023). Contesting policy categories using intersectionality: Reflections for studying migration governance. *Ethnic and Racial Studies 46*(14): 3014–3036.

Cobarrubias, S., Cuttitta, P., Casas-Cortés, M., et al. (2023). Interventions on the concept of externalisation in migration and border studies. *Political Geography 105*(3): 102911.

Coddington, K., Conlon, D., & Martin, L.L. (2020). Destitution economies: Circuits of value in asylum, refugee, and migration control. *Annals of the American Association of Geographers 110*(5): 1425–1444.

Coleman, M. (2007). Immigration geopolitics beyond the Mexico–US border. *Antipode 39*(1): 54–57.

Coleman, M. (2009). What counts as the politics and practice of security, and where? Devolution and immigrant insecurity after 9/11. *Annals of the Association of American Geographers 99*(5): 904–913.

Coleman, M. (2012). The "local" migration state: The site-specific devolution of immigration enforcement in the U.S. South. *Law and Policy 34*(2): 159–190.

Collins, F.L. (2022). Geographies of migration II: Decolonising migration studies. *Progress in Human Geography 46*(5): 1241–1251.

Cuban, S. (2022). *Mapping Southern Routes of Migrant Women.* Abingdon: Routledge.

Darling, J. (2020). Refugee urbanism: Seeing asylum "like a city". *Urban Geography 42*(7): 894–914.

Darling, J., & Bauder, H. (2019). *Sanctuary Cities and Urban Struggles: Rescaling Migration, Citizenship, and Rights.* Manchester: Manchester University Press.

Davis, M.F. (2020). The limits of local sanctuary initiatives for immigrants. *Annals of the American Academy of Political and Social Science 690*(1): 100–116.

De Genova, N., & Roy, A. (2020). Practices of illegalization. *Antipode 52*(2): 352–364.

De Graauw, E. (2022). Mayoral leadership, immigrant sanctuary, and multilevel policy dynamics in San Francisco. *Territory, Politics, Governance 10*(3): 389–406.

de Haas, H., Natter, K., & Vezzoli, S. (2018). Growing restrictiveness or changing selection? The nature and evolution of migration policies. *International Migration Review 52*(2): 324–367.

Delano, A. (2009). From "shared responsibility" to a migration agreement? The limits for cooperation in the Mexico-United States case (2000–2008). *International Migration 50*: e42–e59.

Dennison, J., & Geddes, A. (2019). A rising tide? The salience of immigration and the rise of anti-immigration political parties in Western Europe. *Political Quarterly 90*: 107–116.

Dikeç, M. (2009). The "where" of asylum. *Environment and Planning D: Society and Space 27*(2): 183–189.

Dikeç, M. (2012). Space as a mode of political reasoning. *Geoforum 43*(4): 669–676.

El Qadim, N. (2014). Postcolonial challenges to migration control: French-Moroccan cooperation practices on forced returns. *Security Dialogue 45*(3): 242–261.

Elias, J. (2018). Governing domestic worker migration in Southeast Asia: Public–private partnerships, regulatory grey zones and the household. *Journal of Contemporary Asia 48*(2): 278–300.

Ellerman, A. (2021). *The Comparative Politics of Immigration*. Cambridge: Cambridge University Press.

Favell, A. (2023). The state of migration theory: Challenges, interdisciplinarity, and critique. In Hollifield, J.F., & Brettell, C.B. (eds), *Migration Theory: Talking across Disciplines*. London: Routledge, 341–358.

Feldman, D. (2022). Global crisis and militarized migration management, a world-historic perspective. In Payne, C., Korzeniewicz, R.P., & Silver, B.J. (eds), *World-Systems Analysis at a Critical Juncture*. New York: Routledge.

Feldman, G. (2012). *The Migration Apparatus*. Stanford, CA: Stanford University Press.

Fiddian-Qasmiyeh, E., Loescher, G., Long, K., & Sigona, N. (eds) (2014). *The Oxford Handbook of Refugee and Forced Migration Studies*. Oxford: Oxford University Press.

Foley, L. (2023). The gendered governance of migration. In Pécoud, A., & Thiollet, H. (eds), *Research Handbook on the Institutions of Global Migration Governance*. Cheltenham: Edward Elgar.

Geddes, A., Espinoza, M.V., Abdou, L.H., & Brumat, L. (2019). Introduction: The dynamics of regional migration governance. In Geddes, A., Espinoza, M.V., Abdou, L.H., & Brumat, L. (eds), *The Dynamics of Regional Migration Governance*. Cheltenham: Edward Elgar.

Geiger, M., & Pécoud, A. (eds) (2010). *The Politics of International Migration Management*. Basingstoke: Palgrave Macmillan.

Gill, N. (2010). New state theoretic approaches to asylum and refugee geographies. *Progress in Human Geography 34*(5): 626–645.

Gomberg-Munoz, R., & Wences, R. (2024). Fight for the city: Policing, sanctuary, and resistance in Chicago. In Hillmann, F., & Samers, M. (eds), *Cities, Migration, and Governance*. London: Routledge.

Guiraudon, V., & Lahav, G. (2000). The state sovereignty debate revisited: The case of migration control. *Comparative Political Studies 33*(2): 163–195.

Hammerstad, A. (2012). Securitisation from below: The relationship between immigration and foreign policy in South Africa's approach to the Zimbabwe crisis. *Conflict, Security and Development 12*(1): 1–30.

Hampshire, J. (2013). *The Politics of Immigration: Contradictions of the Liberal State*. London: Polity Press.

Heller, C., & Pécoud, A. (2020). Counting migrants' deaths at the border: From civil society counterstatistics to (inter)governmental recuperation. *American Behavioral Scientist 64*(4): 480–500.

Heller, C., & Pezzani, L. (2018). The Mediterranean mobility conflict: Violence and anti-violence at the borders of Europe. *Humanity*, November 9. http://humanityjournal.org/blog/heller-and-pezzani (accessed November 30, 2024).

Hennebry, J.L., & Petrozziello, A.J. (2019). Closing the gap? Gender and the global compacts for migration and refugees. *International Migration 57*(6): 115–138.

Higuchi, N. (2024). Immigration and nationalism in Japan. In Samers, M., & Rydgren, J. (eds), *Migration and Nationalism: Theoretical and Empirical Perspectives*. Cheltenham: Edward Elgar.

Hillmann, F., & Samers, M. (2024). Introduction: Cities, migration, and governance: Beyond scales and levels. In Hillmann, F., & Samers, M. (eds), *Cities, Migration, and Governance: Beyond Scales and Levels*. London: Routledge.

Hollifield, J. (1992). *Immigrants, Markets, and States: The Political Economy of Postwar Europe*. Cambridge, MA: Harvard University Press.

Hollifield, J. (2004). The emerging migration state. *International Migration Review 38*(3): 885–912.

Hollifield, J., Martin, P.L., Orrenius, P.M., & Héran, F. (eds) (2022). *Controlling Immigration: A Comparative Perspective*, 4th edn. Stanford, CA: Stanford University Press.

Hyndman, J. (2019). Unsettling feminist geopolitics: Forging feminist political geographies of violence and displacement. *Gender, Place and Culture 26*(1): 3–29.

İşleyen, B. (2018). Turkey's governance of irregular migration at European Union borders: Emerging geographies of care and control. *Environment and Planning D: Society and Space 36*(5): 849–866.

İşleyen, B., & El Qadim, N. (2023). Border and im/mobility entanglements in the Mediterranean: Introduction to the special issue. *Environment and Planning D: Society and Space 41*(1): 3–13.

Jacobsen, M.H. (2022). Precarious (dis)placement: Temporality and the legal rewriting of refugee protection in Denmark. *Annals of the American Association of Geographers 112*(3): 819–827.

Jeffries, F., & Ridgeley, J. (2020). Building the sanctuary city from the ground up: Absolutionist solidarity and transformative reform. *Citizenship Studies 24*(4): 548–567.

Jones, R. (2016). *Violent Borders: Refugees and the Right to Move*. London: Verso.

Jones, R. (2021). *White Borders: The History of Race and Immigration in the United States from Chinese Exclusion to the Border Wall*. Boston, MA: Beacon Press.

Joppke, C. (2024). Immigration and nationalism in the neoliberal order. In Samers, M., & Rydgren, J. (eds), *Migration and Nationalism: Theoretical and Empirical Perspectives*. Cheltenham: Edward Elgar.

Joseph, J., Henderson, S., Withers, M., & Shivakoti, R. (2022). Regulation through responsibilisation: Gendered exit policies and precarious migration from India and Sri Lanka. *International Migration*: 1–14. doi:10.1111/imig.13074

Kerr, J. (2019). Cities shaping migration policy. In: Barcelona Institute for International Affairs Global Cities Programme. In *Cities in World Politics: Local Responses to Global Challenges*. Barcelona: CIDOB, 45–52.

Kibria, N. (2023). Decolonizing migration studies: A brief introduction. *Sociological Forum 38*(3): 852–855.

Kocher, A., & Stuesse, A. (2021). Undocumented activism and minor politics: Inside the cramped political spaces of deportation defense campaigns. *Antipode 53*(2): 331–354.

Kofman, E., & Raghuram, P. (2015). *Gendered Migrations and Global Social Reproduction*. New York: Springer.

Kovacs, E., Ramakrishnan, K., & Thieme, T. (2023). Provincializing European responses to the refugee "crisis" through a Hungarian lens. *Political Geography 98*: 102708.

Kreichauf, R., & Glorius, B. (2021). Introduction: Displacement, asylum and the city – theoretical approaches and empirical findings. *Urban Geography 42*(7): 869–893.

Kuusisto-Arponen, A.K., & Gilmartin, M. (2015). The politics of migration. *Political Geography 48*: 143–145.

Lacroix, T. (2022). The transnational state and migration: Reach, flows and policies. *Political Geography 94*: 1–8.

Lafleur, J.M. (2013). *Transnational Politics and the State: The External Voting Rights of Diasporas*. London: Routledge.

Lan, P.C. (2022). Shifting borders and migrant workers' im/mobility: The case of Taiwan during the COVID-19 pandemic. *Asian and Pacific Migration Journal 31*(3): 225–246.

Leitner, H., Sheppard, E., & Sziarto, K.M. (2008). The spatialities of contentious politics. *Transactions of the Institute of British Geographers 33*(2): 157–172.

Leitner, H., & Strunk, C. (2014). Spaces of immigrant advocacy and liberal democratic citizenship. *Annals of the Association of American Geographers 104*(2): 348–356.

Mahon, R. (2020). Gendering migration management. In Geiger, M., & Pécoud, A. (eds), *The International Organization for Migration*. Basingstoke: Palgrave Macmillan, 53–73.

Mayer, M. (2018). Cities as sites of refuge and resistance. *European Urban and Regional Studies 25*(3): 232–249.

McGregor, J. (2011). Contestations and consequences of deportability: Hunger strikes and the political agency of non-citizens. *Citizenship Studies 15*(5): 597–611.

McMahon, S., & Sigona, N. (2021). Death and migration: Migrant journeys and the governance of migration during Europe's "migration crisis". *International Migration Review 55*(2): 605–628.

Meer, N., Dimaio, C., Hill, E., Angeli, M., Oberg, K., & Emilsson, H. (2021). Governing displaced migration in Europe: Housing and the role of the "local". *Comparative Migration Studies 9*(2): 1–17.

Mescoli, E., & Roblain, A. (2021). The ambivalent relations behind civil society's engagement in the "grey zones" of migration and integration governance: Case studies from Belgium. *Political Geography 91*: 102477.

Minca, C. (2015). Geographies of the camp. *Political Geography 49*: 74–83.

Mitchell, K., Jones, R., & Fluri, J. (2019). Introduction. In Mitchell, K., Jones, R., & Fluri, J. (eds), *Handbook on Critical Geographies of Migration*. Cheltenham: Edward Elgar, 1–17.

Mora, M.D.J., Rodriguez, R., Zermeño, A., & Almeida, P. (2018). Immigrant rights and social movements. *Sociology Compass 12*(8): e12599.

Mountz, A. (2020). *The Death of Asylum: Hidden Geographies of the Enforcement Archipelago*. Minneapolis, MN: University of Minnesota Press.

Müller-Funk, L., Fröhlich, C., & Bank, A. (2023). Disentangling forced migration governance: Actors and drivers along the displacement continuum. *International Migration Review 58*(3): 1462–1488.

Müller-Funk, L., & Natter, K. (2024). Forced migration governance in Tunisia: Balancing risks and assets for state-making during independence and democratization. *Mediterranean Politics 29*(5): 617–641.

Nagel, C. (2023). Review essay: Moving beyond the migration state. *International Migration Review 57*(2): 846–852.

Natter, K. (2024). The il/liberal paradox: Conceptualising immigration policy trade-offs across the democracy/autocracy divide. *Journal of Ethnic and Migration Studies 50*(3): 680–701.

Naylor, L., Daigle, M., Zaragocin, S., Ramírez, M.M., & Gilmartin, M. (2018). Interventions: Bringing the decolonial to political geography. *Political Geography 66*: 199–209.

Nicholls, W., & Uitermark, J. (2017). *Cities and social movements: Immigrant rights activism in the United States, France and the Netherlands, 1970–2015*. Chichester: Wiley-Blackwell.

Obradovic-Wochnik, J. (2018). Urban geographies of refugee journeys: Biopolitics, neoliberalism and contestation over public space in Belgrade. *Political Geography 67*: 65–75.

Pearlman, W. (2014). Competing for Lebanon's diaspora: Transnationalism and domestic struggles in a weak state. *International Migration Review 48*(1): 34–75.

Pécoud, A., & Thiollet, H. (2023). *Research Handbook on the Institutions of Global Migration Governance*. Cheltenham: Edward Elgar.

Peterson, V.S. (2017). Towards queering the globally intimate. *Political Geography 56*: 114–116.

Pottie-Sherman, Y., & Graham, N. (2024). Live, work, and stay? Geographies of immigrant receptivity in Atlantic Canada's aspiring gateways. In Hillmann, F., & Samers, M. (eds), *Cities, Migration and Governance: Beyond Scales and Levels*. London: Routledge, 121–141.

Raghuram, P. (2021). Democratizing, stretching, entangling, transversing: Four moves for reshaping migration categories. *Journal of Immigrant & Refugee Studies 19*(1): 9–24.

Roy, A. (2019). The city in the age of Trumpism: From sanctuary to abolition. *Environment and Planning D: Society and Space 37*(5): 761–778.

Samers, M. (2015a). Migration. In Agnew, J., Mamadouh, V., Secor, A., & Sharp, J. (eds), *The Companion to Political Geography*. Chichester: Wiley-Blackwell, 478–492.

Samers, M. (2015b). Regional integration and migration in North America. In Talani, L.S., & McMahon, S. (eds), *Handbook of the International Political Economy of Migration*. Cheltenham: Edward Elgar, 373–398.

Samers, M., & Collyer, M. (2017). *Migration*, 2nd edn. London: Routledge.

Samers, M., & Rydgren, J. (2024). *Migration and Nationalism: Theoretical and Empirical Perspectives*. Cheltenham: Edward Elgar.

Savatic, F., Thiollet, H., Mesnard, A., Senne, J.-N., & Jaulin, T. (2024). Borders start with numbers: How migration data create "fake illegals". *International Migration Review*. doi:10.1177/01979183231222169

Scholten, P., Caponio, T., & Zapata-Barrero, R. (eds) (2019). *The Routledge Handbook of the Governance of Migration and Diversity in Cities*. London: Routledge.

Squire, V. (2021). Unruly migrations, abolitionist alternatives. *Behemoth: A Journal on Civilization 14*(3): 14–24.

Stürner-Siovitz, J. (2023). "All the world's a stage?" A role theory analysis of city diplomacy in global migration governance. *International Migration Review 57*(4): 1329–1361.

Swerts, T. (2017). Creating space for citizenship: The liminal politics of undocumented activism. *International Journal of Urban and Regional Research 41*(3): 379–395.

Szary, A.L.A., & Giraut, F. (2015). Borderities: The politics of contemporary mobile borders. In Szary, A.L.A., & Giraut, F. (eds), *Borderities and the Politics of Contemporary Mobile Borders*. Basingstoke: Palgrave Macmillan.

Tazioli, M. (2019). *The Making of Migration: The Biopolitics of Mobility at Europe's Borders*. London: Sage.

Togral-Koca, B. (2019). Local bordering practices, refugees, and civil society: The case of Berlin. *Geographical Review 109*(4): 544–561.

Triadafilopoulos, T., & Taylor, Z. (2023). The domestic politics of selective permeability: Disaggregating the Canadian migration state. *Journal of Ethnic and Migration Studies 50*(3): 702–725.

Triandafyllidou, A. (2022). Decentering the study of migration governance: A radical view. *Geopolitics 27*(3): 811–825.

Tudor, A. (2022). Ascriptions of migration: Racism, migration and Brexit. *European Journal of Intercultural Relations 26*(2): 230–248.

Turam, B. (2024). The geopolitics of fear: Pro-refugee resistance to Europe's racial security. *Political Geography 109*: 103047.

Tyler, I., & Marciniak, K. (2013). Immigrant protest: An introduction. *Citizenship Studies 17*(2): 143–156.

Varsanyi, M. (2011). Neoliberalism and nativism: Local anti-immigrant policy activism and an emerging politics of scale. *International Journal of Urban and Regional Research 35*(2): 295–311.

Varsanyi, M., Lewis, P., Provine, M., & Decker, S. (2012). A multilayered jurisdictional patchwork: Immigration federalism in the United States. *Law and Policy 34*(2): 138–158.

Wakisaka, D. (2023). Unraveling migration policy-making in the land of "no immigrants": Japanese bureaucracy and the discursive gap. *International Migration Review 58*(3): 1507–1531.

Wallace, S.J. (2014). Papers please: State-level anti-immigrant legislation in the wake of Arizona's SB 1070. *Political Science Quarterly 129*(2): 261–292.

Walters, W. (2010). Anti-political economy: Cartographies of "illegal immigration" and the displacement of the economy. In Best, J., & Paterson, M. (eds), *Cultural Political Economy*. London: Routledge.

Walton-Roberts, M., Rajan, S.I., & Joseph, J. (2022). Gendered mobility and multi-scalar governance models: Exploring the case of nurse migration from India to the Gulf. In Ennis, C.A., & Blarel, N. (eds), *The South Asia to Gulf Migration Governance Complex*. Bristol: Bristol University Press, 35–54.

Watson, J. (2019). Welcoming refugees and the cultural wealth of cities: Intersections of urban development and refugee humanitarianism. *International Journal of Urban and Regional Research 43*: 983–999.

Winders, J. (2012). Seeing immigrants: Institutional visibility and immigrant incorporation in new immigrant destinations. *Annals of the American Academy of Political and Social Science 641*(1): 58–78.

Wong, T.K. (2012). 287(g) and the politics of interior immigration control in the United States: Explaining local cooperation with federal immigration authorities. *Journal of Ethnic and Migration Studies 38*(5): 737–756.

Xiang, B., & Lindquist, J. (2014). Migration infrastructures. *International Migration Review 48*(S1): S122–S148.

Xiang, B., & Lindquist, J. (2018). Postscript: Infrastructuralization: Evolving sociopolitical dynamics in labour migration from Asia. *Pacific Affairs 91*(4): 759–773.

Yeoh, B.S.A. (2021). Conceptual contours of migration studies in and from Asia. *International Migration 59*(6): 225–233.

Yüksel, K. (2023). Ambiguity and migration governance in the satellite city of Duzce, Türkiye. *Asian and Pacific Migration Journal* 32(4): 636–659.

Material and Environmental Political Geographies

Chapter 29

Material Political Geographies

Violante Torre, Clotilde Trivin, and Martin Müller
Department of Geography and Sustainability, University of Lausanne, Lausanne, Switzerland

What Exactly Is "Material"?

Since the 2000s, the social sciences and humanities have been undergoing a "material turn." Political geography is no exception. From the "cultural turn" in the 1990s, with its focus on discourses, text, representation, and symbolic meaning-making, political geography has become more and more interested in questions of matter from the late 2000s onward – things one can touch, feel, and smell. This interest has been reinforced by an increasing awareness of the finiteness of our planet in the Anthropocene, which has produced attention to the material support structures of the Earth and the urgency of finding modes of co-inhabiting this world with all its other species (Kohn 2013; Tsing et al. 2017). The so-called ontological turn, originating from debates in anthropology, has further fueled the attention to the material world, as it questions the nature/culture dualism and advocates for the existence of different lived worlds, rather than just different *views* of the same world (Holbraad & Pedersen 2017).

The theoretical inspirations feeding into the material turn are diverse and eclectic. Among the most prominent are contributions on hybridity and vital materialism (Haraway 1991; Latour 2005; Bennett 2010), affect (Spinoza 1677; Deleuze 1988; Massumi 2002; Barad 2007; Stewart 2007), assemblage thinking (Deleuze & Guattari 1987), phenomenology (Merleau-Ponty 1962), feminist theory (Sedgwick & Frank 2003; Ahmed 2004) and practice theory (de Certeau 1984).

The material turn has resulted in a rehabilitation of the material world in political geography. However, this does not mean a "return" to an earlier idea of

The Wiley Blackwell Companion to Political Geography, Second Edition.
Edited by Virginie Mamadouh, Natalie Koch, Chih Yuan Woon, and John Agnew.

human agents acting on and appropriating the material world. Quite the contrary, this turn seeks to emancipate the material world from being a mere object, a simple tool, put in the hands of powerful humans. It foregrounds ideas of "entanglement" between the human and the material (Barad 2007; Haraway 2008) rather than a neat and clear separation. The term "world" is important when speaking of the "material world." It refers to the continuous enactment of human and more-than-human elements that produce the relations making up our different worlds. Materials become co-constitutive of action and act as much on humans as humans act on them.

Materials, for political geographers, come in different shapes and sizes: materials can be the hardware of borders, with their security cameras, biometric devices, and barriers (Leese, Noori, & Scheel 2022; Pallister-Wilkins 2022; Ozguc & Burridge 2023); they can be the infrastructures of roads, railways, or airports (Usher 2019a; Haughton & McManus 2022); they can be resources such as water, oil, and gas (Koch & Perreault 2019; Barry & Gambino 2020; Dittmer 2021a); and they can be found in the corporeality of human bodies (Militz & Schurr 2016; Merriman & Jones 2017; Gökarıksel & Secor 2020) or in seemingly inert materials like strata and brute rock (Jones & Clark 2019; Lobo & Hine 2023).

This chapter reviews how political geographers have engaged with the material world. With a focus on research produced since 2015, it constitutes an update to the chapter published in the first edition of this companion (Müller 2015). Since then, research on the material world has much increased in volume and breadth. While themes of affect and assemblage continue to be significant, new ones have emerged since the previous edition. Infrastructure and technologies have become important strands of debate, as have natural resources of various kinds, perhaps indicating the increased awareness of their finiteness. That the update has become, with the exception of a few sentences here and there, a complete rewrite is one of the best indicators that debate in the field is advancing at a rapid speed.

Research on the material world in political geography can be clustered around three main foci, which constitute the three main parts of this chapter. First, *sensing the material* considers how materials become felt and experienced, with a focus on questions of affect and emotion. Second, *assembling the material* examines how materials come together to make action possible, highlighting the role of assemblage thinking. Third, *making the material political* is devoted to looking at how the material world is made political or becomes political by itself. The conclusion sketches three issues neglected in the field so far that deserve attention to advance the debate. In so doing, it suggests connections with animal and plant geographies, the discipline of international relations (IR), and Indigenous geographies.

Sensing the Material

In the past two decades, geographers inspired by the actor-network approach to objects, infrastructures, and buildings (Jacobs 2006) and geographers inspired by affect theory (Thrift 2004; Pile 2010) have investigated how human subjectivity is influenced by material things. Studies on affects, emotions, and the material have evoked the ways in which material things bring feelings to human subjectivities. In the case of buildings, for example, this implies "the feel *of* buildings, feelings *in* buildings, and feelings *about* buildings" (Degen, Rose, & Basdas 2010: 334). This renewed engagement with the material invites thinking about material things not only as used and mobilized, but also as sensed, felt, and experienced. Crucially for political geography, this implies an important shift from existing debates about affects and emotions: it is not just that politics works *through* emotions and affects, but that sensing and feeling *themselves* have strong political implications. The material becomes a crucial element in the encounter of politics and affects because it has the capacity to invite and generate affective politics, or prevent it from taking place. This, in turn, challenges "the idea that the material world does little more than provide the stage for political dramas to unfold" (Depledge 2015: 91).

Debates on affects and emotions in the early 2000s, particularly in human geography, focused on the question of ontologies of affects and emotions and their subsequent delimitations (e.g., McCormack 2003; Thrift 2004; Thien 2005; Anderson & Harrison 2006; Tolia-Kelly 2006; Lorimer 2008). Proponents of affect theory defined affect as emergent and fleeting emotional intensities that influence behavior. This implied that affect is precognitive, preconscious, and irrational: it operates in, through, and even beyond the body and bodily experience (Lorimer 2008: 552). Feminist scholars have objected to the conceptualization of affect as precognitive and preconscious, arguing that it is oblivious to the unequal differentiations that position bodies and subjects through race, gender, sexual orientation, or class (e.g., Thien 2005; Tolia-Kelly 2006; Colls 2012; see also Joanne Sharp 2009; Pile 2010). Using the concept of emotions instead, feminist geographers have discussed the interplay between the symbolic positioning of subjects and the intensities of feeling it affords or denies (cf. Anderson & Smith 2001; Davidson, Bondi, & Smith 2016).

The increasingly prominent role of the material world in geographical considerations sheds new light on the political significance of affects and emotions. Recent contributions seek to move beyond the earlier debate about the ontological difference between affects and emotions, acknowledging that both play a role in how politics are produced and lived, and valuing the connections between the different scales at which they operate (Sharp 2021). This renewed interest in the material thus invites geographers to update the vocabulary describing how affects, emotions, and politics relate to each other. Works on "affective atmospheres," for example, extend the question of the capacities of bodies to affect each other and include reflections on the intensities and propagations of feelings across scales (Fregonese 2017). Contributions on "affective spatialities," on the other hand, discuss how space has the capacity to generate and operate such affective intensities and emotional negotiations (Miller & Laketa 2019; Ho 2023). Territories (e.g., Usher 2019b), infrastructure (e.g., Barua 2021), and borders

(Pallister-Wilkins 2022) are understood in more than symbolic terms and contributions show that these are not only produced through elite geopolitical discourses or encounters between bodies, but are also emotionally charged through human and material entanglements. Thus, in her study of the Alpine border between France and Italy in the Susa Valley/Hautes-Alpes, Pallister-Wilkins (2022: 2) shows that the materiality of borders, as made by weather, forests, roads, tunnels, and other infrastructures, plays a crucial role in "the making of the border and migrant (im)mobility." Borders become intimate and embodied spaces filled with more than human entanglements that mark migrants as "out of place" (2022: 4).

An affective approach to the material world acknowledges and even takes as a premise that big categories of political geography are already constructed, embodied, and experienced. Feminist scholars have been at the forefront of these reflections, arguing not only that geopolitics are emotionally constructed and enacted every day, but also that geopolitical issues can be experienced intimately (Hall 2020; Klosterkamp 2023), whether directly through the body and bodily sensations or in intimate spaces usually considered outside of the realm of politics, such as the home. Research centered on "affective geopolitics," for instance, explores the intricate relationship between the material world and geopolitical discourses, practices, and atmospheres. Noteworthy contributions have demonstrated how everyday experiences constitute and embody geopolitical relations, shedding light on lived perceptions of security, peace, and conflicts entwined with material realities (Pain 2009; Sharp 2009; Fregonese 2017). Several ethnographic works examine the state and its material apparatus, encompassing controls, surveillance through drones, police measures, and gunfire. They unveil how the state, through a diverse array of material mechanisms, both regulates and instigates emotions such as anxiety and fear, celebrating enclosure (Mostafanezhad, Farnan, & Loong 2023) and extending even into the realm of everyday domestic life (Jackman & Brickell 2022). By managing this intricate interplay of materialities and human subjectivities, the state itself acquires an "affective" dimension (Laszczkowski & Reeves 2018).

The inquiries on the synergies between state formation and "the material" have not limited themselves to how the state operates but have extended to exploring the synergies between the material and nationalism. These show the often ambivalent relationship between affects, material, and political powers: while the material generates affects to reinforce a sense of belonging to a nation (Merriman & Jones 2017), political powers also mobilize material elements to shape national affective atmospheres (Brady 2021). Tracing how national sentiments and affects have gathered around two infrastructures in Wales, for example, Merriman and Jones (2017) show how national sentiments operate through the assemblage of particular configurations of bodies and materials, thus enabling the emergence of physical infrastructures.

The emotional dimension of politics and geopolitics has been once again highlighted by feminist geographers (Dixon 2015; Smith 2020; Sharp 2021), with nationalism as a core topic in several feminist works. Militz and Schurr (2016), for example, mobilised the term "affective nationalism" to show how bodies are not only sorted and arranged by political powers to orchestrate national sentiment, but actively participate in the creation of such sentiments through the encounter with objects. Gökariksel and Secor (2020) delve more deeply into the question of the materialization of geopolitical imaginaries, discussing how embodied and affective encounters between subjects and objects enable the materialization of symbolic yet highly political issues, foremost among them "the differences between self and other, 'us' and 'them'" (Gökariksel and Secor 2020: 1240). Following the suggestion Dixon (2015) makes for feminist geopolitics, urging to set aside the question of what the field "is" to instead contemplate what it "can do," these reflections on the interconnections among affect, materiality, and politics propose a novel avenue for political geography – one that reconceptualizes affects not merely as an approach to the material in political geography and geopolitics, but rather as constitutive elements of space within geopolitics, forming affective entanglements.

Assembling the Material

By acknowledging that the material is constitutive of worlds, objects, institutions, discourses, and practices, political geographers interested in the material interrogate how it constitutes the political and how it becomes political. Inspired by assemblage thinking (Deleuze & Guattari 1987), political geographers have examined heterogenous arrangements of human and more-than-human elements. Departing from debates on the definition of assemblages (see Anderson & McFarlane 2011), they investigate how more-than-human arrangements cohere to create actions. The material world is not only acted on but also generates the capacity to act. If assemblage thinking is productive for geographical thought in questioning how the material produces new realities (Müller & Schurr 2016), it has proven particularly inspiring for political geography in disassembling big political phenomena, expanding the range of political objects considered and, ultimately, questioning how material things come to matter politically.

Rallied behind the clarion call that "matter matters," political geographers have sought to disassemble big political things such as diplomacy (e.g., Dittmer 2015), democracy (e.g., Eadson & Van Veelen 2021), migration (e.g., Wiertz 2021), and sovereignty (e.g., Mason & Khawlie 2016). They argue for taking sociomaterial entanglements seriously in the (re)production of space and power beyond practices and discourses. Various engagements with the materiality of "territories" deconstruct this analytical object: for example, they argue, on the one hand, that the Earth is a dynamic material substrate with the capacity to shape (geo)political discourses in ways that can reconfigure the (geo)political stage itself (Depledge 2015; Usher 2019b); and, on the other hand, that "political concepts of sovereignty and territoriality" are referred to and produce the digital arrangements of different cloud systems (Baur 2023: 20). Studies of the materiality and territoriality of the digital crucially show how it intersects with space and power, in particular detailing how the military origins of the internet still shape its current technical and territorial configurations (Amoore 2018) and exploring how cloudfrastructures and algorithms open up new political spaces that are not reducible to their digital or material geographies (Amoore 2020; Furlong 2021). It has now become quite commonly accepted to think of such analytical objects as comprising elements drawn from a variety of sites in both space and time (Depledge 2015) or, to put it differently, that temporary sociomaterial assemblages cohere into territories "tying together orders, species, and heterogeneous qualities" (Deleuze & Guattari 1987: 330).

Notably, assemblage thinking has enabled important contributions to debates on the production of scale and on proximity and distance as effects of networked relations: "Scale is what actors achieve by *scaling*, *spacing*, and *contextualizing* each other through the transportation in some specific vehicles of some specific traces" (Latour 2005: 183–184, emphasis in original). With this theorizing of topological spaces of relations, one can observe a popular concern with tracing, identifying, and listing the elements that constitute a growing range of political "things." For example, studies flag the importance of novel worlds of interest such as digital assemblages (e.g., Kitchin 2014; Baur 2023), of analyzing understudied spaces such as airports, bunkers, and hotels (e.g., Adey 2008; Fregonese & Ramadan 2015; Garrett & Klinke 2019), or of tending to geonatural elements such as strata, water, and ice (e.g., Klinger 2018; Marston 2019; Childs 2020; Dodds 2021).

Overall, a search for empirical precision – through the disassembling of material things and the multiplication of the material things considered – has guided research concerns in political geography over the past decade. Some research has enabled the emergence of new subfields of political geography such as the "New Statecraft" (see Dittmer 2021b) that understands the state as performed, multiple, and emergent at a range of scales and temporalities. This literature frames the state as a relational space that depends on the encounters and connections between embodied performances and materials of the state – bureaucratic offices, paperwork, military equipment, and so on (e.g., Rech 2020) – ontologically flattening the actors, elements, and politics that constitute the state.

Political geographers have also called to move further beyond anthropocentric studies of the material world (Barua 2023; Ozguc & Burridge 2023). In his proposal for new ontologies of infrastructures, Barua (2021) argues that, while the material lives of infrastructure have been thoroughly examined, the same cannot be said for the relations between infrastructure and other-than-human lives. Studies seeking to decenter the anthropocentric focus on material things mobilize other theoretical frameworks, drawing for example on political ecology (Pallister-Wilkins 2022), more-than-human (Lobo & Hine 2023), and posthumanist geographies (Oliveras-González 2023) to further examine the agency of material assemblages. Indeed,

Xavier Oliveras-Gonzales's (2023) analysis of the Rio Bravo/Grande is not limited to how it is part of the more-than-human assemblage of objects, actions, and actors that cohere in the US–Mexican border. By showing how the river affects human bodies and consequently shapes border territories, he vividly illustrates how more-than-human actors can transform assemblages and produce new spatialities.

Attending to the material in political geography has key implications for thinking about the agency and the processual qualities of more-than-human assemblages. Studies show that political agency is distributed among both the human and the more-than-human elements of assemblages, offering a framework to analyze multiple realities, open-ended relations, and possibilities of becoming. "'Heterogeneity' is more than a catchword" (Ozguc & Burridge 2023: 476), as these assemblages have the potential to generate politics whether by extending state power (e.g., Meehan, Shaw, & Marston 2013), enforcing (im)mobility at borders (e.g., Squire 2015), or enabling civic participation (e.g., Lachmund 2022). The issue of the heterogeneity of political objects and phenomena is connected to the question of processes, through which assemblages are made, remade, and maintained. Andrew Barry's key study of the Baku–Tbilisi–Ceyhan oil pipeline in *Material Politics* (2013) demonstrated how a sociomaterial assemblage is mutable and assumes different forms, transforming what issues could be bound up with it, what could be known about it, and what could be contested.

By mobilizing the concept of *dispositif*, McGuirk and Dowling (2021) mapped the energy governance system in Sydney to first detail and describe this system and then analyze how processes and devices cohere to produce governance capacity around complex urban problems. These capacities to govern also cohere through political imaginations, prompting Hanna Knox to reflect on another key question: "Just how is it that materials become the grounds for politics?" (2017: 363). In her study of the 70-year struggle over the construction of the Iquitos–Nauta Road in Peru, Knox illustrates how politics are realized through engagements entailing the affective materialities that break through or rupture understandings of sociomaterial relations, wherein the grounds for future politics are remade. Amid the rapidly growing interest in the materiality of infrastructure (see Larkin 2013; Barry 2020), particular attention has been paid to the relationship between infrastructure and power, notably the entanglements of materiality and colonialism (Cowen 2020; Curley 2021; Mason 2023). Understanding material assemblages as processual is therefore important not only to inquire about what political objects and phenomena are made of but also to question how material things become political.

Making the Material Political

The material world can be considered through different theoretical lenses, which are not necessarily competing or exclusive. Scholarship about how politics is felt and circulates through bodies and materials (Ahmed 2004) echoes studies about the agency of more-than-human assemblages (Whatmore 2006). In turn, socio-material and affective assemblages are relational and processual: they tie together bodies, objects, and practices across spaces and temporalities, producing, for example, politics of belonging (Militz & Schurr 2016) and differentiation (Gökarıksel & Secor 2020; Pallister-Wilkins 2022). From how we feel and assemble the material emerges the question of its production, inviting us to think more closely about the articulations between material things and the production of knowledge, discourses, territories, and practices. This, in turn, requires examining how the material comes to matter politically.

Literature in the fields of actor-network theory (ANT) and science and technology studies (STS) has paved the way for analyzing how materials, structures, and knowledge co-produce networks in which agency is a distributed achievement, emerging from associations between human and nonhuman entities (the actor-network). Starting from the notion that social and technical realms are enmeshed, the material can also be conceived in terms of sociotechnical systems shaped by interactions between human and nonhuman elements (Law 2004; Latour 2005). This perspective opens up investigations of how material things become controversial; that is, "matters of concern" (Latour 2004: 225). Rather than representing matters of "fact," they are matters of concern as they can create, catalyze, and maintain human and nonhuman associations and networks.

Drawing on ANT and STS theoretical lineages, articulations between the material, knowledge production, and decision-making have focused on sociotechnical assemblages (de Landa 2006). In this context, the proliferation of technologies has attracted scholarly attention and generated studies about how the digital reconfigures everyday spaces (Cheney-Lippold 2011; Walker & Winders 2021), politics (Wolfe 2021), sovereignty (Amoore 2018), and infrastructure (Bouzarovski, Bradshaw, & Wochnik 2015). These studies converge around how human and nonhuman entities cohere in structures that categorize, organize, and enable different realities. Dorota Golańska (2023) notably shows how the science-based politics of drawing has been mobilized for colonial policy in the Naqab/Negev Desert.

Technologies, and sociotechnical assemblages more generally, can guide decision-making, requiring scrutiny of the politics underlying their workings. For example, in the case of data and border policies, it is necessary to research how algorithms produce "regimes of proof" (Sriraman 2018) and their ensuing recommendations for action, such as stopping and further questioning someone. Knowledge is produced through the analysis of data, but it then dissolves into opaque processes that are distributed between humans and machines. This makes it "increasingly difficult to create accountability" and understand how authority is (co-)produced (Leese, Noori, & Scheel 2022: 13). Tending to the growing complexity and speed of sociotechnical assemblages has become important in the literature to understand how these produce political realities and to open spaces for their possible contestation or subversion.

On the other hand, it is equally necessary to explore how political forces – whether institutions or state actors – mobilize the material to create and reinforce knowledge production. First, different understandings and mobilizations of the material produce and orient scientific framings and concepts at the heart of the development of disciplinary subfields (Braun 2000; McGowran & Donovan 2021; Bobbette 2023). Yusoff (2021), for example, discusses how the materiality of geological confluences linked to colonial extractivism, and how the related geography of racialization and ecological transformation contributes to the construction of the humanities and social sciences. Second, the production and resulting articulations of material things and knowledge shape political practices, the art of governing, and the constitution of sovereignty in often unexpected ways. Such political practices take multiple forms, rely on different objects – technologies, technocratic devices, resources (Koch & Perreault 2019; Lin 2022; Golańska 2023) – and work through different processes by visibilizing and invisibilizing people, objects, and causes (Fleischmann 2023; Minca 2023). If we take the example of resources, Koch and Perreault (2019) highlight how resources and their materiality are inherently political. The controversies, struggles, and clashes of discourses over their management encompass wider political concerns directly linked to national identity, citizenship, and political economy: "Far from being merely incidental features of national identity, resources are fundamentally constitutive of the material and ideological nature of nations and states: the material basis for state power" (Koch & Perreault 2019: 616). Resources are thus visibilized and invisibilized to create narratives and build a story of national power or to capture spaces fallen beyond the gaps of centralized power, with devastating damage to lives and the environment (Klinger 2018).

Finally, discourses, narratives, and story-telling become crucial in the making and dismantling of political questions (Haughton & McManus 2022; Lobo & Hine 2023). While much emphasis is placed on how different actors mobilize material things in often conflictive ways to create political realities, less consideration is given to how the material can also shape and contribute to advancing political struggles. Lobo and Hine delve into this question, demonstrating the power of geological strata in sparking and directing political controversies. In their research, it is the materiality of coal itself that seeds "more-than-human framings of justice" and fuels controversies over energy futures in Australia and India, "opening up geopolitical possibilities for more-than-human climate justice" (Lobo & Hine 2023: 490).

Future Paths

If the swirl of new publications since the first edition of this chapter appeared (Müller 2015) is any indication, material geographies will continue to be a major influence in political geography. This field of research has already opened up novel lines of inquiry and shone new light on established ones. This last section sketches in broad strokes three issues that have been neglected in the field so far and that deserve attention to advance the debate. In so doing, it suggests connections with animal and plant geographies, the discipline of IR, and Indigenous geographies.

Toward Intermaterialism

For the time being, most research examines the material world in relation to humans. It focuses on what the material world does to humans, as well as what it allows them to do or prevents them from doing. Although animal geographies and vegetal geographies are growing fields (Gibbs 2020; Barua 2023), the anthropocentric focus makes humans still central in research. This is also evident in the terminology used, which identifies the material world in relation to humans as more-than-human, other-than-human, nonhuman, and so on. But what about accounts that place relations among the material world front and center in what could be called intermaterialism? One such account is Maan Barua's (2021) study of how infrastructure and other-than-human life interact. Taking inspiration from animal geographies, Barua shows how infrastructure is not just crucial for making animals circulate, for example when using roads, but has also become a medium of nonhuman life, such as for rats living in sewers. For political geography, attention to intermaterialism would entail increased attention to how materials relate not just to humans but among each other.

Toward a More Refined Grammar of the Material

The "material world" is a sweeping category. It encompasses animate materials such as human bodies, animals, and plants; human-made technologies such as technical devices and software; and inanimate – but far from inert – materials such as water, air, and rock. Future research could push toward more nuanced theorizing of the distinctive capacities of these diverse materials to arrive at a more refined grammar for material political geographies. Political capacities and potentials are distributed unevenly across different materials and work through different channels. Inspiration for developing a more refined grammar of the material could come from two edited volumes in IR, which consider the differential ways in which things make relations international (Salter 2015, 2016). Imitating assemblages in their content, the volumes discuss very different, heterogenous objects ranging from corpses to tanks, and from containers to http cookies, showing how these can and must work together to produce political effects, but how each object does so in a different way. The recent emergence of distinct subfields of material geographies, such as political geology or vegetal geographies, is a step toward this more nuanced appreciation of differential agencies and power.

Toward Material Ethics

The contributions discussed in this chapter hint in many ways at two important questions: How do our practices of researching and theorizing the material have political underpinnings? How can the knowledge that we produce participate in reproducing or changing structures of power? A disciplinary engagement with the material invites us not to stop at the empirical sensitivity to the material, but to include reflections on the ethics of material political geographies. In contrast to research with human participants (e.g., Henn et al. 2021), the ethics of research on and with materials, in particular inanimate materials, is less well developed. But if political geography considers materials as lively and agentic, and humans as performatively entangled, this requires a reconsideration of the ethical stance with which materials are approached. Indigenous geographies, with their attentiveness to Indigenous ethics of respect toward other-than-human beings, offer a generative potential here. Rosiek, Snyder, and Pratt (2020), for example, draw inspiration from Indigenous cosmologies to suggest a practice of "ethical reciprocity" in which humans attend to their interdependence with a network of relations of other-than-humans.

The big task ahead for material political geographies is perhaps not so much to push forward, for much has already been achieved there. Rather, it would be well advised to look around and look back. Where can connections with other subfields of geography, other disciplines, or theoretical paradigms be made? What has political geography left behind in the euphoria of "becoming material" that could be useful along the more-than-human journey?

References

Adey, P. (2008). Architectural geographies of the airport balcony: Mobility, sensation and the theatre of flight. *Geografiska Annaler: Series B, Human Geography* 90(1): 29–47. doi:10.1111/j.1468-0467.2008.00274.x

Ahmed, S. (2004). *The Cultural Politics of Emotion*, 2nd edn. Edinburgh: Edinburgh University Press.

Amoore, L. (2018). Cloud geographies: Computing, data, sovereignty. *Progress in Human Geography* 42(1): 4–24. doi:10.1177/0309132516662147

Amoore, L. (2020). *Cloud Ethics: Algorithms and the Attributes of Ourselves and Others.* Durham, NC: Duke University Press.

Anderson, B., & Harrison, P. (2006). Questioning affect and emotion. *Area* 38(3): 333–335.

Anderson, B., & McFarlane, C. (2011). Assemblage and geography. *Area* 43(2): 124–127. doi:10.1111/j.1475-4762.2011.01004.x

Anderson, K., & Smith, S.J. (2001). Editorial: Emotional geographies. *Transactions of the Institute of British Geographers* 26(1): 7–10. doi:10.1111/1475-5661.00002

Barad, K. (2007). *Meeting the Universe Halfway: Quantum Physics and the Entanglement of Matter and Meaning*. Durham, NC: Duke University Press.

Barry, A. (2013). *Material Politics: Disputes along the Pipeline*. Oxford: Blackwell.

Barry, A. (2020). The material politics of infrastructure. In Maasen, S., Dickel, S., & Schneider, C. (eds), *TechnoScienceSociety: Technological Reconfigurations of Science and Society*. Cham: Springer International, 91–109. doi:10.1007/978-3-030-43965-1_6

Barry, A., & Gambino, E. (2020). Pipeline geopolitics: Subaquatic materials and the tactical point. *Geopolitics 25*(1): 109–142. doi:10.1080/14650045.2019.1570921

Barua, M. (2021). Infrastructure and non-human life: A wider ontology. *Progress in Human Geography 45*(6). doi:10.1177/0309132521991220

Barua, M. (2023). Plantationocene: A vegetal geography. *Annals of the American Association of Geographers 113*(1): 13–29. doi:10.1080/24694452.2022.2094326

Baur, A. (2023). European dreams of the cloud: Imagining innovation and political control. *Geopolitics 29*(3): 796–820. doi:10.1080/14650045.2022.2151902

Bennett, J. (2010). *Vibrant Matter: A Political Ecology of Things*. Durham, NC: Duke University Press.

Bobbette, A. (2023). *The Pulse of the Earth: Political Geology in Java*. Durham, NC: Duke University Press.

Bouzarovski, S., Bradshaw, M., & Wochnik, A. (2015). Making territory through infrastructure: The governance of natural gas transit in Europe. *Geoforum 64*: 217–228. doi:10.1016/j.geoforum.2015.06.022

Brady, D. (2021). Between nation and state: Boundary infrastructures, communities of practice and everyday nation-ness in the Chinese rail system. *Environment and Planning C: Politics and Spaces 39*(7): 1436–1452. doi:10.1177/2399654420986810

Braun, B.P. (2000). Producing vertical territory: Geology and governmentality in late Victorian Canada. *Geography, Environment and Society 7*(1): 7–46.

Cheney-Lippold, J. (2011). A new algorithmic identity: Soft biopolitics and the modulation of control. *Theory, Culture and Society 28*(6): 164–181. doi:10.1177/0263276411424420

Childs, J. (2020). Extraction in four dimensions: Time, space and the emerging geo(−) politics of deep-sea mining. *Geopolitics 25*(1): 189–213. doi:10.1080/14650045.2018.1465041

Colls, R. (2012). Feminism, bodily difference and non-representational geographies. *Transactions of the Institute of British Geographers 37*(3): 430–445. doi:10.1111/j.1475-5661.2011.00477.x

Cowen, D. (2020). Following the infrastructures of empire: Notes on cities, settler colonialism, and method. *Urban Geography 41*(4): 469–486. doi:10.1080/02723638.2019.1677990

Curley, A. (2021). Infrastructures as colonial beachheads: The central Arizona project and the taking of Navajo resources. *Environment and Planning D: Society and Space 39*(3): 387–404. doi:10.1177/0263775821991537

Davidson, J., Bondi, L., & Smith, M. (eds) (2016). *Emotional Geographies*. London: Routledge. doi:10.4324/9781315579245

de Certeau, M. (1984). *The Practice of Everyday Life*. Berkeley, CA: University of California Press.

de Landa, M. (2006). *A New Philosophy of Society: Assemblage Theory and Social Complexity*. London: Continuum.

Degen, M., Rose, G., & Basdas, B. (2010). Bodies and everyday practices in designed urban environments. *Science & Technology Studies 23*(2): 60–76. doi:10.23987/sts.55253

Deleuze, G. (1988). *Spinoza: A Practical Philosophy*. San Francisco, CA: City Lights.

Deleuze, G., & Guattari, F. (1987). *A Thousand Plateaus: Capitalism and Schizophrenia*. Minneapolis, MN: University of Minnesota Press.

Depledge, D. (2015). Geopolitical material: Assemblages of geopower and the constitution of the geopolitical stage. *Political Geography 45*: 91–92. doi:10.1016/j.polgeo.2013.05.001

Dittmer, J. (2015). Everyday diplomacy: UKUSA intelligence cooperation and geopolitical assemblages. *Annals of the Association of American Geographers 105*(3): 604–619. doi: 10.1080/00045608.2015.1015098

Dittmer, J. (2021a). The state, all at sea: Interoperability and the global network of navies. *Environment and Planning C: Politics and Spaces 39*(7): 1389–1406. doi:10.1177/23996 54418812469

Dittmer, J. (2021b). The state of this: Introduction to the special issue. *Environment and Planning C: Politics and Spaces 39*(7): 1313–1318. doi:10.1177/23996544211042380

Dixon, D.P. (2015). *Feminist Geopolitics: Material States*. Burlington, VT: Ashgate.

Dodds, K. (2021). Geopolitics and ice humanities: Elemental, metaphorical and volumetric reverberations. *Geopolitics 26*(4): 1121–1149. doi:10.1080/14650045.2019.1697240

Eadson, W., & Van Veelen, B. (2021). Assemblage-democracy: Reconceptualising democracy through material resource governance. *Political Geography 88*: 102403. doi:10.1016/j.polgeo.2021.102403

Fleischmann, L. (2023). More-than-human political geographies: Abjection and sovereign power. *Political Geography 107*: 102949. doi:10.1016/j.polgeo.2023.102949

Fregonese, S. (2017). Affective atmospheres, urban geopolitics and conflict (de)escalation in Beirut. *Political Geography 61*: 1–10. doi:10.1016/j.polgeo.2017.04.009

Fregonese, S., & Ramadan, A. (2015). Hotel geopolitics: A research agenda. *Geopolitics 20*(4): 793–813. doi:10.1080/14650045.2015.1062755

Furlong, K. (2021). Geographies of infrastructure II: Concrete, cloud and layered (in)visibilities. *Progress in Human Geography 45*(1): 190–198. doi:10.1177/0309132520923098

Garrett, B., & Klinke, I. (2019). Opening the bunker: Function, materiality, temporality. *Environment and Planning C: Politics and Spaces 37*(6): 1063–1081. doi:10.1177/2399654418816316

Gibbs, L.M. (2020). Animal geographies I: Hearing the cry and extending beyond. *Progress in Human Geography 44*(4): 769–777. doi:10.1177/0309132519863483

Gökarıksel, B., & Secor, A.J. (2020). Affective geopolitics: Anxiety, pain, and ethics in the encounter with Syrian refugees in Turkey. *Environment and Planning C: Politics and Spaces 38*(7–8): 1237–1255. doi:10.1177/2399654418814257

Golańska, D. (2023). Technologies of drawing (On) a desert: Reading land art and colonial politics in the Naqab/Negev diffractively. *Geopolitics 29*(1): 230–260. doi:10.1080/1465 0045.2022.2160710

Hall, S.M. (2020). The personal is political: Feminist geographies of/in austerity. *Geoforum 110*: 242–251. doi:10.1016/j.geoforum.2018.04.010

Haraway, D. (2008). Encounters with companion species: Entangling dogs, baboons, philosophers, and biologists. *Configurations 14*(1): 97–114. doi:10.1353/con.0.0002

Haraway, D.J. (1991). *Simians, Cyborgs, and Women: The Reinvention of Nature*. New York: Routledge.

Haughton, G., & McManus, P. (2022). Becoming Westconnex – Becoming Sydney: Object-oriented politics, contested storylines and the multi-scalar imaginaries of building a motorway network in Sydney, Australia. *Environment and Planning C: Politics and Spaces 40*(4): 913–932. doi:10.1177/23996544211050941

Henn, S., Miggelbrink, J., & Hörschelmann, K. (2021). *Research Ethics in Human Geography*. London: Routledge.

Ho, E.L.-E. (2023). Social geography III: Emotions and affective spatialities. *Progress in Human Geography 48*(1): 030913252311741. doi:10.1177/03091325231174191

Holbraad, M., & Pedersen, M.A. (2017). *The Ontological Turn: An Anthropological Exposition*. Cambridge: Cambridge University Press.

Jackman, A., & Brickell, K. (2022). "Everyday droning": Towards a feminist geopolitics of the drone-home. *Progress in Human Geography 46*(1): 156–178. doi:10.1177/0309132 5211018745

Jacobs, J.M. (2006). A geography of big things. *Cultural Geographies 13*(1): 1–27. doi:10. 1191/1474474006eu354oa

Jones, A., & Clark, J. (2019). Performance, emotions, and diplomacy in the United Nations assemblage in New York. *Annals of the American Association of Geographers 109*(4): 1262–1278. doi:10.1080/24694452.2018.1509689

Kitchin, R. (2014). Big data, new epistemologies and paradigm shifts. *Big Data & Society 1*(1): 205395171452848. doi:10.1177/2053951714528481

Klinger, J.M. (2018). *Rare Earth Frontiers. From Terrestrial Subsoils to Lunar Landscapes*. Ithaca, NY: Cornell University Press.

Klosterkamp, S. (2023). Unpacking the "global" and the "intimate" of anti-terrorism trials. *Gender, Place and Culture 30*(11): 1638–1642. doi:10.1080/0966369X.2022.2064835

Knox, H. (2017). Affective infrastructures and the political imagination. *Public Culture 29*(2): 363–384. doi:10.1215/08992363-3749105

Koch, N., & Perreault, T. (2019). Resource nationalism. *Progress in Human Geography 43*(4). doi:10.1177/0309132518781497

Kohn, E. (2013). *How Forests Think: Toward an Anthropology Beyond the Human*. Berkeley, CA: University of California Press.

Lachmund, J. (2022). Stewardship practice and the performance of citizenship: Greening tree-pits in the streets of Berlin. *Environment and Planning C: Politics and Spaces 40*(6): 1290–1306. doi:10.1177/23996544211070204

Larkin, B. (2013). The politics and poetics of infrastructure. *Annual Review of Anthropology 42*(1): 327–343. doi:10.1146/annurev-anthro-092412-155522

Laszczkowski, M., & Reeves, M. (eds) (2018). *Affective States: Entanglements, Suspensions, Suspicions*. New York: Berghahn.

Latour, B. (2004). Why has critique run out of steam? From matters of fact to matters of concern. *Critical Inquiry 30*(2): 225–248. doi:10.1086/421123

Latour, B. (2005). *Reassembling the Social*. Oxford: Oxford University Press.

Law, J. (2004). *After Method: Mess in Social Science Research*. London: Routledge.

Leese, M., Noori, S., & Scheel, S. (2022). Data matters: The politics and practices of digital border and migration management. *Geopolitics 27*(1): 5–25. doi:10.1080/14650045.20 21.1940538

Lin, W. (2022). Atmospheric conditioning: Airport automation, labour and the COVID-19 pandemic. *Transactions of the Institute of British Geographers 47*(1): 214–228. doi:10.1111/tran.12499

Lobo, M., & Hine, A. (2023). Storying strata: Entangling coal and energy futures in Australia and India. *Geopolitics 28*(2): 490–512. doi:10.1080/14650045.2022.2040485

Lorimer, H. (2008). Cultural geography: Non-representational conditions and concerns. *Progress in Human Geography 32*(4): 551–559. doi:10.1177/0309132507086882

Marston, A. (2019). Strata of the state: Resource nationalism and vertical territory in Bolivia. *Political Geography 74*: 102040. doi:10.1016/j.polgeo.2019.102040

Mason, M., & Khawlie, M. (2016). Fluid sovereignty: State–nature relations in the Hasbani Basin, Southern Lebanon. *Annals of the American Association of Geographers 106*(6): 1344–1359. doi:10.1080/24694452.2016.1213155

Mason, O. (2023). The geographies of colonial infrastructures: Mobility, im/materiality, and politics on walking trails in the Middle East. *Transactions of the Institute of British Geographers 48*(3): 506–524. doi:10.1111/tran.12599

Massumi, B. (2002). *Parables for the Virtual: Movement, Affect, Sensation*. Durham, NC: Duke University Press. doi:10.1515/9780822383574

McCormack, D.P. (2003). An event of geographical ethics in spaces of affect. *Transactions of the Institute of British Geographers 28*(4): 488–507. doi:10.1111/j.0020-2754.2003.00106.x

McGowran, P., & Donovan, A. (2021). Assemblage theory and disaster risk management. *Progress in Human Geography 45*(6): 1601–1624. doi:10.1177/03091325211003328

McGuirk, P., & Dowling, R. (2021). Urban governance dispositifs: Cohering diverse ecologies of urban energy governance. *Environment and Planning C: Politics and Spaces 39*(4): 759–780. doi:10.1177/2399654420957329

Meehan, K., Shaw, I.G.R., & Marston, S.A. (2013). Political geographies of the object. *Political Geography 33*: 1–10. doi:10.1016/j.polgeo.2012.11.002

Merleau-Ponty, M. (1962). *The Phenomenology of Perception*. London: Routledge.

Merriman, P., & Jones, R. (2017). Nations, materialities and affects. *Progress in Human Geography 41*(5). doi:10.1177/0309132516649453

Militz, E., & Schurr, C. (2016). Affective nationalism: Banalities of belonging in Azerbaijan. *Political Geography 54*: 54–63. doi:10.1016/j.polgeo.2015.11.002

Miller, J.C., & Laketa, S. (2019). The "magic of the mall" revisited: Malls and the embodied politics of life. *Progress in Human Geography 43*(5): 910–926. doi:10.1177/0309132518794274

Minca, C. (2023). Of werewolves, jungles, and refugees: More-than-human figures along the Balkan route. *Geopolitics 28*(2): 550–569. doi:10.1080/14650045.2021.1931840

Mostafanezhad, M., Farnan, R.A., & Loong, S. (2023). Sovereign anxiety in Myanmar: An emotional geopolitics of China's belt and road initiative. *Transactions of the Institute of British Geographers 48*(1): 132–148. doi:10.1111/tran.12571

Müller, M. (2015). More-than-representational political geographies. In Agnew, J., Mamadouh, V., Secor, A.J., & Sharp, J. (eds), *The Wiley Blackwell Companion to Political Geography*. Chichester: John Wiley & Sons., 407–423. doi:10.1002/9781118725771.ch30

Müller, M., & Schurr, C. (2016). Assemblage thinking and actor-network theory: Conjunctions, disjunctions, cross-fertilisations. *Transactions of the Institute of British Geographers 41*(3): 217–229. doi:10.1111/tran.12117

Oliveras-González, X. (2023). Beyond natural borders and social bordering: The political agency of the lower Rio Bravo/Grande. *Geopolitics 28*(2): 533–549. doi:10.1080/14650045.2021.2016706

Ozguc, U., & Burridge, A. (2023). More-than-human borders: A new research agenda for posthuman conversations in border studies. *Geopolitics 28*(2): 471–489. doi:10.1080/14650045.2023.2169879

Pain, R. (2009). Globalized fear? Towards an emotional geopolitics. *Progress in Human Geography 33*(4): 466–486. doi:10.1177/0309132508104994

Pallister-Wilkins, P. (2022). Whitescapes: A posthumanist political ecology of alpine migrant (im)mobility. *Political Geography 92*: 102517. doi:10.1016/j.polgeo.2021.102517

Pile, S. (2010). Emotions and affect in recent human geography: Emotions and affect in recent human geography. *Transactions of the Institute of British Geographers 35*(1): 5–20. doi:10.1111/j.1475-5661.2009.00368.x

Rech, M.F. (2020). Ephemera(l) geopolitics: The material cultures of British military recruitment. *Geopolitics 25*(5): 1075–1098. doi:10.1080/14650045.2019.1570920

Rosiek, J.L., Snyder, J., & Pratt, S.L. (2020). The new materialisms and indigenous theories of non-human agency: Making the case for respectful anti-colonial engagement. *Qualitative Inquiry* 26(3–4): 331–346. doi:10.1177/1077800419830135

Salter, M.B. (ed.) (2015). *Making Things International 1: Circuits and Motion.* Minneapolis, MN: University of Minnesota Press.

Salter, M.B. (ed.) (2016). *Making Things International 2: Catalysts and Reactions.* Minneapolis, MN: University of Minnesota Press.

Sedgwick, E.K., & Frank, A. (2003). *Touching Feeling: Affect, Pedagogy, Performativity.* Durham, NC: Duke University Press.

Sharp, J. (2021). Materials, forensics and feminist geopolitics. *Progress in Human Geography 45*(5): 990–1002. doi:10.1177/0309132520905653

Sharp, J. (2009). Geography and gender: What belongs to feminist geography? Emotion, power and change. *Progress in Human Geography 33*(1): 74–80. doi:10.1177/0309132508090440

Smith, S. (2020). *Intimate Geopolitics: Love, Territory, and the Future on India's Northern Threshold.* New Brunswick, NJ: Rutgers University Press.

Spinoza, B. (1677). *Ethica, Ordine Geometrica Demonstrata.* Mimeo.

Squire, V. (2015). *Post/Humanitarian Border Politics between Mexico and the US.* London: Palgrave Macmillan. doi:10.1057/9781137395894

Sriraman, T. (2018). *In Pursuit of Proof: A History of Identification Documents in India.* Oxford: Oxford University Press.

Stewart, K. (2007). *Ordinary Affects.* Durham, NC: Duke University Press.

Thien, D. (2005). After or beyond feeling? A consideration of affect and emotion in geography. *Area 37*(4): 450–454.

Thrift, N. (2004). Intensities of feeling: Towards a spatial politics of affect. *Geografiska Annaler: Series B, Human Geography 86*(1): 57–78. doi:10.1111/j.0435-3684.2004.00154.x

Tolia-Kelly, D.P. (2006). Affect: An ethnocentric encounter? Exploring the "universalist" imperative of emotional/affectual geographies. *Area 38*(2): 213–217.

Tsing, A.L., Bubandt, N., & Gan, E. (2017). *Arts of Living on a Damaged Planet: Ghosts and Monsters of the Anthropocene.* Minneapolis, MN: University of Minnesota Press.

Usher, M. (2019a). Desali-nation: Techno-diplomacy and hydraulic state restructuring through reverse osmosis membranes in Singapore. *Transactions of the Institute of British Geographers 44*(1): 110–124. doi:10.1111/tran.12256

Usher, M. (2019b). Territory incognita. *Progress in Human Geography 44*(6). doi:10.1177/0309132519879492

Walker, M., & Winders, J. (2021). Where is artificial intelligence? Geographies, ethics, and practices of AI. *Space and Polity 25*(2): 163–166. doi:10.1080/13562576.2021.1985869

Whatmore, S. (2006). Materialist returns: Practising cultural geography in and for a more-than-human world. *Cultural Geographies 13*(4): 600–609. doi:10.1191/1474474006cgj377oa

Wiertz, T. (2021). Biopolitics of migration: An assemblage approach. *Environment and Planning C: Politics and Spaces 39*(7): 1375–1388. doi:10.1177/2399654420941854

Wolfe, S.D. (2021). Blogging the virtual: New geographies of domination and resistance in and beyond Russia. *Antipode 53*(4): 1251–1269. doi:10.1111/anti.12709

Yusoff, K. (2021). The inhumanities. *Annals of the American Association of Geographers 111*(3): 663–676. doi:10.1080/24694452.2020.1814688

Chapter 30

Popular Culture

Silvia Binenti[1] *and Jason Dittmer*[2]

[1]*Department of Conflict and Development Studies, Ghent University, Ghent, Belgium*

[2]*Department of Geography, University College London, London, UK*

What is popular culture? The term itself is contested, and differences over its meaning can be seen to ripple through political and cultural debates over centuries (Dittmer & Bos, 2019). In the term's origination it was meant to contrast with "high culture," or the culture of the elite classes (for instance opera). Popular culture, in this formulation, is the entertainment of the working classes (for instance blues music, or more contemporarily rock and roll or hip-hop). However, another iteration of popular culture can be seen in a contrast with folk culture or "local" traditions. In this iteration, popular culture can be equated with modernity and mass media (for instance Hollywood films, K-pop music). It is this latter formulation that has been the focus of most work in political geography, as the potentially global reach of this type of popular culture has made it seem more salient to scholars. Currently, the study of popular culture in political geography is metamorphosing into a focus on a third definition of popular culture, which draws from both of the preceding. Rather than drawing on an "other" (i.e., high culture, folk culture), this vision of popular culture is one that focuses on the everyday experiences and doings of regular people, and how those experiences and doings might be political.

Popular culture has largely featured in political geography through its examination in the subfield of popular geopolitics. After reviewing the foundation of popular geopolitics, the chapter examines its coevolution with conceptual trends in human geography and the broader social sciences that bring into view a greater appreciation for the ways in which geopolitical agency is produced in/through assemblages consisting of media, objects, and other elements of everyday life. We

The Wiley Blackwell Companion to Political Geography, Second Edition.
Edited by Virginie Mamadouh, Natalie Koch, Chih Yuan Woon, and John Agnew.

then consider two areas that can be pushed further in research on the political geographies of popular culture – consumerism and humor, exemplars of everyday popular culture that can bring more clarity to future trends in the field.

Popular Geopolitics

Political geographers saw the opportunity for new approaches to the study of geopolitical processes with the emergence of the "new cultural geography" in the 1990s (see Dalby 1991; Agnew 1994; Ó Tuathail 1996). The foundation of "critical geopolitics" was laid, with the aim of disrupting hegemonic and imperial practices of statecraft, and a focus on the discursive processes that produced the geographies underpinning various foreign policies. Critical geopolitics has typically been described as heuristically divided into formal geopolitical discourse (the abstract discourse of theorists), practical geopolitical discourse (the official discourse of politicians and the state), and popular geopolitics (geopolitical discourse found in various media texts). The earliest efforts here remained focused on the elite texts of formal and practical geopolitics, with popular geopolitics not seen as worth studying. For example, Ó Tuathail and Agnew (1992: 194) argue that "Our second thesis is that most geopolitical reasoning in world politics is of a practical and not a formal type." At that time, the idea that we might all be marinating in geopolitical discourse of a more popular variety was still considered to be too far from the world of foreign relations.

However, by the early 2000s, political geographers started becoming more interested in everyday life as a site and source of politics, and that included practices of cultural consumption that had previously not featured in political geography. In this context, the field of popular geopolitics emerged from the encounter between critical geopolitics and cultural studies. Its pioneers were paying attention to how geopolitical representations suffused everyday life, shaping consumers' views of foreign affairs. For example, Joanne Sharp (1993, 2000b) studied the way *Reader's Digest*, a US general-interest magazine, contributed to geopolitical imaginaries of the Cold War. Klaus Dodds (2003, 2005, 2006) explored how James Bond audiences were invited to frame geopolitical understandings of post-imperial Britain. Jason Dittmer (2005a, 2007) followed with an analysis of the evolution of the superhero genre and the way the comic book *Captain America* contributed to the configuration of the US national identity and the country's symbolic place in the world map. This body of work asserted that the ways in which place and space were represented in media contributed to the common-sense understandings of foreign policy and the legitimation of violence, even if their conceptual focus in this early work still tended to fall on propagandistic media imparted from above to largely passive consumers (e.g., Dittmer 2005b).

Since then, scholars in popular geopolitics have come to meaningfully engage with the mundane places, practices, and discourses through which geopolitical imaginaries, communities, and identities are formed and performed at the level of the everyday. While still remaining a relatively small area of research, interdisciplinary studies in popular geopolitics flourished in the past two decades to the point that Sidaway (2022: 1622) defined the subdiscipline as "a success story." Scholars have analyzed films (Power & Crampton 2005; Carter & McCormack 2006; Der Derian 2009; Carver 2010; Dodds 2014; Kirby 2015, 2020; Saunders 2019), comic books (Dunnett 2009), television series (Buzan 2010; Glynn & Cupples 2015) including animation (Thorogood 2016), news media outlets (McFarlane & Hay 2003; Debrix 2008; Bernazzoli 2010), social media (Pinkerton & Benwell 2014), radio shows (Pinkerton & Dodds 2009), toys (Carter, Kirby, & Woodyer 2016; Woodyer & Carter 2020), video games (Power 2007; Shaw 2010; Salter 2011; Bos 2018), political cartoons (Dodds 2007, 2010; Holland 2012), and stand-up comedy (Purcell, Brown, & Gokmen 2010). Themes range from issues of gender (Emad 2006; Shepherd 2013; Funnell & Dodds 2017) and nationalism (Dittmer & Larsen 2011) to processes of securitization (MacDonald, Hughes, & Dodds 2010; Dodds 2011) and cultural othering (Semmerling 2006; Dunnett 2009; AlAwadhi & Dittmer 2022). However, it has been argued that the empirical examples scholars have tended to focus on have been media artifacts, be it in screen or textual forms (Kiersey & Neumann 2015), rather than examination of the everyday cultural doings of regular people.

This focus has perpetuated (or resulted from) the ongoing anxiety of scholars around studying popular culture. While high culture is seen as innately important, the fear of being seen to study something unimportant has led scholars to focus on mass media, which can be justified with reference to its extensive reach (Dittmer 2015a). Thus the focus on mass media has maintained geopolitics' traditional focus on elites and the public sphere. Even for popular representations on "screen," Glynn and Cupples (2015) argue that television, which arguably represents a more "everyday" medium of popular geopolitics than blockbuster movies, has been overlooked due to its association with the domestic sphere. As Dittmer and Gray noted in 2010 (1664), this focus resulted in "a popular geopolitics that has been somewhat inexplicably still focused on the elite visions of media moguls, movie directors, and lower-level yet still relatively empowered media functionaries like writers and reporters."

Popular Geopolitics 2.0

By moving away from deconstructions of media artifacts, Dittmer and Gray (2010) have called on scholars to adopt a research agenda that engages with the performative, embodied, and affective dimensions that shape experiences of the geopolitical in everyday life. One way to conceptualize Popular Geopolitics 2.0 is to say that it takes the focus from elite production of media and their discourses and instead foregrounds the way everyday life is suffused by relations that enmesh everyday people in assemblages that are geopolitical in nature. This is now the third definition of popular culture offered in the introduction to this chapter, where popular culture is the experiences and doings of everyday people, through which they both are subjectified and also exert their own imprint on the broader assemblage.

Popular Geopolitics 2.0 shifts the frame of analysis from blockbuster production (that is, elite practices) to everyday practices of mundane consumption that bring to the fore viewers, fans, readers, consumers, players, and users: "By focusing on the interaction of the material body with the discursive world, phenomenological accounts of popular geopolitics could provide a resolutely 'grounded but translocal' perspective of what everyday geopolitical life is, whether linked to the media or otherwise" (Dittmer & Gray 2010: 1667). Indeed, the so-called Popular Geopolitics 2.0 was an attempt to bring together work in feminist geography, nonrepresentational theory, and audience studies. Feminist accounts did and can still further help counteract the inherent elitism and masculinity in scholarship of popular geopolitics (for a critique, see Sharp 2000a) and question boundaries between the private and the public (Koopman 2011) or the international and the personal (Dowler & Sharp 2001). Building on this scholarship, feminist geographers in critical geopolitics have raised important questions about intimate geopolitics (the intersection of geopolitics with reproduction, intimate partnering, and the family; see Smith 2009, 2012; Pain & Staeheli 2014; Pain 2015; Christian, Dowler, & Cuomo 2016; Brickell & Cuomo 2019; Culcasi 2019). Brickell (2012), for example, reflects on how the space of the home is not just affected by geopolitics but also the other way around. Along this line of thought, in a special issue edited by Carter and Woodyer (2020), contributing authors suggest reframing future research around the domestication of geopolitics. Woodyer and Carter (2020) further argue that war toys and children's play, as a significant but neglected mode of the geopolitical, can represent a fruitful lens to explore everyday geopolitics.

Nonrepresentational theory contributed an emphasis on embodiment and affect that had been lacking in the textual deconstruction of early popular geopolitics. By giving voice to how popular culture made people *feel*, and how it produced somatic markers that informed subsequent decision-making at a precognitive time scale, nonrepresentational theory opened new modes of analysis. Carter and McCormack (2006), in their analysis of *Black Hawk Down* (2002), focused on the cultivation of affects by film directors, drawing on the language of film studies to understand the relationship between cinematographic tactics and techniques and how they resonate with "real-world" events and conflicts. However, scholars soon took the concept of affect and everyday life beyond the context of media–audience assemblages, instead focusing on geopolitical encounters in the city, or other geopolitical relations that skip popular media entirely (e.g., Laketa 2016).

From the field of audience studies, scholars of popular geopolitics have drawn the essential insight that audiences are crucial to practices of meaning-making with regard to texts. Klaus Dodds (2006) produced the first effort to look at audiences in popular geopolitics by using Internet Movie Database (IMDb) reviews of

James Bond movies to highlight that "everyday" viewers of these films do not always interpret the text in the deconstructive and critical modes that scholars frequently bring to them. Therefore, understanding the phenomenon of a film requires looking beyond the film itself to its reception in everyday and yet always-social encounters. Still, conceptualizing the audience as the endpoint in a cultural economy of production and consumption would be mistaken, especially in today's world of social media where everyday people are the audience and also the producer. Woon's (2014: 679) study of the *Philippines Daily Inquirer* found that "audiences have the agency and power to animate and rework popular representations, thereby reconfiguring the assemblage within which they are situated, in novel and creative ways." In short, looking at audiences complicates the armchair/disembodied analysis that had been a staple of popular geopolitics.

Despite the emergence of Popular Geopolitics 2.0 and more recent calls for the domestication of popular geopolitics, there is still an overwhelming focus on media texts in the literature, with generally little attention to the actual experience and active role of audiences (there are of course exceptions, see Woon 2014; Dittmer 2015b; Bos 2016; Carter, Kirby, & Woodyer 2016; Mason 2020; Woodyer & Carter 2020; An, Zhang, & Wang 2023). The very definition of popular geopolitics in a recent book maintains that the focus in the literature is toward media production (Saunders 2017). Similarly, Saunders and Strukov (2018) suggest in a disciplinary survey that there is little appreciation for more methodologically ambitious bottom-up approaches to the study of popular geopolitics. They rightly point out that new developments in the field should come to include popular artifacts from non-Anglophone contexts, yet the broader effort to "decolonize" the field still heavily focuses on the screening of geopolitical ideology and messages: the geopolitics of Russian movies (Holland 2012; Saunders & Strukov 2017), screen geopolitics of TV shows in northern European countries (Chow et al. 2020; Saunders & Bruun 2021; Saunders & Halink 2021), spy thrillers on Cold War geopolitics in East Pakistan (Mukharji 2015), and geopolitical representations in Bollywood (Devasundaram 2018; Saxena 2022).

The decolonial intent to move past the West can be paired with another decolonial instinct, which sees agency among everyday people rather than solely with elite producers. Recent postcolonial additions to the scholarship offer fresher and valuable new insights into popular geopolitics, drawing on affect- and audience-focused approaches (Micieli-Voutsinas 2021; Hastie 2023) and moments of excess and counter-hegemonic action, such as Hernandez's (2022) highly original research on pornography set on the US–Mexican border. In many cases, this work takes for granted the very notion of the popular in popular geopolitics. For example, Sidaway (2022: 1622) notes that the lack of attention to classical music or opera is "enmeshed with international politics," because these forms of art are not considered popular "enough." And as Ingram (2019: 8) argues in his book about artists' responses to the war in Iraq, there is also a neglect of aesthetics in geopolitical studies. In addition to taking aesthetics more seriously, new studies that go beyond media objects, screen, literary or theatre productions, or art exhibitions would advance the literature in popular geopolitics. To this end, an assemblage approach to subject formation requires attention to not just a single aesthetic experience (like an art exhibition) but also the relationship of bodies to an array of sensory inputs over a lifetime (or at least a subject's formative years).

Assemblage points to the individual subject as formed out of multitudinous relations with other bodies, discourses, and forces. Indeed, rather than considering a single text or aesthetic experience as being formative, focusing on myriad micro experiences is more likely to shape the formation of geopolitical subjects. As the remainder of this chapter shows, a greater focus on the everyday highlights how mundane places of everyday life are a geopolitical carnival of competing objects, opinions, and styles. This will be further explored in the next section focusing on geopolitical consumption and the consumption of geopolitics, and in the following section examining humor and geopolitics.

Political Consumerism, Fandom, and Resistance

An assemblage approach to popular culture includes within it an interest in materials, objects, and their political agency. That is, people's geopolitical agency is co-constituted with objects, and often operates through objects. If geopolitics is emergent from the myriad everyday interactions that people have with each other and with material objects, it makes sense to look at the fields where such interactions occur, such as the marketplace and public spaces. For that reason, consumption ought to be a key site for considering popular geopolitics, as the communities that form around various geopolitical objects can exert geopolitical force, or at least are imagined doing so by those acting within consumerist frames of politics.

Human geographers have become more interested in consumption and its relationship with politics, underscoring that geographical knowledge could contribute to our understandings of, and reactions to, processes of commodification and globalization (for a review, see Clarke 2008). Scholars largely met with cynicism consumer practices linking politics, ethics, and the market: they mostly seemed to amount to fetishized forms of consumption by virtue-signaling and/or short-sighted consumers. In response to a similar attitude within and outside academia, Mort (2000: 276) encouraged the Left to stop the "rash of socialist moralism" in reaction to the market and acknowledge that for most people consumption represents a source of pleasure and a ground for self-expression:

> For these are the ones we really love to hate: personifications of yuppiedom, the docklands interlopers, those who deal in slick images pulled into the service of capital. The whole fall-out over so-called "designer socialism" has produced a knee-jerk response, branding those who raise its profile as middle-class trendies, hypnotized by the glare of Next and Katherine Hamnett. (Mort 2000: 276)

For Mort, branders and advertisers should not be dismissed as the evil culmination of capitalism but taken seriously as "intermediaries who construct a dialogue between the market on the one hand and consumer culture on the other" (Mort 2000: 276–277). The question of what type of dialogue (if any at all) is still subjected to (not unfounded but often unproductive) anti-capitalist biases from an analytical point of view. In his book treating the geopolitics of spectacle in consumer culture, Jacob Miller (2020: 18) suggests that the looming false binary between the "duped" versus "sovereign" consumer can be overcome through an assemblage approach: "Theories of embodied assemblages help resolve this problem as the idea is to conceptualize consumer subjectivity as emerging through complex relationships with the objects, spaces and indeed practices of consumption."

While cultural geographies of consumption moved away from overly structuralist ideas of consumption (Goss 2006) and even appreciated more positive takes on consumption, in the literature on ethical consumption, the commodification of politics and activism, or corporate social responsibility the idea of "co-option" is very popular. This is even more prominent in the literature on the politics of politically charged commodities, as it serves as fertile ground for their own version of double commodity fetishisms:

> Punk is on the catwalks, Che Guevara t-shirts are everywhere, the Palestinian keffiyeh has become "as ubiquitous as leopard print" and "Liberty" is a shop not a slogan. Nothing is sacred; everything is up for replication . . . There is an almost numbing inevitability to co-option. Punk, whilst a genuine movement, was nonetheless propelled by shrewd entrepreneurs who were key to both defining and marketing the Punk look. (Hoskins 2015: 159)

An ample body of literature has examined how brands strategically align with political values, increasingly targeted at younger generations of fashionably progressive consumers (e.g., Manfredi-Sánchez 2019). However, generally little progress has been made to empirically contribute to our understanding of the multiple, embodied relationships between political identities and political things.

Rather than solely focusing on progressive branding, a more productive approach would entail in-depth engagement with producers, consumers, and the actual objects being consumed as part of the same assemblage. One area where political geographers and others have focused is the commodification of religious and secular symbols, as a useful entryway to explore the political battle between secularists and Islamists in Istanbul in the 1990s and 2000s (Navaro-Yashin 2002; Gökarıksel & Secor 2010, 2012, 2015). On one side of the debate, commodities such as pins and posters with the image of the secularist leader, Kemal Atatürk, gained popularity: "Secularists began to decorate excessively their offices, their homes and themselves with images (in object form) of Ataturk" (Navaro-Yashin 2002: 89). On the other side, Islamists significantly altered their approach to the market by rethinking the products they were buying and selling, the clothes they were wearing, and the food they were eating. In this case, a focus on the commodification of politics does not just reveal something about the politics of consumption, but more crucially it can tell us something about politics per se.

It is important to note that, as Dean (2017: 409) writes, "fandom is now an established feature of contemporary politics deserving greater analytical attention." While the literature has traditionally focused on the way fandom communities become politicized (e.g., Stanfill 2011; Gittinger 2018; Greenwalt & McVey 2022), increasingly scholars turned their attention to the way political support turns into fandom (e.g., Sandvoss 2012; Hinck 2019; Reinhard, Stanley, & Howell 2022). To borrow Sandvoss's (2005: 8) definition, fandom can be interpreted as "regular, emotionally involved consumption of a given popular narrative or text." From this definition, both consumption and affect emerge as key terms in fandom scholarship, which has traditionally emphasized the active, creative, and even resistant role of consumers. In his seminal account of fandom, Hills (2002) rejects dualistic paradigms on consumption between co-option and resistance by also admitting that fans may not exhibit their fandom at all times or perform their fandom in a coherent manner. According to Dittmer and Dodds (2008: 447), this "frees theorists from arguing that either subjects always consciously perform their preferred identity or that their preferred identity is performative."

Dittmer and Gray (2010) maintain that popular geopolitics has tended to neglect the role of the consumer, despite a growing interest in notions of consumer citizenship in political science (see Daunton & Hilton 2001; Micheletti & Stolle 2012; cf. Koch 2020). The predominant focus on branding rather than consumption in the literature on political objects can be compared to the propaganda model of media studies, wherein audiences were seen as passive and fairly homogenous. This largely outdated approach was first replaced by active audience models (see Dittmer & Dodds 2008), which emphasized the way subjects actually interpret, elaborate, and experience objects in popular culture. In this vein, consumers can be depicted as *bricoleurs* who operate in the gaps and the contradictions of dominant, mass-produced consumer culture. The concept is taken from Lévi-Strauss's (1966) idea of *bricolage* as "the re-ordering and re-contextualization of objects to communicate fresh meanings, within a total system of significances, which already includes prior and sedimented meanings attached to the objects used" (Clarke 2008: 149). Bricoleurs thus appropriate and reelaborate commodities in a way that determines the performance of everyday life, including the performance of politics in popular culture. Such creative processes can contribute to cultures of everyday resistance (Scott 1985) and to, perhaps less romanticized, interpretations of "making do" (see Fleming 2005; Gallagher 2008).

More pessimistically than advocates of *bricolage* and "making do," some have argued that political consumption lacks the vigor of protest and more radical lifestyle changes. Indeed, "slacktivism" is often understood as just another way to be captured by the system you are trying to resist: feel-good activism that might help relieve internalized guilt but has no real outside impact (see Morozov 2009). While some scholars are positive about the way slacktivism can shape political consciousness (Dennis 2018), it is believed to give individuals the impression of having fulfilled their political responsibility with little commitment or no follow-up action. However, as James Scott (1985) argues, when a consistent pattern emerges from these scattered acts, they can suddenly become noticeable and acquire political salience – this resonates with the idea of assemblages and networked accounts of power currently at the forefront in debates on resistance (see Bratich 2005, 2008). The point here is not about establishing whether forms of slacktivism amount to "true" activism. Rather, it is to look at how political power emerges from different forms of social participation, including consumerism.

Affective Registers of Popular Geopolitics: Humor

An assemblage approach to popular geopolitics highlights the importance of affect, embodiment, and the event. Significant advances in understanding how geopolitical subjects relate to one another and are transformed by the encounter are found in the literature on geopolitical humor. Early research on the complicated relationship of humor, laughter, and the political (Flint 2001) offered new perspectives on the everyday and distributed nature of agency that is inherent to assemblage thought. After all, any comedian will tell you it is easier to tell a joke when the audience is already laughing; humor is not found only in the joke but is distributed throughout the room.

Humor has its politics, as Clark (2019: 133) provocatively underscores: "What of 'conversational' humor which rolls around the dinner table and living room every day, that expresses discontent by finding a crack in the non-sensical politics that has disadvantaged some in the favour of others?" Laughter as a performative and embodied practice has gained attention in the study of critical geopolitics by opening up debates on its emancipatory and dissenting spirit on one side and its hegemonic and alienating power on the other (see Purcell, Brown, & Gokmen 2010; Dodds & Kirby 2013; Ridanpää 2014; Holland & Levy 2018). Increasingly, in the age of TikTok and tabloidization, geopolitical knowledge tends to be anecdotal, fragmented, and accidental. In this regard, Särmä (2020: 113) argues that "[b]ecause humour and laughter play a role in the circulation process (what is shared, how much, how fast), parodies can sometimes remain our sole connection to an event or issue."

While there is no magic spell to determine what has the potential to become virally funny, political humor is closely linked to structures of power and understandings of power relations (Aaltola 2009). Särmä (2020: 115) speaks of "hegemonic laughter," referring to the role of laughter in shaping the polarized and imperialist spectacle of Western geopolitics, while Fluri (2019) argues for humor as a means of "making do" in precarious geopolitical circumstances. More optimistically, other scholars have engaged with the way satire can represent a counter-hegemonic force, or at least provocatively shake up understandings of the political status quo. Ridanpää (2009, 2014) examines the way the cartoon crisis over the publication of images of Muhammad in a Danish newspaper triggered international debates and turmoil around issues of tolerance, integration, and freedom of speech (on the *Charlie Hebdo* incident, see Issa 2016). Similarly, Hammett (2010) explores the work of Jonathan Shapiro, a South African cartoonist, as a geopolitical tool with real-world effects. In this regard, Van Ramshorst (2019: 899) maintains that "insights into humor's role as a tool of geopolitical resistance underscore the significance of taking humor seriously." The author thus maintains that popular humor not only mirrors geopolitics, but also sets the tone for geopolitical attitudes, understandings, and actions: "As a technique of communication, it serves as an effective tool of dissent and resistance, while other times, it is responsible for setting global geopolitical events in motion" (Van Ramshorst 2019: 900). This, in turn, helps build a bridge between the public and intimate dimensions of geopolitics.

While the debate swings between the hegemonic and counter-hegemonic potential of satire, through an analysis of scatological humor in the show *South Park* Thorogood (2016: 215) argues that satirical media can be "an effective commentator on political issues." His work argues against two common trends in the literature: the elitist assumption that popular culture – and humor more specifically – is one way to make politics comprehensible to a duped audience that would not otherwise comprehend; and the tendency to focus on representations that fit neatly into hegemonic boxes of progressive versus conservative politics. He maintains that there is "a lack of study of satirical humor that generates laughter that is politically ambivalent, vulgar and messy in geopolitical scholarship" (Thorogood 2016: 217). By emphasizing multiple readings and overcoming the binary interpretations of sociopolitical life as commonly channeled by satire, however, *South Park* does not provide viewers with a coherent political narrative with which political subjects can comfortably identify.

Within this body of work, once again the audience perspective has largely been neglected. Thorogood (2016: 231) acknowledges this by stating that "it is hopefully clear that politically ambivalent humor extends beyond the world of television, media and satire into the world of affective, mundane engagements." There is also little focus on right-wing humor, perhaps due to a widespread conviction of an anecdotally Left-leaning scholarship that "The Right Can't Meme" (as termed by a popular internet social media account) and the perception of satire as "progressive" and therefore a prerogative of the Left. In this regard, Dittmer (2013: 498) maintains that "the emphasis in popular geopolitics on 'large' media events such as Hollywood film and newspapers' political cartoons reflects this notion of power as resultant from the hegemonic messages of mass media and the ability of the Left to stage equally large anti-geopolitical spectacles."

By shifting attention away from mass-mediated moods dominant in the literature and focusing on students during a Model United Nations simulation, Dittmer

(2013) shows the way humor as a situated, embodied, and relational practice contributes to geopolitical assemblages. He argues that there is little evidence empirically showing that humor has the ability to shift or alter the convictions that people already hold: "people find things funny (or not) in connection with their already-existing political attitudes" (Dittmer 2013: 498). Instead of taking a definite position within the debate on the relationship between power and humor, therefore, it seems wise to explore the way humor – among other possible affective registers – contributes to larger geopolitical assemblages that ultimately generate power.

While the bodily performance of laughing or not laughing may be seen as an intrinsically political act (Dodds & Kirby 2013) – giving away people's dispositions toward certain matters – its assembled, political power will remain highly labile and context dependent. This leads us to our final point of this section, namely Billig's (2005) notion of "unlaughter" – the absence of laughter where it is expected to occur. While laughter can at times represent a way to exorcise tensions (see Sanchez 2016), the intentional abstinence from laughter can also be a way of expressing disagreement or disapproval. Building on this, Dodds and Kirby (2013: 54) argue that "unlaughter requires, at the very least, an appreciation of the role of audiences." Expectations of people's reactions to jokes and assessments over when and where certain humor may be considered inappropriate can be telling about people's assumptions on the dominant values of certain places, events, and other people (Macpherson 2008).

Conclusion

Popular culture has been a feature of political geography for over three decades as a result of its embedding in popular geopolitics. Its evolution over that time has mirrored (and co-produced) shifts in our understanding of what popular culture is, from a media "thing" to a social "doing." Two fruitful strands of research for future research emphasize the "doing" of geopolitics by people in their everyday lives, consumerism, and humor. There are of course many other directions that could be fruitfully mined by scholars, and of course they should be. Nevertheless, consumerism and humor remain powerful engines of everyday geopolitics and they indicate the way "popular culture" has taken on new meanings in the twenty-first century.

References

Aaltola, M. (2009). *Western Spectacle of Governance and the Emergence of Humanitarian World Politics*. New York: Palgrave Macmillan.

Agnew, J. (1994). The territorial trap: The geographical assumptions of international relations theory. *Review of International Political Economy* 1(1): 53–80.

AlAwadhi, D., & Dittmer, J. (2022). The figure of the refugee in superhero cinema. *Geopolitics* 27(2): 604–628.

An, N., Zhang, J., & Wang, M. (2023). The everyday Chinese framing of Africa: A perspective of tourism-geopolitical encounter. *Geopolitics 28*(4): 1422–1441.

Bernazzoli, R. (2010). "The end of autocracy": Analysing representations of the Austro-Hungarian dissolution as the foundation of US hegemonic discourse. *Geopolitics 15*(4): 643–666.

Billig, M. (2005). *Laughter and Ridicule: Towards a Social Critique of Humour*. London: Sage.

Bos, D. (2016). Critical methodologies for researching military videogames. In Williams, A., Woodward, R., Jenkings, N., & Rech, M.F. (eds), *The Routledge Companion to Military Research Methods*. London: Routledge, 332–344.

Bos, D. (2018). Answering the call of duty: Everyday encounters with the popular geopolitics of military-themed videogames. *Political Geography 63*: 54–64.

Bratich, J. (2008). Activating the multitude: Audience powers and cultural studies. In Goldstein, P., & Machor, J. (eds), *New Directions in American Reception Study*. Oxford: Oxford University Press, 33–56.

Bratich, J.Z. (2005). Amassing the multitude: revisiting early audience studies. *Communication Theory 15*(3): 242–265.

Brickell, K. (2012). Geopolitics of home. *Geography Compass 6*(10): 575–588.

Brickell, K., & Cuomo, D. (2019). Feminist geolegality. *Progress in Human Geography 43*(1): 104–122.

Buzan, B. (2010). America in space: The international relations of Star Trek and Battlestar Galactica. *Millennium: Journal of International Studies 39*(1): 175–180.

Carter, S., Kirby, P., & Woodyer, T. (2016). Ludic – or playful – geopolitics. In Benwell, M., & Hopkins, P. (eds), *Children, Young People and Critical Geopolitics*. Farnham: Ashgate, 61–73.

Carter, S., & McCormack, D. (2006). Film, geopolitics and the affective logics of intervention. *Political Geography 25*(2): 228–245.

Carter, S., & Woodyer, T. (2020). Introduction: Domesticating geopolitics. *Geopolitics 25*(5): 1045–1167.

Carver, T. (2010). Cinematic ontologies and viewer epistemologies: Knowing international politics as moving images. *Global Society 24*(3): 421–431.

Chow, P., Waade, A.M., & Saunders, R.A. (2020). Geopolitical television drama within and beyond the Nordic region. *Nordicom Review 41*(s1): 11–27.

Christian, J., Dowler, L., & Cuomo, D. (2016). Fear, feminist geopolitics and the hot and banal. *Political Geography 54*: 64–72.

Clark, J.H. (2019). "The state kills, we kill, everyone kills": Cracking and framing the field with humor. *Political Geography 68*: 131–138.

Clarke, N. (2008). From ethical consumerism to political consumption. *Geography Compass 2*(6): 1870–1884.

Culcasi, K. (2019). "We are women and men now": Intimate spaces and coping labour for Syrian women refugees in Jordan. *Transactions of the Institute of British Geographers 44*(3): 463–478.

Dalby, S. (1991). Critical geopolitics: Discourse, difference, and dissent. *Environment and Planning D: Society and Space 9*(3): 261–283.

Daunton, M., & Hilton, M. (2001). *The Politics of Consumption: Material Culture and Citizenship in Europe and America*. Oxford: Berg.

Dean, J. (2017). Politicising fandom. *British Journal of Politics and International Relations 19*(2): 408–424.

Debrix, F. (2008). *Tabloid Terror: War, Culture, and Geopolitics.* London: Routledge.

Dennis, J. (2018). *Beyond Slacktivism: Political Participation on Social Media.* London: Palgrave Macmillan.

Der Derian, J. (2009). *Virtuous War: Mapping the Military-Industrial-Media-Entertainment Network.* New York: Routledge.

Devasundaram, A. (2018). Alternative articulations of geopolitical discourse in new Indian films. In Saunders, R.A., & Strukov, V. (eds), *Popular Geopolitics: Plotting an Evolving Interdiscipline.* Abingdon: Routledge.

Dittmer, J. (2005a). Captain America's empire: Reflections on identity, popular culture and geopolitics. *Annals of the Association of American Geographers 95*(3): 626–643.

Dittmer, J. (2005b). NATO, the EU and Central Europe: Differing symbolic shapes in newspaper accounts of enlargement. *Geopolitics 10*(1): 76–98.

Dittmer, J. (2007). The tyranny of the serial: Popular geopolitics, the nation and comic book discourse. *Antipode 39*(2): 247–268.

Dittmer, J. (2013). Humour at the model United Nations: The role of laughter in constituting geopolitical assemblages. *Geopolitics 18*(3): 493–513.

Dittmer, J. (2015a). On Captain America and "doing" popular culture in the social sciences. In Caso, F., & Hamilton, C. (eds), *Popular Culture and World Politics: Theories, Methods, Pedagogies.* Bristol: E-International Relations Publishing, 45–50.

Dittmer, J. (2015b). Playing geopolitics: Utopian simulations and subversions of international relations. *GeoJournal 80*(6): 909–923.

Dittmer, J., & Bos, D. (2019). *Popular Culture, Geopolitics, and Identity*, 2nd edn. Lanham, MD: Rowman and Littlefield.

Dittmer, J., & Dodds, K. (2008). Popular geopolitics past and future: Fandom, identities and audiences. *Geopolitics 13*(3): 437–457.

Dittmer, J., & Gray, N. (2010). Popular geopolitics 2.0: Towards new methodologies of the everyday. *Geography Compass 4*(11): 1664–1677.

Dittmer, J., & Larsen, S. (2011). Aboriginality and the Arctic North in Canadian nationalist superhero comics, 1940–2004. *Historical Geography 38*: 52–69.

Dodds, K. (2003). Cold war geopolitics. In Agnew, J., Mitchell, K., & Toal, G. (eds), *A Companion to Political Geography.* Malden, MA: Blackwell, 204–218.

Dodds, K. (2005). Screening geopolitics: James Bond and the early Cold War films (1962–1967). *Geopolitics 10*(2): 266–289.

Dodds, K. (2006). Popular geopolitics and audience dispositions: James Bond and the Internet Movie Database (IMDb). *Transactions of the Institute of British Geographers NS31*: 116–130.

Dodds, K. (2007). Steve Bell's eye: Cartoons, geopolitics and the visualization of the "war on terror". *Security Dialogue 38*(2): 157–177.

Dodds, K. (2010). Popular geopolitics and cartoons: Representing power relations, repetition and resistance. *Critical African Studies 2*(4): 113–131.

Dodds, K. (2011). Gender, geopolitics, and geosurveillance in *The Bourne Ultimatum. Geographical Review 101*(1): 88–105.

Dodds, K. (2014). Shaking and stirring James Bond: Age, gender, and resilience in *Skyfall* (2012). *Journal of Popular Film and Television 42*(2): 116–130.

Dodds, K., & Kirby, P. (2013). It's not a laughing matter: Critical geopolitics, humour and unlaughter. *Geopolitics 18*: 45–59.

Dowler, L., & Sharp, P.J. (2001). A feminist geopolitics? *Space and Polity 5*(3): 165–176.

Dunnett, O. (2009). Identity and geopolitics in Hergé's *Adventures of Tintin. Social & Cultural Geography 10*(5): 583–598.

Emad, M. (2006). Reading Wonder Woman's Body: Mythologies of gender and nation. *Journal of Popular Culture 39*: 954–984.

Fleming, P. (2005). Metaphors of resistance. *Management Communication Quarterly 19*(1): 45–66.

Flint, C. (2001). The geopolitics of laughter and forgetting: A world-systems interpretation of the post-modern geopolitical condition. *Geopolitics 6*(3): 1–16.

Fluri, J.L. (2019). What's so funny in Afghanistan? Jocular geopolitics and the everyday use of humor in spaces of protracted precarity. *Political Geography 68*: 125–130.

Funnell, L., & Dodds, K. (2017). *Geographies, Genders and Geopolitics of James Bond*. London: Palgrave Macmillan.

Gallagher, M. (2008). "Power is not an evil": Rethinking power in participatory methods. *Children's Geographies 6*(2): 137–150.

Gittinger, J.L. (2018). Hijabi cosplay: Performances of culture, religion, and fandom. *Journal of Religion and Popular Culture 30*(2): 87–105.

Glynn, K., & Cupples, J. (2015). Negotiating and queering US hegemony in TV drama: Popular geopolitics and cultural studies. *Gender, Place and Culture 22*(2): 271–287.

Gökarıksel, B., & Secor, A. (2010). Between fashion and *tesettür*: Marketing and consuming women's Islamic dress. *Journal of Middle East Women's Studies 6*(3): 118–148.

Gökarıksel, B., & Secor, A. (2012). "Even I was tempted": The moral ambivalence and ethical practice of veiling-fashion in Turkey. *Annals of the Association of American Geographers 102*(4): 847–862.

Gökarıksel, B., & Secor, A. (2015). Post-secular geographies and the problem of pluralism: Religion and everyday life in Istanbul, Turkey. *Political Geography 46*: 21–30.

Goss, J. (2006). Geographies of comsumption: The work of consumption. *Progress in Human Geography 30*(2): 237–249.

Greenwalt, D.A., & McVey, J.A. (2022). Get Gritty with it: Memetic icons and the visual ethos of antifascism. *Communication and Critical/Cultural Studies 19*(2): 158–179.

Hammett, D. (2010). Political cartoons, post-colonialism, and critical African studies. *Critical African Studies 2*(4): 1–26.

Hastie, A. (2023). Postcolonial geopolitics: Reading contemporary geopolitics in Maghrebi-French war films. *Geopolitics 28*(1): 239–256.

Hernandez, Y.G. (2022). Border trash: The markings of Latinas as foreign bodies. *Women & Performance: A Journal of Feminist Theory 32*(1): 8–30.

Hills, M. (2002). *Fan Cultures*. London: Routledge.

Hinck, A. (2019). *Politics for the Love of Fandom: Fan-Based Citizenship in a Digital World*. Baton Rouge, LA: LSU Press.

Holland, E.C., & Levy, A. (2018). *The Onion* and the geopolitics of satire. *Popular Communication 16*(3): 182–195.

Holland, T. (2012). "To think and imagine and see differently": Popular geopolitics, graphic narrative, and Joe Sacco's "Chechen War, Chechen Women". *Geopolitics 17*(1): 105–129.

Hoskins, T.E. (2015). *Stitched Up: The Anti-Capitalist Book of Fashion*. London: Pluto Press.

Ingram, A. (2019). *Geopolitics and the Event: Rethinking Britain's Iraq War through Art*. Chichester: Wiley-Blackwell.

Issa, S. (2016). Picturing the *Charlie Hebdo* incident in Arabic political cartoons. *Arabic Studies Quarterly 38*(3): 562–585.

Kiersey, N.J., & Neumann, I.B. (2015). Worlds of our making in science fiction and international relations. In Caso, F., & Hamilton, C. (eds), *Popular Culture and World Politics: Theories, Methods, Pedagogies*. Bristol: E-International Relations Publishing, 74–82.

Kirby, P. (2015). The girl on fire: *The Hunger Games*, feminist geopolitics and the contemporary female action hero. *Geopolitics 20*(2): 460–478.

Kirby, P. (2020). Battle scars: Wonder Woman, aesthetic geopolitics and disfigurement in Hollywood film. *Geopolitics 25*(4): 916–936.

Koch, N. (2020). The corporate production of nationalism. *Antipode 52*(1): 185–205.

Koopman, S. (2011). Alter-geopolitics: Other securities are happening. *Geoforum 42*(3): 274–284.

Laketa, S. (2016). Geopolitics of affect and emotions in a post-conflict city. *Geopolitics 21*(3): 661–685.

Lévi-Strauss, C. (1966). *The Savage Mind*. Chicago, IL: University of Chicago Press.

MacDonald, F., Hughes, R., & Dodds, K. (eds) (2010). *Observant States: Geopolitics and Visual Culture*. London: I. B. Tauris.

Macpherson, H. (2008). "I don't know why they call it the Lake District they might as well call it the rock district!" The workings of humour and laughter in research with members of visually impaired walking groups. *Environment and Planning D: Society and Space 26*(6): 1080–1095.

Manfredi-Sánchez, J.L. (2019). Brand activism. *Communications Society 32*(4): 343–359.

Mason, O. (2020). Walking the line: Lines, embodiment and movement on the Jordan Trail. *Cultural Geographies 27*(3): 395–414.

McFarlane, T., & Hay, I. (2003). The battle for Seattle: Protest and popular geopolitics in *The Australian* newspaper. *Political Geography 22*(2): 211–232.

Micheletti, M., & Stolle, D. (2012). Sustainable citizenship and the new politics of consumption. *Annals of the American Academy of Political and Social Science 644*(1): 88–120.

Micieli-Voutsinas, J. (2021). *Affective Heritage and the Politics of Memory after 9/11: Curating Trauma at the Memorial Museum*. London: Routledge.

Miller, J.C. (2020). *Spectacle and Trumpism: An embodied assemblage approach*. Bristol: Bristol University Press.

Morozov, E. (2009). The brave new world of slacktivism. *Foreign Policy*, May 19. https://foreignpolicy.com/2009/05/19/the-brave-new-world-of-slacktivism (accessed May 7, 2021).

Mort, F. (2000). The politics of consumption. In Lee, M.J. (ed.), *The Consumer Society Reader*. Oxford: Blackwell, 271–281.

Mukharji, P.B. (2015). Technospatial imaginaries: Masud Rana and the vernacularization of popular Cold War geopolitics in East Pakistan, 1966–1971. *History and Technology 31*(3): 324–340.

Navaro-Yashin, Y. (2002). *Faces of the State: Secularism and Public Life in Turkey*. Princeton, NJ: Princeton University Press.

O'Tuathail, G. (1996). *Critical Geopolitics*. Minneapolis, MN: University of Minnesota Press.

O' Tuathail, G., & Agnew, J. (1992). Geopolitics and discourse: Practical geopolitical reasoning in American foreign policy. *Political Geography 11*(2): 190–204.

Pain, R. (2015). Intimate war. *Political Geography 44*: 64–73.

Pain, R., & Staeheli, L. (2014). Introduction: Intimacy-geopolitics and violence. *Area 46*: 344–347.

Pinkerton, A., & Benwell, M. (2014). Rethinking popular geopolitics in the Falklands/Malvinas sovereignty dispute: Creative diplomacy and citizen statecraft. *Political Geography 38*: 12–22.

Pinkerton, A., & Dodds, K. (2009). Radio geopolitics: Broadcasting, listening and the struggle for acoustic spaces. *Progress in Human Geography 33*(1): 10–27.

Power, M. (2007). Digitized virtuosity: Video war games and post-9/11 cyber-deterrence. *Security Dialogue 38*(2): 271–288.

Power, M., & Crampton, A. (2005). Reel geopolitics: Cinemato-graphing political space. *Geopolitics 10*(2): 193–203.

Purcell, D., Brown, M.S., & Gokmen, M. (2010). Achmed the dead terrorist and humor in popular geopolitics. *GeoJournal 75*(4): 373–385.

Reinhard, C.D., Stanley, D., & Howell, L. (2022). Fans of Q: The stakes of QAnon's functioning as political fandom. *American Behavioral Scientist 66*(8): 1152–1172.

Ridanpää, J. (2009). Geopolitics of humour: The Muhammed cartoon crisis and the Kaltio comic strip episode in Finland. *Geopolitics 14*(4): 729–749.

Ridanpää, J. (2014). Geographical studies of humor. *Geography Compass 8*(10): 701–709.

Salter, M.B. (2011). The geographical imaginations of video games: Diplomacy, civilization, America's Army and Grand Theft Auto IV. *Geopolitics 16*(2): 359–388.

Sanchez, A. (2016). Profane relations: The irony of offensive jokes in India. *History and Anthropology 27*(3): 296–312.

Sandvoss, C. (2005). *Fans: The Mirror of Consumption*. Cambridge: Polity Press.

Sandvoss, C. (2012). Enthusiasm, trust and its erosion in mediated politics: On fans of Obama and the Liberal Democrats. *European Journal of Communication 27*(1): 68–81.

Särmä, S. (2020). Collage: An art-inspired methodology for studying laughter in world politics. In Caso, F., & Hamilton, C. (eds), *Popular Culture and World Politics: Theories, Methods, Pedagogies*. Bristol: E-International Relations Publishing, 110–119.

Saunders, R.A. (2017). *Popular Geopolitics and Nation Branding in the Post-Soviet Realm*. London: Routledge.

Saunders, R.A. (2019). (Profitable) imaginaries of Black Power: The popular and political geographies of Black Panther. *Political Geography 69*: 139–149.

Saunders, R.A., & Bruun, H. (2021). "Radio Free Sweden": Satire, national identity, and the un-PC (geo)politics of Jonatan Spang. *Global Society 35*(1): 84–101.

Saunders, R.A., & Halink, S. (2021). Geysers, *Game of Thrones*, and geopolitics: Iceland as a zone of strategic tourism. In Cordoba Azcarate, M., Mostafanezhad, M., & Norum, R. (eds), *Tourism Geopolitics: Assemblages of Infrastructure, Affect, and Imagination*. Tucson, AZ: University of Arizona Press, 113–135.

Saunders, R.A., & Strukov, V. (2017). The popular geopolitics feedback loop: Thinking beyond the "Russia versus the West" paradigm. *Europe-Asia Studies 69*(2): 303–324.

Saunders, R.A., & Strukov, V. (eds) (2018). *Popular Geopolitics: Plotting an Evolving Interdiscipline*. Abingdon: Routledge.

Saxena, C. (2022). Re/presenting Afghans in Hindi cinema: The popular geopolitics of India-Afghanistan relations. *Space and Polity* 26(3): 145–164.

Scott, J. (1985). *Weapons of the Weak: Everyday Forms of Peasant Resistance*. New Haven, CT: Yale University Press.

Semmerling, T. (2006). *Evil Arabs in American Popular Film*. Austin, TX: University of Texas Press.

Sharp, J.P. (1993). Publishing American identity: Popular geopolitics, myth and *The Reader's Digest*. *Political Geography* 12(6): 491–503.

Sharp, J.P. (2000a). Remasculinising geo-politics? Comments on Gearoid O' Tuathail's *Critical Geopolitics*. *Political Geography* 19(3): 361–364.

Sharp, J.P. (2000b). *Condensing the Cold War: Reader's Digest and American Identity*. London: University of Minnesota Press.

Shaw, I. (2010). Playing war. *Social and Cultural Geography* 11(8): 789–803.

Shepherd, L.J. (2013). *Gender, Violence and Popular Culture: Telling Stories*. New York: Routledge.

Sidaway, J.D. (2022). Popular Geopolitics 3.0? Deconstructing the boundaries of popular geopolitics. *Geopolitics* 27(5): 1622–1628.

Smith, S. (2009). The domestication of geopolitics: Buddhist–Muslim conflict and the policing of marriage and the body in Ladakh, India. *Geopolitics* 14(2): 197–218.

Smith, S. (2012). Intimate geopolitics: Religion, marriage, and reproductive bodies in Leh, Ladakh. *Annals of the Association of American Geographers* 102(6): 1511–1528.

Stanfill, M. (2011). Doing fandom, (mis)doing whiteness: Heteronormativity, racialization, and the discursive construction of fandom. *Transformative Works and Cultures* 8. doi:10.3983/twc.2011.0256

Thorogood, J. (2016). Satire and geopolitics: Vulgarity, ambiguity and the body grotesque in *South Park*. *Geopolitics* 21(1): 215–235.

Van Ramshorst, J.P. (2019). Laughing about it: Emotional and affective spaces of humour in the geopolitics of migration. *Geopolitics* 24(4): 896–915.

Woodyer, T., & Carter, S. (2020). Domesticating the geopolitical: Rethinking popular geopolitics through play. *Geopolitics* 25(5): 1050–1074.

Woon, C.-Y. (2014). Popular geopolitics, audiences, and identities: Reading the "war on terror" in the Philippines. *Geopolitics* 19(3): 656–683.

Chapter 31

Natural Resources

Andrea Marston

Department of Geography, Rutgers University, New Brunswick, NJ, USA

If there is one central tenet in the subfield of resource geography, it is that natural resources are not natural. Although a rock, a tree, or a river may appear as a natural thing, none of them is a resource until they have been socially categorized as such. Even more, resources are not usable until they have been scientifically assessed, laboriously extracted and refined, and made available on the market as commodities. This is true regardless of whether these commodities are meant to be consumed (like a bottle of water) or held as an investment (like a tract of land). The epistemological and physical labor of "resource-making" not only muddies the distinction between nature and society, but also is the condition of possibility for contemporary political formations, economic production, and life itself (Huber 2019; Kama 2020; Kuchler & Bridge 2023).

For all these reasons, geographers frequently recite the line that "resources are not, they become" (Zimmermann 1951: 15). Building on this intervention, Gavin Bridge (2009: 1219) describes resources as possessing a "dual quality" insofar as they are "part physical entity, part social entity" and exist "at an interface between the physical and the human." Joe Bryan (2022: 449) similarly writes that resources are "never immutable qualities of the world that they make, so much as they are expressions of a particular understanding of the world." By extension, the naming and creation of resources are fundamentally contentious processes, since there is no universal understanding of the world, either across or within communities.

What qualifies as a resource – and what kind of resource is it – changes over time and space, such that a mineral like lithium can go from being primarily a

The Wiley Blackwell Companion to Political Geography, Second Edition.
Edited by Virginie Mamadouh, Natalie Koch, Chih Yuan Woon, and John Agnew.

psychiatric treatment for mood stabilization to a crucial component of energy storage solutions within a renewable energy transition (Köppel 2020; Soto Hernandez & Newell 2023). Indeed, the concept of a "resource" itself is "vexing," given that it involves overt objectification and "is tied to the colonial-capitalist subjugation of people and environments" (Valdivia, Himley, & Havice 2022: 10). Nevertheless, geographers continue to ask questions about *how* and *with what effects* resources are made, often paying particular attention to how resources shape and are shaped by economic relations, political formations, and embodied subjectivities.

A Burgeoning Field

In the first edition of *The Wiley Blackwell Companion to Political Geography* (2015), Kathryn Furlong and Emma S. Norman note that the term "natural resources" is attributed to Adam Smith in *The Wealth of Nations* (1776), who defines them as "those materials or substances of a place which can be used to sustain life or for economic exploitation" (cited in Furlong & Norman 2015: 424). This utilitarian definition is also geographical, as resources are rooted in and defined by their locations. Nevertheless, the subfield of resource *geography* has a much shorter history than natural resources writ large. The Google Books Ngram Viewer, a tool that charts the appearance of words or phrases in books over time, demonstrates that "resource geography" did not appear in the literature until shortly after World War II. It showed up increasingly until 1973 – the year of the first oil shock – and then dropped away. Its appearances fluctuated wildly thereafter, with the highest subsequent peak recorded in 2004, the height of the twenty-first-century commodity boom. The 2008 recession marked the beginning of the end of this boom, and new contributions to the field of resource geography similarly entered a period of decline.

Resource geography is now undergoing a resurgence of interest. The recent upward trajectory can be partially attributed to a collectively sharpened awareness of both the urgency of climate change and the projected material intensiveness of a global energy transition. In this context, mineral prospecting and extraction are expected to increase in the coming years, even if fossil fuel consumption decreases (which is not its current trend). Meanwhile, the financialization of nature has deepened in the realms of conservation, climate adaptation, green energy, and "green" mining ventures (Dempsey & Suarez 2016; Dafermos, Gabor, & Michell 2021; Knuth et al. 2022; Franz & McNelly 2023).

All these trends are further shaped by geopolitical tensions and anxieties, as exemplified by proliferating discourses around resource and energy security. In the United States, the 2022 Inflation Reduction Act passed by President Joe Biden was hailed as the country's most ambitious climate change policy to date. Yet it has also been critiqued for being too protectionist and automobile centric (Kleimann et al. 2023; Riofrancos et al. 2023). In this context, critical minerals are becoming anchors for new geopolitical trading blocs such as the Mineral Security Partnership, an initiative launched in 2022 by the United States and its Western allies that aims to counterbalance China's mineral dominance (Vivoda, Matthews, & McGregor 2024).

At the same time, and adding fuel to growing interests in resource geography, conflicts around resources are eliciting new political formations and raising questions about what even counts as politics (Huber 2019). Tensions emerge around access to and control over resources, but they also emerge around the defensibility of treating fragments of the Earth as commodities. Resource-making often involves the violent imposition of one way of seeing and interacting with the world on top of another, and what one group of people perceives as a natural resource might be understood by another group as constitutive of a sacred place or being. This is demonstrated by Marisol de la Cadena's (2010) discussion of the potential impact of open-pit gold mining on mountainous Earth beings in Peru, as well as by Megan Ybarra's (2018) reflection on the role of sacred caves in Q'eqchi' social and political life in Guatemala. In such contexts, resource access justifies new rounds of colonial appropriation. The infrastructures that reinforce capitalist-colonial modernity, such as highways, energy grids, and circuit boards, are then constructed using these same resources (LaDuke & Cowen 2020; Hamouchene & Sandwell 2023).

Environmental actions are thus animated by an increasing awareness of the uneven geographies of climate, energy, and environmental injustices, as well as their origins in colonialism and racism. For example, Purifoy and Seamster (2021) highlight the relationship between environmental degradation in nonwhite spaces and development in white spaces through the notion of "creative extraction"; Ranganathan (2022) argues that "environmental unfreedoms" in contemporary urban India are the historical products of caste, labor, and property; and Davies (2022) demonstrates that the apparent invisibility of environmental toxicity in Louisiana's "cancer valley" is politically constructed. Investigations such as these often emphasize the difficulties associated with mounting "resistance" campaigns in environmentally contaminated contexts. For instance, Valdivia (2018) draws on work with residents of the oil refinery city of Esmeraldas, Ecuador, to explore the coexistence of apparently contradictory responses to lived toxicities associated with oil capitalism, while Murphy (2017: 495) reflects on how decolonial efforts in Canada must reckon with an "alterlife" in which persistent organic pollutants have become a part of the "molecular fabric of our bodies." Most poignantly, widespread adoption among geographers of the Latin American Indigenous feminist method of *cuerpo-territorio* (body-territory) has facilitated new understandings of how environmental injustices are experienced at multiple and inextricable scales simultaneously (Zaragocín & Caretta 2021).

There is also a growing focus on the populist dimensions of environmental politics and governance, as countries around the world react to decades of neoliberalism, climate change, and land-use change (McCarthy 2019). Sometimes populist sentiment can be hinged to progressive environmental action, as demonstrated in the late 2010s in the United States by the alliance between Indigenous nations and mostly white landowners against new oil pipelines in the Dakotas (Bosworth 2022). In other instances, an authoritarian variety of environmental populism has reasserted elite power and market domination, particularly in the context of agricultural production (McKay, Oliveira, & Liu 2020; Ortiz Gallego 2024). In these cases, governments can mobilize populist sentiment to expand extractivist dynamics under the auspices of resource security (Graybill 2019; Adaman, Arsel, & Akbulut 2021; İnal 2022; Ofstehage, Wolford, & Borras 2022).

Finally, resource geographers have become increasingly interested in the topic of waste, both as an undesirable byproduct of resource production and as a resource in its own right. Waste is often imagined as a residual category of production, or a material substance whose disposal must be efficiently managed, but geographers emphasize the dialectical relationship between waste and value (Goldstein 2013; Gidwani & Reddy 2011). They also explore the dual meanings of the word "waste" itself, which can be used not only to describe the discarded byproducts of production but also as a place or resource that has been inadequately exploited (Dillon 2014).

The production of wastelands is a process entwined with processes of race and racism, as colonial powers used the improper transformation of nature as evidence of degeneracy, or at least of an insecure claim to land (Voyles 2015; Kopack 2019; Koch 2022; Meché 2022). Geographies of waste thus intersect with environmental justice scholarship, which has long been concerned with the disproportionate presence of toxic waste in poor communities of color (Kuletz 2016 [1998]; Pulido 2017). Waste is unevenly distributed globally, as both a byproduct of resource production (Vasudevan 2021; Hecht 2023) and as part of global recycling industries (Lepawsky 2015; Morinville 2021). Moreover, most waste management strategies involve indefinite containment, which is a colonial approach in that it presupposes ongoing access to stolen land (Liboiron 2021). How to imagine waste disposal differently, particularly when the waste is unlikely to be repurposed as resources and its toxic properties stretch out in perpetuity, is the subject of several recent geographical analyses (Ureta 2016; Beckett & Keeling 2019; Cram 2023).

Amid these trends, the remainder of this chapter addresses just two areas of recent interest for resource geographers: extractivism and plantations. Scholars invoke both words to describe forms of export-oriented resource production, but the terms are often divided by resources. "Extractivism" is more frequently deployed with respect to fossil fuels, minerals, and ocean commodities, while "the plantation" is used to describe agriculture, forestry, and other forms of resource cultivation (but see Labban 2014; Perreault 2018). However, this chapter focuses on the slightly different dynamics bundled with each term. Extractivism emphasizes the appropriation of nature and unequal trade relations at multiple political scales, whereas the plantation draws attention to the socioecological transformations and frequently racialized forms of exploitation that occur at sites of resource production. Read alongside one another, the two sets of literature expose one another's lacunae even as they elucidate many of the shared characteristics of contemporary resource geography.

Extractivism

The definition of extractivism varies widely, but it broadly describes an economic regime in which "the extractive sectors contribute a large share of total value added, the export of primary commodities is among the most important sources of revenue, and labor and natural resources are exploited beyond their ability to reproduce themselves" (Schaffartzik et al. 2016: 102). An extractivist economy typically relies on exporting raw materials with little or no refining or processing that would (through the application of more technical human labor) add value to the exports.

Uruguayan sociologist Eduardo Gudynas (2009) distinguishes between neoliberal extractivism, in which wealth as well as raw materials accrue in an imperial or capitalist metropole, and "neo-extractivism," in which the state becomes more directly involved in extractive sectors, absorbs a comparatively large percentage of resource rent as the ultimate resource landlord, and uses this wealth to fund social programming. Gudynas draws this distinction to highlight continuities and disjunctures between the neoliberal governments that controlled most of Latin America in the 1990s and the comparatively progressive governments that surged regionally in the early 2000s. He argues that extractivist practices persisted across both decades despite differences in how resource rents were distributed and spent. Of course, the state is implicated even in the most neoliberal of extractivist regimes, since the creation of a "free" market requires active and ongoing policymaking. Indeed, Arboleda (2020) argues that theories of extractivism can reify the role of the state over and above the dynamics of the market, given that scholars tend to ascribe extractivist characteristics to states rather than people or social relations.

Neo-extractivism often coexists with resource nationalism, a term used to describe either state-led efforts to profit more from its resource stock (Childs 2016) or the belief that citizens of a nation should benefit directly from resources located within their national territory (Young 2017). This belief has a history, and it is usually connected to a particular imaginary of state-led development within territorially defined limits (Koch & Perreault 2019). The resurgence of resource nationalism during and after the commodity boom has made some analysts nervous, while others celebrate the return of the developmentalist state. Resource geographers, however, highlight the ambiguity of resource nationalism, which is just as easily tied to authoritarian populism as it is to redistributive politics – sometimes at the same time.

The communities most often excluded from the benefits of resource nationalism are those for whom the nation-state fits uneasily, such as Indigenous nations and ethnic minorities. On the other hand, as Andrew Curley (2023) demonstrates in the case of coal mining in the Navajo Nation, resource nationalism is a sentiment and set of policies that can also be wielded by Indigenous nations, and it can sometimes support sovereignty struggles – albeit for time frames constrained by resource horizons. Penelope Anthias (2018) similarly explores how Guaraní people in the Bolivian Chaco are bypassing the nation-state and building direct alliances with natural gas companies in an arrangement she calls "hydrocarbon citizenship." Such arrangements are possible because of the continued global economic dependence on fossil fuels, which are increasingly derived from "unconventional" resources like oil shale, tar sands, and coalbed methane. Extracting fuels from these sources involves the deployment of new knowledges, technologies, and infrastructures (Kama 2016; Kuchler 2017; Labussiere 2021), not to mention new regulatory frameworks and interest groups, including "citizen science" activists committed to documenting unintended harms (Wylie 2018; Baka et al. 2019). The frontiers of fossil fuel extraction have also shifted: fracking has resituated the fossil fuel sector within the United States, and offshore drilling is increasingly the norm, despite disasters like the BP spill of 2010 (Andrews et al. 2021).

Meanwhile, protests around pipelines, which are a primary circulatory infrastructure of oil, have helped to precipitate larger and even cross-border protests (McCreary & Milligan 2014; Awasis 2023). Workers on oil rigs and in other greenhouse gas–producing corporations are not yet protagonists in the struggle against climate change, even if this is a role they might yet be persuaded to fulfill (Huber 2022). Yet people whose land and livelihoods are directly threatened by pipelines have become (sometimes unlikely) participants in the global climate movement (Bosworth 2022). Their actions, however, have thus far failed to shake the hegemonic hold of fossil fuels, and countries like Mexico and Ecuador have expanded their oil sectors despite commitments to climate accords (Fry & Delgado 2018).

The persisting fossil fuel–based economy has diverse dependent sectors, but mining companies are rising to the forefront of global environmental consciousness, both because critical minerals are purportedly key to renewable energy generation and storage, and because of the growth of mining-related controversies. Such controversies are exemplified by lithium, which is used to produce the lithium-ion batteries that power electric vehicles and may provide grid-scale energy storage. Brine-based lithium mining consumes vast quantities of water and is the subject of much debate in Latin America's "lithium triangle" (Argentina, Bolivia, and Chile) (Bustos-Gallardo et al. 2021; Jerez et al. 2021; Blair et al. 2023).

Contemporary efforts toward an energy transition are also shaped by global geopolitical struggles, as China controls numerous mineral supply chains and

Western countries fear losing access to metals like cobalt, rare earths, and lithium (Klinger 2020; DeBoom 2021). In response, both North American and European countries have been "onshoring" their metal production (Riofrancos 2023). This move is controversial, as demonstrated by Indigenous protests against a new lithium mine in Thacker's Pass, Nevada, as well as by Canada's plan to mine critical minerals in the northern Ontario "ring of fire" region, which is home to several Indigenous nations. Like oil and gas companies, miners are also attempting to move *actually* offshore – that is, into maritime space. Although not yet operational, deep sea mining projects promise to access untold mineral wealth while bypassing the complications posed by competing sovereignty claims (Childs 2020; Zalik 2022). Under the United Nations Convention on the Law of the Sea, the seabed in areas beyond national jurisdiction is considered the common heritage of humanity, but in practice the ocean is treated as a resource frontier, with companies and states competing to generate knowledge and technologies that can justify privatized claims and enable their exploitation (Havice & Zalik 2018).

The expansion of mining infrastructure into the ocean could be understood as part of the expansion of the "planetary mine." This term was coined by Labban (2014), who uses it to describe the deterritorialization of mining through two processes: mining metals from waste, which allows mining operations to "transcend geological space-time and extend at the planetary scale" (2014: 563), and using microbes in the metallic leaching process, which extends mining processes to the molecular scale while blurring the line between biologically and nonbiologically based forms of extraction. In this analysis, new technologies transform not only the scale of mining operations but also their spatial arrangements and operational logics. As reworked by Arboleda (2020), the planetary mine traces the geography of extraction that has emerged at the intersection of late industrialization (helmed by China) and the fourth machine age, which has transformed how minerals are mined, transported, and consumed. Shared across both visions of the planetary mine, however, is the assertion that the geography of mining is becoming both more expansive and more pervasive.

For many geographers, the forms of expertise that underpin this changing geography of mining constitute crucial areas of inquiry. Extractive economies rely on natural sciences like geology, hydrology, and chemistry for the practical purposes of determining where to dig and how to refine, but they are also morally justified by the ontological distinction between life and nonlife – a distinction whose origins are entangled with the emergence of modern science itself. As geographers have recently argued, geological knowledge in particular is bound up with histories of racism (Yusoff 2018; Himley 2020), colonial territorialization (Braun 2000; Simpson 2019), and national in/exclusions (Kneas 2018; Marston 2024). The logics of extractivism are "more than simply capitalist" as they involve sexualized and racialized objectifications of both laborers and resources (Murrey & Mollett 2023: 762). Yet they are also deeply capitalist, as they render land knowable, interchangeable, and investable (Goldstein 2016; Le Billon & Sommerville 2017). By interrogating the proximity of science and economics, however, resource geographers create space for alternative ways of narrating the world. Dominant forms of expertise cannot erase knowledges that offer less validation for extractive practices, as is the case for many North American Indigenous knowledges (Cameron 2015; Hoogeveen 2016).

Nevertheless, extractive processes often overflow their borders, and part of the power of the concept of extractivism is that it suggests a set of relationships that might flow from a specific oil rig or government policy, but that are not contained therein. Large-scale resource extraction infrastructure also supports the proliferation of small-scale extractive endeavors such as small-scale mining or even smallholding agriculture that is market rather than subsistence oriented (Holley et al. 2020). Physical infrastructures such as roads, ports, and electrical grids might be extended to serve the needs of a large-scale company, but small-scale operators also benefit from their presences (De Theije & Heemskerk 2009). Similarly, the technologies and knowledges required to produce specific resources tend to circulate as discarded machinery and underemployed workers.

Even outside the world of resource extraction, extractive dynamics are at play in social relationships characterized by unequal exchange. Tellingly, researchers and activists now use the term "extractive" to describe academic research that relies on cannibalizing local knowledges without consent, attribution, and/or compensation (Johnson et al. 2021). The term resonates in part because the people from whom *knowledge* is extracted are often also experiencing the negative impacts of *resource* extraction. In this sense, extractivism describes the old (but ongoing) colonial dynamic of appropriation, typically associated with land and nature but also potentially applicable to unwaged labor (Moore 2015; Nichols 2020). This dynamic undergirds capitalist economies, even if it is not the exclusive domain of capitalism.

Plantations

For the last several decades, geographers have been concerned with "land grabbing," understood as the concentration of enormous tracts of land in the hands of agribusinesses, foreign states, and speculators, among others. More recently, geographers and others have been pushing the land-grab conversation in new directions, especially as it has converged with parallel theoretical discussions around the plantation (Oliveira, McKay, & Liu 2021). Retaining the attention on scale and finance afforded by the land-grabbing framework, the plantation concept draws analytical focus to processes of racialization, labor exploitation, unequal exchange, and the persistence of a socionatural formation rooted in the early colonial era. Many resources are produced in contexts that might be described as plantations, and resource geographers pursue this topic analytically to amplify the political salience of their findings.

Plantation scholarship has a long history in Black studies and remains an important focus of Black geographies. Whether in the Caribbean, Brazil, or the

United States, shared features of the plantation include its extensive land requirements, intensive capital and labor requirements, and internal forms of governance that differ across racialized and sexualized hierarchies (Woods 2017 [1998]). The word "plantation" indexes the historical period in which Europeans began relying on stolen land and forced labor in the Americas, Africa, and Asia to finance industrialization in Europe, a reliance that not only facilitated the transition to capitalism but also constituted it (Robinson 2000 [1983]). Yet the plantation is also very much a contemporary phenomenon. Most obviously, it manifests as agribusinesses (Hetherington 2020; Chao 2022), but it also appears in the prison complex and in destroyed city sectors (McKittrick 2013).

These insights have shaped the emerging conversation between resource geography and plantation studies, pushing the former to attend more closely to the historical manipulation of socioenvironmental infrastructure as well as to the spatialization of race. For instance, Kay, Knudson, and Cantor (2024) argue that the expansion of irrigation infrastructure for sugar plantations in Maui has created water shortages for Native Hawaiians, while Stock (2023) considers how solar park development in India resembles an "energy plantation." Even urban municipalities "are accomplished through the plantation tradition of using racial social structure and natural resources – particularly land and water – as instruments of development" (Purifoy 2022: 115), which means that contemporary environmental injustices are expressions of old plantation logics. In all these examples, the authors approach the plantation not just as an export-oriented tract of land but as a set of spatial logics, infrastructures, and unevenly valued people.

Recently, thinking around the plantation has coincided with attempts to name and define the current geological era. While this latter debate has resulted in an unwieldy proliferation of names, "Plantationocene" proponents underscore how "large-scale, export-oriented agriculture dependent on forced labor has played a dominant role in structuring modern life since the insertion of European power in the Americas, Asia, and Africa" (Wolford 2021: 1622). If the concept of extractivism directs attention to the appropriation of nature, the Plantationocene centers racialized enslavement "as 'the' original sin" (Wolford 2021: 1622), neither a precapitalist remnant nor a temporary feature of capitalist expansion. It also raises the question of scale, suggesting that there is something unique about plantations' geographical expansiveness and modular interchangeability.

The Plantationocene, however, is not without its detractors. The term was originally coined in a forum about the Anthropocene (Haraway et al. 2016) and a troubling series of quips in this conversation were subsequently critiqued for equating Black and Native enslavement with ecological simplification, and even definitionally privileging the latter. These early discussions also elided historical and conceptual work on plantations within Black studies (for a full critique, see Davis et al. 2019). More recent scholars of the Plantationocene have taken care to foreground both the racial logics of the plantation and the rich history of plantation-focused scholarship

(Chao et al. 2024; Li 2023), but this does not mean that the term is universally embraced. For example, in a consideration of the term's applicability in and around the Colorado River, Curley and Smith (2023) argue that the Plantationocene remains limited by its tendency to reproduce Native absence.

Regardless of how useful one finds the "-cene" suffix, however, the plantation – as concept, idea, place, organizing principle, and lived reality – is shaping thinking around the geography of resources, and even inflecting ongoing land-grab conversations. In a recent volume on land grabbing (Neef et al. 2023), several authors engage the plantation analytically. Goffe (2023) argues that land grabbing in Jamaica is accomplished in part through the criminalization of squatting, a practice rooted in the policing of plantations, while Correia (2023) explores the relationship between recent land grabs and the longue durée of plantation violence in Paraguay. Still attentive to the new flows of financial capital and legal dealings that facilitate land grabs, these authors and others draw on the plantation as a concept to highlight the persistence of historically sedimented injustices.

Conclusion

Resources are the outcome of capitalist-colonial ideas about and interventions within nature, yet they are also the condition of possibility for the capitalist-colonial present. They are the material building supplies and the fuels, and their production entails forms of dispossession and dehumanization that are arguably better indicators of capitalism than the wage relation. At present, however, capitalism threatens to undermine its own conditions of possibility, given global environmental crises such as soil depletion, water shortages and contamination, deforestation, ocean acidification, and climate change, among others. These planetary changes threaten many resource industries even in the absence of widespread social protest. The future of resources – including which ones could "save" the planet, where they will be produced, and who or what will exert control over them – is very much up for debate.

In this context, resource geographers work to ground socioenvironmental concerns in specific places and tangible materials. Recent research on extractivism and plantations brings slightly different dynamics into focus: extractivist frameworks tend to feature global flows of resources and money, whereas the language of the plantation emphasizes land, labor, and the perpetuation of racial hierarchies. Yet both concepts ultimately guide resource geographers toward expansive and even planetary analyses, as authors trace close connections between extractive sites, investments, and global markets. Even when empirically focused on a particular tract of land, ocean, or subsoil, extractivist and plantation geographies underscore the economic and social interdependence of world regions, as well as the proliferation of extractive and plantation logics far beyond political borders.

Resource geographers also demonstrate how resources are integral to the emergence and expression of territory, sovereignty, and collective identities. The widespread social and environmental transformations wrought by resource extraction – including dispossessions, policing, new labor relations, diverted water resources, chemical drift, simplified ecologies, watershed or aquifer contamination, and so on – necessarily reshape existing social relations and even social narratives. While this sometimes manifests in public rejection of resource extraction, it can also result in demands for a share of the profit, and conflicts over the distribution of resource rent can be as inflammatory as conflicts over extraction itself. In both scenarios, however, resources play a key role in producing and challenging hegemonic relationships between nations, subnational regions, and supranational blocs. From a political geographical perspective, no territory can be understood in isolation from its resource histories, industries, and trade relations.

References

Adaman, F., Arsel, M., & Akbulut, B. (2021). Neoliberal developmentalism, authoritarian populism, and extractivism in the countryside: The soma mining disaster in Turkey. In Scoones, I., Edelman, M., Borras, S.M., Jr., et al. (eds), *Authoritarian Populism and the Rural World*. New York: Routledge, 154–176.

Andrews, N., Bennett, N.J., Le Billon, P., et al. (2021). Oil, fisheries and coastal communities: A review of impacts on the environment, livelihoods, space and governance. *Energy Research & Social Science 75*: 102009.

Anthias, P. (2018). *Limits to Decolonization: Indigeneity, Territory, and Hydrocarbon Politics in the Bolivian Chaco*. Ithaca, NY: Cornell University Press.

Arboleda, M. (2020). *Planetary Mine: Territories of Extraction Under Late Capitalism*. London: Verso.

Awasis, S. (2023). Decolonial process tracing: Indigenous rights and pipeline resistance movements. *Environment and Planning F 2*(1–2, 162): 144.

Baka, J., Hesse, A., Weinthal, E., & Bakker, K. (2019). Environmental knowledge cartographies: Evaluating competing discourses in U.S. hydraulic fracturing rule-making. *Annals of the American Association of Geographers 109*(6): 1941–1960. doi:10.1080/24694452.2019.1574549

Beckett, C., & Keeling, A. (2019). Rethinking remediation: Mine reclamation, environmental justice, and relations of care. *Local Environment 24*(3): 216–230. doi:10.1080/13549839.2018.1557127

Blair, J.J.A., Balcázar, R.M., Barandiarán, J., & Maxwell, A. (2023). The "alterlives" of green extractivism: Lithium mining and exhausted ecologies in the Atacama Desert. *International Development Policy 16*. doi:10.4000/poldev.5284

Bosworth, K. (2022). *Pipeline Populism: Grassroots Environmentalism in the Twenty-First Century*. Minneapolis, MN: University of Minnesota Press.

Braun, B. (2000). Producing vertical territory: Geology and governmentality in late Victorian Canada. *Cultural Geographies 7*(1): 7–46.

Bridge, G. (2009). Material worlds: Natural resources, resource geography and the material economy. *Geography Compass 3*(3): 1217–1244. doi:10.1111/j.1749-8198.2009.00233.x

Bryan, J. (2022). Mapping resources: Mapping as method for critical resource geographies. In Himley, M., Havice, E., & Valdivia, G. (eds), *The Routledge Handbook of Critical Resource Geography*. London: Routledge, 441–452.

Bustos-Gallardo, B., Bridge, G., & Prieto, M. (2021). Harvesting lithium: Water, brine and the industrial dynamics of production in the Salar de Atacama. *Geoforum 119*: 177–189.

Cameron, E. (2015). *Far off Metal River: Inuit Lands, Settler Stories, and the Making of the Contemporary Arctic*. Vancouver: UBC Press.

Chao, S. (2022). *In the Shadow of the Palms: More-Than-Human Becomings in West Papua*. Durham, NC: Duke University Press.

Chao, S., Wolford, W., Ofstehage, A., Guttal, S., Gonçalves, E., & Ayala, F. (2024). The Plantationocene as analytical concept: A forum for dialogue and reflection. *Journal of Peasant Studies 51*(3): 541–563. doi:10.1080/03066150.2023.2228212

Childs, J. (2016). Geography and resource nationalism: A critical review and reframing. *The Extractive Industries and Society 3*(2): 539–546.

Childs, J. (2020). Extraction in four dimensions: Time, space and the emerging geo(−) politics of deep-sea mining. *Geopolitics 25*(1): 189–213.

Correia, J.E. (2023). Ruptures and continuities: How the global land "rush" (re)produces slow violence on Latin America's resource frontiers. In Neef, A., Ngin, C., Moreda, T., & Mollett, S. (eds), *Routledge Handbook of Global Land and Resource Grabbing*. London: Routledge, 47–62.

Cram, S. (2023). *Unmaking the Bomb: Environmental Cleanup and the Politics of Impossibility*. Oakland, CA: University of California Press.

Curley, A. (2023). *Carbon Sovereignty: Coal, Development, and Energy Transition in the Navajo Nation*. Tempe, AZ: University of Arizona Press.

Curley, A., & Smith, S. (2023). Unruly river and plantation logics. *Annals of the American Association of Geographers 114*: 2182–2110, 2188.

Dafermos, Y., Gabor, D., & Michell, J. (2021). The Wall Street consensus in pandemic times: What does it mean for climate-aligned development? *Canadian Journal of Development Studies / Revue canadienne d'études du développement 42*(1–2): 238–251. doi:10.1080/02255189.2020.1865137

Davies, T. (2022). Slow violence and toxic geographies: "Out of sight" to whom? *Environment and Planning C: Politics and Spaces 40*(2): 409–427.

Davis, J., Moulton, A.A., Van Sant, L., & Williams, B. (2019). Anthropocene, Capitalocene, . . . Plantationocene? A manifesto for ecological justice in an age of global crises. *Geography Compass 13*(5): e12438. doi:10.1111/gec3.12438

de la Cadena, M. (2010). Indigenous cosmopolitics in the Andes: Conceptual reflections beyond politics. *Cultural Anthropology 25*(2): 334–370.

De Theije, M., & Heemskerk, M. (2009). Moving frontiers in the Amazon: Brazilian small-scale gold miners in Suriname. *European Review of Latin American and Caribbean Studies | Revista Europea de Estudios Latinoamericanos y del Caribe* (87): 5. doi:10.18352/erlacs.9600

DeBoom, M.J. (2021). Climate necropolitics: Ecological civilization and the distributive geographies of extractive violence in the Anthropocene. *Annals of the American Association of Geographers 111*(3): 900–912. doi:10.1080/24694452.2020.1843995

Dempsey, J., & Suarez, D.C. (2016). Arrested development? The promises and paradoxes of "selling nature to save it. *Annals of the American Association of Geographers 106*(3): 653–671. doi:10.1080/24694452.2016.1140018

Dillon, L. (2014). Race, waste, and space: Brownfield redevelopment and environmental justice at the Hunters Point Shipyard. *Antipode 46*(5): 1205–1221. doi:10.1111/anti.12009

Franz, T., & McNelly, A. (2023). *The "Finance-Extraction-Transitions Nexus": Towards a Critical Research Agenda Exploring the Scramble for Transition Minerals.* London: SOAS Department of Economics.

Fry, M., & Delgado, E. (2018). Petro-geographies and hydrocarbon realities in Latin America. *Journal of Latin American Geography 17*(3): 10–14. doi:10.1353/lag.2018.0039

Furlong, K., & Norman, E.S. (2015). Resources. In Agnew, J., Secor, A.J., & Sharp, J. (eds), *The Wiley Blackwell Companion to Political Geography.* Chichester: John Wiley & Sons, 424–437.

Gidwani, V., & Reddy, R.N. (2011). The afterlives of "waste": Notes from India for a minor history of capitalist surplus. *Antipode 43*(5): 1625–1658. doi:10.1111/j.1467-8330.2011.00902.x

Goffe, R. (2023). Capture land: Anti-squatting policy as processual land grab in Jamaica. In Neef, A., Ngin, C., Moreda, T., & Mollett, S. (eds), *Routledge Handbook of Global Land and Resource Grabbing.* London: Routledge, 65–80.

Goldstein, J.E. (2016). Knowing the subterranean: Land grabbing, oil palm, and divergent expertise in Indonesia's peat soil. *Environment and Planning A: Economy and Space 48*(4): 754–770. doi:10.1177/0308518X15599787

Goldstein, J. (2013). *Terra economica*: Waste and the production of enclosed nature. *Antipode 45*(2): 357–375. doi:10.1111/j.1467-8330.2012.01003.x

Graybill, J.K. (2019). Emotional environments of energy extraction in Russia. *Annals of the American Association of Geographers 109*(2): 382–394.

Gudynas, E. (2009). Diez tesis urgentes sobre el nuevo extractivismo: Contextos y demandas bajo el progresismo sudamericano actual. In Dávila, F.R. (ed.), *Extractivismo, Política y Sociedad.* CAAP/CLAES: Quito, 187–225.

Hamouchene, H., & Sandwell, K. (eds) (2023). *Dismantling Green Colonialism: Energy and Climate Justice in the Arab Region.* London: Pluto Press.

Haraway, D., Ishikawa, N., Gilbert, S.F., Olwig, K., Tsing, A.L., & Bubandt, N. (2016). Anthropologists are talking – about the Anthropocene. *Ethnos 81*(3): 535–564. doi:10.1080/00141844.2015.1105838

Havice, E., & Zalik, A. (2018). Ocean frontiers: Epistemologies, jurisdictions, commodifications. *International Social Science Journal 68*(229–230): 219–235. doi:10.1111/issj.12198

Hecht, G. (2023). *Residual Governance: How South Africa Foretells Planetary Futures.* Durham, NC: Duke University Press.

Hetherington, K. (2020). *Beans Before the Law: Knowledge Practices, Responsability, and the Paraguayan Soy Boom.* Durham, NC: Duke University Press.

Himley, M. (2020). Underground geopolitics: Science, race, and territory in Peru during the late nineteenth century. In O'Lear, S. (ed.), *A Research Agenda for Environmental Geopolitics.* Cheltenham: Edward Elgar, 74–87. doi:10.4337/9781788971249.00013

Holley, E.A., Smith, N.M., Jimenez, J.A.D., Cabezas, I.C., & Restrepo-Baena, O.J. (2020). Socio-technical context of the interactions between large-scale and small-scale mining in Marmato, Colombia. *Resources Policy 67*: 101696. doi:10.1016/j.resourpol.2020.101696

Hoogeveen, D. (2016). Fish-hood: Environmental assessment, critical Indigenous studies, and posthumanism at Fish Lake (Teztan Biny), Tsilhqot'in territory. *Environment and Planning D: Society and Space 34*(2): 355–370.

Huber, M. (2019). Resource geography II: What makes resources political? *Progress in Human Geography 43*(3): 553–564. doi:10.1177/0309132518768604

Huber, M.T. (2022). *Climate Change as Class War: Building Socialism on a Warming Planet*. London: Verso.

İnal, O. (2022). Authoritarianism, populism, and the environment in Turkey. *Environmental History 27*(4): 634–641.

Jerez, B., Garcés, I., & Torres, R. (2021). Lithium extractivism and water injustices in the Salar De Atacama, Chile: The colonial shadow of green electromobility. *Political Geography 87*: 102382.

Johnson, A., Zalik, A., Mollett, C.S., et al. (2021). Extraction, entanglements, and (im)materialities: Reflections on the methods and methodologies of natural resource industries fieldwork. *Environment and Planning E: Nature and Space 4*(2): 383–428. doi:10.1177/2514848620907470

Kama, K. (2016). Contending geo-logics: Energy security, resource ontologies, and the politics of expert knowledge in Estonia. *Geopolitics 21*(4): 831–856. doi:10.1080/14650045.2016.1210129

Kama, K. (2020). Resource-making controversies: Knowledge, anticipatory politics and economization of unconventional fossil fuels. *Progress in Human Geography 44*(2): 333–356. doi:10.1177/0309132519829223

Kay, K., Knudson, C., & Cantor, A. (2024). Plantation pasts, plantation futures: Resisting zombie water infrastructures in Maui, Hawai'i. *Journal of Peasant Studies 51*(1): 111–134. doi:10.1080/03066150.2023.2185140

Kleimann, D., Poitiers, N., Sapir, A. et al. (2023). How Europe should answer the US Inflation Reduction Act. Bruegel Policy Contribution, No. 04/2023. Brussels: Bruegel.

Klinger, J.M. (2020). *Rare Earth Frontiers: From Terrestrial Subsoils to Lunar Landscapes*. Ithaca, NY: Cornell University Press.

Kneas, D. (2018). From dearth to El Dorado: Andean nature, plate tectonics, and the ontologies of Ecuadorian resource wealth. *Engaging Science, Technology, and Society 4*: 131–154. doi:10.17351/ests2018.214

Knuth, S., Behrsin, I., Levenda, A., & McCarthy, J. (2022). New political ecologies of renewable energy. *Environment and Planning E: Nature and Space 5*(3): 997–1013. doi:10.1177/25148486221108164

Koch, N. (2022). Wastelanding Arabia: America's "Garden of Eden" in Al Kharj, Saudi Arabia. *Journal of Historical Geography 77*: 13–24.

Koch, N., & Perreault, T. (2019). Resource nationalism. *Progress in Human Geography 43*(4): 611–631. doi:10.1177/0309132518781497

Kopack, R.A. (2019). Rocket wastelands in Kazakhstan: Scientific authoritarianism and the Baikonur cosmodrome. *Annals of the American Association of Geographers 109*(2): 556–567.

Köppel, J. (2020). Lithium transformations: An unfinished story. *Transformation 33*: 27–47.

Kuchler, M. (2017). Post-conventional energy futures: Rendering Europe's shale gas resources governable. *Energy Research & Social Science 31*: 32–40. doi:10.1016/j.erss.2017.05.028

Kuchler, M., & Bridge, G. (2023). Speculating on shale: Resource-making and the "politics of possibility" in Poland and the UK. *Political Geography 107*: 102978. doi:10.1016/j.polgeo.2023.102978

Kuletz, V. (2016 [1998]). *The Tainted Desert: Environmental Ruin in the American West.* Abingdon: Routledge.

Labban, M. (2014). Deterritorializing extraction: Bioaccumulation and the planetary mine. *Annals of the Association of American Geographers 104*(3): 560–576.

Labussière, O. (2021). A coalbed methane "stratum" for a low-carbon future: A critical inquiry from the Lorraine Basin (France). *Political Geography 85*: 102328. doi:10.1016/j.polgeo.2020.102328

LaDuke, W., & Cowen, D. (2020). Beyond Wiindigo infrastructure. *South Atlantic Quarterly 119*(2): 243–268. doi:10.1215/00382876-8177747

Le Billon, P., & Sommerville, M. (2017). Landing capital and assembling "investable land" in the extractive and agricultural sectors. *Geoforum 82*: 212–224. doi:10.1016/j.geoforum.2016.08.011

Lepawsky, J. (2015). The changing geography of global trade in electronic discards: Time to rethink the e-waste problem. *Geographical Journal 181*(2): 147–159. doi:10.1111/geoj.12077

Li, T.M. (2023). Indonesia's Plantationocene. *Annals of the American Association of Geographers 114*(10): 2194–2198. doi:10.1080/24694452.2023.2201633

Liboiron, M. (2021). *Pollution Is Colonialism.* Durham, NC: Duke University Press.

Marston, A. (2024). *Subterranean Matters: Cooperative Mining and Resource Nationalism in Plurinational Bolivia.* Durham, NC: Duke University Press.

McCarthy, J. (2019). Authoritarianism, populism, and the environment: Comparative experiences, insights, and perspectives. *Annals of the American Association of Geographers 109*(2): 301–313. doi:10.1080/24694452.2018.1554393

McCreary, T.A., & Milligan, R.A. (2014). Pipelines, permits, and protests: Carrier Sekani encounters with the Enbridge Northern Gateway Project. *Cultural Geographies 21*(1): 115–129. doi:10.1177/1474474013482807

McKay, B.M., Oliveira, G.d.L.T., & Liu, J. (2020). Authoritarianism, populism, nationalism and resistance in the Agrarian South. *Canadian Journal of Development Studies / Revue canadienne d'études du développement 41*(3): 347–362. doi:10.1080/02255189.2020.1814707

McKittrick, K. (2013). Plantation futures. *Small Axe: A Caribbean Journal of Criticism 17*(3): 1–15. doi:10.1215/07990537-2378892

Meché, B. (2022). Black as drought: Arid landscapes and ecologies of encounter across the African diaspora. *Environment and Society 13*(1): 60–77.

Moore, J.W. (2015). *Capitalism in the Web of Life: Ecology and the Accumulation of Capital.* New York: Verso.

Morinville, C. (2021). Mining the waste stream: Value and disposability in the global e-waste economy. PhD dissertation, University of Toronto.

Murphy, M. (2017). Alterlife and decolonial chemical relations. *Cultural Anthropology* *32*(4): 494–503. doi:10.14506/ca32.4.02

Murrey, A., & Mollett, S. (2023). Extraction is not a metaphor: Decolonial and Black Geographies against the gendered and embodied violence of extractive logics. *Transactions of the Institute of British Geographers 48*(4): 761–780. doi:10.1111/tran.12610

Neef, A., Ngin, C., Moreda, T., & Mollett, S. (eds) (2023). *Routledge Handbook of Global Land and Resource Grabbing*. London: Routledge. doi:10.4324/9781003080916

Nichols, R. (2020). *Theft Is Property! Dispossession and Critical Theory*. Durham, NC: Duke University Press.

Ofstehage, A., Wolford, W., & Borras, S.M. (2022). Contemporary populism and the environment. *Annual Review of Environment and Resources 47*(1): 671–696. doi:10.1146/annurev-environ-012220-124635

Oliveira, G.d.L.T., McKay, B.M., & Liu, J. (2021). Beyond land grabs: New insights on land struggles and global agrarian change. *Globalizations 18*(3): 321–338. doi:10.1080/14747731.2020.1843842

Ortiz Gallego, D. (2024). The "agropopulist shift" toward authoritarianism: The conflict between the MAS government and coca growers from the Bolivian Yungas. *Journal of Peasant Studies*. doi:10.1080/03066150.2024.2309297

Perreault, T. (2018). The plantation and the mine: Comment on "After the land grab: Infrastructural violence and the 'mafia system' in Indonesia's oil palm plantation zone" by Tania Li. *Geoforum 96*: 345–347. doi:10.1016/j.geoforum.2018.02.025

Pulido, L. (2017). Geographies of race and ethnicity II: Environmental racism, racial capitalism and state-sanctioned violence. *Progress in Human Geography 41*(4): 524–533. doi:10.1177/0309132516646495

Purifoy, D.M. (2022). The plantation town: Race, resources, and the making of place. In Himley, M., Havice, E., & Valdivia, G. (eds), *The Routledge Handbook of Critical Resource Geography*. London: Routledge, 114–125.

Purifoy, D.M., & Seamster, L. (2021). Creative extraction: Black towns in white space. *Environment and Planning D: Society and Space 39*(1): 47–66. doi:10.1177/0263775820968563

Ranganathan, M. (2022). Caste, racialization, and the making of environmental unfreedoms in urban India. *Ethnic and Racial Studies 45*(2): 257–277. doi:10.1080/01419870.2021.1933121

Riofrancos, T. (2023). The security–sustainability nexus: Lithium onshoring in the Global North. *Global Environmental Politics 23*(1): 20–41.

Riofrancos, Thea, Kendall, A., Dayemo, K. K. et al. (2023). Achieving zero emissions with more mobility and less mining. Climate and Community Project. http://www.climateandcommunity.org/more-mobility-less-mining (accessed December 1, 2024).

Robinson, C.J. (2000 [1983]). *Black Marxism: The Making of the Black Radical Tradition*. Chapel Hill, NC: University of North Carolina Press.

Schaffartzik, A., Mayer, A., Eisenmenger, N., & Krausmann, F. (2016). Global patterns of metal extractivism, 1950–2010: Providing the bones for the industrial society's skeleton. *Ecological Economics 122*: 101–110. doi:10.1016/j.ecolecon.2015.12.007

Simpson, M. (2019). Resource desiring machines: The production of settler colonial space, violence, and the making of a resource in the Athabasca Tar Sands. *Political Geography 74*: 102044. doi:10.1016/j.polgeo.2019.102044

Soto Hernandez, D., & Newell, P. (2023). Oro blanco: Assembling extractivism in the lithium triangle. In Scoones, I., Edelman, M., Borras, S.M., Jr., et al. (eds), *Climate Change and Critical Agrarian Studies*. Abingdon and New York: Routledge, 248–271.

Stock, R. (2023). Power for the Plantationocene: Solar parks as the colonial form of an energy plantation. *Journal of Peasant Studies 50*(1): 162–184. doi:10.1080/03066150.2022.2120812

Ureta, S. (2016). Caring for waste: Handling tailings in a Chilean copper mine. *Environment and Planning A: Economy and Space 48*(8): 1532–1548. doi:10.1177/0308518X16645103

Valdivia, G. (2018). "Wagering life" in the petro-city: Embodied ecologies of oil flow, capitalism, and justice in Esmeraldas, Ecuador. *Annals of the American Association of Geographers 108*(2): 549–557. doi:10.1080/24694452.2017.1369389

Valdivia, G., Himley, M., & Havice, E. (2022). Critical resource geography: An introduction. In Matthew Himley, H., & Valdivia, G. (eds), *The Routledge Handbook of Critical Resource Geography*. New York: Routledge, 1–20.

Vasudevan, P. (2021). An intimate inventory of race and waste. *Antipode 53*(3): 770–790. doi:10.1111/anti.12501

Vivoda, V., Matthews, R., & McGregor, N. (2024). A critical minerals perspective on the emergence of geopolitical trade blocs. *Resources Policy 89*: 104587. doi:10.1016/j.resourpol.2023.104587

Voyles, T.B. (2015). *Wastelanding: Legacies of Uranium Mining in Navajo Country*. Minneapolis, MN: University of Minnesota Press.

Wolford, W. (2021). The Plantationocene: A lusotropical contribution to the theory. *Annals of the American Association of Geographers 111*(6): 1622–1639. doi:10.1080/24694452.2020.1850231

Woods, C. (2017 [1998]). *Development Arrested: The Blues and Plantation Power in the Mississippi Delta*. London: Verso.

Wylie, S.A. (2018). *Fractivism: Corporate Bodies and Chemical Bonds*. Durham, NC: Duke University Press.

Ybarra, M. (2018). *Green Wars: Conservation and Decolonization in the Maya Forest*. Oakland, CA: University of California Press.

Young, K.A. (2017). *Blood of the Earth: Resource Nationalism, Revolution, and Empire in Bolivia*. Austin, TX. University of Texas Press.

Yusoff, K. (2018). *A Billion Black Anthropocenes or None*. Minneapolis, MN: University of Minnesota Press.

Zalik, A. (2022). World-making and the deep seabed: Mining the *area* beyond national jurisdiction. In Himley, M., Havice, E., & Valdivia, G. (eds), *The Routledge Handbook of Critical Resource Geography*. London: Routledge, 412–424.

Zaragocin, S., & Caretta, M.A. (2021). *Cuerpo-territorio*: A decolonial feminist geographical method for the study of embodiment. *Annals of the American Association of Geographers 111*(5): 1503–1518. doi:10.1080/24694452.2020.1812370

Zimmermann, E.W. (1951). *World Resources and Industries: A Functional Appraisal of the Availability of Agricultural and Industrial Resources*, 2nd edn. New York: Harper and Brothers.

Chapter 32

Geopolitical Ecologies

Hilary Faxon[1] *and Kendra Kintzi*[2]

[1]*W.A. Franke College of Forestry and Conservation, University of Montana, Missoula, MT, USA*

[2]*Cornell Atkinson Center for Sustainability, Cornell University, Ithaca, NY, USA*

As global temperatures rise and glaciers melt, it is obvious that the environment is a problem for all of us, and for political geography. In recent years, political geography has embraced political ecology as an approach to understanding governance, power, and resistance across forests, fields, and atmospheres (Benjaminsen et al. 2017; Sultana 2022; Collins 2024; Paredes, Gianella, & Olivera 2024). With a long, if heterogenous, tradition of attention to how access and control over resources shape social identities and power relations across scales, political ecology offers analytical and methodological tools that enhance political geography's ability to tackle environmental problems.

Thinking with political geography and political ecology together underscores both the inherently political nature of the environment and the increasingly environmental nature of politics. Previous accounts of the two fields' intersection have taken the state as their object (Robbins 2008; Harris 2017; Loftus 2020). In contrast, this chapter explores environment, space, and power at the edge of the state and beyond. We offer a "geopolitical ecology" that centers both state and nonstate actors, as well as the socioecological implications of intensive resource extraction to fuel decarbonization and digitalization. As Graddy-Lovelace and Ranganathan (2023: 1) write, "Geopolitical ecology reveals the ecological formations driving state-led territorial and economic expansion via militarized security, corporate-led resource extraction, and the conquest of land." In other words, geopolitical ecologies move beyond resource determinism and the "territorial trap" (Agnew 1994) to question how resources are made and mobilized in particular ways to advance particular interests, across local and global scales.

The Wiley Blackwell Companion to Political Geography, Second Edition.
Edited by Virginie Mamadouh, Natalie Koch, Chih Yuan Woon, and John Agnew.

This chapter shows how a geopolitical ecology approach can be applied to a specific issue: the extraction and commercialization of critical minerals. As Julie Klinger writes, critical minerals – including rare earth minerals – are not actually rare. Their production and circulation are shaped not only by geology, but also by geopolitics: "The twin desires to isolate the hazards [of mining] while capturing the geopolitical benefits of keeping the production of these strategically vital elements within a particular set of borders drives production to the frontiers of empire, state, and capital" (Klinger 2017: 11). This tension shapes geographies of production with profoundly uneven political, social, and ecological effects. How particular materials become prized for extraction, and how they are used, controlled, and consolidated, are mediated through existing relationships across national boundaries. These unfolding processes take place at multiple scales, from grassroots environmental mobilizations to transnational investments (Klinger 2020).

The contemporary critical mineral rush has brought challenges familiar from previous waves of resource extraction, while also raising new questions of global climate governance. Materials like lithium, copper, cobalt, and nickel are key components of renewable technologies like wind and solar, the electricity grid infrastructures they connect to, and the handheld devices, utility-scale batteries, and data-storage facilities that comprise smart systems. But framings that emphasize the importance of these materials for decarbonization and digitalization can be used to justify militarization, extend spatial inequalities, and entrench extractive paradigms. While the name suggests a finite amount, a geopolitical ecology approach impels us to consider how scarcity and value are socially constructed, including through state and nonstate projects of securitization. Critical minerals are not discovered but made.

In this chapter we first provide a brief introduction to political ecology, tracing the genealogy and key concerns of the field. We then turn to recent work at the intersection of political geography and political ecology that examines the role of resource frontiers in state-making and how political contestation reshapes extractive landscapes across scales. We apply these insights to examine copper mining in Jordan and rare earth extraction in Myanmar as illustrative sites of making and maintaining critical mineral frontiers. Finally, we demonstrate how a geopolitical ecology approach to critical mineral frontiers reveals how new forms of value and authority emerge at edge of the state.

The Roots of Political Ecology

Environmental problems are not inevitable, but they are always political. The interdisciplinary field of political ecology is characterized by a theoretical commitment to critical social theory, a methodological commitment to in-depth, direct observation, and a political commitment to questions of social equity and socioenvironmental governance (Perreault, Bridge, & McCarthy 2020). Political ecologists are interested in both material resources and meanings, exploring environmental knowledge, change, governance, identities, and politics (Perreault, Bridge, & McCarthy 2020). They insist on the fundamental interrelation of power, nature, and society.

Political ecology emerged in the 1970s and 1980s in critical dialogue with cultural ecology, natural hazards research, and ecological anthropology. These subfields were concerned with how groups adapted to their environments and how environmental research could contribute to understanding human resilience and social change. As Michael Watts argues in his (1983) treatise "On the poverty of theory," the central failing of these earlier frameworks is that they employ a "billiard ball" approach to human–environment interaction, which posits an atomized view of society as made up of individual agents in closed, functional social systems who rationally respond to external environmental changes and threats.

This framing presents a limited and limiting view of the social relations that mediate how resources are produced, distributed, and used. It fails, crucially, to understand how nature and society are fundamentally linked through everyday practices, and how the boundaries around any social group continually change through interaction with broader social and economic systems. Social ills – be they poverty, famine, or underdevelopment – are recoded as the adaptive failures of overpopulation, ignorance, or inappropriate technology. Instead, political ecology focuses on how nature is transformed, extracted, and appropriated to support the maintenance and reproduction of larger power relations (Svarstad, Benjaminsen, & Overå 2018). While paying close attention to materiality and ecology, the field emphasizes that natural hazards are not just natural but are made through broader national, regional, and global politics (Walker 2005, 2007). Rather than a simple, replicable "resource curse," uneven power relations *produce* uneven access to resources and vulnerability to environmental hazards, shaping the winners and losers in negotiations over environmental benefits and burdens (Peluso & Watts 2001; Watts 2004).

Today, political ecology is an unruly and energetic set of scholarly conversations that both criticize existing injustices and nurture alternative futures, twinned tendencies that Paul Robbins (2012) calls the "hatchet" and the "seed." While still true to its Marxist roots, different epistemologies and approaches – including feminist and decolonial theories – have reshaped the field's concerns and methodologies (Rocheleau & Edmunds 1997; Elmhirst 2011; Sultana 2021). This has given rise to a productive tension between "critical deconstruction" and "radical advocacy," roughly aligned with North American and European schools of political ecology (Desvallées, Arnauld De Sartre, & Kull 2022). Over the past two decades, "ecology" as a topic of inquiry has undergone a transformation. While canonical concerns with land degradation (Fairhead & Leach 1995; Blaikie & Brookfield 2015) and community-based natural resource management (Li 2002) remain relevant, recent years have seen growing interest in new topics such as urban metabolisms (Heynen 2014), renewable energy transitions (Avila et al. 2022; Knuth et al. 2022), and data (Akhter 2017; Goldstein & Nost 2022).

Despite these evolutions, the field remains characterized by several key analytical moves. First, this is *political*, not apolitical, ecology. In Robbins's (2012: 13) terms: "This is the difference between identifying broader systems rather than blaming proximate and local forces; between viewing ecological systems as power-laden rather than politically inert; and between taking an explicitly normative approach rather than one that claims the objectivity of disinterest." Unlike the earlier models of isolated social–environmental interaction that Watts critiqued, political ecology is premised on the existence of unequal distribution. Second, resource conflicts are not universal but rather historical and place based. They take shape through dynamic interactions between human and nonhuman natures, through both material and discursive processes that stretch from local to global scales. Finally, political ecologists think with paradox. By investigating surprises, challenging assumptions, and probing contradictions, they push beyond the party line to generate theory from the margins.

Resource Frontiers and State-Making

Political ecology and political geography have many conceptual synergies. To date, much of the scholarship that explicitly bridges these fields has focused on the state (Robbins 2003, 2008; Meehan & Molden 2015; Harris 2017; Loftus 2020). For example, building on and departing from James Scott's (1998) account of state tendencies toward simplification in land management as a strategy for legibility and control, Robbins (2008) shows how analytical tools from political ecology and political geography reveal the alternative roles of the state in controlling territory, brokering economic relations, and producing knowledge. A 2017 special issue of *Political Geography* extended this inquiry, demonstrating how water infrastructures and forest spirits both materialize and trouble landscapes, ontologies, and authority (Clarke-Sather 2017; Kelly-Richards & Banister 2017; Theriault 2017). Recent research has investigated how modern states frame the natural world (Whitehead, Jones, & Jones 2007) and diverse actors mobilize narratives of resource nationalism (Koch & Perreault 2019), underscoring the geopolitical dimensions of natural – and national – resource struggles (Andreucci 2017; Bigger & Neimark 2017; Dongol & Neumann 2021).

Both political geography and political ecology investigate how resources and landscapes become pivotal terrains of state power. Canonical work in political ecology examines how colonial and postcolonial efforts to control forests produced revenue, disciplined people, and reified the state (Vandergeest & Peluso 1995; Peluso & Vandergeest 2001). This work has inspired accounts that examine how institutions and officials have territorialized land and resources through processes both violent and mundane, and in doing so make up what we know as the state, its subjects, and the value of forest landscapes (Ybarra 2012; Devine & Baca 2020; Faxon et al. 2022). Other accounts have probed the dynamic relation between water infrastructures and state formation (Alatout 2008; Meehan 2014; Akhter 2015; Sneddon 2015; Mason 2020) and investigated the militarization of conservation (Lunstrum 2014; Adams 2020; Massé & Margulies 2020; Marijnen, de Vries, & Duffy 2021). This theoretically and empirically diverse set of studies all show how making resources also makes the state (Bridge 2014; Perreault 2020).

States are also involved in making extractive resource landscapes. This often involves *frontierization*, the dissolution of established orders to serve particular political-economic interests and make way for new regimes of resource control (Klinger 2017; Rasmussen & Lund 2018). Frontiers are crucial to the expansion of state power, opening up avenues to profit from extraction and extend state territoriality beyond existing boundaries (Black 2018). Frontierization involves twin dynamics of devaluation and revaluation, generating new sites of contestation over land rights and livelihood practices (Peluso & Lund 2011; Knuth, Potts, & Goldstein 2019; Sanchez-Lopez 2021). These processes are exemplified through mining and mineral extraction.

As mining frontiers become sites of value extraction, they also become powerful sites for shaping political imaginaries of nationhood, homeland, and belonging (Perreault 2018; Swann-Quinn 2019, 2020). Resource production goes hand in hand with knowledge production, as scientific exploration transforms subterranean strata into the realm of the state (Marston 2019). These processes raise questions about sovereignty and state power, particularly in (post)colonial and settler-colonial contexts (Marston & Perreault 2017; Marston & Kennemore 2019; Bjørst & Rodon 2022; Curley 2023). Mineral extraction also creates cascading socioecological impacts that permeate land and water, compounding dispossession (Perreault 2013) and generating archipelagos of extraction with ever-shifting borders (Swann-Quinn 2021).

Extractive frontiers do not end at the mine, but create new metabolic relations that unevenly distribute the benefits and burdens of subterranean exploitation. Research across resource geographies highlights how the valuation, extraction, circulation, and exchange of subterranean resources shape the geopolitics of empire and global pathways of accumulation (Bridge 2008, 2011; Huber 2009; Mitchell 2011; Barak 2020). In dialogue with political ecologies of mining (Peluso 2017; Côte & Korf 2018; Klein 2022), this work raises important questions about how metals, minerals, and hydrocarbon resources are both embedded within complex material ecologies and constitutive of broader political-economic and sociospatial relations (Perreault 2013; Marston & Perreault 2017; Marston 2019).

New research on "geopolitical ecologies" links insights from political ecology and political geography to highlight the vital role of resources in establishing and contesting authority across scales. Geopolitical ecologies examine state and national security discourses as they intersect with the interests of transnational corporations and nonstate authorities (Bigger & Neimark 2017; Graddy-Lovelace & Ranganathan 2023). Drawing insights from feminist political ecology

(Rocheleau, Thomas-Slayter, & Wangari 1996; Elmhirst 2011; Harcourt & Nelson 2015), this work reframes the state not as a monolithic entity but as a porous, dynamic, evolving set of relations. As Andrea Nightingale writes, state power is not fixed but always in formation, as the "contested, shifting, emergent boundaries of the state contain the possibilities for transformative change" (Nightingale 2018: 704). This work brings us back to foundational questions: Who benefits and who suffers when resources are made and managed? How are power and authority exercised and contested? How do everyday struggles over land and water reshape political authority at local, national, and global scales?

Making Critical Mineral Frontiers

Across the Global South and North, critical mineral frontiers are reshaping state and local politics, refashioning land and sea, and delivering returns to transnational private capital (Zalik 2018; Childs 2019). Political ecology has excelled at diving deep into the place-based microdynamics of particular field sites, but methodological and conceptual challenges remain to link pathways of exploitation and circulation at the new frontiers of "smart" or "green" development. In some ways, the contemporary rush for critical minerals resembles the land rush that followed the 2008 financial crisis, with key concerns at the intersection of global investment flows, state sovereignty, and resource materiality (Peluso & Lund 2011; Wolford et al. 2013). But the processing, refining, and circulation of critical minerals raise new questions. Critical mineral extraction transforms landscapes into sites of value and harm that reshape intimate spaces and everyday practices, even as they are enrolled in larger political-economic processes of establishing authority and accelerating accumulation (Klinger 2017; Dorn & Gundermann 2022; Blair et al. 2023).

What counts as a "critical mineral" is a political question. In 2023, the US Department of Energy published a Critical Materials List that included the "electric eighteen," materials critical to clean energy transition, as well as 50 additional critical minerals (USDOE 2023). The International Energy Agency (IEA), an intergovernmental organization based in Paris and composed predominantly of high-income, Global North countries to coordinate global energy policy and trade, tracks 37 critical minerals (IEA 2024). Mined at particular sites, including conflict areas with multiple, competing authorities in Myanmar and the Democratic Republic of Congo, extraction is an expensive operation that attracts international capital, whether from Chinese conglomerates or electric vehicle (EV) manufacturers in Europe, the United States, and Korea. Refining and processing operations are often more concentrated than extraction, with more than half of planned lithium chemical plants located in China and 90% of planned nickel refineries in Indonesia (IEA 2023). Critical minerals are key to building renewable energy infrastructures, but also vital for everyday technologies such as smartphones and hard drives as well as for military technologies such as lasers, radar, and guidance systems. Some of the best known are lithium, cobalt, nickel, and graphite, used in batteries. Rare earth elements (REEs) are a subgroup of elements used in lighting and digital screens as well as wind turbines and EVs. Critical minerals also include metals that have long been features of global commodity markets, such as copper, which is essential for the wiring needed to transmit electricity.

Struggles over critical minerals play out across complex political geographies. The case of Jordan is exemplary. Jordan's southern deserts and upland mountain ranges, long prized for their natural beauty and ecological diversity, were slated for industrial extraction in 2021. Along the corridor of the Jordan Rift Valley, stretching south toward the Gulf of Aqaba, the landscape is marked by rose-hued cliffs and sunlit ridges that draw tourists and naturalists alike to the protected nature reserves of the Dana Biosphere and Wadi Rum. These deserts and mountains are rich with metals and minerals, including copper, that are seen by investors and state officials as increasingly vital to powering global digitalization and decarbonization (Kintzi 2023).

In 2021, the Jordanian government announced ambitious new plans to begin mining copper inside the Dana Biosphere Reserve (Vidal 2021). The announcement drew swift, sharp criticism from environmentalists, community-based organizations, and local residents living in and around the reserve (Jordan Times 2021). Contestation over the future of copper mining inside one of Jordan's most prized nature reserves reveals two key questions that increasingly mark new critical mineral extraction sites: Who derives value, and how are social and ecological impacts distributed in spatially uneven ways? As one of the flagship national environmental reserves in the country, the Dana Biosphere plays a pivotal role in maintaining biodiversity, providing refuge for thousands of migratory bird species, and preserving the lifeways of Bedouin tribes that have called these hills home for centuries. The entry of copper mining into the Dana Biosphere brings the architectures of extraction deeper into the countryside, transforming the relationship between local residents, the state, and transnational investors.

A geopolitical ecology perspective not only embeds critical minerals in particular landscapes, but also brings scrutiny to competing claims about value, asking who stands to benefit from mineral extraction. In the Dana Biosphere case, a local hydrocarbon magnate and an array of transnational investors stand to profit spectacularly from the possibility of a global bull market for copper (Farchy & Attwood 2024). State officials' framing of this situation as a reasonable tradeoff between investment and conservation obscures the powerful ways in which copper extraction will fortify incumbent hydrocarbon interests and entrench uneven power dynamics at the state and regional scales.

Copper mining in the Dana Biosphere also reveals the dubiousness of claims that critical minerals are vital for decarbonization. Investment seeks profit, not climate change mitigation, and carries real and often damaging consequences for the places where metals are mined and the people who live there. Political ecologists have sounded the alarm about the potential for the global energy transition to create "green sacrifice zones" (Zografos & Robbins 2020; Hamouchene 2023). This scholarship has revealed the neocolonial dynamics of extractivism under the guise of green transitions (Engels & Dietz 2017; Andreucci et al. 2023; Hamouchene 2023; Stock 2023; Ulloa 2023), with a particular focus on lithium mining (Voskoboynik & Andreucci 2022; Blair et al. 2023). These developments threaten to replicate broader inequities by generating socioenvironmental damages in the Global South in the name of sustainability in the Global North, a phenomenon that Jerez, Garcés, and Torres (2021) call "the colonial shadow of green electromobility." Analyzing the global geographies of nickel extraction for energy transition, Andreucci et al. (2023) likewise warn of "decarbonization by dispossession." By revealing how processes of extraction are embedded in ecologies and communities and exposing processes of enclosure and dispossession at the frontiers of decarbonization, recent scholarship asks whose lives and what places are made expendable in green extractivism (Baka 2017; Brock, Sovacool, & Hook 2021; Andreucci & Zografos 2022).

Maintaining Critical Mineral Frontiers

Critical minerals are increasingly framed by powerful states as an object of concern for national security, driving protectionist policies and expansionist measures. In 2024, the United States passed the Critical Minerals Security Act, raising new questions about how the colonial and imperial history of US extractivism might shape critical mineral futures (Black 2018). While China has framed much of its mineral resource development efforts around the world as "South–South" cooperation, critical scholarship has exposed the longer extractive relations and uneven socioecological benefits and burdens that have emerged at these new resource frontiers (Wilson 2011; Lafitte & French 2013; Andrews-Speed, Qiu, & Len 2016; DeBoom 2018, 2020; Kinzley 2018; Zhang et al. 2023).

Dozens of other countries have adopted new policies that frame critical minerals as a strategic priority, which range from tax credits for exploration in Canada to the European Union's 2024 European Critical Materials Act. Governments are making concerted efforts to capture the revenue from resource extraction, for example by nationalizing the lithium industry in Mexico and banning the export of nickel ore in Indonesia to shore up the country's powerful role in nickel processing (IEA 2023). As a result of new demand and new policies, markets for critical minerals have boomed. The global lithium market, for example, grew almost sevenfold between 2017 and 2022 (IEA 2023). Yet lithium politics are embedded within national histories and local memories of exploitation, which shape the forms of contestation and repertoires of resistance that are mobilized at these new frontiers (Jerez, Garcés, & Torres 2021; Sanchez-Lopez 2021; Dorn & Gundermann 2022; Saleth & Varov 2023). Attending to the longer histories of frontier-making is therefore essential to understand the geopolitical ecologies of critical minerals.

This is vividly illustrated in northern Myanmar, where rare earth extraction has come to a landscape long marked by conflict and competing authorities. Myanmar's resource frontiers are shaped by decades of colonialization, militarization, and civil war, even as they powerfully reshape everyday lives and landscapes (Sarma, Faxon, & Roberts 2023). This takes on a particular dynamic in the Myanmar–China borderlands, where ethnic nationalism, markets, and marriages have long cut across official international boundaries (Dean 2005). Home to multiple ethnic minority groups, Myanmar's northernmost administrative zone – Kachin State – is rich in resources, with significant gold deposits and tropical hardwood timber, as well as two-thirds of the world's jadeite. From the jade mines blasted into hillsides to banana plantations leaking pesticides into the waterways, resource extraction, often capitalized with Chinese investment, brings dangerous, grueling work and environmental harms that irrevocably change the landscapes of Kachin State (Sarma, Rippa, & Dean 2023).

Resource frontiers are also shaped by generations of armed conflict and intermittent peace. Ethnic armed organizations, notably the Kachin Independence Organization (KIO), have been engaged in an armed struggle for greater political autonomy for generations, most recently since the collapse of a 17-year ceasefire agreement in 2011. While the KIO controls swaths of territory and functions as a government, providing state-like services to constituents, other armed actors are guilty of extortion, controlling checkpoints, trade routes, and opium fields, sometimes in alliance with the Myanmar military. In the absence of firm or singular state control, a series of uneasy alliances between armed actors facilitate resource extraction, a dynamic that Kevin Woods (2011) dubs "ceasefire capitalism."

Myanmar's critical mineral frontier takes shape in this contentious context, across a topography of jungles and mountains controlled by different armed authorities. Mining of heavy REEs has exploded in the past decade, with dire environmental and social consequences. In 2022 Global Witness commissioned a satellite study that found over 2700 mining collection pools at almost 300 separate locations in Myanmar, fueling exports of an estimated $781 million in REEs to China annually, up from just $1.5 million in 2014 (Global Witness 2022). The quantity of REEs exported to China is difficult to estimate, not least because the minerals cross the Myanmar–China border along with timber and jadeite in convoys long used for both licit and illicit trade (Sadan et al. 2022).

Patterns of resource exploitation are also shaped by distinct patterns of governance on either side of the border. As greater environmental regulation inside China has slowed the expansion of extraction, REE mining has grown across the Myanmar border. One study using satellite imagery between 2005 and 2020 found that mining activities increased 130% on China's side and 327% on Myanmar's side (Chinkaka et al. 2023). Like previous waves of resource extraction in the region, mining is driven by Chinese demand and capitalized by Chinese investments. China is often assumed to dominate the global production of REEs, but rare earth extraction in Myanmar directly feeds China's processing and manufacturing industries. Meanwhile, extractive economies in Myanmar rely on Chinese investment and technological expertise, as well as established cross-border ethnic kinship and brokerage ties.

Linked to global demand and Chinese investment and channeled through informal, often violent, relations of exchange, REEs in Myanmar's borderlands represent a new iteration of longer processes of frontierization. But they also represent new arenas for negotiating authority and legitimacy. In 2023, the KIO announced it was canceling plans for new REE mines due to public protest against environmental damage, especially water contamination (Fishbein et al. 2023). This cancellation did not stop the expansion of extraction, as mines in areas controlled by Myanmar military-allied groups have continued operations. However, it demonstrates how battles over critical minerals are entangled in struggles for local environmental health and political legitimacy. Efforts to extract REEs reflect longstanding, violent patterns of pollution and profit linked to the commodification of resources in an area of fragmented authority and brisk border trade, yet they also provide opportunities to assert sovereignty and solidarity amid ongoing war.

Toward a Geopolitical Ecology

This chapter has brought together insights from political geography and political ecology to illustrate how a geopolitical ecology perspective might extend our view of environmental issues beyond the state. Political ecology offers tools to interrogate the materiality of resources, as well as the local environments and social relations in which they are embedded. Uniting these insights with research in political geography has served not only to highlight the environment as a key domain of inquiry, but also to enrich both fields by thinking about the relationships between resources and political power across scales. We see this in the case of critical minerals, where, around the world, new frontiers are embedded in ongoing struggles over resources and power, even as they are driven by renewed forms of transnational investment motivated by discourses of decarbonization, digitalization, and security. This is further complicated in contested landscapes and conflict settings, like Myanmar and Jordan, where local interests are subordinated to corporate interests linked to global supply chains. For ordinary people who live on and work these landscapes, whether tropical jungles or arid deserts, the arrival and expansion of extraction threaten local livelihoods and bodies. As both the relative abundance and economic value of certain materials fluctuate, future analysis must consider the shifting political discourses, practices of resource control, and relations of authorities that are reshaping local ecologies and global connections.

Geopolitical ecologies offer key tools for future research into how diverse actors negotiate, access, and build infrastructures to extract, refine, process, and recycle a wide range of natural and physical materials, and how locals living and working in these extractive landscapes negotiate the daily realities of frontier life.

Bringing together key insights from political geography and political ecology to transcend the "territorial trap," such an approach examines spatial patterns of control in complex borderlands, resource ecologies, place-based power relations, and linkages to broader political economy along the shifting contours of global investment. In the time of climate crisis, a geopolitical ecology is especially needed to ask how new efforts toward decarbonization and digitalization are grafted onto landscapes already marked by multiple forms of extraction, exploitation, and exclusion, and how they are reshaping socioecological and geopolitical dynamics within and beyond the state.

References

Adams, W.M. (2020). Geographies of conservation III: Nature's spaces. *Progress in Human Geography 44*(4): 789–801. doi:10.1177/0309132519837779

Agnew, J. (1994). The territorial trap: The geographical assumptions of international relations theory. *Review of International Political Economy 1*(1): 53–80. doi:10.1080/09692299408434268

Akhter, M. (2015). Infrastructure nation: State space, hegemony, and hydraulic regionalism in Pakistan. *Antipode 47*(4): 849–870. doi:10.1111/anti.12152

Akhter, M. (2017). Desiring the data state in the Indus Basin. *Transactions of the Institute of British Geographers 42*(3): 377–389. doi:10.1111/tran.12169

Alatout, S. (2008). "States" of scarcity: Water, space, and identity politics in Israel, 1948–59. *Environment and Planning D: Society and Space 26*: 959–982. doi:10.1068/d1106

Andreucci, D. (2017). Resources, regulation and the state: Struggles over gas extraction and passive revolution in Evo Morales's Bolivia. *Political Geography 61*: 170–180. doi:10.1016/j.polgeo.2017.09.003

Andreucci, D., López, G.G., Radhuber, I.M., et al. (2023). The coloniality of green extractivism: Unearthing decarbonisation by dispossession through the case of nickel. *Political Geography 107*: 102997. doi:10.1016/j.polgeo.2023.102997

Andreucci, D., & Zografos, C. (2022). Between improvement and sacrifice: Othering and the (bio)political ecology of climate change. *Political Geography* 92: 102512. doi:10.1016/j.polgeo.2021.102512

Andrews-Speed, P., Qiu, M., & Len, C. (2016). Chinese engagement in Southeast Asian energy and mineral resources: Motivations and outlook. *Eurasian Geography and Economics 57*(3): 316–342. doi:10.1080/15387216.2016.1211025

Avila, S., Deniau, Y., Sorman, A.H., & McCarthy, J. (2022). (Counter)mapping renewables: Space, justice, and politics of wind and solar power in Mexico. *Environment and Planning E: Nature and Space 5*(3): 1056–1085. doi:10.1177/25148486211060657

Baka, J. (2017). Making space for energy: Wasteland development, enclosures, and energy dispossessions. *Antipode 49*(4): 977–996. doi:10.1111/anti.12219

Barak, O. (2020). *Powering Empire: How Coal Made the Middle East and Sparked Global Carbonization*. Oakland, CA: University of California Press.

Benjaminsen, T.A., Buhaug, H., McConnell, F., et al. (2017). Political geography and the environment. *Political Geography 56*: A1–A2. doi:10.1016/j.polgeo.2016.11.008

Bigger, P., & Neimark, B.D. (2017). Weaponizing nature: The geopolitical ecology of the US Navy's biofuel program. *Political Geography 60*: 13–22. doi:10.1016/j.polgeo.2017.03.007

Bjørst, L.R., & Rodon, T. (2022). Progress stories and the contested making of minerals in Greenland and northern Québec. *The Extractive Industries and Society 12*: 100941. doi:10.1016/j.exis.2021.100941

Black, M. (2018). *The Global Interior: Mineral Frontiers and American Power*. Cambridge, MA: Harvard University Press.

Blaikie, P., & Brookfield, H. (2015). *Land Degradation and Society*. London: Routledge.

Blair, J.J.A., Balcázar, R.M., Barandiarán, J., & Maxwell, A. (2023). The "alterlives" of green extractivism: Lithium mining and exhausted ecologies in the Atacama desert. *International Development Policy | Revue internationale de politique de développement 16*. doi:10.4000/poldev.5284

Bridge, G. (2008). Global production networks and the extractive sector: Governing resource-based development. *Journal of Economic Geography 8*(3): 389–419. doi:10.1093/jeg/lbn009

Bridge, G. (2011). Resource geographies 1: Making carbon economies, old and new. *Progress in Human Geography 35*(6): 820–834. doi:10.1177/0309132510385524

Bridge, G. (2014). Resource geographies II: The resource-state nexus. *Progress in Human Geography 38*(1): 118–130. doi:10.1177/0309132513493379

Brock, A., Sovacool, B.K., & Hook, A. (2021). Volatile photovoltaics: Green industrialization, sacrifice zones, and the political ecology of solar energy in Germany. *Annals of the American Association of Geographers 111*(6): 1756–1778. doi:10.1080/24694452.2020.1856638

Childs, J. (2019). Greening the blue? Corporate strategies for legitimising deep sea mining. *Political Geography 74*: 102060. doi:10.1016/j.polgeo.2019.102060

Chinkaka, E., Klinger, J.M., Davis, K.F., et al. (2023). Unexpected expansion of rare-Earth element mining activities in the Myanmar–China border region. *Remote Sensing 15*(18): 4597. doi:10.3390/rs15184597

Clarke-Sather, A. (2017). State power and domestic water provision in semi-arid Northwest China: Towards an aleatory political ecology. *Political Geography 58*: 93–103. doi:10.1016/j.polgeo.2015.12.003

Collins, Y.A. (2024). A political ecology of atmospheres. A voluminous case study of the Guiana Shield. *Political Geography 109*: 103048. doi:10.1016/j.polgeo.2023.103048

Côte, M., & Korf, B. (2018). Making concessions: Extractive enclaves, entangled capitalism and regulative pluralism at the gold mining frontier in Burkina Faso. *World Development 101*: 466–476. doi:10.1016/j.worlddev.2016.11.002

Curley, A. (2023). *Carbon Sovereignty: Coal, Development, and Energy Transition in the Navajo Nation*. Tucson, AZ: University of Arizona Press.

Dean, K. (2005). Spaces and territorialities on the Sino–Burmese boundary: China, Burma and the Kachin. *Political Geography 24*(7): 808–830. doi:10.1016/j.polgeo.2005.06.004

DeBoom, M.J. (2018). Nuclear (geo)political ecologies: A hybrid geography of Chinese investment in Namibia's uranium sector. *Journal of Current Chinese Affairs 46*(3): 53–83.

DeBoom, M.J. (2020). Climate necropolitics: Ecological civilization and the distributive geographies of extractive violence in the anthropocene. *Annals of the American Association of Geographers 111*(3): 900–912. doi:10.1080/24694452.2020.1843995

Desvallées, L., Arnauld De Sartre, X., & Kull, C. (2022). Epistemic communities in political ecology: Critical deconstruction or radical advocacy? *Journal of Political Ecology 29*(1). doi:10.2458/jpe.4702

Devine, J.A., & Baca, J.A. (2020). The political forest in the era of green neoliberalism. *Antipode 52*(4): 911–927. doi:10.1111/anti.12624

Dongol, Y., & Neumann, R.P. (2021). State making through conservation: The case of post-conflict Nepal. *Political Geography 85*: 102327. doi:10.1016/j.polgeo.2020.102327

Dorn, F.M., & Gundermann, H. (2022). Mining companies, indigenous communities, and the state: The political ecology of lithium in Chile (Salar de Atacama) and Argentina (Salar de Olaroz-Cauchari). *Journal of Political Ecology 29*(1). doi:10.2458/jpe.5014

Elmhirst, R. (2011). Introducing new feminist political ecologies. *Geoforum 42*(2): 129–132. doi:10.1016/j.geoforum.2011.01.006

Engels, B., & Dietz, K. (eds) (2017). *Contested Extractivism, Society and the State*. London: Palgrave Macmillan. doi:10.1057/978-1-137-58811-1

Fairhead, J., & Leach, M. (1995). False forest history, complicit social analysis: Rethinking some West African environmental narratives. *World Development 23*(6): 1023–1035. doi:10.1016/0305-750X(95)00026-9

Farchy, J., & Attwood, J. (2024). Copper market grapples with a crucial question. Bloomberg.com, April 18. https://www.bloomberg.com/news/newsletters/2024-04-18/copper-market-grapples-with-a-crucial-question (accessed May 17, 2024).

Faxon, H.O., Goldstein, J.E., Fisher, M.R., & Hunt, G. (2022). Territorializing spatial data: Controlling land through One Map projects in Indonesia and Myanmar. *Political Geography 98*: 102651. doi:10.1016/j.polgeo.2022.102651

Fishbein, E., Brang, H. P., Awng, Z. M. et al. (2023). How the Kachin public overturned a rare earth mining project in KIO territory. *Frontier Myanmar*, May 2. https://www.frontiermyanmar.net/en/how-the-kachin-public-overturned-a-rare-earth-mining-project-in-kio-territory (accessed May 17, 2024).

Global Witness (2022). Myanmar's poisoned mountains. *Global Witness*, August 9. https://www.globalwitness.org/en/campaigns/natural-resource-governance/myanmars-poisoned-mountains (accessed May 17, 2024).

Goldstein, J., & Nost, E. (2022). *The Nature of Data: Infrastructures, Environments, Politics*. Lincoln, NE: University of Nebraska Press.

Graddy-Lovelace, G., & Ranganathan, M. (2023). Geopolitical ecology for our times. *Political Geography 108*: 103034. doi:10.1016/j.polgeo.2023.103034

Hamouchene, H. (2023). *Dismantling Green Colonialism: Energy and Climate Justice in the Arab Region*. London: Pluto Press.

Harcourt, W., & Nelson, I.L. (eds) (2015). *Practising Feminist Political Ecologies: Moving Beyond the "Green Economy"*. London: Zed Books.

Harris, L.M. (2017). Political ecologies of the state: Recent interventions and questions going forward. *Political Geography 58*: 90–92. doi:10.1016/j.polgeo.2017.03.006

Heynen, N. (2014). Urban political ecology I: The urban century. *Progress in Human Geography 38*(4): 598–604. doi:10.1177/0309132513500443

Huber, M.T. (2009). Energizing historical materialism: Fossil fuels, space and the capitalist mode of production. *Geoforum 40*(1): 105–115. doi:10.1016/j.geoforum.2008.08.004

IEA (2023). *Critical Minerals Market Review 2023 – Analysis*. Paris: International Energy Agency. https://www.iea.org/reports/critical-minerals-market-review-2023 (accessed June 11, 2024).

IEA (2024). *Critical Minerals Data Explorer – Data Tools*. Paris: International Energy Agency. https://www.iea.org/data-and-statistics/data-tools/critical-minerals-data-explorer (accessed June 11, 2024).

Jerez, B., Garcés, I., & Torres, R. (2021). Lithium extractivism and water injustices in the Salar de Atacama, Chile: The colonial shadow of green electromobility. *Political Geography 87*: 102382. doi:10.1016/j.polgeo.2021.102382

Jordan Times (2021). RSCN reiterates rejection of copper mining at Dana reserve. *Jordan Times*, September 6. https://jordantimes.com/news/local/rscn-reiterates-rejection-copper-mining-dana-reserve#google_vignette (accessed May 17, 2024).

Kelly-Richards, S.H., & Banister, J.M. (2017). A state of suspended animation: Urban sanitation and water access in Nogales, Sonora. *Political Geography 58*: 104–113. doi:10.1016/j.polgeo.2015.04.002

Kintzi, K. (2023). Desert power: Wind farms and the new frontiers of renewable energy finance in Jordan. *Commodity Frontiers* 6: 24–31. doi:10.26300/WS81-3A59

Kinzley, J.C. (2018). *Natural Resources and the New Frontier: Constructing Modern China's Borderlands*. Chicago, IL: University of Chicago Press.

Klein, B.I. (2022). Mineral commons: Collective claims to territory in the goldfields of Madagascar. *Political Geography 99*: 102783. doi:10.1016/j.polgeo.2022.102783

Klinger, J.M. (2017). *Rare Earth Frontiers: From Terrestrial Subsoils to Lunar Landscapes*. Ithaca, NY: Cornell University Press.

Klinger, J.M. (2020). Environment, development, and security politics in the production of Belt and Road spaces. *Territory, Politics, Governance 8*(5): 657–675. doi:10.1080/21622671.2019.1582358

Knuth, S., Behrsin, I., Levenda, A., et al. (2022). New political ecologies of renewable energy. *Environment and Planning E: Nature and Space 5*(3): 997–1013. doi:10.1177/25148486221108164

Knuth, S., Potts, S., & Goldstein, J.E. (2019). In value's shadows: Devaluation as accumulation frontier. *Environment and Planning A: Economy and Space 51*(2): 461–466. doi:10.1177/0308518X18806112

Koch, N., & Perreault, T. (2019). Resource nationalism. *Progress in Human Geography 43*(4): 611–631. doi:10.1177/0309132518781497

Lafitte, G., & French, P. (2013). *Spoiling Tibet: China and Resource Nationalism on the Roof of the World*. London: Zed Books.

Li, T.M. (2002). Engaging simplifications: Community-based resource management, market processes and state agendas in upland Southeast Asia. *World Development 30*(2): 265–283. doi:10.1016/S0305-750X(01)00103-6

Loftus, A. (2020). Political ecology II: Whither the state? *Progress in Human Geography 44*(1): 139–149. doi:10.1177/0309132518803421

Lunstrum, E. (2014). Green militarization: Anti-poaching efforts and the spatial contours of Kruger National Park. *Annals of the Association of American Geographers 104*(4): 816–832. doi:10.1080/00045608.2014.912545

Marijnen, E., de Vries, L., & Duffy, R. (2021). Conservation in violent environments: Introduction to a special issue on the political ecology of conservation amidst violent conflict. *Political Geography 87*: 102253. doi:10.1016/j.polgeo.2020.102253

Marston, A. (2019). Strata of the state: Resource nationalism and vertical territory in Bolivia. *Political Geography 74*: 102040. doi:10.1016/j.polgeo.2019.102040

Marston, A., & Kennemore, A. (2019). Extraction, revolution, plurinationalism: Rethinking extractivism from Bolivia. *Latin American Perspectives 46*(2): 141–160. doi:10.1177/0094582X18781347

Marston, A., & Perreault, T. (2017). Consent, coercion and cooperativismo: Mining cooperatives and resource regimes in Bolivia. *Environment and Planning A: Economy and Space 49*(2): 252–272. doi:10.1177/0308518X16674008

Mason, M. (2020). Hydraulic patronage: A political ecology of the Turkey-Northern Cyprus water pipeline. *Political Geography 76*: 102086. doi:10.1016/j.polgeo.2019.102086

Massé, F., & Margulies, J.D. (2020). The geopolitical ecology of conservation: The emergence of illegal wildlife trade as national security interest and the re-shaping of US foreign conservation assistance. *World Development 132*: 104958. doi:10.1016/j.worlddev.2020.104958

Meehan, K., & Molden, O.C. (2015). Political ecologies of the state. In Agnew, J., Mamadouh, V., Secor, A.J., et al. (eds), *The Wiley Blackwell Companion to Political Geography*. Chichester: John Wiley & Sons, 438–450. doi:10.1002/9781118725771.ch32

Meehan, K.M. (2014). Tool-power: Water infrastructure as wellsprings of state power. *Geoforum 57*: 215–224. doi:10.1016/j.geoforum.2013.08.005

Mitchell, T. (2011). *Carbon Democracy: Political Power in the Age of Oil*. London: Verso.

Nightingale, A.J. (2018). The socioenvironmental state: Political authority, subjects, and transformative socionatural change in an uncertain world. *Environment and Planning E: Nature and Space 1*(4): 688–711. doi:10.1177/2514848618816467

Paredes, M., Gianella, C., & Olivera, S. (2024). Corrupting climate change institutions from the inside Systemic Collusion in the Peruvian Forest Governance. *Political Geography 108*: 102991. doi:10.1016/j.polgeo.2023.102991

Peluso, N.L. (2017). Plantations and mines: Resource frontiers and the politics of the smallholder slot. *Journal of Peasant Studies 44*(4): 834–869. doi:10.1080/03066150.2017.1339692

Peluso, N.L., & Lund, C. (2011). New frontiers of land control: Introduction. *Journal of Peasant Studies 38*(4): 667–681. doi:10.1080/03066150.2011.607692

Peluso, N.L., & Vandergeest, P. (2001). Genealogies of the political forest and customary rights in Indonesia, Malaysia, and Thailand. *Journal of Asian Studies 60*(3): 761–812. doi:10.2307/2700109

Peluso, N.L., & Watts, M. (eds) (2001). *Violent Environments*. Ithaca, NY: Cornell University Press.

Perreault, T. (2013). Dispossession by accumulation? Mining, water and the nature of enclosure on the Bolivian Altiplano. *Antipode 45*(5): 1050–1069. doi:10.1111/anti.12005

Perreault, T. (2018). Mining, meaning and memory in the Andes. *Geographical Journal 184*(3): 229–241. doi:10.1111/geoj.12239

Perreault, T. (2020). State of nature: on the co-constitution of resources, state and nation. In Moisio, S., Koch, N., Jonas, A.E.G., et al. (eds), *Handbook on the Changing Geographies of the State*. Cheltenham: Edward Elgar, 228–239.

Perreault, T., Bridge, G., & McCarthy, J. (2020). *The Routledge Handbook of Political Ecology*. London: Routledge.

Rasmussen, M.B., & Lund, C. (2018). Reconfiguring frontier spaces: The territorialization of resource control. *World Development 101*: 388–399. doi:10.1016/j.worlddev.2017.01.018

Robbins, P. (2003). Political ecology in political geography. *Political Geography 22*(6): 641–645.

Robbins, P. (2008). The state in political ecology: A postcard to political geography from the field. In Cox, K.R., Low, M., & Robinson, J. (eds), *The SAGE Handbook of Political Geography*. London: Sage, 205–218. doi:10.4135/9781848607880

Robbins, P. (2012). *Political Ecology: A Critical Introduction*, 2nd edn. Chichester: Wiley-Blackwell.

Rocheleau, D., & Edmunds, D. (1997). Women, men and trees: Gender, power and property in forest and agrarian landscapes. *World Development 25*(8): 1351–1371. doi:10.1016/S0305-750X(97)00036-3

Rocheleau, D.E., Thomas-Slayter, B.P., & Wangari, E. (eds) (1996). *Feminist Political Ecology: Global Issues and Local Experiences*. London: Routledge.

Sadan, M., Yü, D.S., Lawn, D.S., Brown, D., & Zhou, R. (2022). *Rare Earth Elements, Global Inequalities, and the "Just Transition"*. London: British Academy. doi:10.5871/just-transitions-s-i/M-S

Saleth, L.A., & Varov, I. (2023). Anticipating lithium extraction in northern Portugal: A Sacrifice Zone in the making? *Journal of Political Ecology 30*(1). doi:10.2458/jpe.4849

Sanchez-Lopez, M.D. (2021). Territory and lithium extraction: The Great Land of Lipez and the Uyuni Salt Flat in Bolivia. *Political Geography 90*: 102456. doi:10.1016/j.polgeo.2021.102456

Sarma, J., Faxon, H.O., & Roberts, K.B. (2023). Remaking and living with resource frontiers: Insights from Myanmar and beyond. *Geopolitics 28*(1): 1–22. doi:10.1080/14650045.2022.2041220

Sarma, J., Rippa, A., & Dean, K. (2023). "We don't eat those bananas": Chinese plantation expansions and bordering on Northern Myanmar's Kachin borderlands. *Eurasian Geography and Economics 64*(7–8): 842–868. doi:10.1080/15387216.2023.2215802

Scott, J.C. (1998). *Seeing Like a State: How Certain Schemes to Improve the Human Condition Have Failed*. New Haven, CT: Yale University Press.

Sneddon, C. (2015). *Concrete Revolution: Large Dams, Cold War Geopolitics, and the US Bureau of Reclamation*. Chicago, IL: University of Chicago Press.

Stock, R. (2023). Power for the plantationocene: Solar parks as the colonial form of an energy plantation. *Journal of Peasant Studies 50*(1): 162–184. doi:10.1080/03066150.2022.2120812

Sultana, F. (2021). Political ecology 1: From margins to center. *Progress in Human Geography 45*(1): 156–165. doi:10.1177/0309132520936751

Sultana, F. (2022). The unbearable heaviness of climate coloniality. *Political Geography 99*: 102638. doi:10.1016/j.polgeo.2022.102638

Svarstad, H., Benjaminsen, T.A., & Overå, R. (2018). Power theories in political ecology. *Journal of Political Ecology 25*(1). doi:10.2458/v25i1.23044

Swann-Quinn, J. (2019). Mining the homeland: Imagining resources, nation, and territory in the Republic of Georgia. *Eurasian Geography and Economics 60*(2): 119–151. doi:10.1080/15387216.2019.1605921

Swann-Quinn, J. (2020). Digging deep: Crossing scale in the Georgian mining industry. In O'Lear, S. (ed.), *A Research Agenda for Environmental Geopolitics*. Cheltenham: Edward Elgar, 136–150.

Swann-Quinn, J. (2021). Toward extractive archipelagos. *Territory, Politics, Governance 9*(2): 180–202. doi:10.1080/21622671.2019.1687322

Theriault, N. (2017). A forest of dreams: Ontological multiplicity and the fantasies of environmental government in the Philippines. *Political Geography 58*: 114–127. doi:10.1016/j.polgeo.2015.09.004

Ulloa, A. (2023). Aesthetics of green dispossession: From coal to wind extraction in La Guajira, Colombia. *Journal of Political Ecology 30*(1). doi:10.2458/jpe.5475

USDOE (2023). Notice of Final Determination on 2023 DOE Critical Materials List. Federal Register. https://www.federalregister.gov/documents/2023/08/04/2023-16611/notice-of-final-determination-on-2023-doe-critical-materials-list (accessed May 31, 2024).

Vandergeest, P., & Peluso, N.L. (1995). Territorialization and state power in Thailand. *Theory and Society 24*(3): 385–426.

Vidal, M. (2021). Copper mine threatens Jordan's largest nature reserve. *Al Jazeera*, October 3. https://www.aljazeera.com/news/2021/10/3/copper-mine-threatens-jordans-largest-nature-reserve (accessed August 5, 2022).

Voskoboynik, D.M., & Andreucci, D. (2022). Greening extractivism: Environmental discourses and resource governance in the "Lithium Triangle". *Environment and Planning E: Nature and Space 5*(2): 787–809. doi:10.1177/25148486211006345

Walker, P.A. (2005). Political ecology: Where is the ecology? *Progress in Human Geography 29*(1): 73–82. doi:10.1191/0309132505ph530pr

Walker, P.A. (2007). Political ecology: Where is the politics? *Progress in Human Geography 31*(3): 363–369. doi:10.1177/0309132507077086

Watts, M. (1983). On the poverty of theory: Natural hazards research in context. In Hewitt, K. (ed.), *Interpretations of Calamity*. London: Routledge, 231–262.

Watts, M. (2004). Resource curse? Governmentality, oil and power in the Niger Delta, Nigeria. *Geopolitics 9*(1): 50–80. doi:10.1080/14650040412331307832

Whitehead, M., Jones, R., & Jones, M. (2007). *The Nature of the State: Excavating the Political Ecologies of the Modern State*. Oxford: Oxford University Press.

Wilson, J.D. (2011). Resource nationalism or resource liberalism? Explaining Australia's approach to Chinese investment in its minerals sector. *Australian Journal of International Affairs 65*(3): 283–304. doi:10.1080/10357718.2011.563779

Wolford, W., Borras, S.M., Jr., Hall, R., Scoones, I., & White, B. (2013). Governing global land deals: The role of the state in the rush for land. *Development and Change 44*(2): 189–210. doi:10.1111/dech.12017

Woods, K. (2011). Ceasefire capitalism: Military–private partnerships, resource concessions and military–state building in the Burma–China borderlands. *Journal of Peasant Studies 38*(4): 747–770. doi:10.1080/03066150.2011.607699

Ybarra, M. (2012). Taming the jungle, saving the Maya Forest: Sedimented counterinsurgency practices in contemporary Guatemalan conservation. *Journal of Peasant Studies 39*(2): 479–502. doi:10.1080/03066150.2012.666974

Zalik, A. (2018). Mining the seabed, enclosing the area: Ocean grabbing, proprietary knowledge and the geopolitics of the extractive frontier beyond national jurisdiction. *International Social Science Journal 68*(229–230): 343–359. doi:10.1111/issj.12159

Zhang, C., Chen, F., Huang, S., & Shen, S. (2023). Resource nationalism and platinum group metals: Chinese strategies in a hydrogen-powered future. *The Extractive Industries and Society 15*: 101317. doi:10.1016/j.exis.2023.101317

Zografos, C., & Robbins, P. (2020). Green sacrifice zones, or why a green new deal cannot ignore the cost shifts of just transitions. *One Earth 3*(5): 543–546. doi:10.1016/j.oneear.2020.10.012

Chapter 33

Water Politics

Chris Sneddon[1,2] and Coleen Fox[1]

[1]Department of Geography, Dartmouth College, Hanover, NH, USA
[2]Department of Environmental Studies, Dartmouth College, Hanover, NH, USA

Questions surrounding water politics are of critical importance to the human and nonhuman collectives inhabiting the planet, to the diverse array of governments striving to balance ecological and economic goals within water governance, and to scholars attempting to interpret water–society relations in the twenty-first century. Water dynamics and water politics are also central concerns of geography. There is a lengthy tradition within the discipline of "using water to broker disciplinary 'unity' and to conceptualize matters of history, politics and theory" (Liao & Schmidt 2023: 242). We approach water in a similar vein, bringing politics to center stage. However, capturing with any degree of depth the incredible diversity of work on water politics within political geography, political ecology, Indigenous geographies, science and technology studies (STS), and many neighboring fields is a colossal task. This chapter therefore focuses on lines of inquiry into water politics that are enduring, as well as ones that signal novel pathways of thinking about water and politics.

Given the diversity of research on water politics, this chapter can hardly be considered comprehensive. We have therefore sidestepped a number of themes that political geographers have recently explored, including the politics of the world's oceans and coasts in all their complexity and vital biophysical importance in terms of climate dynamics, navigation, fisheries, and many other dimensions (but see Hung & Lin, this volume). Other scholarship that deserves more space includes the politics of groundwater (Birkinshaw 2022; Kroepsch & Clifford 2022; Lawless 2023), how climate change is related to altered hydrologies and draught (Dinar et al. 2015; Melville-Rea 2022), and the complex politics of aridity in the

The Wiley Blackwell Companion to Political Geography, Second Edition.
Edited by Virginie Mamadouh, Natalie Koch, Chih Yuan Woon, and John Agnew.

world's desert regions (Sternberg 2016; Koch 2021; Sizek 2021; Voyles 2021). Geographers have further worked to reconcile the physical and human traditions of the field through specific engagement with water–society relations and the politics of these relations (Lave 2012; Ashmore 2015). Despite being partial, however, the remainder of the review amply demonstrates that we are witnessing an efflorescence of geographical thinking on water politics, one that serves as a vital marker of what it means to think geographically.

Water, Politics, and Power

Over a decade ago, Agnew (2011) argued the need to recover a more broadminded understanding of politics – whether formal or informal – in the realm of water–society relations and embrace it as a tool for social change where and when the conditions suggest it would be effective. Agnew (2011) cites multiple challenges embedded within water politics, including how to transcend the territorial limitations of traditional political communities, how to legitimize multiple voices within debates over governance, and how to effectively and justly incorporate different knowledges in debates over policies or other kinds of water–society interventions. In many ways, the work reviewed here has responded to analysis of these challenges. However, Agnew's bluntest message is that "(w)e have politics available to us and we need to invigorate it" (2011: 474). We do see in the following review the many ways in which scholars of water politics are carefully identifying and interrogating the politics we have available to "us," while not assuming that the same kinds of politics are available to all people and waters and their relations everywhere in the same fashion.

Geographical understandings of water politics see power relations as central conceptual and empirical themes. At one level, water is a biophysical presence of enormous consequence for nearly every human subsistence and economic activity imaginable and one that holds complex and varied meanings among human societies and cultures. Relatedly, as a nonhuman entity, water exercises agency in multiple ways, not only as one of the organizing and mediating substances around which the planet is given structure and evolves, but as a powerful agent of both human flourishing (e.g., water for existence, food production, electricity generation) and human disaster (e.g., flooding, hurricanes, habitat for disease vectors). Water is, as Bakker (2012) astutely notes, "political, biopolitical [and] material." Thus, water politics is simultaneously about the power to govern, control, and manipulate water access, and about the power to confer meaning over a wide spectrum of cosmologies regarding water's place in the world.

Geographical thinking also takes water politics – and hence politics itself – in a number of fruitful and critical directions. First, as work in the hydrosocial tradition has shown, water politics are never only about water in isolation. As with other socio-natures, water is always bound up in a wide range of material and cultural practices, biophysical processes, ideological positions, and institutional networks (Swyngedouw 2004). In addition, material struggles over water are constantly filtered through the different social meanings attached to water. A brilliant example is Bakker's (2013) work showing how water's entanglement within capitalist economic relations in the context of private water supply networks across a range of locales has shifted during the transition from a neoliberal to a postneoliberal orientation within dominant water governance models.

Second, geographers raise questions about who is and should be considered a political actor, particularly outside more formal water governance arrangements. A case in point, Sultana's (2007) seminal work notes that water politics must always be considered in relation to axes of social difference. The everyday *pani* ("water") politics in rural Bangladesh, where drinking wells contaminated with arsenic have created enormous hardships and conflicts within and among communities, is enacted through individuals' gender, social standing, and age. In a similar vein, Barnes's (2017) keen ethnography of the everyday politics of irrigation systems in the Nile Valley exemplifies how power relations are constantly negotiated not only through the daily interactions involving farmers and government technicians, but also through international forums involving state and international aid officials. Thus, if politics is about power, it is not only about the more abstract modes of power relations seen from the "heights" of state bureaucracies or water management corporations, but also the more everyday politics of obtaining safe drinking water, or critical irrigation flows, for oneself and one's family under often difficult conditions.

Third, geographical work is not unexpectedly concerned with the spaces of politics: Where does water politics occur? Menga (2017: 102) draws on the work of Hannah Arendt and others to highlight the *polis* (taken from the ancient Greek term for city-state, and the root of "political") as the "arena within which politics is conducted." While the *polis* can imply a physical space of some kind, at whatever spatial relief, it also refers to the socially bounded political community that might be contesting an issue or proposal. This more metaphorical "space" of political interactions is where "speech and action can . . . be manipulated" by a ruling elite or other powerful actor to narrow the possible terms and stakes of political debates and hence "foreclose political encounter" (2017: 102–103). This curtailing of more egalitarian political encounters across a wide range of actors within a political community is the essence of what Swyngedouw (2011) labels the postpolitical condition of numerous state–society relations in the twenty-first century. Applied to water, the "political terrain" (*hydropolis* for Menga) in a postpolitical era is radically restricted to conventional debates over water governance, or "the sphere of consensual governing and policy-making, centered on the technical, managerial and consensual administration (policing) of environmental, social, economic or other domains" (Swyngedouw 2011: 266). Transformative or revolutionary water politics thus becomes at best highly implausible and at worst unimaginable. Yet many geographers today argue that a more radical and transformative water politics can be clawed back from a postpolitical purgatory.

Another theme of recent geographical research, albeit to varying degrees, is a radical decentering of state actors as protagonists and/or antagonists in understanding water politics. There remain many compelling reasons to retain an analytical eye on state actors in both their regulatory and everyday formulations (Ranganathan & Balazs 2015), not least because of their centrality in water governance regimes (Menga & Swyngedouw 2018). But scholars today have demonstrated the power of paying more direct attention to a range of actors, many previously unrecognized in water governance scholarship, beyond the state in the study of water politics. The remainder of this chapter highlights this work, reviewing efforts of scholars in political geography and related fields who are defining and extending the boundaries of how water politics is conceived and enacted in intriguing and politically cogent ways. Specific themes we consider include spatializing water politics, critical hydropolitics, the politics of water infrastructure, and the political ontologies of water. These are not isolated lines of inquiry but are reinforcing one another continuously and creatively.

Politics, Water, Spaces

Geographical work on water politics fundamentally interrogates the spatiality of political dynamics. For example, research on the hydrosocial cycle in general and hydrosocial territories more specifically has enlivened interpretations of water politics in recent years. By zeroing in on how territories and other spatial units both constrain and are shaped by struggles over water, this work emphasizes the power imbalances within existing governance institutions and infrastructures, and how different social groups throughout the world are mounting place-based political and conceptual challenges to hegemonic modes of water governance. Accordingly, whether conceptualized as hydrosocial territories (Boelens et al. 2016), waterscapes (Karpouzoglou & Vij 2017; Flaminio, Rouillé-Kielo, & Le Visage 2022), or communities of riverhood (Boelens et al. 2023), geographers are at the forefront of rethinking how water–society relations are conceptualized and prove meaningful within and across dynamic sets of spatial relations.

By foregrounding how water politics occurs through everyday practices as well as through more formal social movements, thinking on what collectively can be called hydrosocial geographies (Liao & Schmidt 2023) invites us to reconsider conventional ways of spatially organizing water–society relations, such as the watershed or the river basin. It is not that the basin or watershed should be abandoned as part of our analytical toolkit, but rather that these units have often been coopted by state actors working within technical and managerial mindsets perceiving water exclusively as a "resource" for developmental and geopolitical aims (see Molle 2009; Sneddon 2015). For example, Sarna-Wojcicki et al. (2019) show how the Karuk tribe's efforts to assert their traditional water rights in the Klamath River basin of the northwestern United States are hampered by a watershed focus that fails to recognize Karuk notions of water's spatiality. From the perspective of water politics, scalar narratives that speak of the "global," "regional," "national," or "local" as common-sense things are always in flux and often contested by multiple actors (Perreault 2005). A good example of this is found in the work of Diné geographer Andrew Curley (2021), who shows how Western water laws and governance arrangements in the southwestern United States were predicated on European notions of space, nature, and property that violently erased and "enclosed" Indigenous territories.

The notion of the hydrosocial territory emerged as the spatial expression of earlier work on the hydrosocial cycle that was associated with urban political ecology and efforts to rethink the ontological, epistemological, and social status of water (Linton 2010; Budds et al. 2014). Hydrosocial territories refer to the "socially, naturally and politically constituted spaces that are (re)created through the interactions amongst human practices, water flows, hydraulic technologies, biophysical elements, socio-economic structures and cultural-political institutions" (Boelens et al. 2016: 1) and, crucially, are where water politics takes place. Work centering hydrosocial territories examines how struggles over water control, access, and meaning alter power relations among actors and redefine their tactics (Flaminio, Rouillé-Kielo, & Le Visage 2022). In sum, geographers see waterscapes and hydrosocial territories as sites of politics, where the biophysical and political-economic processes that shape and alter these sociohydrological spaces over time are also deeply influenced by how these spaces have been previously defined and imagined by the long-term inhabitants of specific locales.

A majority of the pioneering work within the hydrosocial territories tradition has highlighted how rural and Indigenous communities throughout Latin America have invoked novel territorial framings as a means of defending their livelihoods and cultural identities in the face of state-led extractive activities such as mining and hydropower development (e.g., Boelens 2014; Duarte-Abadia, Boelens, & Roa-Avendaño 2015; Lopez et al. 2019; Manosalvas, Hoogesteger, & Boelens 2021; Artiga-Purcell 2022). The hydrosocial territories framework has proven fungible, as evidenced by recent studies applying it to the Salween River basin in Southeast Asia (Götz & Middleton 2020) and urban settings (Hommes & Boelens 2017; Cantor 2021). Scholars also point out that state actors themselves – a series of water bureaucracies and their economic allies in urban southern California (Cantor 2021) and a river basin targeted for hydropower development in the Salween (Götz & Middleton 2020) – are actively constructing hydrosocial territories according to a managerial and developmental logic. Hydrosocial territories are thus intensive zones of political struggle, where largely Eurocentric understandings of water governance undergirding government perspectives on water resource development confront more diverse notions of human–water relations.

Hydropolitics

There is a commonly held view that material conflicts over water, at the heart of a great deal of water politics scholarship, are on an inevitable path toward escalation in the remaining decades of the twenty-first century. The argument goes something like this: rising human populations and consumption patterns combined with more extreme and frequent climatic events will exacerbate struggles over freshwater in regions of the planet already suffering from scarcity. Even in water-abundant regions, competing uses (e.g., agricultural, industrial, urban) will place increasing strains on rivers and other freshwater sources and stoke conflicts. These crisis discourses reach their acme in the "water wars" thesis, positing that nation-states that share transnational river basins – particularly those in water-scarce regions – are inevitably on a path toward overt conflict.

As years of research by political geographers and scholars of hydropolitics have shown, this thesis rests on empirically shaky ground, and it obscures the numerous ways in which political dynamics among riparian states are more likely to lead to cooperative interstate arrangements (Wolf 2007; Brochmann & Gleditsch 2012). This, however, is not necessarily reason to celebrate: for many of the riparian states that share transnational river basins, cooperation implies joint efforts to rapidly develop basin "resources," wherein water, or Modern Water in Linton's (2010) phrasing, is valued primarily for its capacity to generate electricity, irrigate agricultural lands, and in general to be "useful." Roughly two decades of critical hydropolitics, in addition to what Liao and Schmidt (2023) refer to as critical water studies, have clearly shown that a myopic fixation on cooperation and conflict among state actors that share transnational river basins occludes the multiple ways in which nonstate actors, hydraulic infrastructure, knowledge claims and counterclaims, and biophysical networks participate in the politics of water concerning international river basins (Rogers et al. 2023).

In addition, scholarship in critical hydropolitics criticizes the simplistic assumptions regarding territory emanating from conventional international relations theory that perceive territorial spaces as "containers" where state actors exercise unchallenged sovereignty and effectively separate domestic and international political concerns. Furlong (2006) noted some time ago, via an investigation of transboundary water dynamics in southern Africa that include the Okavango Basin and the Lesotho Highlands Water Project, that these assumptions are largely unwarranted since they fail to account for the important roles played by nonstate actors and the often unclear lines between domestic and international territories and political dynamics. In conventional hydropolitical studies, the river basin scale is naturalized and immutable and riparian states hold a monopoly over (geo)political negotiations concerning things like the equitable utilization of a basin's total flows or prior notification of development projects.

Critical hydropolitics thoroughly disrupts these assumptions, demonstrated in a range of theoretical and empirical studies on transnational basins of vastly different hydrological, political-economic, and social circumstances such as the Mekong (Sneddon & Fox 2006; Middleton & Allouche 2016), Ganges-Brahmaputra (Hanasz 2017), Nile (Menga 2017), Indus (Akhter 2015, 2019), Salween (Suhardiman & Middleton 2020), the Great Lakes basin (Norman 2014), and the Jordan (Allouche 2019), among others. Moreover, this work collectively suggests that the roots of water conflicts within transboundary basins are often fueled through the discursive practices of government agencies and the experts they engage to "rationally" advance river basin development while willfully ignoring the needs and voices of marginalized communities and civil society organizations opposed to wanton infrastructure development.

Studies in critical hydropolitics have also explored how disputes over transnational waterways serve specific state regimes in a variety of ways related to nation-building and enhancing political legitimacy through large-scale, visible water development projects. For example, Allouche (2019) interrogated how postcolonial societies in Israel and India – using conflicts over the Jordan and Indus basins, respectively – engaged in forms of "water nationalism" to prop up identities pegged to the nation-state at the expense of other identities not founded on ethnicity. In this instance, state efforts to bolster external territorialization were themselves supported through assertion of rights to develop a "national" resource (e.g., river flows) made possible through large-scale infrastructure projects. Similarly, Haines (2016) situates negotiations between Pakistan and India over governing the Indus River within the complex history of decolonization in the region and each state's territorial ambitions, and ultimately makes a case for alternative river–society relations that transcend the limitations of state-to-state transboundary approaches. As these case studies illustrate, critical hydropolitics remains a vital framework for analyzing how both state and nonstate actors engage in a complex politics of spatial scale in transboundary waters, highlighting in particular how hydrosocial geographies are imagined, altered through infrastructure development, and shift over time.

Water, Technology, Infrastructure

Over the past two decades, geographers have also made a compelling case for broadening out the politics of water to include a wide range of nonhumans – prominently infrastructure and water entities themselves – arguing that these are inextricably enmeshed in how water circulates and how different social actors influence these circulations. As a precursor to the politics of various forms of water infrastructure, scholars in geography and STS traditions argue that water and infrastructure are themselves part of heterogenous assemblages of human practices, biophysical processes, financial flows, technological artifacts and expertise, and many more entities and institutions (Sneddon 2015; Rogers & Crow-Miller 2017). This perspective implies that the politics of water is never solely about water as an abstraction or dilution: water politics also encompasses all the things, ideas, and processes to which hydrologies sit in relation. This is also the essence of thinking hydrosocially.

Indeed, a hallmark of recent research on geographical engagement with water is the integration of social theories highlighting the multiple and strong connections that make water, human societies, technologies and infrastructure, and biophysical networks virtually inseparable. Oftentimes the politics of water infrastructure feature state power, in multiple forms, as a central theme. In a seminal work on the connections between state power and infrastructure, Meehan (2014) deftly shows that hydraulic infrastructure – in this case, in the form of flood control and drinking water supply networks in Mexico – can serve as a crucial extension of state power. But it can also serve to subvert state power in unexpected ways, particularly through the creative actions of households in Tijuana that draw on alternative infrastructures to maintain consistent water access. Infrastructure is thus not only the purview of state actors, but must be theorized in relation to what people actually do in their daily lives to enhance water security, even when the water infrastructure in question is more modest than large dams or irrigation structures.

In hydrosocial worlds, infrastructures influence how water is moved and distributed, treated, and transformed in myriad ways. Indeed, some of the most violent water politics concerns the construction of large dams (Del Bene, Scheidel, & Temper 2018; Han & Webber 2020) and the damaging water impacts of large-scale energy infrastructure (Whyte 2017; Yazzie & Baldy 2018; Estes 2019). As numerous thinkers point out, however, dam-affected communities around the world are contesting infrastructure in a range of intriguing ways. For example, Fung and Lamb (2023) introduce the innovative concept of "slow resistance" – a counterpart to Nixon's (2011) concept of "slow violence" – to highlight the many subtle and long-term methods that communities living within the Salween River basin have used to contest interbasin transfer projects initiated by successive authoritarian regimes in Thailand.

Geographers have also highlighted the contentious politics of more prosaic water infrastructure, where conflicts over water provision, in urban, periurban, and rural settings, shed light on the often subtle and less visible ways in which water infrastructure produces and sustains social inequities (e.g., Sultana 2007; Colven 2020; Millington & Scheba 2021; Diaz-Combs 2023). Other work focuses on the particular types of technopolitics that emerge out of hydraulic infrastructures such as water pipelines (Mason 2020) and desalination projects (Usher 2019), as state agents prioritize technological interventions in hydrosocial worlds as a means of bolstering national security and geopolitical interests.

A focus on infrastructure within these watery assemblages also reveals the complex power relations circulating through water's uses and meanings. For example, scholars of water conflicts and politics have in recent years investigated the variable ways in which water infrastructure – perhaps most obviously in the form of large dams, irrigation networks, and water supply systems – both shapes and reflects power relations. Going back to Wittfogel's (in)famous thesis on the connections between large-scale irrigation works and state power in the form of "despotic regimes," a multitude of studies have tracked how the monopoly of state agents over large-scale infrastructure tends in the long run to lead to a concentration of power over water access and usage within the hands of state bureaucracies and their allies (Strang 2016; Chien & Hong 2018). This is the essence of Curley's (2021) research on the Central Arizona Project, a network of water diversion infrastructure that served as a colonial "beachhead" for the colonial settler state in its long-term efforts to expropriate Diné lands and waters during the twentieth century in the service of territorial control. In a different vein, Mullenite (2019) aptly describes the concentration of governmental power in the case of Guyana, where state-sponsored water control systems – ostensibly for irrigation and flood control – have fomented ethnic divisions in ways that have legitimized authoritarian rule.

In many instances, infrastructural water politics are not as straightforward as the Arizona and Guyana cases might suggest. In a study of the highly variable and contested nature of water supply networks in Delhi, India, Truelove (2019) emphasizes that the politics that flow out of governance practices are not characterized by straightforward power relations over control and access. Rather, power over water–society relations is exercised through "gray zones" that include, for example, "political assemblages . . . that defy dualisms such as legal-illegal, formal-informal, and public private" (2019: 1760). Other work highlights how the maintenance of infrastructure is equally culpable in how state power is expressed over time as space. For example, Barnes (2017) highlights the subtle ways in which the upkeep of irrigation infrastructure by government agencies throughout Egypt's Nile Valley is not simply technical expertise, but a means for state agents to exert control over water access. Famers utilizing irrigation works, by contrast, perceive the maintenance of irrigation canals as expressions of communal relations constructed around water sharing.

One of the key lessons of work on the linkages between hydraulic infrastructure and water politics is that the role of infrastructure in altering and controlling human interactions with water, although all too common, does not necessarily lead to domination and social injustice. Water infrastructure, like water itself, is embedded within social relations. Although infrastructure has historically been a means for states to assert their power and rights over water, there is no inevitable path toward despotism. Water infrastructure can "equally support desires to change political arrangements and the human and nonhuman relationship that these compose" (Strang 2016: 311). Cautious support for this position can perhaps be found in the multiple, significant efforts to remove dams in the United States and elsewhere (Fox, Magilligan, & Sneddon 2016), as well as the numerous cases of Indigenous communities in diverse geographical regions reasserting alternate understandings of water worlds.

Water, Worlds, Politics

The so-called ontological turn in the social sciences has contributed to a flourishing of thinking across studies of nature–society relations and is increasingly being integrated into a range of geographical work in critical water studies. These engagements involve the "political processes of ontological entanglement and disentanglement" that characterize the many contemporary struggles over water (Yates, Harris, & Wilson 2017: 798). Ontological politics refers to how different perspectives on what constitute reality overlap and intersect, and what this means for how politics are strategized and enacted. Water governance is one of the key arenas of friction between national governments and Indigenous peoples (Curran 2019), and nearly all water politics concerning Indigenous peoples and conflicts with settler societies involves a "deeper disconnect over what water is, what it represents, and the place of humans within the wider world" (Neville & Coulthard 2019: 4). A great deal of critical water research has recently focused on Indigenous ontologies of water, recognizing the challenges of Indigenous communities across the planet that must navigate the conventional norms and institutions of water governance propped up within settler-colonial states.

Scholars of Indigenous studies point out that Indigenous water ontologies are diverse, yet many approach water – whether as river, lake, or other form – as a force that exercises powerful and often unknowable agencies through and with humans and other nonhuman persons. From an Indigenous perspective, human relations with water are often characterized in terms of relationships, reciprocities (both positive and negative), and kinship (McGregor 2014; Fox et al. 2017; Daigle 2018; Wilson & Inkster 2018; Wilson 2019; Leonard et al. 2023). More directly, an Indigenous cosmology is "water view," meaning "a view *from* the river not a view *of* the river" (Yazzie & Baldy 2018: 2). Importantly, this is a "theoretical standpoint . . . that foregrounds water view, (re)claiming knowledges not just for the people, but also for the water" (Yazzie & Baldy 2018: 2). From this ontological position, threats to water and the contentious responses to these threats are deep struggles between different worlds of water, enacted through the humans who inhabit them. For many if not all Indigenous communities, engaging in water politics with long-dominant societies starts with a basic assumption: "Western ontologies of water-as-a-resource are not entirely capable of acknowledging water as a living entity" (Parsons & Fisher 2020: 130). Moreover, much Indigenous water politics is not so much about asserting a "right" to water in the more conventional sense, but about asserting a *relationship* with water that conforms with Indigenous views of reality. In these cases, the phrase water politics, or even ontological politics, may not do full justice to the complex ways in which water and humans shape each other's worlds.

Another key concern of research on Indigenous water politics is how to both contest and work within settler water governance frameworks and political institutions. For example, Diné communities in the southwestern United States have for decades simultaneously sought recognition of their "rights" to water under colonial legal regimes while also forwarding an Indigenous and decolonial view of waterscapes such as the Little Colorado River (Curley 2019). In a comparable vein, Laborde and Jackson (2022) describe how Aboriginal Australians are using the concept of "Living Waters," a stark contrast to conventional ontologies of water as a resource, to discuss hybrid water governance approaches at a national level, while Mora et al. (2022) illuminate a similar effort by the Mistawasis Nêhiyawak First Nation in Saskatchewan, Canada, advocating an alternative methodology for governing water grounded in Indigenous perspectives. The goal, in both these cases and many others, is to insert Indigenous ontologies into dominant water governance institutions in ways that acknowledge the presence of multiple ontologies as a starting point, and that eventually lead to more inclusive, just, and ecologically sustainable water governance for all (Yates, Harris, & Wilson 2017).

A significant challenge to these integrative efforts is the capacity of Indigenous peoples to engage in shared governance arrangements and subsequent uncertainties over what political consequences might emerge from such hybrid institutions. For example, Wilson and Inkster (2018) forefront the multiple ways in which Yukon First Nation ontologies of water – its essence as a living entity and a relative – clash with conventional, state-driven ontologies of water as a "resource." Indeed, even conceptualizing water as a resource is an act of colonialism, one that "violently" removes water "from complex and interconnected ecological processes [including humans] for purposes of extraction and exploitation" (Curley 2021: 79). As conceived within dominant water governance models, resolutions to water access or other problems through hybrid approaches expressed as co-management, participation, or decentralization frequently serve to depoliticize water–society relations and cut off the potential means through which marginalized groups might genuinely engage in water governance on their own terms (Wilson 2019: 8). This point comes through clearly in how Yukon First Nations conceptualize the notion of "respect" for water, which is a form of "politics that insists on the viability of Indigenous forms of governance and refuses to subsume their forms of governance within that of settler-colonial society" (Wilson & Inkster 2018: 17). Indigenous ontologies focus on a concern of governance *with* water (a reciprocal partnership) in a fashion that settler-colonial norms of governance *over* water cannot comprehend and rarely acknowledge.

While advocates of Indigenous water ontologies are increasingly encountering dominant legal frameworks that devalue Indigenous ways of relating to water, several recent works underscore the creative advocacy of Indigenous communities in promoting alternative water ontologies and how these hold lessons for improved water governance more generally. Viaene (2021), for example, argues persuasively that international legal institutions and human rights advocates would benefit from thinking through "plurilegal water realities" such as those of the Maya Q'eqchi' communities in Guatemala that are opposing hydropower development. Relatedly, O'Donnell (2023) reviews efforts by Indigenous communities in Aotearoa New Zealand, Canada, Australia, and the United States to contest settler water laws. She concludes that forceful proclamations of Indigenous understandings of justice and rights have brought to light the glaring flaws in many current water governance arrangements in these societies and laid the groundwork for transformative water governance. One such case in Aotearoa New Zealand followed a long struggle headed by the Whanganui Iwi community, which granted the Whanganui River the legal status of "personhood" in the eyes of the national legal system (Parsons & Fisher 2020). Relatedly, ongoing work concerning the Lummi Nation's efforts to simultaneously bolster treaty rights and sustain Indigenous ways of interacting with water in the Salish Basin region in the coastal Pacific (near the Canadian–US border) clearly demonstrates the tenacity and ingenuity of Indigenous activists in working both within and around dominant institutions of water governance (Norman 2019). Clearly, scholarship on the many and heterogenous forms of Indigenous water ontology have blossomed in recent years, and are enlivening debates over water politics within geography and related fields.

Water Futures

As this chapter has shown, research on water politics within geography is highly interdisciplinary, drawing from Indigenous studies, critical geopolitics, STS, anthropology, and many other traditions of inquiry. However we categorize it, this work is vibrant, attuned to novel ideas and voices, and highly relevant to conversations concerning the past, present, and future of water–society relations. As with other socio-natures, water is always bound up in a wide range of material and cultural practices, biophysical processes, ideological positions, and institutional networks (Swyngedouw 2004). Given this water–society hybridity, water is never only the "*object* of politics," but instead "internalizes and expresses politics" through how it is utilized and conceptualized by different social groups (Linton & Budds 2014: 176). Our hope is that this brief overview of water politics will inspire even more creative and transformative thinking and scholarship in the years to come.

Acknowledgment

We dedicate this chapter to Karen Bakker, whose work on water–society relations will be immediately familiar to readers interested in critical scholarship on water politics. In many ways, this entire entry could be organized as a reflection on Karen's prolific and momentous contributions. Her intellectual curiosity and graciousness have inspired and will continue to inspire generations of scholars and advocates of water justice.

References

Agnew, J. (2011). Waterpower: Politics and the geography of water provision. *Annals of the Association of American Geographers 101*(3): 463–476.

Akhter, M. (2015). The hydropolitical cold war: The Indus waters treaty and state formation in Pakistan. *Political Geography 46*: 65–75.

Akhter, M. (2019). Adjudicating infrastructure: Treaties, territories, hydropolitics. *Environment and Planning E: Nature and Space* 2(4): 831–849.

Allouche, J. (2019). State building, nation making and post-colonial hydropolitics in India and Israel: Visible and hidden forms of violence at multiple scales. *Political Geography* 75: 102051.

Artiga-Purcell, J.A. (2022). Hydrosocial extractive territories: Gold, sugarcane and contested water politics in El Salvador. *Geoforum* 131: 93–104.

Ashmore, P. (2015). Towards a sociogeomorphology of rivers. *Geomorphology* 251: 149–156.

Bakker, K. (2012). Water: Political, biopolitical, material. *Social Studies of Science* 42(4): 616–623.

Bakker, K. (2013). Neoliberal versus postneoliberal water: Geographies of privatization and resistance. *Annals of the Association of American Geographers* 103(2): 253–260.

Barnes, J. (2017). States of maintenance: Power, politics, and Egypt's irrigation infrastructure. *Environment and Planning D: Society and Space* 35(1): 146–164.

Birkinshaw, M. (2022). Grabbing groundwater: Capture, extraction and the material politics of a fugitive resource. *Geoforum* 136: 32–45.

Boelens, R. (2014). Cultural politics and the hydrosocial cycle: Water, power and identity in the Andean highlands. *Geoforum* 57: 234–247.

Boelens, R., Escobar, A., Bakker, K., et al. (2023). Riverhood: Political ecologies of socionature commoning and translocal struggles for water justice. *Journal of Peasant Studies* 50(3): 1125–1156.

Boelens, R., Hoogesteger, J., Swyngedouw, E., Vos, J., & Wester, P. (2016). Hydrosocial territories: A political ecology perspective. *Water International* 41(1): 1–14.

Brochmann, M., & Gleditsch, N.P. (2012). Shared rivers and conflict – a reconsideration. *Political Geography* 31(8): 519–527. doi:10.1016/j.polgeo.2012.11.001

Budds, J., Linton, J., & McDonnell, R. (2014). The hydrosocial cycle. *Geoforum* 57: 167–169.

Cantor, A. (2021). Hydrosocial hinterlands: An urban political ecology of Southern California's hydrosocial territory. *Environment and Planning E: Nature and Space* 4(2): 451–474.

Chien, S.S., & Hong, D.L. (2018). River leaders in China: Party-state hierarchy and transboundary governance. *Political Geography* 62: 58–67.

Colven, E. (2020). Subterranean infrastructures in a sinking city: The politics of visibility in Jakarta. *Critical Asian Studies* 52(3): 311–331.

Curley, A. (2019). "Our winters' rights": Challenging colonial water laws. *Global Environmental Politics* 19(3): 57–76.

Curley, A. (2021). Resources is just another word for colonialism. In Himley, M., Havice, E., & Valdivia, G. (eds), *The Routledge Handbook of Critical Resource Geography*. London: Routledge, 79–90.

Curran, D. (2019). Indigenous processes of consent: Repoliticizing water governance through legal pluralism. *Water* 11(3): 571.

Daigle, M. (2018). Resurging through Kishiichiwan: The spatial politics of Indigenous water relations. *Decolonization: Indigeneity, Education & Society* 7(1): 159–172.

Del Bene, D., Scheidel, A., & Temper, L. (2018). More dams, more violence? A global analysis on resistances and repression around conflictive dams through co-produced knowledge. *Sustainability Science* 13: 617–633.

Díaz-Combs, C. (2023). "People are no longer quiet": Ordinary environmental citizenship in Lago Agrio, Ecuador. *Environment and Planning E: Nature and Space* 7(1): 25148486231160642.

Dinar, S., Katz, D., De Stefano, L., & Blankespoor, B. (2015). Climate change, conflict, and cooperation: Global analysis of the effectiveness of international river treaties in addressing water variability. *Political Geography 45*: 55–66.

Duarte-Abadía, B., Boelens, R., & Roa-Avendaño, T. (2015). Hydropower, encroachment and the re-patterning of hydrosocial territory: The case of Hidrosogamoso in Colombia. *Human Organization* 74(3): 243–254.

Estes, N. (2019). *Our History Is the Future: Standing Rock versus the Dakota Access Pipeline, and the Long Tradition of Indigenous Resistance.* London: Verso Books.

Flaminio, S., Rouillé-Kielo, G., & Le Visage, S. (2022). Waterscapes and hydrosocial territories: Thinking space in political ecologies of water. *Progress in Environmental Geography 1*(1–4): 33–57.

Fox, C., Reo, N., Turner, D., et al. (2017). "The river is us; the river is in our veins": Re-defining river restoration in three Indigenous communities. *Sustainability Science 12*: 521–533.

Fox, C.A., Magilligan, F.J., & Sneddon, C.S. (2016). "You kill the dam, you are killing a part of me": Dam removal and the environmental politics of river restoration. *Geoforum 70*: 93–104.

Fung, Z., & Lamb, V. (2023). Dams, diversions, and development: Slow resistance and authoritarian rule in the Salween River Basin. *Antipode 55*(6): 1662–1685.

Furlong, K. (2006). Hidden theories, troubled waters: International relations, the "territorial trap", and the Southern African Development Community's transboundary waters. *Political Geography 25*: 438–458.

Götz, J.M., & Middleton, C. (2020). Ontological politics of hydrosocial territories in the Salween River basin, Myanmar/Burma. *Political Geography 78*: 102115.

Haines, D. (2016). *Rivers Divided: Indus Basin Waters in the Making of India and Pakistan.* New York: Oxford University Press.

Han, X., & Webber, M. (2020). From Chinese dam building in Africa to the Belt and Road Initiative: Assembling infrastructure projects and their linkages. *Political Geography 77*: 102102.

Hanasz, P. (2017). Muddy waters: International actors and transboundary water cooperation in the Ganges-Brahmaputra problemshed. *Water Alternatives 10*(2): 459–474.

Hommes, L., & Boelens, R. (2017). Urbanizing rural waters: Rural-urban water transfers and the reconfiguration of hydrosocial territories in Lima. *Political Geography 57*: 71–80.

Karpouzoglou, T., & Vij, S. (2017). Waterscape: A perspective for understanding the contested geography of water. *Wiley Interdisciplinary Reviews: Water 4*(3): e1210.

Koch, N. (2021). The political lives of deserts. *Annals of the American Association of Geographers 111*(1): 87–104.

Kroepsch, A.C., & Clifford, K.R. (2022). On environments of not knowing: How some environmental spaces and circulations are made inscrutable. *Geoforum 132*: 171–181.

Laborde, S., & Jackson, S. (2022). Living waters or resource? Ontological differences and the governance of waters and rivers. *Local Environment 27*(3): 357–374.

Lave, R. (2012). *Fields and Streams: Stream Restoration, Neoliberalism, and the Future of Environmental Science.* Atlanta, GA: University of Georgia Press.

Lawless, C. (2023). How water features: Negotiating and reassembling the sociomaterial politics of central Californian groundwater. *Space and Polity 27*(2): 218–233.

Leonard, K., David-Chavez, D., Smiles, D., et al. (2023). Water back: A review centering rematriation and Indigenous water research sovereignty. *Water Alternatives 16*(2): 374–428.

Liao, Y.K., & Schmidt, J.J. (2023). Hydrosocial geographies: Cycles, spaces and spheres of concern. *Progress in Environmental Geography 2*(4): 240–265.

Linton, J. (2010). *What Is Water? The History of a Modern Abstraction.* Vancouver: UBC Press.

Linton, J., & Budds, J. (2014). The hydrosocial cycle: Defining and mobilizing a relational-dialectical approach to water. *Geoforum 57*: 170–180.

Lopez, R.R., Boelens, R., Vos, J., & Rap, E. (2019). Hydrosocial territories in dispute: Flows of water and power in an interbasin transfer project in Bolivia. *Water Alternatives 12*(1): 267–284.

Manosalvas, R., Hoogesteger, J., & Boelens, R. (2021). Contractual reciprocity and the re-making of community hydrosocial territories: The case of La Chimba in the Ecuadorian páramos. *Water 13*(11): 1600.

Mason, M. (2020). Hydraulic patronage: A political ecology of the Turkey-Northern Cyprus water pipeline. *Political Geography 76*: 102086.

McGregor, D. (2014). Traditional knowledge and water governance: The ethic of responsibility. *AlterNative: An International Journal of Indigenous Peoples 10*(5): 493–507.

Meehan, K.M. (2014). Tool-power: Water infrastructure as wellsprings of state power. *Geoforum 57*: 215–224.

Melville-Rea, H. (2022). Parched and impatient: Political engagement after drought. *Political Geography 96*: 102516.

Menga, F. (2017). Hydropolis: Reinterpreting the polis in water politics. *Political Geography 60*: 100–109.

Menga, F., & Swyngedouw, E. (eds) (2018). *Water, Technology and the Nation-State.* London: Routledge.

Middleton, C., & Allouche, J. (2016). Watershed or powershed? Critical hydropolitics, China and the "Lancang-Mekong cooperation framework". *International Spectator 51*(3): 100–117.

Millington, N., & Scheba, S. (2021). Day zero and the infrastructures of climate change: Water governance, inequality, and infrastructural politics in Cape Town's water crisis. *International Journal of Urban and Regional Research 45*(1): 116–132.

Molle, F. (2009). River-basin planning and management: The social life of a concept. *Geoforum 40*(3): 484–494.

Mora, M.F., Johnston, A.B., Watson, M., & Bharadwaj, L. (2022). Honouring water: The Mistawasis Nêhiyawak water governance framework. *International Indigenous Policy Journal 13*(3): 1–27.

Mullenite, J. (2019). Infrastructure and authoritarianism in the land of waters: A genealogy of flood control in Guyana. *Annals of the American Association of Geographers 109*(2): 502–510. doi:10.1080/24694452.2018.1490635

Neville, K.J., & Coulthard, G. (2019). Transformative water relations: Indigenous interventions in global political economies. *Global Environmental Politics 19*(3): 1–15.

Nixon, R. (2011). *Slow Violence and the Environmentalism of the Poor.* Cambridge, MA: Harvard University Press.

Norman, E.S. (2014). *Governing Transboundary Waters: Canada, the United States, and Indigenous Communities*. New York: Routledge.

Norman, E.S. (2019). Finding common ground: Negotiating downstream rights to harvest with upstream responsibilities to protect—dairies, berries, and shellfish in the Salish Sea. *Global Environmental Politics 19*(3): 77–97.

O'Donnell, E. (2023). Water sovereignty for Indigenous peoples: Pathways to pluralist, legitimate and sustainable water laws in settler colonial states. *PLOS Water 2*(11): e0000144.

Parsons, M., & Fisher, K. (2020). Indigenous peoples and transformations in freshwater governance and management. *Current Opinion in Environmental Sustainability 44*: 124–139.

Perreault, T. (2005). State restructuring and the scale politics of rural water governance in Bolivia. *Environment and Planning A 37*(2): 263–284.

Ranganathan, M., & Balazs, C. (2015). Water marginalization at the urban fringe: Environmental justice and urban political ecology across the North–South divide. *Urban Geography 36*(3): 403–423.

Rogers, S., & Crow-Miller, B. (2017). The politics of water: A review of hydropolitical frameworks and their application in China. *Wiley Interdisciplinary Reviews: Water 4*(6): e1239.

Rogers, S., Fung, Z., Lamb, V., et al. (2023). Beyond state politics in Asia's transboundary rivers: Revisiting two decades of critical hydropolitics. *Geography Compass 17*(4): e12685.

Sarna-Wojcicki, D., Sowerwine, J., Hillman, L., Hillman, L., & Tripp, B. (2019). Decentring watersheds and decolonising watershed governance: Towards an ecocultural politics of scale in the Klamath Basin. *Water Alternatives 12*(1): 241–266.

Sizek, J. (2021). Zombie infrastructure: A legal geography of railroad monstrosity in the California desert. *Environment and Planning D: Society and Space 39*(4): 758–775. doi:10.1177/02637758211023504

Sneddon, C. (2015). *Concrete Revolution: Large Dams, Cold War Geopolitics, and the US Bureau of Reclamation*. Chicago, IL: University of Chicago Press.

Sneddon, C., & Fox, C. (2006). Rethinking transboundary waters: A critical hydropolitics of the Mekong basin. *Political Geography 25*(2): 181–202.

Sternberg, T. (2016). Water megaprojects in deserts and drylands. *International Journal of Water Resources Development 32*(2): 301–320. doi:10.1080/07900627.2015.1012660

Strang, V. (2016). Infrastructural relations: Water, political power and the rise of a new "despotic regime". *Water Alternatives 9*(2): 292–318.

Suhardiman, D., & Middleton, C. (2020). The Salween River as a transboundary commons: Fragmented collective action, hybrid governance and power. *Asia Pacific Viewpoint 61*(2): 301–314.

Sultana, F. (2007). Water, water everywhere, but not a drop to drink: Pani politics (water politics) in rural Bangladesh. *International Feminist Journal of Politics 9*(4): 494–502.

Swyngedouw, E. (2004). *Social Power and the Urbanization of Water: Flows of Power*. Oxford: Oxford University Press.

Swyngedouw, E. (2011). Depoliticized environments: The end of nature, climate change and the post-political condition. *Royal Institute of Philosophy Supplement 69*: 253–274.

Truelove, Y. (2019). Gray zones: The everyday practices and governance of water beyond the network. *Annals of the American Association of Geographers 109*(6): 1758–1774.

Usher, M. (2019). Desali-nation: Techno-diplomacy and hydraulic state restructuring through reverse osmosis membranes in Singapore. *Transactions of the Institute of British Geographers 44*(1): 110–124.

Viaene, L. (2021). Indigenous water ontologies, hydro-development and the human/more-than-human right to water: A call for critical engagement with plurilegal water realities. *Water 13*(12): 1660.

Voyles, T.B. (2021). *The Settler Sea: California's Salton Sea and the Consequences of Colonialism*. Lincoln, NE: University of Nebraska Press.

Whyte, K. (2017). The Dakota access pipeline, environmental injustice, and US colonialism. *Red Ink: An International Journal of Indigenous Literature, Arts, & Humanities 19*(1). https://ssrn.com/abstract=2925513 (accessed December 2, 2024)

Wilson, N.J. (2019). "Seeing water like a state?" Indigenous water governance through Yukon First Nation self-government agreements. *Geoforum 104*: 101–113.

Wilson, N.J., & Inkster, J. (2018). Respecting water: Indigenous water governance, ontologies, and the politics of kinship on the ground. *Environment and Planning E: Nature and Space 1*(4): 516–538.

Wolf, A.T. (2007). Shared waters: Conflict and cooperation. *Annual Review of Environment and Resources 32*: 241–269.

Yates, J.S., Harris, L.M., & Wilson, N.J. (2017). Multiple ontologies of water: Politics, conflict and implications for governance. *Environment and Planning D: Society and Space 35*(5): 797–815.

Yazzie, M., & Baldy, C.R. (2018). Introduction: Indigenous peoples and the politics of water. *Decolonization: Indigeneity, Education and Society 7*(1): 1–18.

Chapter 34

Health Geographies

Brian King and Halie Kampman
Department of Geography, Pennsylvania State University, University Park, PA, USA

It is nearly impossible to engage with current events and fail to observe the inextricable links between politics and human health. The Covid-19 pandemic made plain the ways in which political, economic, and cultural systems shape disease exposure and health management. The rapid development of a vaccine collided with biobehavioral practices that proved stubbornly resistant to change. The rollout of vaccines and treatment protocols were not evenly experienced, as the pandemic generated divergent effects within resource-constrained environments including within the Global South (Brunn & Gilbreath 2022).

The rapid expansion of Covid-19 underscores the co-occurrence of health challenges in the twenty-first century. And it did not occur in isolation. Prior to the onset of Covid-19, the United States was already contending with a severe opioid epidemic, now the leading cause of injury-related death in the nation (Rudd et al. 2016; Ruhm 2017). And previous epidemics, such as the HIV/AIDS epidemic, have not disappeared, as HIV has transitioned into a chronic condition. These challenges extend beyond infectious disease. Diet-related diseases fueled by huge shifts in agrarian systems and increasing levels of food insecurity compound the co-occurrence of existing health issues. The intersecting landscape of health challenges exacerbated by Covid-19 exposes how the production, management, experience, and potential for human health and wellbeing are inherently political.

While health has not been a central area of focus for political geography, it is fundamentally related to the subfield's contributions, and the subfield has much to contribute to how health is understood and studied. In this chapter, we provide a brief history of geographical research on human health, focusing on the subfields

The Wiley Blackwell Companion to Political Geography, Second Edition.
Edited by Virginie Mamadouh, Natalie Koch, Chih Yuan Woon, and John Agnew.

of medical geography and health geography. This is followed by a review of some key scholarship on human health, including a focus on food systems and care theory. We conclude by identifying some key challenges and future directions for scholarship on the politics of human health.

Medical and Health Geography

Medical geography emerged in the 1950s, in part from scholarship on human disease ecology.

Meade and Earickson (2000) argued that the addition of the Commission on Medical Geography (Ecology) of Health and Disease to the International Geographical Union in 1952 represented an early disciplinary commitment to medical geography. Jacques May's publication of *The Ecology of Disease* in 1958 focused on the interactions between human populations and the natural environment, including the production of infectious disease (May 1958; Mayer 1996). In a landmark review, Mayer (1996: 441) detailed that disease ecology considers how "humanity, including culture, society and behavior; the physical world, including topography, vegetation and climate; and biology, including vector and pathogen ecology, interact together in an evolving and interactive system, to produce foci of disease." Recent volumes surveying the fields of medical and health geography emphasize the continued influence of disease ecology (Brown, McLafferty, & Moon 2009; Emch, Root, & Carrell 2017).

Medical geography has a longstanding interest in patterns of disease diffusion and the location of healthcare services. As Meade and Earickson (2000: 9) explain, "When the epidemiological and statistical study of the spread of contagious disease within a population and its subgroups is considered spatially, settlement forms and urban systems and transportation networks and trade flows and global connections and population mobility become the objects of modeling." The spatial analytical tradition provided a foundation from which health geography would emerge in the 1990s.

A broader interest in human health in addition to disease patterns contributed to the emergence of the health geography subfield in the 1990s. Central to this shift were a series of papers by Robin Kearns, Graham Moon, and others who articulated a "post-medical geography of health" in order to emphasize broader notions of health and wellbeing in place (Kearns 1993: 144). Some key themes to health geography include interest in therapeutic places and landscapes, referring to concern with the places and processes that contribute to health (Gesler 1992; Williams 1999; Gesler & Kearns 2002; Kearns & Moon 2002). Reflecting an early critique of the biomedical model that continues to drive hegemonic understandings of health, Dyck (1995) advocated for health geography to challenge the positivist paradigm that had dominated traditional inquiry. In a nonpositivist approach, understandings of space and place are instead seen as relational.

A relational approach continues to prevail in health geography today, as scholars focus on the production of place and how place-specific factors shape the possibilities for health and wellbeing (Gatrell & Elliott 2014; Evans et al. 2015; Philo 2016; Prior, Manley, & Sabel 2019). Diverse understandings and definitions of health and wellbeing are thus central to health geography. Rather than considering health as the absence of disease, health geographers consider broader physical, mental, and social factors that contribute to wellbeing. As medical and health geography shift the focus toward the intricate interplay between health and broader spatial and social forces, this exposes how health becomes a political concern, shaping and being shaped by political agendas, governance strategies, and social responses.

Health, Politics, and Political Geography

Human disease and health have not been traditional foci of political geography but are increasingly seen as relevant to the subfield. The work of Alan Ingram (2009) was instrumental in defining disease as a geopolitical concern, for the way in which it implicates power, economy, and the mechanisms by which healthcare is distributed. In a review of the intersections between (geo)politics and disease, Ingram (2009) identifies three themes as particularly relevant: the spatialization of governance, biopolitics, and transnational political economies. Ingram's focus on biopolitics (Foucault 1976) raises important questions around the role of the modern state in governing (and inciting self-governing of) the health of its subjects. Some scholars suggest that biopolitics has given rise to new forms of self-responsibility (Lemke 2007) or *responsibilization* (King & Rishworth 2022).

Public health policies, and the broader political contexts and histories in which they operate, are central to scholarship in this area. Michael Brown (2006, 2008, 2009; Brown & Knopp 2010) has been influential in connecting public health, urban geographies, and human health. Brown's work has been instrumental in leveraging frameworks within political geography to address health interventions by state agencies. As he notes: "Those who conceptualize 'urban politics' rarely consider public-health departments as an interesting apparatus of the state . . . Public health in all its wide and shifting forms is an important part of urban political geography that we should consider more fully" (Brown 2009: 23). Public health must also be considered in the context of broader political histories. For instance, even in cases when governments explicitly address public health crises, unless they can also address politically created social inequities, public health will be slow to improve (King, Burka, & Winchester 2018). King (2017) illustrates this with the case of HIV/AIDS in South Africa. While medications are made widely available, disease management requires a much broader attention to social ecological processes, specifically the inequities left by apartheid that cause vulnerable populations to lack access to resources such as healthy foods needed to supplement medication. Other geographers have similarly examined the intersections between urban geography, health, and political control (Craddock 2000a), looking also at intersections with social identity (Craddock 2000b; Cooper 2006).

Political geography has recently addressed Covid-19 to expose different forms of governance that emerged during the pandemic. Covid-19 limited movement, which in some cases eroded hard-fought rights for free movement (Cole & Dodds 2021). Travel restrictions in response to Covid-19 are analyzed as just one layer in a broader landscape of highly uneven global mobility (Häkli 2021). Reflecting this concern, Liu and Bennett (2020) examine the case of border control in mainland China and Hong Kong to critique how national governments' seemingly functionalist efforts to set borders intersect with biopower, surveillance, and identity. Using her own family's experience of borders closing between Switzerland and France, Fall (2020) considers how new borders link the intimate and the global, asserting that standard state-sanctioned safety measures invoked feelings of emotional disconnect and restriction. On the other hand, Leonard (2020) argues that the maintenance of borders around Indigenous lands enabled a type of compassionate community care during the pandemic, as well as helping to prevent and track the spread of disease.

Covid-19 brought about a diversity of policy reactions around the globe, which have been used to gain insight into the relationship between disease and governance. Considering the spatial differentiation in governance, Covid-19 offered a lens with which to explore and evaluate state policies explicitly designed to address the diversity of landscapes across the United States (Agnew 2022). A popular perspective within political geography considers how Covid-19 reinforced various forms of strong-arm politics. According to Murphy (2022), the pandemic amplified state-sanctioned surveillance and various data-gathering projects went largely uncontested. At the same time, populist movements gained ground to emphasize territorial sovereignty, in some cases making the proposed safety of authoritarianism attractive to citizens (Lim 2022; Murphy 2022).

Covid-19 also demonstrated that the discourses about disease, medicine, and health shape broader geopolitical imaginations and advance political agendas. Global conspiracy theories around medicine and health – like that of the "plandemic" – assert that the pandemic was planned as part of a larger political project of surveillance (Mercille et al. 2021). These can serve to deny the actual epidemiological complexities of a disease while advancing politics based on fear and inward-looking behavior (Sturm & Albrecht 2021).

Political geography helps to illuminate the implications of health governance and disease management. Political contexts and histories are fundamental in shaping landscapes of health. We can see this in the ways in which HIV/AIDS management remains spatially unequal in South Africa, or in how Covid-19 exposed different forms of governance. Relatedly, geographies of food and diet build on perspectives in political geography to show how food landscapes and dietary choices are political unto themselves.

Politics of Food and Diet

How food is produced, policed, distributed, and consumed is inherently political, and health geographies extend to explore the broader systems that shape landscapes of health, including food systems. Spatial approaches to food access and insecurity show how socioeconomic disenfranchisement is related to limited access. The concept of "food deserts" – spaces that are characterized by a lack of food and grocery options – has been useful to geographers seeking to unpack underlying structural factors (Walker, Keane, & Burke 2010; Mayer et al. 2014). Many studies focus on the presence or absence of supermarkets, but can also explore the interaction between a diversity of spaces where residents purchase food (restaurants, corner stores, gas stations, etc.) (Walker, Keane, & Burke 2010). Food desert frameworks can invite a policy angle, considering the root causes of income disparities related to food access and the distribution of supermarkets in respect to a neighborhood's economics. The concept of "food apartheid," coined by social activist Karen Washington, builds on this by invoking a more critical angle on structural factors behind inequitable access (Brones 2018). It specifically references systemic issues of racism and discrimination that shape spatial differentiation (Reese 2019).

Inequities undergirding food access are also studied and addressed in the context of food- and health-oriented social movements. This includes movements to improve the quality and quantity of food available to various groups, such as the "local" food movement and movements centered around race and equity. Starting in the 1990s, conversations developed around the rise in "local" and "sustainable" food movements, generally based in middle- and upper-class communities throughout the Global North. Local food, celebrated as an alternative to increasingly globalizing foodways, began to be promoted as an alternative form of food consumption that involves connecting with local producers, nourishing yourself, and investing in your community.

Geographers apply a critical lens to food movements by considering how they may also have defensive mechanisms that narrow the scope of what should be valued and can create an "us versus them" mindset. Viewed thus, "local" is not an innocent term, as it can house nativist sentiments leading to alternative standards of purity and perfection vulnerable to corporate cooption (Allen et al. 2003; Allen 2010; Alkon 2012; Goodman, DuPuis, & Goodman 2012). Early research on local food movements has since been applied to contexts throughout the Global South, where the cultural and historical mechanisms of "localism" mean different things (Miller 2019; Montefrio 2020; Kampman forthcoming). Studies of food movements have also become more attentive to the subject of race, challenging the assumption that movements are white and highlighting everyday resistances among communities of color (Sbicca 2018; Reese 2019).

The racialized nature of food geographies is foundational to food politics, particularly around spaces of activism. Alkon's (2012) work was instrumental in questioning and challenging the whiteness of alternative food movements while also examining how race intersects with understandings of local and organic food systems. She explores how the "green economy" – referring to sustainable and local movements – functions on the ground and how these meanings are raced and classed. Other approaches aim to further expand the category of alternative food, or food justice work, by rethinking the ways in which communities of color fall outside initial norms set by early (white) food movement evangelists (Garth & Reese 2020). In her book on Black food geographies, Reese (2019) shows how and where Black people create spaces for nourishing food landscapes, despite food inequities created by anti-Black racism.

Beyond race, other social constructs shape the politics of food, which can be explored through the discursive meanings and measures of dietary choice. The politics of dietary choice addresses the ways in which dietary choices, and associated dietary health issues, both shape and are shaped by society. The determinants of human health are, of course, social and political. The concept of "deaths of despair" (Case & Deaton 2015, 2017, 2020) explores how lifestyle choices – specifically referencing alcohol consumption and drug use – have skyrocketed in the United States (and beyond) throughout the twenty-first century. Contemporary health problems at the scale of the body are intricately interwoven with societal ills, related to education, social standing, or race (Gennuso et al. 2019; Zheng & Choi 2024).

Relatedly, there is a significant body of work on diet-related diseases (like type 2 diabetes, heart disease, or hypertension) and the ways in which these diseases are socially constructed (Guthman 2011; Biltekoff 2013; Kimura 2013). Diet-related diseases, as well as risk factors for disease (e.g., being overweight), carry heavy stigmas that shape how they are experienced and treated across multiple scales. For instance, Guthman (2011, 2012) argues for scholarship and activism around obesity and food justice. Rather than considering obesity to be a personal problem, it is necessary to explore a diversity of causes (like epigenetics, race, and class) that go beyond personal choice. An overexaggerated focus on personal choice redirects attention away from the broader context of causal factors and normalizes social biases. These biases are also often racialized and, in particular, fat phobias have racial origins. As Strings (2019: 6) explains, "phobia about fatness and preference for thinness have not, principally or historically been about health. Instead they have been one way the body has been used to craft and legitimate race, sex and class hierarchies." Black bodies have historically been othered to sharpen and legitimize racial distinctions.

The politics of food and health take on a new dimension when considered globally, including topics of international food aid and international movements for food sovereignty. Food aid invites a complex negotiation of power between donor and recipient, which implicates geopolitics, economics, and cultural norms. Contemporary food aid refers to not only donated goods but also more long-term structural development projects focused on improving food systems across the

Global South. Recently, these projects have taken a more explicitly health-focused turn, seeking to improve nutrition by advancing the economic, social, and environmental health of agricultural production. For instance, one such project across the continent of Africa, known as the New Green Revolution for Africa, launched in the early 2000s and directs donor funding to improve the production of staple crops throughout the continent. Though structured as a more sustainable approach to food aid, critical geographers have argued that in practice it is highly abstracted and inattentive to local realities (Kerr 2012; Gengenbach et al. 2018). Based on the logic of neoliberal developmentalism, which advances a supposedly scalable markets-based approach, such programs can be blind to the material or cultural realities of a place.

Neoliberal approaches to international food aid have been countered by movements for food sovereignty, referring to systems in which the people who produce and distribute food have full control over these processes. Other frameworks have made particular contributions to this area of study, incorporating Indigenous perspectives centered on resistance to the colonial-capitalist legacies that threaten sovereignty (Trauger 2014; Grey & Patel 2015; Daigle 2019). Normalized neoliberal logics can be countered through the pursuit of food sovereignty, informed by Indigenous beliefs. Scholars consider the potential of non-commodified food exchanges, the role of civil disobedience in the face of extractive food policies, and the reframing of property rights as "use" rights (Trauger 2014). Just as access to nutritious food reflects broader social, economic, and spatial disparities, so too does access to healthcare: how care work is organized, experienced, and distributed across different regions.

Geographies of (Health)Care

Integral to research on health is *care* – referring to practical or emotional support. Health geographers approach care as a complex concept (which is neither inherently good nor bad) and explore how it is organized, experienced, and spatially distributed. As in other subthemes of health geographies, the politics of care emerged from early studies of spatial inequalities, in this case in regard to access to (health)care. In other words, geographies of care research initially focused on how healthcare facilities are distributed across different regions and how these spatial differentiations overlap with other factors like race and class (Staeheli & Brown 2003).

Branching out from a focus on spatial inequalities, the concept of "landscapes of care" is useful to theorize the myriad factors – spatial and otherwise – that play a role in care. The term emerged from the concept of landscapes of despair (Gesler 1992) and therapeutic landscapes (Williams 1999) as a less prescriptive way to think about the "complex embodied and organizational spatialities that emerge from and through the relationships of care" (Milligan 2000: 740). Scale has always been important to care work, particularly because of the temptation to think about care at the level of the individual. Theorizations of care push against this temptation: while relations of care at the individual level can be an important launching point, the politics of care is ignited when it is understood at the structural level (how it operates in institutions, governments, etc.) (Tronto 1993).

The related concept of "therapeutic citizenship" can be applied to theorize care at the structural level. This term refers to the ways in which individuals or communities actively engage with health-related practices, policies, and systems to claim rights, access resources, and promote social justice. It encompasses the idea that individuals not only passively receive healthcare services but also actively participate in shaping their own health outcomes and advocating for their health needs. This has been applied to the case of HIV treatment across Africa. In the early days of the epidemic in West Africa, Nguyen (2010: 9) described the importance of testimonials that "trained people to talk about themselves, and talking about yourself came to be about staying alive." This underscored the political entanglements associated with accessing healthcare as the epidemic expanded across the continent.

In research on the geographies of (health)care, there is particular focus on gendered and intersectional care. In a sense, all theorizations of care trace back to an attention to gender, as feminist theory is central to care work. Feminist theorists who trailblazed care work, such as Gilligan (1982), challenge canonical theories of moral development, asserting that justice must be complemented by care. More recent work emphasizes how care is a relational interdependent experience and is not just individual and innate to women (Tronto 2015; Rishworth & King 2021). Attention to care serves as resistance to liberal economy, not to be confused with an inherently feminine trait (Lawson 2007; Bartos 2019). Furthermore, understanding *intersectional* care can extend to Indigenous notions of care that focus on community-level relations and in care between human and nonhuman organisms (Tallbear 2022). Intersectional care considers wider understandings of what it means to care, who has access, who is excluded from good care, and how all this is immersed in structural power relations (Hankivsky 2014). For instance, care has been applied as an analytical framework by which to explore colonialism; Narayan (1995) argues that colonialism and slavery were made palatable to the enslaved and their captors through a rhetoric of care and responsibility.

Gendered approaches to care are related to embodied geographies of emotion and care, a relatively contemporary turn in care work. These approaches are based on the conviction that care exposes the epistemological power of emotions, extending sites of geographical concern to other scales, like that of the body or touch (Lawson 2007; Dixon 2015). Broadly speaking, this area of study draws from emotional geographies, which seek to understand emotion in terms of social spatial mediation (Parr 2003; Davidson 2005; Bondi 2016). As health geographers are focused on issues of power, emotion itself can be seen as a site of power

(Lawson 2007). The focus on emotions broadens the scope of what is considered power and how power operates. Embodiment can also be explored through ways in which political geographies shape people's everyday experience and lives. For instance, Sarma, Faxon, and Roberts (2023) consider the embodied political geographies of extractive resource frontiers in Myanmar by showing how politics exists beyond the scale of the state, manifesting also in the intimate and mundane.

The study of care within health geographies illuminates its multifaceted nature and its pivotal role in shaping individual wellbeing. From analyzing spatial inequalities to exploring intersectional perspectives, this examination underscores the intricate dynamics that inform care provision and its broader societal implications.

Challenges and Future Directions

Geographical scholarship on human disease and health is rich and diverse. The value of this research is especially clear in the wake of the Covid-19 pandemic and the persistent challenges of opioid epidemics and chronic diseases. As these cases show and as geographers have long argued, health outcomes are deeply entwined with political, economic, and cultural systems.

The interdisciplinary nature of health geographies has shed light on the multifaceted dimensions of health and disease. From the foundational principles of medical geography to the emergence of health geography as a distinct subfield, scholars have grappled with diverse understandings of health, ranging from disease patterns to broader notions of wellbeing in place. Political geography has similarly engaged with questions of human disease and health, emphasizing the geopolitical dimensions of disease governance, biopolitics, and transnational political economies. The politics of food and diet have emerged as critical areas of inquiry within health geographies, elucidating how food systems are political constructs shaped by power relations, socioeconomic disparities, and cultural norms. From the analysis of food deserts and food apartheid to the examination of local food movements and international food aid, scholars have interrogated the structural factors behind inequitable access to nutritious food and the implications for public health. The discourses surrounding dietary choices and diet-related diseases further highlight the societal constructions of health and the ways in which intersecting axes of identity, such as race and class, shape health outcomes and experiences.

By situating health within broader political and spatial contexts, researchers can better understand the root causes of health disparities and advocate for more equitable and inclusive policies and interventions. Moving forward, it is imperative to continue exploring the politics of human health, engaging with diverse perspectives and voices to foster a more holistic understanding of health and wellbeing in an increasingly interconnected world.

References

Agnew, J. (2022). Failing federalism? US dualist federalism and the 2020–22 pandemic. *Regional Studies, Regional Science* 9(1): 149–171. doi:10.1080/21681376.2022.2045214

Alkon, A. (2012). *Black, White, and Green: Farmers Markets, Race, and the Green Economy*. Athens, GA: University of Georgia Press.

Allen, P. (2010). Realizing justice in local food systems. *Cambridge Journal of Regions, Economy and Society* 3(2): 295–308. doi:10.1093/cjres/rsq015

Allen, P., Fitzsimmons, M., Goodman, M., & Warner, K. (2003). Shifting plates in the agrifood landscape: The tectonics of alternative agrifood initiatives in California. *Journal of Rural Studies* 19(1): 61–75. doi:10.1016/S0743-0167(02)00047-5

Bartos, A.E. (2019). Introduction: Stretching the boundaries of care. *Gender, Place and Culture* 26(6): 767–777. doi:10.1080/0966369X.2019.1635999

Biltekoff, C. (2013). *Eating Right in America: The Cultural Politics of Food and Health*. Durham, NC: Duke University Press.

Bondi, L. (2016). *Emotional Geographies*. New York: Routledge.

Brones, A. (2018). Food apartheid: The root of the problem with America's groceries. *The Guardian*, May 15. https://www.theguardian.com/society/2018/may/15/food-apartheid-food-deserts-racism-inequality-america-karen-washington-interview (accessed March 15, 2024).

Brown, M. (2006). Sexual citizenship, political obligation and disease ecology in gay Seattle. *Political Geography* 25(8): 874–898. doi:10.1016/j.polgeo.2006.05.004

Brown, M. (2008). Working political geography through social movements theory: The case of gay and lesbian Seattle. In Cox, K.R., Low, M., & Robinson, J. (eds), *The SAGE Handbook of Political Geography*. London: Sage, 285–304. doi:10.4135/9781848607880.n17

Brown, M. (2009). 2008 Urban geography plenary lecture—Public health as urban politics, urban geography: Venereal biopower in Seattle, 1943–1983. *Urban Geography* 30(1): 1–29. doi:10.2747/0272-3638.30.1.1

Brown, M., & Knopp, L. (2010). Between anatamo- and bio-politics: Geographies of sexual health in wartime Seattle. *Political Geography* 29(7): 392–403. doi:10.1016/j.polgeo.2010.10.001

Brown, T., McLafferty, S., & Moon, G. (2009). *A Companion to Health and Medical Geography*. Hoboken, NJ: John Wiley & Sons.

Brunn, S.D., & Gilbreath, D. (eds) (2022). *Covid-19 and a World of Ad Hoc Geographies*. Cham: Springer Nature.

Case, A., & Deaton, A. (2015). Rising morbidity and mortality in midlife among white non-Hispanic Americans in the 21st century. *Proceedings of the National Academy of Sciences of the United States of America* 112(49): 15078–15083. doi:10.1073/pnas.1518393112

Case, A., & Deaton, A. (2017). Mortality and morbidity in the 21st century. *Brookings Papers on Economic Activity* 2017(1): 397–443. doi:10.1353/eca.2017.0005

Case, A., & Deaton, A. (2020). *Deaths of Despair and the Future of Capitalism*. Princeton, NJ: Princeton University Press.

Cole, J., & Dodds, K. (2021). Unhealthy geopolitics? Bordering disease in the time of coronavirus. *Geographical Research* 59(1): 169–181. doi:10.1111/1745-5871.12457

Cooper, D. (2006). Active citizenship and the governmentality of local lesbian and gay politics. *Political Geography 25*: 921–943. doi:10.1016/j.polgeo.2006.08.009

Craddock, S. (2000a). *City of Plagues: Disease, Poverty, and Deviance in San Francisco*. Minneapolis, MN: University of Minnesota Press.

Craddock, S. (2000b). Disease, social identity, and risk: Rethinking the geography of AIDS. *Transactions of the Institute of British Geographers 25*(2): 153–168. doi:10.1111/j.0020-2754.2000.00153.x

Daigle, M. (2019). Tracing the terrain of Indigenous food sovereignties. *Journal of Peasant Studies 46*(2): 297–315. doi:10.1080/03066150.2017.1324423

Davidson, J. (2005). Contesting stigma and contested emotions: Personal experience and public perception of specific phobias. *Social Science & Medicine 61*: 2155–2164. doi:10.1016/j.socscimed.2005.04.030

Dixon, D.P. (2015). *Feminist Geopolitics: Material States*. Burlington, MA: Ashgate. doi:10.4324/9781315582221

Dyck, I. (1995). Transforming the relations of research: Towards culturally safe geographies of health and healing. *Health and Place 1*(3): 137–147. doi:10.1016/1353-8292(95)00020-M

Emch, M., Root, E.D., & Carrel, M. (2017). *Health and Medical Geography*. New York: Guilford Press.

Evans, J., Semogas, D., Smalley, J.G., & Lohfeld, L. (2015). This place has given me a reason to care: Understanding "managed alcohol programs" as enabling places in Canada. *Health & Place 33*: 118–124.

Fall, J. (2020). Fenced in. *Environment and Planning C: Politics and Spaces 38*(5): 771–794. doi:10.1177/2399654420933900

Foucault, M. (1976). *The History of Sexuality, Vol. 1: An Introduction*, trans. Robert Hurley. New York: Vintage.

Garth, H., & Reese, A. (2020). *Black Food Matters: Racial Justice in the Wake of Food Justice*. Minneapolis, MN: University of Minnesota Press.

Gatrell, A.C., & Elliott, S.J. (2014). *Geographies of Health: An Introduction*. Hoboken, NJ: John Wiley & Sons.

Gengenbach, H., Schurman, R.A., Bassett, T.J., Munro, W.A., & Moseley, W.G. (2018). Limits of the new green revolution for Africa: Reconceptualising gendered agricultural value chains. *Geographical Journal 184*(2): 208–214. doi:10.1111/geoj.12233

Gennuso, K.P., Blomme, C.K., Givens, M.L., Pollock, E.A., & Roubal, A.M. (2019). Deaths of despair(ity) in early 21st century America: The rise of mortality and racial/ethnic disparities. *American Journal of Preventive Medicine 57*(5): 585–591. doi:10.1016/j.amepre.2019.06.018

Gesler, W.M. (1992). Therapeutic landscapes: Medical issues in light of the new cultural geography. *Social Science & Medicine 34*(7): 735–746.

Gesler, W.M., & Kearns, R.A. (2002). *Culture/Place/Health*. London: Routledge.

Gilligan, C. (1982). *In a Different Voice: Psychological Theory and Women's Development*. Cambridge, MA: Harvard University Press.

Goodman, D., DuPuis, M., & Goodman, M. (2012). *Alternative Food Networks, Knowledge, Practice, and Politics*. London: Routledge.

Grey, S., & Patel, R. (2015). Food sovereignty as decolonization: Some contributions from Indigenous movements to food system and development politics. *Agriculture and Human Values 32*: 431–444. doi:10.1007/s10460-014-9548-9

Guthman, J. (2011). *Weighing In: Obesity, Food Justice and the Limits of Capitalism*. Berkeley, CA: University of California Press.

Guthman, J. (2012). Opening up the black box of the body in geographical obesity research: Toward a critical political ecology of fat. *Annals of the Association of American Geographers 102*(5): 951–957. doi:10.1080/00045608.2012.659635

Häkli, J. (2021). COVID-19 certificates as a new form of mobility control. *European Journal of Risk Regulation* 12(2): 362–369. doi:10.1017/err.2021.21

Hankivsky, O. (2014). Rethinking care ethics: On the promise and potential of an intersectional analysis. *American Political Science Review 108*(2): 252–264. doi:10.1017/S0003055414000094

Ingram, A. (2009). The geopolitics of disease. *Geography Compass 3*(6): 2084–2097. doi:10.1111/j.1749-8198.2009.00284.x

Kampman, H. forthcoming. The culinary cosmopolitanism of a Ghanaian local food movement.

Kearns, R. (1993). Place and health: Towards a reformed medical geography. *Professional Geographer 45*(2): 139–147.

Kearns, R., & Moon, G. (2002). From medical to health geography: Novelty, place and theory after a decade of change. *Progress in Human Geography 26*(5): 605–625. doi:10.1191/0309132502ph389oa

Kerr, R.B. (2012). Lessons from the old Green Revolution for the new: Social, environmental and nutritional issues for agricultural change in Africa. *Progress in Development Studies 12*(2–3): 213–229. doi:10.1177/146499341101200308

Kimura, A.H. (2013). *Hidden Hunger: Gender and the Politics of Smarter Foods*. Ithaca, NY: Cornell University Press.

King, B. (2017). *States of Disease: Political Environments and Human Health*. Berkeley, CA: University of California Press.

King, B., Burka, M., & Winchester, M.S. (2018). HIV citizenship in uneven landscapes. *Annals of the American Association of Geographers 108*(6): 1685–1699.

King, B., & Rishworth, A. (2022). HIV responsibilisation: Stigma, disclosure, and care in the age of 90-90-90. *Transactions of the Institute of British Geographers 47*(2): 499–513. doi:10.1111/tran.12527

Lawson, V. (2007). Geographies of care and responsibility. *Annals of the Association of American Geographers 97*(1): 1–11. doi:10.1111/j.1467-8306.2007.00520.x

Lemke, T. (2007). Susceptible individuals and risky rights: Dimensions of genetic responsibility. In Burri, R.V., & Dumit, J. (eds), *Biomedicine as Culture*. London: Routledge, 159–174.

Leonard, K. (2020). Medicine lines and COVID-19: Indigenous geographies of imagined bordering. *Dialogues in Human Geography 10*(2): 164–168. doi:10.1177/2043820620934941

Lim, K.F. (2022). Authoritarianism recalibrated: The COVID-19 crisis and the reproduction of one-party dominance in Singapore. In Koch, N. (ed.), *Spatializing Authoritarianism*. Syracuse, NY: Syracuse University Press, 338–355.

Liu, X., & Bennett, B. (2020). Viral borders: COVID-19's effects on securitization, surveillance, and identity in Mainland China and Hong Kong. *Dialogues in Human Geography 10*(2): 158–163. doi:10.1177/2043820620933828

May, J. (1958). *The Ecology of Human Disease*. New York: M.D. Publications.

Mayer, J.D. (1996). The political ecology of disease as one new focus for medical geography. *Progress in Human Geography 20*: 441–456. doi:10.1177/030913259602000401

Mayer, V.L., Hillier, A., Bachhuber, M.A., et al. (2014). Food insecurity, neighborhood food access, and food assistance in Philadelphia. *Journal of Urban Health 91*: 1087–1097. doi:10.1007/s11524-014-9887-2

Meade, M.S., & Earickson, R.J. (2000). *Medical Geography*, 2nd edn. New York: Guilford Press.

Mercille, J.T., Albrecht, J., Cole, K., & Dodds, A.L. (2021). Interventions in critical health geopolitics: Borders, rights, and conspiracies in the COVID-19 pandemic. *Political Geography 91*: 201445. doi:10.1016/j.polgeo.2021.102445

Miller, B.S. (2019). Food and nationalism in an independent Ghana. In Miller, B.S. (ed.), *The Emergence of National Food*. London: Bloomsbury, 61–70. doi:10.5040/9781350074163.0010

Milligan, C. (2000). Bearing the burden: Towards a restructured geography of caring. *Area 32*(1): 49–58.

Montefrio, M.J.F. (2020). Cosmopolitan translations of food and the case of alternative eating in Manila, the Philippines. *Agriculture and Human Values 37*(2): 479–494. doi:10.1007/s10460-019-10000-z

Murphy, A.B. (2022). Taking territory seriously in a fluid, topologically varied world: Reflections in the wake of the populist turn and the COVID-19 pandemic. *Geografiska Annaler: Series B, Human Geography 104*(1): 27–42. doi:10.1080/04353684.2021.2022987

Narayan, U. (1995). Colonialism and its others: Considerations on rights and care discourses. *Hypatia 10*(2): 133–140.

Nguyen, V.K. (2010). *The Republic of Therapy: Triage and Sovereignty in West Africa's Time of AIDS*. Durham, NC: Duke University Press.

Parr, H. (2003). Medical geography: Care and caring. *Progress in Human Geography 27*: 212–221. doi:10.1191/0309132503ph423pr

Philo, C. (2016). "Healthy debate" and "healthy ferment": Medical and health geographies. *Progress in Human Geography*: 1–21. doi:10.1177/0309132516678343

Prior, L., Manley, D., & Sabel, C.E. (2019). Biosocial health geography: New "exposomic" geographies of health and place. *Progress in Human Geography 43*(3): 531–552. doi:10.1177/0309132518772644

Reese, A. (2019). *Black Food Geographies*. Chapel Hill, NC: University of North Carolina Press.

Rishworth, A., & King, B. (2021). "There is a secret in love": Gender, care and HIV management in South Africa. *Gender, Place and Culture 28*: 1561–1583. doi:10.1080/0966369X.2020.1819209

Rudd, R.A., Seth, P., David, F., & Scholl, L. (2016). Increases in drug and opioid-involved overdose deaths – United States, 2010–2015. *Morbidity and Mortality Weekly Report 65*(50–51): 1445–1452. doi:10.15585/mmwr.mm655051e1

Ruhm, C.J. (2017). Geographic variation in opioid and heroin involved drug poisoning mortality rates. *American Journal of Preventative Medicine 53*: 745–753. doi:10.1016/j.amepre.2017.06.009

Sarma, J., Faxon, H.O., & Roberts, K.B. (2023). Remaking and living with resource frontiers: Insights from Myanmar and beyond. *Geopolitics 28*(1): 1–22. doi:10.1080/14650045.2022.2041220

Sbicca, J. (2018). *Food Justice Now! Deepening the Roots of Social Struggle*. Minneapolis, MN: University of Minnesota Press.

Staeheli, L.A., & Brown, M. (2003). Where has welfare gone? Introductory remarks on the geographies of care and welfare. *Environment and Planning A: Economy and Space 35*(5): 771–777. doi:10.1068/a35132

Strings, S. (2019). *Fearing the Black Body: The Racial Origins of Fat Phobia*. New York: NYU Press.

Sturm, T., & Albrecht, T. (2021). Constituent COVID-19 apocalypses: Contagious conspiracism, 5G, and viral vaccinations. *Anthropology & Medicine 28*(1): 122–139.

Tallbear, K. (2022). Making love and relations beyond settler sex and family. In Rayter, S., & Zisman Halpern, L. (eds), *Queerly Canadian: An Introductory Reader in Sexuality Studies*. Toronto: Women's Press, 145–166.

Trauger, A. (2014). Toward a political geography of food sovereignty: Transforming territory, exchange and power in the liberal sovereign state. *Journal of Peasant Studies 41*(6): 1131–1152. doi:10.1080/03066150.2014.937339

Tronto, J. (1993). *Moral Boundaries: A Political Argument for an Ethic of Care*. New York: Routledge.

Tronto, J. (2015). *Who Cares? How to Reshape a Democratic Politics*. Ithaca, NY: Cornell University Press.

Walker, R.E., Keane, C.R., & Burke, J.G. (2010). Disparities and access to healthy food in the United States: A review of food deserts literature. *Health & Place 16*(5): 876–884. doi:10.1016/j.healthplace.2010.04.013

Williams, A. (ed.) (1999). *Therapeutic Landscapes*. New York: Ashgate.

Zheng, H., & Choi, Y. (2024). Reevaluating the "deaths of despair" narrative: Racial/ethnic heterogeneity in the trend of psychological distress-related death. *Proceedings of the National Academy of Sciences 121*(8): e2307656121. doi:10.1073/pnas.230765612

Chapter 35

Ocean Geographies

Po-Yi Hung and Yu-Chia Lin

Department of Geography, National Taiwan University, Taipei, Taiwan

In 1986, geographer Martin Ira Glassner called for a "new political geography of the sea" to examine its "spatial or territorial component, including such matters as the territorial sea and the contiguous zone, the exclusive economic zone, international straits, the regimes of islands and archipelagos, freedom of transit for land-locked states, deep seabed mining and protection and preservation of the marine environment" (Glassner 1986: 6). For him, the aim of "new" political geography at the time was to investigate the politics of the spatial or territorial component under the Law of the Sea and its framework, in particular the Third United Nations Conference on the Law of the Sea (UNCLOS III, 1973–1982). In the four decades since Glassner's publication, studies regarding the spatial or territorial component under the framework of the Law of the Sea have built up the scholarship of political geography of the ocean.

Political geographers and related scholars continue to study how UNCLOS served as the foundation for maritime zoning and bordering procedures (Foley 2022). Critically, UNCLOS offers a general legal framework to configure the operation of maritime boundaries, even though it does not explicitly address territorial disputes or border conflicts over the ocean. For instance, in 2023 UNCLOS came to a consensus to safeguard high seas marine life through the establishment of Marine Protected Areas and environmental impact assessments. In centering such issues, political geographies of the ocean continue to trace the border forces that still exist over the ocean, which may not be "legible" from a land-based perspective on a global atlas.

The Wiley Blackwell Companion to Political Geography, Second Edition.
Edited by Virginie Mamadouh, Natalie Koch, Chih Yuan Woon, and John Agnew.

Yet the "new" in Glassner's 1986 article should also be renewed. For example, in addition to the visible borders on land and ocean depicted in the world atlas, countries are competing to establish their authority over less visible arenas, like maritime areas, undersea continental plates, melting ice in the Arctic, and coral reefs. In addition to persistent questions about state authority and territoriality, scholars are now giving more attention to the physical characteristics of the ocean when examining the political geography of the ocean (e.g., Billé 2020; Hung & Lien 2022). This involves emphasizing the spatial nature of the ocean, while avoiding the conventional approach of viewing it as a two-dimensional map on land.

Oceans encompass an alternative framework of space and time (Duara 2021) and water offers a rich setting for reconsidering how our political geographies are shaped by and exert influence on a dynamic and abundant material existence. Steinberg and Peters (2015) accordingly suggest a "wet ontology" approach to challenge the idea of a fixed and controlled terrestrial ontology based on confined areas and linear structures. The wet ontology approach has emphasized the interaction between the physical nature, time, and movement of the ocean, as well as the importance of ocean water's diverse states as gaseous vapor, liquid water, and solid ice.

Fluid oceanic ontologies have compelled scholars to rethink notions related to political geography, such as maritime borders and state territoriality (Hung & Lien 2022). This perspective prompts us to reassess the ocean as a dynamic and ever-changing realm characterized by movements, drifts, and the constant reshaping of ocean borders. The aim of this chapter is thus to highlight the emerging political geographies of the ocean from a wet ontology perspective. In so doing, we integrate Glassman's classic political geography "spatial or territorial components" with a cross-disciplinary dialogue about the mobility, materiality, and changing form of state governance over the ocean. New exchanges between political geography and other subfields, including island geography, migration geography, infrastructural geography, and science studies, provide diverse approaches to the spatial politics of the ocean.

To introduce the emerging geographies of the ocean, this chapter focuses on four themes: disputed islands and the oceanic gray zone, asylum seekers and migration, fisheries and infrastructure, and high seas governance and science studies. Since we cannot comprehensively review the many political geographical issues pertaining to oceans, we use these four themes as the starting point to complicate the relationship between political geography and critical ocean studies. And though they are separated here for clarity, these themes are all interconnected. The ocean is, after all, a vast waterbody connecting the whole planet. As a result, the political geographies of disputed islands, asylum seekers, fisheries, and high seas governance spotlight the many interconnections that it facilitates.

Disputed Islands and the Oceanic Gray Zone

The Chagos Archipelago in the Indian Ocean, the Falkland Islands/Islas Malvinas in the Atlantic Ocean, the South China Sea Islands, Dokdo/Takeshima Island, and Senkaku/Diaoyu Island in the western Pacific are among the many disputed islands around the world. These islands have been a subject of geopolitical disputes regarding state territorial claims over the ocean, which escalated after World War II. For example, China has bolstered its territorial assertions in the South China Sea by enhancing its military facilities on the islands and even the coral reefs (Hung & Lien 2022). As such, China's activities on the South China Sea Islands have heightened tensions about the territorial conflicts involving China, Taiwan, Vietnam, the Philippines, and Malaysia.

Effective border practices are essential in addressing the challenges posed by marine gray zones. The Chinese government has deployed its domestic fishing fleet in the South China Sea as a means to assert its dominance and establish its claim to sovereignty over the disputed "nine-dash line" border (Roszko 2016). This move is seen as a plan to legitimize China's exercise of control in the region (Zhang 2016; Kennedy 2018; Patalano 2018). The Chinese government not only considers its fishing fleet as part of its border infrastructure, but also actively encourages "patriotic tourism" in the South China Sea – aimed at promoting feelings of nationalism among citizens toward maritime territories, through a purposeful manipulation of emotions and spatial dynamics (Rowen 2018; Huang & Suliman 2020; Mostafanezhad 2020). In addition to creating emotional connections to establish control over maritime areas, state actors can also build new facilities that might strengthen their claims. For example, in responding to perceived threats from China, the Taiwanese government has redefined the Taiwan Strait as a region for the development of green energy, officially designating the neighboring offshore region for wind farm development. By extending offshore windfarm infrastructures into the sea, the government has reterritorialized the oceanic space of the Taiwan Strait (Hung 2020).

A "maritime gray zone" illustrates the intricate connection between sovereignty and territory. According to Billé (2020), maritime borders and the extension of a state's sovereign claims do not necessarily reflect an expansion of sovereign territory. Therefore, conversations about the power of states in the water take place in a realm of political intricacy where territory and sovereignty do not always align, and challenge the current international legal framework of oceanic governance (Rolf & Agnew 2016; Braverman & Johnson 2020). This development has compelled border studies to reconsider matters concerning maritime borders, without automatically assuming that sovereignty implies a defined state area. Studies of maritime gray zones must thus be sensitive to the manner in which these behaviors at ocean borders actively disturb, alter, and restructure relationships among states, sovereignty, territories, and borders (Koch 2022).

Maritime gray zones can be a purposeful tactic employed by a state to occupy contested ocean areas without provoking an armed response. In doing so, government leaders strategically aim to stay below the threshold that would prompt a military reaction (Harold et al. 2017). In East Asia, the South China Sea, contested among China and countries of Southeast Asia (Rolf & Agnew 2016; Rice et al. 2016; Yorgason 2017; Gaynor 2019; Diener & Hagen 2022; Rodenbiker 2023), and the Diaoyu/Senkaku Islands, contested among Japan, Taiwan, and China (Wirth 2016; Harold et al. 2017) are two vivid examples. For the Diaoyu/Senkaku Islands in particular, the 2012 fisheries agreement between Taiwan and Japan establishes a defined region of 74,000 km^2 surrounding the contested islands. Taiwan and Japan possess the authority to fish, police, and enforce laws in the maritime gray zone (Yeh et al. 2015). However, they do not grant these powers to other countries, including China. The fisheries agreement between Taiwan and Japan establishes a bounded gray zone where fishing activities occur. This zone produces a situation where the territorial rights for fishing overlap, and both governments have flexible sovereign control over it.

Another example of a maritime gray zone is the Arctic. The frozen ice of the Arctic makes the Arctic Ocean a "quasi-island," but it raises different territorial issues to the Diaoyu/Senkaku Islands case. Dodds and Benwell (2010) employ the notion of "resource-led nationalism" to examine how various Arctic nations assert their sovereignty by means of flag-planting and map-making (see also Steinberg 2010; Bridge 2014; Drewniak et al. 2018). In addition to asserting their sovereignty, governments in the Arctic region are facing environmental issues, such as global warming, that affect the area. With rapidly melting sea ice, the Arctic has transformed into an area of state governance that is marked by uncertainty (Dodds 2010, 2012; Dittmer et al. 2011). As a marine area with ambiguous boundaries, the changing region offers scholars a chance to explore the connection between state territorial claims and ocean borders, free from the limitations of rigid land-based perspectives (Steinberg & Kristoffersen 2017; Bennett et al. 2020).

Overall, the overlapping territorial and resource claims in maritime gray zones highlight the flexibility of state governance in defining borders in oceanic areas. Political geography scholarship on these issues offers a valuable opportunity to question the fixed and frequently abstract ideas about borders that come from the traditional understanding of state territories (Hung & Lien 2022) and ocean governance (Suarez de Vivero & Rodriguez-Mateos 2023).

Asylum Seekers and Migration

Political geographers have shown that the distribution of mobilities and immobilities among people is not equal (Sparke 2006; Cresswell 2011). For political geographies of the ocean, these issues are related to a broader debate about migration of people and state territoriality over the ocean. Refugees, asylum seekers in particular, who are traveling by sea, as well as those who are escaping severe human rights abuses and environmental catastrophes by ocean routes, are trapped within an inadequate legal system. Their security is contingent not just on the benevolence of the ocean but also on the dynamic interplay between funding, state security, and border interests. These linkages infiltrate the holes in the legal system, introducing policies and programs that change according to political interests (MacIntosh 2019). Attempting to address a refugee crisis through the management of maritime transit routes is inevitably doomed to be unsuccessful, as evidenced by the predictable occurrence of large-scale drownings (Albahari 2018; MacIntosh 2019). Paradoxically, it is these large-scale drownings that initiate the most significant moments of humanitarian action, which appear to persist until a substantial number of individuals are saved at sea, at which point other political agendas become prominent.

As a result of the increased frequency of pushbacks and sinking of refugee boats, migration by sea has been the topic of attention, despite the fact that it is not a new phenomenon. Refugee boats have become the public face of the refugee crisis worldwide because they highlight the tension between sovereign rights and refugee rights, as well as the shift in focus from terrestrial boundaries to oceans as alternative borders (Joseph & Balakrishnan 2023). The political geography of the ocean and migration studies have become closely associated, united around key issues of immigration restrictions, refugees, and humanitarian borders (Richardson 2013; Loyd & Mountz 2014; Dickson 2015; Jones 2016, 2019; Kahn 2019). These studies have indicated that the phenomenon of migration, particularly the flow of refugees and asylum seekers, has presented states with significant difficulties in managing their border policy and maintaining territorial control.

Migration scholars have examined these shifting flows through the concept of "border regimes," which situates borders as a means to govern mobile populations and distinguish them as migrants. State border regimes can include the management of refugees traveling across the ocean – and often this includes tactics designed to "push out" border enforcement to meet migrant refugees on the ocean (Stierl 2016, 2018; Tazzioli 2016). In this way, political arguments over the rescue of migrants traveling by water become entangled with the politics of border displacement. Tazzioli (2016: 12–13) thus reconceptualizes borders as "apparatuses of capture" by examining "the spatial and political transformations in the government of migration at sea and beyond the sea." Such bordering practices not only produce migrants and citizens, but also "*act on* subjects' freedom in order to govern it" (2016: 12–13, emphasis in original). From this perspective, maritime borders are no longer determined by linear territorial governance. Instead, they are shaped by the blending of heterogenous forms of governance and the ever-changing nature of humanitarian borders.

The arrival of asylum seekers by sea has led academics to reconsider the complex and varied ways in which borders are formed by focusing on the concept of mobility (Cresswell 2010; Sheller 2018; Minca et al. 2022). In the new global border system for the purpose of controlling migration from the ocean, the state continues to play a significant role. As such, political geographers have closely observed the new ways in which global borders are being managed for asylum seekers from the sea, both within and beyond the territory of the receiving country (Johnson et al. 2011; Burridge et al. 2017; Vives 2017). Crucially, state mobility constraints imposed on asylum seekers from the ocean do not completely prohibit their migration. Rather, refugee governance oscillates between movement and cessation, between migrants in sending and receiving groups, and between mobility and immobility in the ocean.

Political geographies of the ocean have thus pushed scholars to reassess the notion of "mobile borders" to account for the movement of migrants (Tazzioli 2018) as well as the spatial organization of how migrants are intercepted, arrested, and forcibly moved through the marine domain (Dickson 2021). Building from these insights, the related concept of "mobile sovereignty" has also helped to explain the expansion of zones of state sovereignty into the ocean through the movements of sovereign agents and populations under external control (Rolf & Agnew 2016; Everuss 2020). In the broader concerns of political geography of the ocean, human mobility will continue to shed light on the techniques that states use to de- and reterritorialize the sea, which are then incorporated into governance policies regarding migration control over the ocean.

Fisheries and Infrastructure

Recent scholarship on the political geographies of the ocean has also given special attention to the physical characteristics of the ocean, approaching it as a dynamic, voluminous, liquid, drifting, and nonlinear temporal space that resists human ordering, bordering, and territorializing processes. Emphasizing the ocean's materialities can challenge traditional terrestrial-based approaches to political geography. For some scholars, the ocean's material constraints are transformed into a profitable frontier space (Tsing 2011; Neilson 2013; Hepworth 2014; Havice & Zalik 2019). Recent technological and infrastructural advances have made it possible to explore and exploit the ocean space in ways that were not possible before (Squire 2022). For example, new technologies have made it possible to drill for oil, gas, and minerals in deep water (Höhler 2002; Zalik 2015, 2018; Phillips 2018; Childs 2019), to generate electricity from wind and waves (Hung 2020), and even to project power and governance onto the sea (Hung & Lien 2022).

These activities rely on, cooperate with, and are interrelated within the operation of infrastructure built on/in/over the ocean. Infrastructure is a collective term for those material components supporting higher-order goals that reshape the human condition (Carse 2017; Barua 2021). It transcends material form and serves as a complex mediator that both facilitates and problematizes state power (Meehan 2014). So too does infrastructure shape the entanglement between human and nonhuman actors (environment, resources, species, space)

(Furlong 2011; Carse 2012; Nelson & Bigger 2022). Political geographies of the ocean are part of this entanglement, leading scholars to examine how people overcome the obstacles presented by the ocean and its multiple materialities, and how marine infrastructure mediates between multiple materialities and governance to obtain resources from it (Bustos-Gallardo 2021; Campbell et al. 2021; Probyn 2021; Zalik 2021).

Fisheries are exemplary of the relevance of infrastructure for the emerging political geography of the ocean. Fishing infrastructures operate at the interface between human and environment, and they manage resource flows and exert influence on environmental practices (Monstadt 2009). In the maritime context, both the production and governance of marine resources rely on infrastructure to mediate the relationship between humans and the materialities of the ocean. Fishing vessels, fishing and freezing technology, port facilities, and fishers' bodies in the water all serve as crucial infrastructure that facilitates human access to resources.

In addition, both the movement of fishing vessels and the catches they harvest are major concerns for state actors, whose efforts to govern ocean spaces are challenged by their lack of information from these seacraft. As such, making invisible actions in distant waters visible renders the ocean more governable (Toonen & Mol 2016; Song, Johnsen, & Morrison 2018). Industrial fisheries increasingly rely on the assembly of digital technologies to make the vast ocean visible. New technologies aim to make mobile infrastructure on the sea, such as moving fishing vessels, transparent, accountable, and subject to governance (Boucquey et al. 2019; Drakopulos 2022; Drakopulos, Havice, & Campbell 2022). For example, Taiwan, as one of the major fishing states, has taken steps to install vessel monitoring systems (VMS) on fishing vessels, which allow for hourly tracking of their positions. Additionally, the government implemented electronic logbook (e-logbook) systems, trained fishers to use them properly, and established penalties for falsifying reports. As such, by projecting fishing vessels on a screen located in the state building away from the ocean, the VMS and e-logbook put state governance into practice and generate real-time management of the ocean (Maxwell et al. 2015).

Ports, as the critical infrastructure for ocean governance, are located between land and sea, and constitute a chokepoint in the circulation of people, materials, resources, and commodities between terrestrial and ocean spaces (Jensen 2023). The impact of ports is obvious in that logistics infrastructure strengthens the ability of inland areas to obtain resources and cash out commodities, including fish (Monios, Bergqvist, & Woxenius 2018; Iftikhar & Zhan 2020). As a result, state governments often regard seaports as crucial infrastructure to connect the international market and conduct policy to guide their construction and investment (Hall 2009; Jenss & Schuetze 2021). For example, Akhavan (2017) emphasizes the role of container ports, which served as the connection to the world economy, in the process of building Dubai from a village to a global city.

As globalization has evolved in the past decades, the shipping and transportation industry has thrived and ports have taken on ever more significance for commercial activities, city competition, and military strategy. Ports today are not only commercial facilities but the infrastructure for projecting geopolitics. The Belt and Road Initiative (BRI), for example, is a twenty-first-century Chinese strategic policy that combined the Silk Road Economic Belt and the Maritime Silk Road and is aimed at enhancing connectivity around Afro-Eurasia (Blanchard & Flint 2017; Sidaway & Woon 2017; Alff 2020; Jensen 2023). Iftikhar and Zhan (2020) suggest that Chinese overseas port investments in Greece and Pakistan offer a great position for the Chinese government to expand its trading network in energy, materials, and goods while simultaneously projecting military force. However, while increasing numbers of studies label these scenarios as part of the so-called rise of China, political geographers also proceed with the idea of a rising China with caution. Sidaway and Woon (2017) propose a viewpoint beyond the Eurocentric or US-centric perspective, especially when the representation of China could be a Western strategy for developing its own geopolitical deployment. As such, the building up of maritime infrastructures under the BRI, including ports, has led to "important geopolitical questions that pose thorny analytical challenges" (Sidaway & Woon 2017: 600).

Since infrastructure mediates the tension between the seeming invisibility of the ocean and the governance of oceanic resources, including fishing activities, "it flags the modernist desire to render social and environmental heterogeneity manageable and amenable to standardized solutions" (Carse 2017: 28). That is, modern states exercise spatial governance through the processes of measuring, visualizing, and demarcating property rights (Scott 2016; Steinberg & Kristoffersen 2017; Campbell et al. 2021). Thinking about the broader issue concerning resource governance over the ocean, infrastructure plays a crucial role in overcoming the challenges of the ocean and mediating the relationship between humans and marine resources. Since infrastructure facilitates resource extraction and shapes state governance efforts, a larger question for future scholarship on the political geography of the ocean will be how state power and the function of infrastructure have both stabilized and destabilized each other in terms of making and remaking the ocean a resource frontier to be governed (see Faxon and Kintzi, this volume).

High Seas Governance and Science Studies

Questions of ocean governance and its resources inherently involve questions of expertise. For example, the worldwide depletion of marine resources, such as fish, has been acknowledged as a global issue and as leading to disputes between states (Havice 2018; Fabinyi 2020). Indigenous communities around the world, on the other hand, are well aware of these challenges and have either maintained or struggled over their own forms of science and expertise to manage marine resources (Sakakibara 2020; Blair 2023). However, the contemporary capitalist state has sought to regulate marine resources to ensure political and commercial profits (Dupont & Baker 2014; Kim 2019). As a result, the management of fishing resources in the ocean has been advanced through modern fisheries science, involving governments, environmental nongovernmental organizations, and fishing enterprises (Dupont et al. 2020). Pursuing scientific governance has served as the means by which interstate government over the ocean is justified (Polacheck 2012). United around implementing science-based governance of marine resources, governmental and nongovernmental entities work to expand their authority over the ocean in order to ensure fairness, justice, and sustainability based on scientific correctness.

Yet claims to scientific precision, when used as a means to negotiate control over marine resources, become quickly entangled in political debates. Although it is rare for modern data collecting to achieve complete scientific precision, the pursuit of increasingly accurate numerical values or scientific outcomes is an ongoing and endless endeavor with political negotiations among states. This is not to deny the scientific data or conclusions generated by fishery scientists. Rather, it says that the supposedly "nonpolitical" discourses of marine science are in fact highly political, especially when they are used in governing maritime areas between states (Boucquey 2020). One way to transition from an apolitical to a political perspective in science-driven ocean governance is to focus on the territorial statecraft enabled by scientific data creation on marine resources, such as fish populations.

The high seas have historically been subject to political dynamics of ocean appropriation rather than being seen as a shared resource (Lambach 2021; Teo 2023). Yet fisheries science and fishing operations have served as important sites for the state to exercise its authority by way of interstate governance, specifically through the governance of Regional Fisheries Management Organizations (RFMOs). RFMOs are interstate organizations that enforce regulatory measures to establish territorial orders for the resource frontier of the high seas. In order to establish regulatory measures, member states must engage in negotiations with one another under the framework of RFMOs. Rather than exerting their own territorial power, each member state has chosen to collaborate with nonstate actors, especially fisheries scientists, within RFMOs to establish shared territorial authority over the high seas. As such, RFMOs have been crucial in promoting interstate collaboration for the conservation and management of fisheries resources, particularly for migratory fish species (Havice 2021; Lambach 2021; Rosello 2022; Haas et al. 2023).

Crucially, for states to achieve the territorialization of the high seas by RFMOs, they must depend on nonstate fisheries scientists to generate scientific data. The scientific evidence the latter produce is crucial for RFMOs and member states to establish the legitimacy of their scientifically regulated fishing areas (Ortuño Crespo et al. 2020). While scientific data on fish stocks has remained unknown, it has allowed the state to collaborate with fisheries scientists to protect the state's best interests (Hubbard 2014). As geographers and scholars in science and technology studies have long shown, science is consistently imbricated with state power and it can easily become an "anti-politics machine" (Ferguson 1990; Jasanoff 1990, 2004; Akhter & Ormerod 2015; Barnes 2022). Fisheries science has done precisely this, as it generates data for studying fish populations but also contributes to the establishment of RFMO regions, where states actively participate in creating governance for the open ocean. The case of RFMOs suggests that future scholarship on the political geographies of the ocean can continue to interrogate the dynamic relationship between science, state power, and ocean governance.

Conclusion

In reconsidering Glassner's (1986) call for a "new political geography of the sea," this chapter has pointed to persistent themes in ocean geographies alongside new ways of thinking through fluid oceanic ontologies. States continue to maintain strategic concerns regarding the maritime areas that surround their territories, and international frameworks like UNCLOS continue to delineate the ocean into distinct areas and legal principles (Braverman & Johnson 2020; Hung & Lien 2022). The resulting global normative framework serves as an expansion of the state's sovereign rights over the ocean. Additionally, it establishes a contested governance regime for humanitarian borders for refugees from the sea, the exploitation of marine resources, and the conservation of the environment. Although recent scholarship has emphasized the distinct physical characteristics of seascape and landscape, state power remains central to the emerging political geographies of the ocean. This is because state sovereignty continues to serve as the basis for determining oceanic territoriality and exercising authority over the ocean.

Although the political technology of state territoriality (Elden 2010) and the development of oceanic space (Lambach 2022) are ongoing processes, the methods of exerting state authority have changed. The concept of mobile state governance has played a crucial role in examining the evolving connection between a state's sovereign rights and the maritime domain (Hung & Lien 2022). Specifically, when it comes to governing the ocean, the state's control differs from its control over land-based territory. The state responds to various aspects of the ocean in a subtle manner and also exerts its power in a more mobile manner. A view from the ocean shows how states are forced to modify regulations and institutions regarding marine spaces, human mobility, natural resources, and migrating fish populations.

These forms of mobile state governance explored in this chapter do not imply an unrestricted extension of state power. Instead, for future studies on the political geography of the ocean, it is critical to investigate how a law of the sea might establish the legitimacy of ocean borders, which would serve as the definitive boundaries for a state's sovereign rights to regulate oceanic phenomena. But just as the movement of fish and asylum seekers challenges the way the government operates and its authority, the ongoing scenarios also demonstrate the need for future research on the nonhuman elements of the ocean, especially in the face of increasingly severe effects of climate change.

Acknowledgments

Completion of this chapter was supported by the National Science and Technology Council in Taiwan (111-2410-H-002-099-MY3; 110-2410-H-002-186) and the Taiwan Social Resilience Research Center (Grant no. 112L900303) from the Higher Education SPROUT Project by the Ministry of Education in Taiwan.

References

Akhavan, M. (2017). Development dynamics of port-cities interface in the Arab Middle Eastern world – the case of Dubai global hub port-city. *Cities 60*: 343–352.

Akhter, M., & Ormerod, K.J. (2015). The irrigation technozone: State power, expertise, and agrarian development in the U.S. West and British Punjab, 1880–1920. *Geoforum 60*: 123–132.

Albahari, M. (2018). From right to permission: Asylum, Mediterranean migrations, and Europe's war on smuggling. *Journal on Migration and Human Security* 6(2): 121–130.

Alff, H. (2020). Belts and roads every- and nowhere: Conceptualizing infrastructural corridorization in the Indian Ocean. *Environment and Planning C: Politics and Spaces 38*(5): 815–819.

Barnes, T.J. (2022). The discipline that came in from the cold: American human geography becomes a Cold War social science. *Environment and Planning F 1*(2–4): 145–167.

Barua, M. (2021). Infrastructure and non-human life: A wider ontology. *Progress in Human Geography 45*(6): 1467–1489.

Bennett, M.M., Stephenson, S.R., Yang, K., Bravo, M.T., & De Jonghe, B. (2020). The opening of the transpolar sea route: Logistical, geopolitical, environmental, and socioeconomic impacts. *Marine Policy 121*: 104178.

Billé, F. (2020). *Voluminous States: Sovereignty, Materiality, and the Territorial Imagination*. Durham, NC: Duke University Press.

Blair, J. (2023). *Salvaging Empire: Sovereignty, Natural Resources, and Environmental Science in the South Atlantic*. Ithaca, NY: Cornell University Press.

Blanchard, J.M.F., & Flint, C. (2017). The geopolitics of China's maritime silk road initiative. *Geopolitics 22*(2): 223–245.

Boucquey, N. (2020). The nature of fisheries governance: Narratives of environment, politics, and power and their implications for changing seascapes. *Journal of Political Ecology 27*(1): 169189.

Boucquey, N., Martin, K.S., Fairbanks, L., Campbell, L.M., & Wise, S. (2019). Ocean data portals: Performing a new infrastructure for ocean governance. *Environment and Planning D: Society and Space 37*(3): 484–503.

Braverman, I., & Johnson, E. (2020). *Blue Legalities: The Life and Laws of the Sea*. Durham, NC: Duke University Press.

Bridge, G. (2014). Resource geographies II: The resource-state nexus. *Progress in Human Geography 38*(1): 118–130.

Burridge, A., Gill, N., Kocher, A., & Martin, L. (2017). Polymorphic borders, territory, politics. *Governance 5*(3): 239–251.

Bustos-Gallardo, B. (2021). Anadromous frontiers: Reframing citizenship in extractive regions. The salmon industry in Los Lagos, Chile. In Himley, M., Havice, E., & Valdivia, G. (eds), *The Routledge Handbook of Critical Resource Geography*. New York: Routledge, 369–380.

Campbell, L.M., Gray, N.J., Zigler, S.B.J., Acton, L., & Gruby, R. (2021). World-making through mapping: Large-scale marine protected areas and the transformation of global oceans. In Himley, M., Havice, E., & Valdivia, G. (eds), *The Routledge Handbook of Critical Resource Geography*. New York: Routledge, 425–440.

Carse, A. (2012). Nature as infrastructure: Making and managing the Panama Canal watershed. *Social Studies of Science 42*(4): 539–563.

Carse, A. (2017). Keyword: Infrastructure – how a humble French engineering term shaped the modern world. In Harvey, P., Jensen, C.B., & Morita, A. (eds), *Infrastructures and Social Complexity: A Routledge Companion*. London: Routledge, 27–39.

Childs, J. (2019). Greening the blue? Corporate strategies for legitimising deep sea mining. *Political Geography 74*: 102060.

Cresswell, T. (2010). Towards a politics of mobility. *Environment and Planning D: Society and Space 28*(1): 17–31.

Cresswell, T. (2011). Mobilities I: Catching up. *Progress in Human Geography 35*(4): 550–558.

Dickson, A. (2015). Distancing asylum seekers from the state: Australia's evolving political geography of immigration and border control. *Australian Geographer 46*(4): 437–454.

Dickson, A.J. (2021). The carceral wet: Hollowing out rights for migrants in maritime geographies. *Political Geography 90*: 102475.

Diener, A.C., & Hagen, J. (2022). Bordering the South China Sea: Maritime claims, contested sovereignty, and novel territorialities. In Diener, A.C., & Hagen, J. (eds), *Invisible Borders in a Bordered World: Power, Mobility, and Belonging*. New York: Routledge, 82–104.

Dittmer, J., Moisio, S., Ingram, A., & Dodds, K. (2011). Have you heard the one about the disappearing ice? Recasting Arctic geopolitics. *Political Geography 30*(4): 202–214.

Dodds, K. (2010). A polar Mediterranean? Accessibility, resources and sovereignty in the Arctic Ocean. *Global Policy 1*(3): 303–311.

Dodds, K. (2012). Graduated and paternal sovereignty: Stephen Harper, Operation Nanook 10, and the Canadian Arctic. *Environment and Planning D: Society and Space 30*(6): 989–1010.

Dodds, K., & Benwell, M.C. (2010). More unfinished business: The Falklands/Malvinas, maritime claims, and the spectre of oil in the South Atlantic. *Environment and Planning D: Society and Space 28*(4): 571–580.

Drakopulos, L. (2022). Privatizing the fisheries observer industry: Neoliberal science and policy in the U.S. West Coast fisheries. *Geoforum 131*: 116–125.

Drakopulos, L., Havice, E., & Campbell, L. (2022). Architecture, agency and ocean data science initiatives: Data-driven transformation of oceans governance. *Earth System Governance 12*: 100140.

Drewniak, M., Dalaklis, D., Kitada, M., Ölçer, A., & Ballini, F. (2018). Geopolitics of Arctic shipping: The state of icebreakers and future needs. *Polar Geography 41*(2): 107–125.

Duara, P. (2021). Oceans as the paradigm of history. *Theory, Culture and Society 38*(7–8): 143–166.

Dupont, A., & Baker, C.G. (2014). East Asia's maritime disputes: Fishing in troubled waters. *Washington Quarterly 37*(1): 79–98.

Dupont, C., Blanc-Petitjean, P., Cortet, M., et al. (2020). Dissatisfaction of women with induction of labour according to parity: Results of a population-based cohort study. *Midwifery 84*: 102663.

Elden, S. (2010). Land, terrain, territory. *Progress in Human Geography 34*(6): 799–817.

Everuss, L. (2020). Mobile sovereignty: The case of "boat people" in Australia. *Political Geography 79*: 102162.

Fabinyi, M. (2020). Maritime disputes and seafood regimes: A broader perspective on fishing and the Philippines–China relationship. *Globalizations 17*(1): 146–160.

Ferguson, J. (1990). *The Anti-politics Machine: "Development", Depoliticization, and Bureaucratic Power in Lesotho*. New York: Cambridge University Press.

Foley, P. (2022). Proximity politics in changing oceans. *Maritime Studies 21*: 53–64.

Furlong, K. (2011). Small technologies, big change: Rethinking infrastructure through STS and geography. *Progress in Human Geography 35*(4): 460–482.

Gaynor, J.L. (2019). Liquid territory, shifting sands: Property, sovereignty, and space in Southeast Asia's tristate maritime boundary zone. In Braverman, I., & Johnson, E. (eds), *Blue Legalities: The Life and Laws of the Sea*. Durham, NC: Duke University Press, 107–127.

Glassner, M.I. (1986). The new political geography of the sea. *Political Geography Quarterly 5*(1): 6–8.

Haas, B., Goodman, C., Sinan, H., & Davis, R. (2023). Fact or fiction? Unpacking the terminologies used in fisheries allocation discussions. *Marine Policy 152*: 105630.

Hall, P.V. (2009). Container ports, local benefits and transportation worker earnings. *GeoJournal 74*: 67–83.

Harold, S.W., Nakagawa, Y., Fukuda, J., et al. (2017). *The U.S.–Japan Alliance and Deterring Gray Zone Coercion in the Maritime, Cyber, and Space Domains*. Santa Monica, CA: RAND Corporation.

Havice, E. (2018). Unsettled sovereignty and the sea: Mobilities and more-than-territorial configurations of state power. *Annals of the American Association of Geographers 108*(5): 1280–1297.

Havice, E. (2021). The environmental geopolitics of allocation: State power and institutional stability in Eastern Atlantic bluefin tuna management. *Political Geography 88*: 102395.

Havice, E., & Zalik, A. (2019). Ocean frontiers: Epistemologies, jurisdictions, commodifications. *International Social Science Journal 68*(229–230): 219–235.

Hepworth, K. (2014). Enacting logistical geographies. *Environment and Planning D: Society and Space 32*(6): 1120–1134.

Höhler, S. (2002). Depth records and ocean volumes: Ocean profiling by sounding technology, 1850–1930. *History and Technology 18*: 119–154.

Huang, Y., & Suliman, S. (2020). Geopolitics, (re)territorialisation, and China's patriotic tourism in the South China Sea. *Geopolitics 28*(4): 1465–1488.

Hubbard, J. (2014). In the wake of politics: The political and economic construction of fisheries biology, 1860–1970. *Isis 105*(2): 365–378.

Hung, P.Y. (2020). Placing green energy in the sea: Offshore wind farms, dolphins, oysters, and the territorial politics of the intertidal zone in Taiwan. *Annals of the American Association of Geographers 110*(1): 56–77.

Hung, P.Y., & Lien, Y.H. (2022). Maritime borders: A reconsideration of state power and territorialities over the ocean. *Progress in Human Geography 46*(3): 870–889.

Iftikhar, M., & Zhan, J.V. (2020). The geopolitics of China's overseas port investments: A comparative analysis of Greece and Pakistan. *Geopolitics 27*(3): 826–851.

Jasanoff, S. (1990). *The Fifth Branch: Science Advisers as Policymakers*. Cambridge, MA: Harvard University Press.

Jasanoff, S. (2004). *States of Knowledge: The Co-production of Science and Social Order*. New York: Routledge.

Jensen, F. (2023). State capitalism and Spanish port development along the maritime silk road. *Environment and Planning A: Economy and Space 55*(3): 636–654.

Jenss, A., & Schuetze, B. (2021). Rethinking authoritarian power: The logistics space and authoritarian practices in and between secondary port cities of the global South. *International Studies Quarterly 65*(1): 82–94.

Johnson, C., Jones, R., Paasi, A., et al. (2011). Interventions on rethinking "the border" in border studies. *Political Geography 30*: 61–69.

Jones, R. (2016). *Violent Borders: Refugees and the Right to Move*. Brooklyn, NY: Verso.

Jones, R. (2019). *Open Borders. Defense of Free Movement*. Athens, GA: University of Georgia Press.

Joseph, C.E., & Balakrishnan, V. (2023). Through the eyes of the boat people: Redefining oceans in the twenty-first century. In Menon, D.M., & Zaidi, N. (eds), *Cosmopolitan Cultures and Oceanic Thought*. Delhi: Routledge India, 168–178.

Kahn, J.S. (2019). *Islands of Sovereignty: Haitian Migration and the Borders of Empire*. Chicago, IL: University of Chicago Press.

Kennedy, C. (2018). The struggle for blue territory: Chinese maritime militia grey-zone operations. *RUSI Journal 163*(5): 8–19.

Kim, H.J. (2019). South Korea's use of force against Chinese illegal fishing in the course of law enforcement in the Yellow Sea. *Marine Policy 99*: 148–156.

Koch, N. (2022). Planting flags in water. *Dialogues in Human Geography 12*(2): 302–306.

Lambach, D. (2021). The functional territorialization of the high seas. *Marine Policy 130*: 104579.

Lambach, D. (2022). The territorialization of the global commons: Evidence from ocean governance. *Politics and Governance 10*(3): 41–50.

Loyd, J.M., & Mountz, A. (2014). Managing migration, scaling sovereignty on islands. *Island Studies Journal 9*(1): 23–42.

MacIntosh, C. (2019). Refugees at sea. In International Ocean Institute – Canada (Ed.). In *The Future of Ocean Governance and Capacity Development: Essays in Honor of Elisabeth Mann Borgese (1918–2002)*. Nijhoff: Brill, 426–430.

Maxwell, S.M., Hazen, E.L., Lewison, R.L., et al. (2015). Dynamic ocean management: Defining and conceptualizing real-time management of the ocean. *Marine Policy 58*: 42–50.

Meehan, K.M. (2014). Tool-power: Water infrastructure as wellsprings of state power. *Geoforum 57*: 215–224.

Minca, C., Rijke, A., Pallister-Wilkins, P., et al. (2022). Rethinking the biopolitical: Borders, refugees, mobilities. . .. *Environment and Planning C: Politics and Spaces 40*(1): 3–30.

Monios, J., Bergqvist, R., & Woxenius, J. (2018). Port-centric cities: The role of freight distribution in defining the port city relationship. *Journal of Transport Geography 66*: 5364.

Monstadt, J. (2009). Conceptualizing the political ecology of urban infrastructures: Insights from technology and urban studies. *Environment and Planning A: Economy and Space 41*(8): 1924–1942.

Mostafanezhad, M. (2020). Tourism frontiers: Primitive accumulation, and the "free gifts" of (human) nature in the South China Sea and Myanmar. *Transactions of the Institute of British Geographers 45*(2): 434–447.

Neilson, B. (2013). Life and code: Logistics, culture and economy. In Bennett, T. (ed.), *Challenging (the) Humanities*. Melbourne: Australian Scholarly Publishing, 99–116.

Nelson, S.H., & Bigger, P. (2022). Infrastructural nature. *Progress in Human Geography 46*(1): 86–107.

Ortuño Crespo, G., Mossop, J., Dunn, D., et al. (2020). Beyond static spatial management: Scientific and legal considerations for dynamic management in the high seas. *Marine Policy 122*: 104102.

Patalano, A. (2018). When strategy is "hybrid" and not "grey": Reviewing Chinese military and constabulary coercion at sea. *Pacific Review 31*(6): 811–839.

Phillips, J. (2018). Order and the offshore: The territories of deep-water oil production. In Peters, K., Steinberg, P., & Stratford, E. (eds), *Territory Beyond Terra*. London: Rowman and Littlefield, 51–67.

Polacheck, T. (2012). Assessment of IUU fishing for southern bluefin tuna. *Marine Policy 36*(5): 1150–1165.

Probyn, E. (2021). Extracting fish. In Himley, M., Havice, E., & Valdivia, G. (eds), *The Routledge Handbook of Critical Resource Geography*. New York: Routledge, 361–388.

Rice, S., Tyner, J., Munro-Stasiuk, M., Kimroy, S., & Sirik, S. (2016). Making land to make life: Island-building in the South China Sea and the biopolitics of geophysical transformation. *Geographical Journal 182*(4): 444–448.

Richardson, T. (2013). Borders and mobilities: Introduction to the special issue. *Mobilities 8*(1): 1–6.

Rodenbiker, J. (2023). Ecological militarization: Engineering territory in the South China Sea. *Political Geography 106*: 102932.

Rolf, S., & Agnew, J. (2016). Sovereignty regimes in the South China Sea: Assessing contemporary Sino-US relations. *Eurasian Geography and Economics 57*(2): 249–273.

Rosello, M. (2022). Regional fishery management organization measures and the imposition of criminal and administrative sanctions in respect of high seas fishing. *Marine Policy 144*: 105213.

Roszko, E. (2016). Fishers and territorial anxieties in China and Vietnam: Narratives of the South China Sea beyond the frame of the nation. *Cross-Currents: East Asian History and Culture Review 21*: 19–46.

Rowen, I. (2018). Tourism as a territorial strategy in the South China Sea. In Spangler, K. (ed.), *Enterprises, Localities, People, and Policy in the South China Sea*. London: Palgrave Macmillan, 61–74.

Sakakibara, C. (2020). *Whale Snow: Iñupiat, Climate Change, and Multispecies Resilience in Arctic Alaska*. Tempe, AZ: University of Arizona Press.

Scott, K. (2016). Insular territories: US colonial science, geopolitics, and the (re)mapping of the Philippines. *Geographical Journal 182*(1): 2–14.

Sheller, M. (2018). *Mobility Justice: The Politics of Movement in an Age of Extremes*. London: Verso Books.

Sidaway, J.D., & Woon, C.Y. (2017). Chinese narratives on "one belt, one road" (一带一路) in geopolitical and imperial contexts. *Professional Geographer 69*(4): 591–603.

Song, A.M., Johnsen, J.P., & Morrison, T.H. (2018). Reconstructing governability: How fisheries are made governable. *Fish and Fisheries 19*(2): 377–389.

Sparke, M. (2006). Political geography: Political geographies of globalization (2) – governance. *Progress in Human Geography 30*(3): 357–372.

Squire, R. (2022). Depth: Discovering, "mastering", exploring the deep. In Peters, K., Anderson, J., Davies, A., & Steinberg, P. (eds), *The Routledge Handbook of Ocean Space*. London: Routledge, 347–361.

Steinberg, P. (2010). You are (not) here: On the ambiguity of flag planting and finger pointing in the Arctic. *Political Geography 29*(2): 81–84.

Steinberg, P., & Kristoffersen, B. (2017). "The ice edge is lost . . . nature moved it": Mapping ice as state practice in the Canadian and Norwegian North. *Transactions of the Institute of British Geographers 42*(4): 625–641.

Steinberg, P., & Peters, K. (2015). Wet ontologies, fluid spaces: Giving depth to volume through oceanic thinking. *Environment and Planning D: Society and Space 33*(2): 247–264.

Stierl, M. (2016). A sea of struggle–activist border interventions in the Mediterranean Sea. *Citizenship Studies 20*(5): 561–578.

Stierl, M. (2018). A fleet of Mediterranean border humanitarians. *Antipode 50*(3): 704–724.

Suarez de Vivero, J.L., & Rodríguez-Mateos, J.C. (2023). Facing a new ocean decade: Geopolitical change and challenges in ocean governance. *Geopolitics 29*(1): 148–173.

Tazzioli, M. (2016). Border displacements. Challenging the politics of rescue between Mare Nostrum and Triton. *Migration Studies 4*(1): 1–19.

Tazzioli, M. (2018). The temporal borders of asylum: Temporality of control in the EU border regime. *Political Geography 64*: 13–22.

Teo, J. (2023). Ocean materialities and the ontologies of ocean conservation in areas beyond national jurisdiction: The Costa Rica thermal dome. *Marine Policy 153*: 105646.

Toonen, H.M., & Mol, A.P.J. (2016). Governing the marine environment through information: Fisheries, shipping, and tourism. In MacDonald, B.H., Soomai, S.S., De Santo, E.M., & Wells, P.G. (eds), *Science, Information and Policy Interface for Effective Coastal and Ocean Management*. Boca Raton, FL: CRC Press, 125–151.

Tsing, A. (2011). Frontiers of capitalism. In:. In Tsing, A. (ed.), *Friction: An Ethnography of Global Connection*. Princeton, NJ: Princeton University Press, 27–50.

Vives, L. (2017). Unwanted sea migrants across the EU border: The Canary Islands. *Political Geography 61*: 181–192.

Wirth, C. (2016). Securing the seas, securing the state: Hope, danger and the politics of order in the Asia-Pacific. *Political Geography 53*: 76–85.

Yeh, Y.H., Tseng, H.S., Su, D.T., & Ou, C.H. (2015). Taiwan and Japan: A complex fisheries relationship. *Marine Policy 51*: 293–301.

Yorgason, E. (2017). Eastern Asia's revitalization of the state ideal through maritime territorial disputes. *Political Geography 61*: 203–214.

Zalik, A. (2015). Trading on the offshore: Territorialization and the ocean grab in the international seabed. In Ervine, K., & Fridell, G. (eds), *Beyond Free Trade*. London, 190: Palgrave Macmillan, p. 173.

Zalik, A. (2018). Mining the seabed, enclosing the area: Ocean grabbing, proprietary knowledge and the geopolitics of the extractive frontier beyond national jurisdiction. *International Social Science Journal 68*(229–230): 343–359.

Zalik, A. (2021). World-making and the deep seabed: Mining the area beyond national jurisdiction. In Himley, M., Havice, E., & Valdivia, G. (eds), *The Routledge Handbook of Critical Resource Geography*. New York: Routledge, 425–440.

Zhang, H. (2016). Chinese fishermen in disputed waters: Not quite a "people's war". *Marine Policy 68*: 65–73.

Index

The Wiley Blackwell Companion to Political Geography, Second Edition.
Edited by Virginie Mamadouh, Natalie Koch, Chih Yuan Woon, and John Agnew.